# Lecture Notes in Computer Science 11340

*Commenced Publication in 1973*
Founding and Former Series Editors:
Gerhard Goos, Juris Hartmanis, and Jan van Leeuwen

More information about this series at http://www.springer.com/series/7407

Paola Flocchini · Giuseppe Prencipe
Nicola Santoro (Eds.)

# Distributed Computing by Mobile Entities

## Current Research in Moving and Computing

*Editors*
Paola Flocchini
University of Ottawa
Ottawa, ON, Canada

Giuseppe Prencipe
University of Pisa
Pisa, Italy

Nicola Santoro
Carleton University
Ottawa, ON, Canada

ISSN 0302-9743 ISSN 1611-3349 (electronic)
Lecture Notes in Computer Science
ISBN 978-3-030-11071-0 ISBN 978-3-030-11072-7 (eBook)
https://doi.org/10.1007/978-3-030-11072-7

Library of Congress Control Number: 2018965922

LNCS Sublibrary: SL1 – Theoretical Computer Science and General Issues



Cover illustration: The drawing on the front cover was created by the editors. Used with permission.

This Springer imprint is published by the registered company Springer Nature Switzerland AG
The registered company address is: Gewerbestrasse 11, 6330 Cham, Switzerland

# Preface

The past two decades have seen the rapid growth and development of the field of distributed computing by mobile entities, whose concern is the study of the computational and complexity issues arising in systems of decentralized computational entities operating in a spatial universe $\mathcal{U}$. The entities can move in $\mathcal{U}$ (their movement is constrained by the nature of $\mathcal{U}$), and are autonomous in their actions (e.g., they are not directed by an external controller). Depending on the nature of $\mathcal{U}$, two different settings are usually identified.

The first setting, called *discrete universe* (sometimes, *netscape* or *graph world*), is when $\mathcal{U}$ is a simple graph and the entities, usually called *agents*, move from node to neighboring node. This setting has a long history. The pioneering work of Shannon in 1951 on `Exploration` [6] was the first of a prolific line of algorithmic investigations on computations by a *single* agent. It was, however, only later that *multiple* agents (two, to be precise) were first considered [2]. The distributed computing concerns started in full in 2001 with the study of the `Black-Hole Search` [3, 4] and `Intruder Capture` [1] problems. Interestingly, in parallel and independently of this theoretical work, the notion and use of mobile agents were being intensively investigated in other fields, noticeably AI and software engineering. Indeed, the use of mobile software agents has been very popular in networked environments, ranging from the Internet to the Data Grid, both as a theoretical paradigm and as a system-supported programming platform; in these fields, the theoretical research had focused mainly on the descriptive and semantic concerns.

The other setting, called *continuous universe*, is when $\mathcal{U}$ is a Euclidean space that the entities, usually called *robots*, can perceive and move in. The continuous setting had long been investigated in the traditional fields of AI, robotics, and control. Within the distributed computing community, the setting was introduced in 1996 with the study of the `Pattern Formation` problem [7, 8]. The investigations in the field were initially sporadic; this situation has drastically changed over the years, and a continuously increasing number of algorithmic investigations examine this setting. A comprehensive snapshot of the status of the research in this setting at that time appeared in 2012 [5].

In both settings, the research concern is on determining what tasks can be performed by the entities, under what conditions, and at what cost. In particular, the central question is to determine what minimal hypotheses allow a given problem to be solved.

Encompassing and modeling a large variety of application environments and systems, from robotic swarms to networks of mobile sensors, from software mobile agents in communication networks to crawlers and viruses on the Web, the theoretical research in this area intersects distributed computing with the fields of computational geometry (especially for continuous spaces), control theory, graph theory, and combinatorics (especially for discrete spaces).

In spite of the apparent differences between the two settings, and of the distinct technical tools required in each, at a higher level, the basic principles are similar; the vision and mind frame required to examine questions and solve problems share many commonalities. This fact has allowed ideas, problems, and questions to be transferred from one setting to the other. Indeed, it is quite common for researchers in the field to do research in both settings.

As mentioned, over the years, the community of researchers investigating the field of distributed computing by mobile entities has become quite large. An important aggregation point of this community has been the series of research meetings, called MOVING AND COMPUTING (MAC). Since their inception in 2004, these events have comprised scheduled tutorials and lectures and, more importantly, open sessions for the presentations of open problems, new results, discussions, and free research. Regardless of their length (micro-MAC, mini-MAC, big-MAC), the goals of these meetings have been first of all to create a clearer picture of the state of the art in the field, then to identify outstanding open problems and research directions, and finally to foster a collaborative attack of some problems. While so much has been done to advance our knowledge of the field, the larger part of the territory is still unexplored and remains uncharted.

Since the field has expanded so rapidly in recent years, and the many results are spread through a large number of conferences and journals, these meetings have provided a snapshot of the state of the art. However, since the MAC events are without proceedings, this benefit has been limited mostly to the participants. The need for a comprehensive description of the state of the art, to be available to all researchers already working or interested in the area, as well as to those (e.g., PhD students and postdocs) entering the field for the first time, is clearly felt. The main motivation of this book is to satisfy this need.

This book is based on the lectures and tutorial presented at the MAC meeting held in La Maddalena Island in 2017. Greatly expanded, revised, and updated, each of the lectures forms an individual chapter. Together, they provide a map of the current knowledge about the boundaries of "moving and computing."

This has been a truly collective project, and we would like to thank all the authors for the enthusiasm with which they have embarked this project and carried it through.

The book is organized in five parts. Part 1 contains two chapters focusing on "moving and computing" models in the two spatial settings; Part 2 and Part 3 are devoted to robots in the classic Look–Compute–Move model and in a continuous time variation, respectively; Part 4 contains chapters on agents in discrete settings; finally, Part 5 is dedicated to novel subjects.

December 2018

Paola Flocchini
Giuseppe Prencipe
Nicola Santoro

## References

1. L. Barrière, P. Flocchini, P. Fraignaud, and N. Santoro. Capture of an intruder by mobile agents. In *Proceedings of 14th Symposium on Parallel Algorithms and Architectures (SPAA)*, pages 200–209, 2002.
2. M. Blum and D. Kozen. On the power of the compass (or, why mazes are easier to search than graphs). In *19th Symposium on Foundations of Computer Science (FOCS)*, pages 132–142, 1978.
3. S. Dobrev, P. Flocchini, G. Prencipe, and N. Santoro. Mobile search for a black hole in an anonymous ring. In *Proceedings of 15th International Conference on Distributed Computing, (DISC)*, pages 166–179, 2001.
4. S. Dobrev, P. Flocchini, G. Prencipe, and N. Santoro. Searching for a black hole in arbitrary networks: optimal mobile agents protocols. In *Proceedings of 21st ACM Symposium on Principles of Distributed Computing (PODC)*, pages 153–161, 2002.
5. P. Flocchini, G. Prencipe, and N. Santoro. *Distributed Computing by Oblivious Mobile Robots.* Morgan & Claypool, 2012.
6. C. E. Shannon. Presentation of a maze-solving machine. In *8th Conference of the Josiah Macy Jr. Found. (Cybernetics)*, pages 173–180, 1951.
7. K. Sugihara and I. Suzuki. Distributed algorithms for formation of geometric patterns with many mobile robots. *Journal of Robotics Systems*, 13:127–139, 1996.
8. I. Suzuki and M. Yamashita. Distributed anonymous mobile robots. In *Proceedings of 3rd International Colloquium on Structural Information and Communication Complexity (SIROCCO)*, pages 313–330, 1996.

# Contents

## Continuous Time Robots



## Agents



## Other Computational Settings

# Models

# Moving and Computing Models: Robots

Paola Flocchini[1], Giuseppe Prencipe[2](✉), and Nicola Santoro[3]

[1] University of Ottawa, Ottawa, Canada
pflocchi@uottawa.ca
[2] University of Pisa, Pisa, Italy
prencipe@di.unipi.it
[3] Carleton University, Ottawa, Canada
santoro@scs.carleton.ca

**Abstract.** This chapter provides a map of the current knowledge about the boundaries of Moving and Computing in Continuous Spaces, describing the "models" under which the results known so far have been obtained.

## 1 Introduction

Within the field of *Distributed Computing by Mobile Entities*, or *Moving and Computing*, a prominent role is played by the theoretical investigations on the computational and complexity issues related to systems of mobile computational entities that operate and move in continuous spaces. Because the original inspiration and motivation for these studies arose from the application field of swarm robotics, the computational entities are usually called *robots*.

The last two decades have seen a flurry of studies on the computational and complexity issues related to distributed computing by such robots, and a large number of results have been established. Each individual study, with its results derived under specific set of assumptions, has contributed to forming and detailing the map of the "computational universe" of Moving and Computing in Continuous Spaces, enlarging our understanding of this universe, of its nature, its limits, its boundaries, of its critical factors. On the other hand, the specific assumptions under which each one of these results is established constitutes the restricted view point from which the universe is observed and described; this obviously limits what can be seen, observed and found. Furthermore, the set of all the assumptions of all these results is very large and varied, encompassing many elements and factors of very different nature and whose interaction is often unknown. In other words, there is no such a thing as a single "Model" of Moving and Computing in Continuous Spaces. Rather, there are parts of the universe that are defined and delimited by particular sets of assumptions, "models" which have been well-studied and clearly defined. To these, as a result of recent investigations, new "models" have been added to describe other parts of the universe newly discovered.

Aim of this chapter is to provide a map of the current knowledge about the boundaries of Moving and Computing in Continuous Spaces, describing the


P. Flocchini et al. (Eds.): Distributed Computing by Mobile Entities, LNCS 11340, pp. 3–14, 2019.
https://doi.org/10.1007/978-3-030-11072-7_1

"models" under which the known results so far (topic of Chaps. 3–15) have been obtained.

## 2 Standard Model: $\mathcal{OBLOT}$

Perhaps the most general, well-known and investigated model is the one where the robots, silent and anonymous, are *oblivious*, that is, every time they are activated, they have no recollection of past activities. This de-facto standard model, which we shall call $\mathcal{OBLOT}$, is defined by a set of fundamental features, which are are presented first.

The $\mathcal{OBLOT}$ model covers a large spectrum of settings and situations, each defined by specific choices, among a range of possibilities, with respect to a fundamental component, *time*, as well as two less crucial but still important elements, *orientation* and *mobility*; these will be described next.

### 2.1 Fundamental Features

The system is composed of a set $\mathcal{R} = \{r_1, \ldots, r_n\}$ of $n$ computational entities, called *robots*, that live and operate in a connected spatial universe $\mathcal{U} \subseteq \mathbb{R}^d$, $d \geq 1$, in which they can move. The robots are viewed as points in $\mathbb{R}^d$; let $r(t)$ denote the position of robot $r$ at time $t$. More than one robot can occupy the same location at the same time; when this occurs, we say that there is *multiplicity*.

Each robot is provided with its own local memory and is capable of performing local computations with (infinite precision) real arithmetic. Each robot is endowed with motorial capabilities; it can turn and move in any direction.

The robots are externally *identical*; that is, they are indistinguishable by their appearance. They are *anonymous*, that is, the do not have distinct identities that can be used during the computation. The robots are *autonomous*; that is, they operate without a central control or external supervision. They are *homogeneous*; that is, they all have and execute the same protocol, or algorithm.

Each robot has its own local coordinate system, whose *origin* always coincides with the robot's position (hence it follows the robot as it moves), while its *unit of length* as well as the *orientation* and *handedness* are fixed. The coordinate systems of different robots might be different in all their aspects.

A robot is capable of observing the universe $\mathcal{U}$ within its visibility range $\nu$ (the same for all robots); as a result, it determines the positions (expressed in its local coordinate system) of all the robots within distance $\nu$. The visibility is said to be *limited* if $\nu \neq \infty$, *unlimited* otherwise. We say that there is *multiplicity detection* if the observation distinguishes if a point is occupied by a one, or more than one robot; the multiplicity detection is said to be *strong* if it allows to detect the exact number of robots on the same point.

The robots are *silent*: they have no means of direct communication of information to other robots. Thus the only means of interaction between robots are observations and movements: that is, communication is *stigmergic*.

Each robot operates in `Look-Compute-Move` (LCM) cycles; in each cycle, it observes its surroundings, it computes a destination, and it moves towards it. Specifically:

(i) `Look`: The robot observes $\mathcal{U}$. This operation is instantaneous and the result is a *snapshot* indicating the positions of all robots within its radius of visibility expressed in its own coordinate system.
(ii) `Compute`: The robot executes the algorithm (the same for all robots), using the snapshot of the `Look` operation as input. The result of the computation is a destination point.
(iii) `Move`: The robot moves towards the computed destination; if the destination is the current location, the robot stays still, performing a *null movement*.

The sequence `Look-Compute-Move` forms an operational *cycle* of the robot. At the end of a cycle, a robot may either start a new cycle or become *inactive*; in the latter case, it does not perform any operation ("sleeps") until it becomes *active* again, and then starts a new cycle.

The robots are *oblivious*: at the end of a cycle, all obtained information (observations, computations, and move) are erased. In other words, at the beginning of a cycle, the robots have *no* memory of past actions and computations ("every time is the first time"), and the computation is based solely on what determined in the current cycle. The importance of this property, sometimes called *memoryless* or *stateless*, comes from its link to reliability and self-stabilization.

This concludes the description of the fundamental features of the $\mathcal{OBLOT}$ model. Any system exhibiting these features (i.e., meeting these specifications) is an $\mathcal{OBLOT}$ system. This definition does not impose any specific restriction on several important component of a system: e.g., the activation schedule, the synchronization between cycles of different robots, the mobility range, etc. The range of possibilities, all still within the $\mathcal{OBLOT}$ model, are discussed in the next subsections.

## 2.2 Temporal Features: Time, Activation and Synchronization

The temporal dimension is perhaps the most important, as any assumption made on this regards greatly impact the computational capabilities of the robots.

Although *time* is easily defined in terms of an external observer, this notion might generally not be available to the robots; even if each robot were to be endowed with a local clock, the clocks might not sign the same value, nor run at the same rate, nor be otherwise synchronized. The *duration* of a cycle, i.e. the time spent by a robot computing and moving, might be different for different cycles, and different for different robots. The *delay* between two successive cycles of the same robot (i.e., the duration of inactivity), might not be the same, and could be different for different robots.

Assumptions made on the last two aspects of the temporal dimension, the *activation* schedule of the robots and of timing of the *operations* within their cycles, define specific instances of the general $\mathcal{OBLOT}$ model. The two most important models, intensively and extensively studied in the literature, are $\mathcal{A}$SYNC and $\mathcal{S}$SYNC.

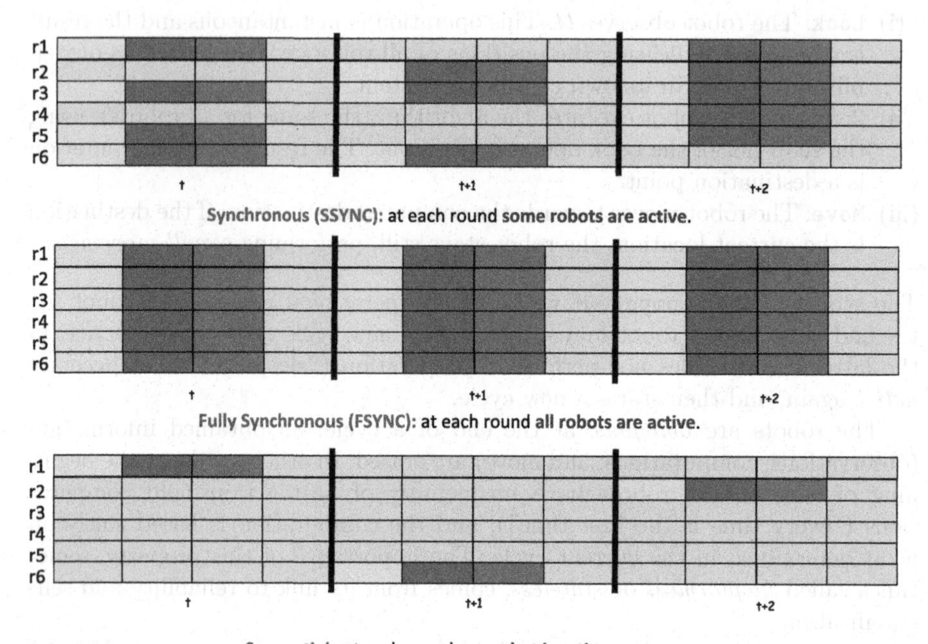


**Fig. 1.** $\mathcal{S}$SYNC

In the *synchronous* model ($\mathcal{S}$SYNC) (originally called *semi-synchronous*), the activations of the robots is logically divided into global rounds; in each round, one or more robots are activated and obtain the same snapshot; based on that snapshot, they compute and perform their move, ending their cycle by the next round. Note that such a system is computationally *equivalent* to a synchronous system in which the chosen robots are activated simultaneously and all operations are instantaneous. The choice of which robots are activated in a given round is assumed to be made by *the activation scheduler* (or daemon) (Fig. 1).

An extreme case of activation scheduler in $\mathcal{S}$SYNC defines the *fully-synchronous* ($\mathcal{F}$SYNC) model: *all* the robots are activated in every round; introduced in [20], used e.g. in [10,13,21].

Another extreme case of daemon is the so-called *centralized* activation scheduler, which defines the *sequential* model ($\mathcal{S}$EQUENTIAL) where *only one* robot is active at any time.

In all cases, in $\mathcal{S}$SYNC the activation scheduler is *fair*: for every robot $r$ and time $t$, there exists a time $t' \geq t$ at which $r$ is activated; that is, every robot is activated infinitely often.

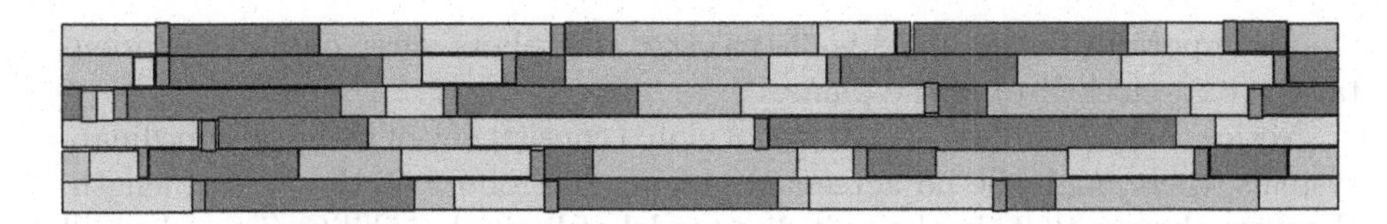


**Fig. 2.** $\mathcal{A}$SYNC

In the *asynchronous* ($\mathcal{A}$SYNC) model, there is no common notion of time, each robot is activated asynchronously and independently from the other robots. Furthermore, the duration of each `Compute`, `Move` and `Sleep` are finite but otherwise unpredictable. As a result, computations can be made based on totally obsolete observations, taken arbitrarily far in the past. Also, robots can be seen while moving, creating further inconsistencies in the robots' understanding of the universe. Also in $\mathcal{A}$SYNC the activation scheduler is *fair*: for every robot $r$ and time $t$, there exists a time $t' \geq t$ at which $r$ is activated; that is, every robot is activated infinitely often (Fig. 2).

Although the activation scheduler is fair both in $\mathcal{A}$SYNC and $\mathcal{S}$SYNC, between two successive activations of the same robot $r$, the number of activations of the other robots is possibly unbounded. A stronger requirement is that offered by a *k-bounded* (or *k-fair*) scheduler: For every $r \in \mathcal{R}$, between two successive activations of $r$, every other robot has been activated at most $k$ times [9]. The bound $k$ may be not known to the robots, in which case the daemon is called just *bounded*. Note that, in $\mathcal{F}$SYNC, the scheduler is 1-bounded by definition.

An example of 2-bounded centralized scheduler is the *slicing* scheduler: starting from time $t = 0$, after $n$ successive rounds (a slice), all the robots in the system have been activated exactly once [7,9]. A particular slicing daemon is the classical *round-robin* scheduler: In each slice, all robots are activated always in the same order.

### 2.3 Orientation Features

Each robot $r$ has its own unit of length, and a local Cartesian coordinate system defining the directions of the coordinate axes, together with their orientations. This local coordinate system of $\mathbb{R}^d$ is self-centric, i.e. the origin is the position of the robot.

The absence of any a-priori assumption on consistency of the local coordinate systems is called `Disorientation`, in which case we say that the robots are *disoriented.*

Depending on the level of (a-priori known) consistency among the local coordinate systems, different levels of *global* geometric agreement can be identified. The strongest form is `GlobalConsistency`: all robots agree on the direction and the orientation of all axes; a weaker form is `k-Axes`: all robots agree on the direction and orientation of $k$ axes ($1 \leq k < d$).

An important factor in $\mathbb{R}^2$ is `Chirality`: the robots agree on a cyclic orientation (e.g., clockwise) of the plane.

Notice that, regardless of the level of global consistency of the local coordinate systems, there might be no agreement among the robots on the unit of length. However observe that, in the unit-disc model of limited visibility (i.e., when all robots have the same visibility radius), the robots have a *de facto* agreement on the unit of length.

### 2.4 Mobility Features

#### Mobility Scheduler

The actual movement of a mobile robot is controlled by an external *mobility scheduler*. The scheduler decides how fast the robot moves toward its destination point, and it may even interrupt its movement before the destination point is reached.

In a $\mathcal{S}$SYNC system, the speed of the robot in a given `Move` is not important as all movements terminate before the next global round starts; in $\mathcal{A}$SYNC, the speed determines the duration of the `Move` operation of the robot in that cycle.

Regardless of the speed, a move may stop before the robot reaches its destination, e.g. because of energy limit. In this regards, two main sub-models can be defined:

- **rigid** (or *unlimited*) mobility: All robots always reach their destinations when performing `Move`; this type of mobility is also called *undisturbed-motion* [2].
- **fixed** mobility: There exists a constant $\gamma > 0$ such that every robot performing `Move` will move exactly $Min\{dest, \gamma\}$, where *dest* is the distance to the destination point. The quantity $\gamma$ might not be known *a priori* to the robots. Fixed mobility is usually assumed when working in $\mathcal{S}$SYNC.
- **non-rigid** mobility: The scheduler may stop a move before the robot reaches its destination; the robot is not notified that an interruption has occurred (but it may be able to infer it from its next observation). However, the distance traveled in a move by $r$ is not infinitesimally small (unless it brings the robot to its destination): there exists an (arbitrarily small) constant $\delta > 0$, such that, if the destination is closer than $\delta$, $r$ will reach it; otherwise, 5 will move towards it by at least $\delta$. This guarantees, for example, that if a robot keeps computing the same destination point, it will reach it in a finite number of iterations; without this assumption, it would be impossible for $r$ to ever reach its destination, following a classical Zenonian argument. The quantity $\delta$ might not be known to the robots.

#### Trajectories

In the standard model, robots move towards their destination in straight line movement. A stronger assumption is for the robots to capable of moving along a specified curve; in this variant, which we shall call **guided** movement, the result of the `Compute` operation is not only the destination point but also the trajectory to be followed. Note that the concepts of rigid, fixed and non-rigid mobility can be extended to the variant when movement is guided.

**Collisions**
Multiplicity can be accidentally created during the `Move` operation; indeed moving robot might non-intentionally be at the same place of other robots, moving or stationary. More precisely, a robot $r$ is said to *collide* with robot $s$ at time $t$ if $r(t) = s(t)$ and at time $t$, $r$ is performing `Move`, and its destination is not $r(t)$. With respect to collisions, there are three main approches.

- **immaterial** collisions: Since robots are viewed as points, a moving robot causing a collision proceeds in its movement unaware that the collision occurred (e.g., [21]).
- **fail-stop** collisions: A moving robot stops moving when it collides with another robot (e.g., [5]).
- **intolerable** collisions: Collisions are undesirable events (with possibly negative consequences), and thus to be avoided. Since the only times a robots is aware of other robots is during `Look`, and while moving it cannot detect the position of other robots, collision avoidance must be done algorithmically, designing protocols that are *collision-free* in every feasible execution.

## 3 Weaker Variants: Opacity and Extent

In the standard model, the robots are viewed as points; i.e., they are *dimensionless.* Furthermore, a snapshot contains the set of the positions of all the robots within the visibility range. In other words, the visibility of a robot is considered to be *unobstructed*: if three robots $r$, $s$ and $z$ are collinear with $s$ in the middle, $s$ does not prevent $r$ from viewing $z$; i.e., the robots are *transparent.* That is, in the standard model, the robots have neither **extent** (because they are dimensionless) nor **opacity** (because they are transparent).

Three weaker models have been defined with respect to these two qualities: extent and opacity.

### 3.1 Opaque Robots with No Extent

Even with robots with no extent, one can consider the case when the line of sight of a robot is *obstructed* by the closest robot on that line; that is, they are **opaque**. Specifically, two robots $r_1$ and $r_2$ within visibility range of each other can see each other if and only if there is no other robot on the segment $\overline{r_1 r_2}$. Clearly opacity restricts visibility; collisions are defined as in the standard model. This weaker model has been considered e.g., in [3,15,17,18].

### 3.2 Opaque Robots with Extent

Opacity can be a natural consequence of robots that have a physical dimension; that is, entities with an *extent.*

Opaque robots with a physical dimension, also called *solid*, are viewed as opaque circular disks of a fixed diameter (hence they are are assumed to have a common unit distance). The distance between a robot and some point (or some other robot) is defined as the distance from (or to) the centre of the corresponding circular disks, and the perimeter of the disk is called *contour*.

The robots' visibility is clearly affected by their opacity and extent. Specifically, a point $p$ is visible by a robot $r_1$ (or equivalently, $r_1$ can see $p$) if $p$ is within the visibility range of $r_1$ and if there is a point $p_1$ on the contour of $r_1$ such that the straight segment $\overline{p_1 p}$ does not contain any point occupied by other robot. Robot $r_1$ can see another robot $r_2$ if there is at least one point on the contour of $r_2$ that is visible by $r_1$. Note that if a robot $r_1$ can see robot $r_2$, it can see some non-zero arc of its bounding circle and thus it can always compute its centre.

In addition to restricted visibility, extent clearly restricts mobility because robots with extent cannot cross each other. Collisions occur when two (or more) robots touch. It is sometimes assumed that, when a robot collides with another, its movement stops (fail-stop collision).

This much weaker model has been considered e.g., in [4,5].

### 3.3 Transparent Robots with Extent

Opacity does not need to be implied by the fact that robots have an extent. In fact, it is possible that the snapshot contains the positions of all the robots within visibility range even if they have an extent; e.g., the snapshot is provided by a drone. In this model, the drawbacks due to the physical dimension of the robots are all present; however, visibility is not affected. This model has been considered e.g. in [1].

Notice that the models *transparent with extent* and *opaque with no extent* are computationally orthogonal, and a solution in one model cannot generally be transformed into a solution in the other.

## 4 Stronger Variants: Memory and Communication

In the standard model, robots are *oblivious* and *silent*, i.e., they have no memory of past cycles and no explicit means of communication. Stronger variants of the standard model have been studied in the literature, where little memory and communication capabilities are provided by the presence of *lights*, introduced in [6,7]. The variants described below are discussed in detail in Chap. 11.

### 4.1 Finite-State Robots $\mathcal{F}$-STATE

In addition to the algorithm, each robot has a local working memory, or *workspace*, used for computations and to store various information (e.g., regarding the location of other robots) obtained during the cycles. In the standard $\mathcal{OBLOT}$ model, this workspace is volatile and it is initialized to be empty at every new activation. Consider now the situation when robots are equipped with some persistent memory. In this case, part (or all) of the workspace is *legacy*: unless explicitly erased by the robot, it will persist throughout the robot's cycles.

In this model, an important parameter is the *size* of the persistent workspace. One extreme is the *unbounded memory* case, where no information is ever erased; hence robots can remember all past computations and actions (e.g., see [19,21]).

On the opposite side is the case when the size of the persistent workspace is *constant*, that is, is contains only a constant number of bits. In this case, the model is called $\mathcal{F}$-STATE (e.g., [8,12,14]).

In between these two extremes one may consider different sizes of the persistent workspace: with $f(n)$-STATE we indicate robots with a persistent workspace of size $O(f(n))$ bits.

### 4.2 $\mathcal{LUMINOUS}$ Robots

Robots are endowed with a persistent and externally visible state variable, called *visible light*, that can assume values from a finite set of colors. The light can be set in each cycle by the robot at the end of its Compute operation. It is externally visible in the sense that its color at time $t$ is visible to all robots in its visibility radius that perform a Look operation at that time. It is persistent in the sense that, while the robot is oblivious and forgets all information from previous cycles, the variable is not automatically reset at the end of a cycle. The color a robot sees is used as input during the computation. In other words, $\mathcal{LUMINOUS}$ robots are finite-state robots where the state variable is visible to the other robots within range during their Look phase, which means that this persistent information can be used, not only to remember, but also to communicate. Also in this case, we can generalize the concept to larger sets of colors by indicating with $f(n)$-$\mathcal{L}$UMINOUS robots whose lights have $O(f(n))$ colors available.

Luminous robots have been studied, for example, in [7,12,16]; see also [11].

### 4.3 Finite-Communication Robots $\mathcal{F}$-COMM

In the $\mathcal{F}$-COMM model, robots are still endowed with an externally visible light that can be set during the Compute operation and that persists through cycles, unless erased or changed by the robots. However, the light of a robot is visible *only* to the other robots within that are performing the Look operation. In other words, when a robot is activated, it cannot see the current color of its light, but it can reset it. In this model, the light variable cannot be used to remember information from past activations, but it can be used as a communication tool. With $f(n)$-COMM we indicate robots whose external lights have $O(f(n))$ colors available. $\mathcal{F}$-COMM robots have been studied, for example, in [12].

## 5 Germane Models

### 5.1 Robots in Discrete Spaces

The oblivious robots model has been employed also in *discrete* spaces, i.e., when the spatial universe $\mathcal{U}$ in which the robots operate is a graph. Oblivious robots in graphs are discussed in Chap. 8 for the gathering problem, and in Chap. 9 for the exploration problem.

Let $G = (V, E)$ be an undirected connected simple graph where $V$ is the set of vertexes and $E$ the set of edges, with $|V| = n$; $E(x)$ be the edges incident to $x \in V$. The vertexes are unlabeled (i.e., the graph is anonymous), while the edges might or might not be labeled. If the graph is edge-labeled, let $\lambda_x : E(x) \to \mathcal{L}$ be an injective local labeling function that associates a label from a set $\mathcal{L}$ to each edge incident on $x$.

Let $\lambda = \{\lambda_x : x \in V\}$ be the global labeling function and let $(G, \lambda)$ denote the resulting edge-labeled graph. Let $\Psi$ be the placement function describing the position of the robots in the graph. Let $(G, \Psi)$ (resp. $(G, \lambda, \Psi)$) denote the graph (reps. edge-labeled graph) with the placement of the robots. More than one robot could be on the same node; $\mu(v)$ denotes the number of robots present in node $v$. The description of the graph, together with the indication of the exact number of robots located on each vertex, is called a *configuration* and is denoted by $(G, \mu)$.

The robots move from node to neighbouring node still operating in Look-Compute-Move cycles. The $\mathcal{A}$SYNC, $\mathcal{S}$SYNC, $\mathcal{F}$SYNC models can be defined similarly to the continuous space case, with some distinctions due to discreteness. In particular, the Look operation provides a snapshot of the graph (or part of it in case of limited visibility), showing the presence or absence of robots on the nodes. The Move operation is considered instantaneous in all three models. In $\mathcal{A}$SYNC the robots look in arbitrarily different moments and the time to complete both Compute and Move operations is finite but unpredictable, in $\mathcal{F}$SYNC all robots simultaneously perform each of the operations, while in $\mathcal{S}$SYNC in each cycle a subset of the robots is active, and those robots simultaneously perform each of the operations.

Like in the case of continuous spaces, the robots might have *limited visibility* (i.e., see only up to a certain distance given in terms of number of hops), they may have *multiplicity detection* capabilities and be able to distinguish nodes containing more than one robots (often called *towers*). In this setting also *local multiplicity* detection has been considered, where a robot can detect multiplicity (or strong multiplicity) only on the node it resides.

### 5.2 Continuous Time Robots

In the *continuous time* model, all robots continuously observe and at the same time measure and adjust their movement paths. This causes the trajectories of the robots, that are assumed to have some constant maximum movement speed, to become (continuous) curves. Robots might be able to change their speed and movement direction, resulting in trajectories that are continuous but not necessarily differentiable; usually, however, trajectories are restricted to be right-differentiable.

Continuous time robots are discussed in Chap. 13 for the gathering problem, in Chap. 14 for Search and Evacuation, and in Chap. 15 for patrolling problems.

## 6 Fault Tolerance

Robots are typically assumed to be *correct*, that is to operate without faults. More realistically, failures may occur. The robots' faults that have been investigated fall in two categories: *crash* faults (i.e. a faulty robot stops executing its cycle forever) and *Byzantine* faults (i.e. a faulty robot may exhibit arbitrary behavior and movement). Of course, the Byzantine fault model encompasses the crash fault model, and is thus harder to address. Chapter 10 is fully devoted to fault-tolerant robots.

## References

1. Gan Chaudhuri, S., Mukhopadhyaya, K.: Gathering asynchronous transparent fat robots. In: Janowski, T., Mohanty, H. (eds.) ICDCIT 2010. LNCS, vol. 5966, pp. 170–175. Springer, Heidelberg (2010). https://doi.org/10.1007/978-3-642-11659-9_17
2. Cohen, R., Peleg, D.: Convergence properties of the gravitational algorithm in asynchronous robot systems. SIAM J. Comput. **34**, 1516–1528 (2005)
3. Cohen, R., Peleg, D.: Local spreading algorithms for autonomous robot systems. Theor. Comput. Sci. **399**, 71–82 (2008)
4. Cord-Landwehr, A., et al.: Collisionless gathering of robots with an extent. In: Černá, I., et al. (eds.) SOFSEM 2011. LNCS, vol. 6543, pp. 178–189. Springer, Heidelberg (2011). https://doi.org/10.1007/978-3-642-18381-2_15
5. Czyzowicz, J., Gasieniec, L., Pelc, A.: Gathering few fat mobile robots in the plane. Theor. Comput. Sci. **410**(6–7), 481–499 (2009)
6. Das, S., Flocchini, P., Prencipe, G., Santoro, N., Yamashita, M.: The power of lights: synchronizing asynchronous robots using visible bits. In: Proceedings of 32nd IEEE International Conference on Distributed Computing Systems (ICDCS), pp. 506–515 (2012)
7. Das, S., Flocchini, P., Prencipe, G., Santoro, N., Yamashita, M.: Autonomous mobile robots with lights. Theor. Comput. Sci. **609**, 171–184 (2016)
8. Barrameda, E.M., Das, S., Santoro, N.: Deployment of asynchronous robotic sensors in unknown orthogonal environments. In: Fekete, S.P. (ed.) ALGOSENSORS 2008. LNCS, vol. 5389, pp. 125–140. Springer, Heidelberg (2008). https://doi.org/10.1007/978-3-540-92862-1_11
9. Défago, X., Gradinariu, M., Messika, S., Raipin-Parvédy, P.: Fault-tolerant and self-stabilizing mobile robots gathering. In: Dolev, S. (ed.) DISC 2006. LNCS, vol. 4167, pp. 46–60. Springer, Heidelberg (2006). https://doi.org/10.1007/11864219_4
10. Dieudonné, Y., Dolev, S., Petit, F., Segal, M.: Deaf, dumb, and chatting asynchronous robots. In: Abdelzaher, T., Raynal, M., Santoro, N. (eds.) OPODIS 2009. LNCS, vol. 5923, pp. 71–85. Springer, Heidelberg (2009). https://doi.org/10.1007/978-3-642-10877-8_8
11. Flocchini, P.: Computations by luminous robots, pp. 238–252 (2015)
12. Flocchini, P., Santoro, N., Viglietta, G., Yamashita, M.: Rendezvous with constant memory. Theor. Comput. Sci. **621**, 57–72 (2016)
13. Ganguli, A., Cortes, J., Bullo, F.: Visibility-based multi-agent deployment in orthogonal environments, pp. 3426–3431 (2007)

14. Hsiang, T.-R., Arkin, E.M., Bender, M.A., Fekete, S.P., Mitchell, J.S.B.: Algorithms for rapidly dispersing robot swarms in unknown environments. In: Boissonnat, J.-D., Burdick, J., Goldberg, K., Hutchinson, S. (eds.) Algorithmic Foundations of Robotics V. STAR, vol. 7, pp. 77–93. Springer, Heidelberg (2004). https://doi.org/10.1007/978-3-540-45058-0_6
15. Di Luna, G.A., Flocchini, P., Gan Chaudhuri, S., Santoro, N., Viglietta, G.: Robots with lights: overcoming obstructed visibility without colliding. In: Felber, P., Garg, V. (eds.) SSS 2014. LNCS, vol. 8756, pp. 150–164. Springer, Cham (2014). https://doi.org/10.1007/978-3-319-11764-5_11
16. Di Luna, G.A., Flocchini, P., Gan Chauduri, S., Poloni, F., Santoro, N., Viglietta, G.: Mutual visibility by luminous robots without collisions. Inf. Comput. **254**, 392–418 (2017)
17. Sharma, G., Busch, C., Mukhopadhyay, S.: Bounds on mutual visibility algorithms (2015)
18. Sharma, G., Busch, C., Mukhopadhyay, S.: Brief announcement: complete visibility for oblivious robots in linear time. In: 29th ACM Symposium on Parallelism in Algorithms and Architectures (SPAA), pp. 325–327 (2017)
19. Sugihara, K., Suzuki, I.: Distributed algorithms for formation of geometric patterns with many mobile robots. J. Robotics Syst. **13**, 127–139 (1996)
20. Suzuki, I., Yamashita, M.: Distributed anonymous mobile robots. In: Proceedings of $3^{rd}$ International Colloquium on Structural Information and Communication Complexity (SIROCCO), pp. 313–330 (1996)
21. Suzuki, I., Yamashita, M.: Distributed anonymous mobile robots: formation of geometric patterns. SIAM J. Comput. **28**(4), 1347–1363 (1999)

# Moving and Computing Models: Agents

Shantanu Das[1(✉)] and Nicola Santoro[2(✉)]

[1] Aix-Marseille University, CNRS, LIS, Marseille, France
shantanu.das@lis-lab.fr
[2] Carleton University, Ottawa, Canada
santoro@scs.carleton.ca

**Abstract.** This chapter introduces and discusses the existing computational models employed in the literature for studying the feasibility and complexity of computations by *mobile agents*: computational mobile entities that operate and move in discrete spaces, modeled as graphs.

While almost all models share some fundamental features, making basic common assumptions, their fundamental differences depend on the assumptions made on the capabilities of the agents, in particular on the means of interaction with the environment and of inter-agent communication. Clearly, there are many variations of the models, depending on the assumed level of synchrony, anonymity, persistent memory, and topological knowledge. This Chapter aims to provide an overview of these models and assumptions.

**Keywords:** Mobile agents · Graph · Communication
Coordination · Synchronization · Whiteboards · Tokens
Face-to-Face · Wireless · Memory

## 1 Introduction

Distributed computing by mobile computational entities operating in a discrete space, i.e. a graph, is a research field with a long research history, antecedent the one of mobile entities in continuous spaces.

Its original inspiration derives from the research on *mobile agents* (MA), an area extensively studied for a long time by investigators especially in the fields of Artificial Intelligence and of Software Engineering. Indeed, MA offer a simple and natural way to describe systems where mobility is inherent, and an explicit and direct way to model the entities of those systems, such as mobile code, software agents, viruses, web crawlers, etc. Hence their success as a programming and software design paradigm for a variety of networked systems (e.g., [13,57,61,69]).

From a *computational* point of view, the settings of autonomous mobile entities operating in graphs are an immediate and natural extension of the traditional message-passing network settings studied in distributed computing, as first observed by David Wall [68] in his pioneering vision of *messages* as the *active* agents of the distributed computation.


P. Flocchini et al. (Eds.): Distributed Computing by Mobile Entities, LNCS 11340, pp. 15–34, 2019.
https://doi.org/10.1007/978-3-030-11072-7_2

This setting can be described as a collection of autonomous mobile computational entities, called *agents*, operating in a graph $G$, called *netscape*. The agents have limited computing capabilities and private storage, can move from node to neighboring node in the graph, and perform computations at each node, according to a predefined set of behavioral rules called *protocol*, the same for all agents. They may be *synchronous*, i.e acting simultaneously at each step, or they may be *asynchronous*, in the sense that every action they perform (computing, moving, etc.) takes a finite but otherwise unpredictable amount of time and there is no synchronization between the actions of different agents. The communication and interaction between agents can be explicit and direct (e.g., *wireless*) although possibly limited in range (e.g., *face-to-face*); explicit but indirect, through shared memory areas provided by each node (i.e., *whiteboard*); or implicit, through the positioning of identical pebbles (e.g., *tokens*) that can be carried, dropped and picked up by the agents.

The algorithmic focus has been on how to develop efficient protocols that will allow a team of such identical simple agents to cooperatively perform (possibly complex) tasks. Examples of basic tasks are `Search`, `Traversal`, `Exploration`, `Rendezvous`, `Election`. The coordination of the agents necessary to perform these tasks is not necessarily simple or easy to achieve. In fact, the computational problems related to these operations are definitely non trivial, and a great deal of theoretical research is devoted to the study of conditions for the solvability of these problems and to the discovery of efficient algorithmic solutions; e.g., see [1,2,7–9,31,36,37,41,43].

The main research concern has been on determining what tasks can be performed by such entities, under what conditions, and at what cost. In particular, a central question is to determine what minimal hypotheses allow a given problem to be solved.

Different computational models exist, each based on different assumptions about the capabilities of the agents, mainly with respect to communication and synchronization. Aim of this Chapter is to provide an overview of these models. We will first start with the fundamental features, that is, the basic assumptions common to (almost) all models.

## 2 Fundamental Features

The setting is a simple, finite, connected undirected[1] graph $G = (V, E)$, where $V$ is the set of nodes (or sites) and $E \subseteq V \times V$ is the set of edges (or links).

The nodes of $G$ may or may not have distinguished names or labels; For generality, we assume the nodes to be *anonymous* (i.e. without labels) unless otherwise specified. The incident links at any node are however locally labeled, providing a total order on the set of incident links. Let $V(u) = \{v \in V : (u, v) \in E\}$ denote the set of neighbours of $u$; $E(u)$ denote the set of edges incident to node $u \in V$; and let $\lambda_u : E(u) \rightarrow \mathcal{L}$ be an injective function that associates to

[1] Although directed graphs may be more natural for certain applications, the problems of agents moving in directed graphs are largely unexplored.

each incident edge a distinct label, sometimes called *port number*, from a set of labels $\mathcal{L}$. Note that for each edge $e = (u, v)$ there are two associated labels, $\lambda_u(e)$ and $\lambda_v(e)$, which are possibly different. The set $\lambda = \{\lambda_u : u \in V\}$ constitutes the labeling of $G$, and by $(G, \lambda)$ we shall denote the corresponding edge-labeled graph.

The edge-labeled graph $(G, \lambda)$, also called *graph world* or *netscape*, defines the discrete spacial universe under consideration.

Operating in $(G, \lambda)$, is a finite set $\mathcal{A} = \{a_1, ..., a_k\}$ of autonomous mobile computational entities called *agents*. The agents have computing capabilities and bounded storage, execute the same protocol, and can move from node to neighbouring node in $G$; the initial location of an agent is usually called *home-base*. Each agent has a personal persistent memory (bounded number of bits) which defines the *state* of the agent (c.f. Sect. 4.2).

The *lifetime* of an agent $a \in \mathcal{A}$ is a sequence of *activity stages*, possibly separated by periods of inactivity.

At the beginning of an activity stage, an agent $a \in \mathcal{A}$ is at a node $u \in V$; it observes its surrounding and interacts with it. The nature of the observation and interaction differs depending on the specific model, and will be discussed later; in any case, the agent can see the outgoing label $\lambda_u(u, v)$ of each edge $(u, v) \in E(u)$ incident to $u$.

The agent then executes a protocol (the same for all agents) to determine what to do; in particular whether or not it will move and, if so, to which neighboring node $v \in V(u)$ (by specifying $\lambda_u(u, v)$). If the decision is not to move, described as a move to a *null* port, the current activity stage terminates, and the agent may become inactive. If the decision is to move (e.g., to $v \in V(u)$) $a$ performs the move (e.g., the system transports $a$ to the node at the other end of the specified edge). In general, upon arriving to $v$, the agent has available the label $\lambda_v(v, u)$ of the edge from which it arrived; this allows the agent to backtrack if needed. Once the move is completed, the current activity stage terminates, and the agent may become inactive.

When the next activity stage starts, the agent is reactivated in the same state in which it terminated the previous stage; in other words, the personal memory of the agent is persistent from stage to stage.

In general it is assumed that the agents have no prior knowledge of the graph $G$ as well as the parameters $|V| = n$, $|E| = m$, and $k$. In some cases, the agents may have partial knowledge of the network (c.f. Sect. 4.4).

There are several cost measures used to express the complexity of a problem and the efficiency of a solution protocol. The main ones are (i) the *team size*: the number of agents employed to solve the problem; and (ii) the number of *moves* made by the agents in total to solve the problem. For synchronous environments (i.e., under the assumption of a synchronous scheduler, discussed in Sect. 4.1), another important one is *time*: the amount of time elapsed until the problem is solved. Other important measures are those related to memory and to the communication and interaction mechanisms, which will be discussed next.

# 3 Communication and Interaction

The communication and interaction between mobile agents, referred to as *coordination* in the MA community [14], is perhaps the most crucial element for computing in such distributed environments.

There are several models of communication and interaction between mobile agents that have been studied in the distributed computing literature. They can be classified depending on the temporal and spatial requirements imposed on the sender (the "writer") and the receiver (the "reader") of the information, and on whether or not the communication is explicit.

Consider an agent $a \in \mathcal{A}$ communicating some information $I$ at node $v \in V$ at time $t$. With an abuse of the MA terminology, that communication is said to be *temporally coupled* ($T$) if it can only be received at time $t$, and *spatially coupled* ($S$) if it can only be received at node $v$. Hence, there are four possible categories: $ST$, $S\bar{T}$, $\bar{S}T$, $\bar{S}\bar{T}$, where $\bar{X}$ indicates absence of requirement $X \in \{S, T\}$. The communication is said to be *explicit* if the content of the communication is precisely $I$, *implicit* otherwise.

The main models of communication and interaction studied in the literature can be categorized using these four classes; see Fig. 1, where the darker boxes indicate models in which communication is implicit.

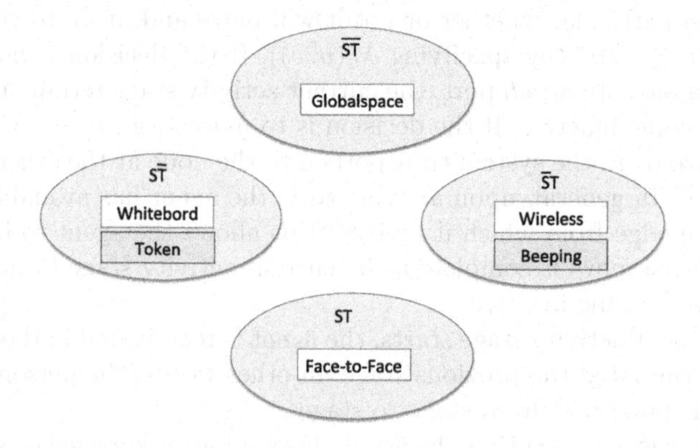


**Fig. 1.** Communication model classes.

## 3.1 $\bar{S}\bar{T}$: $\mathcal{G}$lobalspace

A model where communication is fully uncoupled both spatially and temporally is the powerful $\mathcal{G}$LOBALSPACE model. In this model, there is a global shared memory area, called *globalspace*, with concurrent-read and exclusive-write, that every agent can access from any node; reading the globalspace is an unlocked operation, while writing is in fair mutual exclusion. In MA, this model corresponds to the *associative blackboard* of Linda-like models [14]. The interaction between agents is only through the globalspace; in particular, agents are not visible to each other even when at the same node.

Summarizing, in the $\mathcal{G}$LOBALSPACE model, each agent is capable of explicit communication with all other agents, regardless of their location and without need of synchronization. This overly powerful communication model can be used to show impossibility results, i.e. to prove that certain distributed tasks are impossible even under full exchange of information.

### 3.2 $S\bar{T}$: $\mathcal{W}$hiteboard

A spatially coupled (but temporally uncoupled) explicit form of interaction and communication is offered by the $\mathcal{W}$HITEBOARD model. In this model, each node $v \in V$ of the network, also called *host*, provides a local shared memory area, called *whiteboard*, where any agent visiting $v$ can read and write (and erase) information; access to a whiteboard is restricted by fair mutual exclusion so that at most one agent can access the whiteboard of a node at the same time, and any requesting agent will be granted access within finite time.

In this model, the observation and interaction by agents is only through the whiteboards; in particular, agents are not visible to each other even when at the same node. Starting an activity stage, agent $a \in \mathcal{A}$ at node $v \in V$ asks for access to the whiteboard of node $v$. On obtaining access, it reads the contents of the whiteboard; based on this information (and its own state), it performs computations according to the protocol (the same for all agents), possibly updates the content of the whiteboard (by erasing and writing), and determines its destination (possibly the null port). Finally the agent relinquishes access to the whiteboard and performs its (possibly null) move.

If the agents are *anonymous*, the author(s) of the information written on the whiteboard might not be identifiable. In general, no limitations are assumed on the size of the whiteboards; however, reducing the required size for whiteboards is one of the optimization issues for the design of efficient algorithms in this model.

This model has been extensively used in the literature and solutions developed for a variety of problems (e.g., see [9,18,31,32,43,67]).

Notice that, under the $\mathcal{W}$HITEBOARD model, the problems of `Election` and `Rendezvous` are *computationally equivalent*; that is, any solution protocol for one can be easily modified to solve the other. If a leader is elected, the leader can easily make all the other agents gather in a node of its choice, say node $v$, simply by writing in the whiteboard of each node a sequence of edge-labels corresponding to a path to the node $v$. Conversely, if the agents gather in a node, a leader can easily be elected by exploiting the mutual-exclusion access to the whiteboard of that node; e.g., the first agent to access it becomes the leader. Hence, the answer to any computational question for either problem is valid for both.

It has also been shown [19,33] that a system of mobile agents in a graph $(G, \lambda)$ under the $\mathcal{W}$HITEBOARD model can solve the same set of problems as a traditional message-passing distributed network having the same underlying topology as the graph $(G, \lambda)$.

### 3.3 $S\bar{T}$ (Implicit): $\mathcal{T}$okens

The $\mathcal{T}$OKEN model is one of the oldest ones, introduced for graph explorations and rendezvous search [11,12]. Although spatially coupled and temporally uncoupled like $\mathcal{W}$HITEBOARD, in this model communication is *implicit.* It is achieved by means of identical *tokens* (sometimes called *pebbles*) that can be held by the agents, placed on nodes, and (in the case of *removable* tokens) picked-up from nodes.

In this model, the observation and interaction by agents is only through the tokens. In particular, agents are not visible to each other even when at the same node, while tokens placed on a node are visible to the agents at that node, once they are given access. Initially, the tokens are held by (some of) the agents and/or placed at (some of) the nodes; it is usually assumed that initially an agent has only a small constant number of tokens (typically one).

Starting an activity stage, agent $a \in \mathcal{A}$ at node $v \in V$ asks for access to the tokens at node $v$; access is granted in fair mutual exclusion. On obtaining access, it counts the number of tokens in $v$. Based on this information (and its own state), it performs computations according to the protocol (the same for all agents) and decides whether (1) to place a subset of its own tokens at the node, (2) to pick up a subset of the tokens from the node (in the case of removable tokens), or (3) do nothing; it then determines its destination (possibly the null port). Finally the agent relinquishes access to the token area and performs its (possibly null) move. This model has been used e.g. in [4,12,42,49,62].

A more restrictive model is that of $\mathcal{U}$UNMOVABLE $\mathcal{T}$OKEN in which Action (2) is not allowed: once a token is placed at a node, it cannot be picked up (and thus moved) by any agent (e.g. [17]).

A more powerful model is the $\mathcal{E}$NHANCED $\mathcal{T}$OKEN model introduced in [42]. This model allows the agent to place tokens on a node to mark specific ports (i.e. specific incident edges) of the node; an agent having access to the tokens at node $v$ is then able to see how many tokens are placed in correspondence of each port of $v$. Note that $h$ tokens in the $\mathcal{E}$NHANCED $\mathcal{T}$OKEN model correspond to at most $h\Delta$ tokens in the $\mathcal{T}$OKEN model, where $\Delta$ is the maximum degree in $G$.

There are some obvious interesting computational relationships between the $\mathcal{T}$OKEN and $\mathcal{W}$HITEBOARD models. In fact, any protocol which uses at most $\tau$ tokens in $\mathcal{T}$OKEN, can be directly implemented in $\mathcal{W}$HITEBOARD using whiteboards of size at most $\lceil \log \tau \rceil$ bits. This measure $\tau$, called the *token load*, is clearly important in that it determines the usability of token-protocols in the whiteboard model, and provides a simple mechanism to transform complexity results from the token setting to the whiteboard one. Importantly, the transformation preserves also other costs (such as total number of moves by agents and time). Note that an automatic transformation exists also in the other direction:

any solution that uses whiteboards of size $s$ bits can be implemented with a token load $\tau \leq 2^s n$.

### 3.4 $\bar{S}T$: $\mathcal{W}$ireless

An explicit form of interaction and communication that is spatially uncoupled but temporally coupled is offered by the $\mathcal{W}$IRELESS model.

In this model, agents are not visible to each other even when at the same node. Each agent is capable of explicit and direct communication with all other agents. More precisely, every agent is capable of *transmitting* information; whenever an agent transmits, all other agents active at that time receive the information regardless of their location in the graph $G$. An agent, not active at the time of the transmission, will not receive the information.

To guarantee that an agent is active when a transmission takes place, temporal coupling (i.e., synchronization) is required. This model is usually associated with the assumption of a *fully synchronous* scheduler (discussed later), so that all agents are active whenever a transmission takes place. In such a case, each agent has possibly full information about the actions taken by other agents during the execution of a distributed algorithm. This abstraction helps to separate the problem of communication from the problem of solving the actual tasks such as `Searching`, `Gathering`, `Scattering` or `Pattern Formation`.

In the model it is assumed that more than one transmission can take place at the same time, and that each active agent will receive all the messages. This model has been used e.g. in [30, 54].

A weaker form of this model is when reception of a transmission by an active agent is limited to those who are not transmitting at that time. In other words, an active agent can either transmit or listen but not both. An even weaker form of the model considers the effect of *collisions* (i.e. multiple simultaneous transmissions) to all active agents (not only the transmitting ones); different sub-models are considered depending on the assumed outcome on reception after a collision (e.g., no information, the information of the closest transmitter, information of an arbitrary transmitter, etc.).

A different type of restriction is the one described by the $l - \mathcal{H}$OP $\mathcal{W}$IRELESS model, $1 \leq l \leq diam(G)$, in which the *transmission range* of the agents is limited to the $l$-neighbourhood of the node they are currently in, where $diam(G)$ is the diameter of $G$. Clearly, when $l \geq diam(G)$, the two models coincide. Obviously weaker forms of this model can be defined by considering *collisions* as we did for the general model.

### 3.5 $\bar{S}T$ (Implicit): $\mathcal{B}$eeping

The most recent model of inter-agent communication and coordination is the $\mathcal{B}$EEPING model, introduced in [22]; like in $\mathcal{W}$IRELESS, the communication is spatially uncoupled and temporally coupled; however, unlike in $\mathcal{W}$IRELESS, communication is *implicit* and thus severely limited.

In this model, agents are not visible to each other even when at the same node. Every agent is capable of transmitting a simple signal, called *beep*. Whenever a beep is transmitted, every agent active at that time hears a beep regardless of the current location, unless it is also transmitting a beep; that is, an active agent can either transmit or listen but not both. Agents that are not active at the time of a transmission will not hear it. This model is usually associated with the assumption of a *fully synchronous* scheduler (discussed in Sect. 4.1), so that all agents are active whenever a transmission takes place. This model has been used e.g. in [23,46,53].

Clearly one can define the restricted model $l - \mathcal{H}$OP $\mathcal{B}$EEPING in a way analogous to what done in the $l - \mathcal{H}$OP $\mathcal{W}$IRELESS model.

### 3.6 *ST*: $\mathcal{F}$ace-to-$\mathcal{F}$ace

The simplest model of communication is the direct and explicit exchange of information between agents at the same node at the same time. Both spatially and temporally coupled, this model is called $\mathcal{F}$ACE-TO-$\mathcal{F}$ACE ($\mathcal{F}2\mathcal{F}$).

In this model, when at a node $v$, an agent can see the other agents present at $v$, it can distinguish one from another (even if they are anonymous), and communicate with each one of them. Communication happens in a pair-wise manner, with each agent in each pair reading the content of the memory of the other agent.

This model is usually associated with the assumption of a *fully synchronous* scheduler (discussed in Sect. 4.1), so that the activity stages of all agents take place simultaneously. Starting an activity stage, each agent at node $v \in V$ communicates with all other agents at node $v$. Each agent then executes the protocol (the same for all agents) to determine, based on the information received, from which port (possibly null) to leave $v$. Finally, they all execute their (possibly null) move. This model has been used e.g. in [37,40].

Computationally, $\mathcal{F}$ACE-TO-$\mathcal{F}$ACE is a very weak model for communication between agents. Being the model both spatially and temporally coupled, to communicate two agents need to be at the same node at the same time; this means that the two agents will never be able to direcly exchange information unless they rendezvous at a node. Since `Rendezvous` is in itself a difficult and sometime unsolvable problem, many tasks become impossible in this model of communication.

## 4 Fundamental Dimensions

### 4.1 Time

Although *time* is easily defined in terms of an external observer, this notion might generally not be available to the agents; even if each agent were to be endowed with a local clock, the clocks might not sign the same value, nor run at the same rate, nor be otherwise synchronized. The *duration* of an activity stage, i.e. the time spent by a agent observing and interacting with its surrounding, computing, and moving, might be different for different stages, and different for different agents. The *delay* between two successive stages of the same agent (i.e., the duration of inactivity), might not be the same, and could be different for different agents.

Assumptions made on the last two aspects of the temporal dimension, the activation schedule of the agents and the duration of the operations within their stages, define specific instances, or models, of the computational universe in which the agents operate. The two most important models, intensively and extensively studied in the literature, are $\mathcal{A}$SYNC and $\mathcal{S}$SYNC.

In the *semi-synchronous* (or simply *synchronous*) model ($\mathcal{S}$SYNC), time is logically divided into global rounds; in each round, some agents are activated and perform their activities (interact with the environment, compute and move), ending them by the end of the round; the other agents are inactive for the entire round. The choice of which agents are activated in a given round is made by an adversarial *activation scheduler* (or daemon), which is however assumed to be *fair*: for every agent $a \in \mathcal{A}$ and time $t$, there exists a time $t' \geq t$ at which $a$ is activated; that is, every agent is activated infinitely often.

An extreme case of activation scheduler in $\mathcal{S}$SYNC defines the *fully-synchronous* ($\mathcal{F}$SYNC) model: *all* the agents are activated in every round. Another extreme case of daemon is the so-called *centralized* activation scheduler, which defines the *sequential* model ($\mathcal{S}$EQUENTIAL) where *only one* agent is active in each round.

In the *asynchronous* ($\mathcal{A}$SYNC) model, there is no common notion of time, each agent is activated asynchronously and independently from the other agents. Furthermore, the duration of each operation is finite but otherwise unpredictable. As a result, computations can be made based on totally obsolete observations, taken arbitrarily far in the past. Also in $\mathcal{A}$SYNC the activation scheduler is *fair*: for every agent $a$ and time $t$, there exists a time $t' \geq t$ at which $r$ is activated; that is, every agent is activated infinitely often.

Although the activation scheduler is fair both in $\mathcal{A}$SYNC and $\mathcal{S}$SYNC, between two successive activations of the same agent $a$, the number of activations of the other agents is possibly unbounded. A stronger requirement rs that offered by a *p-bounded* (or *p-fair*) scheduler: For every $a \in \mathcal{A}$, between two successive activations of $a$, every other agent has been activated at most $p$ times. Note that, in $\mathcal{F}$SYNC, the scheduler is 1-bounded by definition.

## 4.2 Memory

The computational capabilities of the agents and the solvability of tasks depend on the amount of *persistent memory* (bits of information) available to the agents. By persistent we mean that the information the agent carries when moving from one node to another is preserved during the inactivity periods (i.e., it is the same at the end of a stage and at the beginning of the next). This does not include the working memory used by the agent when performing computations at a node for which there are, in general, no restrictions. An optimization issue is minimizing the amount of persistent memory used by the agent.

### 4.2.1 Finite-State

A *finite-state* mobile agent has $O(1)$ bits of persistent memory i.e. the amount of memory is a constant independent of the size of the problem (the parameters $n$, $m$, $k$ and $\Delta$); e.g., see [10,17,29,62]. There are several limitations with finite-state agents. Note that such an agent cannot identify a node, even in a labelled graph (since any unique identifiers assigned to nodes must have $O(\log n)$ bits). Moreover, it is known that any such agent (or any constant-size team of such agents) cannot explore an arbitrary graph of degree 3 or more [55,66]. Finite-state agents can navigate only in special topologies such as ring networks, trees, complete graphs and other specific topologies. A finite-state agent equipped with a few tokens can explore a larger set of graphs including e.g. unoriented grids and tori.

A special case of the finite-state agent is the *oblivious* agent which has zero persistent memory c.f. Chaps. 8 and 9.

### 4.2.2 Bounded

A *bounded* memory agent has an amount of persistent memory either logarithmic (i.e. $O(\log n)$) or polynomial (i.e. $O(f(n))$ for some polynomial function $f$) in the size of the graph. The former is called *log-Bounded*, while the latter is called *poly-Bounded*. Note that any agent having $\Omega(\log n)$ memory can explore all graphs of size $n$ with termination, even if the nodes of the graph are unlabelled and the agents have no means of marking the nodes. Thus many distributed tasks can be performed by log-bounded agents, given sufficient time. However, there is often a trade-off between time complexity and the memory complexity of a problem, so poly-bounded agents can often perform a task much faster than log-bounded agents. In particular a poly-bounded agent can store complete information of the graph $G$ in its memory. In particular, this allows such an agent to solve the problem of `Map-Construction`—to explore and build a map of an unknown graph. Most results concerning mobile agents assume poly-bounded agents, unless otherwise specified.

### 4.2.3 Unbounded

The most powerful agent in terms of memory is the *Unbounded* memory agent, which does not have any limits of the amount of information it can store in its memory. Such agents are mainly used for proving impossibility results, i.e to show that certain tasks can not be solved irrespective of the amount of memory available to an agent. Unbounded memory agents are also used for solving distributed tasks in infinite graphs, see e.g. [28].

## 4.3 Visibility

### 4.3.1 Local Visibility

Mobile agents are generally assumed to have only *local visibility* of the graph structure; that is, an agent is only aware of the port numbers, and hence the degree, of the node where it currently is. Depending on the amount of persistent memory, the agent can keep some knowledge of the nodes and edges that it has visited so far; clearly this knowledge may not be sufficient for the agent to build a coherent image of the subgraph that has visited.

### 4.3.2 $q$-hop Visibility

In some cases, the mobile agent may have a larger visibility of the graph structure allowing it to see the subgraph containing all nodes at a distance of at most $k$ from the current node, for some small constant integer $q \geq 1$. This model is called $q$-hop visibility and studied e.g. in [10,29,59]. It was recently shown [20] agents with 1-hop visibility are more powerful than agents with local visibility for the tasks of `Exploration` and `Map Construction` in unlabelled graphs. Note that the visibility range $q$ here, is a constant independent of the size or diameter of the graph.

At the other extreme, when the visibility range is as large as the diameter of the graph, the agents are said to have *global visibility.* This feature has been assumed, for example in the "*Robots in Graphs*" model studied in Chaps. 8 and 9.

## 4.4 Knowledge

Some tasks require the mobile agents to have some prior knowledge about its environment. For example, it is well known that visiting all the nodes of an unknown and unlabelled graph is impossible without some knowledge about the size of the graph, under the face-to-face model of communication. We list below some types of knowledge that may be provided to the agent as input in order to facilitate its task.

- $n$ **(number of nodes of** $G$**):** Each agent is provided as input the value $n$ of the number of nodes in the graph. Sometimes, the exact value of $N$ may not be needed and instead the agent may be provided with an upper-bound $N$ on the number of nodes in the graph.
- $m$ **(number of edges of** $G$**):** Each agent is provided as input the value of $M$, the number of edges in the graph. This is sometimes equivalent to knowing the number of nodes of the graph (e.g. ring or tree). In other cases, this information can be used to deduce the value of $n$ during exploration. Note that the value of $m+1$ is an upper bound on the value of $n$ for any connected graph.
- $k$ **(number of agents):** Each agent is provided as input the value $n$, the total number of agents. This knowledge is particularly useful for tasks such as gathering all agents at a node, or scattering the agents uniformly in a given

space. When the homebases of agents are marked, the knowledge of $k$ can be used to deduce the size $n$ of the graph. However, when the agents are not allowed to mark their homebases (e.g. under the $\mathcal{F}2\mathcal{F}$ model), the knowledge of $k$ is less powerful than the knowledge of $n$.
- **Topology of $G$:** In some cases the agents may be provided with information about the topology of $G$. Using this information, the agents may execute distinct algorithms for different families of graphs, e.g. trees, grids, rings, hypercubes etc.
- **Sense of direction:** In many cases the agents do not need information about the complete topology of the graph; instead it is sufficient for the agents to have a sense of direction allowing it to navigate in the graph. As an example, in an oriented grid graph where the outgoing edges at each node are marked consistently as North, East, South and West, an agent following any given sequence of edges can determine whether it will arrive at an already visited node or a new one. For an arbitrary graph, this concept can be generalized by labelling the edges appropriately and then providing the agents with a "Sense of direction" encoding that maps any sequence of edge labels to a unique integer such that the following property holds: The encoding for any two sequences of edge labels are equal if and only if the two corresponding paths from the homebase of the agent lead to the same node. When the agents are provided with such an encoding function, we say that the agents have *Sense of Direction* (SoD) information [50,52].
- **Map of the graph:** We say that an agent is provided with a *map* of the graph, if the input to the agent is a copy of the underlying graph, each edge is labelled with the port-numberings on its end-points, and where the node corresponding to the current location of the agent is specially marked.
- **Full Knowledge:** We say that the agents have *full knowledge* if each agent is provided with a map of the graph as well as the starting location of all the agents present in the graph. With this knowledge, each agents can independently compute the optimal solution strategy (if it exists) to any given distributed task, without the need to move from its starting location. This corresponds to executing a centralized algorithm for precomputing the strategy of each agent.

In general, an algorithm to solve a distributed task must correctly solve the problem in any graph of any size. Such an algorithm is said to work for *arbitrary topology.* On the other hand, if an algorithm works for a specific family of graphs, then we say that the algorithms works for *fixed topology* and for such an algorithm, the knowledge of topology is the usual assumption.

### 4.4.1 Quantitative Model for Knowledge

Instead of providing knowledge as input to the mobile agent, another possibility is to provide knowledge in an online fashion as and when needed by the agent. Several research investigation on mobile agents have considered the so-called *Oracle model* (e.g., see [56,64]). Under this model, a mobile agent can ask questions to an "oracle" which has full knowledge about the environment as well as the current locations of the the agents. Typically the questions must have a "YES -or- NO" answer. The optimization issue is to minimize the amount of information (in bits) needed by any agent during the execution of an algorithm, assuming a one-bit answer to each question asked by the agent.

# 5 Security and Fault Tolerance

In systems supporting mobile agents, *security* is the most pressing concern and possibly the most difficult to address. Among the severe security threats faced in these systems, two are particularly troublesome. The first is the threat to the network caused by harmful mobile agents (e.g., viruses); this problem is particularly acute in unregulated noncooperative settings such as the Internet. The second is the threat to the agents posed by harmful network components, hosts or links; this problem exists also in environments with regulated access and where agents cooperate toward common goals (e.g., sharing of resources or distribution of a computation on the grid); in fact, a local (hardware or software) failure might render a component harmful. In the context of distributed computing, security problems of both types have been considered and studied.

## 5.1 Dangerous Mobile Agents

Dangerous agents are typically extraneous to the system, e.g. intruders or viruses entering the system from outside. On the other hand, even internal agents, which are by definition cooperative, might create problems due to failures.

### 5.1.1 External Agents: Intruders and Viruses

Consider the presence of an *external* harmful mobile agent in the system. A type of harmful agents that has been well investigated is that of a *mobile virus* that infects any visited network site. In this case, the crucial task is clearly to decontaminate the infected network; this task, called `Network Decontamination` is to be carried out by a team of anti-viral system agents (the *cleaners*), able to decontaminate visited sites, avoiding any recontamination of decontaminated areas. This problem is equivalent to the one of `Intruder Capture` [7], and is related to the classical problem known as `Graph Search` (e.g., see [63,65]), which is in turn closely related to standard graph parameters. A detailed description of this problem and the existing solutions can be found in Chap. 19.

The intruders considered in the above results do not cause any direct harm or affect in any way the functioning of the mobile agents. A more powerful adversarial entity, simply called the *malicious agent*, was considered recently [34,35] in the context of the `Gathering` problem. The malicious agent prevents access to the node that it occupies, any mobile agent that attempts to move to the same node receives a failure notification. Further, the malicious agent is assumed to move arbitrarily fast and has full knowledge of the graph, the positions of all the agents and even the algorithm followed by the mobile agents.

Thus, any single mobile agent can be blocked forever by the malicious agent. However, a team of multiple mobile agents may be able to defeat the malicious agent by surrounding it, and thus bypass it. The connectivity of the graph is important in this context, since the malicious agent can block any articulation point in a graph, potentially dividing the graph into disconnected components.

#### 5.1.2 Internal Agents: Crashes and Byzantine Faults

System agents, also called *internal* agents, are typically assumed to be *correct*, that is to operate without faults. More realistically, failures may occur, leading some internal agents to behave in a possibly harmful way with respect to the execution of the common protocol.

The agents' faults that have been investigated fall in two categories: *crash* faults (i.e. a faulty agent stops executing its cycle forever) and *Byzantine* faults (i.e. a faulty agent may exhibit arbitrary behavior and movement). Of course, the Byzantine fault model encompasses the crash fault model, and is thus harder to address. When agents have unique identities, *weak* byzantine agents may not lie about their identities, while *strong* Byzantine agents may assume fake identities.

Agent crashes in graphs were studied in [24,26,33], and Byzantine agents have been studied in [25,40].

### 5.2 Dangerous Graphs

The term *dangerous graph* is usually used in reference to networks where there are (possibly several) harmful components, called *black holes* and *black edges*. These components, as well as the related harmful presence of a *black virus*, are described next. The problems and solutions related to this topic are treated in Chap. 18.

#### 5.2.1 Black Holes and Edges

A severe harmful component is a *black hole*, a node that disposes of any incoming agent, leaving no observable trace of its destruction (see e.g. [27,43–45]). The task, called `BlackHoleSearch` is for a team of system agents to unambiguously determine and report the location of the black hole within finite time. The task is clearly dangerous for the agents, as any agent entering the black hole is destroyed without trace. The research concern is to determine under what conditions and at what cost mobile agents can successfully accomplish this task.

Similarly harmful components are the *black edges*, links that dispose of any agent traversing them, leaving no observable trace of the destruction [18].

#### 5.2.2 Black Virus

The fact that a node is black hole can be seen as of the presence at that node of a harmful process disposing of the incoming agents; such a process is static, i.e. it does not move from the node where it resides. A harmful process that has some of the destructive aspects of a black holes and some of the mobility of an intruder is the *black virus* [15]. Like a black hole, a black virus disposes of the system agents arriving at the note; when such an event occurs, the process migrates to all neighboring nodes installing itself there (unless a system agent is already there). The migration consists of the original black virus sending clones of itself along all incident edges and then becoming inactive. The task, called `BlackVirusDecontamination` is for a team of system agents to remove all black viruses from the network.

### 5.2.3 Other Dangerous Nodes

While black holes are extremely dangerous nodes that always destroy any agent that enters it, there are less malicious but still faulty nodes. A *gray hole* is a dangerous node that may or may not destroy an agent entering this node [3,6]. Such a node acts sometimes as a black hole and sometimes acts as a normal node, which makes it difficult to locate it. The gray hole fault may also change the contents of the whiteboard of the node after an agent has successfully left the node. Another less dangerous fault is the *Repairable hole* [21] which acts like a black hole until an agent visits this node. The first agent visiting such a node is destroyed after repairing the fault. On any subsequent visits the node acts like a fault-free node. Naturally, repairable holes are easier to locate than black holes.

# 6 Agents in Other Discrete Spaces

## 6.1 Agents in Infinite Graphs

Most results on mobile agents assume that the agents are moving in a finite graph.

There are also some results on infinite graphs [28], in particular infinite lines and infinite grids of finite dimensions [5]. Some tasks on infinite graphs, such as `Exploration` and `Gathering`, require the agents to explore a finite region around their homebases. For instance, to solve the rendezvous of two agents which are a distance $D$ apart, each agent may need to visit nodes at most distance $D$ from its starting location.

A large body of research on agents in infinite graphs is constituted by the investigations on the geometric $\mathcal{A}$MOEBOT model, discussed in Chap. 22: the entities (called particles) are finite-state machines communicating in the 1$-\mathcal{H}$OP $\mathcal{W}$IRELESS model, operating and moving in the infinite graph of the hexagonal tasselation of the Euclidean plane.

For solving tasks on infinite graphs, the complexity is clearly measured in terms of the finite parameters of the problems, e.g. maximum degree of the graph, distance between agents, number of agents etc.

## 6.2 Agents in Dynamic Graphs

Mobile agents have been recently studied also when they operate on dynamic graphs; that is, on graphs whose structure varies over time. Various models have been defined to describe dynamic graphs (time-varying graphs, evolving graphs, temporal graphs, etc.) each with a slightly different focus, and each using different terminology. In the following we define the most general one, that encompasses most of the others [16].

A *time-varying graph* (*TVG*) is defined as a quintuple $\mathcal{G} = (V, E, \mathcal{T}, \rho, \zeta)$, where $V$ is a finite set of nodes; $E \subseteq V \times V$ is a finite set edges, $\mathcal{T} \subseteq \mathbb{T}$ is the time-span of the graph (or lifetime of the system), and the temporal domain $\mathbb{T}$ is generally assumed to be $\mathbb{N}$ for discrete-time systems or $\mathbb{R}^+$ for continuous-time systems; $\rho : E \times \mathcal{T} \rightarrow \{0, 1\}$ is the edge presence function, which indicates whether a given edge is available at a given time; $\zeta : E \times \mathcal{T} \rightarrow \mathbb{T}$, is the latency function, which indicates the time it takes for an agent to cross a given edge if starting at a given time. The footprint of $\mathcal{G}$ is a static graph composed by the union of all nodes and edges ever appearing during the lifetime $\mathbb{T}$.

Typically, the study of mobile agents in dynamic graphs has considered fully synchronous or semi-synchronous activation schedules; in these cases, time is discrete (i.e., $\mathbb{T} = \mathbb{N}$), the agents operate at discrete rounds, and $\zeta$ is constant (usually $\zeta = 1$). Under these assumptions, it is common to view the time-varying graph as a sequence of static graphs $G_1, G_2, \ldots$, called *Evolving Graph*, where $G_t$ corresponds to the static snapshot of $\mathcal{G}$ at time $t$ [48,58].

Problems studied in dynamic graphs are, for example, `Exploration` and `Gathering`, for which solutions under particular restrictions on the graph's connectivity and on its footprint are made. Solutions have been proposed for decentralized settings (e.g., see [38,39,51]), as well for centralized ones (e.g., see [47,60]). Chapter 20 is devoted to this topic and contains details on the model, as well as on the existing results.

## References


1. Albers, S., Henzinger, M.: Exploring unknown environments. In: Proceedings of 29th ACM Symposium on Theory of Computing (STOC), pp. 416–425 (1997)
2. Alpern, S., Gal, S.: The Theory of Search Games and Rendezvous. Chapman & Hall, Kluwer (2003)
3. Královič, R., Miklík, S.: Periodic data retrieval problem in rings containing a malicious host. In: Patt-Shamir, B., Ekim, T. (eds.) SIROCCO 2010. LNCS, vol. 6058, pp. 157–167. Springer, Heidelberg (2010). https://doi.org/10.1007/978-3-642-13284-1_13
4. Balamohan, B., Dobrev, S., Flocchini, P., Santoro, N.: Exploring an unknown dangerous graph with a constant number of tokens. Theor. Comput. Sci. **610**, 169–181 (2016)
5. Bampas, E., Czyzowicz, J., Gąsieniec, L., Ilcinkas, D., Labourel, A.: Almost optimal asynchronous rendezvous in infinite multidimensional grids. In: Lynch, N.A., Shvartsman, A.A. (eds.) DISC 2010. LNCS, vol. 6343, pp. 297–311. Springer, Heidelberg (2010). https://doi.org/10.1007/978-3-642-15763-9_28
6. Bampas, E., Leonardos, N., Markou, E., Pagourtzis, A., Petrolia, M.: Improved periodic data retrieval in asynchronous rings with a faulty host. Theor. Comput. Sci. **608**, 231–254 (2015)

7. Barrière, L., Flocchini, P., Fraigniaud, P., Santoro, N.: Capture of an intruder by mobile agents. In: Proceedings of 14th ACM Symposium on Parallel Algorithms and Architectures (SPAA), pp. 200–209 (2002)
8. Barrière, L., Flocchini, P., Fraigniaud, P., Santoro, N.: Can we elect if we cannot compare? In: Proceedings of 15th ACM Symposium on Parallel Algorithms and Architectures (SPAA), pp. 324–332 (2003)
9. Barrière, L., Flocchini, P., Fraigniaud, P., Santoro, N.: Rendezvous and election of mobile agents: impact of sense of direction. Theor. Comput. Syst. **40**(2), 143–162 (2007)
10. Barrière, L., Flocchini, P., Mesa-Barrameda, E., Santoro, N.: Uniform scattering of autonomous mobile robots in a grid. Int. J. Found. Comput. Sci. **22**(3), 679–697 (2011)
11. Baston, V., Gal, S.: Rendezvous search when marks are left at the starting points. Naval Res. Logist. **38**, 469–494 (1991)
12. Bender, M., Fernandez, A., Ron, D., Sahai, A., Vadhan, S.: The power of a pebble: exploring and mapping directed graphs. In: Proceedings of 30th ACM Symposium on Theory of Computing (STOC), pp. 269–287 (1998)
13. Braun, P., Rossak, W.: Mobile Agents. Morgan Kaufmann, Burlington (2005)
14. Cabri, G., Leonardi, L., Zambonelli, F.: Mobile-agent coordination models for internet applications. Computer **33**(2), 82–89 (2000)
15. Cai, J., Flocchini, P., Santoro, N.: Network decontamination from a black virus. Int. J. Netw. Comput. **4**(1), 151–173 (2014)
16. Casteigts, A., Flocchini, P., Quattrociocchi, W., Santoro, N.: Time-varying graphs and dynamic networks. Int. J. Parallel Emergent Distrib. Syst. **27**(5), 387–408 (2012)
17. Chalopin, J., Das, S., Labourel, A., Markou, E.: Tight bounds for black hole search with scattered agents in synchronous rings. Theor. Comput. Sci. **509**, 70–85 (2013)
18. Chalopin, J., Das, S., Santoro, N.: Rendezvous of mobile agents in unknown graphs with faulty links. In: Pelc, A. (ed.) DISC 2007. LNCS, vol. 4731, pp. 108–122. Springer, Heidelberg (2007). https://doi.org/10.1007/978-3-540-75142-7_11
19. Chalopin, J., Godard, E., Métivier, Y., Ossamy, R.: Mobile agent algorithms versus message passing algorithms. In: Shvartsman, M.M.A.A. (ed.) OPODIS 2006. LNCS, vol. 4305, pp. 187–201. Springer, Heidelberg (2006). https://doi.org/10.1007/11945529_14
20. Chalopin, J., Godard, E., Naudin, A.: Anonymous graph exploration with binoculars. In: Moses, Y. (ed.) DISC 2015. LNCS, vol. 9363, pp. 107–122. Springer, Heidelberg (2015). https://doi.org/10.1007/978-3-662-48653-5_8
21. Cooper, C., Klasing, R., Radzik, T.: Locating and repairing faults in a network with mobile agents. Theor. Comput. Sci. **411**(14–15), 1638–1647 (2010)
22. Cornejo, A., Kuhn, F.: Deploying wireless networks with beeps. In: Lynch, N.A., Shvartsman, A.A. (eds.) DISC 2010. LNCS, vol. 6343, pp. 148–162. Springer, Heidelberg (2010). https://doi.org/10.1007/978-3-642-15763-9_15
23. Czumaj, A., Davies, P.: Communicating with beeps. In: Proceedings of 20th International Conference on Principles of Distributed Systems (OPODIS), pp. 1–16 (2016)
24. Czyzowicz, J., Gasieniec, L., Kosowski, A., Kranakis, E., Krizanc, D., Taleb, N.: When patrolmen become corrupted: monitoring a graph using faulty mobile robots. Algorithmica **7916**(3), 925–940 (2017)
25. Czyzowicz, J., et al.: Search on a line by Byzantinerobots. In: Proceedings of 27th International Symposium onAlgorithms and Computation (ISAAC), pp. 27:1–27:12 (2016)

26. Czyzowicz, J., Godon, M., Kranakis, E., Labourel, A., Markou, E.: Exploring graphs with time constraints by unreliable collections of mobile robots. In: Tjoa, A.M., Bellatreche, L., Biffl, S., van Leeuwen, J., Wiedermann, J. (eds.) SOFSEM 2018. LNCS, vol. 10706, pp. 381–395. Springer, Cham (2018). https://doi.org/10.1007/978-3-319-73117-9_27
27. Czyzowicz, J., Kowalski, D., Markou, E., Pelc, A.: Searching for a black hole in synchronous tree networks. Comb. Probab. Comput. **16**, 595–619 (2007)
28. Czyzowicz, J., Pelc, A., Labourel, A.: How to meet asynchronously (almost) everywhere. ACM Trans. Algorithms **8**(4), 37:1–37:14 (2012)
29. Barrameda, E.M., Das, S., Santoro, N.: Deployment of asynchronous robotic sensors in unknown orthogonal environments. In: Fekete, S.P. (ed.) ALGOSENSORS 2008. LNCS, vol. 5389, pp. 125–140. Springer, Heidelberg (2008). https://doi.org/10.1007/978-3-540-92862-1_11
30. Das, S., Dereniowski, D., Karousatou, C.: Collaborative exploration of trees by energy-constrained mobile robots. Theor. Comput. Syst. **62**(5), 1223–1240 (2018)
31. Das, S., Flocchini, P., Kutten, S., Nayak, A., Santoro, N.: Map construction of unknown graphs by multiple agents. Theor. Comput. Sci. **385**(1–3), 34–48 (2007)
32. Das, S., Flocchini, P., Nayak, A., Santoro, N.: Effective elections for anonymous mobile agents. In: Asano, T. (ed.) ISAAC 2006. LNCS, vol. 4288, pp. 732–743. Springer, Heidelberg (2006). https://doi.org/10.1007/11940128_73
33. Das, S., Flocchini, P., Santoro, N., Yamashita, M.: Fault-tolerant simulation of message-passing algorithms by mobile agents. In: Prencipe, G., Zaks, S. (eds.) SIROCCO 2007. LNCS, vol. 4474, pp. 289–303. Springer, Heidelberg (2007). https://doi.org/10.1007/978-3-540-72951-8_23
34. Das, S., Focardi, R., Luccio, F., Markou, E., Squarcina, M.: Gathering of robots in a ring with mobile faults. Theor. Comput. Sci. (2018, in press). https://doi.org/10.1016/j.tcs.2018.05.002
35. Das, S., Luccio, F.L., Markou, E.: Mobile agents rendezvous in spite of a malicious agent. In: Bose, P., Gąsieniec, L.A., Römer, K., Wattenhofer, R. (eds.) ALGOSENSORS 2015. LNCS, vol. 9536, pp. 211–224. Springer, Cham (2015). https://doi.org/10.1007/978-3-319-28472-9_16
36. Deng, X., Papadimitriou, C.H.: Exploring an unknown graph. J. Graph Theor. **32**(3), 265–297 (1999)
37. Dessmark, A., Fraigniaud, P., Pelc, A.: Deterministic rendezvous in graphs. Algorithmica **46**, 69–96 (2006)
38. Di Luna, G., Dobrev, S., Flocchini, P., et al.: Distrib. Comput. (2018, in press). https://doi.org/10.1007/s00446-018-0339-1
39. Di Luna, G.A., Flocchini, P., Pagli, L., Prencipe, G., Santoro, N., Viglietta, G.: Gathering in dynamic rings. In Proocedings of the 24th International Colloquium Structural Information and Communication Complexity (SIROCCO), pp. 339–355 (2017)
40. Dieudonné, Y., Pelc, A., Peleg, D.: Gathering despite mischief. ACM Trans. Algorithms **11**(1), 1–28 (2014)
41. Diks, K., Fraigniaud, P., Kranakis, E., Pelc, A.: Tree exploration with little memory. J. Algorithms **51**(1), 38–64 (2004)
42. Dobrev, S., Flocchini, P., Kralovic, R., Santoro, N.: Exploring an unknown dangerous graph using tokens. Theor. Comput. Sci. **472**, 28–45 (2013)
43. Dobrev, S., Flocchini, P., Prencipe, G., Santoro, N.: Searching for a black hole in arbitrary networks: optimal mobile agents protocols. Distribut. Comput. **19**(1), 1–19 (2006)

44. Dobrev, S., Flocchini, P., Prencipe, G., Santoro, N.: Mobile search for a black hole in an anonymous ring. Algorithmica **48**(1), 67–90 (2007)
45. Dobrev, S., Královič, R., Santoro, N., Shi, W.: Black hole search in asynchronous rings using tokens. In: Calamoneri, T., Finocchi, I., Italiano, G.F. (eds.) CIAC 2006. LNCS, vol. 3998, pp. 139–150. Springer, Heidelberg (2006). https://doi.org/10.1007/11758471_16
46. Dufoulon, F., Burman, J., Beauquier, J.: Beeping a deterministic time-optimal leader election. In: Proceedings of 32nd International Symposium on Distributed-Computing (DISC) (2018)
47. Erlebach, T., Hoffmann, M., Kammer, F.: On temporal graphexploration. In: Proceedings of the 42nd International Colloquium on Automata, Languages, and Programming (ICALP), pp. 444–455 (2015)
48. Ferreira, A.: Building a reference combinatorial model for MANETs. IEEE Netw. **18**(5), 24–29 (2004)
49. Flocchini, P., Ilcinkas, D., Santoro, N.: Ping pong in dangerous graphs: optimal black hole search with pebbles. Algorithmica **62**(3–4), 1006–1033 (2012)
50. Flocchini, P., Mans, B., Santoro, N.: Sense of direction in distributed computing. Theor. Comput. Sci. **291**, 29–53 (2003)
51. Flocchini, P., Mans, B., Santoro, N.: On the exploration of time-varying networks. Theor. Comput. Sci. **469**, 53–68 (2013)
52. Flocchini, P., Roncato, A., Santoro, N.: Backward consistency and sense of direction in advanced distributed systems. SIAM J. Comput. **32**(2), 281–306 (2003)
53. Förster, K.-T., Seidel, J., Wattenhofer, R.: Deterministic leader election in multi-hop beeping networks. In: Kuhn, F. (ed.) DISC 2014. LNCS, vol. 8784, pp. 212–226. Springer, Heidelberg (2014). https://doi.org/10.1007/978-3-662-45174-8_15
54. Fraigniaud, P., Gasieniec, L., Kowalski, D., Pelc, A.: Collective tree exploration. Networks **48**(3), 166–177 (2006)
55. Fraigniaud, P., Ilcinkas, D., Peer, G., Pelc, A., Peleg, D.: Graph exploration by a finite automaton. Theor. Comput. Sci. **345**(2–3), 331–344 (2005)
56. Gorain, B., Pelc, A.: Deterministic graph exploration with advice. ACM Trans. Algorithms **15**(1), 8 (2018, to appear)
57. Gray, R., Kotz, D., Nog, S., Rus, D., Cybenko, G.: Mobile agents: the next generation in distributed computing. In: Proceedings of 2nd AIZU International Symposium on Parallel Algorithms/Architecture Synthesis (PAS) (1997)
58. Harary, F., Gupta, G.: Dynamic graph models. Math. Comp. Model. **25**(7), 79–88 (1997)
59. Hsiang, T.-R., Arkin, E.M., Bender, M.A., Fekete, S.P., Mitchell, J.S.B.: Algorithms for rapidly dispersing robot swarms in unknown environments. In: Boissonnat, J.-D., Burdick, J., Goldberg, K., Hutchinson, S. (eds.) Algorithmic Foundations of Robotics V. STAR, vol. 7, pp. 77–93. Springer, Heidelberg (2004). https://doi.org/10.1007/978-3-540-45058-0_6
60. Ilcinkas, D., Wade, A.M.: Exploration of the T-interval-connected dynamic graphs: the case of the ring. Theor. Comput. Syst. **62**(5), 1144–1160 (2018)
61. Johansen, D., van Renesse, R., Schneider, F.B.: Operating system support for mobile agents. In: Proceedings of 5th Workshop Hot Topics in Operating Systems (HotOS), pp. 42–45 (1995)
62. Kranakis, E., Krizanc, D., Markou, E.: Deterministic symmetric rendezvous with tokens in a synchronous torus. Discrete Appl. Math. **159**(9), 896–923 (2011)
63. Megiddo, N., Hakimi, S., Garey, M., Johnson, D., Papadimitriou, C.: The complexity of searching a graph. J. ACM **35**(1), 18–44 (1988)

64. Nisse, N., Soguet, D.: Graph searching with advice. Theor. Comput. Sci. **410**(14), 1307–1318 (2009)
65. Parson, T.: The search number of a connected graph. In: Proceedings of 9th Southeastern Conference on Combinatorics, Graph Theory and Computing, pp. 549–554 (1978)
66. Rollik, H.A.: Automaten in planaren graphen. Acta Informatica **13**(3), 287–298 (1980)
67. Sudo, Y., Baba, D., Nakamura, J., Ooshita, F., Kakugawa, H., Masuzawa, T.: An agent exploration in unknown undirected graphs with whiteboards. In: Proceedings of 3rd Workshop on Reliability, Availability, and Security (WRAS) (2010)
68. Wall, D.: Messages as active agents. In: Proceedings of 9th ACM Symposium on Principles of Programming Languages (POPL), pp. 549–554 (1978)
69. Zambonelli, F., Jennings, N.R., Wooldridge, M.: Developing multiagent systems: the Gaia methodology. ACM Trans. Softw. Eng. Methodol. **12**(3), 317–370 (2003)

# Robots in Look-Compute-Move

# Pattern Formation

Giuseppe Prencipe(✉)

University of Pisa, Pisa, Italy
giuseppe.prencipe@unipi.it

**Abstract.** The PATTERN FORMATION problem is one of the most important coordination problem for robotic systems. Initially the entities are in arbitrary positions; within finite time they must arrange themselves in the space so to form a pattern given in input. In this chapter, we will mainly deal with the problem in the $\mathcal{OBLOT}$ model.

**Keywords:** Pattern formation · Agreement · Multiplicity detection

## 1 Introduction

In this chapter, we will describe the PATTERN FORMATION problem, where the robots are required to form, in a not predetermined area of the plane where they operate, a pattern they receive in input. The pattern can be given as a set of points in the plane (expressed in their Cartesian coordinates), or as a geometric predicate (e.g., "form a *circle*").

The standard requirements are that, initially, no two entities are in the same position (i.e., there are no dense points), and that the number of points prescribed in the pattern and the number of robots are the same. The robots are said to *form the pattern* if, at the end of the computation, the positions of the robots coincide, in everybody's local view, with the points of the pattern (or satisfy the predicate). Depending on the application, the formed pattern may be *translated*, and/or *rotated*, and/or *scaled*, and/or *flipped* into its mirror position with respect to the initial pattern. If dense points are allowed in the robots configurations and in the pattern, the problem is called *pattern formation with multiplicity.*

The PATTERN FORMATION problem is practically relevant because, if the robots can form a given pattern, they can agree on their respective roles in a subsequent, coordinated action.

The more general and difficult version of this problem is the ARBITRARY PATTERN FORMATION problem, where the robots must be able to form *any* arbitrary pattern $\mathbb{P}$ they are given in input, starting from *any* arbitrary initial configuration where the robots occupy distinct location. The pattern formation problem, in its general as well as in the more specific versions, has been extensively investigated (e.g., see [1–11]).


P. Flocchini et al. (Eds.): Distributed Computing by Mobile Entities, LNCS 11340, pp. 37–62, 2019.
https://doi.org/10.1007/978-3-030-11072-7_3

## 2 Views and Symmetricity

Useful tools to study what patterns are formable by oblivious robots are based on the notion of *view* [9,12]: this notion is strictly related to that of a *symmetricity* of a set of entities in the plane (either point or robots). In this chapter we will just give a quick overview of these concepts, that will be detailed in Chap. 6.

Let $Z_i$ denote the local coordinate system of robot $r_i$. The *global view* $GV_i(t)$ from robot $r_i$ at time $t$ is the infinite rooted tree defined as follows (refer also to Fig. 1):

1. The root $v_i$ of $GV_i(t)$ corresponds to $r_i$.
2. Node $v_i$ has $n-1$ children, one for each robot $r_j$, with $j \neq i$. The edge from node $v_i$ to node $v_j$ corresponding to $r_j$ is labeled $((a,b),(c,d))$, where $(a,b)$ is the position of $r_j$ with respect to $Z_i$, and $(c,d)$ is the position of $r_i$ with respect to $Z_j$.
3. Node $v_j$, with $j \neq i$ has $n-1$ children, one for each robot $r_l$, with $j \neq l$; the edge from $v_j$ to $v_l$ is labeled $((a',b'),(c',d'))$, where $(a',b')$ is the position of $r_j$ with respect to $Z_i$, and $(c',d')$ is the position of $r_i$ with respect to $Z_j$.

Since in general a robot does not know the coordinate systems of the other robots, which are integral part of the definition of global view, the global view of a configuration is in general not available to the robots and, in most cases, impossible to derive.

Something that the robots can locally compute in absence of any other information is the *local view.* The *local view* $LV_i(t)$ *of* robot $r_i$ at time $t$ is the set of vectors $vec(r_i, r_j)$ for all $j \neq i$ with respect to $Z_i$. In other words, the local view $LV_i(t)$ corresponds to the information that $r_i$ obtains when performing `Look` at time $t$. Two local views $LV_i(t)$ and $LV_j(t)$ are said to be *equivalent* ($LV_i(t) \equiv LV_j(t)$) if they are equal up to rotations, mirroring, and scaling.

An important property of the equivalence classes defined by the views (both in the case of global views, and of local views) is that they all have the same size.

**Lemma 1** ([9]). *Given a configuration* $\mathbb{E}$ *at time* $t$*, all the equivalence classes of robots with the same global (resp., local) view, have the same cardinality* $m$*.*

Moreover, the robots can be partitioned into $\frac{n}{m}$ groups of $m$ robots each, such that two robots have an equivalent view if and only if they belong to the same group. Note that, in the case of global view, the equivalence relationship is equality.

**Lemma 2** ([9]). *If the system has* `Chirality`*, given a configuration* $\mathbb{E}$ *at time* $t$*, the robots in the same equivalence class form a regular* $m$*-gon, and the regular* $m$*-gons formed by all the groups have a common center.*

The size $m$ of the equivalence classes, called *symmetricity,* is denoted by $\sigma(\mathbb{E})$ in the case of global views, and by $\rho(\mathbb{E})$ in the case of local views.

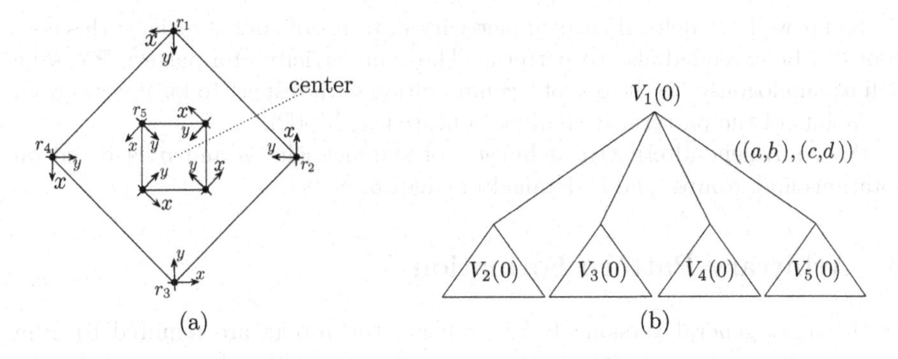


**Fig. 1.** (a) A configuration of robots. (b) The global view of $r_1$.

For example, in the configuration of robots depicted in Fig. 1(a) there are two classes of symmetry, each containing 4 robots, both when considering global and local views: In this case $\sigma(\mathbb{E}) = \rho(\mathbb{E}) = 4$. Moreover, since in this example there is chirality, Lemma 2 holds and the robots can be partitioned into 2 groups of 4 robots each group forming a 4-gon with a common centre.

Note that Lemma 2 does not hold when there is no chirality, i.e., when the axis of the coordinate systems of the robots are not rotationally symmetric. Consider, for example, the configuration depicted in Fig. 1(b). It is clear that all robots have the same global and local views, thus belong to the same equivalence class; they however do not form a single $n$-gon, but rather 2 distinct $\frac{n}{2}$-gons.

More examples are shown in Fig. 2 where, in all cases, the robots have the same local views; thus $\rho(\mathbb{E}_1) = \rho(\mathbb{E}_2) = \rho(\mathbb{E}_3) = n$. On the other hand, the global views are not always the same. More precisely, in Fig. 2(a), we have that $\sigma(\mathbb{E}_1) = 1$ because all global views are different; in Fig. 2(b), $\sigma(\mathbb{E}_2) = n$ because the global views are all identical; finally, in Fig. 2(c), $\sigma(\mathbb{E}_3) = \frac{n}{2}$.

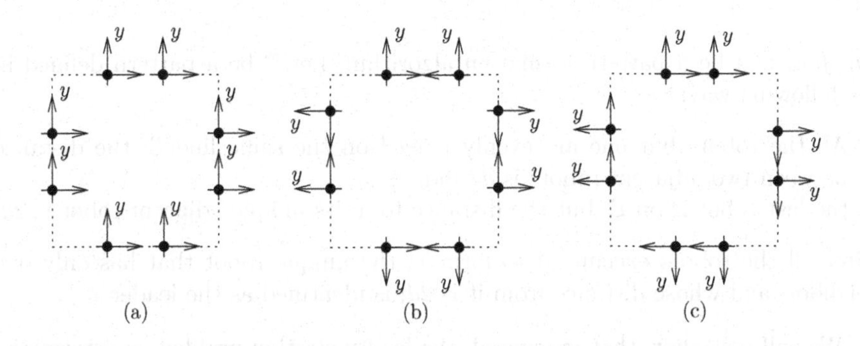


**Fig. 2.** Three configurations of robots: (a) $\mathbb{E}_1$, with $\rho(\mathbb{E}_1) = n$ and $\sigma(\mathbb{E}_1) = 1$; (b) $\mathbb{E}_2$, with $\rho(\mathbb{E}_2) = \sigma(\mathbb{E}_2) = n$; and (c) $\mathbb{E}_3$, with $\rho(\mathbb{E}_3) = n$ and $\sigma(\mathbb{E}_3) = \frac{n}{2}$.

So far we have defined the symmetricity in terms of configurations; this concept can be extended also to patterns. The symmetricity of a pattern $\mathbb{P}$ can be defined analogously to the one of a configuration with respect to local views from the points of the pattern. It shall be indicated with $\rho(\mathbb{P})$.

An equivalent alternative definition of symmetricity $\rho$ and $\sigma$ is based on rotations and groups, and is detailed in Chap. 6.

# 3 Arbitrary Pattern Formation

In the most general version of the problem, the robots are required to form any *arbitrary* pattern $\mathbb{P}$ they are given in input, starting from any *arbitrary* plain initial configuration; that is, they are required to solve the ARBITRARY PATTERN FORMATION problem. We note that, since rotation is allowed, two robots always form the desired pattern. Therefore we will assume to have at least three robots in the system. Also, we will assume that the robots operate under the $\mathcal{OBLOT}$ model.

## 3.1 Arbitrary Pattern Formation and Leader Election

A problem related to the ARBITRARY PATTERN FORMATION problem is the LEADER ELECTION problem: the robots in the system are said to *elect a leader* if, after a finite number of cycles, all the robots deterministically agree on (i.e., choose) the same robot $l$, called the leader. A deterministic algorithm that lets the robots in the system elect a leader in a finite number of cycles, given any initial configuration, is called a *leader election algorithm.*

The relationship between the arbitrary pattern formation problem and the leader election problem, is as follows:

**Theorem 1** ([5]). *If it is possible to solve the* ARBITRARY PATTERN FORMATION *problem for $n \geq 3$ robots, then the* LEADER ELECTION *problem is solvable too.*

*Proof.* Let $\mathcal{A}$ be a pattern formation algorithm. Let $\mathbb{P}$ be a pattern defined in the following way:

1. All the robots but one are evenly placed on the same line $\mathcal{L}$; the distance between two adjacent robots is $d$; and
2. the last robot is on $\mathcal{L}$, but the distance from its unique adjacent robot is $2d$.

After all the robots execute $\mathcal{A}$ to form $\mathbb{P}$, the unique robot that has only one neighbor, and whose distance from it is $2d$, is identified as the leader.

We will now show that in general, the leader election problem is deterministically unsolvable.

**Theorem 2** ([5]). *There exists no deterministic algorithm that solves the* LEADER ELECTION *problem, even in* $\mathcal{F}$SYNC *with* `Chirality`.

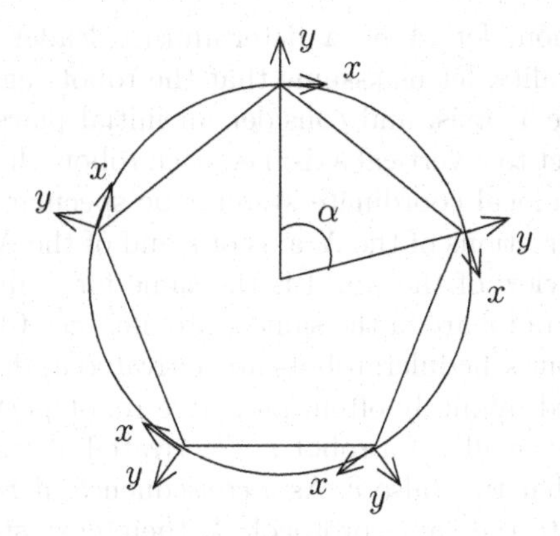


**Fig. 3.** Theorem 2: The unbreakable symmetry of a 5-gon.

*Proof.* By contradiction, let $\mathcal{A}$ be a deterministic algorithm for solving the LEADER ELECTION problem, and let us assume that the robots have no agreement on the local compasses (i.e., `Disorientation` holds). Consider any pattern different from a regular $n$-gon or a single point, and let the initial positions be such that the robots form a regular $n$-gon. Let $\alpha = 360°/n$ be the characteristic angle of the $n$-gon, and let the local coordinate system of each robot be rotated by $\alpha$ with respect to its neighbor on the polygon (see Fig. 3). In this situation, all the robots have the same (local) view of the world. Now, for any move that any one robot can make in its local coordinate system by executing algorithm $\mathcal{A}$, we know that each robot can make the same move in its local coordinate system. If all of them move in the exact same way at the same time (i.e., they move according to a synchronous schedule), they again end up in a regular $n$-gon or a single point. Therefore, by letting all the robots move at the same time in the same way, we always proceed from one regular $n$-gon or single point to the next. Hence, no leader can be elected. The same argument applies even if `Chirality` holds.

Thus, by Theorem 1, we can state the following:

**Corollary 1.** *In a system with $n > 2$ robots, the* ARBITRARY PATTERN FORMATION *problem is unsolvable.*

Furthermore, even if the robots agree on the direction and direction of one axis (agreement `k-Axes`, with $k = 1$), the LEADER ELECTION problem is still unsolvable when $n$ is even: in the following, we will refer to this kind of agreement as `OneAxis`.

**Theorem 3** ([5]). *Let the robots agree only on the direction and orientation of one axis; there exists no deterministic algorithm that solves the* LEADER ELECTION *problem, hence the* ARBITRARY PATTERN FORMATION *problem, when $n$ is even.*

*Proof.* By contradiction, let $\mathcal{A}$ be a deterministic leader election algorithm. Without loss of generality, let us assume that the robots agree on the direction and orientation of the $Y$ axis, and consider an initial placement of the robots symmetric with respect to a vertical axis; i.e., each robot $r$ has a *specular partner* $\widehat{r}$. In addition, let the local coordinate systems be specular with respect to the symmetry axis: the directions of the $X$ axis of $r$ and of the $X$ axis of $\widehat{r}$ are opposite; thus the (local) view of the world is the same for $r$ and $\widehat{r}$. In this setting, at time $t = 0$, both $r$ and $\widehat{r}$ are in the same state; i.e., $\tau(r, 0) = \tau(\widehat{r}, 0)$. Consider now a semi-synchronous scheduler: robots are activated at discrete time instants; each robot is activated infinitely often; an active robot performs its operations instantaneously. Additionally, if a robot $r$ is activated at time $t \geq 0$, the scheduler will activate at that time also $\widehat{r}$. As a consequence, if $\tau(r, t) = \tau(\widehat{r}, t)$, since the two robots execute the same protocol $\mathcal{A}$, their next state will still be the same: if $r$ moves to $d$, $\widehat{r}$ moves to the point $\widehat{d}$ specular to $d$ with respect to the symmetry axis. In other words, in this execution of protocol $\mathcal{A}$, $\tau(r, t) = \tau(\widehat{r}, t)$ for all $t \geq 0$. On the other hand, since $\mathcal{A}$ is an election protocol, it must exist a time $t' > 0$ such that a robot, say $r'$ becomes leader. Since the leader is unique, $\tau(r', t') \neq \tau(r, t')$ for all $r \neq r'$, contradicting the fact that $\tau(r', t') = \tau(\widehat{r'}, t')$.

Let us consider now the converse relationship between the ARBITRARY PATTERN FORMATION problem and the LEADER ELECTION problem. Assume that all robots share a common protocol LEADER($\mathbb{E}$) that, given any configuration $\mathbb{E}$, deterministically returns a unique leader in $\mathbb{E}$. We can now employ such a protocol to form an arbitrary target pattern $\mathbb{P}$, i.e., to solve the ARBITRARY PATTERN FORMATION problem, assuming that the robots agree on a common chirality. The overall idea of the algorithm consists of three main steps [13]: (1) the robots move to some appropriate positions, and build a kind of global coordinate system; (2) next, they compute the final positions to occupy in order to form the input pattern; (3) finally, the robots move towards these final positions, paying attention to maintain unchanged the global coordinate system.

In particular, given a set of points $P$ and its $SEC(P)$, we call the *concentric enclosing circles* of $SEC(P)$ all the circles having the same center of $SEC(P)$ and passing through at least one point in $P$. Starting from a *leader configuration* (i.e., a configuration where a leader can be located), the robots first move to an *agreement configuration*:

**Definition 1 (Agreement Configuration).** *A configuration* $\mathbb{T}$ *is an agreement configuration if and only if both following conditions hold:*

1. *There exists a robot* $r_l$ *in* $\mathbb{T}$ *such that* $r_l$ *is the unique robot located on the smallest concentric enclosing circle of* $SEC(\mathbb{T})$*;*
2. *There is no robot at the center of* $SEC(\mathbb{T})$*.*

In order to achieve an agreement configuration from a leader configuration $\mathbb{E}$, the robots act as follows. If there is a robot $r$ that is located at the center $c$ of $SEC(\mathbb{E})$, let $s$ be the closest robot to $c$ among the robots in $\mathbb{E} \setminus \{r\}$, and $p$ the median point on the segment $\overline{rs}$. Then, by moving $r$ towards $p$, an agreement configuration is achieved. Otherwise (no robot is at $c$), we consider the smallest concentric enclosing circle of $SEC(\mathbb{E})$, call it $\mathcal{C}$; if there is only one robot on this circle, then the robots are already in an agreement configuration. Thus, let us assume there is more than one robot on $\mathcal{C}$. Now, the availability of protocol LEADER() is exploited: let $r^* = \text{LEADER}(\mathbb{E})$, and let $r$ be the first robot on $\mathcal{C}$, according to the clockwise orientation, with respect to the half-line $\overrightarrow{cr^*}$ (recall that Assumption `Chirality` holds). By moving $r$ towards the median point of the segment $\overline{rc}$, an agreement configuration is obtained. Note that the previous strategy works also when all robots are on $SEC(\mathbb{E})$ (i.e., when $\mathcal{C} \equiv SEC(\mathbb{E})$): the only difference is in the way robot $r$ is chosen. In fact, in this case, $r$ is the first *non-critical* robot on $\mathcal{C}$, i.e. the first robot on $\mathcal{C}$ whose movement would not change $SEC(\mathbb{E})$ (in this case $r$ might coincide with $r^*$).

Once the robots are in an agreement configuration $\mathbb{T}$, they can also agree on their final positions: in particular, the center $c$ of $SEC(\mathbb{P})$ is mapped onto the center $o$ of $SEC(\mathbb{T})$; the pattern is rotated so that $\overrightarrow{orl}$ is mapped onto $\overrightarrow{cs}$, with $s$ the first non-critical point located on the smallest concentric enclosing circle of $\mathbb{P}$; and $\mathbb{P}$ is scaled with respect to the radius of $SEC(\mathbb{T})$ so that all the distances are expressed according to the radius of $SEC(\mathbb{T})$ (in particular $SEC(\mathbb{T}) = SEC(\mathbb{P})$).

Then, the robots occupy these positions, starting from those situated on $SEC$, and then on all the circles concentric to $SEC$ from the largest to the smallest. During this phase, the final positions are maintained unchanged, by making sure that the robots remain in an agreement configuration until the pattern is formed. In particular, the protocol makes sure that no angle above 180° is created on $SEC$ (otherwise the smallest enclosing circle changes), and that the leader of the agreement configuration remains the unique closest robot from the center of $SEC$ and does not leave the radius where it is located. In other words,

**Theorem 4** ([13]). *In* $\mathcal{A}$SYNC, *assuming* `Chirality`, *for any* $n \geq 4$ *if the* LEADER ELECTION *problem is solvable, then the* ARBITRARY PATTERN FORMATION *problem is solvable.*

### 3.2 Arbitrary Pattern Formation and Compasses

The solvability of the ARBITRARY PATTERN FORMATION problem, and in general which patters can be formed regardless of the starting configuration, strictly depend on the level of agreement that the robots have about their local coordinate systems.

Following the ideas of the proof of previous Theorem 2, it is possible to show a *necessary* condition for the solvability of the ARBITRARY PATTERN FORMATION problem: the absence of common agreement on the coordinate system, leads to the inability to form arbitrary patterns.

**Theorem 5** ([5]). *Without any agreement on the local compasses,* ARBITRARY PATTERN FORMATION *is* impossible, *even in* $\mathcal{F}$SYNC *with chirality.*

As a consequence, some agreement is necessary.

*Sense of Direction.* Total agreement on the coordinate system (Assumption ConsistentCompass) is indeed *sufficient* to solve the ARBITRARY PATTERN FORMATION problem even in $\mathcal{A}$SYNC. To see how, consider the following protocol [5]:

1. Each robot establishes the (lexicographic) total order of the points of the local pattern (Fig. 4(a)).
2. Each robot establishes the (lexicographic) total order of the robots' positions retrieved in the last Look (Fig. 4(b)). As we will see, this order will be the same for all robots.
3. The first and second robots move to the positions matching the first and second pattern points. This movement can be performed in such a way that the order of the robots does not change (Fig. 4(c) and (d)). Once this is done, the first two robots' positions will determine the translation and scaling of the pattern (Fig. 4(e)).
4. All other robots go to the other points of the pattern. This can be done by moving the robots sequentially to the pattern's points. The sequence is chosen in such a way to guarantee that, after one robot has made even only a small move towards its destination, no other robot will move before that one has reached its destination (Fig. 4(f)).

We note that the final positions of the robots are not rotated w.r.t. the input positions; in other words the algorithm keeps the "orientation" given by the input pattern. Moreover, in this case Theorem 1 holds also for $n = 2$, since the rightmost and topmost robot in the system can always be identified as the leader.

**Theorem 6** **([5]).** *With* ConsistentCompass*,* ARBITRARY PATTERN FORMATION *is solvable in* $\mathcal{A}$SYNC*.*

*Partial Agreement: Odd Number of Robots.* Let us now consider the case when the robots have partial agreement: they agree only on the orientation of one axis, say $Y$; that is, there is common agreement also on the direction of the $X$ axis, but not on its orientation (assumption OneAxis). Note that this case, if there is also chirality, would trivially coincide with the total agreement one.

As stated by Corollary 1, the ARBITRARY PATTERN FORMATION problem is unsolvable in general; furthermore, by Theorem 3, it is also unsolvable by an even number of robots when the Assumption OneAxis considered in this section holds, since symmetric initial configuration can impede the formation of arbitrary patterns. However, for breaking the symmetry, it is sufficient to know that the number $n$ of robots is odd: in this case, in fact, either the robots are in a symmetric initial situation, in which there is a unique middle robot that will move in order to break the symmetry; or the initial situation is not symmetric, and this asymmetry can be used to identify an orientation of the $X$ axis.

In more detail, let us define some *references* related to a set of points $\mathbb{E}$ that will be used in the following:

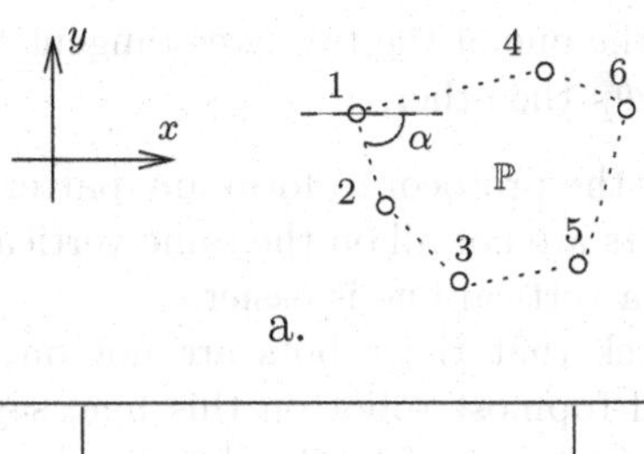


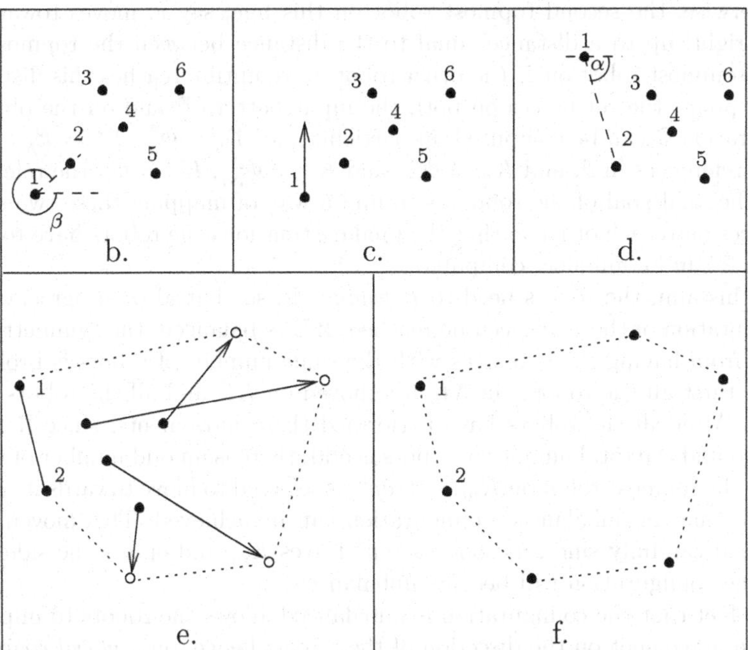


**Fig. 4.** An example of the arbitrary pattern formation protocol in presence of `ConsistentCompass`. (a) The input pattern $\mathbb{P}$. The robots have complete knowledge on the local coordinate systems. The numbers represent the lexicographical ordering the robots give to the points of $\mathbb{P}$, and $\alpha = \texttt{Angle}(p_1, p_2)$. (b) The robots sort the robots' positions retrieved in the last `Look` state, and compute $\beta = \texttt{Angle}(r_1, r_2)$. (c) $r_1$ moves in such a way that $\texttt{Angle}(r_1, r_2) = \alpha$. (d) The relative positions of $r_1$ and $r_2$ are such that $\texttt{Angle}(r_1, r_2) = \alpha$. (e) At this point, all the robots can translate and scale the input pattern according to $\overline{r_1 r_2}$. Then, all the robots, one at a time, reach the final positions of the pattern to form. (f) The final configuration.

- The two vertical lines that are tangent to the convex hull of $\mathbb{E}$, and the vertical axis $\Phi_m^{\mathbb{E}}$ that is in the middle between them.
- These three vertical lines delimit two regions (or *sides*): one to the left of $\Phi_m^{\mathbb{E}}$ and one to its right. Let $\mathcal{M}_{\mathbb{E}}$ and $\mathcal{L}_{\mathbb{E}}$ denote the side in $\mathbb{E}$ with more and less points, respectively. If the two sides have the same number of points, then $\mathcal{M}_{\mathbb{E}}$ is the rightmost side. If $|\mathcal{M}_{\mathbb{E}}| \neq |\mathcal{L}_{\mathbb{E}}|$, then $\mathbb{E}$ is said to be *unbalanced*; otherwise, we will call it *balanced*.

- Finally, $\Phi^{\mathbb{E}}_{\mathcal{M}}$ denotes the one of the two axes tangent to the convex hull of $\mathbb{E}$ that lies in $\mathcal{M}_{\mathbb{E}}$, and $\Phi^{\mathbb{E}}_{\mathcal{L}}$ the other.

We will describe now the protocol to form any pattern with an odd number of points, where the points are not all on the same vertical line. The case, where the robots have to form a vertical line is easier.

First, the robots check that the robots are not on the same vertical line $\Xi$; otherwise, the second topmost robot on this line, say $r$, moves towards its (local) right, up to a distance equal to the distance between the topmost and the bottommost robot on $\Xi$ (no other robot move until $r$ reaches this distance). At this point, the references on both the input pattern $\mathbb{P}$ and on the observed configuration $\mathbb{D}$ can be computed: in particular, let $\Upsilon_m = \Phi^{\mathbb{P}}_m$, $\Upsilon^+ = \Phi^{\mathbb{P}}_{\mathcal{M}}$, $\Upsilon^- = \Phi^{\mathbb{E}}_{\mathcal{L}}$ the references in $\mathbb{P}$, and $K_m = \Phi^{\mathbb{D}}_m$, and $K^+ = \Phi^{\mathbb{D}}_{\mathcal{M}}$, $K^- = \Phi^{\mathbb{E}}_{\mathcal{L}}$ the references in $\mathbb{D}$. The final goal of the robots is to find a way of mapping these two sets of references onto each other so that the final destinations the robots have to reach to form $\mathbb{P}$ can be uniquely computed.

To this aim, the robots need to *unbalance* $\mathbb{D}$, so that also an agreement on the orientation of the $x$ axis cen be reached. If $\mathbb{D}$ is balanced, the symmetry that derives from having the two sides with the same number of robots is broken as follows. First all the robots[1] in $\mathcal{M}_{\mathbb{D}}$ are moved on $K^+$ and all the robots in $\mathcal{L}_{\mathbb{D}}$ on $K^-$. After all the robots have performed these movements, since $\mathbb{D}$ is still balanced and the total number of robots is odd, there is an odd number of robots on $K_m$: the topmost robot on $K_m$, say $top^*$, is selected to move towards its (local) right, so that an unbalanced configuration can be achieved. This movement is performed carefully since, as soon as $top^*$ leaves $K_m$ and enters the side to its right, the configuration will become unbalanced.

The fact that the configuration is unbalanced allows the robots to implicitly reach an agreement on the direction of the $x$ axis; hence, on a *global coordinate system* (*GCS*): the common orientation of the $x$ axis is given by mapping $\mathcal{M}_{\mathbb{P}}$ onto $\mathcal{M}_{\mathbb{D}}$.

Once the *GCS* has been established, the topmost robots on $K^+$ and on $K^-$ ($top^+$ and $top^-$, respectively) move vertically on $K^+$ and on $K^-$, respectively, until they reach positions corresponding to the two topmost points on $\Upsilon^+$ and $\Upsilon^-$ in $\mathbb{P}$. Once $top^+$ and $top^-$ place themselves in the correct positions, they will never move again. At this point, the set of final positions of the robots can be easily computed, by scaling the pattern according to these mappings. Note that here the pattern does not need to be rotated.

Now, all robots are ready to reach their final destinations. Note that at this point it might be possible that the unbalancing process is not completed yet; i.e., $top^*$ is still moving towards its destination. Should this be the case, the other robots can however detect it, and will not start their move until $top^*$ stops (again, details can be found in [5]). The robots reach their final destinations sequentially:

[1] Note that, since at this time the robots still do not have a common agreement on the direction of the $X$ axis, for some robots $\mathcal{M}_{\mathbb{D}}$ and $\mathcal{L}_{\mathbb{D}}$ might be different. All of them, however, agree on $K_m$.

- First, the robots in $\mathcal{S}^-$ (side of $\mathbb{D}$ where $K^-$ lies) sequentially fill the final positions that are in $\mathcal{S}^-$. If there are more robots than available final positions, the "extra" robots are sequentially moved towards $K_m$, starting from the topmost robots that is closest to $K_m$.
- Second, the robots in $\mathcal{S}^+$ (side of $\mathbb{D}$ where $K^-$ lies), except for the bottommost on $K^+$, sequentially fill the final positions in $\mathcal{S}^+$. If there are more robots than available final positions, the "extra" robots are sequentially moved towards $K_m$, starting from the topmost robots that is closest to $K_m$.
- Third, if there are still unfilled final positions in $\mathcal{S}^+$ (that is, there were not enough robots in $\mathcal{S}^+$ in the second step), the robots on $K_m$ are sequentially moved in $\mathcal{S}^+$, starting from the topmost, to fill the final positions occupied by no robots.
- Fourth, if there are still unfilled final positions in $\mathcal{S}^-$ (that is, there were not enough robots in $\mathcal{S}^-$ in the first step), the robots on $K_m$ are sequentially moved in $\mathcal{S}^-$, starting from the topmost, to fill the final positions still available.

At this point, all the robots not on $K_m$ occupy the correct positions except one: the bottommost robot on $K^+$, say $r$.

- If there is an available destination in $\mathcal{S}^+$, then $r$ goes there. At this point, all the robots but those on $K_m$ are in correct positions. Note that now all available destinations are also on $K_m$: thus, the robots on $K_m$ move sequentially (and only vertically on $K_m$) towards the available final destinations.
- If there are no available final positions inside $\mathcal{S}^+$ and $\mathcal{S}^-$, $r$ moves towards $K_m$. Once it reaches the median axis, all the robots but those on $K_m$ are in correct positions, and again the algorithm proceeds as in the previous case.
- If there is an available destination in $\mathcal{S}^-$, $r$ first moves towards $K_m$. Then, the topmost robot on $K_m$ moves in $\mathcal{S}^-$ on the last unfilled final position. Once also this position becomes occupied, only the robots on $K_m$ must be adjusted, as in the two previous cases.

Thus, the above plus Theorem 3 imply the following:

**Theorem 7** ([5]). *With* `OneAxis`, Arbitrary Pattern Formation *is solvable only if $n$ is odd, and this can be done in* $\mathcal{A}$sync.

*Partial Agreement: Even Number of Robots.* By Theorem 3, an arbitrary pattern can not be formed by an even number of robots with `OneAxis`. In this section, we are interested in determining which class of patterns, if any, can be formed in this case starting from any initial position. Again, we will assume that the robots in the system have common agreement on the direction and orientation of only the $Y$ axis, and that the number $n$ of robots in the system is even.

We say that $\mathbb{P}$ is a *symmetric pattern* if it has at least one axis of symmetry $\Lambda$; that is, for each $p \in \mathbb{P}$ there exists exactly another point $p' \in \mathbb{P}$ such that $p$ and $p'$ are symmetric with respect to $\Lambda$ (see Figs. 5(b), (c) and (d)).

The proof of the unsolvability result of Theorem 3 is useful to better understand what kind of patterns can not be formed, hence what kind of pattern

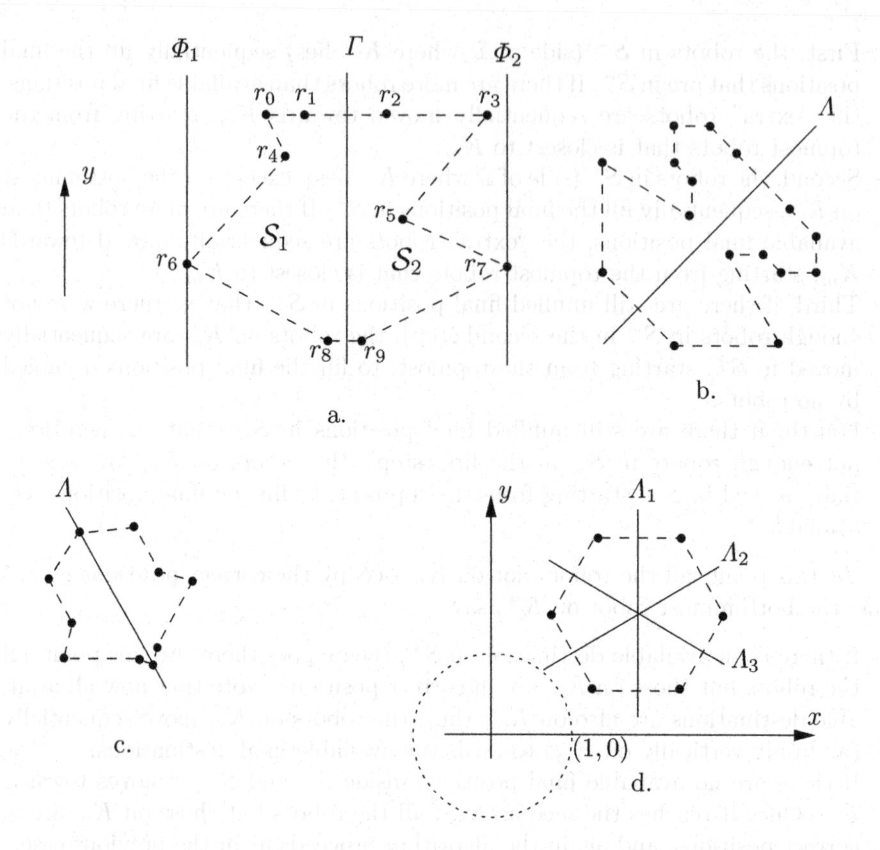


**Fig. 5.** (a) An unachievable asymmetric pattern. (b) An achievable pattern with one axis of symmetry not passing through any vertex. (c) An unachievable pattern. (d) An achievable pattern that has three axes of symmetry not passing through any vertex. Note that this pattern has also axes of symmetry passing through vertexes.

formation algorithms can not be designed. In fact, the ability to form a particular type of patterns would imply the ability to elect a robot in the system as the leader. Formally,

**Theorem 8** ([5]). *If an algorithm $\mathcal{A}$ lets the robots form (a) an asymmetric pattern, or (b) a symmetric pattern that has all its axes of symmetry passing through some vertex, then $\mathcal{A}$ is a leader election algorithm.*

From Theorems 3 and 8, it follows that:

**Corollary 2.** *There exists no pattern formation algorithm that lets the robots in the system form (a) an asymmetric pattern, or (b) a symmetric pattern that has all its axes of symmetry passing through some vertex.*

Let us call $\mathfrak{T}$ the class containing all the arbitrary patterns, and $\mathfrak{P} \subset \mathfrak{T}$ the class containing only patterns with at least one axis of symmetry not passing through any vertex (e.g., see Figs. 5(b) and (d)); let us call *empty* such an axis. Corollary 2 states that if $\mathbb{P} \in \mathfrak{T} \setminus \mathfrak{P}$, then $\mathbb{P}$ can not be in general formed; hence, according to Part (b), the only patterns that might be formed are symmetric ones with at least one *empty axis*.

The idea behind the algorithm that solves the ARBITRARY PATTERN FORMATION problem with partial agreement and an even number of robots is as follows. First, the robots compute locally an *empty axis* of the input pattern $\mathbb{P}$, say $\Lambda$, and then rotate $\mathbb{P}$ so that $\Lambda$ is parallel to the common understanding of the orientation of $y$; let us denote by $\mathbb{P}_R$ the rotated pattern.

If the robots lie all on the same line, the algorithm forces them to place on at least two distinct vertical lines, $\Gamma$ and $\Gamma'$ (this is achieved as for the odd case). Then, the topmost robot on $\Gamma$, say *Out*, and the topmost robot on $\Gamma'$, say *Out'*, move so that they place themselves in the correct position: in particular, since $\mathbb{P}_R$ is symmetric with respect to $\Lambda$, *Out* and *Out'* must place themselves to the same height. This is because, by Corollary 2, the input pattern can not be a vertical line.

At this point, the set of final positions can be computed, by scaling the input pattern with respect to $\overline{\Gamma\Gamma'}$, and by translating it so that the topmost point on the rightmost vertical axis tangent to $\mathbb{P}$ is mapped onto *Out*, and the topmost point on the leftmost vertical axis tangent to $\mathbb{P}$ is mapped onto[2] *Out'*.

At this point, the robots move to reach a balanced configuration, with each side containing half of the robots. The balancing is obtained as follows. Let $\mathcal{S}$ and $\mathcal{S}'$ be the two sides determined by $\Gamma_m$, the vertical median axis between $\Gamma$ and $\Gamma'$.

- In the side that has more than $n/2$ robot (if any), the robots are moved sequentially (starting from the topmost with the smallest horizontal distance from $\Gamma_m$) towards $\Gamma_m$, using a path that avoids collisions, until there are exactly $n/2$ robots in that side.
- In a side that has $\leq n/2$ robots, the robots are moved towards the final positions in that side.
- The robots that are on $\Gamma_m$ wait until $|\mathcal{S}| \leq n/2$ and $|\mathcal{S}'| \leq n/2$, and all the robots in the two sides are on a final position. At this point, sequentially (from the topmost) they move towards the final positions still available in the two sides. In fact, by the way the input pattern has been rotated, no final positions can be on $\Gamma_m$.

Thus, we can state the following:

**Theorem 9** ([5]). *With* `OneAxis`*, when $n$ is even only patterns in $\mathfrak{P}$ can be formed, and this can be done in* $\mathcal{A}$SYNC.

[2] Note that, since $\mathbb{P}_R$ is symmetric, nothing changes if the topmost point on the leftmost vertical axis tangent to $\mathbb{P}$ is mapped onto *Out*, and the topmost point on the rightmost vertical axis tangent to $\mathbb{P}$ is mapped onto *Out'*.

*No Agreement.* In absence of any additional assumption, and in particular in absence of any agreement on the compasses, Theorem 2 implies that no asymmetric pattern can be formed from all arbitrary initial configurations. Furthermore, as discussed later in Sect. 4, a symmetric pattern $\mathbb{A}$ with symmetricity $\sigma(\mathbb{A})$, can be formed only with the same or lower symmetricity. This means that the only patterns that might (possibly) be formed from all initial configurations are either an *n-gon*, or the *uniform circle* (i.e. a circle along which the robots are placed at equal distance), or the *point* (i.e., all robots are gathered at the same location). Note that this fact holds regardless of the synchronicity (i.e., even in $\mathcal{F}$SYNC).

The problem of forming a point (a.k.a. *Gathering*) and forming a uniform circle are important in their own right, and are analyzed respectively in Chap. 4 and Chap. 5, respectively. Interestingly, to date it is not known whether these problems can be solved in $\mathcal{A}$SYNC without additional assumptions; in the case of $\mathcal{S}$SYNC, they are both solvable.

## 3.3 Landmarks Covering: Formation of Visible Patterns

An interesting problem related to ARBITRARY PATTERN FORMATION (APF) is the LANDMARKS COVERING problem: in the space there are $n$ points, the *landmarks*, visible to all robots[3]; the problem is for the robots to reach a configurations where at each landmark there is precisely one robot. A solution protocol must enable the robots to cover the landmarks, regardless of the location of the landmarks and of the initial location of the robots.

In other words, the LANDMARKS COVERING problem is precisely the ARBITRARY PATTERN FORMATION problem when the points of the input pattern are globally visible. Clearly, any solution to APF under some conditions, will solve also LANDMARKS COVERING under those conditions. The research interest is whether LANDMARKS COVERING can be solved more efficiently than APF, or with fewer conditions than APF, or in situations where APF is not (known to be) solvable. In terms of efficiency, the main goal of any LANDMARKS COVERING solution protocol is that of minimizing the robots movements, i.e., the total amount traveled by the robots to reach the final configurations in which all landmarks are covered.

Interestingly, unlike the ARBITRARY PATTERN FORMATION problem, the LANDMARKS COVERING problem can always be solved in $\mathcal{A}$SYNC, provided there is `Chirality`. Furthermore this can be done always with minimal travel costs and without collisions [14].

The solution strategy consists in the robots computing a unique perfect *matching* between robots and landmarks which minimizes the total travel costs from each robot to the landmark assigned by the matching; each robot then moves until it reaches the assigned landmark, avoiding collisions. The clear difficulty is to perform this process obliviously; to do so, the determined matching must be *invariant* to the movements of the robots towards their destination, so

[3] Equivalently, the position of the landmarks is known *a priori* to all robots.

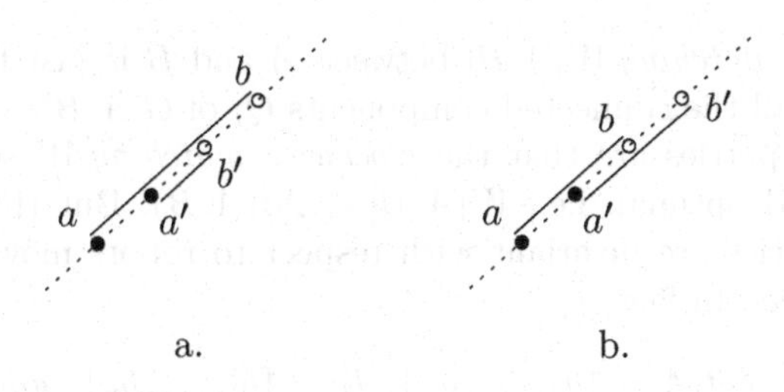


**Fig. 6.** Examples of matchings.

that each robot, every time it becomes active, can determine which landmark was initially assigned it, regardless of the progress made by the other robots towards their assigned landmarks.

Consider the initial configuration of the robots $\mathbb{A}$ and let $\mathbb{B}$ denote the pattern of the landmarks. We can view a perfect matching $M$ from $\mathbb{A}$ to $\mathbb{B}$ as a set of pairs $\{(a, b)\}$ where $a$ is a robot location in $\mathbb{A}$ and $b$ is a landmark in $\mathbb{B}$, and its *cost* is the sum $\sum_{(a,b)\in M} |\boldsymbol{ab}|$ of the Euclidean distances between the matched points. Let $\mathcal{M}(\mathbb{A}, \mathbb{B})$ denote the set of all perfect matchings $M$ of minimum cost between $\mathbb{A}$ to $\mathbb{B}$ such that for all distinct pairs $(a, b), (a', b') \in M$, the points $a, a', b', b$ do not reside on the same line in that specific order. For example, $M$ may not include the match shown in Fig. 6(a), but may include the pairs shown in Fig. 6(b). We call the matchings in this set *optimal*. It is easy to verify that $\mathcal{M}(\mathbb{A}, \mathbb{B}) \neq \emptyset$; note that there might be more than one optimal matching between $\mathbb{A}$ and $\mathbb{B}$.

We can compute a unique optimal matching, called *clockwise matching*, between two set of $n$ distinct points, $A$ and $B$, as follows:

(1) First consider the bipartite graph $G[A, B] = (V, E)$ whose vertex set $V = A \cup B$ comprises of the points of $A$ and $B$, and where the edge set $E = \cup_{M\in\mathcal{M}(A,B)} M$ contains all pairs matched in at least one optimal matching.
(2) Consider now the connected components $G_1, G_2, \ldots, G_k$ of $G[A, B]$, and the periphery[4] $C_i$ of component $G_i$; let $A_i$ and $B_i$ be the points of $A$ in $G_i \setminus C_i$ and the points of $B$ in $G_i \setminus C_i$, respectively.
(3) Consider next the subgraph $\hat{G}[A, B]$ of $G[A, B]$ recursively defined as follows: $\hat{G}[A, B] = \emptyset$ if $A = B = \emptyset$; otherwise $\hat{G}[A, B] = \cup_{1\leq j\leq k}(C_i \cup \hat{G}[A_i, B_i])$. Note that each connected component $Q$ of $\hat{G}[A, B]$ is either a cycle or a single edge.
(4) Finally, for each connected component $Q_i$ of $\hat{G}[A, B]$, construct the matching $W_i$ where $W_i = Q_i$ if $Q_i$ is a single edge, otherwise, $W_i$ is a clockwise tour $(a_1, b_1), (a_2, b_2), \ldots, (a_m, b_m)$ of $Q_i$.

[4] For a plane graph, the periphery is the boundary of the exterior face.

The *clockwise matching* $W[A, B]$ between $A$ and $B$ is just the union of the matchings $W_i$ of all the connected components $Q_i$ of $\hat{G}[A, B]$.

Important properties are that the *clockwise matching* $W$ so determined is unique and indeed optimal; i.e., $W[A, B] \in \mathcal{M}(A, B)$. But the crucial fact is that these properties are invariant with respect to robots moving towards the matched landmarks. In fact,

**Lemma 3** **([14]).** *Let* $A = \{a_1, \ldots, a_n\}$, $B = \{b_1, \ldots, b_n\}$, *and* $= \{c_1, \ldots, c_n\}$ *be set of points which satisfy following:*

1. $\{(a_1, b_1), (a_2, b_2), \ldots, (a_n, b_n)\} \in W[A, B]$
2. $c_i \in \overline{a_i b_i}$
3. *if there exists* $j \neq i$ *such that* $a_j \in \overline{a_i b_i}$ *then* $c_i = a_i$

*Then* $\{(c_1, b_1), (c_2, b_2), \ldots, (c_n, b_n)\} \in W[C, B]$.

Thus, the (collision avoiding) solution protocol is simply [14]:

Algorithm LandmarkCover
Assumptions: Visible Landmarks; Chirality.

1. Let $A = \{a_1, \ldots, a_n\}$ be the position of the robots (as returned by Look) and let $B = \{b_1, \ldots, b_n\}$ be the positions of the landmarks in my coordinate system.
2. Compute the clockwise matching $W[A, B]$. Let $a \in A$ be my position and $b \in B$ the landmark assigned to me in $W[A, B]$.
3. If $\forall a' \in A \setminus \{a\}, a' \notin \overline{ab}$, then move towards $b$.

**Theorem 10** **([14]).** *The* Landmarks Covering *problem can be solved in* $\mathcal{A}$sync *with minimal travel costs, provided there is* Chirality.

In other words, with Chirality, every *visible pattern* can be formed in $\mathcal{A}$sync.

# 4 Pattern Formation and Initial Configuration

The proof of Theorem 2 shows that, with no agreement on the local coordinate systems, the Arbitrary Pattern Formation problem cannot be solved. Thus, an interesting question is what patterns could be formed, in absence of common coordinate system, starting from a *specific* configuration $\mathbb{E}$. Once again, we will assume the $\mathcal{OBLOT}$ scenario.

## 4.1 Impossibility

The patterns that the robots can or cannot form starting from configuration $\mathbb{E}$ at time $t = 0$ are strictly related to the classes of equivalence derived from the definition of views (seen in Sect. 2).

If the views of two or more robots are identical, in some executions (e.g., under a scheduler that activates them always at the same time) those robots will always perform the same actions, without being able to break their symmetry; so, the patterns that can be possibly formed must have the same or higher symmetricity, but always a multiple of the original one.

**Theorem 11** ([9]). *Starting from a configuration* $\mathbb{E}$ *with symmetricity* $\sigma(\mathbb{E})$, *it is* impossible *to form any pattern* $\mathbb{P}$ *with* $\sigma(\mathbb{P}) < \sigma(\mathbb{E})$, *or* $\sigma(\mathbb{P}) \neq k \cdot \sigma(\mathbb{E})$ *for some integer* $k > 1$.

In other words, if $\mathbb{E}$ is totally asymmetric (i.e., $\sigma(\mathbb{E}) = 1$), all patterns are potentially formable; on the other hand, if $\sigma(\mathbb{E}) = m > 1$, only patterns with the same symmetricity or with a symmetricity that is a multiple of $m$ are candidate to be formable. Notice that this impossibility holds even if the robots are not oblivious. In case of systems with chirality, by Lemma 2, we obtain the inability to form a pattern that cannot be partitioned, as the initial configuration, in $\frac{n}{m}$ regular $m$-gons.

**Theorem 12** ([9]). *In systems with* `Chirality`, *starting from a configuration* $\mathbb{E}$ *with symmetricity* $\sigma(\mathbb{E}) = m$, *it is* impossible *to form any pattern unless it is the union of* $\frac{n}{m}$ *regular* $m$*-gons all having the same center.*

## 4.2 Possibility

Once we know which are the only patterns that could be formed starting from a configuration $\mathbb{E}$, the questions become whether those patterns can be formed, and how. In $\mathcal{A}$SYNC, no answers are known. In the case of $\mathcal{S}$SYNC there are some conditional answers.

If the robots are *not oblivious* (recall that the impossibility holds even in this case), they can record all the snapshots in which they are active; the change of coordinates in two successive snapshots allows to detect movement and to measure it; hence information can be communicated by moving appropriate distances [9]. In particular, they can communicate their own coordinate systems and unit of measures, so that the complete views can be locally constructed and examined; once this is done, forming the pattern is straightforward.

We are however interested in oblivious robots, for which there is *no* memory, and hence no tool to record information, to detect and measure movement, and thus to communicate. Interestingly, it is possible for oblivious robots to form all the formable patterns [15], if the robots have `Chirality`, move with *fixed mobility* (possibly different for each robot) and know the *maximum movement* $\hat{\delta}$.

**Theorem 13** ([15]). *A team of oblivious robots in* $\mathcal{S}$SYNC *with* `Chirality`, fixed mobility, *and* known maximum movement, *starting from configuration* $\mathbb{E}$ *with* $\sigma(\mathbb{E}) = m$ *can form any pattern* $\mathbb{P}$ *decomposable into* $\frac{n}{m}$ *regular* $m$*-gons all having the same center.*

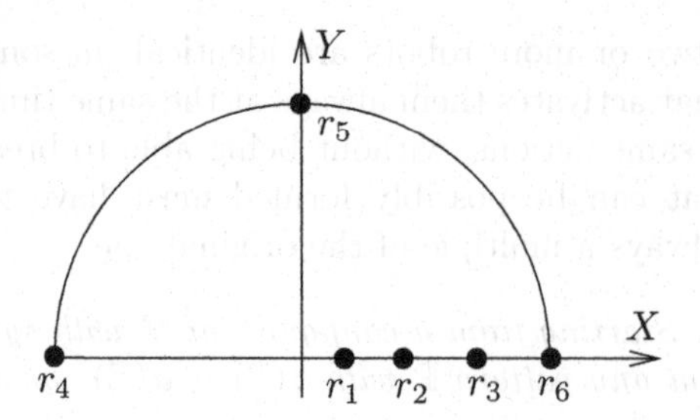


**Fig. 7.** A $T$-stable configuration with six robots.

Notice that, since the robots do not agree on a common coordinate system, the level of symmetry perceived by the robots (given by their local views) might not correspond to the actual level of symmetry of the global view which, as defined earlier, take into account also the coordinate systems.

Let $\mathbb{P} = p_1, \ldots, p_n$ be the pattern to be formed and let us assume that the robots start in $n$ distinct positions. For simplicity we describe only the case when the pattern does not contain dense points. Moreover, we assume (again for simplicity) that each robot knows the origin of its own coordinate system, which does not change throughout the algorithm. The result still holds with some modifications also when these assumptions are removed. Also for simplicity we assume the unit distance of a robot coincides with $\hat{\delta}$.

The algorithm distinguishes the case when $\rho(\mathbb{E}) = 1$, and thus the initial configuration is totally asymmetric, even without considering the coordinate systems, from the case when $\rho(\mathbb{E}) > 1$.

*Case* $\rho(\mathbb{E}) = 1$. In this case the initial configuration $\mathbb{E}$ is perceived as asymmetric. This is the simplest case and also a building block possibly used in the other cases.

Since the symmetricity is 1 and there is chirality, a total order can be imposed on the robots, even in absence of a common coordinate system. The robots are in fact ordered in non decreasing order of their radii with respect to the centre $c$ of the smallest enclosing circle $SEC(\mathbb{E})$ (for points with the same distance, ties are broken by using chirality). Let this order correspond to $r_1, \ldots, r_n$, where the robots are aware of their own index. The algorithm is designed in such a way that $SEC$ will never change until the pattern is "almost" formed.

Intuitively, the robots move from $\mathbb{E}$ to a special configuration, called a *$T$-stable configuration*, where $SEC$ contains exactly three robots on the circumference: two opposite on a diameter and the third at 90° from both, and no robots occupy the center (see Fig. 7). The robots can then agree on a common coordinate system by selecting as $X$ the line passing through the two robots positioned opposite on the diameter of $SEC$, and as $Y$ the line passing through $c$ and through the third robot placed at 90° on the circumference.

The unit distance of this common coordinate system is chosen in a very specific way as $\frac{Rad}{2^l}$ where $Rad$ is the radius of $SEC$, and $l$ is the smallest positive integer such that $|r_j| < |p_j|$ for each $1 \leq j \leq n$, where $|r_j|$ (resp., $|p_j|$) indicates the distance from point $r_j$ (resp., vertex $p_j$) to its own origin. This choice is made for the unit distance to be sufficiently small so that robots never move away from $c$ while going towards their position to form the pattern. The robots now move one by one to their final destination following their order (which implies that robots closer to $c$ move to their destination first). This order, combined with the fact that no robot has to move away from $c$ in the process, guarantees that the magnitude of the unit distance does not change in the formation process, and that a robot that has reached its final position does not have to move anymore. The movements are performed without destroying the $T$-stable configuration, paying particular attention to the movements of the last three robots.

*Case* $\rho(\mathbb{E}) > 1$. When the robots perceive $\rho(\mathbb{E}) > 1$, it does not necessarily mean that $\sigma(\mathbb{E}) > 1$, because the different coordinate systems might induce more asymmetry. In this general case, the robots perform two procedures. First they try to move from $\mathbb{E}$ to a configuration that reflects a symmetry $m$ that divides $\rho(\mathbb{P})$. Once/if such a situation is reached, they proceed to form the pattern. If, while changing symmetricity, they happen to form a configuration $\mathbb{E}'$ with $\rho(\mathbb{E}') = 1$, they instead form the pattern using the algorithm described in the previous case.

Let us describe the first procedure that allows the robots to appropriately reduce the perceived symmetricity $\rho$ until it divides the symmetricity of the pattern to be formed.

The idea is the following. First the centre of the smallest enclosing circle $c$ is identified. Point $c$ is also the centre of symmetry of $\mathbb{E}$; that is, the unique point such that the robots can be divided in $\frac{n}{\rho(\mathbb{E})}$ groups each forming a regular $\rho(\mathbb{E})$-gon with centre $c$. Then each robot moves away from $c$ in a straight line according to its coordinate system of a small amount. The amount is very carefully computed so to guarantee that: (1) it is smaller than the robot's unit distance and thus can be reached instantaneously in one step, (2) if two robots are located symmetrically with respect to $c$ and have non symmetrical local coordinate systems, they will move of a *different* amount.

Depending on the activation schedule of the robots, the above procedure is shown to either break completely the symmetry in one step reaching a configuration $\mathbb{E}'$ where $\rho(\mathbb{E}') = 1$, or to reduce the symmetry eventually reaching, after repeated applications of the procedure, a configuration $\mathbb{A}$ such that $m = \rho(\mathbb{A})$ divides $\rho(\mathbb{P})$.

Now both the pattern and the configuration can be partitioned into $k = \frac{n}{m}$ regular $m$-gons all having the same center so to have a correspondence between each $m$-gon with a group of $m$ robots. Let $\mathcal{R}_1, \ldots, \mathcal{R}_k$ be the $k$ sets of robots and let $\mathcal{R}_k = \{r_1, \ldots, r_m\}$. Set $\mathcal{R}_k$ is special and it is used to create consistent coordinate systems. In fact, in this case it is not possible for the robots to agree on a common coordinate system based on a $T$-stable configuration (like for the asymmetric case). Because of the rotational symmetry induced by the $\frac{n}{m}$ regular

$m$-gons around $c$, each robot in one class of symmetry decides its destination individually. Each robot $s_j$ in $\mathcal{R}_i$ chooses as $X$-axis the line passing through the common centre $c$ and the closest among the robots in $\mathcal{R}_k$; while the unit distance is chosen as described earlier. As for the destination point: each robot chooses the closest location among the $m$ possible locations and ties are broken, for example, by chirality. In this way, the coordinate systems of robots belonging to the same class are rotational symmetric with respect to $c$ and intervals $\frac{2\pi}{m}$, and the destinations form a regular $m$-gon with the same centre that matches the one to be formed.

Notice that the algorithm described above works also for configurations and patterns with dense points, provided the robots have strong multiplicity detection. Indeed it allows to form in $\mathcal{S}$SYNC all patterns formable according to the strong *global* symmetricity $\sigma(\mathbb{E})$ of the initial configuration $\mathbb{E}$.

*Possibility in $\mathcal{A}$*SYNC. If we restrict ourselves to just plain patterns and initial configurations, and consider the weaker *local* symmetricity $\rho(\mathbb{E})$ of the initial configuration $\mathbb{E}$, than it is possible to form the patterns with symmetricity divisible by $\rho(\mathbb{E})$, even in $\mathcal{A}$SYNC:

**Theorem 14 ([16]).** *A team of oblivious robots in $\mathcal{A}$SYNC with* `Chirality`*, starting from a plain configuration $\mathbb{E}$ with $\rho(\mathbb{E})$ can form any pattern $\mathbb{P}$ such that $\rho(\mathbb{E})$ divides $\rho(\mathbb{P})$.*

It is unknown whether this can be done without chirality.

## 5 Forming a Sequence of Patterns in $\mathcal{S}$SYNC

In this chapter we have discussed, under a variety of assumptions on the robots' capabilities and features, how to form a (possibly arbitrary) pattern given in input. A natural question is whether the robots can form not just a single pattern but a *series of distinct patterns*, given in a particular order, or, more generally of characterizing the series that can be formed. To enable a series of pattern to be formed, a protocol must guarantee that a robot that wakes up in an arbitrary configuration can, in spite of its obliviousness, figure out what pattern in the sequence is being formed so to join the others in performing the required tasks. In other words, a solution must provide, through the robots' movement some form of memory in an otherwise memoryless system.

In this section we consider $\mathcal{OBLOT}$ robots with `Chirality`, in $\mathcal{S}$SYNC under unlimited mobility (i.e., all robots always reach their destinations when performing their move). The focus is on *infinite* series: *periodic* (or cyclic) series $\mathbb{S}^\infty = \langle \mathbb{P}_1, \mathbb{P}_2, \ldots, \mathbb{P}_m \rangle^\infty$, i.e. the periodic repetition of a finite series $\mathbb{S}$ of distinct patterns. The results are then generalizable to infinite *aperiodic* series. Three different scenarios are analyzed, depending on the level of anonymity of the robots: completely anonymous robots, visibly indistinguishable but ordered set of robots and distinctly labeled robots.

Before describing the three scenarios, we introduce some special patterns needed in the rest of the section: (1) `POINT` is the pattern consisting of a single point; (2) TWO-POINTS is the pattern consisting of exactly two points; (3) POLYGON($k$), for any $k \geq 3$, is the pattern consisting of points $p_1, p_2, \ldots, p_k$ that are vertexes of a regular convex polygon of $k$ sides.

### 5.1 Anonymous Robots

Consider $n$ identical robots starting from distinct locations. Central to the anonymous case is the notion of symmetry in a configuration, which is quantified using the concept of *centered view*, and *centered symmetricity*, $\hat{\rho}$, a slight modification of the notion of local view and of symmetricity $\rho$ discussed in Sect. 4.

If $r_i$ is not located at the centre of the smallest enclosing circle, its centered view $CV_i(t)$ contains the coordinates of all the other robots considering as origin $(0, 0)$ its own position and as $(1, 0)$ the position of the center. On the other hand, if $r_i$ is in the centre of the smallest enclosing circle, the origin is still the location of $r$, but any robot $r_j$ whose view $CV_j(t)$ is minimum among all the other robots is thought to be at coordinate $(1, 0)$. Finally, no information about the coordinate system of the robots is available in these views because they are assumed unknown and not necessarily consistent.

Notice that, given any arbitrary configuration $\mathbb{E}$, there is a total order of the distinct centered views of the robots in $\mathbb{E}$, in spite of their anonymity. The elements of $CV_i$ can be ordered lexicographically to obtain an ordered sequence $Q(CV_i)$, for each robot $r_i \in \mathbb{E}$. For any two robots $r_i$ and $r_j$, the ordered sequences $Q(CV_i)$ and $Q(CV_j)$ contain the same number of elements and these sequences can be ordered lexicographically. So, $CV_i < CV_j$ if and only if $Q(CV_i)$ is lexicographically smaller than $Q(CV_j)$.

An obvious consequence of anonymity is that from a configuration $\mathbb{E}$ consisting of anonymous robots at $w$ distinct locations, a configuration $\mathbb{E}'$ where the robots occupy more than $w$ distinct locations might not be reachable, which restricts the size of patterns in any formable series of patterns. To form repetitively any series $\mathbb{S}$ of patterns, all the patterns in $\mathbb{S}$ should be of the same size. Thus, only patterns of size $n$ are considered, where $n$ is the number of robots. Each robot starts from a distinct location and during the pattern formation algorithm, no two robots should occupy the same location (i.e. no dense points are allowed). Moreover, those patterns are indeed formable.

The formation algorithm is based on the identification of special configurations: the *bi-circular* and the *q-symmetric-circular* configurations. Before giving an intuition of the technique employed, we define these special configurations (see Fig. 8 for an example of a bi-circular configuration).

**Definition 2 (BCC).** *A configuration is called* bi-circular *(denoted by BCC) if: (i) there is a unique location (called the* pivot*), such that the smallest enclosing circle SEC containing all the robots, has diameter more than three times the diameter of the circle C containing all robots except those at the pivot; (ii) SEC and C intersect at exactly one point: the point directly opposite the pivot (called the* base-point*).*

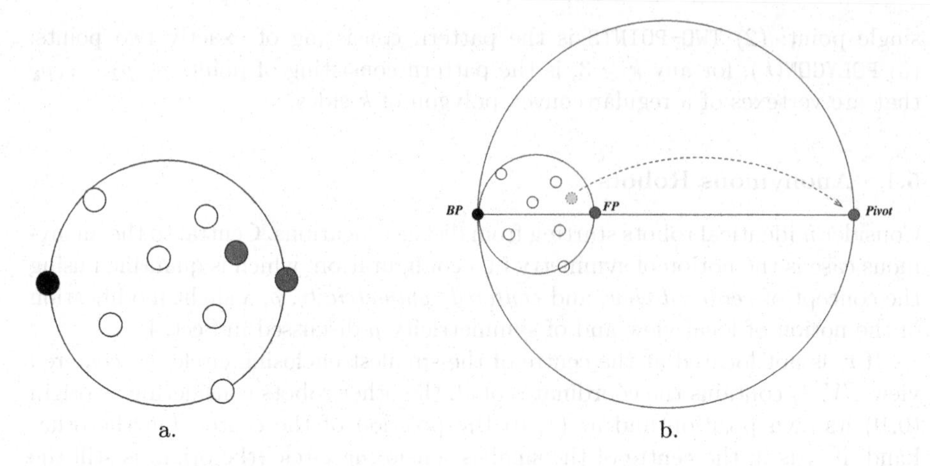


**Fig. 8.** (a) An arbitrary configuration of robots and the smallest enclosing circle. (b) A bi-circular configuration.

**Definition 3** (SCC). *A configuration containing n robots is called* q-symmetric-circular *or, SCC(q), $1 < q < n$, if: (i) the smallest enclosing circle SEC has exactly q points on its circumference that are occupied by robots; (ii) all the other robots lie on or in the interior of a smaller circle C that is concentric to SEC such that $Diameter(SEC) \geq (5 + \sin^{-1}(\pi/q)) \cdot Diameter(C)$; (iii) there are no robots in the center of SEC.*

In both configurations, the former circle (*SEC*) is called the *primary enclosure* while the latter (*C*) is called the *secondary enclosure*. The point on the secondary enclosure directly opposite the base-point is called the *frontier-point*. The ratio of the diameter of the primary enclosure over the diameter of the secondary enclosure is called the *stretch* of the configuration.

An interesting property of the bi-circular configuration is that in such a configuration the robots can agree on a coordinate system and define a unique way to order the robots. It can also be shown that from an arbitrary initial configuration either a particular type of BCC configuration or a particular type of SCC($q$) configuration can always be formed. More precisely:

**Lemma 4** ([17]). *Starting from any configuration $\mathbb{E}$ with symmetricity $\hat{\rho}(\mathbb{E}) = q$, and for any $k \geq (5 + \sin^{-1}(\pi/q))$ we can reach a configuration $\mathbb{E}'$ such that either (i) $\mathbb{E}'$ is SCC($q'$) having stretch k, where $q' > 1$ is a factor of q, or, (ii) $\mathbb{E}'$ is BCC having stretch $k' = (k+1)/2$.*

It can also be shown that, once a bi-circular configuration containing $n$ robots is formed, any pattern $\mathbb{P}$ of size $n$ can be formed.

**Lemma 5** ([17]). *(i) In any bi-circular configuration, the robots can agree on a unique coordinate system. (ii) Starting from a bi-circular configuration with $n \geq 4$ robots in distinct locations, any pattern $\mathbb{P}$ of size n can be formed.*

Similarly, it can be shown that:

**Lemma 6** ([17]). *Starting from a configuration of type SCC(q), $q > 1$, with n robots occupying distinct locations any pattern $\mathbb{P}$ such that the symmetricity $\hat{\rho}(\mathbb{P}) = q \cdot a$, $a \geq 1$ and size($\mathbb{P}$) = n can be formed.*

Based on the above properties, the idea of the algorithm for forming a cyclic series of distinct patterns $\langle \mathbb{P}_1, \mathbb{P}_2, \ldots, \mathbb{P}_m \rangle^\infty$ by $n$ anonymous robots is the following.

Let $F$ be a function that maps each pattern $\mathbb{P}_i$ to a real number $t_i = F(\mathbb{P}_i)$ that satisfies the condition of Lemma 6. To signal the formation of pattern $\mathbb{P}_i$, one of the following configurations is unambiguously used: either SCC($x$) with stretch $k_i$, where $x$ is any factor of $q$ or, configuration BCC with stretch $k'_i = (k_i + 1)/2$. Due to Lemma 4 it is possible to form one of these configurations starting from an arbitrary configuration of symmetricity $q$. By computing the stretch of the configuration, the robot can then identify which pattern $\mathbb{P}_i$ is being formed. The robots can then form, by Lemmas 5 and 6, pattern $\mathbb{P}_i$. During the formation of pattern $\mathbb{P}_i$, at each intermediate configuration, each robot can uniquely identify which pattern is being formed. Once the pattern has been completed the resulting configuration has symmetricity $q$. Hence, by Lemma 4, it is again possible to form a SCC or BCC configuration having the appropriate stretch for the next pattern $\mathbb{P}_{i+1}$ in the sequence. Using this technique, the robots can move from one pattern to the next, and thus they can form the required sequence of patterns.

**Theorem 15** ([17]). *In* $\mathcal{S}$SYNC *with* unlimited mobility *and* chirality, *n anonymous robots starting from distinct locations in an arbitrary configuration $\mathbb{E}$, can form a cyclic series of distinct patterns $\langle \mathbb{P}_1, \mathbb{P}_2, \ldots, \mathbb{P}_m \rangle$, each of size n, if and only if $\hat{\rho}(\mathbb{P}_i) = \hat{\rho}(\mathbb{P}_j) \geq \hat{\rho}(\mathbb{E})$ $\forall i, j \in \{1, 2, \ldots m\}$.*

The condition imposed by the previous theorem on the kind of patterns in the sequence can be relaxed if the robots are equipped with lights: this scenario will be analyzed in Chap. 11.

## 5.2 Robots with Distinct Visible Identities

Let us consider now the case when each robot $r_i$ has a unique identity $ID_i$ (w.l.g, $ID_i = i$) and any other robot can see this identity. During the Look operation, a robot $r_i$ obtains a snapshot containing $(j, x_j, y_j)$ tuples where $j \neq i$ and $(x_j, y_j)$ is the location of the $j$-th robot, with respect to the local coordinate system of robot $r_i$. In this case, even in absence of agreement on directions, the symmetry among the robots can be broken by the use of distinct labels. The view of each robot is unique as it contains information about both the identities and locations of the other robots. Thus, there are no symmetric configurations. Moreover, as opposed to the anonymous case, robots can be allowed to form dense points, since the robots can be separated later, if required.

When there is only one robot, the only pattern that can be formed is obviously POINT. With $n = 2$ robots, only two patterns can be formed: POINT and TWO-POINTS and it is easy to form the sequence (POINT, TWO-POINTS)$^\infty$, by movement of a single robot (say $r_2$). The more interesting cases occur when there are at least three robots (i.e., $n \geq 3$), in this case any sequence of distinct patterns $\mathbb{S} = \langle \mathbb{P}_1, \mathbb{P}_2, \ldots \mathbb{P}_m \rangle$ can be formed, with the only restriction that each pattern $\mathbb{P}_i$ has at most $n$ points. A description of the algorithm is given below.

Robots $r_1$, $r_2$ and $r_n$ have special roles. In particular, $r_1$ and $r_2$ remain fixed in distinct locations for the entire algorithms serving as fixed points of reference for the other robots. The idea is to apply a known function $F$ to each pattern $\mathbb{P}_j$ so to obtain a real number $w_j = F(\mathbb{P}_j)$, $w_j \in (1, \infty)$ (distinct for every pattern). Before forming pattern $\mathbb{P}_j$, robot $r_k$ moves to a location between $r_1$ and $r_2$ such that the ratio of distances $dist(r_1, r_2)/dist(r_1, r_n)$ is equal to $w_j$. This is the signal for the other robots to indicate which pattern is being formed. Each robot $r_i$, $2 < i < n$ can compute the location where it should move to in order to form pattern $\mathbb{P}_j$. Once each of these robots has moved into the correct positions, robot $r_n$ moves to complete the pattern. During the execution of the algorithm every configuration of the robots (excluding at most the first two configurations) either corresponds to some pattern $\mathbb{P}_l \in \mathbb{S}$, or is an intermediate configuration which signals the formation of $\mathbb{P}_l$ (i.e. where $r_1$, $r_2$, and $r_n$ maintain a ratio of $w_j = F(\mathbb{P}_l)$). The function $F$ must be chosen in such a way that the ratio $dist(r_1, r_2)/dist(r_1, r_n)$ in an actual pattern never matches any values in the range of $F$. Thus, each robot can unambiguously determine the location that it needs to move to, by looking at the current configuration.

This algorithm works for any sequence of patterns not containing the `POINT` pattern. In order to include the `POINT` pattern in the sequence of patterns formed, small modifications must be done to the algorithm in the behaviour of robots $r_2$ and $r_n$. Based on the algorithm above, the author conclude that:

**Theorem 16** **([17]).** *In* $\mathcal{S}$SYNC *with* unlimited mobility *and* chirality, $n \geq 2$ *robots having distinct visible identities, can form any cyclic sequence of distinct patterns* $\langle \mathbb{P}_1, \mathbb{P}_2, \ldots, \mathbb{P}_m \rangle$ *provided that* $\forall i$, $size(\mathbb{P}_i) = n_i \leq n$.

### 5.3 Robots with Invisible Distinct Identities

In this case the identities of the robots are not visible to other robots. The robots are assumed to be ordered with labels $1, 2, 3, \ldots, n$ and each robot $r_i$ knows its own label $i$, but it can not visibly identify the label of other robots. In this case, the information contained in the views of the robots is similar to the anonymous case. Thus, two robots may have identical views (in particular, robots at the same location have identical views). However, since the robots have distinct identities, they can execute different algorithms depending on their own labels.

Consider first the case when there are at least four robots. The `BCC` configuration, defined for the anonymous case, is used here as well to signal the formation of specific patterns in a series. As already mentioned, dense points are allowed and the algorithm must ensure that there is at least one robot at the pivot and one at the base-point of the bi-circular configuration.

From any arbitrary configuration $\mathbb{E}$ with more than 3 robots, a bi-circular configuration of any given stretch $k > 3$, can be formed by the movement of a single robot (this single robot will place itself in a pivot position).

The technique for forming any given pattern $\mathbb{P}$ starting from a bi-circular configuration of stretch $k_i$ is as follows: As mentioned before, the bi-circular configuration can be formed by robot $r_n$ jumping to the pivot location. Once the robots are in bi-circular configuration BCC with stretch $k_i$, robot $r_1$ and robot $r_{n-1}$ occupy the base-point and the frontier-point. These three robots remain in their location while the other robots move to the required positions for forming pattern $\mathbb{P}$. The positions are assigned in the following manner. The points in the pattern $\mathbb{P}$ are mapped to locations in the bi-circular configuration such that the smallest enclosing circle of pattern $\mathbb{P}$ coincides with the secondary enclosure of the configuration and the base-point coincides with the lexicographically smallest point $p_i$ on the smallest enclosing circle of $\mathbb{P}$, i.e., $p_i \in SEC(\mathbb{P})$ and $p_i \leq p_j$, for any $p_j \in SEC(\mathbb{P})$. Notice that this mapping is unique. Let $\Gamma(\mathbb{P})$ be the unique mapping obtain by each robot (i.e., the locations that correspond to points in the pattern $\mathbb{P}$). The elements of $\Gamma(\mathbb{P})$ are sorted in such a way that the first point is the base-point of the current BCC configuration of the robots, and all points which lie on the secondary enclosure $\mathcal{C}$ precede those that are located in the interior of $\mathcal{C}$. For $1 \leq i \leq size(\mathbb{P}_i)$ robot $r_i$ is assigned the $i$th location in $\Gamma(\mathbb{P})$ and for $size(\mathbb{P}_i) < j \leq n$ robot $r_j$ is assigned the $n$-th location in $\Gamma(\mathbb{P})$.

During the formation of a pattern $\mathbb{P}_i$ of size $size(\mathbb{P}_i)$, the algorithm ensures that the BCC configuration is maintained by keeping robots $r_1$, $r_{n-1}$ and $r_n$ stationary at the base-point, at the frontier-point and at the pivot positions respectively. Only when all the other robots have moved to their assigned location, robot $r_{n-1}$ moves to its own assigned location, and also this is done ensuring that BCC is preserved with the appropriate stretch so that robot $r_n$ can unambiguously move to the required position to complete the pattern.

The remaining cases are when there are exactly 2 or 3 robots. For $n = 2$, the case of invisible identities is same as that of visible identities. The case of $n = 3$ has been studied in [18] and an algorithm for forming any sequence of patterns of at most three points has been given. As mentioned before, the transformations between any two patterns of size 3 is straightforward and requires the movement of a single robot (say $r_3$). The only challenging scenario involves the formation of POINT and TWO-POINTS, where the intermediate configurations before and after forming POINT must be distinguished from the configuration forming TWO-POINTS. In conclusion:

**Theorem 17** ([17]). *In* $\mathcal{S}$SYNC *with* unlimited mobility *and* chirality, *n robots having distinct invisible identities can form any cyclic sequence of distinct patterns* $\langle \mathbb{P}_1, \mathbb{P}_2, \ldots, \mathbb{P}_m \rangle$ *where* $\forall i$, $size(\mathbb{P}_i) \leq n$.

## References


1. Ando, H., Suzuki, I., Yamashita, M.: Formation and agreement problems for synchronous mobile robots with limited visibility. In: Proceedings of the 1995 IEEE Symposium on Intelligent Control, pp. 453–460 (1995)
2. Chatzigiannakis, I., Markou, M., Nikoletseas, S.: Distributed circle formation for anonymous oblivious robots. In: Ribeiro, C.C., Martins, S.L. (eds.) WEA 2004. LNCS, vol. 3059, pp. 159–174. Springer, Heidelberg (2004). https://doi.org/10.1007/978-3-540-24838-5_12

3. Défago, X., Konagaya, A.: Circle formation for oblivious anonymous mobile robots with no common sense of orientation. In: Proceedings of 2nd Workshop on Principles of Mobile Computing, pp. 97–104 (2002)
4. Défago, X., Souissi, S.: Non-uniform circle formation algorithm for oblivious mobile robots with convergence toward uniformity. Theor. Comput. Sci. **396**(1–3), 97–112 (2008)
5. Flocchini, P., Prencipe, G., Santoro, N., Widmayer, P.: Arbitrary pattern formation by asynchronous oblivious robots. Theor. Comput. Sci. **407**(1–3), 412–447 (2008)
6. Kasuya, M., Ito, N., Inuzuka, N., Wada, K.: A pattern formation algorithm for a set of autonomous distributed robots with agreement on orientation along one axis. Syst. Comput. Jpn. **37**(10), 89–100 (2006)
7. Katreniak, B.: Biangular circle formation by asynchronous mobile robots. In: Pelc, A., Raynal, M. (eds.) SIROCCO 2005. LNCS, vol. 3499, pp. 185–199. Springer, Heidelberg (2005). https://doi.org/10.1007/11429647_16
8. Sugihara, K., Suzuki, I.: Distributed algorithms for formation of geometric patterns with many mobile robots. J. Robot. Syst. **13**, 127–139 (1996)
9. Suzuki, I., Yamashita, M.: Distributed anonymous mobile robots: formation of geometric patterns. SIAM J. Comput. **28**(4), 1347–1363 (1999)
10. Tanaka, O.: Forming a circle by distributed anonymous mobile robots. Master's thesis, Department of Electrical Engineering (1992)
11. Wang, P.K.C.: Navigation strategies for multiple autonomous mobile robots moving in formation. J. Robot. Syst. **8**(2), 177–195 (1991)
12. Yamauchi, Y., Uehara, T., Kijima, S., Yamashita, M.: Plane formation by synchronous mobile robots in the three-dimensional euclidean space. J. ACM **64**(3), 16:1–16:43 (2017)
13. Dieudonné, Y., Petit, F., Villain, V.: Leader election problem versus pattern formation problem. In: Lynch, N.A., Shvartsman, A.A. (eds.) DISC 2010. LNCS, vol. 6343, pp. 267–281. Springer, Heidelberg (2010). https://doi.org/10.1007/978-3-642-15763-9_26
14. Fujinaga, N., Ono, H., Kijima, S., Yamashita, M.: Pattern formation through optimum matching by oblivious CORDA robots. In: Lu, C., Masuzawa, T., Mosbah, M. (eds.) OPODIS 2010. LNCS, vol. 6490, pp. 1–15. Springer, Heidelberg (2010). https://doi.org/10.1007/978-3-642-17653-1_1
15. Yamashita, M., Suzuki, I.: Characterizing geometric patterns formable by oblivious anonymous mobile robots. Theor. Comput. Sci. **411**(26–28), 2433–2453 (2010)
16. Fujinaga, N.: Oblivious pattern formation algorithms for asynchronous mobile robots based on bipartite matching approach. Master's thesis, Department of Informatics, Kyushu University (2012)
17. Das, S., Flocchini, P., Santoro, N., Yamashita, M.: On the computational power of oblivious robots: forming a series of geometric patterns. In: Proceedings of 29th Annual ACM Symposium on Principles of Distributed Computing (PODC), pp. 267–276 (2010)
18. Bouzid, Z., Lamani, A.: Robot networks with homonyms: the case of patterns formation. In: Défago, X., Petit, F., Villain, V. (eds.) SSS 2011. LNCS, vol. 6976, pp. 92–107. Springer, Heidelberg (2011). https://doi.org/10.1007/978-3-642-24550-3_9

# Gathering

Paola Flocchini[✉]

University of Ottawa, Ottawa, Canada
pflocchi@uottawa.ca

**Abstract.** In this Chapter, we focus on the GATHERING problem: that is, the problem of having the robots, initially located in arbitrary distinct points of the plane, gather in the exact same location. In this Chapter we examine GATHERING in the standard $\mathcal{OBLOT}$ model when robots have unlimited visibility; we also briefly review results about the relaxed problem of CONVERGENCE, where robots only need to move infinitely close to each other, without necessarily reaching the same point.

## 1 Introduction

The previous Chapter treated one of the most important problems for $\mathcal{OBLOT}$: pattern formation. Among specific patterns, a special place is occupied by two classes: `Point` and `Uniform Circle`. The class `Point` is the set consisting of a single point; point formation corresponds to the important GATHERING problem requiring all robots to gather at a same location, not determined in advance. The other important class of patterns is `Uniform Circle`: the points of the pattern form the vertices of a regular $n$-gon, where $n$ is the number of robots.

In addition to their relevance as individual problems, the classes `Point` and `Uniform Circle` play another important role. As seen in the previous Chapter, a crucial observation is that formability of a pattern $P$ from an initial configuration $\Gamma$ depends on the relationship between the *symmetricity* of $P$ and the symmetricity of $\Gamma$. More precisely, a pattern $P$ can be formed from a configuration $\Gamma$ only if the symmetricity of $\Gamma$ divides that of $P$ [20,30]. Based on this observation, it follows that the only patterns that *might* be formable from any arbitrary initial configuration are the ones with maximum symmetricity, that is, `Point` and `Uniform Circle`. It is rather easy to see that both points and uniform circles can be formed in $\mathcal{F}$SYNC, i.e., if the robots are fully synchronous. After a long quest by several researchers, it has finally been shown that `Point`, as well as `Uniform Circle` are indeed formable in $\mathcal{A}$SYNC (and thus also in $\mathcal{S}$SYNC), but with extremely complex solutions. The complexity of the problems is due to the difficulties inherent in the simultaneous presence of asynchrony, obliviousness, and disorientation. As we will see in the next Chapter, `Uniform Circle` can be formed from arbitrary initial configurations without additional assumptions in the standard $\mathcal{OBLOT}$ model; on the other hand, as we will see in this Chapter, the formation of `Point` requires the introduction of either some form of orientation or multiplicity detection.


P. Flocchini et al. (Eds.): Distributed Computing by Mobile Entities, LNCS 11340, pp. 63–82, 2019.
https://doi.org/10.1007/978-3-030-11072-7_4

This Chapter is mainly devoted to the solution to GATHERING in $\mathcal{A}$SYNC [7]. More precisely, in Sect. 4 we show the impossibility result for GATHERING with disoriented robots without multiplicity detection; in Sect. 5 we describe the solution to GATHERING for disoriented robots in $\mathcal{A}$SYNC with multiplicity detection, and in Sect. 6 for oriented robots in absence of multiplicity detection; in Sect. 7 we briefly summarize the current state of the art for the related problem of CONVERGENCE, and for the special case of RENDEZVOUS (where the system contains only two robots), as well as for the case of robots with limited visibility (treated in detail in Chap. 7); in Sect. 8 we summarize the feasibility results.

## 2 The Problems

Let $\mathcal{R} = \{r_1, \ldots, r_n\}$ be a set of $n$ robots in the plane, operating in `Look-Compute-Move` cycles following the $\mathcal{OBLOT}$ model (for the details of the model see Chap. 1).

Let $dist^t(r_i, r_j)$ denote the distance between robots $r_i$ and $r_j$ at time $t$, and $D(t) = \max\{dist^t(r_i, r_j)\}$ the maximum distance between any two robots at that time.

The GATHERING problem is solved if $\exists t : \forall t' \geq t, D(t') = 0$, that is, in finite time all robots are located at the same point, not necessarily determined a priori. When $n = 2$ the problem is usually called RENDEZVOUS.

CONVERGENCE is achieved if $lim_{t \to \infty} D(t) = 0$, that is, if the robots get arbitrarily close to each other.

## 3 Preliminary Observations

**Rendezvous.** RENDEZVOUS occupies a special place in the research on mobile robots because such a simple situation presents more challenges than the general case. The problem is easily solvable in $\mathcal{F}$SYNC, where the simple strategy that lets a robot move to the half point between the robots' positions would achieve rendezvous, but, without introducing agreement on the coordinate system, it is unsolvable in $\mathcal{S}$SYNC.

**Theorem 1** ([30]). *Without any agreement on the local coordinate systems, in* $\mathcal{S}$SYNC, GATHERING *of* $n = 2$ *robots is* impossible, *even with multiplicity detection.*

It is not difficult to see why this is the case. Consider two robots $r$ and $s$ that agree on one axis but not on the orientation of the other. These robots have specular view of the environment. Assume by contradiction that a solution exists and let $\mathcal{A}$ the corresponding protocol. Consider an execution $\mathcal{E}$ of $\mathcal{A}$ where, in the very last round before achieving rendezvous, robot $r$ is activated and moves while robot $s$ sleeps. Such an execution can be shown to exist. Consider now the execution $\mathcal{E}'$ up to (and excluding) the last round; at this point, let both robots be activated. When this happens, $r$ will perform the same move as in $\mathcal{E}$, whose destination is the observed position of $s$ and, since the view of $s$ is specular, $s$ will choose the observed position of $r$ as its destination. The result will be a switch of the positions of the robots. Since they are oblivious, in the same conditions they will repeat the same actions; this means that, if they are both activated in every turn from now on, they will continue to switch without ever gathering.

**Gathering in $\mathcal{F}$SYNC.** In $\mathcal{F}$SYNC, the *move-to-half* solution can be generalized to solve GATHERING of $n > 2$ robots by having the robots *move to the centre of gravity* of their positions. In $\mathcal{F}$SYNC all robots are activated at the same time, they can all easily compute the centre of gravity (which is the same for all robots), and they all move there (Algorithm GO-TO-CoG).

**Theorem 2** ([9]). GATHERING *is solvable in* $\mathcal{F}$SYNC *in the standard* $\mathcal{OBLOT}$ *model.*

This simple algorithm, however, can achieve GATHERING only because of synchronicity. In fact, in $\mathcal{S}$SYNC, where only a subset of robots is activated at each round, the protocol fails. The reason of the failure is that the center of gravity is not invariant with respect to robots' movements towards it, and the robots are oblivious. Once a robot makes a move towards the center of gravity, its position changes; since the robots act independently and have no memory of the past, a robot (even the same one) observing the new configuration will compute and move towards a different point.

## 4 Impossibility of Gathering in $\mathcal{S}$SYNC

If the robots have no agreement on the coordinate system and cannot detect multiplicity, GATHERING is not solvable in $\mathcal{S}$SYNC (and thus in $\mathcal{A}$SYNC) [28].

**Theorem 3** ([28]). *In absence of multiplicity detection and of any agreement on the coordinate systems,* GATHERING *is deterministically unsolvable in* $\mathcal{S}$SYNC.

To see why this is the case, assume that the $n > 2$ robots in the system have no agreement on the coordinate system and cannot detect multiplicity. By contradiction, let $\mathcal{A}_g$ be a deterministic algorithm that correctly solves the GATHERING problem in $\mathcal{S}$SYNC.

Consider the following scenario: (*i*) All robots have the same unit distance; (*ii*) All robots move of at each round of at least $\delta$ (the same for all robots); (*iii*) Robots $r_1, \ldots, r_{n-1}$, called the *black* robots, have the same orientation and direction of the local coordinate system, while $r_n$, f called the *white* robot, has a local coordinate system where both axes have the same direction but opposite orientation with respect to the coordinate system of the black robots.

Note that the information on the unit distance or the common minimum movement, as well as the black and white colouring, are not known to the robots.

Clearly, $\mathcal{A}_g$ must solve gathering also in the scenario described above. By definition, if the robots execute $\mathcal{A}_g$, they will gather on the same point, say P, in finite time, say at time $t_g$. Let $\mathcal{E}$ be an arbitrary execution (defined in terms of the activation of the robots) of $\mathcal{A}_g$ in $\mathcal{S}$SYNC. We now design an alternative activation schedule of the robots, $\mathcal{E}'$, that behaves exactly as $\mathcal{E}$ until time $t_g - 1$.

1. If $r_n$ is not on P at time $t_g - 1$, then in $\mathcal{E}'$ $r_n$ is left inactive at this time and all the other robots behave like in $\mathcal{E}$. At time $t_g$ the robots then reach a configuration $\mathbb{E}_1$ where all the black robots occupy the same position $p_b$, while the white robot does not (see Fig. 1a). At time $t_g$ the robots in the system sense the world as if there were only two robots (they cannot detect multiplicity), and these two robots have the same view of the world. Hence, following the same reasoning of the impossibility proof for $n = 2$, we can conclude that $\mathcal{A}_g$ does not correctly solve the GATHERING problem.
2. Otherwise (i.e., if $r_n$ is on P at time $t_g - 1$), in $\mathcal{E}'$ at time $t_g - 1$, $r_{n-1}$ is left inactive, while all the other robots behave like in $\mathcal{E}$. Also in this case, at time $t_g$ the robots reach a configuration where $n - 2$ of the black robots and the white robot occupy the same position $p_w$, while the last black robot does not: we will denote such a configuration as $\mathbb{E}_2$ (see Fig. 1b).

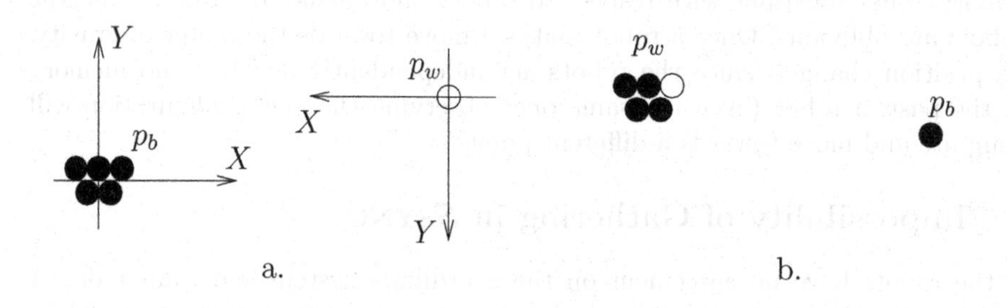


**Fig. 1.** [28]: configurations used in the proof of Theorem 3. (a) A $\mathbb{E}_1$-configuration, (b) a $\mathbb{E}_2$-configuration.

In this case, the situation is more delicate. As before, only two points on the plane are occupied by robots: on one, $p_b$, there is $r_{n-1}$ (a black robot), and on the other one, $p_w$, there are $r_1, \ldots, r_{n-2}$ (black robots) and $r_n$ (white robot). However, the robots on $p_w$ do not all have the same view of the world, hence the argument of the previous case cannot be used.
We build a new activation schedule $\mathcal{E}''$ as follows. $\mathcal{E}''$ is the same as $\mathcal{E}'$ until time $t_g$. After time $t_g$, $\mathcal{E}''$ activates $r_1, \ldots, r_{n-2}$ always together (i.e., as if they were one robot); $r_{n-1}$ and $r_n$ are arbitrarily and fairly activated by $\mathcal{E}''$. Since, by hypothesis, $\mathcal{A}_g$ solves GATHERING, in a finite number of cycles, say at time $\tilde{t}$, the robots gather at P.
Note that, at time $\tilde{t} - 1$, it is not possible that both $r_n$ and $r_{n-1}$ are already at P. In fact, this would imply that $r_n$ (and $r_{n-1}$) would not move between time $\tilde{t} - 1$ and $\tilde{t}$, and that only the robots $r_1, \ldots, r_{n-2}$ would move. However, the view of the world of $r_n$ and of $r_1, \ldots, r_{n-2}$ at time $\tilde{t} - 1$ is the same; hence, since $\mathcal{A}_g$ is assumed to be deterministic, $r_1, \ldots, r_{n-2}$ should take the same decision taken by $r_n$, that is to *not* move, thus not reaching gathering at time $\tilde{t}$.

A similar argument can be applied to show that (1) It is impossible that, at time $\tilde{t}-1$, $r_1,\ldots,r_{n-1}$ are already at P, while $r_n$ is not; (2) It is impossible that, at time $\tilde{t}-1$, $r_n$ and $r_{n-1}$ are already at P, while $r_1,\ldots,r_{n-2}$ are not. At this point, we build a third activation schedule $\mathcal{E}'''$ that is the same as $\mathcal{E}''$ until time $\tilde{t}-1$. The behaviour of $\mathcal{E}'''$ is as follows:

(a) If at time $\tilde{t}-1$ no robot is at P, then in $\mathcal{E}'''$ robots $r_1,\ldots,r_{n-1}$ are activated, while $r_n$ is not. Hence, at time $\tilde{t}$ the robots are in a $\mathbb{E}_1$-configuration.
(b) If at time $\tilde{t}-1$ only $r_{n-1}$ is at P, then in $\mathcal{E}'''$ robots $r_1,\ldots,r_{n-2}$ are activated, while $r_n$ is not. Hence, at time $\tilde{t}$ the robots are in a $\mathbb{E}_1$-configuration.
(c) If at time $\tilde{t}-1$ only robots $r_1,\ldots,r_{n-2}$ are at P, then in $\mathcal{E}'''$ robots $r_{n-1}$ is activated, while $r_n$ is not. Hence, at time $\tilde{t}$ the robots are in a $\mathbb{E}_1$-configuration.
(d) If at time $\tilde{t}-1$ only $r_n$ is at P, then in $\mathcal{E}'''$ robots $r_1,\ldots,r_{n-2}$ are active, while $r_{n-1}$ is not. Hence, the robots are once again in a $\mathbb{E}_2$-configuration.

In (a)–(c), a contradiction is reached by following the previous case. In (d), by iterating the above argument, we have that either the robots keep forming $\mathbb{E}_2$-configuration, or they form an $\mathbb{E}_1$-configuration; in both cases, it is shown that $\mathcal{A}_g$ cannot correctly solve the GATHERING problem.

This impossibility holds even if the adversary is very restricted in the choice of the scheduler. Indeed it holds if the scheduler is not only fair and centralized but also *slicing* (i.e., only one robot is activated in each round, and starting from time $t=0$, after $n$ successive rounds, all the robots in the system have been activated exactly once) [14].

# 5 Gathering in $\mathcal{A}$SYNC with Multiplicity Detection

As seen earlier, gathering is impossible in absence of any form of agreement on the coordinate system and of multiplicity detection (see Theorem 3); this is true even if the robots are rigid (i.e., they always reach their destination in a single round). The question then becomes what additional capabilities/assumptions need to be made for gathering to become possible. The algorithm described in this section achieves GATHERING in $\mathcal{A}$SYNC and uses, as additional assumption, multiplicity detection.

## 5.1 Preliminary Observation and Terminology

**Convergence Based on Invariant Target Points.** Before describing the solution to the GATHERING problem, we make some observation about the simpler CONVERGENCE problem, where the robots are required only to move infinitely close to each other, without necessarily gathering at the same point. CONVERGENCE can be easily solved by having each robot compute some appropriate target point (e.g., the centre of gravity) and move there. The center of gravity, however, is not invariant with respect to robots' movements towards it, which means that after a robot moves in its direction, the position of the center of gravity changes. As a consequence, since the robots are oblivious and they act independently from each other, a robot performing the Look operation will not necessarily compute the same point it computed in the previous cycle. For CONVERGENCE this is shown not to be a problem, because the targets get closer and closer to each other, but GATHERING cannot be achieved.

The natural solution for gathering would be then to choose as destination a point that, unlike the center of gravity, is *invariant* with respect to the robots' movements towards it. The only known point with such a property is the unique point in the plane that minimizes the sum of the distances between itself and all positions of the robots (the *Weber* (or *Fermat* or *Torricelli*) *point*) [25,31]. Unfortunately, the Weber point is not expressible as an algebraic expression involving radicals since its computation requires finding zeroes of high-order polynomials [2]. In other words, the Weber point is *not computable* even by radicals; thus it cannot be used to solve the gathering for $n \geq 5$. Interestingly, even *convergence* towards the Weber point can not be guaranteed due to its instability with respect to changes in the point set [17].

In conclusion, to solve GATHERING a very different strategy has then to be devised.

**Background and Difficulties.** If robots can detect multiplicity, a strategy starting from distinct initial positions of the robots could be to have some of them create a unique point of multiplicity (a dense point). Once such a point is created, all the other robots could simply move there solving the problem. This idea has been employed in the $\mathcal{S}$SYNC model, with the assumption of *fixed mobility* (i.e, a robot always travels the minimum between the distance to its destination and a fixed $\delta$, reaching its destination point) [30]. Depending on the initial configuration, the dense point is formed either in the centre of the smallest enclosing circle containing all the robots, or on the rim of the smallest enclosing circle. Such a solution is however heavily based on instantaneous movements and on synchronicity.

In the asynchronous setting the general strategy of creating a unique dense point can be employed, but in this case the overall gathering algorithm is very complex [7]. In fact, several difficulties have to be overcome because of the combination of asynchrony and obliviousness.

*Collision Avoidance.* Among the various difficulties is the one of avoiding *collisions*: since the robots do not look while moving, and the destination is computed based on possibly outdated information about the position (and moves) of the other robots, to avoid collisions, the computation of a robot $r$ must take into account all possible movements of all the other robots from the time $t$ of the Look to the *unknown* and a-priori *unbounded* time $t' > t$ when $r$ will actually end its move. In other words, collision avoidance, if required, is difficult and it is sole responsibility of the protocol. As we will see, the algorithm of [7] will ensure that collision are never except for the final gathering point.

*Symmetry Breaking.* An additional difficulty due to obliviousness and related to collisions is that if two robots (accidentally or by design) terminate a cycle at the same location, then they become potentially indistinguishable, and from that moment on they might behave exactly in the same way (in fact there is at least one execution in which they will do so); in particular, it might not be possible for them to separate ever again. More generally, due to asynchrony, symmetric configurations are more difficult to break.

*Symmetry Recognition.* Symmetric configuration can be an advantage for gathering. Consider, for example the *equiangular* configuration (i.e., all the robots are the vertexes of a $n$-gon). In such a case the only strategy the robots can apply to possibly solve GATHERING would be to move towards the center of the $n$-gon (GO-TO-CENTER strategy); in fact, any other movement would be symmetrically executed by all the robots by an adversary that would create a perfectly synchronized execution, resulting in yet another equiangular configuration. So, if a robot perceives an equiangular configuration, it must act in this way. On the other hand, if such a configuration is "accidentally" created by the movement of some robots during the execution, a robot starting the LOOK phase might observe the equiangular configuration and decide to apply the go-to-center strategy, while those already moving continue their procedure (possibly destroying the newly formed equiangularity). Any algorithm must ensure that, if a symmetric configuration like this one is formed during the execution, *all* robots become aware of it (recall, however, that the robots are oblivious and do not remember previous observations), so that *all* robots follow the same strategy. As we will see, the algorithm of [7] will ensure this for the class of biangular configuration (of which the equiangular is a particular case).

**Notation and Basic Definition.** We now introduce some important terminology used in the algorithm.

– **succ() and pred().** Given a set $P$ of $n$ distinct points in the plane, a point $c \notin P$ called the *center*, and the set $RadSet(P, c)$, we define the successor of $p \in P$ with respect to $c$, denoted by $\texttt{succ}(p, c)$, as the point $q \in P$ such that (refer to Fig. 2a): either $q$ is the closest point to $p$ on the radius where $p$ lies, with $dist(c, q) > dist(c, p)$ (if any); or $\overrightarrow{cq}$ is the radius following $\overrightarrow{cp}$ in the order implied by the clockwise direction, and $q$ is the closest point to $c$ on $\overrightarrow{cq}$. Symmetrically, given a point $q \in P$, the predecessor of $q$ with respect to $c$, denoted by $\texttt{pred}(q, c)$, is the point $p \in P$ such that $\texttt{succ}(p, c) = q$.

– **Cyclic String of Angles.** The functions succ() and pred() define a unique cyclic order on $P$, which we shall denote by $<p_0, p_1, \ldots, p_{n-1}>$, where $p_{i+1} = \texttt{succ}(p_i)$, and all operations on indices are modulo $n$. This, in turns, defines a cyclic *string of angles* $SA^+(P, c) = <\alpha_0, \alpha_1, \ldots, \alpha_{n-1}>$, where $\alpha_i = \sphericalangle(p_i, c, p_{i+1})$; $p_i$ is called the (clockwise) *start point* of $\alpha_i$. The string of angles in the opposite direction is denoted by $SA^-(P, c) = <\alpha_{n-1}, \ldots, \alpha_0>$.

Associated to the cyclic string of angles $SA^+(P, c)$ there is the set of strings $SA^+(P, c)[i] = <\alpha_i, \alpha_{i+1}, \ldots, \alpha_{i+n-1}>$, with $0 \leq i \leq n-1$ (refer to example depicted in Fig. 2b, where the string of angles are computed with respect to

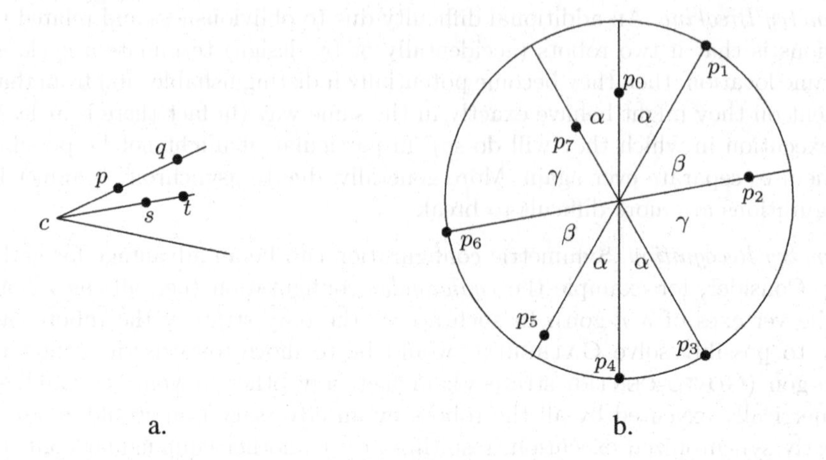


**Fig. 2.** [7]: (a) Example: $\texttt{succ}(p,c) = q$, $\texttt{pred}(q,c) = p$, $\texttt{succ}(q,c) = s$, $\texttt{pred}(s,c) = q$. (b) Example of the string of angles of $P = p_0,\ldots,p_7$, computed with respect to their *SEC*, with a clockwise orientation of the circle. We have $SA^+(P,c)[1] = \langle\alpha,\beta,\gamma,\alpha,\alpha,\beta,\gamma,\alpha\rangle$; $LexMinString(P,c) = \langle\alpha,\alpha,\beta,\gamma,\alpha,\alpha,\beta,\gamma\rangle$; $StartSet^+(P,c) = \{p_3,p_7\}$; and $StartSet^-(P,c) = \emptyset$.

the *SEC* of the 8 points); similarly, associated to $SA^-(P,c)$ there is the set of strings $SA^-(P,c)[i] = <\alpha_{i-1},\ldots,\alpha_i>$; here and in the following, all operations on the indices are modulo $n$. We define the *start point* of $SA^+(P,c)[i]$ as the start point of $\alpha_i$, that is $p_i$. Finally, let $SA(P,c)[i] = SA^+(P,c)[i] \cup SA^-(P,c)[i]$, and $SA(P,c) = \bigcup_i SA(P,c)[i]$.

– **Simple/Mixed String of Angles.** We say that $SA(P,c)$ is *simple* if $SA^+(P,c)$ does not contain any angle of zero degrees; otherwise, at least two points are on the same radius, and we say that $SA(P,c)$ is *mixed*.

– **Lexographically Minimum String of Angles.** We denote by *LexMinString*$(P,c)$ the lexicographically minimum string among all strings in $SA(P,c)$. Let $StartSet^+(P,c) = \{p_i \in P | SA^+(P,c)[i] = LexMinString(P,c)\}$ be the set of start points of $LexMinString(P,c)$ in $SA^+(P,c)$, and let $StartSet^-(P,c)$ be defined similarly. Let $StartSet(P,c) = StartSet^+(P,c) \cup StartSet^-(P,c)$.

– **Biangular Configurations.** We say that a set of $n$ distinct points in the plane $P$ is *biangular* if there exists a point $b$ such that $\forall i \geq 0\ \alpha_i = \alpha_{i+2} > 0$ where $SA^+(P,b) = <\alpha_0,\ldots,\alpha_{n-1}>$; $b$ is then called *center of biangularity* of $P$. Given a set $P$ of $n-1$ points on the plane we say that $P$ is *biangular with one gap* and center $b$ if there exists a point $x_g \notin P$, such that $P \cup \{x_g\}$ is *biangular* with center $b$. Analogous definition holds for a set of points *biangular with two gaps.* Finally, given a set $P$ of $n$ points, we say that $P$ is *irregular biangular* if there exists a point $p \in P$, the center, such that $P \setminus \{p\}$ is *regular biangular with*

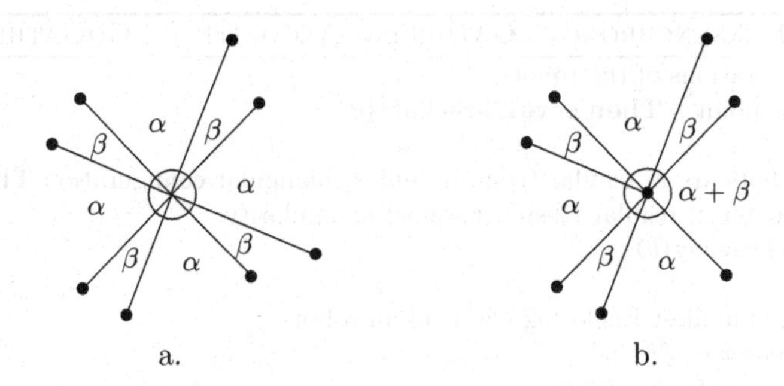


**Fig. 3.** [7]: (a) A regular biangular and (b) irregular biangular set of 8 points.

*one gap* with center $p$. (refer to Fig. 3b). Note that it can be shown that, if $P$ is irregular biangular, then its center is unique.

– **Periodic Configurations.** We say that a set $P$ of $n$ points is regular periodic if $SA^+(P, c)$ is a *periodic* string with period greater than or equal to 3, where $c$ is the center of the SEC of $P$; Similarly to the "gaps" introduced for a biangular set of points, we say that a set $P$ of $n-1$ points is *periodic with one gap* if there exists a string $W$, with $|W| \geq 3$, and $e \geq 2$ such that $SA^+(P, c) = W^{e-1} \circ W'$, with $W = \langle w_0, \ldots, w_{n/e-1} \rangle$ and $W' = \langle w_0, w_1, \ldots, w_{i-1}, \overline{w}, w_{i+2}, \ldots, w_{n/e-1} \rangle$, for some $0 \leq i \leq n/e - 1$, and with $\overline{w} = w_i + w_{i+1}$ (refer to Fig. 5b). Note that, since $n \geq 5$ and $e \geq 2$, if $P$ is periodic with one gap, then $i$ is unique. Furthermore, we say that a set $P$ of $n$ points is *irregular periodic*, if one of the points in $P$ is at $c$, and $P \setminus \{c\}$ is periodic with one gap.

### 5.2 The Strategy

The overall strategy follows the principle of making the robots create a unique dense point within finite time and then have all the remaining robots gather at this point. As already observed, since the robots are disoriented, oblivious, and operate in a totally asynchronous manner, several difficulties are encountered. There are two main difficulties already discussed to consider, the first is ensuring that a unique dense point is created other than the final gathering one (i.e., that the robots never collide during their movement); the other is to have the robot collectively recognize the presence of special symmetric configurations: the biangular configurations.

In particular, Algorithm GoGather works by examining the configuration observed by a robot in the `Look` operation (see also Fig. 4). First of all, a robot checks whether there is already a single dense point, $p$; if so, the robot moves towards $p$. If no dense point is present, the robot verifies whether the current configuration is *biangular* (of which a special case is the equiangular). If the check for a biangular configuration is positive, the robot moves towards the center of biangularity $b$ if no other robots are in the way (routine `moveIfFreeWay`). The

**Algorithm 1.** ASYNCHRONOUS GATHERING ALGORITHM – GOGATHER

$\mathcal{R}$ := Set of positions of the robots;
**If** One dense point $p$ **Then** `moveIfFreeWay(`$p$`)`.
**Else**
  **If** The robots are in regular (resp. irregular) biangular configuration **Then**
    $b$ := Center of regular (resp. irregular) biangularity;
    `moveIfFreeWay(`$b$`)`.
  **Else**
    $SEC$ := Smallest Enclosing Circle of all robots;
    $c$ := Center of $SEC$;
    **If** No robot is at $c$ **Then**
      Compute the set of strings $SA(\mathcal{R})$, $LexMinString(\mathcal{R})$;
      Compute $StartSet^{+}(\mathcal{R}), StartSet^{-}(\mathcal{R})$;
      $s := |StartSet^{+}(\mathcal{R}) \cup StartSet^{-}(\mathcal{R})|$;
      **If** $SA(\mathcal{R})$ is simple **Then** `Case 1.` **Else** `Case 3.`
    **Else** %One robot $r$ is at $c$%
      $\overline{\mathcal{R}} := \mathcal{R} \setminus \{c\}$;
      Compute the set of strings $SA(\overline{\mathcal{R}})$, $LexMinString(\overline{\mathcal{R}})$;
      Compute $StartSet^{+}(\overline{\mathcal{R}}), StartSet^{-}(\overline{\mathcal{R}})$;
      $s := |StartSet^{+}(\overline{\mathcal{R}}) \cup StartSet^{-}(\overline{\mathcal{R}})|$;
      **If** $SA(\overline{\mathcal{R}})$ is simple **Then** `Case 2.` **Else** `Case 4.`

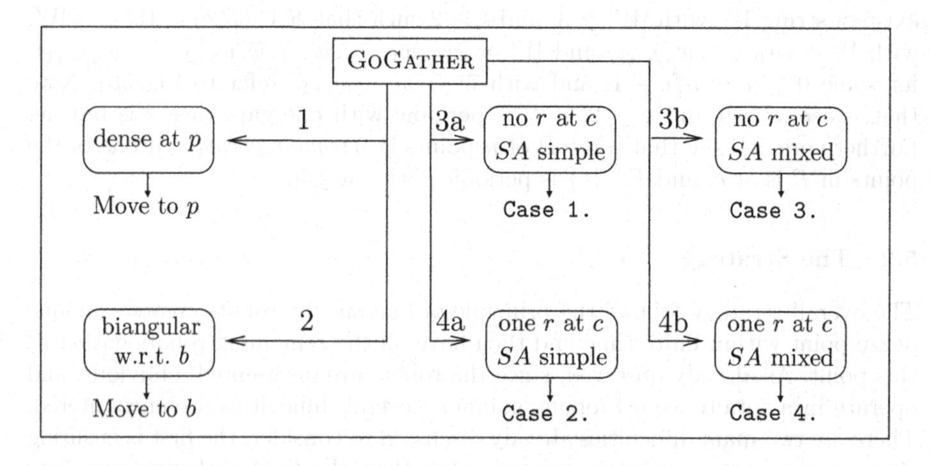


**Fig. 4.** [7]: Schematic overview of our solution; the numbers on the arrows represent the ordering of the tests performed by Algorithm GOGATHER.

algorithm ensures that, if this case is recognized by one robot, then *all* robots will recognize it, and will move towards the same point $b$; in this case, within finite time $b$ will become dense.

If the first two tests fail, the robot analyzes the *string of angles* (SA) of the robots with respect to the center $c$ of the smallest enclosing circle. (See an Example in Fig. 5).

The algorithm distinguishes four cases depending on whether there is one or no robot at the center of *SEC*, and on whether the *SA* is *simple* (i.e., the string does not contain any angle of zero degrees) or *mixed* (i.e., at least one angle of zero degrees is in the string, which implies that at least two robots are on the same radius).

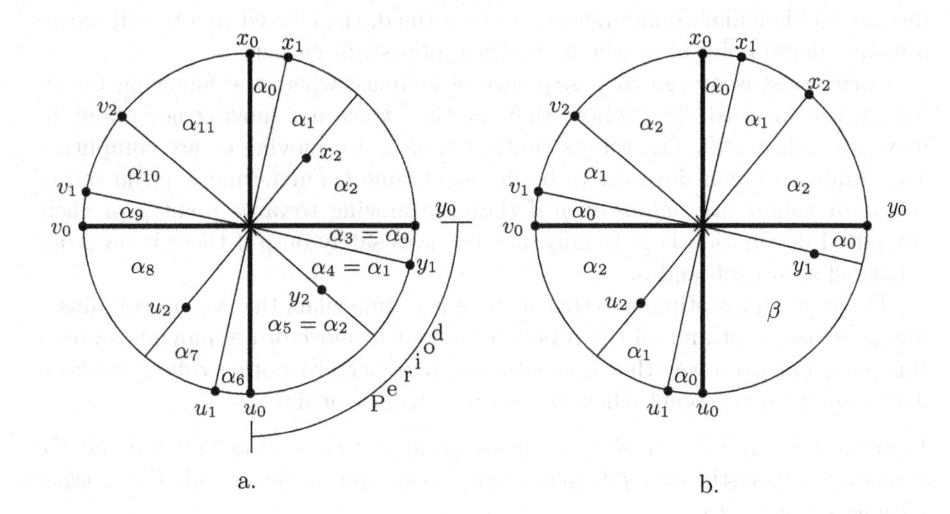


**Fig. 5.** [7]: (a) Example with $|StartSet^+(P, c)| = 4$, $LexMinString(P, c) = \langle\alpha_0, \ldots, \alpha_{11}\rangle$ with period $\langle\alpha_0, \alpha_1, \alpha_2\rangle$. There are $\frac{n}{k} = \frac{12}{3} = 4$ periods, with $\beta = 90°$. and $\beta = \alpha_1 + \alpha_2 + \alpha_3 = 360°/\frac{n}{k} = 360°/\frac{12}{3} = 90°$. The thick lines represent the starting points of each of the four periods. Robots $x_i$, $y_i$, $u_i$, and $v_i$, $0 \le i \le 2$, are *equivalent*. (b) If $y_2$ is removed from $P$, we obtain an example of a set of points that is periodic with one gap, with $\beta = \alpha_1 + \alpha_2$.

In all these cases, the algorithm "elects" a subset of the robots as leaders, on the basis of the string of angles. Several cases arises (see Fig. 6 where $s$ is the size of the elected set) and the algorithm is complicated by various technicalities.

At a high level of description, the simplest situation is when the elected set consists of a single robot, in which case that robot moves in such a way that it maintains the leader status, until it reaches another robot, thus creating a unique dense point. If there are more robots elected, they all move towards the center of the smallest enclosing circle $c$. In doing so they move *cautiously*, that is, ensuring that the smallest enclosing circle stays invariant during their movements, and paying attention to *potential* biangular configuration that might be formed during their movements. Indeed, if there is this possibility, the robots will try to reach a biangular configuration rather than avoiding it. This is achieved as follows: in the `Compute` phase, an elected robot checks if there exists sets of points in the trajectories of all the robots that might render the configuration biangular (*critical* points), and it explicitly computes all those sets. If the elected robot has a critical point on its way towards the destination, it will move towards such a point. the algorithm ensures that the critical point will be reached and that when this happens, no robot is moving or about to move (the configuration is *still*). This is orchestrated by a careful synchronization mechanism collectively performed by the robots. Hence, if a biangular configuration is formed during the movements of the elected robots, all the other robots will observe it in their next Look state, and they will eventually gather in the center of biangularity. If instead no biangular configuration can be formed, the elected robots will create a unique dense point at $c$, where all other robots will gather.

Correctness is proven by a sequence of lemmas, where the following terminology will be used. The robots that, at time $t$ are not moving nor about to move are called *still*. The robots that, at time $t$, are moving or are computing a non-null movement are said to be *acting* at time $t$. Furthermore, c the acting robots at time $t$ are *acting on* $p$ if they are moving towards point $p$ or their computed destination is $p$. Finally, a robot acts safely on $p$ if there is no other robot between itself and $p$.

The first lemma stipulates that if there is a time when the system contains a unique dense point and all the robots are about to move or are moving towards this point in such a way that none of the robots perceives other robots between itself and the dense point, then GATHERING will be achieved.

**Lemma 1** ([7]). *Let $p$ be the only dense point at time $t$. If at that time all the robots are either* still *or safely acting on $p$, then there exists a time $t' > t$ when all robots gather at $p$.*

Another important lemma proves that, if a biangular configuration is formed at a time when all robots are *still*, the robots will reach the center of biangularity.

**Lemma 2** ([7]). *Let at time $t$ the configuration be plain,* still *and biangular (either regular or irregular) with center $b$. Then there exists a time $t' > t$ when all robots gather at $b$.*

The next lemma proves that any arbitrary configuration will reach, within finite time one of the following: ($a$) a biangular configuration where all robots are *still*; ($b$) a configuration with a unique dense point where all robots are *still*; ($c$) a configuration with a unique dense point where all acting robots are acting safely on the unique dense point.

**Lemma 3** ([7]). *From any initial configuration $\mathbb{E}$, within a finite number of cycles, the robots reach a configuration that is either biangular (regular or irregular) and* still*, or dense and* still*, or dense with all acting robots safely acting on the dense point.*

From Lemmas 1, 2, and 3, it follows:

**Theorem 4** ([7]). *In $\mathcal{A}$SYNC, with* multiplicity detection, *$n \geq 5$ robots can solve the* GATHERING *problem within finite time.*

Since the Weber point can be computed for $n < 5$, this fact as well as the availability of multiplicity detection, can be exploited to solve GATHERING in $\mathcal{A}$SYNC also for $n = 3$ and $n = 4$ [8].

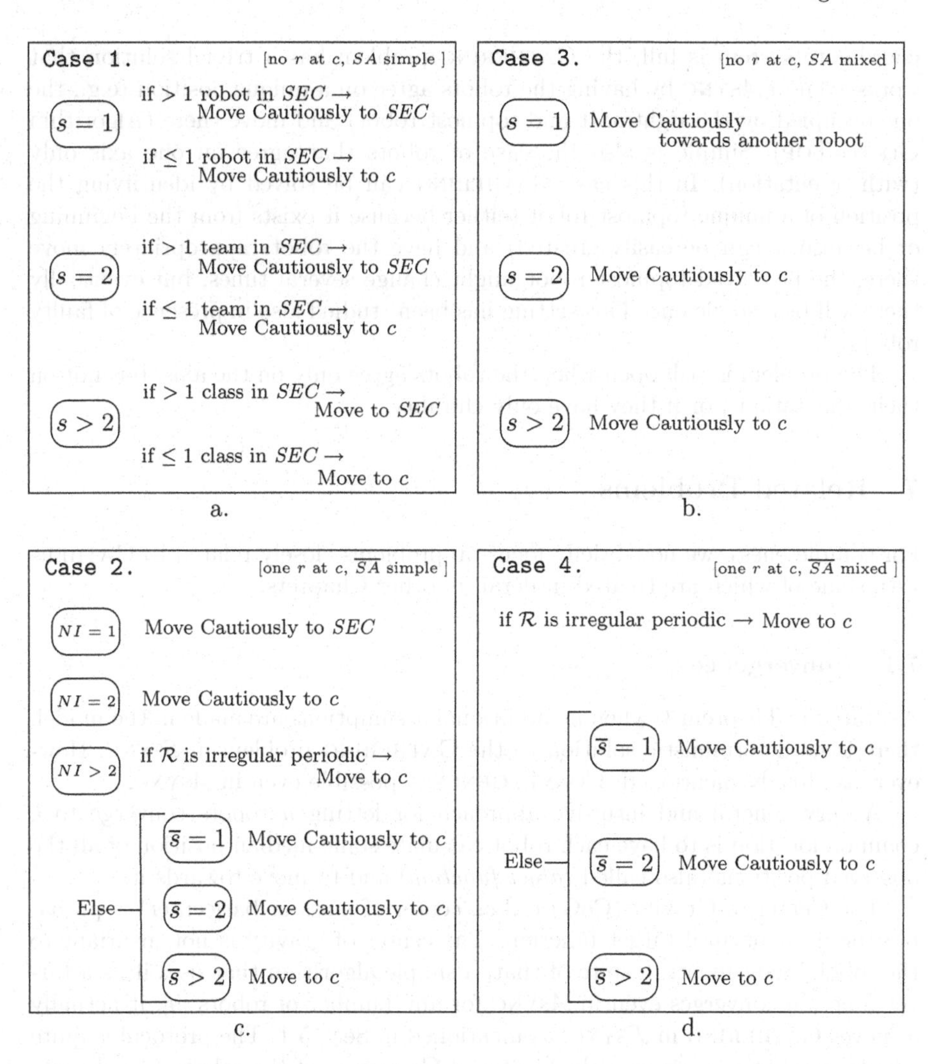


**Fig. 6.** [7]: The four cases of the algorithm when there is no dense point, and the configuration is not biangular. In the figure, $s$ is the number of start points in the string of angles; $\overline{s}$ is the number of start points in the string of angles built not considering $c$ (such string of angles is denoted by $\overline{SA}$ in the figure); and $NI$ is the number of robots inside $SEC$. Note that the figure specifies only the kind of movement that is performed in each case, and not which robot (or subset of robots) is performing it.

## 6 Gathering in $\mathcal{A}$SYNC with Compasses

By Theorem 3, either multiplicity detection or some form of agreement on the coordinate system is necessary for GATHERING to be solvable in $\mathcal{S}$SYNC (and thus in $\mathcal{A}$SYNC). We have seen that with multiplicity detection GATHERING can be solved. Much simpler is the case of axis agreement. If agreement on the coordinate system is full, the GATHERING problem has a trivial solution that works even in $\mathcal{A}$SYNC by having the robots agree on a unique position (e.g., the one occupied by the rightmost and topmost robot) and move there (Algorithm GO-TO-TOP). Simple is also the case of robots that agree on one axis only (with orientation). In this case GATHERING can be solved by identifying the position of a unique topmost robot (either because it exists from the beginning or because it can be easily created) and have the robots appropriately move there; the perceived topmost robot might change several times, but eventually there will be a single one. This setting has been studied also in presence of faulty robots [3].

The problem is still open when the robots agree only on the axis, but not on their orientations, or if they have only chirality.

## 7 Related Problems

For completeness, we now briefly focus on problems closely related to GATHERING, some of which are treated in detail in other Chapters.

### 7.1 Convergence

As stated in Theorem 3, when no additional assumptions are made in the model, there is no deterministic solution to the GATHERING problem in $\mathcal{S}$SYNC. However, as already mentioned, CONVERGENCE is possible even in $\mathcal{A}$SYNC.

A very general and intuitive approach for letting $n$ robots converge to a common location is to have each robot calculate some median position of all the observed positions (also called *target function*) and to move towards it.

The Center-of-Gravity (CoG) (a.k.a. center of mass or baricenter) is probably the most natural target function. The center of gravity is not invariant to the robots' movement; in spite of that, a simple algorithm that uses it as a target function converges even in $\mathcal{A}$SYNC for any number of robots [9]; it actually achieves GATHERING in $\mathcal{F}$SYNC, as mentioned in Sect. 5.1. The protocol is quite simple: a robot $r_i$ computes the Center of Gravity $c_i$ of the robots $c_i = \frac{1}{n}\Sigma_j r_j$, where $r_j$ is the position of robot $r_j$, and moves towards $c_i$.

This strategy has several advantages: in fact, it uses simple calculation, and it can be applied to any number of dimensions and to any number of robots. The crucial property on which convergence is based is that, even if the centre of gravity changes with the movements of the robots, there is still an invariant measure that allows the robots to get closer and closer. Define the destination point $\psi_i(t)$ of robot $r_i$ to be the final point of the movement made by $r_i$ following the last Look performed by $r_i$ before or at time $t$. Let $H(t)$ denote the convex hull of the points $r_i(t)$ and $\psi_i(t)$. Then the convex hull $H(t)$ cannot increase in time. In other words, using protocol CoG, we have that CONVERGENCE is solvable in $\mathcal{A}$SYNC in the standard $\mathcal{OBLOT}$ model.

The *convergence time* of the solution can be studied, based on the notion of *rounds*. Starting at time $t$, a round is said to terminate at the earliest point in time $t'$ when all robots have performed at least one complete cycle (Look–Compute–Move) during the time between $t$ and $t'$. With this definition, the convergence time is define as the number of rounds that are required to halve the convex hull [9]. As expected, the convergence time depends on the synchronicity level of the model; it also depends on the type of mobility, that is whether the usual mobility model is assumed, where the robots move towards the destination of at least a small amount $\delta$, or the rigid model is assumed, where the robots always reach their destination [9]. Other target functions have been also investigated with the main goal of improving the convergence time in $\mathcal{A}$SYNC [11].

Convergence has been studied also in presence of inaccurate measurements deriving impossibility and possibility results depending on the level of inaccuracy [10].

### 7.2 Rendezvous

When $n = 2$, i.e. the system contains only *two* robots, the GATHERING problem is very special, and it is called RENDEZVOUS. As seen in Sect. 5.1, the RENDEZVOUS problem is unsolvable in $\mathcal{S}$SYNC (and thus $\mathcal{A}$SYNC) without a common coordinate system, and easily solvable in $\mathcal{F}$SYNC. The CONVERGENCE problem of two oblivious robots without a common coordinate system, on the other hand, is easily solvable even in $\mathcal{A}$SYNC with he *move-to-half* strategy.

The existing gap between trivial possibility of rendezvous of two oblivious robots in presence of ConsistentCompass and impossibility in presence of Disorientation has lead to the study of additional assumption required to achieve rendezvous in presence of Disorientation. One natural direction is to explore what level of consistency of the coordinate systems (i.e., accuracy of their compasses) would allow the robots to solve rendezvous. This approach has been followed in [23] to study the case of compasses with Chirality, but with axis tilted up to a certain degree. It is shown that the level of inaccuracy (i.e., the amount of tilt) tolerable by the robots depends on the level of synchrony, as well as on whether the tilt is permanent or may change from round to round. Another research direction is the study of feasibility for robots when a little memory is available in the form of visible (or not visible) lights trying to minimize the number of lights employed [13,19,22]. This topic is studied in detail in Chap. 11.

### 7.3 Gathering and Convergence with Limited Visibility

The problems of GATHERING and CONVERGENCE have been investigated also in the context of limited visibility. This setting, together with other forms of restricted visibilities) is considered in detail in Chap. 7; for completeness, we summarize the main results.

**Gathering.** GATHERING can be achieved in $\mathcal{A}$SYNC, when there is agreement on the coordinate systems, and the robots initially form a connected visibility graph [18]. Taking advantage of the common coordinate system, the algorithm prescribes the robots to appropriately move only "down" and/or "right" guaranteeing that the initial visibility graph is preserved and eventually making the robots gather in the bottom-less, right-most point of the initial rectangular space containing all robots. It is not known whether it is possible to solve the problem with weaker assumptions. GATHERING has also been studied when the robots do not initially form a connected visibility graph and do not even agree on a common coordinate system. In this case, however, it is assumed the presence in the system of a "leader", which takes the form of a special configuration of three robots (called *Turing Mobile*) that coordinates the activities leading to Gathering [16].

**Convergence.** CONVERGENCE can be achieved in the standard $\mathcal{OBLOT}$ model in $\mathcal{S}$SYNC [1] by having the robot appropriately move towards the center of their visible smallest enclosing circle. This algorithm, designed to achieve CONVERGENCE, actually solves the GATHERING problem in $\mathcal{F}$SYNC; the reason is that synchronicity makes the robots form a point when they all are within visibility. The behavior of the robots following this algorithm in presence of faults has been studied in a 1-dimensional setting in [5]. CONVERGENCE has been shown to be possible also in $\mathcal{A}$SYNC, under special schedulers: partial $\mathcal{A}$SYNC [26] (where the time spent in the `Move` operation is bounded), and 1-fair $\mathcal{A}$SYNC [24] (where between two successive activations of the same robot, all the other robots have been activated at most 1 time). Finally, CONVERGENCE has been studied also in presence of perception inaccuracies (radial errors in locating a robot) and it has been show how to reach convergence in $\mathcal{F}$SYNC for small inaccuracies [32].

**Collision-less Convergence.** The *Collision-less Convergence* problem (also called, *Near-Gathering*) consists of achieving convergence without any collision among the robots. Slight modifications can make the algorithm of [1] collision-less, thus solving Near-Gathering in $\mathcal{S}$SYNC. Near-Gathering can be achieved also in $\mathcal{A}$SYNC, with two additional assumptions [27]: the robots must partially agree on a common coordinate system (one axis is sufficient) and the initial visibility graph must be well-connected, that is, the subgraph of the visibility graph that contains only the edges corresponding to robots at distance strictly smaller than $V$ must be connected.

**Gathering as a Combination of Algorithms.** To achieve gathering with limited visibility in $\mathcal{S}$SYNC, it might be possible to combine a collision-less convergence version of [1] in $\mathcal{S}$SYNC until all robots can see each other, with the existing gathering algorithm for unlimited visibility [7]. Indeed, this can be done only if the robots can be made aware that they are all within visibility so to be able to start the unlimited visibility algorithm in a coherent way. It is not known whether this can be done in the $\mathcal{OBLOT}$ model. In a stronger model it can be achieved, for example, if the robots know $n$ (the total number of robots), or if they have lights (which can be used to signal when they are all within visibility).

### 7.4 Gathering and Convergence in Other Settings

GATHERING and CONVERGENCE in the $\mathcal{OBLOT}$ model in presence of faults have been extensively studied (e.g., see [15]); fault-tolerant computing by oblivious robots is treated in detail in Chap. 10. GATHERING has also been investigated in the context of robots with extent (e.g., see [4,12,21]); the results are described in Chap. 7 in the context of restricted visibility. A variant of GATHERING has been proposed where the robots must gather only at some predetermined points in the plane (the meeting-points) and solutions have been proposed with the goal of minimizing objective functions based on the traveled distance performed by the robots [6].

## 8 Summary

Tables 1 and 2 summarize the feasibility results for point formation and convergence in the unlimited and in the limited visibility settings, starting from

**Table 1.** Feasibility results in the unlimited visibility setting.

| Unlimited visibility | Assumptions | $\mathcal{F}$SYNC | $\mathcal{S}$SYNC and $\mathcal{A}$SYNC |
|---|---|---|---|
| Convergence | | | |
| ($n = 2$) | No assumptions | GO-TO-HALF | GO-TO-HALF |
| ($n > 2$) | No assumptions | GO-TO-COG | GO-TO-COG [9] |
| Point formation | | | |
| RENDEZVOUS | No assumptions | GO-TO-HALF | Impossible [30] |
| ($n = 2$) | Tilted compasses | GO-TO-HALF | [23] |
| | Eventually consistent | GO-TO-HALF | [29] |
| | With lights | GO-TO-HALF | [13,19,22] |
| Gathering | No assumptions | GO-TO-COG | Impossible [28] |
| ($n > 2$) | Multiplicity only | GO-TO-COG | [7] |
| | 1 axis (and orientation) | GO-TO-COG | GO-TO-TOP [3] |
| | Axis (and no orientation) | GO-TO-COG | ? |

**Table 2.** Feasibility results in the limited visibility setting.

| Limited visibility | Assumptions | $\mathcal{F}$SYNC | $\mathcal{S}$SYNC | $\mathcal{A}$SYNC |
|---|---|---|---|---|
| Convergence | No assumptions | [1] | [1] | ? |
| Gathering | No assumptions | [1] | Impossible [28] | Impossible [28] |
| | Axis | [1,18] | [18] | [18] |
| | Multiplicity and $n$ | [1] | [1]+ [7] | ? |
| | Multiplicity and lights | [1] | [1] + [7] | ? |
| | Multiplicity only | [1] | ? | ? |

arbitrary configurations where robots occupy different positions, and where the visibility graph is connected.

Summarizing: solving the point formation problem (and thus convergence) with unlimited visibility is always trivial in $\mathcal{F}$SYNC: in the case of two robots, each robot moves half-way towards the other (Algorithm GO-TO-HALF), in the case of $n > 2$ robots, each robot moves toward the centre of gravity of the visible robots (Algorithm GO-TO-COG). Solutions become increasingly complex in $\mathcal{S}$SYNC and $\mathcal{A}$SYNC, where some assumptions must be made as it is shown that without multiplicity detection and some form of orientation the problem is unsolvable. Whether Gathering could be achieved without multiplicity detection and with a form of orientation weaker than agreement on the robots' axis and orientation is still an open problem.

Solving gathering and convergence with limited visibility is more challenging. It is still always doable in $\mathcal{F}$SYNC. Solutions become increasingly complex in $\mathcal{S}$SYNC and $\mathcal{A}$SYNC, and several cases are still open. For example, it is not known whether a convergence algorithm exists in $\mathcal{A}$SYNC without any additional assumption, no solution has been proposed yet to solve the gathering problem in $\mathcal{S}$SYNC (and $\mathcal{A}$SYNC) with multiplicity detection only. When multiplicity detection is complemented by knowledge of the number of robots $n$ or by the presence of a constant number of lights, the convergence algorithm in limited visibility can be combined with the gathering algorithm in unlimited visibility to obtain a solution in $\mathcal{S}$SYNC; it is still open whether a solution exists under these assumptions in $\mathcal{A}$SYNC.

## References


1. Ando, H., Oasa, Y., Suzuki, I., Yamashita, M.: A distributed memoryless point convergence algorithm for mobile robots with limited visibility. IEEE Trans. Robot. Autom. **15**(5), 818–828 (1999)
2. Bajaj, C.: The algebraic degree of geometric optimization problems. Discrete Comput. Geom. **3**, 177–191 (1988)
3. Bhagat, S., Gan Chaudhuri, S., Mukhopadhyaya, K.: Fault-tolerant gathering of asynchronous oblivious mobile robots under one-axis agreement. J. Discrete Algorithms **36**, 50–62 (2016)
4. Bhagat, S., Gan Chaudhuri, S., Mukhopadhyaya, K.: Gathering of opaque robots in 3D space. In: 19th International Conference on Distributed Computing and Networking (ICDCN), pp. 1–10 (2018)
5. De Carufel, J.-L., Flocchini, P.: Fault-induced dynamics of oblivious robots on a line. In: Spirakis, P., Tsigas, P. (eds.) SSS 2017. LNCS, vol. 10616, pp. 126–141. Springer, Cham (2017). https://doi.org/10.1007/978-3-319-69084-1_9
6. Cicerone, S., Di Stefano, G., Navarra, A.: Gathering of robots on meeting-points: feasibility and optimal resolution algorithms. Distrib. Comput. **31**(1), 1–50 (2018)
7. Cieliebak, M., Flocchini, P., Prencipe, G., Santoro, N.: Distributed computing by mobile robots: gathering. SIAM J. Comput. **41**(4), 829–879 (2012)
8. Cieliebak, M., Prencipe, G.: Gathering autonomous mobile robots. In: 9th International Colloquium on Structural Information and Communication Complexity (SIROCCO), pp. 57–72 (2002)

9. Cohen, R., Peleg, D.: Convergence properties of the gravitational algorithm in asynchronous robot systems. SIAM J. Comput. **34**, 1516–1528 (2005)
10. Cohen, R., Peleg, D.: Convergence of autonomous mobile robots with inaccurate sensors and movements. In: Durand, B., Thomas, W. (eds.) STACS 2006. LNCS, vol. 3884, pp. 549–560. Springer, Heidelberg (2006). https://doi.org/10.1007/11672142_45
11. Cord-Landwehr, A., et al.: A new approach for analyzing convergence algorithms for mobile robots. In: Aceto, L., Henzinger, M., Sgall, J. (eds.) ICALP 2011. LNCS, vol. 6756, pp. 650–661. Springer, Heidelberg (2011). https://doi.org/10.1007/978-3-642-22012-8_52
12. Czyzowicz, J., Gasieniec, L., Pelc, A.: Gathering few fat mobile robots in the plane. Theor. Comput. Sci. **410**, 481–499 (2009)
13. Das, S., Flocchini, P., Prencipe, G., Santoro, N., Yamashita, M.: Autonomous mobile robots with lights. Theor. Comput. Sci. **609**, 171–184 (2016)
14. Défago, X., Gradinariu, M., Messika, S., Raipin-Parvédy, P.: Fault-tolerant and self-stabilizing mobile robots gathering. In: Dolev, S. (ed.) DISC 2006. LNCS, vol. 4167, pp. 46–60. Springer, Heidelberg (2006). https://doi.org/10.1007/11864219_4
15. Défago, X., Gradinariu, M., Messika, S., Raipin Parvédy, P.: Fault and byzantine tolerant self-stabilizing mobile robots gathering - feasibility study. Technical report (2016)
16. Di Luna, G., Flocchini, P., Santoro, N., Viglietta, G.: Turingmobile: a turing machine of oblivious mobile robots with limited visibility and its applications. In: 32nd International Symposium on Distributed Computing (DISC) (2018)
17. Durocher, S., Kirkpatrick, D.: The projection median of a set of points. Comput. Geom.: Theory Appl. **42**(5), 364–375 (2009)
18. Flocchini, P., Prencipe, G., Santoro, N., Widmayer, P.: Gathering of asynchronous mobile robots with limited visibility. Theor. Comput. Sci. **337**, 147–168 (2005)
19. Flocchini, P., Santoro, N., Viglietta, G., Yamashita, M.: Rendezvous with constant memory. Theor. Comput. Sci. **621**, 57–72 (2016)
20. Fujinaga, N., Yamauchi, Y., Kijima, S., Yamashita, M.: Asynchronous pattern formation by anonymous oblivious mobile robots. SIAM J. Comput. **44**(3), 740–785 (2015)
21. Gan Chaudhuri, S., Mukhopadhyaya, K.: Leader election and gathering for asynchronous fat robots without common chirality. J. Discrete Algorithms **33**, 171–192 (2015)
22. Heriban, A., Défago, X., Tixeuil, S.: Optimally gathering two robots. In: 19th International Conference on Distributed Computing and Networking (ICDCN), pp. 1–10 (2018)
23. Izumi, T., et al.: The gathering problem for two oblivious robots with unreliable compasses. SIAM J. Comput. **41**(1), 26–46 (2012)
24. Katreniak, B.: Convergence with limited visibility by asynchronous mobile robots. In: Kosowski, A., Yamashita, M. (eds.) SIROCCO 2011. LNCS, vol. 6796, pp. 125–137. Springer, Heidelberg (2011). https://doi.org/10.1007/978-3-642-22212-2_12
25. Kupitz, Y., Martini, H.: Geometric aspects of the generalized Fermat-Torricelli problem. Intuitive Geom. **6**, 55–127 (1997)
26. Lin, J., Morse, A.S., Anderson, B.D.O.: The multi-agent rendezvous problem. Part 2: the asynchronous case. SIAM J. Control. Optim. **46**(6), 2120–2147 (2007)
27. Pagli, L., Prencipe, G., Viglietta, G.: Getting close without touching: near-gathering for autonomous mobile robots. Distrib. Comput. **28**(5), 333–349 (2015)
28. Prencipe, G.: Impossibility of gathering by a set of autonomous mobile robots. Theor. Comput. Sci. **384**(2–3), 222–231 (2007)

29. Souissi, S., Défago, X., Yamashita, M.: Using eventually consistent compasses to gather memory-less mobile robots with limited visibility. ACM Trans. Auton. Adapt. Syst. **4**(1), 1–27 (2009)
30. Suzuki, I., Yamashita, M.: Distributed anonymous mobile robots: formation of geometric patterns. SIAM J. Comput. **28**(4), 1347–1363 (1999)
31. Weiszfeld, E.: Sur le point pour lequel la somme des distances de $n$ points donnés est minimum. Tohoku Math. **43**, 355–386 (1936)
32. Yamamoto, K., Izumi, T., Katayama, Y., Inuzuka, N., Wada, K.: The optimal tolerance of uniform observation error for mobile robot convergence. Theor. Comput. Sci. **444**, 77–86 (2012)

# Uniform Circle Formation

Giovanni Viglietta(✉)

JAIST, Nomi, Japan
johnny@jaist.ac.jp

**Abstract.** We treat the second of the two patterns that are formable in the $\mathcal{OBLOT}$ model from every initial configuration of $n$ robots: `Uniform Circle`, i.e., the pattern where the robots are located at the vertices of a regular $n$-gon. The algorithm presented in this chapter solves the `Uniform Circle` Formation Problem in the standard $\mathcal{OBLOT}$ model under the $\mathcal{A}$SYNC scheduler.

**Keyword:** Uniform Circle Formation

## 1 Introduction

In this chapter we treat the second of the two patterns that are formable in $\mathcal{OBLOT}$ from every initial configuration of $n$ robots: `Uniform Circle`, i.e., the pattern where the robots are located at the vertices of a regular $n$-gon. The algorithm presented in this chapter asssumes that no two robots are initially in the same location, and solves the `Uniform Circle` Formation Problem in the standard $\mathcal{OBLOT}$ model under the $\mathcal{A}$SYNC scheduler, and has been published in [12,15].

The main body of the algorithm is described in Sect. 2, and deals with the case of $n > 5$ robots. The cases with fewer robots are solved with ad-hoc algorithms, and are discussed in Sect. 3.

Several related papers have appeared, studying the `Uniform Circle` Formation Problem or its variants, and solving it in special cases or under different assumptions [2,4–8,10,16]. One of the earliest results on the problem appeared in [18], where it is shown that robots can form `Uniform Circle` under the $\mathcal{S}$SYNC scheduler, provided that they can remember past observations. Another early algorithm, given in [5], makes a swarm of oblivious robots converge towards `Uniform Circle` (possibly without ever forming it) under the $\mathcal{S}$SYNC scheduler. Other algorithms from the same period can be combined to prove that `Uniform Circle` is actually formable under the $\mathcal{S}$SYNC scheduler: by concatenating the algorithm in [14], for forming a biangular configuration, with the one in [7], for forming `Uniform Circle` from a biangular starting configuration, it is possible to form `Uniform Circle` from any initial configuration (the case with four robots has been solved separately in [8]). Observe, however, that the two algorithms can be concatenated only because the scheduler is $\mathcal{S}$SYNC: under the $\mathcal{A}$SYNC scheduler, this technique would be ineffective.


P. Flocchini et al. (Eds.): Distributed Computing by Mobile Entities, LNCS 11340, pp. 83–108, 2019.
https://doi.org/10.1007/978-3-030-11072-7_5

Some results for the $\mathcal{A}$SYNC scheduler assume implicit agreements among all the robots. As shown in [10], if the local coordinate systems of all the robots have the same orientation (i.e., the system has chirality), there is a simple algorithm to form `Uniform Circle`. This has later been improved by a general result of [13], which only assumes that all local coordinate systems are right-handed. Then, in [11], a `Uniform Circle` Formation algorithm was given for $\mathcal{A}$SYNC robots with no assumptions on their local coordinate systems, but allowing them to move along circular arcs, as well as straight line segments. A solution for $n \neq 4$ robots under the $\mathcal{A}$SYNC scheduler without extra assumptions (i.e., in the standard $\mathcal{OBLOT}$ model) was finally given in [12], and the solution for the special case $n = 4$ appeared in [15].

Related results about the formation of `Uniform Circle` include a study of the problem of minimizing the maximum distance traveled by a robot, where robots have one bit of internal memory [1], an algorithm for transparent robots with extent that are $\mathcal{S}$SYNC, perform rigid movements, and agree on one axis [17], and an algorithm for opaque $\mathcal{LUMINOUS}$ robots in $\mathcal{F}$SYNC [9].

## 2 General Algorithm for $n > 5$ Robots

If the swarm of robots is not too small, i.e., if $n > 5$, there is a general `Uniform Circle` Formation algorithm that works in all cases [12]. Recall that the goal of the robots is to position themselves on the vertices of a regular $n$-gon and stop moving: we call this type of configuration *Regular*.

A fundamental geometric tool used in the algorithm is the concept of *smallest enclosing circle (SEC)* of the swarm of robots: this is the circle of smallest radius that contains all robots. It is well known that such a circle exists, is unique, and its center and radius can be effectively computed by the robots.

The algorithm can be outlined as follows: the general strategy is to make the robots move to the SEC, determine their final "target points", and then move to such points to form a *Regular* configuration. The robots always move in an orderly fashion, and the ones that are allowed to move at each step are carefully chosen in such a way that the SEC does not change during their movements. The only exception to this protocol is when the robots form, either "intentionally" or "accidentally", a particular type of configuration called *Pre-regular*: in this case, they follow a special procedure that ignores the SEC.

The details of the algorithm are given below.

### 2.1 Special Cases: *Biangular* and *Pre-regular* Configurations

We first consider the *Biangular* configurations, exemplified in Fig. 1(a). In such a configuration, the number of robots $n$ is even, and the set of their locations has exactly $n/2$ axes of symmetry. Note that a *Biangular* configuration can be partitioned into two *Regular* configurations of size $n/2$. This is a particularly interesting situation, because the robots may all have the exact the same view, provided that their axes are oriented symmetrically, as Fig. 1(a) shows. In this scenario, the scheduler may force all robots to perform the same computation and move at the same time, which will make the configuration remain *Biangular* at all times (or become *Regular*). So, any `Uniform Circle` Formation algorithm must ensure that this type of synchronous behavior from a *Biangular* configuration indeed results in the formation of a *Regular* configuration. Nonetheless, the algorithm must also take into account that movements may be asynchronous and non-rigid: while moving toward their destinations, the robots may also form different and possibly asymmetric intermediate configurations, and asynchronously compute new destinations from those configurations. Therefore, it is clearly desirable that the robots preserve some geometric invariant as they move, so that any such intermediate configuration is treated coherently with the *Biangular* case.

A solution to the problem of forming a regular polygon starting from a *Biangular* configuration is given in [7], where the robots identify a "supporting regular polygon" (see Fig. 1(b)), and each robot moves toward the closest vertex of such a polygon. Any intermediate configuration possibly formed while the robots move asynchronously towards the vertices of the supporting polygon is called *Pre-regular* (note that all *Biangular* configurations are also *Pre-regular*). While executing this procedure from a *Pre-regular* configuration, the supporting polygon remains invariant (e.g., see Fig. 1(c)). So, whenever the configuration is perceived as *Pre-regular* by *all* the robots, moving toward the appropriate vertex of the supporting polygon results in the formation of a *Regular* configuration.

Note that a robot can effectively determine whether the observed configuration is *Pre-regular*. Moreover, thanks to the following lemma, all robots in a *Pre-regular* configuration implicitly agree on the same supporting polygon and behave coherently with one another.

**Lemma 1** ([12]). *The supporting polygon of a* Pre-regular *configuration is unique.*

## 2.2 General Strategy: SEC and Analogy Classes

Consider now a starting position of the robots that is not *Pre-regular* (and hence not *Biangular*). Recall that the robots have no common reference frame, and there are no "environmental" elements that can help them orient themselves. This difficulty may prevent the robots from coordinating their movements and act "consistently" from one cycle to another. To overcome this, the SEC of the robots' positions is identified (see Fig. 2(a)), and the algorithm ensures that the robots move in such a way as to keep the SEC fixed. This will hold true so long as the configuration is not *Pre-regular*. If the configuration happens to become *Pre-regular* during the execution, then the procedure of Sect. 2.1 will be executed, and the SEC will no longer be preserved.

As a preliminary step, the general algorithm attempts to make all robots reach the perimeter of the SEC. So, let us consider a configuration that is not *Pre-regular* and in which all robots lie on the perimeter of SEC. In this situation, we identify pairs of robots $r_1$ and $r_2$ that are located in "symmetric" positions, i.e.,

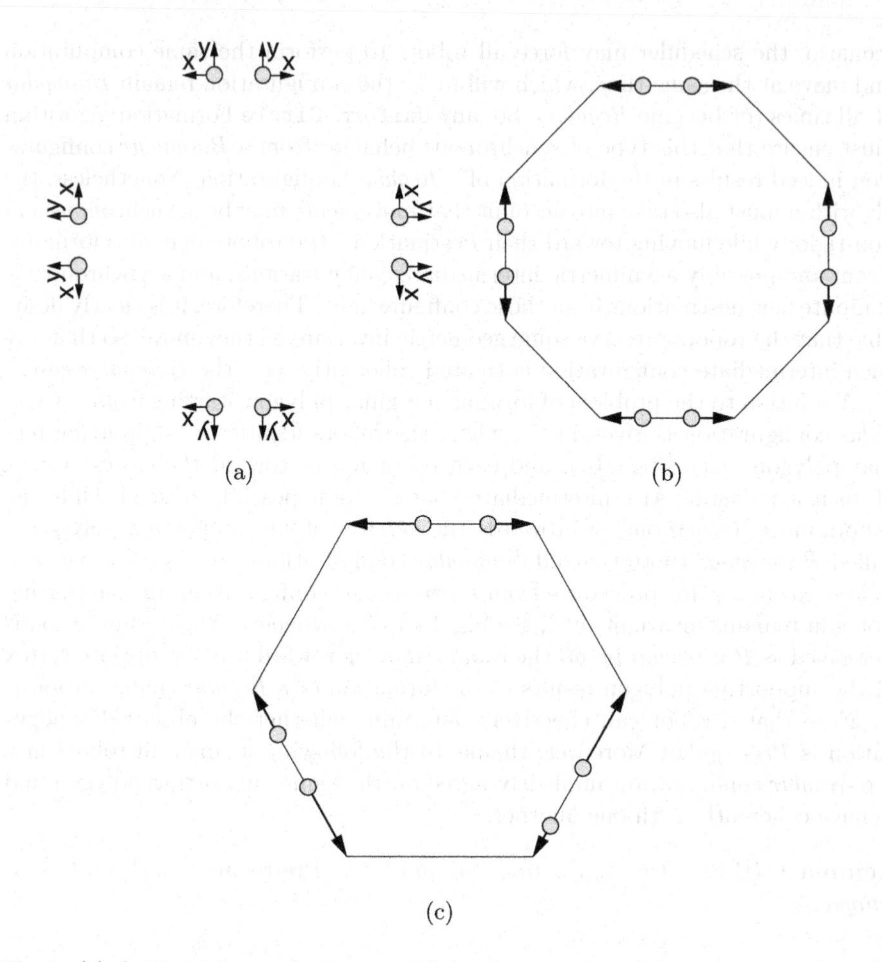


**Fig. 1.** (a) A *Biangular* configuration with local axes oriented in such a way that all robots have the same view. (b) How the algorithm resolves a *Biangular* configuration. (c) A generic *Pre-regular* configuration with its supporting polygon, which remains invariant as the robots move along the arrows. (Source: [12])

such that there is an isometry of the plane that permutes the robots' locations switching the positions of $r_1$ and $r_2$. We call two such robots *analogous*, and the swarm is thus partitioned into *analogy classes* of analogous robots (see Fig. 2(b)). In general, an analogy class has either the shape of a *Regular* set or of a *Biangular* set (with some degenerate cases, such as a single point or a pair of points).

Similarly to the *Biangular* case (cf. Sect. 2.1), the scheduler may force all the robots in an analogy class to perform the same computation and move at the same time, thus occupying symmetric positions again, and potentially forever. To accommodate this, the algorithm incorporates this type of behavior and makes all analogous robots always *deliberately* move together in the same fashion.

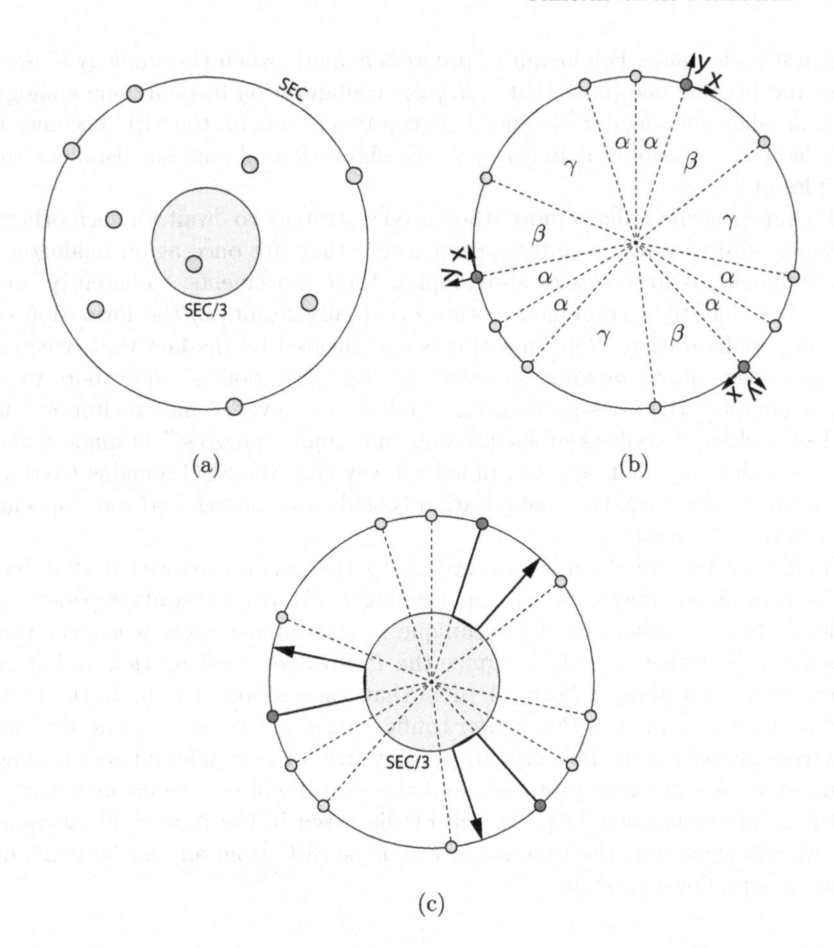


**Fig. 2.** (a) A swarm of robots with its SEC and SEC/3. (b) The three highlighted robots form an analogy class. If their axes are oriented as indicated, the three robots have the same view. (c) The three dark-shaded robots are selected as walkers and move according to the arrows. At the end, for each walker there is a non-walker at an angular distance of $\pi/3$ from it (note that $\pi/3$ is a multiple of $2\pi/n = \pi/6$). (Source: [12])

More specifically, the algorithm lets only one analogy class move at any given time, while all the other robots wait on the SEC (see Fig. 2(c)). The robots in the analogy class that is allowed to move are called *walkers*. When the walkers have been chosen, they move radially to SEC/3, which is the circle whose radius is 1/3 of the radius of the SEC and concentric with the SEC. Once they are all on the perimeter of SEC/3, they move to their so-called *finish set*, while staying within SEC/3 (or in its interior). When they are all in their finish set, they move radially to the SEC again. After that, a new analogy class of walkers is chosen, and so on. The walkers and the finish set are chosen in such a way that, when the walkers are done moving, some kind of "progress" toward a *Regular* configuration is made. For instance, progress is made when two analogy classes merge and become one (note that a *Regular* configuration has only one analogy class), or when the angular distance between two robots on the SEC becomes a multiple of $2\pi/n$ (note that in a *Regular* configuration all angular distances are multiples of $2\pi/n$).

Of course, as the walkers move, they need a strategy to "wait for each other" and make sure to reach a configuration where they are once again analogous. Also, different analogy classes should plan their movements "coherently", in such a way that their combined motion eventually results in the formation of a *Regular* configuration. Note that this is complicated by the fact that, when a class of walkers starts moving, some of the "reference points" the robots were using to compute their destinations may be lost. Moreover, it may be impossible to select a class of walkers in such a way that some "progress" is made when they reach their destinations, and in such a way that the SEC remains fixed as they move. In this case, the configuration is said to be *locked*, and some special moves have to be made.

Finally, as the robots move according to the general algorithm that has just been outlined, they may end up forming a *Pre-regular* configuration "by accident". So, the robots need a technique to stop immediately whenever this happens, so that they can all recognize the *Pre-regular* configuration and start executing the procedure of Sect. 2.1 (note that some robots may be in the middle of a movement when a *Pre-regular* configuration is formed accidentally, and countermeasures have to be taken to prevent this, or else different robots may end up executing different protocols, and the swarm will behave incoherently).

All the aforementioned aspects will be discussed in the rest of this section. Next we will show how the robots can reach the SEC from any initial configuration, as a preliminary step.

### 2.3 Preliminary Step: Reaching the SEC

A simple way to make all robots reach the SEC without colliding is to make each of them move radially, away from the center, as in Fig. 3(a). This protocol only works assuming that no two robots are *co-radial*, i.e., on the same half-line extending from the center of the SEC. A special case is the *Central* configuration, in which one robot lies at the center of the SEC. *Central* configurations are easily resolved by simply making the central robot move to SEC/3, in such a way as not to become co-radial with any other robot.

The *Co-radial* configurations that are not *Central* are handled as follows. First of all, if there is any robot in the interior of SEC/3 that is not co-radial with any other robot, it moves radially to SEC/3 (note how the evolution of a *Central* configuration naturally blends with this protocol). Then, for each maximal set of at least two mutually co-radial robots, the robot that is closest to the center of the SEC moves radially toward the center until it is in SEC/3 (see Fig. 3(b)). Finally, the most internal co-radial robots make a lateral move to become non-co-radial, as in Fig. 3(c). The lateral move is within SEC/3 (or its interior) and it is "sufficiently small", in order to prevent collisions. A sufficiently small move

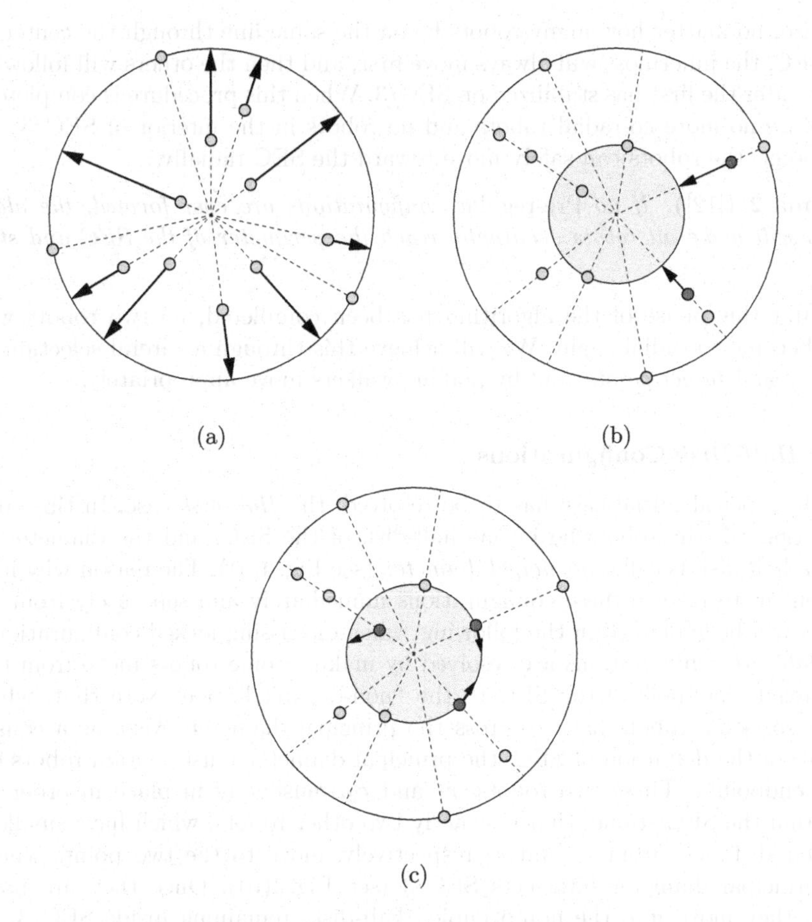


**Fig. 3.** (a) All robots move radially to reach the SEC. (b) The most internal co-radial robots move radially to SEC/3. (c) When they are in SEC/3, they make a small lateral move. (Source: [12])

is, for instance, a move that reduces the angular distance to any other robot by no more than 1/3.

The reason why we make robots reach SEC/3 before performing lateral moves is because we want to prevent the accidental formation of *Pre-regular* configurations. We will discuss this aspect later, in Sect. 2.9.

It is easy to see how this strategy makes the robots coordinate themselves and avoid collisions. Indeed, as soon as a robot $r$ makes a lateral move and stops being co-radial with other robots, it is seen by the other robots as a non-co-radial robot lying in the interior of SEC/3. Hence, no other robot will take initiatives, and will just wait until $r$ has reached SEC/3 and has stopped there. This guarantees that, when a robot decides to perform a lateral move, no other robot is in the middle of a lateral move.

Also, no matter how many robots lie on the same line through the center of the SEC, the innermost will always move first, and then the others will follow in order, after the first has stabilized on SEC/3. When this procedure is completed, there are no more co-radial robots and no robots in the interior of SEC/3. At this point, the robots can safely move toward the SEC radially.

**Lemma 2** ([12]). *If no* Pre-regular *configurations are ever formed, the algorithm will make all robots eventually reach the perimeter of the SEC and stop there.*

After this phase of the algorithm has been completed, no two robots will ever become co-radial again. We will achieve this through a careful selection of *walkers* and *target points*, and by making walkers move appropriately.

### 2.4 *Half-Disk* Configurations

Another special initial case has to be resolved: the *Half-disk* case. In this configuration, all the robots lie in one half-disk of the SEC, and the diameter of such a half-disk is called *principal diameter* (see Fig. 4(a)). The reason why it is convenient to resolve these configurations immediately and separately from all others will be explained in the following, when discussing locked configurations.

*Half-disk* configurations are resolved by making some robots move from the "occupied" half-disk of the SEC to the "non-occupied" one. Note that, while doing so, some robots have to cross the principal diameter. Also, as a consequence of the definition of SEC, the principal diameter must contain robots on both endpoints. These two robots, $r_1$ and $r_2$, must stay in place in order to maintain the SEC stable. Hence, exactly two other robots, which have smallest angular distances from $r_1$ and $r_2$ respectively, move to the two points where the principal diameter intersects SEC/3 (see Fig. 4(b)). Once they are both there, they move into the non-occupied half-disk, remaining inside SEC/3, as in Fig. 4(c). (More precisely, if the principal diameter already contains some robots on or inside SEC/3, such robots do not preliminarily move to the perimeter of SEC/3, because it is unnecessary and it may even cause collisions; in this case, they move into the unoccupied half-disk right away.)

A very special *Half-disk* case is the one where all robots lie on the same line. This case is handled as a generic *Half-disk*, with two robots first moving on SEC/3 (if they are not already on it or in its interior), and then moving away from the principal diameter. If they move in opposite directions, the configuration is no longer *Half-disk*. If they move in the same direction, they form a generic *Half-disk*, which is then resolved normally.

When analyzing the possible evolutions of a *Half-disk* configuration, one has to keep in mind that it transitions into a different configuration while one or two robots are still moving. This turns out to be relatively easy, since the moving robots are inside SEC/3 (like the robots that move laterally in the *Co-radial* case) and move in a very predictable and controlled way. When the configuration ceases to be *Half-disk*, the robots will move on SEC as described before, and they will never form a *Half-disk* configuration again.

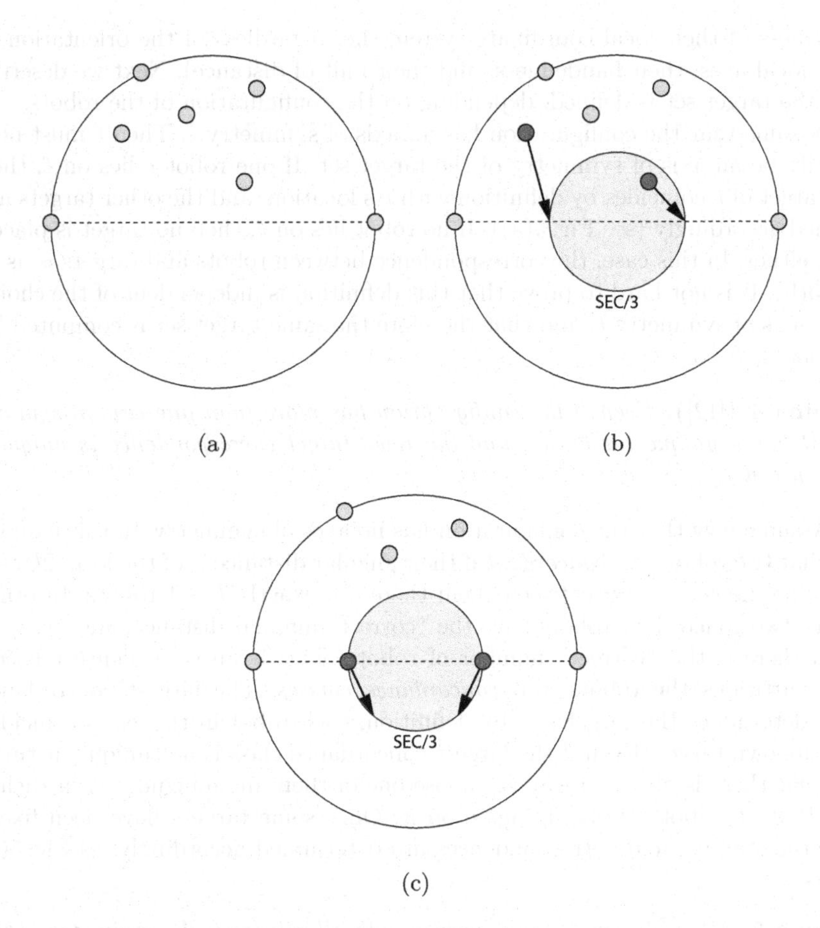


**Fig. 4.** (a) A *Half-disk* configuration and the principal diameter. (b) Two robots move to the intersection between the principal diameter and SEC/3. (c) The same two robots move to the non-occupied half-disk. (Source: [12])

**Lemma 3** ([12]). *If no* Pre-regular *configurations are ever formed, the algorithm will make all robots eventually reach the perimeter of the SEC and stop there, in such a way that in every half-circle of the SEC there is at least one robot.*

## 2.5 Identifying Targets

Suppose now that all robots lie on the perimeter of the SEC, and the configuration is not *Pre-regular* and not *Half-disk.* In this case we can define a *target set*, which represents the final *Regular* configuration that the robots are attempting to form. Each element of the target set is called a *target*, and corresponds to some robot's intended destination. Hence the target set is a *Regular* set of $n$ points arranged on the SEC in such a way that it can be computed by all robots, regardless of their local coordinate system (i.e, regardless of the orientation of their local axes, their handedness, and their unit of distance). Next we describe how the target set is defined, depending on the configuration of the robots.

Assume that the configuration has an axis of symmetry $\ell$. Then $\ell$ must necessarily be an axis of symmetry of the target set. If one robot $r$ lies on $\ell$, then the target of $r$ coincides by definition with $r$'s location, and the other targets are defined accordingly (see Fig. 5(a)). If no robot lies on $\ell$, then no target is placed on $\ell$, either. In this case, the correspondence between robots and targets is as in Fig. 5(b). It is not hard to prove that this definition is independent of the choice of an axis of symmetry $\ell$, and that therefore the same target set is computed by all robots.

**Lemma 4** ([12]). *Even if the configuration has more than one axis of symmetry, it has a unique target set, and the robot-target correspondence is uniquely determined.*

Assume now that the configuration has no axes of symmetry. In this case we say that two robots are *concordant* if their angular distance is of the form $2k\pi/n$, for some integer $k$, and between them there are exactly $k-1$ robots. In other terms, two concordant robots have the "correct" angular distance, and between them there is the "correct" number of robots. This is an equivalence relation that partitions the robots into *concordance classes*. The largest concordance class determines the target set: by definition, each robot in this class coincides with its own target. Even if the largest concordance class is not unique, it turns out that there is always a way to choose one of them unambiguously, in such a way that all robots implicitly agree on it. Once some targets have been fixed, the other targets and correspondences are determined accordingly, as Fig. 5(c) shows.

**Lemma 5** ([12]). *In every configuration with all robots on the perimeter of the SEC, the target set and the robot-target correspondence are uniquely determined.*

### 2.6 Identifying Walkers, Locked Configurations

When the target set has been identified, then the *walkers* can be defined. The walkers are simply the analogy class of robots that are going to move next.

Typically, the algorithm will attempt to move an analogy class of robots to their corresponding targets. The robots that currently lie on their targets are called *satisfied*, and these robots should not move. Moreover, the walkers should be chosen in such a way that, when they abandon the perimeter of the SEC and move into its interior, they do not cause the SEC to change. An analogy class of robots with this property is called *movable*. Finally, no two robots should become co-radial as a result the walkers' movements. This means that the walkers should be chosen in such a way that, as they move toward their targets, they do not become co-radial with other robots. The targets of such robots are said to be *reachable*.

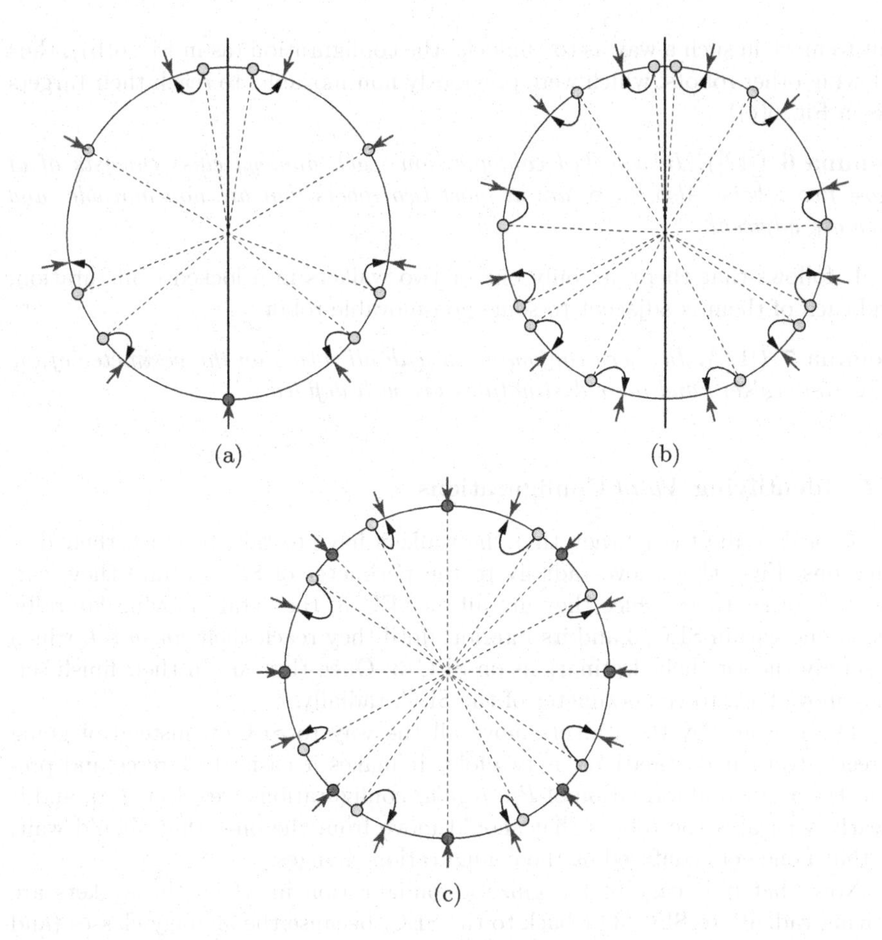


**Fig. 5.** The outer arrows indicate targets, and the inner arrows indicate correspondences between robots and targets. (a) The dark-shaded robot lies on an axis of symmetry. (b) There are some axes of symmetry, none of which contains a robot. (c) There are no axes of symmetry, and the dark-shaded robots form the largest concordance class. (Source: [12])

Therefore, the walkers are selected to be a movable analogy class of robots that are not satisfied and can reach their targets without ever becoming co-radial with other robots. If such a class is not unique, one can always be chosen unambiguously.

There are special cases where no such analogy class or robots exists: these configurations are said to be *locked* (see for instance Fig. 6(a)). In a locked configuration, the walkers are chosen with a different criterion: they are an analogy class that is movable and not satisfied, and that is adjacent to some non-movable analogy class. Such an analogy class is called *unlocking.* The goal of these walkers is not to reach their targets (if they could, the configuration would not be locked), but to move in such a way as to "unlock" the configuration (as in Fig. 6(b)), thus allowing other robots, which were previously non-movable, to reach their targets (as in Fig. 6(c)).

**Lemma 6** **([12]).** *In a locked configuration, each analogy class consists of at most two robots. Also, there are at most two robots that are non-movable, and they are adjacent.*

It follows that there are only one or two walkers in a locked configuration, and each of them is adjacent to some non-movable robot.

**Lemma 7** **([12]).** *In every configuration with all robots on the perimeter of the SEC, the walkers and their destinations are well defined.*

### 2.7 Identifying *Valid* Configurations

Next we describe the journey that the walkers have to take to reach their destinations. First they move radially to the perimeter of SEC/3, and they wait for each other there. Once they are all on SEC/3, they start moving laterally, remaining within SEC/3 and its interior, until they reach their *finish set*, which is simply the set their destinations on SEC/3. Once they are in their finish set, they move back to the perimeter of the SEC, radially.

The reason why the walkers move all the way to SEC/3, instead of going directly to their destinations, is two-fold. It makes it easier to foresee and prevent the accidental formation of *Pre-regular* configurations (see Sect. 2.9), and it clearly separates the robots that should move from the ones that should wait, so that none gets confused as the configuration changes.

Note that it is easy to recognize a configuration in which the walkers are moving radially to SEC/3 or back to the SEC, because the analogy classes (and hence the walkers) are defined only based on angular distances between robots. Thus, if all robots are on the SEC, except for a few analogous robots that are between SEC and SEC/3, then the configuration is recognized as a "consistent", or *Valid* one, in which the walkers are either moving to SEC/3 or are moving back to the perimeter of the SEC (see Fig. 7(a)).

If the walkers have already started moving laterally in SEC/3, then recognizing the configuration as a *Valid* one is more difficult. This can be done by "guessing" where the internal robots were located when they were still on the SEC and they have been selected as walkers. If there is a way to re-position the internal robots within their respective "sectors" of the SEC in such a way as to make them become a full analogy class, then the configuration is considered *Valid*, and the internal robots are considered walkers (see Fig. 7(b)). Otherwise, it means that the execution is in one of the earlier stages, and the robots still have to make their preliminary move to the SEC.

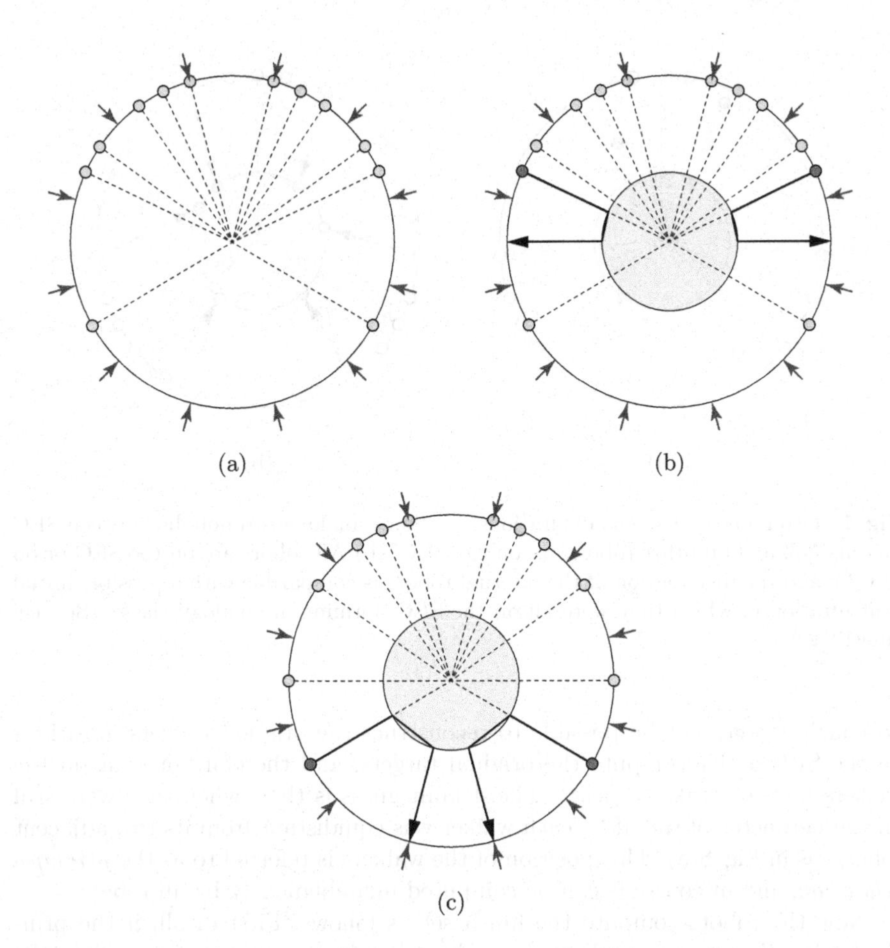


**Fig. 6.** (a) A locked configuration: the topmost robots are satisfied, the bottommost robots are non-movable, and all other robots would become co-radial in the process of reaching their targets. (b) A preliminary move is made to unlock the configuration. (c) When the configuration is unlocked, the bottommost robots become movable. (Source: [12])

## 2.8 Identifying the Finish Set

Once the configuration has been recognized as *Valid* and all walkers are in SEC/3, they compute their *finish set*. Recall that this is the set of their destinations on SEC/3, which they want to reach before moving back to SEC.

In order to understand where they should be going, the walkers have to recompute their targets. Indeed, note that the original targets have been computed when the walkers were on the perimeter of SEC. As they are now in SEC/3, in the process of moving laterally to their destinations, they need a robust way to define targets. This means that different walkers should compute the same target set, and the target set should not change as the walkers move within SEC/3.

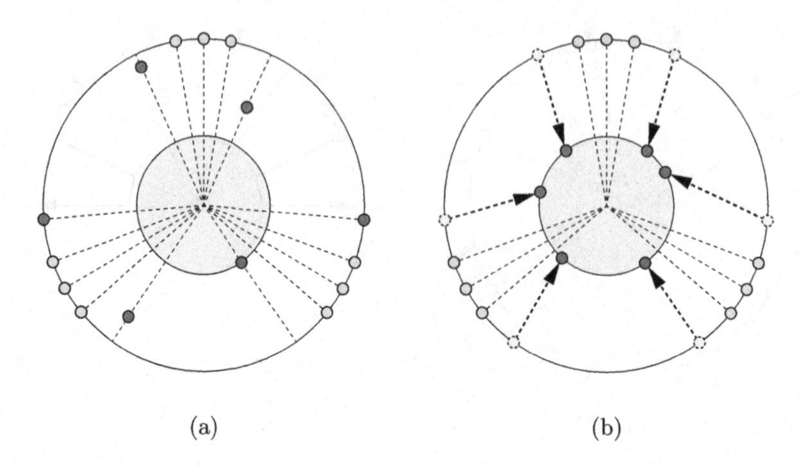

(a) (b)

**Fig. 7.** Two types of *Valid* configurations. (a) Some analogous robots lie between SEC and SEC/3, and all other robots are on the SEC. (b) All robots are on the SEC or on SEC/3, and the distribution of the internal robots is compatible with a possible initial configuration in which they were all on the SEC, forming an analogy class. (Source: citeFPSV17)

Of course it may not be possible to reconstruct the original walkers' positions on the SEC and recompute the original targets, and therefore once again the walkers have to "take a guess". The default guess is that, when they were still on the perimeter of the SEC, each walker was equidistant from its two adjacent robots, as in Fig. 8(a). This position of the walkers is referred to as the *principal relocation*, and of course it can be computed unambiguously by all robots.

Now the robots compute the finish set as follows. First of all, if the principal relocation is not a full analogy class, but just a subset of one, then the walkers know that it could not possibly be their initial position on the SEC (see Fig. 8(b)). In this case, the finish set is defined to be the principal relocation itself. The reason is that, by moving to their principal relocation, the walkers all join some bigger analogy class: this is not an "ineffective" move, because it makes progress toward having a unique analogy class.

On the other hand, if the principal relocation forms in fact an analogy class, then the walkers assume that to be their original position on the SEC. Hence they compute the new targets based on that configuration, with the usual algorithm (see Fig. 8(c)). Now, if the walkers can reach their respective targets from inside SEC/3 (that is, without becoming co-radial with other robots), then the finish set is the set of their targets. Otherwise, the walkers are confused, and by default their finish set is the principal relocation again.

Now that the finish set has been defined, the robots move there, always remaining within SEC/3, and without becoming co-radial with each other. There is only one exception: suppose that the walkers reach their finish set and move radially to the perimeter of the SEC: let $R$ be the set of the final positions of

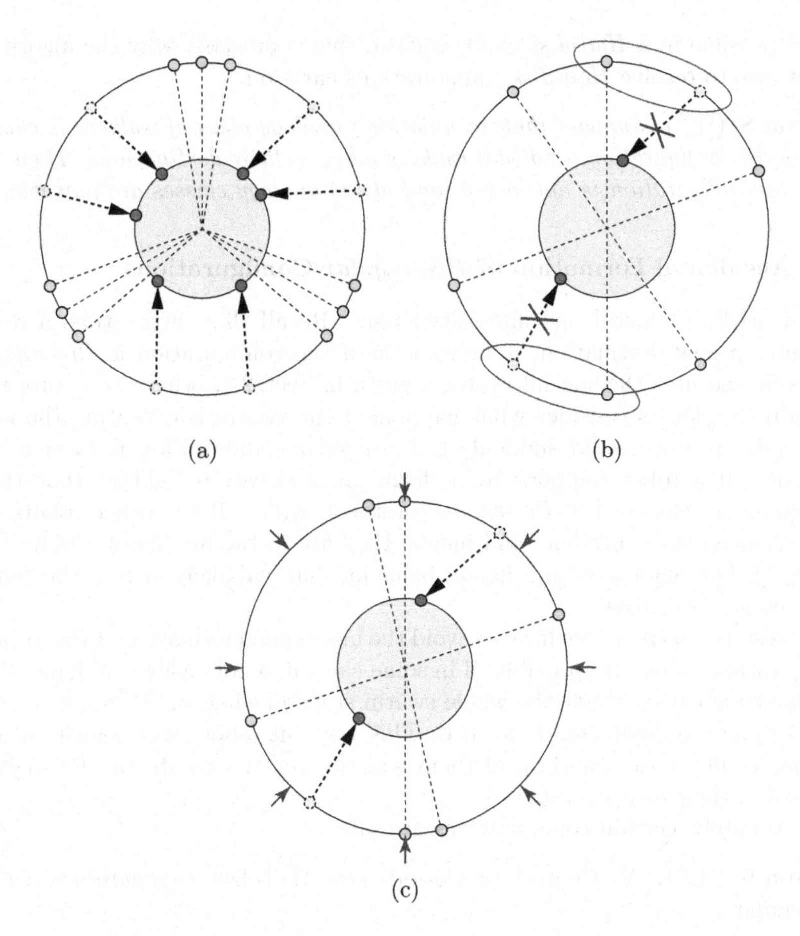


**Fig. 8.** (a) The principal relocation of the internal robots. (b) If the principal relocation is a proper subset of an analogy class, it cannot be the original position of the internal robots, or else a larger set of walkers would have been selected. (c) If the principal relocation forms an analogy class, it is used to determine the target set. Such targets remain fixed as the internal robots move within their respective sectors. (Source: [12])

the walkers on the SEC. If the new configuration is locked, and the robots in $R$ happen to form an unlocking analogy class, then it was not a correct move for the walkers to go to $R$. Indeed, this would cause them to become walkers again (unless there are two unlocking analogy classes and the other one is chosen), and the execution would enter an infinite loop. In this special case, the walkers have to do something to unlock the configuration, instead of reaching $R$. The strategy is as follows: if the walkers are two, they move to two antipodal points (as in Fig. 6(b)); if there is a unique walker, it becomes antipodal with some non-movable robot currently located on the SEC. Note that this type of move would not be possible in a *Half-disk* configuration: this is precisely why the algorithm makes sure to resolve *Half-disk* configurations early on.

**Lemma 8** ([12]). *Suppose that an unlocking analogy class of walkers is chosen in a locked configuration, and said walkers move to their destinations. Then, the resulting configuration is not locked, and all its analogy classes are movable.*

### 2.9 Accidental Formation of *Pre-regular* Configurations

The algorithm has still one unresolved issue. Recall that, every time a robot computes a new destination, it first checks if the configuration is *Pre-regular*. If it is, it executes the special protocol given in Sect. 2.1; otherwise it proceeds normally. So, let us consider what happens if the swarm is executing the non-*Pre-regular* protocol, and suddenly a *Pre-regular* configuration is formed "by accident". If a robot happens to perform an observation right at that time, it is going to execute the *Pre-regular* protocol, while all the other robots are still executing the other one, and maybe they are in the middle of a move (see Fig. 9(a)). This leads to an incoherent behavior that will likely disrupt the "flow" of the entire algorithm.

To resolve this issue, we have to avoid the unintended formation of *Pre-regular* configurations whenever possible. If in some cases it is not easily avoidable, then we have to make sure that the whole swarm stops moving, or *freezes*, whenever a *Pre-regular* configuration is formed. This way, all robots will transition into the new configuration, and all of them will coherently execute the *Pre-regular* protocol in their next cycles.

Fortunately, certain configurations are safe:

**Lemma 9** ([12]). *No* Central *or* Co-radial *or* Half-disk *configuration can be* Pre-regular.

So, in these initial phases, no *Pre-regular* configuration can be formed accidentally. Another important observation is the following:

**Lemma 10** ([12]). *In a* Pre-regular *configuration, no robot can be in SEC/3.*

This explains why we make our walkers move radially to SEC/3 first, and we allow them to move laterally only within SEC/3.

Hence, the only moves that have to be analyzed are the radial ones, which are performed by the walkers between SEC and SEC/3 or by the robots that are reaching the perimeter of the SEC during the preliminary step. We can conveniently simplify the problem if we let only one analogy class of robots move at a time. Note that this is already the case when the moving robots are the walkers, and in the other cases there is always a way to totally order the analogy classes unambiguously. If only one analogy class is moving radially, it is then easier to keep the swarm's behavior under control and analyze all possible outcomes.

The general protocol that is used for radial moves is called *cautious move*, and is described next.

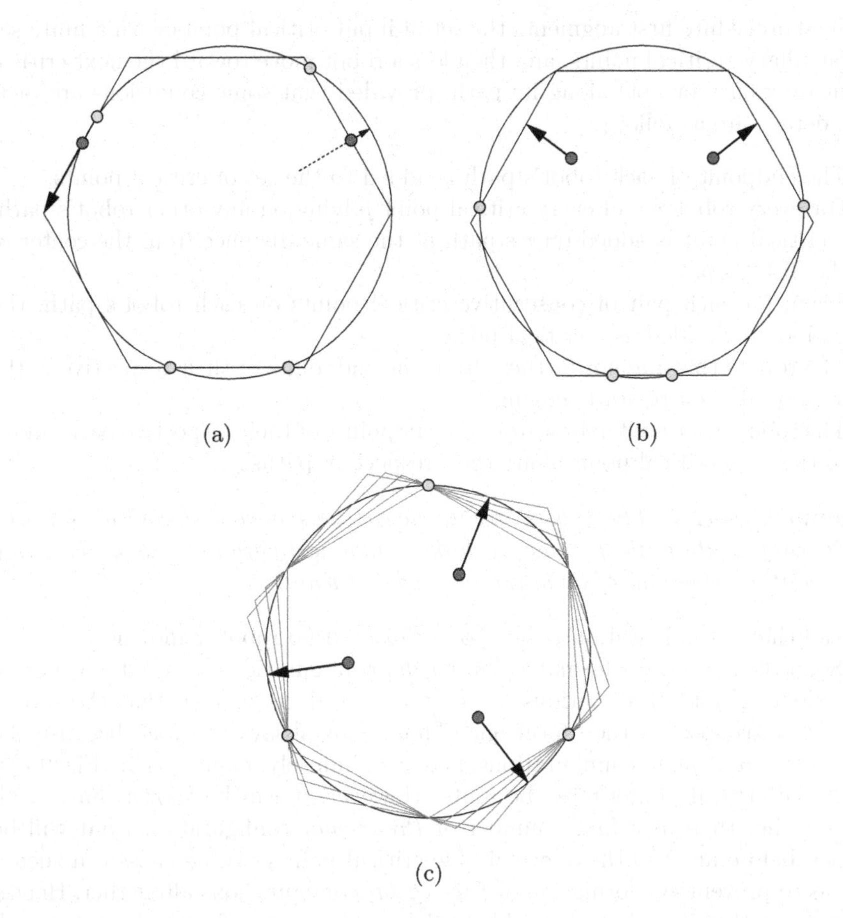


**Fig. 9.** (a) As the robot on the right moves to the SEC, a *Pre-regular* configuration is accidentally formed. The robot on the left recognizes a *Pre-regular* configuration, and starts executing the corresponding protocol, which is inconsistent with the other robot's move. (b) To prevent this behavior, enough critical points are added. Now the swarm is guaranteed to stop as soon as a *Pre-regular* configuration is formed. (c) A case in which infinitely many *Pre-regular* configurations are formable. Still, only the innermost is relevant, because it can be reached before all the others. (Source: [12])

## 2.10 Cautious Moves

In a cautious move, there is an analogy class of robots that have to move radially, either all from SEC/3 to SEC or all from SEC to SEC/3, while the other robots wait. Each moving robot has a final destination and is given as input a finite set of *critical points* along its path. Collectively, the moving robots execute a protocol that makes them move in such a way as to freeze whenever they are all located at a critical point (see for instance Fig. 9(b))[1].

[1] A similar concept has been used in [3], with some technical differences.

The procedure first augments the set of input critical points with a finite set of "auxiliary" critical points, and then lets a robot move toward the next critical point (auxiliary or not) along its path, provided that some conditions are met. The details are as follows.

- The endpoint of each robot's path is added to the set of critical points.
- For every robot $r$ and every critical point $p$ lying on any other robot's path, a critical point is added on $r$'s path at the same distance from the center of the SEC as $p$.
- Then, for each pair of consecutive critical points on each robot's path, the midpoint is added as a critical point.
- The robots that are not farthest from the endpoints of their respective paths are not allowed to start moving.
- The robots that are farthest from the endpoints of their respective paths move to the next critical point along their respective paths.

**Lemma 11** ([12]). *If the robots execute the cautious move protocol from a frozen initial configuration, they either reach their final destinations or they freeze in a configuration where all of them are in a critical point.*

(Recall that a configuration is said to be *frozen* if no robot is moving.)

So, if the potentially formable *Pre-regular* configurations are used to generate the critical points of a cautious move, it is indeed guaranteed that the robots will freeze as soon as they form one. This is not always possible, because the formable *Pre-regular* configurations may be infinitely many (as in Fig. 9(c)), while the critical points must be finite. However, it can be shown that, in all cases, either there is a finite number of *Pre-regular* configurations that will be formed before all the others, or suitable critical points can be chosen in such a way as to prevent the formation of *Pre-regular* configurations altogether. Hence, it turns out that it is always possible to choose a finite set of critical points for all cautious moves, and guarantee that the swam is frozen whenever it transitions into a *Pre-regular* configuration.

**Lemma 12** ([12]). *Let an analogy class of robots perform a radial cautious move from SEC to SEC/3 or vice versa, with suitable critical points. Then, either all robots reach their destinations, or they freeze in a* Pre-regular *configuration.*

### 2.11 Correctness of the Algorithm

All the elements of the `Uniform Circle` Formation algorithm have been presented. When a robot executes the algorithm, it determines the current configuration type and executes the corresponding procedure to compute a destination point. Observe that some configurations fall in more than one category (for instance, the *Central* configurations are also *Co-radial*), and so the order in which such categories are tested matters. The order is the following:

- *Regular*,
- *Pre-regular*,
- *Central*,
- *Half-disk*,
- *Co-radial*,
- *Valid*,
- *Invalid*.

All these classes have already been defined, except *Invalid*, which is the set of configurations that do not fall in any other class. In an *Invalid* configuration, all robots simply perform a cautious move toward the perimeter of the SEC.

The correctness of the *Pre-regular* case of the algorithm, as well as the *Central*, *Co-radial*, and *Half-disk* cases is relatively straightforward to prove. Other parts of the algorithm, however, need a more careful analysis: these include a characterization of the locked configuration and the determination of critical points in every configuration where a radial move is made, in order to avoid the accidental formation of *Pre-regular* configuration.

These theoretical tools allow to finally tackle the *Valid* case, and so analyze the main "loop" of the algorithm. It can be shown that the different phases of the execution "hinge together" as intended: all the walkers reach SEC/3 and freeze there (unless a *Pre-regular* configuration is formed in the process), then they all move to their finish set, freeze again, and finally they move back to the perimeter of the SEC. As the execution continues and more iterations of this phase are made, it is necessary to study how the target set changes, in order to make sure that a *Pre-regular* configuration is eventually formed.

To this end, it can be proven that, at each iteration, some "progress" is made toward a *Regular* or *Biangular* configuration. This could mean that the walkers join another analogy class (thus reducing the total number of analogy classes), or that a new axis of symmetry is acquired, or that more robots become satisfied. Of course the configuration may also be locked: in this case it can be proved that, after one iteration, either the configuration is no longer locked, or some analogy classes have merged, or a previously non-movable analogy class has become movable.

Also, by design, the algorithm never allows an analogy class to split (because the walkers constitute an analogy class when they are selected, and are again all analogous when they reach their finish set), and it never causes a symmetric configuration to become asymmetric from one iteration to the next. However, it is true that the targets may change, and thus the number of satisfied robots may actually decrease. However, this can happen only when some analogy classes merge or when the configuration becomes symmetric, and thus it can happen only finitely many times.

So, either a *Pre-regular* configuration is formed by accident (and this case leads to a quick resolution), or eventually there will be only one analogy class left, and hence the configuration will be *Regular* or *Biangular*. Figure 10 shows the possible transitions between configuration types that the algorithm allows. Observe that every possible flow ends in a *Regular* configuration.

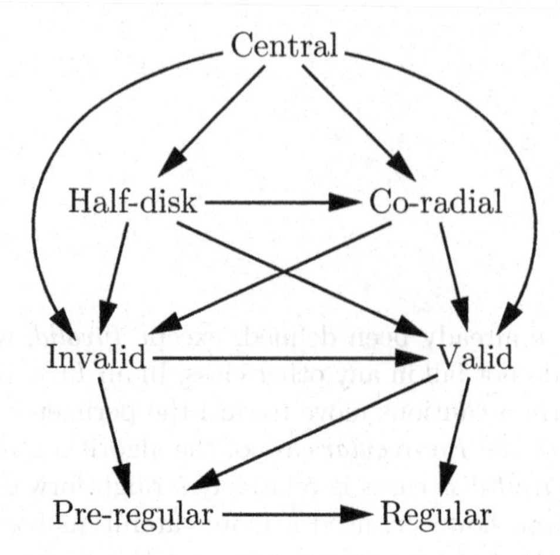


**Fig. 10.** Possible transitions between configurations in the `Uniform Circle` Formation algorithm. (Source: [12])

**Theorem 1** ([12]). *The* `Uniform Circle` *Formation Problem is solvable by* $n > 5$ *robots in the standard* $\mathcal{OBLOT}$ *model under the* $\mathcal{A}$SYNC *scheduler.*

## 3 Special Algorithms for $n \leq 5$ Robots

For small values of $n$, i.e., $n \leq 5$, the `Uniform Circle` Formation algorithm of Sect. 2 fails, and ad-hoc algorithms have been designed for these cases. If $n \leq 3$ the problem is relatively simple, and the case $n = 5$ is obtained by modifying the general algorithm of Sect. 2. The case $n = 4$, on the other hand, requires a special treatment.

### 3.1 $n \neq 4$ Robots

If $n = 1$ or $n = 2$, the pattern `Uniform Circle` is automatically formed, and nothing has to be done. If $n = 3$, the algorithm is as follows:

- if the three distances between pairs of robots are all distinct and robots $r_1$ and $r_2$ are farthest apart, then robot $r_3$ moves parallel to $r_1r_2$ toward the axis of $r_1r_2$;
- otherwise, if $r_1r_3 = r_2r_3$, then $r_3$ moves to the closest point that forms an equilateral triangle with $r_1$ and $r_2$ (in case there are two such points, one is chosen arbitrarily).

The analysis of this algorithm is straightforward.

**Theorem 2** ([12]). *The* `Uniform Circle` *Formation Problem is solvable by* $n \leq 3$ *robots in the standard* $\mathcal{OBLOT}$ *model under the* $\mathcal{A}$SYNC *scheduler.*

If $n = 5$, it turns out that the algorithm of Sect. 2 works as it is, except in one situation: there are locked configurations where an unlocking analogy class cannot be found in the usual way, because all the classes that are adjacent to some unmovable class are satisfied. This cannot happen if $n > 5$, and in this type of configuration the algorithm is undefined.

The algorithm can be modified to encompass these cases as follows. Suppose that the configuration is *Valid*, with all robots on the perimeter of the SEC, and without axes of symmetry. Two robots are said to be *antipodal* if they occupy antipodal points of the SEC.

- If no two robots are antipodal, a movable robot is chosen unambiguously, it becomes a walker, and moves to become antipodal with some other robot. It can be shown that there exists one such robot that can complete its movement without becoming co-radial with any other robot.
- If exactly two robots are antipodal, there is a unique robot that is adjacent to both of them (recall that a *Valid* configuration cannot be *Half-disk*). Such a robot is movable, and it becomes the walker and moves to become antipodal with another robot.
- If two pairs of robots are antipodal, the non-antipodal robot is movable, it becomes the walker, and moves to the midpoint of its adjacent robots, thus creating an axis of symmetry.

When the configuration has an axis of symmetry $\ell$, since $n = 5$ is odd, there must be a robot $r$ on $\ell$. If $\ell$ is not unique, the configuration is *Regular*; so, assume that $\ell$ is unique. If the two analogous robots $s$ and $s'$ that are farthest from $r$ are not satisfied, the other two robots $q$ and $q'$ make a preliminary move to become antipodal (on the diameter of the SEC that is perpendicular to $\ell$), thus making $s$ and $s'$ able to move to their targets. When $s$ and $s'$ are satisfied, $q$ and $q'$ move to their targets to form a *Regular* configuration.

Observe that, since $n = 5$ is an odd number, no *Pre-regular* configuration can be formed, and thus the cautious moves require no ad-hoc critical points.

**Theorem 3** ([12]). *The* `Uniform Circle` *Formation Problem is solvable by* $n = 5$ *robots in the standard* $\mathcal{OBLOT}$ *model under the* $\mathcal{A}$SYNC *scheduler.*

### 3.2 $n = 4$ Robots

The case with $n = 4$ robots presents difficulties that make it unique, and has been approached with radically different techniques. This case has been resolved in [15].

When trying to apply the algorithm of Sect. 2 to $n = 4$ robots, it is immediate to recognize that the *Biangular* configurations are now rectangular, and for such configurations the supporting polygon defined in Sect. 2.1 is not a unique square, but there are infinitely many of them. Even if the robots tried to implicitly agree on one such square with a common criterion, the square would shift as the robots move asynchronously to its vertices, yielding a convergence algorithm at best (as opposed to a formation algorithm).

The approach adopted in [15] is, roughly speaking, to "tilt" the supporting square by 45°. That is, a square is found such that each robot lies on a distinct edge of it (or on the extension of an edge), and the target of each robot is the midpoint of the edge on which it lies. Once again there is more than one such square, but the construction in Fig. 11 unambiguously produces one.

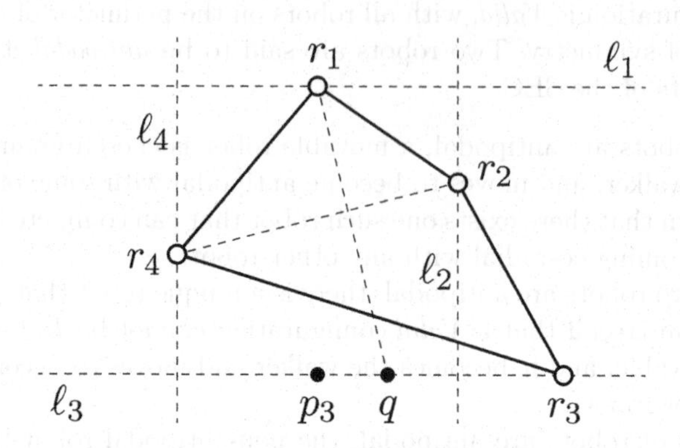


**Fig. 11.** Construction of the supporting square. (Source: [15])

Let $r_1r_2r_3r_4$ be a strictly convex quadrilateral whose diagonals $r_1r_3$ and $r_2r_4$ are not perpendicular. Let $q$ be the unique point such that $r_1q = r_2r_4$, the lines $r_1q$ and $r_2r_4$ are perpendicular, and the ray emanating from $r_1$ and passing through $q$ intersects the line $r_2r_4$. Let $\ell_3$ be the line through $r_3$ and $q$, and let $\ell_1$ be the line through $r_1$ parallel to $\ell_3$. (Since $r_1r_3$ and $r_2r_4$ are not perpendicular, $\ell_3$ is well defined and is distinct from $\ell_1$.) Let $\ell_2$ be the line through $r_2$ perpendicular to $\ell_1$, and let $\ell_4$ be the line through $r_4$ parallel to $\ell_2$. By construction, these four lines intersect at four points that are vertices of a square $Q$, the supporting square. In turn, the midpoints of the edges of $Q$ form a second square $Q'$, whose vertices are the target points of the robots. Referring to Fig. 11, the target of $r_3$ is $p_3$, and the segment $r_3p_3$ is called $r_3$'s *pathway*, etc.

**Lemma 13** ([15]). *All robots agree on the same supporting square, which remains fixed as all robots move to their target points asynchronously. Moreover, no two robots' pathways intersect.*

As the above construction assumes that the robots form a convex quadrilateral with non-perpendicular diagonals, special protocols are needed in these cases. If the quadrilateral is non-convex, there is a unique robot that is contained in the triangle formed by the other three: this robot moves to the foot of an altitude of such triangle, thus forming a loosely convex configuration with perpendicular diagonals. On the other hand, whenever the quadrilateral is (loosely) convex and its diagonals are perpendicular, then the robots that are closest to the intersection point of the diagonals move away from it until the configuration becomes a square.

The above protocol has yet one exception: it does not apply when all four robots are on the same line. In this case, the two non-extremal robots move to either side of the line by a small amount. As they move asynchronously in this fashion, their supporting square changes wildly, and so a "safe region" has to be defined in which the robots do not rely on their supporting square but follow a different protocol. The safe region has the form of a *thin hexagon*, depicted in Fig. 12. The proportions of the hexagon are carefully chosen for reasons that will be clear later.

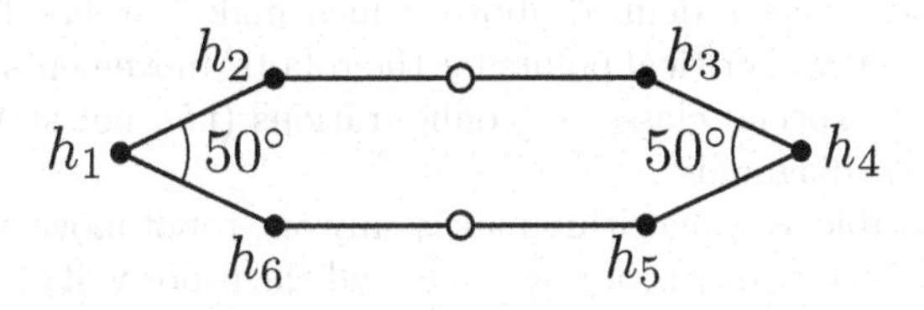


**Fig. 12.** Thin hexagon: the angles at $h_1$ and $h_4$ are 50°, and $h_1h_4 = 4 \cdot h_1h_2$. (Source: [15])

If there are robots in $h_1$ and $h_4$ and two internal robots on different sides of the diagonal $h_1h_4$, then the internal robots move to the white dots of Fig. 12, turning the configuration into one with perpendicular diagonals. If the internal robots are on the same side of the diagonal $h_1h_4$, say above it in Fig. 12, then they move to the vertices $h_2$ and $h_3$ respectively, and they wait for each other. When they both arrive and stop, the supporting square can be computed without ambiguity, and the protocol can safely switch to the normal one, which makes all robots move to their target points on the supporting square.

Of course, if all robots asynchronously move toward their target points, they may accidentally form configurations that are not strictly convex or have perpendicular diagonals or are thin hexagons, much like they could accidentally form *Pre-regular* configurations in the algorithm of Sect. 2. Again, the issue is resolved by limiting the number of robots that move concurrently and by employing a concept of critical point and cautious move similar to that of Sect. 2.10.

To define the critical points, some definitions are needed. Two robots are *concordant* if they have to move around the center of the supporting square in the same "direction" (i.e., clockwise or counterclockwise) as they go toward their targets. They are *discordant* if they move in opposite directions, and a robot is said to be *finished* when it is on its target point. Let the guidelines of two discordant robots $r_i$ and $r_j$ intersect in a point $g$. If $p_i$ lies on the segment $r_ig$ and $p_j$ lies on the segment $r_jg$, then $r_i$ and $r_j$ are said to be *convergent*; otherwise, they are *divergent* (if both $r_i$ and $r_j$ are finished, they are both convergent and divergent). If the pathway of $r_i$ intersects the segment $r_jr_k$ in $v$, then $r_i$ is said

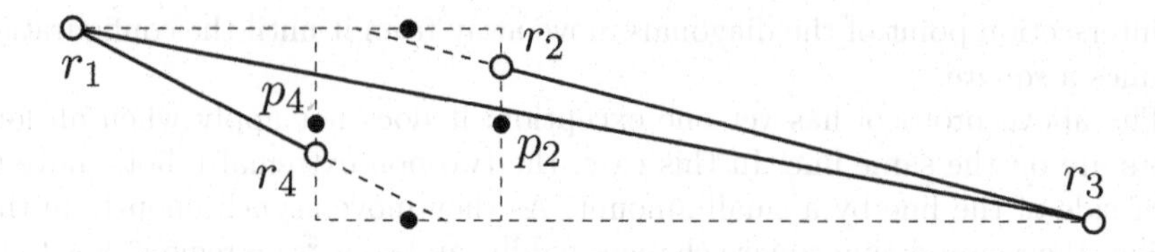


**Fig. 13.** Robot $r_2$ is blocked; robots $r_1$ and $r_3$ are hindered. (Source: [15])

to be *blocked* at $v$. If the pathway of $r_i$ intersects an extension of the segment $r_jr_k$ in $v$, then $r_i$ is said to be *hindered* at $v$ (see Fig. 13).

Of course, the points $v$ defined above, which make a robot blocked or hindered, constitute natural critical points for the robots' movements, as they determine transitions to special classes of configurations (i.e., not strictly convex or with perpendicular diagonals).

Whenever possible, the algorithm makes only one robot move while the others wait. In this case asynchrony is not an issue, and the robot will either move to its target or to the first critical points on its pathway. An example of such a situation is when one robot is discordant with the other three: then, the discordant one will move before the other three. Another example is when all robots are concordant: in this case, the two robots on opposite sides of the supporting square that are closest to each other will move (as they move toward their targets, their distance will keep being the shortest). It can be proven that these two robots cannot be both blocked or both hindered. Thus, if one of them is blocked, it is chosen as the only robot to move; if none of them is blocked but one is hindered, it is the one to move. In all situations in which a single robot cannot be distinguished in a robust way, it is always possible to choose two robots to move while the other two robots wait. Suitable critical points also exist for these situations.

A notable fact is that, when looking for critical points, in almost all cases it is unnecessary to take into account the accidental formation of thin hexagon configurations. Indeed, in most of the relevant configurations, if the robots to move are selected properly, no thin hexagon can ever be formed, as Fig. 14 exemplifies. This is due to the proportions of the thin hexagon detailed in Fig. 12, which have been chosen specifically for this purpose. For instance, suppose that robots $r_1$ and $r_2$ are divergent and robots $r_3$ and $r_4$ are divergent, as well. Further assume that $r_1$ is the only robot that does not lie on an edge of the supporting square, but on the extension of it: in this configuration, $r_1$ is chosen to move toward its target. It is easy to verify that, if $r_2$ is the robot closest to $r_1$, then the two angles $\angle r_1r_3r_4$ and $\angle r_1r_4r_3$ are both greater than 25°. Hence, if $r_1$ is on an acute vertex of a thin hexagon and $r_3$ (respectively, $r_4$) is on the opposite vertex, then $r_4$ (respectively, $r_3$) cannot be contained in the same thin hexagon.

As in Sect. 2, for the algorithm to work properly, it is necessary to identify to which class the observed configuration belongs, by checking them one by one in the correct order, because some classes have intersections. The order is the following:

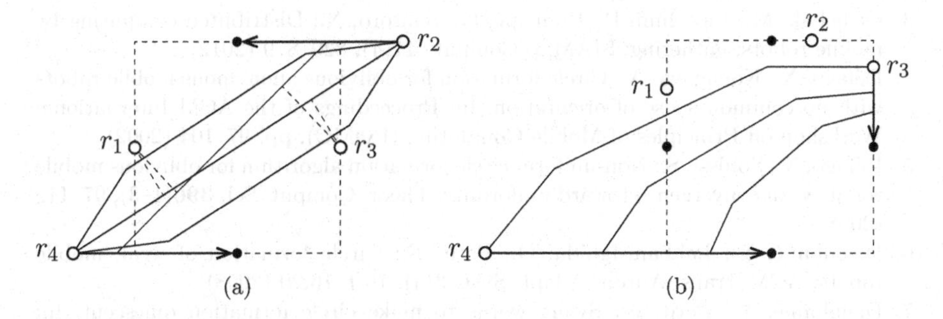


**Fig. 14.** (a) If $r_1$ and $r_3$ are on their targets and $r_2$ and $r_4$ move, no thin hexagon is formed. (b) As $r_3$ and $r_4$ move, no thin hexagon is formed. (Source: [15])

- Perpendicular diagonals,
- Thin hexagon,
- Non-convex,
- All robots concordant,
- Two robots convergent, two other robots divergent,
- Two robots divergent, two other robots divergent,
- Three robots concordant, one discordant.

To each class corresponds a different set of actions. The algorithm is designed in such a way that, whenever a class transition occurs, no robot is moving: thus, all robots will "witness" the transition and will coherently execute the instructions for the new class. Also, when the configuration transitions from class $A$ to class $B$, the class $B$ comes before $A$ in the list above: this ensures that some progress toward the solution is always made. The last rule has only one exception: when all robots are on the same side of a thin hexagon, they will eventually be found on four consecutive vertices of it. This configuration is then interpreted as one with two pairs of divergent robots, which is resolved by the normal algorithm without possibility of forming a thin hexagon again.

We conclude that `Uniform Circle` is formable by $n = 4$ robots, as well.

**Theorem 4** ([15]). *The* `Uniform Circle` *Formation Problem is solvable by* $n = 4$ *robots in the standard* $\mathcal{OBLOT}$ *model under the* $\mathcal{A}$SYNC *scheduler.*

## References


1. Bhagat, S., Mukhopadhyaya, K.: Optimum circle formation by autonomous robots. In: Chaki, R., Cortesi, A., Saeed, K., Chaki, N. (eds.) Advanced Computing and Systems for Security. AISC, vol. 666, pp. 153–165. Springer, Singapore (2018). https://doi.org/10.1007/978-981-10-8180-4_10
2. Chatzigiannakis, I., Markou, M., Nikoletseas, S.: Distributed circle formation for anonymous oblivious robots. In: Ribeiro, C.C., Martins, S.L. (eds.) WEA 2004. LNCS, vol. 3059, pp. 159–174. Springer, Heidelberg (2004). https://doi.org/10.1007/978-3-540-24838-5_12

3. Cieliebak, M., Flocchini, P., Prencipe, G., Santoro, N.: Distributed computing by mobile robots: gathering. SIAM J. Comput. **41**(4), 829–879 (2012)
4. Défago, X., Konagaya, A.: Circle formation for oblivious anonymous mobile robots with no common sense of orientation. In: Proceedings of the ACM International Workshop on Principles of Mobile Computing (POMC), pp. 97–104 (2002)
5. Défago, X., Souissi, S.: Non-uniform circle formation algorithm for oblivious mobile robots with convergence toward uniformity. Theor. Comput. Sci. **396**(1–3), 97–112 (2008)
6. Dieudonné, Y., Labbani-Igbida, O., Petit, F.: Circle formation of weak mobile robots. ACM Trans. Auton. Adapt. Syst. **3**(4), 16:1–16:20 (2008)
7. Dieudonné, Y., Petit, F.: Swing words to make circle formation quiescent. In: Prencipe, G., Zaks, S. (eds.) SIROCCO 2007. LNCS, vol. 4474, pp. 166–179. Springer, Heidelberg (2007). https://doi.org/10.1007/978-3-540-72951-8_14
8. Dieudonné, Y., Petit, F.: Squaring the circle with weak mobile robots. In: Hong, S.-H., Nagamochi, H., Fukunaga, T. (eds.) ISAAC 2008. LNCS, vol. 5369, pp. 354–365. Springer, Heidelberg (2008). https://doi.org/10.1007/978-3-540-92182-0_33
9. Feletti, C., Mereghetti, C., Palano, B.: Uniform circle formation for swarms of opaque robots with lights. In: Izumi, T., Kuznetsov, P. (eds.) Stabilization, Safety, and Security of Distributed Systems. LNCS, vol. 11201, pp. 317–332. Springer, Cham (2018). https://doi.org/10.1007/978-3-030-03232-6_21
10. Flocchini, P., Prencipe, G., Santoro, N.: Self-deployment algorithms for mobile sensors on a ring. Theor. Comput. Sci. **402**(1), 67–80 (2008)
11. Flocchini, P., Prencipe, G., Santoro, N., Viglietta, G.: Distributed computing by mobile robots: solving the uniform circle formation problem. In: Aguilera, M.K., Querzoni, L., Shapiro, M. (eds.) OPODIS 2014. LNCS, vol. 8878, pp. 217–232. Springer, Cham (2014). https://doi.org/10.1007/978-3-319-14472-6_15
12. Flocchini, P., Prencipe, G., Santoro, N., Viglietta, G.: Distributed computing by mobile robots: uniform circle formation. Distrib. Comput. **30**(6), 413–457 (2017)
13. Fujinaga, N., Yamauchi, Y., Kijima, S., Yamashita, M.: Asynchronous pattern formation by anonymous oblivious mobile robots. SIAM J. Comput. **44**(3), 740–785 (2015)
14. Katreniak, B.: Biangular circle formation by asynchronous mobile robots. In: Pelc, A., Raynal, M. (eds.) SIROCCO 2005. LNCS, vol. 3499, pp. 185–199. Springer, Heidelberg (2005). https://doi.org/10.1007/11429647_16
15. Mamino, M., Viglietta, G.: Square formation by asynchronous oblivious robots. In: Proceedings of the 28th Canadian Conference on Computational Geometry (CCCG), pp. 1–6 (2016)
16. Miyamae, T., Ichikawa, S., Hara, F.: Emergent approach to circle formation by multiple autonomous modular robots. J. Robot. Mechatron. **21**(1), 3–11 (2009)
17. Mondal, M., Chaudhuri, S.G.: Uniform circle formation by mobile robots. In: Proceedings of the Workshop Program of the 19th International Conference on Distributed Computing and Networking (ICDCN), pp. 20:1–20:2 (2018)
18. Suzuki, I., Yamashita, M.: Distributed anonymous mobile robots: formation of geometric patterns. SIAM J. Comput. **28**(4), 1347–1363 (1999)

# Symmetry of Anonymous Robots

Yukiko Yamauchi(✉)

Faculty of Information Science and Electrical Engineering, Kyushu University, Fukuoka, Japan
yamauchi@inf.kyushu-u.ac.jp

**Abstract.** Symmetry of anonymous mobile robots imposes many impossibilities. We focus on the formation problem that requires the robots to form a target pattern. We consider the robots moving in the three-dimensional space and the two-dimensional space (3D and 2D space, respectively) and introduce the notion of *symmetricity* of a set of points that represents the set of rotation groups that the robots cannot resolve. However, the symmetricity does not always match the rotational symmetry of geometric positions of the robots. We demonstrate that the robots are capable of breaking symmetry by their movement in some cases. The goal of this chapter is to present the following characterization of formable patterns; anonymous synchronous mobile robots in 3D space or 2D space can form a target pattern from an initial configuration if and only if the symmetricity of an initial configuration is a subset of the symmetricity of the target pattern.

**Keywords:** Symmetry · Rotation group · Pattern formation problem
Plane formation problem · Symmetry breaking

## 1 Introduction

Symmetry is a source of impossibilities of agreement problems in *anonymous* distributed systems, where computing entities are indistinguishable and execute a common deterministic algorithm. For example, consider a traditional distributed message-passing system consisting of two anonymous processes. There is no deterministic algorithm to elect a leader because the two processes have an identical local "view" of the system and they always demonstrate identical behavior. This notion has been formalized as the *symmetricity* of a network of anonymous processes [22].

In this chapter, we introduce the *symmetricity* of anonymous mobile robots moving in the three-dimensional Euclidean space and the two-dimensional Euclidean space (3D and 2D space, respectively). Each robot is a point and repeats a `Look-Compute-Move` cycle with a common deterministic algorithm, i.e., we consider the $\mathcal{OBLOT}$ model introduced in Chap. 1. It cannot access the global coordinate system and use any explicit communication medium. It observes the positions of other robots in its *local coordinate system*, which is a right-handed


P. Flocchini et al. (Eds.): Distributed Computing by Mobile Entities, LNCS 11340, pp. 109–133, 2019.
https://doi.org/10.1007/978-3-030-11072-7_6

orthogonal coordinate system in the specified space with an arbitrary unit distance and arbitrary directions and orientations of coordinate axes. Although sensing is the only means of communication, the robots may not obtain a consistent result (i.e., observation) due to their inconsistent local coordinate systems. In such a distributed robot system, symmetry arises from the geometric positions of the robots and their local coordinate systems. To show the effect of symmetry, we consider fully-synchronous robots equipped with unlimited sensing range.

We focus on the *formation problem* that requires the robots to form a specified pattern and show that the initial symmetry determines the set of formable patterns; thus, it determines the robots' self-organization power. Previous studies [16,19] have pointed out that the formation problem is related to the agreement problem, and the problem is further classified as follows depending on the target pattern.

- The *point formation problem*, which is also known as the *gathering problem*, requires the robots to gather at a single point with no predefined gathering point. The point formation problem is the simplest agreement problem.
- The *circle formation problem* requires the robots to form a circle (i.e., a regular polygon on a plane). Circle formation implies that the robots agree on the center and radius of a circle.
- The *pattern formation problem* requires the robots to form a target pattern (shape), where each robot is given the target pattern as a multiset of point coordinates. Since the robots do not know the global coordinate system, any uniform scaling, translation, rotation, and their combination on the target pattern are allowed.[1] The robots can form an arbitrary pattern when they agree on a common coordinate system consisting of the origin, unit distance, and coordinate axes.

The formation problem was initially considered for the robots moving in 2D space. The point formation problem and the pattern formation problem are easily extended to 3D space. However, the circle formation problem is not directly extended to 3D space because an agreement on a point and radius results in a sphere in 3D space. To solve the circle formation problem, the robots must first agree on a plane on which they form a circle. This problem itself is an important formation problem in 3D space.

- The *plane formation problem* requires the robots in 3D space to land on a common plane; however, multiple robots cannot land on a single point to avoid multiplicity. Thus, point formation is not a solution for the plane formation problem. Plane formation implies that the robots agree on a single direction and a single point.

The equivalent in 2D space is the *line formation problem*, which requires the robots to form a line. We consider the following two examples.

[1] The pattern formation problem for the robots with chirality does not allow any reflection of the target pattern by a mirror plane.

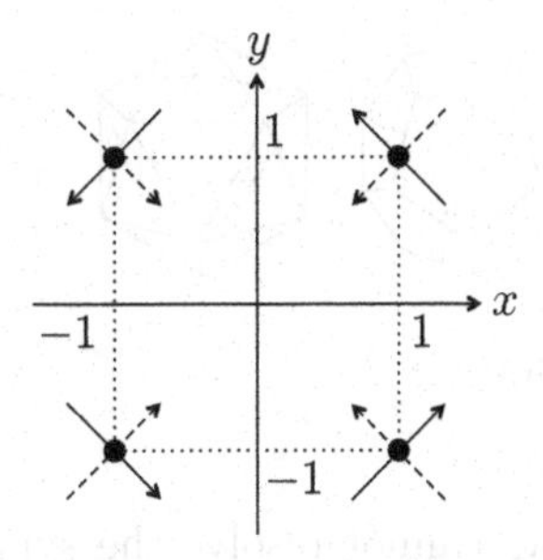


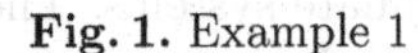
**Fig. 1.** Example 1

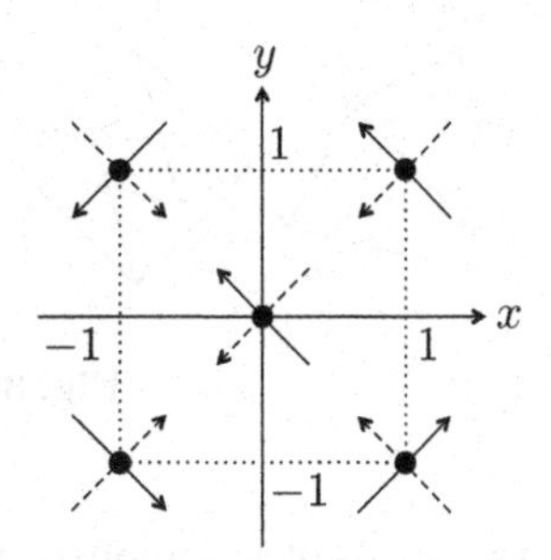


**Fig. 2.** Example 2

**Example 1.** Consider an initial configuration of four robots placed at $(1,1)$, $(-1,1)$, $(-1,-1)$, and $(1,-1)$ in the global coordinate system (Fig. 1). Here, the robots' local coordinate systems are symmetric with respect to the origin $(0,0)$. In Fig. 1, the $x$ and $y$ axes of each local coordinate system are indicated by a solid arrow and a broken arrow, respectively.

If the four robots observe their positions, they obtain identical observations because their local coordinate systems are symmetric with respect to the origin $(0,0)$, which is also the center of the rotational symmetry of their positions. Note that the observation consists of the coordinates of the four points in the robot's local coordinate system. The next positions locally computed by each robot using a common algorithm form another regular square, and the robots never resolve their rotational symmetry. Furthermore, the final lines that they propose are symmetric with respect to the origin, and they cannot form a line.

In 2D space, the symmetry among the robots is recognized by the rotational symmetry. Consider a set $P$ of points in 2D space. We consider a decomposition of $P$ into regular $m$-gons with a common center, where one point forms a regular 1-gon with an arbitrary center, and two points form a regular 2-gon with the center being the midpoint. Then, we consider the maximum value of $m$. For example, $m$ is four in Example 1. The value of $m$ represents the rotational symmetry of $P$, and when $m > 1$, the common center is the center $c(P)$ of the smallest enclosing circle of $P$. As demonstrated in Fig. 1, there exists a set of local coordinate systems for $P$ that is symmetric with respect to $c(P)$ and prevents the robots from breaking the regular $m$-gons.

The crucial difference between the rotational symmetry of $P$ and the value of $m$ arises when $c(P)$ is contained in $P$.

**Example 2.** Five robots are placed at $(0,0)$, $(1,1)$, $(-1,1)$, $(-1,-1)$, and $(1,-1)$ in the global coordinate system (Fig. 2).

In this case, the value of $m$ is one, and the robot on the origin can break the symmetry by leaving its current position. The robot can propose a final line by moving to a point on one of the diagonals of the regular square formed by the other four robots. In [16,19,23], this value $m$ is called the *symmetricity* of a set $P$ of points, denoted by $\rho(P)$.

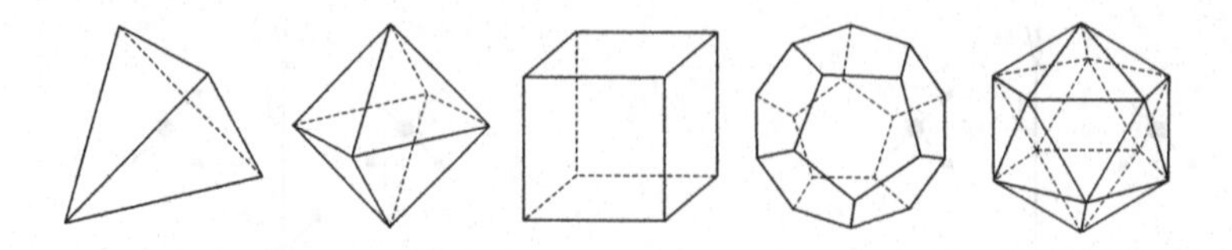

**Fig. 3.** Regular polyhedra

The two examples demonstrate that the robots cannot resolve the symmetry of their initial geometric positions and local coordinate systems. The initial symmetry prevents the robots from forming shapes with lower symmetry. Symmetric configurations are recognized as regular polygons in 2D space. Thus, we consider the symmetry in 3D space in relation to the five regular polyhedra, i.e., the regular tetrahedron, the regular octahedron, the cube, the regular dodecahedron, and the regular icosahedron (Fig. 3). In addition, there are thirteen semi-regular polyhedra that are considered less symmetric than the five regular polyhedra. The symmetry of such a polyhedron is recognized as a set of direct congruent transformations, i.e., a set of symmetry operations that preserve the center and maintain Euclidean distance and handedness. As in 2D space, such operations consist of rotations around axes in 3D space that form the special orthogonal group $SO(3)$. Since we assume a finite number of robots, we consider a set consisting of finite rotation operations. The subgroups of $SO(3)$ with finite order are classified into five types of *rotation groups*, i.e., the *cyclic groups*, the *dihedral groups*, the *regular tetrahedral group*, the *regular octahedral group*, and the *regular icosahedral group*. A cyclic group and a dihedral group are recognized as a set of rotations on a pyramid with a regular polygon base and a set of rotations on a prism with regular polygon bases, respectively. Each of the remaining three rotation groups is recognized as a set of rotations on the corresponding regular polyhedra. Other symmetry operations in 3D space consist of reflections for mirror planes (*bilateral symmetry*), reflections for a point (*central inversion*), and *rotation-reflections* [6]. However, these operations change handedness.[2] In this chapter, we consider rotation groups.

Let us begin with the plane formation problem. We say that the robots *form* a regular polyhedron when they are placed on the vertices of the regular polyhedron. The regular tetrahedron, the regular octahedron, and the regular icosahedron have the corresponding rotational symmetry. The cube is the dual of the regular octahedron, and the regular dodecahedron is the dual of the regular icosahedron in the sense that the centers of the faces of one regular polyhedron form its dual polyhedron. Thus, the cube and the regular dodecahedron have the identical rotational symmetry as their duals. The robots seem to be unable to form a plane when they form a regular polyhedron in an initial configuration. However, the robots can form a plane from an initial configuration where they

[2] In Sect. 6, we discuss generalization of these symmetry operations to the robots without *chirality*, in which model the local coordinate system of a robot is either right-handed or left-handed.

form one of the regular polyhedra, except the regular icosahedron. In addition, the robots can form a plane from an initial configuration where they form an icosidodecahedron; however, they cannot form a plane from the remaining semi-regular polyhedra. This counter-intuitive fact derives from the symmetry of local coordinate systems. For example, if one arranges local coordinate systems whose origins form a cube, these local coordinate systems will be asymmetric or agree on one direction. The eight local coordinate systems are less symmetric than the cube. Similarly, there is no set of local coordinate systems for the vertices of the regular icosahedron that is symmetric regarding the regular icosahedral group. However, there exists a set of local coordinate systems for these 12 vertices that is symmetric regarding the regular tetrahedral group, which is a subgroup of the regular icosahedral group. The 12 robots may be caught in the regular tetrahedral group, which does not act on a set of points on a plane and they cannot form a plane. Consequently, to investigate the rotational symmetry in 3D space, we must consider the structure of the rotation groups.

The goal of this chapter is to characterize the set of formable patterns by synchronous mobile robots in 3D space. We first define the *symmetricity* in 3D space that contains the symmetricity in 2D space as a subclass. For a given set $P$ of points in 3D space, its symmetricity, denoted by $\varrho(P)$, consists of a set of rotation groups that the robots cannot break when they are placed on $P$. Then, we give the following characterization.

**Theorem 1** [25]. *Regardless of obliviousness, fully-synchronous robots can form a target pattern $F$ from an initial configuration $P$ if and only if* $\varrho(P) \subseteq \varrho(F)$.

Intuitively, this necessary and sufficient condition means that the "symmetry" of an initial configuration must be lower than that of a target pattern. The symmetricity is defined such that the subset relation reflects the subgroup relation in group theory.

The chapter begins with the definition of rotation groups and related notions. The impossibility part first considers oblivious robots and then extends the necessary condition to non-oblivious robots. After that, matching sufficient condition is given by a pattern formation algorithm for oblivious robots. Non-oblivious robots can execute the algorithm by ignoring the content of local memory.

The reminder of this chapter is organized as follows. We describe the system model and define the formation problem in Sect. 2. In Sect. 3, we define the symmetricity of a set of points and demonstrate how the robots are caught in their initial symmetricity. Section 4 presents initial configurations that allow the robots to break their symmetry through movement, and Sect. 5 presents an overview of a pattern formation algorithm for solvable instances wherein the robots use their symmetry to coordinate themselves. As a corollary of the main theorem, we give a necessary and sufficient condition for the robots to solve the plane formation problem. Conclusions are presented in Sect. 6. In addition, we extend the main results to weaker robot models.

## 2 Robot Model and Formation Problem

**Robot System.** We consider a set $R = \{r_1, r_2, \ldots, r_n\}$ of $n$ ($n \geq 2$) robots in a specified space (i.e., 2D space or 3D space). Each robot is an anonymous point, and we use the indices to facilitate description. The position of $r_i$ at time $t$ in the global coordinate system is denoted by $p_i(t)$. A *configuration* is a multiset of robots' positions denoted by $P(t) = \{p_1(t), p_2(t), \ldots, p_n(t)\}$. We assume $p_i(0) \neq p_j(0)$ holds for all $i \neq j$ because there is no deterministic algorithm that can separate more than one robots at the same point. All possible multisets of points in 3D space and those in 2D space are denoted by $\mathcal{P}_n^3 \in (\mathbb{R}^3)^n$ and $\mathcal{P}_n^2 \in (\mathbb{R}^2)^n$, respectively.

Each robot $r_i$ repeats a Look-Compute-Move cycle. In the Look phase, $r_i$ obtains the snapshot of other robots' positions. Visibility is $\mathcal{U}$NLIMITED; therefore, $r_i$ can observe all other robots. Robot $r_i$ uses its local coordinate system when it observes the positions of other robots and when it moves to a next position. Each local coordinate system is a right-handed orthogonal coordinate system in the specified space. For example, in 3D space, it is a right-handed $x$-$y$-$z$ coordinate system. The origin of the local coordinate system of $r_i$ is its current position and it changes as $r_i$ moves. On the other hand, the directions and orientations of the axes and the unit distance do not change. Thus, a local coordinate system is represented by a point and a set of vectors of its unit length. Note that a local coordinate system is a uniform scaling, a transformation, a rotation, or their combination of the global coordinate system. Each robot is equipped with the *(strong) multiplicity detection ability,* and when more than one robot is on a single position, each robot can count the number of such robots. In the Compute phase, $r_i$ computes its next position using a common deterministic algorithm. We say that the algorithm is *oblivious* when its input is the observation obtained in the preceding Look phase and *non-oblivious* when its input contains past observations and computations. The robots executing an oblivious algorithm are called *oblivious*, and the robots executing a non-oblivious algorithm are called *non-oblivious.* The output of the Compute phase is the coordinates of the next position in $r_i$'s local coordinate system. In the Move phase, $r_i$ moves toward the next position. We say a movement is *rigid* when a robot reaches its next position. Otherwise, a robot may stop en route, and we say the movement is *non-rigid.* A non-rigid movement guarantees that a robot moves by an unknown minimum moving distance $\delta$; however, after moving $\delta$, it may stop at any point on its route. If the distance between the current position and the next position is less than or equal to $\delta$, the robot reaches the next position.

There are three models for the timing of Look-Compute-Move cycles, i.e., *fully-synchronous* ($\mathcal{F}$SYNC), *semi-synchronous* ($\mathcal{S}$SYNC), and *asynchronous* ($\mathcal{A}$SYNC). In the $\mathcal{F}$SYNC model, the Look, Compute, and Move phases are completely synchronized. We consider that the $t$th Look-Compute-Move cycle starts at time $(t-1)$ and ends by time $t$ for $t = 1, 2, \ldots$. In other words, in the $t$th Look-Compute-Move cycle, each robot observes $P(t-1)$ in its local coordinate system, computes its next position, and the Move phase ends before time $t$. In the $\mathcal{S}$SYNC model, like the $\mathcal{F}$SYNC model, Look-Compute-Move cycles start at a discrete time; however, some robots may skip cycles. To guarantee fairness, we assume that each robot executes a Look-Compute-Move cycle infinitely many times. In the $\mathcal{A}$SYNC model, we only assume that the length of each Look-Compute-Move cycle is finite and each robot executes a cycle infinitely many times. The main difference between the $\mathcal{A}$SYNC model and the other two models is that a robot may be observed while it is moving. However, since a robot can observe the positions of other robots, it cannot recognize which robot is moving. In other words, the duration of each cycle is negligible in the $\mathcal{F}$SYNC model and the $\mathcal{S}$SYNC model; however, it is not negligible in the $\mathcal{A}$SYNC model.

An *execution* of an algorithm starting from an initial configuration $P(0)$ is a sequence $P(0), P(1), P(2), \ldots$ of configurations for discrete time $t = 0, 1, 2, \ldots$ for the $\mathcal{F}$SYNC model and the $\mathcal{S}$SYNC model. For the $\mathcal{A}$SYNC model, we consider the time at which at least one robot takes a snapshot of the positions of the robots. Let $t_0, t_1, t_2, \ldots$ be the sequence of such time instants that satisfies $t_i < t_{i+1}$ for each $i = 0, 1, 2, \ldots$. Then, we consider $t_i$ as time $i$ and we focus on the sequence $P(0), P(1), P(2), \ldots$.

Given an initial configuration $P(0)$ and robots' local coordinate systems, the execution of algorithm $\psi$ from $P(0)$ is uniquely determined in the $\mathcal{F}$SYNC model with rigid movement. On the other hand, there are many executions of $\psi$ from $P(0)$ in the $\mathcal{S}$SYNC model and the $\mathcal{A}$SYNC model with rigid movement. However, any execution of $\psi$ in the $\mathcal{F}$SYNC model occurs in the $\mathcal{S}$SYNC model, and any execution of $\psi$ in the $\mathcal{S}$SYNC model occurs in the $\mathcal{A}$SYNC model. The relation between the rigid movement and the non-rigid movement is the same. As a result, if the robots cannot accomplish a given task in the $\mathcal{F}$SYNC model with rigid movement, there exists an execution where the robots cannot accomplish the task in the $\mathcal{S}$SYNC model and the $\mathcal{A}$SYNC model regardless of movement rigidity. However, if there exists an algorithm $\psi$ by which the robots can accomplish a given task in the $\mathcal{A}$SYNC model with non-rigid movement, $\psi$ also makes the robots in the $\mathcal{S}$SYNC model and the $\mathcal{F}$SYNC model accomplish the task regardless of movement rigidity.

In this chapter, we mainly consider $\mathcal{F}$SYNC robots with rigid movement. Generalization of the results to the $\mathcal{S}$SYNC model and the $\mathcal{A}$SYNC model is discussed in Sect. 6.

**Formation Problem.** The *pattern formation* problem requires the robots to form a *target pattern* from a given initial configuration. The target pattern is given to each robot as a multiset $F$ of points in the global coordinate system. Thus, the pattern formation problem allows any uniform scaling, translation, rotation, and their combination on the target pattern. We say a multiset of points $F'$ is *similar* to $F$ when such a translation exists (denoted by $F' \sim F$). An algorithm *forms* a target pattern $F$ from a given initial configuration $P$ if, regardless of the robots' initial local coordinate systems and the initial local memory content (of non-oblivious robots), for any execution $P(0)(= P), P(1), P(2), \ldots$ under the specified model, there exists a finite $t$ such that $(i)$ $P(t) \sim F$ and (*ii*) for any positive integer $t' \geq t$, $P(t') = P(t)$. A target pattern $F$ is *formable* from a given initial configuration $P$ if there exists an algorithm that forms $F$ from $P$.

The *plane formation problem* requires the robots in 3D space to land on a plane that is not predefined; however, multiple robots cannot land on a single point to avoid multiplicity. Thus, point formation is not a solution for the plane formation problem. We also assume that $n$ is larger than three because fewer than four robots are on one plane. An algorithm *forms a plane* from a given initial configuration $P$ if, regardless of the robots' initial local coordinate systems and the initial local memory content (of non-oblivious robots), for any execution $P(0)(=P), P(1), P(2), \ldots$ under the specified model, there exists finite $t$ such that (*i*) $P(t)$ is contained in a plane, and (*ii*) for any positive integer $t' \geq t$, $P(t') = P(t)$.

These formation problems require the robots to form a target pattern or a plane regardless of their initial local coordinate systems, execution schedule of cycles (in the $\mathcal{S}$SYNC model and the $\mathcal{A}$SYNC model), and movements (in the non-rigid model). In other words, a formation algorithm must defeat the adversary that controls the initial local coordinate systems, schedules, and movements. Regarding non-oblivious robots, an algorithm must accomplish formation regardless of initial local memory content.

We say that a set of points *form* a polyhedron when the points are placed on the vertices of the polyhedron. Thus, we sometimes use a polyhedron and its vertices interchangeably.

Consider a ball $B$ centered at point $b$ in 3D space. The *interior* (*exterior*) of $B$ does not contain the sphere of $B$. The smallest enclosing ball of a set $P$ of points and its center are denoted by $B(P)$ and $b(P)$, respectively.

## 3 Symmetricity

To investigate the symmetry among the robots, we consider the *rotation group* and the *symmetricity* of a set of points, and the *rotation group* of a set of local coordinate systems. These notions represent the set of symmetry operations that can be performed on a set of points and a set of local coordinate systems. As described in Sect. 2, these symmetry operations are classified into five types of rotation groups.

The rotation group $\gamma(P)$ of a set $P$ of points represents the "maximal" rotation group that acts on $P$. Here, the "maximality" addresses the subgroup relation in group theory. Clearly, $\gamma(P)$ does not depend on the coordinate system to observe $P$.

Let $C = \{(p_i, x_i, y_i, z_i) \mid r_i \in R\}$ denote a set of local coordinate systems of the robots, where $p_i$ is the coordinates of the current position of $r_i$ (i.e., the origin), and $x_i$, $y_i$, and $z_i$ are the coordinates of $(1,0,0)$, $(0,1,0)$, and $(0,0,1)$ of $r_i$'s local coordinate system in the global coordinate system. We extract the origins of local coordinate systems from $C$ and $P(C)$ denotes the set $\{p_i \mid r_i \in R\}$. The rotation group $\sigma(C)$ of a set of local coordinate systems $C$ represents the "maximal"

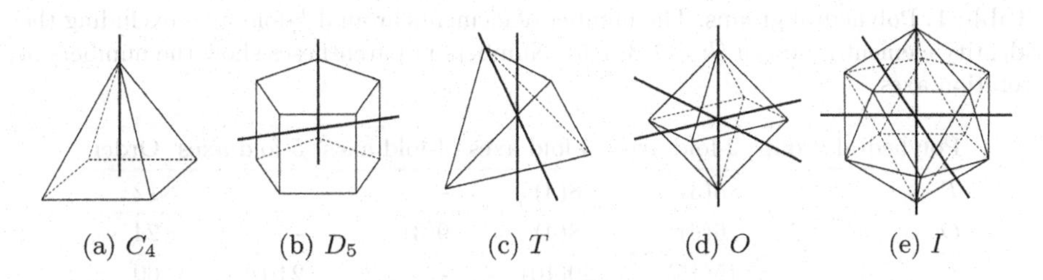

(a) $C_4$ (b) $D_5$ (c) $T$ (d) $O$ (e) $I$

**Fig. 4.** Rotation groups. A bold line indicates a rotation axis of each folding.

rotation group that acts on $C$. Intuitively, the robots cannot resolve $\sigma(C)$ of initial local coordinate systems $C$ (Sect. 1). However, the robots observe only $P(C)$ and they do not know $C$ and $\sigma(C)$.

The symmetricity $\varrho(P)$ of a set $P$ of points is a set of all possible rotation groups of local coordinate systems compatible with $P$. Because the robots cannot observe their local coordinate systems, the set lists all possible symmetries that the robots cannot resolve. Note that $\varrho(P)$ does not depend on the local coordinate system to observe $P$.

These three notions play a central role in characterizing formable target patterns from a given initial configuration $P$. Especially, $\varrho(P)$ shows possible rotation groups that the robots are caught in, and $\gamma(P)$ enables coordination among the robots by decomposing $P$ into its orbits.

### 3.1 Rotation Group of a Set of Points

Consider a regular pyramid that has a regular $k$-gon as its base (Fig. 4). The appearance of the pyramid does not change after the rotation by $2\pi/k$ around the axis containing the apex and the center of the base. There are $k$ such symmetry operations around this axis, i.e., rotations by $2\pi i/k$ for $i = 1, 2, \ldots, k$, which do not change the appearance of the pyramid. We say the rotation axis is *k-fold* because it admits $k$ rotations around it. Let $a^i$ be the rotation by $2\pi i/k$ around the $k$-fold axis with $a^k = e$ where $e$ is the identity element. Then, $a^1, a^2, \ldots, a^k$ form the cyclic group $C_k$. The order of $C_k$ is $k$. The set of all possible symmetry operations by rotations in 2D space (i.e., $SO(2)$) consists of the cyclic groups.

The special orthogonal group $SO(3)$ has five types of subgroups of finite order [2,6], i.e., the *cyclic groups* $C_k$ $(k = 1, 2, \ldots)$, the *dihedral groups* $D_\ell$ $(\ell = 2, 3, \ldots)$, the *regular tetrahedral group* $T$, the *regular octahedral group* $O$, and the *regular icosahedral group* $I$. The groups are identified by the rotations of a regular pyramid with a regular $k$-gon base, a regular prism with regular $\ell$-gon bases, the regular tetrahedron, the regular octahedron, and the regular icosahedron, respectively (Fig. 4). The latter three rotation groups, $T$, $O$, and $I$ are called *polyhedral groups*.

A regular prism (except a cube) has two parallel regular $\ell$-gons as its top and bottom bases and two types of rotation axes, i.e., the $\ell$-fold axis containing the centers of its top and bottom bases and $\ell$ 2-hold axes that exchange the top and

**Table 1.** Polyhedral groups. The number of elements around $k$-fold axes excluding the identity element is shown ($k = 2, 3, 4, 5$). Numbers in parentheses show the numbers of rotation axes.

| Polyhedral group | 2-fold axes | 3-fold axes | 4-fold axes | 5-fold axes | Order |
|---|---|---|---|---|---|
| $T$ | 3(3) | 8(4) | - | - | 12 |
| $O$ | 6(6) | 8(4) | 9(3) | - | 24 |
| $I$ | 15(15) | 20(10) | - | 24(6) | 60 |

the bottom. We call this $\ell$-fold axis the *principal axis* and the remaining $\ell$ 2-fold axes the *secondary axes*. These rotation operations on a regular prism form the *dihedral group* $D_\ell$. The order of $D_\ell$ is $2\ell$. For $\ell = 2$, we can define $D_2$ in the same manner; however, in group theory we do not distinguish the principal axis.

Table 1 shows the number of rotation axes and the number of elements around each type of rotation axes of polyhedral groups. We call the cyclic groups and the dihedral groups *2D rotation groups* because they act on a set of points on a plane, and we call $T$, $O$, and $I$ *3D rotation groups* because they do not act on a set of points on a plane.

Let $\mathbb{S} = \{C_k, D_\ell, T, O, I \mid k = 1, 2, \ldots and\ \ell = 2, 3, \ldots\}$ be the set of rotation groups with finite order, where $C_1$ is the rotation group with order 1 (its unique element is the identity element). The order of $G \in \mathbb{S}$ is denoted by $|G|$.

If $G'$ is a subgroup of $G$ ($G, G' \in \mathbb{S}$), we write $G' \preceq G$. If $G'$ is a proper subgroup of $G$ (i.e., $G' \neq G$), we write $G \prec G'$. For example, we have $T \prec O$, $T \prec I$; however, $O \not\prec I$. If $G \in \mathbb{S}$ contains a $k$-fold axis, then $C_k \preceq G$. Clearly, $C_{k'} \preceq C_k$ if $k'$ divides $k$ (i.e., $k'|k$), which also holds for dihedral groups. Note that the relation $\prec$ is asymmetric and transitive. Figure 5 shows the structure of subgroups of 3D rotation groups.

**Definition 1.** *Let $P \in \mathcal{P}_n^3$ be a set of points. The rotation group $\gamma(P)$ of $P$ is the rotation group in $\mathbb{S}$ that acts on $P$ and none of its proper supergroups in $\mathbb{S}$ act on $P$.*

By definition, we can uniquely determine $\gamma(P)$ regardless of a coordinate system to observe $P$. For example, when $P$ forms a cube, $\gamma(P) = O$. When $\gamma(P) = D_2$, we can recognize the principal axis because of the arrangement of $P$ around the three 2-fold axes. When $\gamma(P) \succ C_1$, all rotation axes of $\gamma(P)$ contain $b(P)$, which is the single intersection of all rotation axes.

A rotation axis of $G \in \mathbb{S}$ is *oriented* if there are no 2-fold rotation axes perpendicular to it, otherwise it is *unoriented.* For example, when $\gamma(P) = C_k$, the rotation axis of $C_k$ is *oriented.* In fact, any set $P$ of points with $\gamma(P) = C_k$ is not symmetric relative to any plane perpendicular to the rotation axis of $\gamma(P)$. The secondary axes of $D_\ell$ are oriented if and only if $\ell$ is odd. The 3-fold rotation axes of $T$ are oriented while the 2-fold rotation axes of $T$ are unoriented. All rotation axes of $O$ and $I$ are unoriented.

Finally, we address the rotation group of a set of points contained in a plane. When we consider a set $P$ of points contained in a plane in 3D space, $\gamma(P)$ is

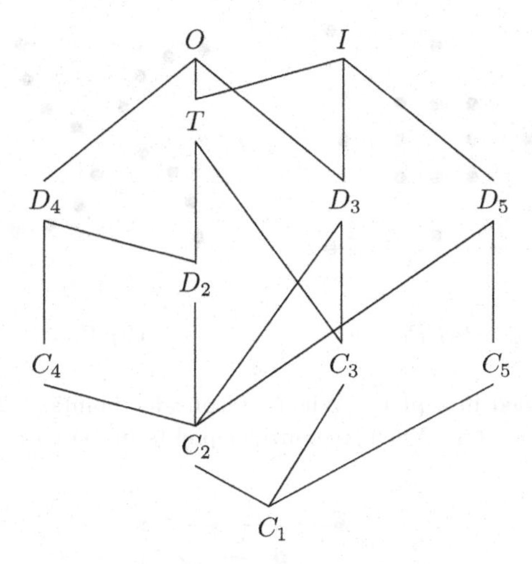


**Fig. 5.** Subgroups of 3D rotation groups. For each edge, the upper rotation group is a proper supergroup of the lower group. We omit some edges because the subgroup relation is transitive.

either a cyclic group or a dihedral group. On the other hand, when we consider a set $P$ of points in 2D space (i.e., a plane), $\gamma(P)$ is a cyclic group. Since we consider $x$-$y$ coordinate systems in 2D space, we do not have an operation that changes the top and bottom of the plane (Fig. 6).[3]

### 3.2 Symmetricity of a Set of Points

A set $P$ of points is *transitive* regarding a rotation group $G$ if it is an orbit of $G$ through some seed $s \in P$, i.e., $P = Orb(s) = \{g * s \mid g \in G\}$, where $*$ denotes the action of $g$ on $s$. Consider the case where seed point $s$ is on a $k$-fold axis ($k > 1$) of $G$. Then, the rotation operations around this $k$-fold axis do not move $s$. We call $\mu(s) = |\{g \in G \mid g * s = s\}|$ the *folding* of $s$. For any transitive set $P$ of points regarding $G \in \mathbb{S}$, $\mu(s) = \mu(s')$ holds for all $s, s' \in P$, and $|P| = |G|/\mu(s)$ holds for any $s \in P$.

By definition, any set $P$ of points is decomposed into orbits of $\gamma(P)$. Let $\{P_1, P_2, \ldots, P_k\} = \{Orb(p) \mid p \in P\}$ be the orbit space of $\gamma(P)$. Clearly, $P_i \cap P_j = \emptyset$ holds for any $i, j \in \{1, 2, \ldots, k\}$ and each $P_i$ is transitive regarding $\gamma(P)$. We call this decomposition the $\gamma(P)$-*decomposition* of $P$. It is worth emphasizing that $P_i$'s may have different sizes (Fig. 7).

We define an embedding of a rotation group to an arrangement of its supergroup. For two groups $G, H \in \mathbb{S}$, an embedding of $G$ to $H$ is an embedding of each rotation axis of $G$ to one of the rotation axes of $H$ such that any $k$-fold axis

[3] We can recognize the robots in 2D space as those that agree on the "top" direction and move on a plane in 3D space.

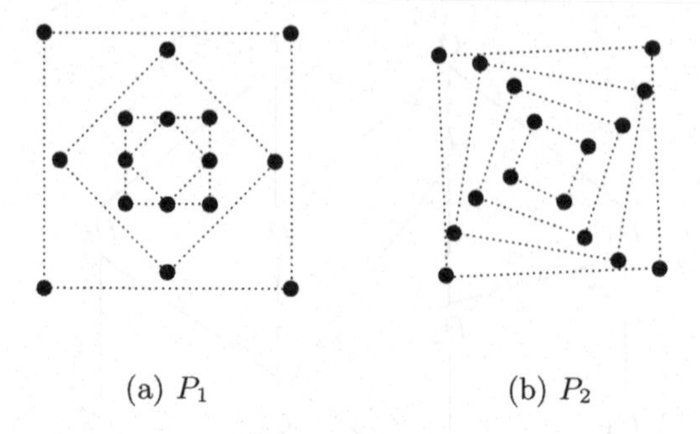

(a) $P_1$ (b) $P_2$

**Fig. 6.** Points contained in a plane. When we consider points in 2D space (i.e., on a plane), $\gamma(P_1) = \gamma(P_2) = C_4$. When we consider points in 3D space, $\gamma(P_1) = D_4$ while $\gamma(P_2) = C_4$.

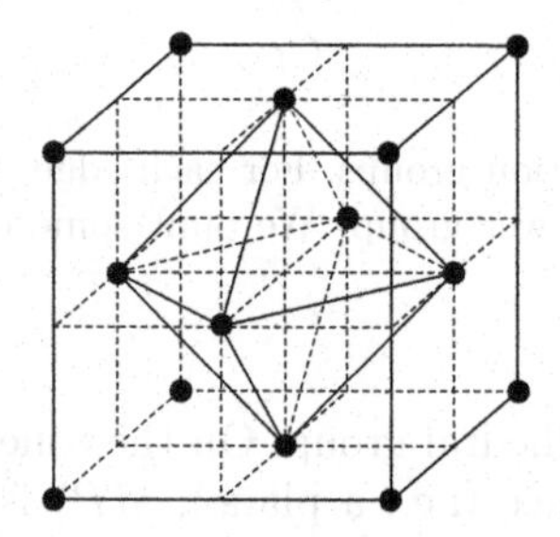

**Fig. 7.** A $\gamma(P)$-decomposition with elements of different sizes. The set $P$ of points consists of 14 points and $\gamma(P) = O$. The $\gamma(P)$-decomposition of $P$ consists of two elements: one forms a regular octahedron and the other forms a cube.

of $G$ overlaps a $k'$-fold axis of $H$ satisfying $k|k'$ while maintaining the arrangement. If a rotation axis of $G$ is oriented, it overlaps either an oriented rotation axis of $H$ with the same orientation or an unoriented rotation axis of $H$. Note that there may be many embeddings of $G$ to $H$. For example, there are three embeddings of $D_4$ to $O$ depending on the choice of the principal axis of $D_4$. Observe that we can embed $G$ to $H$ if and only if $G \preceq H$.

For a set $P$ of points and an embedding of $G \preceq \gamma(P)$ to $\gamma(P)$ ($G \in \mathbb{S}$), we define the decomposition of $P$ into the orbit space of $G$ in the same manner as the $\gamma(P)$-decomposition of $P$. We call the decomposition the *G-decomposition* of $P$ relative to the embedding of $G$.

**Definition 2.** *Let $P \in \mathcal{P}_n^3$ be a set of points. The symmetricity $\varrho(P)$ of $P$ is the set of rotation groups $G \in \mathbb{S}$ satisfying the following conditions:*

1. *$G$ acts on $P$ (thus, $G \preceq \gamma(P)$), and*
2. *there is an embedding of $G$ into $\gamma(P)$ such that every element of the $G$-decomposition of $P$ is a $|G|$-set.*

By definition, if $G \in \varrho(P)$, $\varrho(P)$ contains any $G' \prec G$. We sometimes describe $\varrho(P)$ as a set of such maximal elements rather than listing all its elements.

For example, when $P$ forms a cube, $\varrho(P) = \{D_4\}$. The above definition is rephrased as follows for an initial configuration $P$ without multiplicity; $\varrho(P)$ consists of $C_1$ and all rotation groups formed by the rotation axes of $\gamma(P)$ not containing any point of $P$. Thus, any element $G \in \varrho(P)$ satisfies $G \preceq \gamma(P)$. We say a rotation axis of $\gamma(P)$ is *occupied* when it contains a point of $P$; otherwise the axis is *unoccupied*.

To demonstrate the role of each element of $\varrho(P)$, we first define the rotation group of a set of local coordinate systems. Recall that a local coordinate system is specified by its origin, the directions and orientations of its orthogonal axes, and the unit distance. Let $C = \{(p_i, x_i, y_i, z_i) \mid r_i \in R\}$ be a set of local coordinate systems of $R$. We consider symmetry operations that make each element of $C$ overlap another element of $C$.

**Definition 3.** *Let $C$ be a set of local coordinate systems of $n$ robots. The rotation group $\sigma(C)$ of $C$ is the rotation group in $\mathbb{S}$ that acts on $C$ and none of its proper supergroups in $\mathbb{S}$ act on $C$.*

Intuitively, $\varrho(P)$ of a set $P$ of points lists all possible symmetries of the local coordinate systems for $P$, i.e., $\sigma(C)$ for all sets $C$ of local coordinate systems such that $P(C) = P$. Clearly, $\gamma(P(C)) \succeq \sigma(C)$ holds.

**Lemma 1.** *Let $P$ be an initial configuration of the robots. For each $G \in \varrho(P)$, there is a set $C$ of local coordinate system that satisfies*

*1. $P(C) = P$, and*
*2. $\sigma(C) = G$.*

We demonstrate a construction of the set of local coordinate systems that satisfy Lemma 1. For a set $P$ of points and any $G \in \varrho(P)$, we fix an arbitrary embedding of $G$ to the unoccupied rotation axes of $\gamma(P)$. Such an embedding always exists by definition. Let $\{P_1, P_2, \ldots, P_{k'}\}$ be the $G$-decomposition of $P$, where elements are ordered arbitrarily. For each $P_i$ ($i = 1, 2, \ldots, k'$), we select one point $q_i \in P_i$ and arbitrarily fix its local coordinate system $(q_i, x_i, y_i, z_i)$. Then we apply all elements of $G$ to $(q_i, x_i, y_i, z_i)$ and obtain the remaining points of $P_i$ and their local coordinate systems. With this construction, the resulting local coordinate systems $C$ are symmetric regarding $G$ (i.e., $\sigma(C) = G$). Figure 8 shows an example with $O$.

When we consider the symmetricity of a set of points in 2D space, we check only the cyclic groups in the same manner as the definition of the rotation group of a set of points.[4]

The rotation group of a set of points, the rotation group of a set of local coordinate systems, and the symmetricity of a set of points are easily extended to a multiset of points and a multiset of local coordinate systems via symmetry operations that conform to multiplicities. However, for the $\gamma(F)$-decomposition $\{F_1, F_2, \ldots, F_m\}$ of a given multiset $F$ of points, the robots cannot always agree on the ordering of the elements. For example, the robots cannot agree on an ordering of more than one point at a single position.

[4] The definition of symmetricity in [16,19,23] considers the maximum order of the cyclic groups in the symmetricity.

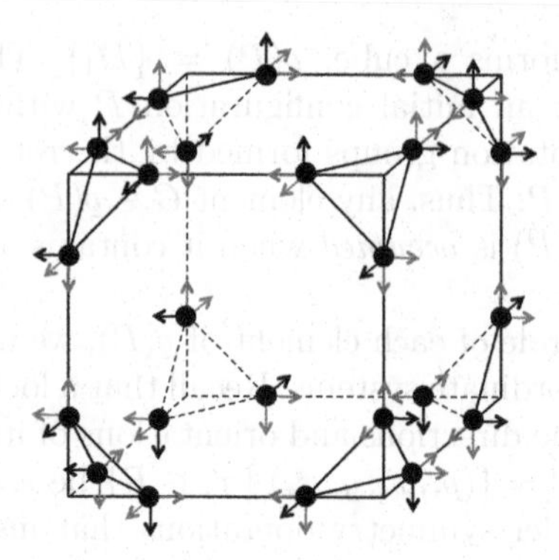

**Fig. 8.** Symmetric local coordinate systems for 24 robots. The rotation group of the local coordinate systems is $O$.

### 3.3 Impossibility Results

We demonstrate that regardless of obliviousness, $\mathcal{F}$SYNC robots cannot break their symmetricity. We first consider an initial configuration $P$ of oblivious $\mathcal{F}$SYNC robots. Consider an execution $P(0)(= P), P(1), \ldots$ where the initial local coordinate systems $C$ for $P$ satisfy $\sigma(C) = G \succ C_1$ for some $G \in \varrho(P)$. Let $\{P_1, P_2, \ldots, P_m\}$ be the $G$-decomposition of $P(= P(C))$ relative to an arbitrary embedding of $G$ to unoccupied rotation axes of $\gamma(P)$ and $R_i$ ($i = 1, 2, \ldots, m$) be the robots in $P_i$. For each $R_i$, the next positions the robots in $R_i$ compute in $P(0)$ are symmetric relative to the embedding of $G$. After the robots move, a new configuration $P(1)$ is obtained, and $P(1)$ is still symmetric relative to the embedding of $G$. Since the axes of the local coordinate systems never change during the execution, the local coordinate systems of the robots are still symmetric relative to the embedding of $G$. Thus, we obtain

$$\gamma(P(1)) \succeq \sigma(P(1)) \succeq \sigma(P(0)) = G.$$

By repeating the above argument, for any $t = 1, 2, \ldots$ we obtain

$$\gamma(P(t)) \succeq \sigma(P(t)) \succeq \sigma(P(0)) = G.$$

Recall that $\varrho(F)$ contains all subgroups of $G'$ if $G' \in \varrho(F)$. This means that if $G'' \notin \varrho(F)$, $\varrho(F)$ contains none of its supergroups. Assume that there exists $G$ that satisfies $G \in \varrho(P)$ and $G \notin \varrho(F)$. By Lemma 1, there exists a set $C$ of local coordinate systems for $P$ that satisfies $\sigma(C) = G$. If the initial local coordinate systems of the robots in $P$ are identical to $C$, $\gamma(P(t)) \succeq \sigma(P(t)) \succeq G$ for any execution $P(0)(= P), P(1), \ldots$. Assume that there exists an algorithm that enables the robots to form $F$ from $P$. In a terminal configuration $P(t)$, $\sigma(P(t)) \in \varrho(F)$ holds because $P(t) \sim F$. Since $G \notin \varrho(F)$, its supergroup $\sigma(P(t))$ satisfies $\sigma(P(t)) \notin \varrho(F)$, which is a contradiction. Consequently, we obtain the following necessary condition.

**Lemma 2.** *Oblivious $\mathcal{F}$SYNC robots can form a target pattern $F$ from an initial configuration $P$ only if $\varrho(P) \subseteq \varrho(F)$.*

When the robots are equipped with local memory, the impossibility holds if the content of local memory is identical in an initial configuration.

**Theorem 2** [25]. *Regardless of obliviousness, $\mathcal{F}$SYNC robots can form a target pattern $F$ from an initial configuration $P$ only if $\varrho(P) \subseteq \varrho(F)$.*

As a corollary, we obtain a necessary condition for the plane formation problem.[5] Recall that a 3D rotation group does not act on a set of points on a single plane. Thus, any 3D rotation group does not act on any terminal configuration of the plane formation problem.

**Corollary 1.** *Regardless of obliviousness, $\mathcal{F}$SYNC robots can form a plane from an initial configuration $P$ only if $\varrho(P)$ consists of 2D rotation groups.*

Note that there are infinitely many sets $P$ of points satisfying $\gamma(P) \in \varrho(P)$. Here, we consider the construction of such $P$. When we fix an arrangement of $G \in \mathbb{S}$, there are infinitely many choices of a seed point to obtain an orbit of size $|G|$. On the other hand, if we ignore uniform scaling, there is a finite number of sets $P$ of points that satisfy $\gamma(P) \notin \varrho(P)$ because such $P$ occupies (a subset of) rotation axes of $\gamma(P)$.

## 4 Symmetry Breaking

Recall that the definition of the symmetricity ignores the occupied rotation axes of $\gamma(P)$ of a set $P$ of points. Example 2 demonstrated the validity of this definition by showing a simple symmetry breaking algorithm in 2D space. The algorithm can be extend to 3D space.

Table 2 shows the polyhedra formed by $G \in \{T, O, I\}$ and a seed point on a rotation axis of $G$. This is a complete list of polyhedra where some axes of the 3D rotation groups are occupied. See also Figs. 9 and 10. When the robots form one of these seven polyhedra in configuration $P$, $\gamma(P)$ coincides with the corresponding rotation group; however, $\varrho(P)$ does not contain $\gamma(P)$. For example, when the robots form a cube, $\gamma(P) = O$ and $\varrho(P) = \{D_4\}$. Thus, the local coordinate systems of the robots agree on some direction (without orientation), and the robots may be able to form, for example, a prism and a plane. However, the robots cannot agree on their actual symmetry by simply observing $P$.

Algorithm 1 enables the robots to break their symmetry in the sense that after the execution of the algorithm, the rotation group of robots' positions is an element of their initial symmetricity. The resulting symmetry is lower than the

[5] Corollary 1 is a rephrasing of the characterization in [24], where the authors compared the order of 3D rotation groups (i.e., 12, 24, and 60) with the size of each element of the $\gamma(P)$-decomposition of an initial configuration $P$ in order to check the symmetricity of $P$.

**Table 2.** Polyhedra with occupied rotation axes

| Rotation group | Folding of a seed | Polyhedron |
|---|---|---|
| $T$ | 3 | Regular tetrahedron |
| $T$ | 2 | Regular octahedron |
| $O$ | 4 | Regular octahedron |
| $O$ | 3 | Cube |
| $O$ | 2 | Cuboctahedron |
| $I$ | 5 | Regular icosahedron |
| $I$ | 3 | Regular dodecahedron |
| $I$ | 2 | Icosidodecahedron |

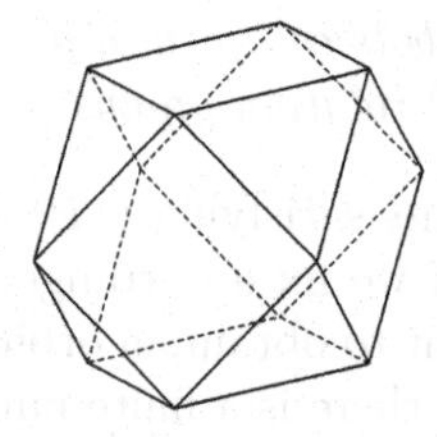

**Fig. 9.** Cuboctahedron

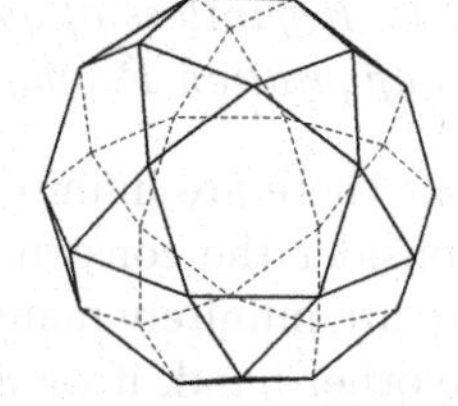

**Fig. 10.** Icosidodecahedron

symmetry of their initial positions. The algorithm makes the robots forming one of the seven polyhedra select an adjacent face and sends them toward the center of the selected face. However, it stops the robots before they reach the center. Thus, it is called the "go-to-center" algorithm. For example, consider a cube consisting of eight vertices and six faces. Any set of selected faces is no more symmetric regarding $O$, and the robots break the symmetry by their movements. The go-to-center algorithm is based on the same strategy as Example 2. That is, the robots on the rotation axes leave their current positions. We obtain the following lemma.

**Lemma 3** [25]. *Let $P$ be an initial configuration of oblivious $\mathcal{F}$SYNC robots where the robots form one of the following polyhedra; a regular tetrahedron, a regular octahedron, a cube, a cuboctahedron, a regular dodecahedron, a regular icosahedron, and an icosidodecahedron. When the robots execute the go-to-center algorithm in $P$, the resulting configuration $P'$ satisfies $\gamma(P') \in \varrho(P)$.*

When $P$ consists of more than one orbits of $\gamma(P) \in \{T, O, I\}$ and $\gamma(P) \notin \varrho(P)$, at least one element of the $\gamma(P)$-decomposition of $P$ forms one of the seven polyhedra. Since the robots can agree on the ordering among the elements of the $\gamma(P)$-decomposition of $P$, they can agree on one of such elements with the smallest order, and the robots forming the element execute the go-to-center algorithm. In this case also, Lemma 3 holds for any resulting configuration.

The "leave-rotation-axes" strategy works for 2D rotation groups. For example, when the robots form a pyramid with a regular polygon base, the robot at

**Algorithm 1.** Go-to-center algorithm for robot $r_i \in R$ [24]

**Notation**

$P$: The positions of the robots forming one of the seven polyhedra.
$p_i$: Current position of $r_i$.
$\epsilon$: $\ell/100$ where $\ell$ is the length of an edge of the polyhedron that $P$ forms.

**Algorithm**

**Switch** ($P$) **do**
  **Case** cuboctahedron:
    Select an adjacent triangle face.
    Destination $d$ is the point $\epsilon$ before the center of the selected face and on the line from $p_i$ to the center.
  **Case** icosidodecahedron:
    Select an adjacent pentagon face.
    Destination $d$ is the point $\epsilon$ before the center of the selected face and on the line from $p_i$ to the center.
  **Default**:
    Select an adjacent face.
    Destination $d$ is the point $\epsilon$ before the center of the selected face and on the line from $p_i$ to the center.
**Enddo**

the apex can break the symmetry simply by leaving its current position. This strategy also works for the principal axis of dihedral groups. Robots on the secondary axes of a dihedral group $D_\ell$ form a regular $\ell$-gon. In this case, each of these $\ell$ robots selects a single direction parallel to the principal axis, and moves slightly in that direction. Note that this procedure works when the robots form at least two orbits of $D_\ell$. Otherwise, the robots form a regular $n$-gon, and their symmetricity in 3D space contains $C_\ell$. In the worst case, the robots always select the same direction and forever maintain a regular $n$-gon.

## 5 Formation Algorithms

The two necessary conditions, i.e., Theorem 2 and Corollary 1, are also sufficient conditions for the pattern formation problem and the plane formation problem, respectively. We first give an overview of a pattern formation algorithm for all solvable instances. Note that we explain the algorithm in English because it is difficult to understand its pseudo-code. Then, we will give an overview of a plane formation algorithm. The difference between the pattern formation algorithm and the plane formation algorithm is how the robots agree on the final plane because after that the problem is reduced to formation of a regular $n$-gon, which is always formable from a solvable instance of the plane formation algorithm.

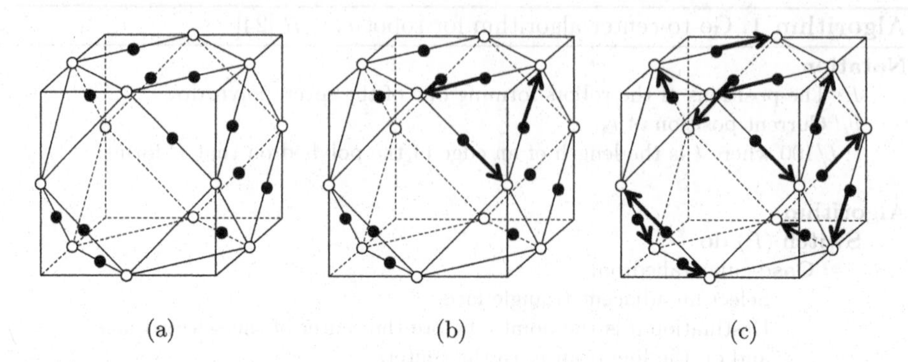


**Fig. 11.** Assignment of the final positions with multiple nearest target points. Black circles represent the robots and white circles represent the points of $\widetilde{F}$. (a) An example. (b) A right-hand screw rule around a 3-fold rotation axis. (c) Whole perfect matching.

### 5.1 Pattern Formation Algorithm

Assume that an initial configuration $P$ satisfies $\gamma(P) \in \varrho(P)$ and $\varrho(P) \subseteq \varrho(F)$, i.e., $\gamma(P) = G \in \varrho(F)$. The robots can agree on $G$; thus, they can agree on the $G$-decomposition $\{P_1, P_2, \ldots, P_m\}$ of $P$ that consists of $|G|$-sets. In the same manner, the robots can agree on an embedding of $G$ to $\gamma(F)$ according to an arbitrary rule; however, they cannot agree on the ordering among the elements of the $G$-decomposition $\{F_1, F_2, \ldots, F_m\}$ of $F$. The robots first embed $F$ to $P$ such that $B(F)$ overlaps $B(P)$ and the rotation axes of $\gamma(F)$ overlap $\gamma(P)$. When $\gamma(F) \succ \gamma(P)$, $\gamma(F)$ is fixed by embedding $G$ into $\gamma(F)$. Let $\widetilde{F}$ denotes this embedding. The formation is completed by sending the robots to the nearest unoccupied points of $\widetilde{F}$ in the order of $P_1, P_2, \ldots, P_m$. These nearest points actually form an element of the $G$-decomposition of $\widetilde{F}$, say $\widetilde{F}_j$. When there is more than one nearest points of $\widetilde{F}_j$ for $q \in P_i$, all $q' \in P_i$ also have more than one nearest points. These matchings form a circle around some rotation axis (Fig. 11). Since the robots agree on the handedness, they can resolve collisions using the right-hand screw rule and by selecting the direction from $b(P)$ as the positive direction. Finally, the robots move to their final positions. These procedures are performed in a single cycle by making each robot compute the final positions of all other robots in the order of $P_1, P_2, \ldots, P_m$.

When $P$ satisfies $\gamma(P) \notin \varrho(P)$, the robots can translate $P$ into another configuration $P'$ satisfying $\gamma(P') \in \varrho(P')$ using the go-to-center algorithm and "leave-the-axes" strategy presented in Sect. 4. The robots can agree on the termination of the algorithm by simply observing $P'$. Then, the robots begin the pattern formation algorithm.

Non-oblivious robots can execute this algorithm by ignoring its memory content. Consequently, we obtain Theorem 1.

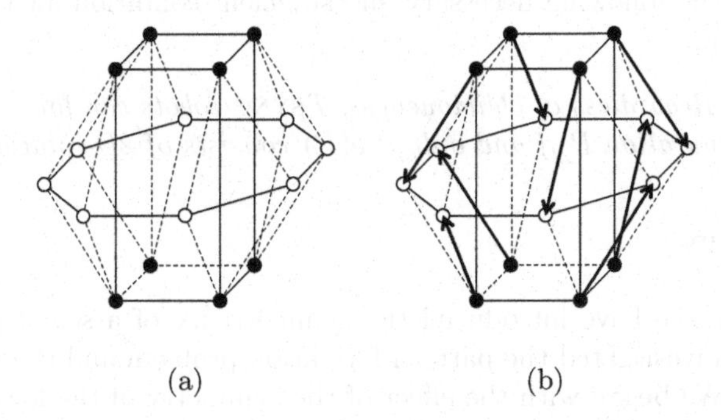


**Fig. 12.** Assignment of final positions with multiple nearest target points on a final plane. Black circles represent the robots and white circles represent the points of regular octagon on the final plane. (a) Current configuration and the final plane. (b) Right-hand screw rule around the principal axis.

### 5.2 Plane Formation Algorithm

The plane formation algorithm does not specify the positions of the robots in a terminal configuration. However, we can assume the terminal configuration requires a regular $n$-gon (i.e., a circle). In fact, a regular $n$-gon is formable from all solvable instances of the plane formation problem; thus, we use the pattern formation algorithm after the robots agree on the final plane.

Let us start with an easy case. Let $P$ be an initial configuration of the robots such that $\gamma(P)$ is a 2D rotation group. The robots can agree on the plane that is perpendicular to the unique rotation axis (or the principal axis) and contains the center $b(P)$ of the smallest enclosing ball of $B(P)$ regardless of the coordinate system to observe $P$. After the robots on the rotation axis perform "leave-rotation-axes" movements, the robots can form a regular $n$-gon in a single cycle by embedding a regular $n$-gon on the agreed plane and using the pattern formation algorithm.

Note that when $P$ forms a regular prism with regular $n/2$-gon bases (except a cube), the robots must use the right-hand screw rule around the principal axis. Figure 12 shows an example for a square prism. Here, the final plane is the one that contains all the secondary axes, and the final regular $n$-gon is embedded so that it is symmetric with respect to the secondary axes. There are two nearest vertices for each robot, and the right-hand screw rule resolves the collision.

When $\gamma(P)$ is a 3D rotation group, from Corollary 1, $\varrho(P)$ consists of 2D rotation groups, which means that there exists at least one element $P_i$ in the $\gamma(P)$-decomposition of $P$ that forms a regular tetrahedron, a cube, a regular octahedron, a regular dodecahedron, or an icosidodecahedron. The robots forming one of such elements execute the "go-to-center" algorithm, and the rotation group $\gamma(P')$ of a resulting configuration $P'$ is a 2D rotation group. In the same manner as the previous case, the robots can agree on a target plane and land on different points on the plane.

We obtain the following necessary and sufficient condition for the plane formation problem.

**Corollary 2.** *Regardless of obliviousness, $\mathcal{F}$SYNC robots can form a plane from an initial configuration $P$ if and only if $\varrho(P)$ consists of 2D rotation groups.*

## 6 Summary

In this chapter, we have introduced the symmetricity of a set of points in 3D space, and we investigated the pattern formation problem and the plane formation problem. We began with the effect of the symmetry of the local coordinate systems of the robots. To formalize the degree of symmetry, we used the rotation groups and defined the symmetricity of a set of points. We then gave the characterization of formable patterns in terms of symmetricity. The necessary condition is based on the impossibility of symmetry breaking while there is a finite number of cases where the robots break the symmetry of their configuration by movement. We have also presented an overview of the pattern formation algorithm to demonstrate matching sufficient condition. We derived the characterization of initial configurations from which the plane formation is accomplished from the main theorem and we have presented the overview of the plane formation algorithm.

In this section, we first present generalization of the symmetricity and the characterization of formable patterns.

**Robots in the $\mathcal{A}$SYNC model.** Impossibility results for a robot system with stronger assumptions hold for other robot systems with weaker assumptions, while algorithms for a robot system with weaker assumptions work correctly for other robot systems with stronger assumptions. Recall that Theorem 1 assumes the $\mathcal{F}$SYNC model. Theorem 1 is generalized to the $\mathcal{A}$SYNC model (thus, the $\mathcal{S}$SYNC model) with a pattern formation algorithm for oblivious $\mathcal{A}$SYNC robots. The same generalization is possible for Corollary 2. Regarding the pattern formation problem in 2D space, it has been shown that oblivious $\mathcal{A}$SYNC robots (thus oblivious $\mathcal{S}$SYNC robots) with non-rigid movement have the same formation power as non-oblivious $\mathcal{F}$SYNC robots with rigid movement except the rendezvous problem [16,19,23]. For the plane formation problem, it has been shown that oblivious $\mathcal{S}$SYNC robots with non-rigid movement have the same formation power as non-oblivious $\mathcal{F}$SYNC robots with rigid movement [21].

**Robots without chirality.** We can consider a weaker robot system by removing the chirality assumption. Thus, the local coordinate system of a robot is either right-handed or left-handed. We must consider symmetry operations that change the handedness, thereby adopting the composite symmetry of rotations around axes and reflections by mirror planes. In 3D space, there are seventeen types of such symmetry, each of which forms a group [6]. A rotation group of a configuration is generalized to a *symmetry group* by considering all these symmetry groups, and symmetricity is also generalized by considering unoccupied rotation axes and unoccupied mirror planes. The formation power of the robots is reduced by the additional symmetry with mirror planes [3,20]. For example, the robots cannot form a plane from an initial configuration where they form a cube because any final plane is a mirror plane and the robots cannot avoid multiplicities.

**Robots with $\mathcal{L}$IMITED visibility.** A robot with $\mathcal{L}$IMITED visibility can observe other robots in its visibility range. In this case, the robots may not be aware of the symmetry of the entire configuration, and there are infinitely many initial configurations of oblivious $\mathcal{F}$SYNC robots in 2D space where slight movements of the robots increase the overall symmetry. Thus, oblivious $\mathcal{F}$SYNC robots with $\mathcal{L}$IMITED visibility have weaker formation power in both 2D space and 3D space [26].

We close this chapter by identifying open problems related to the formation problem and the symmetry among the robots.

1. The formation problem in 3D space.
   (a) Pattern formation algorithms for $\mathcal{S}$SYNC and $\mathcal{A}$SYNC robots.
   (b) The effect of obliviousness.
   (c) The effect of $\mathcal{L}$IMITED visibility.
   (d) The effect of randomization.
2. The formation problem in other space, such as higher-dimensional space, a sphere, a torus, etc.
3. Limited or obstructed visibility in a considered space and its effect on symmetry and the formation problem.
4. New distributed coordination problems, such as the formation of a sequence of patterns [8] and the team assembling problem [18].

## 7 Bibliography

Armstrong provides a good overview of the rotation groups [2], and the book by Cromwell presents an algorithm to determine the rotation group of a given polyhedron [6].

Yamashita and Kameda introduced the notion of symmetricity of a network of anonymous processes [22]. They showed the impossibility of fundamental agreement tasks in anonymous networks such as the leader election problem. Suzuki and Yamashita introduced the notion of *symmetricity* among anonymous mobile robots in 2D space and showed that non-oblivious $\mathcal{S}$SYNC robots can form the patterns that non-oblivious $\mathcal{F}$SYNC robots can form except the point of multiplicity two (called the *rendezvous problem*) [19]. Their definition of symmetricity is based on the decomposition of a set of points into concentric congruent regular polygons addressed in Sect. 1. After that, a series of papers discussed the effect of obliviousness and synchrony on the formation power of mobile robots. Yamashita and Suzuki showed that obliviousness does not affect the formation power of $\mathcal{F}$SYNC robots and $\mathcal{S}$SYNC robots [23]. Flocchini et al. first introduced the $\mathcal{A}$SYNC model [14] and showed that an agreement on the directions and orientations of $x$ and $y$ axes enables oblivious $\mathcal{A}$SYNC robots to form any target pattern [14]. Dieudonné et al. showed that oblivious $\mathcal{A}$SYNC robots can form an arbitrary pattern if and only if they can solve the leader election problem [11]. Fujinaga et al. presented a pattern formation algorithm for oblivious $\mathcal{A}$SYNC robots and showed that synchrony has no effect on the formation power except the rendezvous problem [16]. Their algorithm is based on the techniques for the *embedded pattern formation problem*, where the target pattern is drawn on the plane as landmarks [15]. While the pattern formation problem is not always solvable, the embedded pattern formation problem is always solvable by oblivious $\mathcal{A}$SYNC robots by the "clockwise matching" strategy. In terms of symmetricity, a point or a circle is formable from any initial configuration in 2D space. Cieliebak presented a point formation algorithm for oblivious $\mathcal{A}$SYNC robots without chirality [4]. Flocchini et al. showed that oblivious $\mathcal{A}$SYNC robots without chirality can from a circle from any initial configuration [12]. Das et al. considered *formation of a sequence of geometric patterns* by oblivious $\mathcal{S}$SYNC robots and characterized formable sequences in terms of symmetricity [8]. These papers consider the robots in 2D space. Yamauchi et al. extended the notion of symmetricity to 3D space using rotation groups and characterized the set of formable patterns in 3D space [25].

Deterministic symmetry breaking in 2D space has been considered [19,23]. For example, non-oblivious $\mathcal{S}$SYNC robots can show their local coordinate systems by their movement [19]. Each robot first moves along its $x$ axis and then along its $y$ axis. After that, it moves to show its unit distance. Other robots remember these movements to obtain the local coordinate system of the robot. In another study [23], an oblivious $\mathcal{S}$SYNC robot moves away from the center of the smallest enclosing circle of the robots by some distance that encodes its local coordinate system. The go-to-center algorithm in 3D space can be seen as an equivalent of these symmetry breaking algorithms.

Randomization enables the robots to break symmetry; however, it is difficult to make oblivious $\mathcal{A}$SYNC robots show the result of their random choices. Yamauchi et al. proposed a randomized pattern formation algorithm for oblivious $\mathcal{A}$SYNC robots in 2D space that enables arbitrary pattern formation with probability 1 [27]. As mentioned previously, initial multiplicities cannot be resolved because the robots at the same point with the same local coordinate systems move in the same way. Dieudonné and Petit proposed the *scattering problem*, which requires the robots to resolve initial multiplicity [10]. They proposed a randomized scattering algorithm for oblivious $\mathcal{S}$SYNC robots with the weak multiplicity detection ability. Clement et al. showed the expected time of the algorithm [5]. Izumi et al. proposed a randomized scattering algorithm for oblivious $\mathcal{S}$SYNC robots with $\mathcal{L}$IMITED visibility and the local-weak multiplicity detection ability and showed that its execution time depends on the diameter of the initial visibility graph [17].

Ando et al. proposed a convergence algorithm for oblivious $\mathcal{S}$SYNC robots with $\mathcal{L}$IMITED visibility in 2D space [1]. Flocchini et al. considered the convergence of oblivious $\mathcal{A}$SYNC robots with $\mathcal{L}$IMITED visibility and a consistent compass that assumes the robots agree on the direction and/or orientation of the $x$ and $y$ axes of local coordinate systems [13].

Cicerone et al. considered the embedded pattern formation problem in 2D space by oblivious $\mathcal{A}$SYNC robots without chirality [3]. The lack of chirality results in unsolvable instances due to axes of symmetry, i.e., mirror planes perpendicular to the plane. Tomita et al. investigated the plane formation problem by oblivious $\mathcal{F}$SYNC robots without chirality [20]. The lack of chirality also reduces the solvable instances. They presented a characterization in terms of symmetricity considering symmetry operations by mirror planes, rotation axes, and their combinations.

The notion of symmetricity is extended to other systems of anonymous mobile computing entities by adopting the symmetry operations of the considered system. Das et al. considered the formation of a sequence of patterns by the robots with externally visible lights [7]. Each robot can change the color of its light at the end of a `Compute` phase, and the color is persistent in the next cycle. Symmetry operations are required to maintain the positions of the robots and the colors of their lights, and the *chromatic symmetricity* is defined. Liu et al. considered the *team assembling problem* by "colored" robots that requires the robots to form teams specified by the number of robots for each color [18]. The color of a robot does not change and the robots of the same color are indistinguishable. The requirement for the symmetry operations is the same as that for the robots with lights. Liu et al. characterized formable teams and provided a team assembling algorithm for oblivious $\mathcal{A}$SYNC colored robots. Di Luna et al. considered the *unbreakable symmetry* in the *amoebot* model consisting of a set of anonymous particles moving in a triangular grid [9]. Each particle can sense other particles in neighboring grid vertices, communicate with them, and move by repeating an expansion and a contraction. The particles do not agree on the clockwise-direction. In a triangular grid, possible symmetry operations are $C_2$, $C_3$, and a reflection by a mirror plane. Based on the unbreakable symmetry, Di Luna et al. characterized formable shapes and proposed a shape formation algorithm for sufficiently many particles.

**Acknowledgment.** This work was supported by JSPS KAKENHI Grant Number JP18H03202.

## References

1. Ando, H., Oasa, Y., Suzuki, I., Yamashita, M.: Distributed memoryless point convergence algorithm for mobile robots with limited visibility. IEEE Trans. Robot. Autom. **15**(5), 818–828 (1999). https://doi.org/10.1109/70.795787
2. Armstrong, M.A.: Groups and Symmetry. Springer, New York (1988). https://doi.org/10.1007/978-1-4757-4034-9
3. Cicerone, S., Di Stefano, G., Navarra, A.: Asynchronous embedded pattern formation without orientation. In: Gavoille, C., Ilcinkas, D. (eds.) DISC 2016. LNCS, vol. 9888, pp. 85–98. Springer, Heidelberg (2016). https://doi.org/10.1007/978-3-662-53426-7_7

4. Cieliebak, M., Flocchini, P., Prencipe, G., Santoro, N.: Distributed computing by mobile robots: gathering. SIAM J. Comput. **41**(4), 829–879 (2012). https://doi.org/10.1137/100796534
5. Clement, J., Défago, X., Potop-Butucaru, M.G., Izumi, T., Messika, S.: The cost of probabilistic agreement in oblivious robot networks. Inf. Process. Lett. **110**(11), 431–438 (2010). https://doi.org/10.1016/j.ipl.2010.04.006
6. Cromwell, P.R.: Polyhedra. University Press, Cambridge (1997)
7. Das, S., Flocchini, P., Prencipe, G., Santoro, N.: Synchronized dancing of oblivious chameleons. In: Ferro, A., Luccio, F., Widmayer, P. (eds.) FUN 2014. LNCS, vol. 8496, pp. 113–124. Springer, Cham (2014). https://doi.org/10.1007/978-3-319-07890-8_10
8. Das, S., Flocchini, P., Santoro, N., Yamashita, M.: Forming sequence of geometric patterns with oblivious mobile robots. Distrib. Comput. **28**, 131–145 (2015). https://doi.org/10.1007/s00446-014-0220-9
9. Di Luna, G.A., Flocchini, P., Santoro, N., Viglietta, G., Yamauchi, Y.: Shape formation by programmable particles. In: Proceedings of the 21st International Conference on Principles of Distributed Systems (OPODIS 2017), pp. 31:1–31:16. Schloss Dagstuhl-Leibniz-Zentrum fuer Informatik (2017). https://doi.org/10.4230/LIPIcs.OPODIS.2017.31
10. Dieudonné, Y., Petit, F.: Robots and demons (the code of the origins). In: Crescenzi, P., Prencipe, G., Pucci, G. (eds.) FUN 2007. LNCS, vol. 4475, pp. 108–119. Springer, Heidelberg (2007). https://doi.org/10.1007/978-3-540-72914-3_11
11. Dieudonné, Y., Petit, F., Villain, V.: Leader election problem versus pattern formation problem. In: Lynch, N.A., Shvartsman, A.A. (eds.) DISC 2010. LNCS, vol. 6343, pp. 267–281. Springer, Heidelberg (2010). https://doi.org/10.1007/978-3-642-15763-9_26
12. Flocchini, P., Prencipe, G., Santoro, N., Viglietta, G.: Distributed computing by mobile robots: uniform circle formation. Distrib. Comput. **30**(6), 413–457 (2017). https://doi.org/10.1007/s00446-016-0291-x
13. Flocchini, P., Prencipe, G., Santoro, N., Widmayer, P.: Gathering of asynchronous robots with limited visibility. Theor. Comput. Sci. **337**, 147–168 (2005). https://doi.org/10.1016/j.tcs.2005.01.001
14. Flocchini, P., Prencipe, G., Santoro, N., Widmayer, P.: Arbitrary pattern formation by asynchronous, anonymous, oblivious robots. Theor. Comput. Sci. **407**, 412–447 (2008). https://doi.org/10.1016/j.tcs.2008.07.026
15. Fujinaga, N., Ono, H., Kijima, S., Yamashita, M.: Pattern formation through optimum matching by oblivious CORDA robots. In: Lu, C., Masuzawa, T., Mosbah, M. (eds.) OPODIS 2010. LNCS, vol. 6490, pp. 1–15. Springer, Heidelberg (2010). https://doi.org/10.1007/978-3-642-17653-1_1
16. Fujinaga, N., Yamauchi, Y., Ono, H., Kijima, S., Yamashita, M.: Pattern formation by oblivious asynchronous mobile robots. SIAM J. Comput. **44**(3), 740–785 (2015). https://doi.org/10.1137/140958682
17. Izumi, T., Potop-Butucaru, M.G., Tixeuil, S.: Connectivity-preserving scattering of mobile robots with limited visibility. In: Dolev, S., Cobb, J., Fischer, M., Yung, M. (eds.) SSS 2010. LNCS, vol. 6366, pp. 319–331. Springer, Heidelberg (2010). https://doi.org/10.1007/978-3-642-16023-3_27
18. Liu, Z., Yamauchi, Y., Kijima, S., Yamashita, M.: Team assembling problem for asynchronous heterogeneous mobile robots. Theor. Comput. Sci. **721**, 27–41 (2018). https://doi.org/10.1016/j.tcs.2018.01.009

19. Suzuki, I., Yamashita, M.: Distributed anonymous mobile robots: formation of geometric patterns. SIAM J. Comput. **28**(4), 1347–1363 (1999). https://doi.org/10.1137/S009753979628292X
20. Tomita, Y., Yamauchi, Y., Kijima, S., Yamashita, M.: Plane formation by synchronous mobile robots without chirality. In: Proceedings of the 21st International Conference on Principles of Distributed Systems (OPODIS 2017), pp. 13:1–13:17 (2017). https://doi.org/10.4230/LIPIcs.OPODIS.2017.13
21. Uehara, T., Yamauchi, Y., Kijima, S., Yamashita, M.: Plane formation by semi-synchronous robots in the three dimensional Euclidean space. In: Bonakdarpour, B., Petit, F. (eds.) SSS 2016. LNCS, vol. 10083, pp. 383–398. Springer, Cham (2016). https://doi.org/10.1007/978-3-319-49259-9_30
22. Yamashita, M., Kameda, T.: Computing on anonymous networks: part I-characterizing the solvable cases. IEEE Trans. Parallel Distrib. Syst. **7**(1), 69–89 (1996). https://doi.org/10.1109/71.481599
23. Yamashita, M., Suzuki, I.: Characterizing geometric patterns formable by oblivious anonymous mobile robots. Theor. Comput. Sci. **411**, 2433–2453 (2010). https://doi.org/10.1016/j.tcs.2010.01.037
24. Yamauchi, Y., Uehara, T., Kijima, S., Yamashita, M.: Plane formation by synchronous mobile robots in the three dimensional Euclidean space. J. ACM **64**(3), 16:1–16:43 (2017). https://doi.org/10.1145/3060272
25. Yamauchi, Y., Uehara, T., Yamashita, M.: Brief announcement: pattern formation problem for synchronous mobile robots in the three dimensional Euclidean space. In: Proceedings of the 35th ACM Symposium on Principles of Distributed Computing (PODC 2016), pp. 447–449. ACM (2016). https://doi.org/10.1145/2933057.2933063
26. Yamauchi, Y., Yamashita, M.: Pattern formation by mobile robots with limited visibility. In: Moscibroda, T., Rescigno, A.A. (eds.) SIROCCO 2013. LNCS, vol. 8179, pp. 201–212. Springer, Cham (2013). https://doi.org/10.1007/978-3-319-03578-9_17
27. Yamauchi, Y., Yamashita, M.: Randomized pattern formation algorithm for asynchronous oblivious mobile robots. In: Kuhn, F. (ed.) DISC 2014. LNCS, vol. 8784, pp. 137–151. Springer, Heidelberg (2014). https://doi.org/10.1007/978-3-662-45174-8_10

# Computation Under Restricted Visibility

Subhash Bhagat[1(✉)], Krishnendu Mukhopadhyaya[1], and Srabani Mukhopadhyaya[2]

[1] Indian Statistical Institute, Kolkata, Kolkata, India
subhash.bhagat.math@gmail.com, krishnendu@isical.ac.in
[2] Birla Institute of Technology Mesra, Lalpur Extension Centre, Ranchi, India
srabanim@gmail.com

**Abstract.** In a swarm of robots, each robot has certain capabilities to perform their computations to achieve a global objective. One such capability is the sensing capability, known as *vision*. This enables a robot to sense the positions of the other robots in the system. The sensing capability may be restricted by two factors: (i) the sensing range and (ii) the opacity of the robots. The sensing range may be limited or unlimited and the robots may be transparent or opaque. When robots have limited sensing range, a robot can sense all the robots within a fixed radius around it. If three opaque robots are collinear, the middle robot obstructs the vision of the two other robots. This chapter deals with these two constraints on the vision of the robots. A model with such a constraint is referred to as the *restricted visibility* model. This chapter presents different geometric formation problems for swarm robots under the *restricted visibility* model.

**Keywords:** Swarm robots · Limited visibility · Opaque robots

## 1 Introduction

A distributed system of swarm robots provides efficient solutions to many real life problems. A major focus of research in this branch is to identify minimal sets of capabilities which help robots to accomplish some predefined goal. Each set of assumptions stands for a new model of computation for swarm of robots. Restrictions on the vision of the robots yield one such computational model, the *restricted visibility* model. Limited visibility range and opacity of the robots are two such restrictions. These two restrictions impair the robots from obtaining complete view of the robot positions in the system. This setting is more complex than the traditional model in which robots are transparent and have unlimited visibility range. Thus, computations with such restrictions are more involved and challenging.

The assumptions of limited visibility range and opacity are natural and make such system more suitable in large number of real life applications. Researchers have considered different combinations of these two assumptions


P. Flocchini et al. (Eds.): Distributed Computing by Mobile Entities, LNCS 11340, pp. 134–183, 2019.
https://doi.org/10.1007/978-3-030-11072-7_7

in the *look-compute-move* model and studied different geometric formation problems therein. The robots with limited visibility range may be transparent [2,15,17,18,21,22,25,26,28,30] and opaque robots may have unlimited visibility range [1,3–5,10,12,19]. Researchers have also considered opaque robots with limited visibility range [6].

## 2 Chapter Organization

In Sect. 3, general model and notations used in this chapter are presented. The rest of the chapter is divided into two basic blocks: the *limited visibility* model in Sect. 4 and the *obstructed visibility* model in Sect. 5.

The organization of the works done under the *limited visibility* model is as follows:

- Section 4.1 considers the *convergence* problem. First, in Sect. 4.1.1, the problem is considered under the $\mathcal{S}$SYNC model. In Sect. 4.1.2, the problem is considered for a set of robots which can not measure distances. Section 4.1.3 presents a solution to problem under the $\mathcal{A}$SYNC model with *1-fair schedule.*
- In Sect. 4.2, the *near gathering* problem is considered which is closely related to the *convergence* problem.
- The *gathering* problem is considered in Sect. 4.3. Section 4.3.1 considers the *gathering* problem under the $\mathcal{A}$SYNC model when robots have agreement on both coordinate axes. Section 4.3.2 presents a solution to the *gathering* problem under the $\mathcal{S}$SYNC model when robots have eventually consistent compasses. An optimal solution to the *gathering* problem in this setting is provided in Sect. 4.3.3.
- Section 4.4 contains a randomized distributed algorithm for the *scattering* problem.
- A study on the *pattern formation* problem under *limited visibility* model is presented in Sect. 4.5.

Following is the organization of the works done under *obstructed visibility* model in Sect. 5.

- Section 5.1 contains the works done for the *gathering* problem for opaque robots with no extent. Sections 5.1.1 and 5.1.2 describe the solutions to the *gathering* problem in the Euclidean plane and three dimensional Euclidean space respectively.
- Section 5.2 considers gathering of opaque robots with extent. First, in Sect. 5.2.1, solutions to the problem with three and four opaque robots with extent under the $\mathcal{A}$SYNC model are presented. Then, a gathering algorithm of an arbitrary number of asynchronous robots is described. Finally, in Sect. 5.2.2, a solution to the problem under the $\mathcal{F}$SYNC model is described, which assumes that robot are opaque and have limited visibility range.
- In Sect. 5.3, two algorithms for the *mutual visibility* problems are described.

Section 6 concludes the chapter.

## 3 General Model and Notations

Consider a set $\mathcal{R} = \{r_1, r_2, \ldots, r_n\}$ of $n$ homogeneous, autonomous robots. The robots repeatedly execute the computational cycle consisting of three phases *look-compute-move*. This chapter presents works done under the $\mathcal{OBLOT}$ model.

- **Robot configuration:** A robot configuration, $\mathcal{R}(t) = \{r_1(t), \ldots, r_n(t)\}$ is the multi-set of positions occupied by the robots in $\mathcal{R}$ at time $t$. Let $\mathcal{CH}(t)$ denote the convex hull of the points in $\mathcal{R}(t)$.
- **Measurement of angles:** If not stated otherwise, the angle between two line segments refers to the angle which is less than or equal to $\pi$.
- **Limited Visibility:** If robots have limited visibility range, a robot can only sense the other robots lying at a distance less than or equal to $V$ (Fig. 1). The value of $V$ is assumed to be the same for all the robots and this value is known to the robots. The model in which robots have limited visibility range is known as *limited visibility* model.

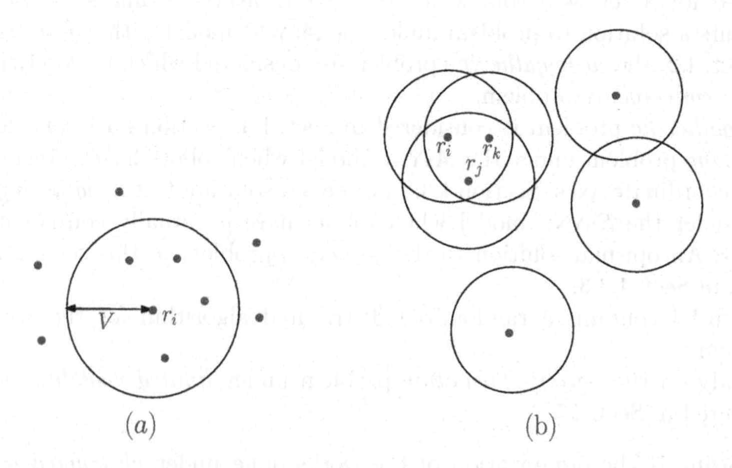


**Fig. 1.** An illustration of limited visibility: (a) robot $r_i$ can see all robots within radius V around it and (b) only robots $r_i, r_j$ and $r_k$ are mutually visible

- **Obstructed Visibility:** If robots are non-transparent and three robots are collinear, the middle robot obstructs the vision of the two other robots. Let $r_i$, $r_j$ and $r_k$ be three collinear robots such that $r_k$ lies between $r_i$ and $r_j$. In such a case, $r_k$ blocks the vision of $r_i$ and $r_j$, i.e., $r_i$ and $r_j$ are mutually invisible (Fig. 2). This visibility model is known as *obstructed visibility*.

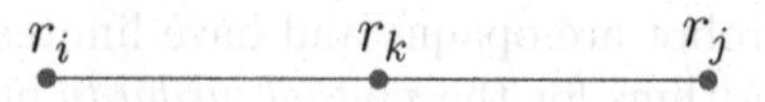


**Fig. 2.** An illustration of obstructed visibility: robot $r_k$ blocks the vision of $r_i$ and $r_j$

- **Vision of a robot:** The vision of a robot $r_i$ at time $t$ is the set of robot positions visible to $r_i$ (excluding $r_i$). This set is denoted by $\mathcal{V}_i(t)$. Let $\mathcal{S}_i(t)$ denote the smallest enclosing circle of the points in $\mathcal{V}_i(t) \cup \{r_i(t)\}$. $\mathcal{O}_i(t)$ is the centre of $S_i(t)$. The convex hull of $\mathcal{V}_i(t) \cup \{r_i(t)\}$ is denoted by $\mathcal{CH}_i(t)$.
- **Visibility graph:** Due to limited visibility, a robot may not have locations of all the robots in the system. The *visibility graph* depicts the mutual visibility status among the robots. The *visibility graph* is the graph $G = (\mathcal{R}, E)$ where $\forall r_i, r_j \in \mathcal{R}$, $(r_i, r_j) \in E$ iff $dist(r_i(t), r_j(t)) \leq V$. If the visibility graph is complete, then all the robots in the system are mutually visible. In this chapter, all the works under the *limited visibility* model assume that the initial visibility graph of a robot configuration is connected.
- The *visibility polygon* of $r_i$ at time $t$, denoted by $\mathcal{VP}_i(t)$, is defined as follows: sort the points in $\mathcal{V}_i(t)$ angularly in counter-clockwise direction w.r.t. $r_i(t)$, starting from any robot position in $\mathcal{V}_i(t)$. Then connect them in that order to generate the polygon $\mathcal{VP}_i(t)$ (Fig. 3).

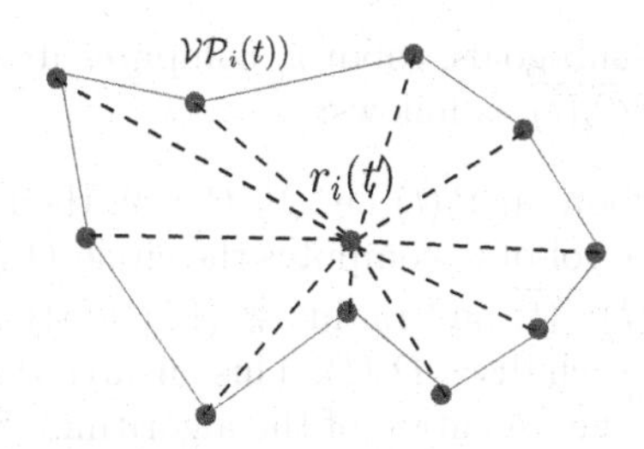


**Fig. 3.** An example of visibility polygon

- Consider two points $p$ and $q$. Let $\overline{pq}$ denote the closed line segment joining two points $p$ and $q$, including the end points $p$ and $q$, and $|\overline{pq}|$ denote the length of $\overline{pq}$.

# 4 Limited Visibility Model

In this model, a robot takes decisions based on the robot positions lying within its visibility range. During the execution of an algorithm, robots move in such a way that the visibility graph remains connected. All the works described in this section, assume transparent point robots in the Euclidean plane.

## 4.1 The Convergence Problem

The *convergence* problem is one of the fundamental geometric formation problem for a swarm of robots. It requires the robots to become increasingly closer to each other, without necessarily reaching the same point. More precisely, let $d_{max}(t)$ denote the maximum distance between the robots at time $t$. The robots are said to *converge* (to a point) if $d_{max}(t)$ is monotonically non-increasing and, for any $\epsilon > 0$ and $t > 0$, there exists a time $t' \geq t$ such that $d_{max}(t') < \epsilon$. The *convergence* problem requires the swarm to converge. The converge point is not known to the robots a priori. The problem has been studied for the robots with limited visibility range under the $\mathcal{S}$SYNC and $\mathcal{A}$SYNC models [2,11,17,18,22].

#### 4.1.1 Convergence in the $\mathcal{S}$SYNC Model

Ando et al. [2] presented a solution to the *convergence* problem for a set of semi-synchronous robots with limited visibility, represented as points in the Euclidean plane.

The outline of the convergence algorithm is as follows: an active robot $r_i$ moves towards the centre $\mathcal{O}_i(t)$ of $S_i(t)$, in such a way that the following are satisfied:

(i) the mutually visible robots remain visible after the movements and
(ii) the geometric span of the robot positions reduces within finite time.

To achieve these two sub-goals, robot $r_i$ computes its destination point $p_i(t)$ on the line segment $\overline{r_i(t)\mathcal{O}_i(t)}$ as follows:

- If $r_i$ finds no other robots in $\mathcal{V}_i(t)$ i.e., $\mathcal{V}_i(t) = \emptyset$, then it does not move.
- Otherwise, $\forall r_j \in \mathcal{V}_i(t)$, robot $r_i$ computes the circle $D_j(t)$ having centre at the midpoint $m_j(t)$ of $\overline{r_i(t)r_j(t)}$ and radius $\frac{V}{2}$ (Fig. 4(a)). The point $p_i(t)$ lies in the intersection of all such discs $D_j(t)$. This ensures that the robots $r_i$ and $r_j$ remain visible during the execution of the algorithm. The maximum distance $l_j(t)$ that $r_i$ can move toward $\mathcal{O}_i(t)$, without leaving $D_j(t)$, is depicted in Fig. 4(b). A suitable choice for the point $p_i(t)$ is provided in the pseudo-code *CONVERGE-$\mathcal{S}$SYNC()*.

Correctness of the above algorithm is based on the following two lemmas.

**Lemma 1.** *For two robots $r_i, r_j \in \mathcal{R}$ at time $t \geq t_0, r_i, r_j \in E(t)$ implies $r_i, r_j \in E(t+1)$.*

**Lemma 2.** *For $t \geq t_0, \mathcal{CH}(t) \subset \mathcal{CH}(t+1)$.*

**Theorem 1** [2]. *A set of n point robots with limited visibility can converge within finite time under the $\mathcal{S}$SYNC model.*

For fully synchronous robots, the convergence time of the above algorithm was analyzed by Degener et al. [11] and they provided the following result.

**Theorem 2** [11]. *A set of n fully synchronous robots with limited visibility, executing the algorithm CONVERGE-$\mathcal{S}$SYNC(), can converge in $O(n^2)$ rounds and this bound is tight.*

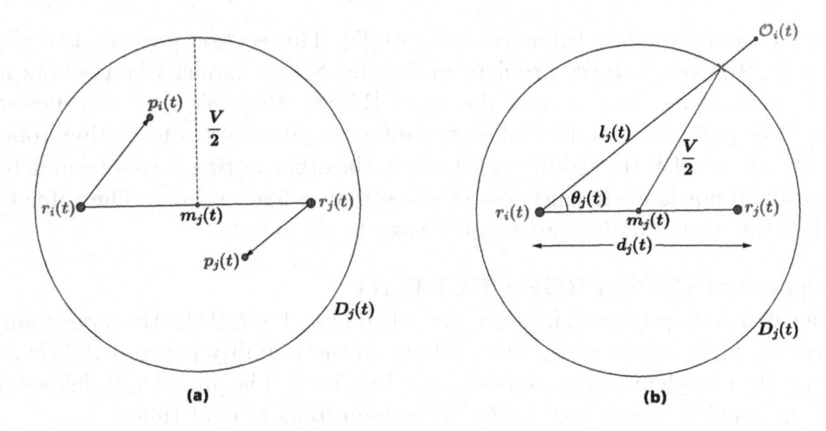


**Fig. 4.** (a) $r_i$ and $r_j$ remain visible to each other even if both of them move simultaneously. (b) $l_j(t)$ is the maximum distance that $r_i$ can move without leaving $D_j(t)$.

---

**Algorithm 1.** CONVERGE-$\mathcal{S}$SYNC()

---

**Input**: $r_i(t) \in R(t), \mathcal{V}_i(t)$.
**Output**: A destination point for robot $r_i$.

1 **if** $|\mathcal{V}_i(t)| == 0$ **then**
2 $\quad p_i(t) \leftarrow r_i(t)$;
3 **else**
4 $\quad$ **for** $\forall r_j \in \mathcal{V}_i(t)$ **do**
5 $\quad\quad d_j(t) \leftarrow dist(r_i(t), r_j(t))$;
6 $\quad\quad \theta_j(t) \leftarrow \angle \mathcal{O}_i(t) r_i(t) r_j(t)$;
7 $\quad\quad l_j(t) \leftarrow \frac{d_j}{2}\cos\theta_j(t) + \sqrt{(\frac{V}{2})^2 - ((\frac{d_j}{2})\sin\theta_j(t))^2}$;
8 $\quad limit \leftarrow min\{l_j(t) : \forall r_j \in \mathcal{V}_i(t) \backslash \{r_i\}$;
9 $\quad goal \leftarrow dist(r_i(t), \mathcal{O}_i(t))$;
10 $\quad d_i(t) \leftarrow min\{goal, limit\}$;
11 $\quad p_i(t) \leftarrow$ point on $\overline{r_i(t)\mathcal{O}_i(t)}$ at a distance $d_i(t)$ from $r_i(t)$;
12 $r_i$ moves towards $p_i(t)$;

---

### 4.1.2 Convergence with Crude Distance Sensing

In the traditional model, it is assumed that robots can measure their mutual distances accurately. However, the measurements by the robots may have some non-negligible inaccuracies. The inaccuracies may occur both in distance and angle measurements. Researchers have studied the gathering and the *convergence* problem under the models in which robots may obtain inaccurate measurements [7,17,18,24].

It was proved that the gathering of $n = 2$ robots with inaccurate distance measurements is impossible under the $\mathcal{F}$SYNC model even with consistent compasses and strong multiplicity detection [7]. This section presents two algorithms for the *convergence* problem under the $\mathcal{S}$SYNC model when robots are unable to measure their mutual distances [17,18]. However they can measure angles. The problem is studied under two different settings. In one setting, robots have knowledge of $\delta$, the rigidity constant. In the other setting, a robot can detect if a visible robot is at distance less or more than a fixed value $l$. The value of $l$ is called the *near-visibility* range and known to the robots.

- **Algorithm CONVERGE-CRUDE-1()**
  The algorithm presented in [18] is described here. Let $\xi_i(t)$ be the largest angle made at $r_i(t)$ by two consecutive robots on the visibility polygon $\mathcal{VP}_i(t)$ and $\psi_i(t)$ the complementary angle of $\xi_i(t)$ (Fig. 5(a)). The angle $\psi_i(t)$ defines the of the smallest wedge containing all visible robots of $r_i$ at time $t$.

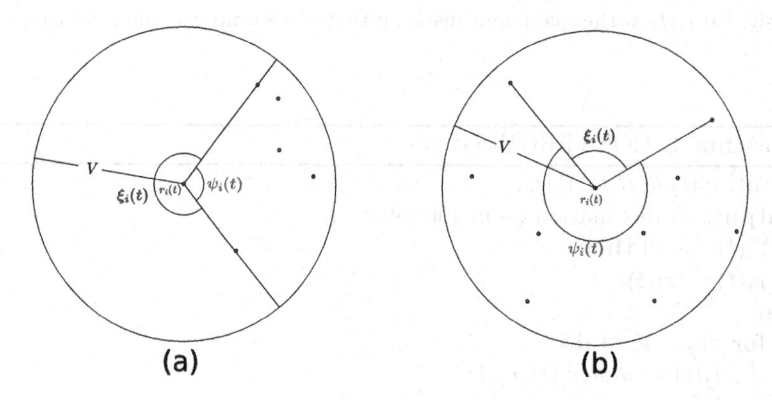


**Fig. 5.** (a) $\xi_i(t) > \pi$ and $\psi_i(t)$; the destination of robot $r_i$ should lie in the wedge defined by $\psi_i(t)$ to remain visible to the other robots in $\mathcal{V}_i(t)$ when it moves. (b) $\xi_i(t) < \pi$ and robot $r_i$ cannot move.

The basic idea of the algorithm is to move the robots on the vertices of the local convex hull of the visible robot positions and gradually shrink it. A robot does not move until it lies on some hull vertex (Fig. 5(b)). Robot $r_i$ moves only when it finds $\psi_i(t) < \pi$ and it moves along the bisector of $\psi_i(t)$. Robot $r_i$ moves a distance $d_i(t)$ defined as follows:

$$d_i(t) = min\{\frac{V}{2}, Vcos(\frac{\psi_i(t)}{2}), \delta\}.$$

By moving a distance $d_i(t)$, robot $r_i$ maintains the connectivity of the visibility graph (Fig. 6). The pseudo-code of the algorithm is given in *CONVERGE-CRUDE-1()*.

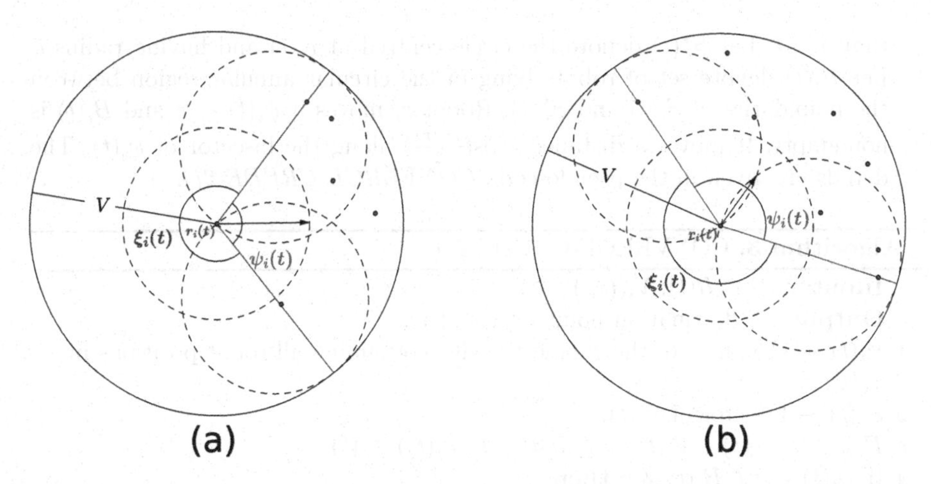


**Fig. 6.** Illustrations showing how to choose allowable movement distance for $r_i$, when $\delta > \frac{V}{2}$ (a) the distance is $\frac{V}{2}$ when $\psi_i(t) < \frac{2\pi}{3}$ (b) the distance is $Vcos(\frac{\psi_i(t)}{2})$ when $\psi_i(t) > \frac{2\pi}{3}$. Dashed circles have radius $\frac{V}{2}$ and there common intersection is the allowable movement area for $r_i$.

---

**Algorithm 2.** CONVERGE-CRUDE-1()

---

**Input**: $r_i(t) \in R(t)$, $\mathcal{V}_i(t), V$.
**Output**: A destination for robot $r_i$.

1 $\psi_i(t) \leftarrow$ the angle of the smallest wedge containing all robot positions in $\mathcal{V}_i(t)$;
2 $\mathcal{L}_i(t) \leftarrow$ bisector of $\psi_i(t)$;
3 **if** $\psi_i(t) < \pi$ **then**
4 $\quad d_i(t) \leftarrow min\{\frac{V}{2}, Vcos(\frac{\psi_i}{2}), \delta\}$;
5 $\quad p_i(t) \leftarrow$ the point on $\mathcal{L}_i(t)$ at a distance $d_i(t)$ from $r_i(t)$;
6 **else**
7 $\quad p_i(t) \leftarrow r_i(t)$;
8 $r_i$ moves towards $p_i(t)$;

---

- **Algorithm CONVERGE-CRUDE-2()**
In algorithm *CONVERGE-CRUDE-1()*, the step length in each movement is computed so that the visibility graph remains connected after the robots move. This does not include the mutual distances among the robots. The robots which are closer, move in more conservative ways than needed. The work in [17] studied the *convergence* problem under the $\mathcal{S}$SYNC model when robots have some crude knowledge of the distances that helps them to identify whether a robot is far or near. The robots have some *near visibility* range $l$ and this value is known to them. Robot $r_i$ can identify whether a visible robot $r_j$ is at a distance less than equal to $l$ or more. For simplicity, it is assumed

that $\delta = l$. Let $S_i^*(t)$ denote the circle centred at $r_i(t)$ and having radius $l$. Let $B_i(t)$ denote set of robots lying in the circular annular region between the boundaries of $S_i(t)$ and $S_i^*(t)$. Robot $r_i$ moves if $\psi_i(t) < \pi$ and $B_i(t)$ is non-empty. It moves a distance $l\cos(\frac{\psi_i(t)}{2})$ along the bisector of $\psi_i(t)$. The details are given in the pseudo-code *CONVERGE-CRUDE-2()*.

**Algorithm 3.** CONVERGE-CRUDE-2()

**Input**: $r_i(t) \in R(t)$, $\mathcal{V}_i(t)$, $V$, $l$.
**Output**: A destination point for robot $r_i$.
1 $\psi_i(t) \leftarrow$ the angle of the smallest wedge containing all robot positions in $\mathcal{V}_i(t)$;
2 $\mathcal{L}_i(t) \leftarrow$ bisector of $\psi_i(t)$;
3 $B_i(t) \leftarrow \{r_j(t) \in \mathcal{V}_i(t) : l \leq dist(r_i(t), r_j(t)) \leq V\}$;
4 **if** $\psi_i(t) < \pi \wedge B_i(t) \neq \emptyset$ **then**
5 $\quad d_i(t) \leftarrow lcos(\frac{\psi_i(t)}{2})$;
6 $\quad p_i(t) \leftarrow$ the point on $\mathcal{L}_i(t)$ at a distance $d_i(t)$ from $r_i(t)$;
7 **else**
8 $\quad p_i(t) \leftarrow r_i(t)$;
9 $r_i$ moves towards $p_i(t)$;

The correctness of the above two algorithms is derived from the fact that $\mathcal{CH}(t+1) \subset \mathcal{CH}(t)$.

**Theorem 3** [17,18]. *The convergence problem is solvable for a set of $n$ point robots with limited visibility under the $\mathcal{S}$SYNC model even if the robots can not measure distances.*

#### 4.1.3 Convergence in the $\mathcal{A}$SYNC Model

Convergence is possible in the $\mathcal{A}$SYNC model under a *1-fair scheduler* [22]. A *1-fair* asynchronous scheduler is an asynchronous scheduler such that between two consecutive activations of a robot $r_i$, all other robots can become active at most once. Thus, during a *look-compute-move* cycle of $r_i$, other robots in the system can perform at most one *look-compute-move* cycle.

The main focus of the convergence algorithm under limited visibility model is to preserve the connectivity of the initial visibility graph, when robots make their movements. To achieve this, robots need to compute their destination points very carefully. Due to asynchrony, this becomes more difficult for the robots. The assumption of the 1-fair scheduler makes it easier for the robots to deterministically compute their destination points and to guarantee the finite time convergence. To maintain the connectivity between two robots $r_i$ and $r_j$, robots preserve the following invariant: [I] if $r_j \in \mathcal{V}_i(t)$, then $p_i(t)$ and $p_j(t)$ lie within the circle $S_{ij}(t)$ with radius $\frac{V}{2}$ and centre at $\frac{r_i(t)+r_j(t)}{2}$.

There are allowable regions in which a robot can move by preserving [I]. Two such regions are *move towards* and *move around* (Fig. 7). These two regions are defined considering distances of the robot positions in $\mathcal{V}_i(t)$ from $r_i(t)$ in the following way: consider a robot position $r_j(t) \in \mathcal{V}_i(t)$.

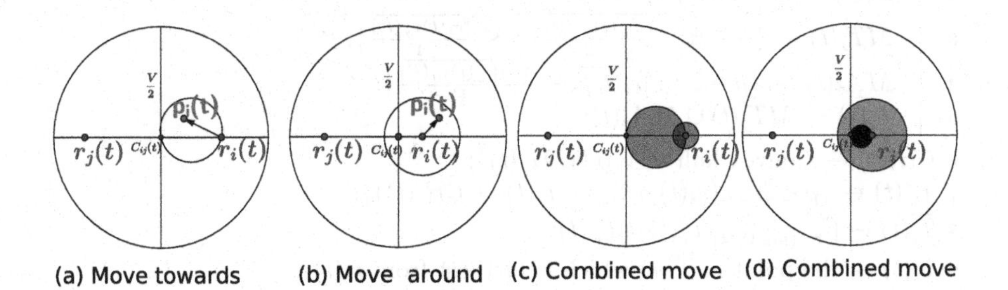


**Fig. 7.** The regions move towards, move around and their combinations

- Let $c_{ij}(t) = (r_i(t) + r_j(t))/2$ and $d = dist(r_i(t), r_j(t))$.
- The region *move towards* contains all the points within the circle having $\overline{c_{ij}(t)r_i(t)}$ as diameter (Fig. 7(a)). A point $p$ in this region satisfies the inequality $|\frac{c_{ij}(t)+r_i(t)}{2}p| \leq \frac{d}{4}$.
- The region *move around* is the set of all points which lie inside the circle having centre at $r_i(t)$ and radius $\frac{V-d}{4}$ (Fig. 7(b)). A point $p$ in this region satisfies the inequality $|r_i(t)p| \leq \frac{V-d}{4}$.
- Let $T_j$ be the union of the two regions *move towards* and *move around* (Fig. 7(c) and (d)).

The robots inside $\mathcal{CH}(t)$ should not move towards the boundary of $\mathcal{CH}(t)$. For the convergence, the robots on the boundary of $\mathcal{CH}(t)$ should move inside the convex hull.

This constraint gives another allowable region for the destination points of the robots. It is denoted by $T_i^*(t)$ and contains all the points in $\mathcal{V}_i(t)$ that are not further than halfway from the boundary of $\mathcal{CH}_i(t)$; i.e., $T_i^*(t) = \{p \in \mathbb{R}^2 : (r_i(t) + 2(p - r_i(t)) \in \mathcal{CH}_i(t)\}$.

The destination point of $r_i$ should lie inside $\overline{T_i}(t) = \bigcap_{r_j(t) \in \mathcal{V}_i(t)} T_j(t) \cap T_i^*(t)$. Robot $r_i$ chooses the furthest point in $\overline{T_i}(t)$ from $r_i(t)$. The movements of the robots according to these rules shrink $\mathcal{CH}(t)$, and within finite time all the robots converge towards a point. The pseudo-code of the algorithm is given in *CONVERGE-1-FAIR()*.

**Theorem 4** [22]. *The convergence problem is solvable for a set of asynchronous robots under a 1-fair scheduler.*

**Algorithm 4.** CONVERGE-1-FAIR()

**Input:** $r_i(t) \in R(t)$, $\mathcal{V}_i(t)$.
**Output:** A destination point for robot $r_i$.
1 **for** $\forall r_j(t) \in \mathcal{V}_i(t)$ **do**
2 $\quad c_{ij}(t) \leftarrow (r_i(t) + r_j(t))/2;$
3 $\quad MT_j(t) \leftarrow \{p \in \mathbb{R}^2 : |\overline{\frac{c_{ij}(t)+r_i(t)}{2}p}| \leq \frac{|\overline{r_i(t)r_j(t)}|}{4}\};$
4 $\quad MA_j(t) \leftarrow \{p \in \mathbb{R}^2 : |\overline{r_i(t)p}| \leq \frac{V-|\overline{r_i(t)r_j(t)}|}{4}\};$
5 $\quad T_j(t) \leftarrow MT_j(t) \cup MA_j(t);$
6 $\mathcal{CH}_i(t) \leftarrow$ Convex hull of $\mathcal{V}_i(t) \cup \{r_i(t)\}$;
7 $T_i^*(t) \leftarrow \{p \in \mathbb{R}^2 : (r_i(t) + 2(p - r_i(t)) \in \mathcal{CH}_i(t))\};$
8 $\overline{T_i}(t) \leftarrow \bigcap_{r_j(t) \in \mathcal{V}_i(t)} T_j(t) \cap T_i^*(t);$
9 $p_i(t) \leftarrow$ the point in $\overline{T_i}(t)$ which is furthest from $r_i(t)$;
10 $r_i$ moves towards $p_i(t)$;

### 4.2 The Near Gathering Problem

Pagli et al. [25] considered the *near gathering* problem which is very close to the *convergence* problem. The *near gathering* problem requires all the robots to coordinate their movements in such a way that within finite time all of them hold positions in a disc of radius $\epsilon > 0$ (known to the robots) and no two robots share same position.

Let $\sigma > 0$ be an arbitrary small constant and $D = V - \sigma$.

**Definition 1.** *(Initial Strong Distance Graph). The initial strong distance graph $I = (\mathcal{R}, E)$ of the robot positions is the graph such that for any two distinct robot $r_i$ and $r_j$ in $\mathcal{R}$, $(r_i, r_j) \in E$ if and only if $dist(r_i(t_0), r_j(t_0)) \leq D$.*

(A) **Assumptions:**
- Initially all robots are stationary and they occupy distinct positions.
- The initial Strong Distance Graph is connected.
- The value of $D$ is the same for all robots and the value is known to the robots.
- The value of $\epsilon$ (required for termination) is known to the robots.
- Robots operate under the $\mathcal{A}$SYNC model and their movements are non-rigid.
- Robots agree on the directions and orientations of the both local coordinate axes.

Under the above assumptions, a distributed algorithm for the *near gathering* problem was presented for a set of asynchronous robots in [25]. It was proved that the same algorithm also works when robots have agreement on the direction and orientation of one local coordinate axis.

(B) **Notations:** To describe the algorithm, the following notations and definitions are used:

- Let $D_i^0(t)$, $D_i^1(t)$ and $D_i^2(t)$ be the closed discs having radii $V$, $V - \frac{\rho}{2}$ and $V - \rho$ respectively, where $\rho = min\{\frac{V}{4}, V - D\}$. All of them have centres at $r_i(t)$ (Fig. 8).
- The point $a_i^1(t)$ is the leftmost intersection point between $D_i^1(t)$ and the horizontal line at a distance $V-\rho$ from $r_i(t)$ and lying above it. The point $a_i^2(t)$ is the bottommost intersection point between $D_i^1(t)$ and the vertical line at a distance $V - \rho$ from $r_i(t)$ and lying right side of it (Fig. 8).
- $\mathcal{SQ}_i(t)$ denotes the closed square circumscribed around $D_i^2(t)$ with sides parallel to the X-axis and Y-axis and $\mathcal{A}_i(t) = D_i^2(t) \cap \mathcal{SQ}_i(t)$.

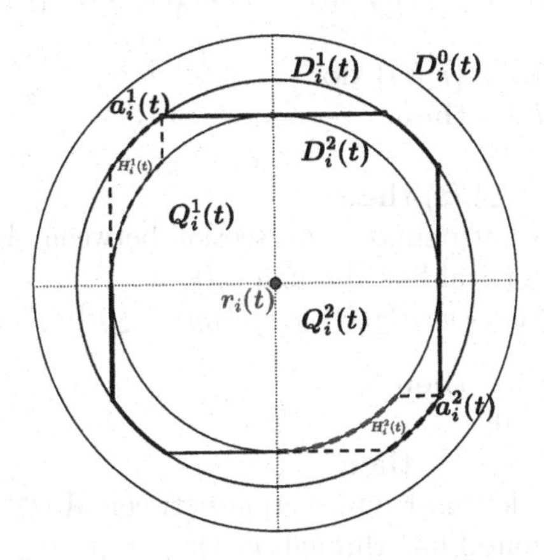


**Fig. 8.** The elements computed by robot $r_i$.

- $Q_i^1(t)$ is the set of all points in $D_i^0(t)$ having positive y-coordinate and non-positive x-coordinate. $Q_i^2(t)$ is the set of all points in $D_i^0(t)$ having positive x-coordinate and non-positive y-coordinate.
- $H_i^1(t)$ is the set of points in $(\mathcal{A}_i(t) \backslash D_i^2(t)) \cap Q_i^1(t)$ which has x-coordinate value less than $p_1.x$, the x-coordinate value of $p_1$. $H_i^2(t)$ is the set of points in $(\mathcal{A}_i(t) \backslash D_i^2(t)) \cap Q_i^2(t)$ which has y-coordinate value less than $p_2.y$, the y-coordinate value of $p_1$ (the area bounded by the dashed segments and arcs in Fig. 8).
- $\mathcal{NW}(t) = \{r_j(t) \in \mathcal{V}_i(t) : r_j(t) \in Q_i^1(t)\}$ and $\mathcal{SE}(t) = \{r_k(t) \in \mathcal{V}_i(t) : r_k(t) \in Q_i^2(t)\}$.

**Definition 2.** ***Move Space:*** *The move space of a robot $r_i$ at time $t$, denoted by $\mathcal{MS}_i(t)$, is the set $\{(x', y') \in \mathbb{R}^2 : x' \geq r_i(t).x \wedge y' \geq r_i(t).y\}$.*

**Algorithm 5.** NEAR-GATHER()

**Input**: $r_i(t) \in R(t), \mathcal{V}_i(t), V, D, \epsilon$.
**Output**: A destination point for robot $r_i$.
1 $\rho \leftarrow min\{\frac{V}{4}, V - D\}$;
2 $\epsilon' \leftarrow min\{\epsilon, \frac{\rho}{2}\}$;
3 $b \leftarrow 1$ if $\forall r_j(t), r_k(t) \in \mathcal{V}_i(t) \cup \{r_i(t)\}, dist(r_j(t), r_k(t)) \leq \epsilon'$, otherwise 0;
4 **if** $b == 1$ **then**
5 $\quad p_i(t) \leftarrow r_i(t)$
6 **else**
7 $\quad p_i(t).x \leftarrow min\{min\{r_j(t).x | r_j(t) \in \mathcal{SE}(t)\}, max\{r_k(t).x | r_k(t) \in \mathcal{V}_i(t)\}, \frac{\rho}{2}\}$;
8 $\quad p_i(t).y \leftarrow min\{min\{r_j(t).y | r_j(t) \in \mathcal{NW}(t)\}, max\{r_k(t).y | r_k(t) \in \mathcal{V}_i(t)\}, \frac{\rho}{2}\}$;
9 $\quad$ **for** $\forall r_j(t) \in \mathcal{V}_i(t) \cup \{r_i(t)\}$ **do**
10 $\quad\quad$ **if** $r_j(t) \in H_i^1(t)$ **then**
11 $\quad\quad\quad p_i(t).x \leftarrow 0$;
12 $\quad\quad\quad$ **if** $r_j(t) \in \mathcal{A}_i(t)$ **then**
13 $\quad\quad\quad\quad s_2 \leftarrow$ bottommost intersection between $\mathcal{A}_i(t) \backslash H_i^2(t)$ and the vertical line through $r_j(t)$;
14 $\quad\quad\quad\quad p_i(t).y \leftarrow min\{p_i(t).y, r_j(t).y - s_2.y\}$;
15 $\quad\quad$ **if** $r_j(t) \in H_i^2(t)$ **then**
16 $\quad\quad\quad p_i(t).y \leftarrow 0$;
17 $\quad\quad\quad$ **if** $r_j(t) \in \mathcal{A}_i(t)$ **then**
18 $\quad\quad\quad\quad s_1 \leftarrow$ leftmost intersection between $\mathcal{A}_i(t) \backslash H_i^1(t)$ and the horizontal line through $r_j(t)$;
19 $\quad\quad\quad\quad p_i(t).x \leftarrow min\{p_i(t).x, r_j(t).x - s_1.x\}$;
20 $\quad$ **if** $p_i(t).x > p_i(t).y$ **then**
21 $\quad\quad p_i(t) \leftarrow (\frac{p_i(t).x}{2}, 0)$
22 $\quad$ **else**
23 $\quad\quad p_i(t) \leftarrow (0, \frac{p_i(t).y}{2})$
24 $r_i$ moves towards $p_i(t)$;

(C) **Computation of destination point:** A robot $r_i$ computes it's destination point as follows:
- $r_i$ has only two possible directions of movements: rightward and upward. It never moves diagonally. The distance traversed by $r_i$ in each move is maximized subject to the following restrictions.
- $r_i$ never enters the *move space* of a robot in $\mathcal{V}_i(t)$, provided it already lies in some other robot's move space.
- $r_i$ never moves to the right of (respectively above) the rightmost (respectively topmost) robot position in $\mathcal{V}_i(t)$.

- If $r_i$ finds a robot in the left halt zone, i.e. in $H_i^1(t)$, it does not move rightward. If $r_i$ finds a robot in the right halt zone i.e., in $H_i^2(t)$, it does not move upward.
- If $r_i$ finds a robot $r_j$ in $\mathcal{A}_i(t)\backslash H_i^1(t)$ (respectively $\mathcal{A}_i(t)\backslash H_i^2(t)$), it remains inside $\mathcal{A}_i(t)\backslash H_i^1(t)$ (respectively $\mathcal{A}_i(t)\backslash H_i^1(t)$).
- In each movement, the length of $r_i$'s displacement must not be greater than $\frac{\rho}{4} \leq \frac{V}{16}$.

The pseudo-code of the algorithm is given in *NEAR-GATHER()*. By executing the above algorithm, robots aggregate in finite time, around the top-right corner of the smallest box that contains all the robot positions. The above algorithm is perfectly symmetric with respect to the X-axis and Y-axis. This implies that above algorithm also works when robots have agreement in one coordinate axis. Also, the algorithm can be modified to solve the *near gathering* problem in the models that use any *p-norm* to measure distances.

**Theorem 5** [25]. *The near gathering problem is solvable for a set of asynchronous robots with agreement in one coordinate axis if the initial strong distance graph is connected.*

### 4.3 The Gathering Problem

The *gathering* problem requires a set of robots to meet at a point, not known a priori. This basic formation problem has been studied extensively in the literature under different computational models [14].

#### 4.3.1 Gathering in the $\mathcal{A}$SYNC Model with Consistent Compasses

The gathering of asynchronous robots with limited visibility becomes difficult since robots may not have complete view of the system. The *gathering* problem for a set of oblivious semi-synchronous robots (thus also for asynchronous robots) is not solvable without multiplicity detection capability or agreements on the local coordinate systems [27].

In this section, a solution to the *gathering* problem for a set of asynchronous robots with limited visibility range is discussed [15]. It is assumed that the robots have agreements in the directions and orientations of both local coordinate axes (*Consistent Compass* model). The robots are assumed to be transparent. The overall idea of the algorithm is to bring all the robots at the rightmost-bottommost point. To achieve this, robots move from left to right and from top to bottom directions. Since they have limited visibility range, during movements, they need to preserve visibility with the current visible robots so that the visibility graph remains connected throughout the execution. Consider a robot $r_i \in \mathcal{R}$ at time $t \geq t_0$. The destination point of $r_i$ depends on its position. Let $VL_i(t)$ and $HL_i(t)$ denote the vertical and horizontal lines through $r_i(t)$. Let $Left_i(t)$ and $Right_i(t)$ be the sets of robot positions in $\mathcal{V}_i(t)$ lying on the left and right open halves of $VL_i(t)$ respectively. $Above_i(t)$ and $Below_i(t)$ are the sets of robot positions in $\mathcal{V}_i(t)$ lying on the upper and lower open halves of $HL_i(t)$ respectively. None of these four sets contains $r_i(t)$.

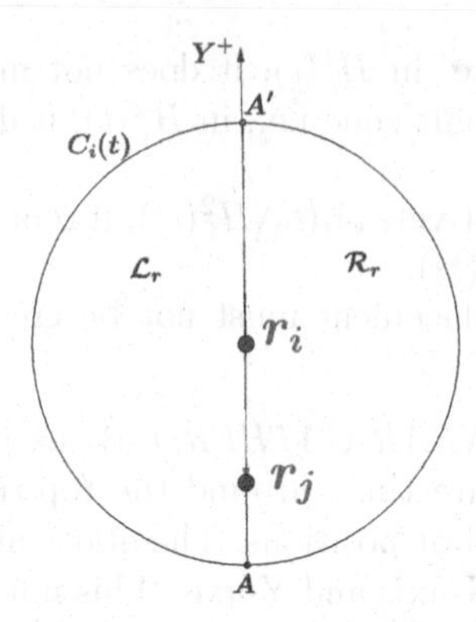


Fig. 9. An illustration of a vertical movement

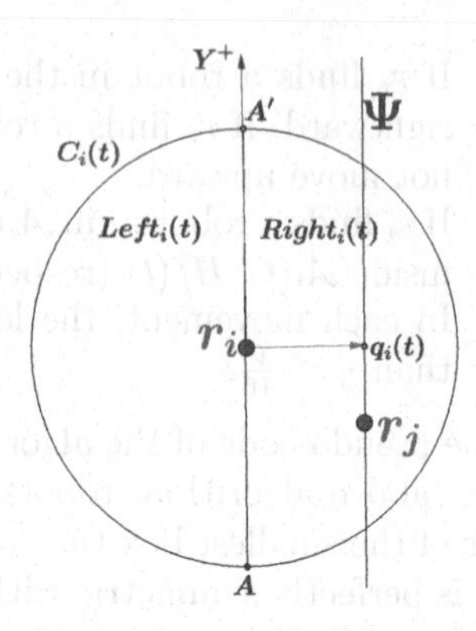


Fig. 10. An illustration of a horizontal movement

- If there is at least one robot position in $Left_i(t)$ or in $Above_i(t)$, robot $r_i$ does nothing.
- If all the robot positions in $\mathcal{V}_i(t)$ lie below on the line $VL_i(t)$, $r_i$ moves towards the nearest visible robot (*vertical move* in Fig. 9).
- If all robot positions in $\mathcal{V}_i(t)$ lie on the right hand side of $r_i$ i.e., $\mathcal{V}_i(t) = Right_i(t)$, $r_i$ makes a *Horizontal move* (Fig. 10 and line no. 15–18 in the pseudo-code *Gathering-*$\mathcal{A}$SYNC()).
- If there are robot positions both below on the line $VL_i(t)$ and in $Right_i(t)$, $r_i$ makes a *Diagonal* move (Figs. 11 and 12 and line no. 20–32 in the pseudo-code *Gathering-*$\mathcal{A}$SYNC()).

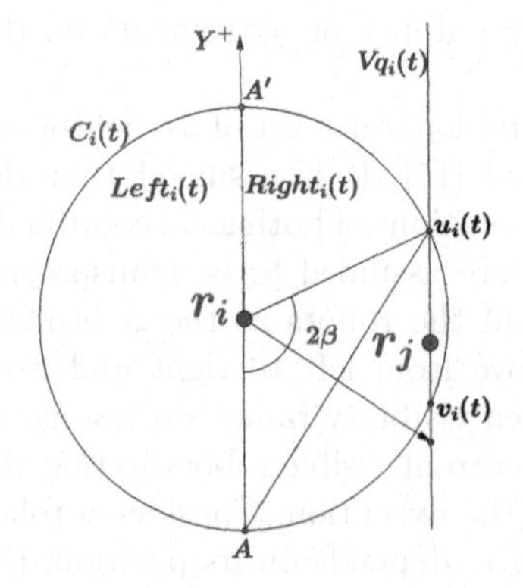


Fig. 11. Case $\beta < 60°$, diagonal move, in *Gathering-*$\mathcal{A}$SYNC()

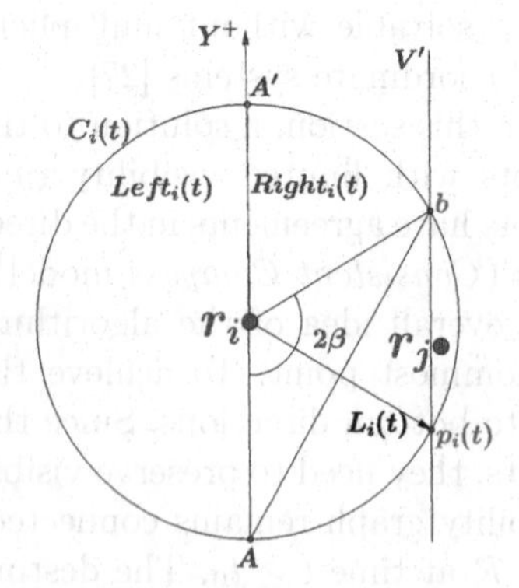


Fig. 12. Case $\beta = 60°$, diagonal move, in *Gathering-*$\mathcal{A}$SYNC()

**Algorithm 6.** Gathering-$\mathcal{A}$SYNC()

```
   Input: r_i(t) ∈ R(t), V_i(t), V.
   Output: A destination point for robot r_i.
 1 VL_i(t) ← vertical line through r_i(t);
 2 HL_i(t) ← horizontal line through r_i(t);
 3 Left_i(t) ← set of all robots on the left of VL_i(t) not lying on VL_i(t);
 4 Right_i(t) ←set of all robots on the right of VL_i(t) not lying on VL_i(t);
 5 Above_i(t) ←set of all robots on the above of HL_i(t) not lying on HL_i(t);
 6 Below_i(t) ←set of all robots on the below of HL_i(t) not lying on HL_i(t);
 7 VL_i^+(t) ← set of all robots in Above_i(t) lying on VL_i(t);
 8 VL_i^-(t) ← set of all robots in Below_i(t) lying on VL_i(t);
 9 if Left_i(t) ∪ VL_i^+(t) ≠ ∅ then
10 |  p_i(t) ← r_i(t);
11 else
12 |  if V_i(t) == VL_i^-(t) then
13 |  |  p_i(t) ←, the nearest robot position in V_i(t) from r_i(t);
14 |  else
15 |  |  if V_i(t) == Right_i(t) then
16 |  |  |  A_i(t) ← projection of all points in Right_i(t) on HL_i(t);
17 |  |  |  p_i(t) ← the nearest point in A_i(t) from r_i(t);
18 |  |  else
19 |  |  |  S_i(t) ← minimum enclosing circle of V_i(t) ∪ {r_i(t)};
20 |  |  |  A_i(t) ← projection of all points in V_i(t) on HL_i(t);
21 |  |  |  q_i(t) ← the nearest point in A_i(t) from r_i(t);
22 |  |  |  Vq_i(t) ← the vertical line through q_i(t);
23 |  |  |  u_i(t) ← the upper intersection point between S_i(t) and Vq_i(t);
24 |  |  |  v_i(t) ← the lower intersection point between S_i(t) and Vq_i(t);
25 |  |  |  A ← the point on negative Y-axis of r_i at a distance V from
   |  |  |    r_i(t);
26 |  |  |  2β ← ∠u_i(t)r_i(t)A;
27 |  |  |  if β < 60° then
28 |  |  |  |  u_i(t) ← the point b on S_i(t) such that ∠u_i(t)r_i(t)b is
   |  |  |  |    minimum and ∠br_i(t)A = 120° ;
29 |  |  |  |  Vq_i(t) ← the vertical line through u_i(t);
30 |  |  |  L_i(t) ← the ray from r_i(t) and perpendicular to Au_i(t) (overline);
31 |  |  |  p_i(t) ← the intersection point between Vq_i(t) and D_i(t);
32 r_i moves towards p_i(t);
```

**Theorem 6** [15]. *The gathering problem is solvable for a set of robots with limited visibility under* $\mathcal{A}$SYNC *model if they have Consistent Compasses.*

### 4.3.2 Gathering in the $\mathcal{S}$SYNC Model with Eventually Consistent Compasses

Consider a set of robots with eventually consistent compasses in the $\mathcal{S}$SYNC model [28]. With eventually consistent compasses, robots agree on a fixed direction, say North direction, after some unknown finite time. An eventually consistent compass has the following properties: (1) the north directions of the robots may change over time; (2) two different robots may disagree on a common north direction; and (3) there exists some time $GST$, unknown to the robots, after which the robots agree on a common north direction for a sufficiently long period. Algorithm *Gathering-*$\mathcal{A}$SYNC*()* does not solve the *gathering* problem for a set of semi-synchronous robots with eventually consistent compasses. When robots with eventually consistent compasses execute algorithm *Gathering-*$\mathcal{A}$SYNC*()*, the initial visibility graph may get disconnected [28].

The solution in [28] assumes common *chirality.* While solving the *gathering* problem, the movements of the robots are coordinated to satisfy three sub-goals at every time instant $t$:

(i) All the mutually visible robots at time $t$ remain visible at time $t+1$.
(ii) The robots move close to each other until they become sufficiently close, as defined below.
(iii) When robots become sufficiently close, they move from left to right to gather at the rightmost-bottommost robot position and, after $GST$, they gather at the rightmost-bottommost robot position within finite time.

An active robot $r_i$ at time $t$ first checks whether the system has been sufficiently converged or not. The convergence criteria is defined as follows. Let $S_{ij}(t)$ be the circle with centre $\frac{r_i(t)+r_j(t)}{2}$ and radius $\frac{V}{2}$, $r_j(t) \in \mathcal{V}_i(t)$. If the intersection $\cap_{r_j(t)\in\mathcal{V}_i(t)} S_{ij}(t)$ contains all robot positions in $\mathcal{V}_i(t)$, then robot $r_i$ concludes that the system has sufficiently converged. Once the system has sufficiently converged, robots start moving from left side towards the rightmost-bottommost robot position as follows. We use terms defined in pseudo-code *Gathering-*$\mathcal{A}$SYNC():

- If there is a robot in $Left_i(t)$ or in $VL_i^+(t)$, then $r_i$ does not move.
- If there are robots neither in $Left_i(t)$ or $VL_i^+(t)$ and $r_i(t)$ is collinear with all robot positions in $\mathcal{V}_i(t)$, then $r_i$ moves linearly to the nearest robot position. In this case, if $r_i$ moves, it must be the topmost or leftmost robot on the line of collinearity.
- If there are robots neither in $Left_i(t)$ or $VL_i^+(t)$ and there are some robots in $Right_i(t)$ or $VL_i^-(t)$, then $r_i$ moves to the closest robot position in $VL_i^-(t)$, if any. Otherwise, $r_i$ moves to the closest robot position in $Right_i(t)$.

When $r_i$ finds that the convergence criteria is not satisfied, it computes its destination point as follows.

- If all the robot positions in $\mathcal{V}_i(t)$ are collinear and $r_i(t)$ does not lie between two other robot positions, $r_i$ moves to the nearest robot position along the line of collinearity.
- If there are robots neither in $Left_i(t)$ or $VL_i^+(t)$ and there are some robot positions in $Right_i(t)$ or $VL_i^-(t)$, then $r_i$ does the following. First, $r_i$ considers the smallest sector $SC_i(t)$ of $S_i(t)$ containing all the robots in $\mathcal{V}_i(t)$ within or on its boundary. Let $r_j(t)$ and $r_k(t)$ be the two farthest robot positions in two different sides of the sector $SC_i(t)$. Let $l_i(t)$ be the foot of a perpendicular from $r_i(t)$ to $\overline{r_j(t)r_k(t)}$. If $l_i(t)$ lies within the triangle $\triangle_{ijk}(t)$ formed by $r_i(t), r_j(t)$ and $r_k(t)$, robot $r_i$ moves to $l_i(t)$. Otherwise, $r_i$ moves to the closest robot position among $r_j(t)$ and $r_k(t)$.

**Theorem 7** [28]. *The gathering problem is solvable for a set of robots with chirality and eventually consistent compasses under the $\mathcal{S}$SYNC model.*

### 4.3.3 Optimal Gathering Algorithm

A new aspect of the classical *gathering* problem has been introduced by Poudel et al. [26]. Most of the earlier papers on the classical *gathering* problem did not discuss about the time complexity of their proposed algorithms, other than their finite time termination. However, in the recent papers [9,11,13,23] runtime of the algorithms has been given importance, while designing a solution. For example, Degener et al. [11] proposed an algorithm for gathering, whose expected runtime is $O(n^2)$ under fully synchronous model.

In some recent papers, the authors have also considered different viewing and connectivity range. The work in [20] proposed a $O(D_G)$ time solution for gathering on plane in $\mathcal{F}$SYNC model, where $D_G$ is the diameter of the initial visibility graph. In this work, robots are assumed to have one-axis agreement. The connectivity range is assumed to be $\frac{1}{\sqrt{2}}$, while visibility range is 1, higher than the connectivity range.

A new aspect of the running time calculation has been brought into consideration by Poudel et al. [26]. Let $D_E$ be the largest Euclidean distance between any pair of nodes in the initial configuration. For the *gathering* problem, a natural lower bound for solutions is $\Omega(D_E)$. For any initial configuration $D_E \leq D_G$ and in the worst case $D_G = \Theta(D_E^2)$. In [26], an attempt was made to bridge the gap between $O(D_G)$ and $O(D_E)$. For any initial configuration having viewing range $\sqrt{10}$ and square connectivity range of $\sqrt{2}$, an algorithm for gathering on a plane has been proposed under $\mathcal{A}$SYNC setting with full axis agreement.

Consider a robot $r_i \in R$. A horizontal line $L_x$ and a vertical line $L_y$ are assumed to pass through $r_i(t)$. Consider an axis-aligned square $ABCD$ with each side of length 2 units, centred at $r_i(t)$ (Fig. 13(a)). Robots having $\sqrt{2}$ square connectivity implies that robot $r_i$ is connected to all other robots placed within the square $ABCD$ (including its boundary), though $r_i$ can see beyond this square. In this solution, it is assumed that robots are oblivious and they take rigid motion only. Throughout the algorithm the robots are allowed to take moves only in the horizontal direction (left or right), vertically downward direction, and diagonally south-east or south-west direction. All these movements are taken on the basis of the positions of the neighbours in the connectivity graph.

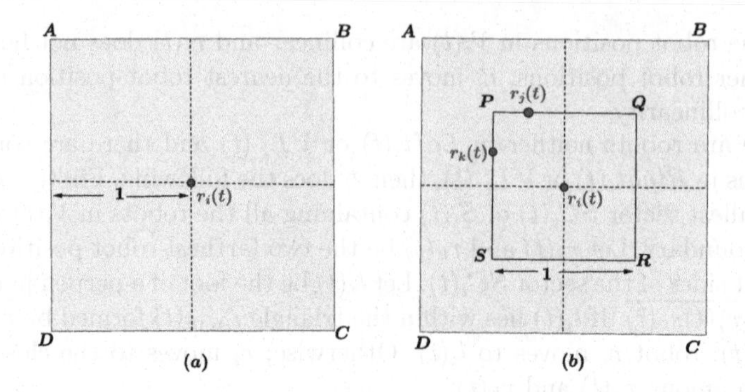


**Fig. 13.** An illustration of squares $ABCD$ and $PQRS$

As shown in Fig. 13(b), let $r_j$ and $r_k$ be the topmost (if any) and the left-most (if any) robot present within the square $ABCD$. While the robot takes its decision about its move, it considers a square $PQRS$ within the square $ABCD$ with its top boundary $R_T$ passing through $r_j(t)$ and left boundary $R_L$ passing through $r_k(t)$ and each side having length 1. The decision of $r_i$ depends not only the position of other robots present in $ABCD$, but also sometimes on the connectivities of other robots present within $PQRS$ and having neighbors outside $ABCD$.

The three types of hops are taken by the robots until all the robots are placed within $1 \times 1$ square (axis-aligned) area and finally a special termination procedure is followed to gather at a single point. Under all the moves, it is established that the visibility is retained. It is shown that the algorithm solves gathering in $O(D_E)$ time, with time measured in terms of look-compute-move (LCM) rounds, in the $\mathcal{F}$SYNC model. For the $\mathcal{S}$SYNC and $\mathcal{A}$SYNC models, $O(D_E)$ epochs solve the gathering under full axis agreement with visibility range of $\sqrt{10}$ and square connectivity range of $\sqrt{2}$. Under one-axis agreement, all other constraints being identical, the robots can form a horizontal line of unit length in $O(D_E)$ epochs in the $\mathcal{A}$SYNC system.

**Theorem 8** [26]. *The gathering problem is solvable in $O(D_E)$ epochs under the $\mathcal{A}$SYNC model for a set of asynchronous robots with both axis agreement and rigid movements when visibility range is $\sqrt{10}$ and square connectivity range is $\sqrt{2}$.*

### 4.4 The Scattering Problem

The *scattering* problem is defined as follows: starting from an arbitrary initial configuration, robots are required to obtain, within finite time, a configuration in which no two distinct robots occupy the same location. This problem can be considered as the dual of the *gathering* problem. To solve the *pattern formation* problem, starting from an arbitrary initial configuration, the *scattering* problem can be used as a preprocessing protocol. Due to the hardness of symmetry breaking problem, it is impossible to design deterministic scattering algorithms.

This implies that there is no deterministic solution to separate two robots, occupying the same location, when they work synchronously. Thus, randomization techniques are used to study the problem.

When robots have limited visibility, a solution for the *scattering* problem also requires to maintain the connectivity among the robots. The work in [21] presented a randomized scattering algorithm with connectivity-preserving property for a set of semi-synchronous robots. The robots have local-weak multiplicity detection capability, which enables them to identify multiple occurrences of the robots at a single point. Robots have no knowledge of $n$. Movements of the robots are rigid. It is assumed that scheduler is a *fair* scheduler, i.e. it activates each robot infinitely often. The number of random bits used by the robots in a cycle is one.

**Definition 3.** *Let $p$ be a point and $\mathcal{S}$ be a circle of unit radius centred at $p$. Let $\mathcal{B} = \{a_1, a_2, \cdots, a_m\}$ be a set of points on the circumference of $\mathcal{S}$. The point $p$ is said to be blocked if no arc of $\mathcal{S}$ with a centre angle less than $\pi$ contains all the points in $\mathcal{B}$ (Fig. 14).*

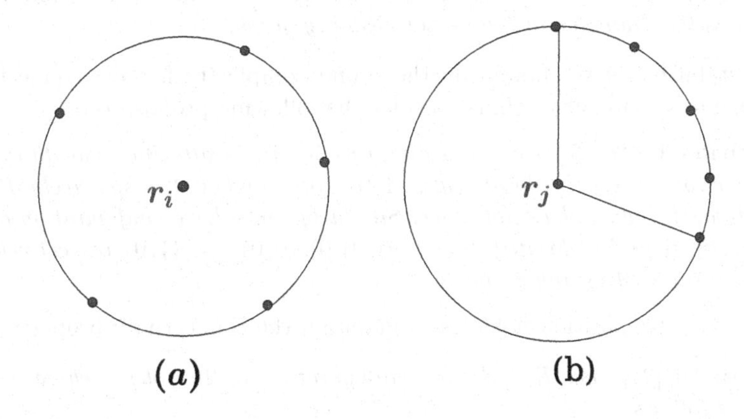


**Fig. 14.** An example of (a) a blocked robot $r_i$, and of (b) a non-blocked robot $r_j$

The outline of the algorithm is as follows [21]. To preserve the connectivity property, a robot at a blocked position does not move. First, non-blocked robots move in such a way that within finite number of rounds all the blocked robots become non-blocked. Then robots at the multiplicity points move in probabilistic way to achieve scattering. Now consider an active non-blocked robot $r_i$ at time $t$. First, $r_i$ computes a point $p$ which represents the direction and maximum amount of movement for $r_i$ and computes two possible destination points on the segment $\overline{r_i(t)p}$. Finally, by using the random bit, it chooses any one of these destination points. Let $S_i^*(t)$ be the circle of unit radius centred at $r_i(t)$. There are two possibilities for the point $p$:

- Suppose the circumference of $S_i^*(t)$ contains no robot position. The possible travel length of $r_i$ is denoted by $d$. The value of $d$ is bounded between two quantities: (i) the minimum distance between the boundary of $S_i^*(t)$ and the

robot positions in $\mathcal{V}_i(t)$ which lies inside $S_i^*(r)$, and (ii) the distance between $r_i(t)$ and its nearest robot position. The value of $d$ is used to compute the destination of $r_i$ which preserves connectivity and avoids creating new point of multiplicity. Robot $r_i$ sets $p$ as the point at a distance $\frac{d}{3}$ from $r_i(t)$ in the direction of the nearest robot from $r_i(t)$.
- Suppose the circumference of $S_i^*(t)$ contains at least one robot position. Let $A$ be the arc of $S_i^*(t)$ which contains all the robots on the boundary of $S_i^*(t)$. Let $\mathcal{L}_c$ be the chord of $A$ and $\mathcal{L}$ be the half line bisecting the centre angle of $A$. Let $\mathcal{L}_c$ and $\mathcal{L}$ intersect at $x$. Then $p$ be the point on $\overline{r_i(t)x}$ at a distance $d$, where $d$ is same as defined in the first case.

Robot $r_i$ takes two candidate destination points on the segment $\overline{r_i(t)p}$ and chooses one of them as its destination, by using the random bit. The robot $r_i$ moves, even if $r_i(t)$ contains exactly one robot, to make a blocked robot a non-blocked one. Details of the strategies are given in the pseudo-code *SCATTER()*.

**Theorem 9** [21]. *Algorithm SCATTER() achieves the connectivity-preserving scattering within* $O(min\{n, D^2+logn\})$ *expected rounds under the* SSYNC *model, where* $D$ *is the diameter of the initial visibility graph.*

To establish a lower bound on the round complexity for any connectivity-preserving scattering algorithm, consider the following proposition.

**Proposition 1.** *Let* $\psi$ *be a connectivity-preserving scattering algorithm. Consider three robots* $r_1, r_2, r_3$ *located at* $(-1,0)$, $(0,0)$ *and* $(1,0)$ *respectively. If robot* $r_2$ *is activated, then it does not move. Similarly, consider a configuration of four robots* $r_1, r_2, r_3$ *and* $r_4$ *located at* $(-1,0), (0,0), (0,0)$ *and* $(1,0)$ *respectively. On activation, none of* $r_2$ *and* $r_3$ *move.*

The above proposition holds as to maintain the connectivity property.

**Theorem 10** [21]. *Let* $\mathcal{R}(t)$ *be the configuration of* $2n$ *robots where* $r_k(t)$ *is defined as follows:*

$$r_k = \begin{cases} (k-1,0) & if\, 1 \leq k \leq n \\ (k,0) & if\, n+1 \leq k \leq 2n \end{cases}$$

*Then, any connectivity-preserving algorithms takes* $\Omega(n)$ *rounds if the initial configuration is* $\mathcal{R}(t)$ *(Fig. 15).*

The proof of the above theorem is done by using Proposition 1 and providing a round-robin activation of the robots. From Proposition 1, robots $r_{k+1}$ and $r_{2n-k+1}$ never move unless robots $r_k$ and $r_{2n-k}$ change their positions. Now consider the round-robin activation schedule of the robots in the order

$$r_n, r_{n-1}, r_{n+1}, r_{n-2}, \cdots, r_{n+j}, r_{n-j-1}, \cdots, r_{2n}, r_1.$$

The robots which can change their positions in round $k$ are $r_k$ and $r_{2n-k}$. This implies that $\Omega(n)$ rounds are needed to scatter $r_n$ and $r_{n-1}$.

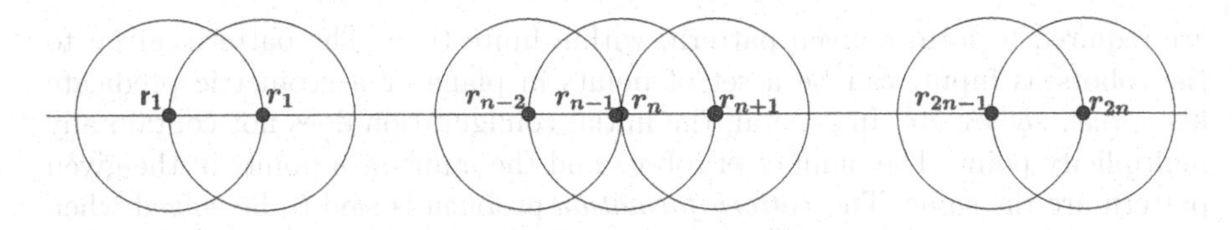


**Fig. 15.** The configuration $\mathcal{R}(t)$ used in Theorem 10

**Algorithm 7.** SCATTER()

**Input**: $r_i(t) \in \mathcal{R}(t), \mathcal{V}_i(t), V$.
**Output**: A destination point for $r_i$.

```
1  S_i^*(t) ← the circle of unit radius centred at r_i(t);
2  R_i^*(t) ← set of robot positions on the boundary of S_i^*(t);
3  m_i(t) ← binary variable indicating multiplicity of r_i(t);
4  Rand() ← random oracle;
5  b_i(t) ← 1 if r_i(t) is non-blocked, otherwise 0;
6  if b_i(t) == 1 then
7      d ← min{min{1 − |r_i(t)r_j(t)|, |r_i(t)r_j(t)|} : r_j(t) ∈
       (V_i(t) − R*(t)) ∪ {r_i(t)}};
8      if R*(t) == ∅ then
9          r_j(t) ← nearest robot position from r_i(t);
10         L_ij(t) ← ray starting from r_i(t) and passing through r_j(t);
11         q ← the point on L_ij(t) at a distance d/3 from r_i(t);
12     else
13         A ← arc of S_i^*(t) containing all robots of R_i^*(t);
14         L_c ← chord of A;
15         L ← the half line bisecting the centre angle of A;
16         x ← intersection point of L_c and L;
17         q ← the point on r_i(t)x at a distance d from r_i(t);
18     l ← dist(r_i(t), q);
19     if Rand() == 1 then
20         p_i(t) ← the point on r_i(t)q at a distance l/4;
21     else
22         p_i(t) ← the point on r_i(t)q at a distance l/2;
23 else
24     p_i(t) ← r_i(t);
25 r_i moves towards p_i(t);
```

### 4.5 The Pattern Formation Problem

The *pattern formation* problem is one of the most important coordination problem for a swarm of robots. Starting from an arbitrary initial configuration, robots are required to form a given pattern, within finite time. The pattern, given to the robots as input, can be a set of points in plane or a geometric predicate like *circle, square* etc. In general, the initial configuration does not contain any multiplicity point. The number of robots and the number of points in the given pattern are the same. The *pattern formation* problem is said to be solved when each point of the pattern is occupied by a robot, within finite time or the robot positions satisfy the predicate. The *pattern formation* problem is mostly investigated for the robots with unlimited visibility range [14]. In this section, a study on the *pattern formation* problem is discussed when robots have limited visibility range [30]. The study reveals the formation power of the robots with limited visibility with respect to the robots with unlimited visibility range.

Consider a set $\mathcal{P}$ of distinct points. The symmetricity $\rho(\mathcal{P})$ of $\mathcal{P}$ is defined to be 1 if the center $c(\mathcal{P})$ of the smallest enclosing circle of $\mathcal{P}$ contains a point of $\mathcal{P}$ (Fig. 16(a)). Otherwise, $\rho(\mathcal{P})$ is the number of angles $\theta \in (0, 2\pi]$ such that rotation of $\mathcal{P}$ by angle $\theta$ yields $\mathcal{P}$ (Fig. 16(b)). For robots with unlimited visibility range, a target pattern $\mathcal{F}$ is formable from an initial robot configuration $\mathcal{I}$ if and only if $\rho(\mathcal{I})$ divides $\rho(\mathcal{F})$ [29]. Let $\mathcal{P}_n$ be the set of all patterns of $n$ points.

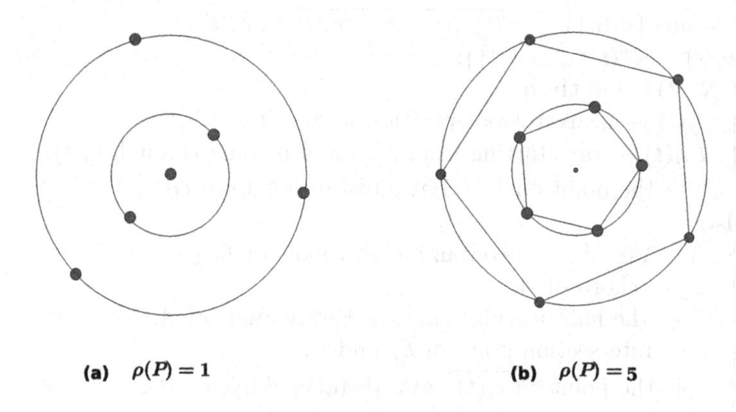


**Fig. 16.** An illustration of symmetricity

Following lemma holds even for the robots with limited visibility.

**Lemma 3** [29]. *A pattern $\mathcal{F} \in \mathcal{P}_n$ is not formable from an initial configuration $\mathcal{I} \in \mathcal{P}_n$ by oblivious $\mathcal{F}$SYNC robots with limited visibility, if $\rho(\mathcal{I}) > \rho(\mathcal{F})$, where $n \geq 3$.*

Following theorem presents a negative result for the *pattern formation* problem under *limited visibility* model.

**Theorem 11** [30]. *Let $\psi$ be an arbitrary pattern formation algorithm for oblivious $\mathcal{F}$SYNC robots with limited visibility. Then there exist $\mathcal{F}, \mathcal{I} \in \mathcal{P}_n$ for $n \geq 3$ such that $\mathcal{F}$ is not formable starting from $\mathcal{I}$ even when $\rho(\mathcal{I})$ divides $\rho(\mathcal{F})$.*

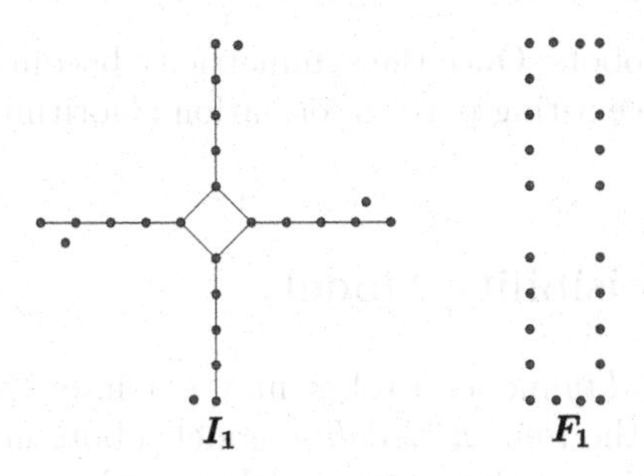


**Fig. 17.** A counter example of non-formable pattern: pattern $\mathcal{F}_1$ is not formable starting from the initial configuration $\mathcal{I}_1$ even if $\rho(\mathcal{I}_1) = \rho(\mathcal{F}_1)$.

The above theorem is proved by means of a counter example. Consider the initial configuration $\mathcal{I}_1$ and the pattern $\mathcal{F}_1$ as shown in Fig. 17. Note that $\rho(\mathcal{I}_1) = \rho(\mathcal{F}_1) = 2$. Let $\psi$ be an arbitrary pattern formation algorithm for oblivious $\mathcal{F}$SYNC robots with limited visibility. $\psi$ forms any target pattern $\mathcal{F}$ from an initial configuration $\mathcal{I}$ when $\rho(\mathcal{I})$ divides $\rho(\mathcal{F})$. It can be proved that there exists an execution of $\psi$ in which robots in $\mathcal{I}_1$ move symmetrically to increase symmetricity to 4 and $\mathcal{F}_1$ is no more formable by Lemma 3.

Theorem 11 implies that limited visibility substantially weakens the formation power of oblivious robots. However, for non-oblivious robots things are different. With limited visibility, non-oblivious $\mathcal{F}$SYNC robots with non-rigid movements and non-oblivious $\mathcal{S}$SYNC robots with rigid movements have same formation power as robots with unlimited visibility range. Thus, we have following two theorems.

**Theorem 12** [30]. *Let $\mathcal{F}, \mathcal{I} \in \mathcal{P}_n$ for $n \geq 3$. Pattern $\mathcal{F}$ is formable from the initial configuration $\mathcal{I}$ by non-oblivious $\mathcal{S}$SYNC robots with limited visibility and rigid moves if and only if $\rho(\mathcal{I})$ divides $\rho(\mathcal{F})$.*

**Theorem 13** [30]. *Let $\mathcal{F}, \mathcal{I} \in \mathcal{P}_n$ for $n \geq 3$. Pattern $\mathcal{F}$ is formable from the initial configuration $\mathcal{I}$ by non-oblivious $\mathcal{F}$SYNC robots with limited visibility and non-rigid moves if and only if $\rho(\mathcal{I})$ divides $\rho(\mathcal{F})$.*

The pattern formation algorithms have two phases. In the first phase, the robots use a convergence [2] algorithm to obtain mutual visibility among each other. Robots can detect termination of this phase as they know $|F| = n$ and $V$. Once, mutual visibility is obtained, the second phase is started. In this phase, robots use existing pattern formation algorithms for robots with unlimited visibility [16,29,30]. However, during any execution of the convergence algorithm, the symmetricity of the configuration may be increased by the movements of the robots (since robots do not have knowledge of the global configuration). Therefore, one need to reduce the symmetricity below $\rho(\mathcal{F})$. This can be done by *symmetricity control* algorithm which uses the local views and local outputs recorded by the non-oblivious robots. Once the symmetricity becomes smaller than $\rho(\mathcal{F})$, the robots execute the existing pattern formation algorithm to achieved the target pattern.

# 5 Obstructed Visibility Model

In the *obstructed visibility* model a robot may not have the complete visibility across the system. In the *limited visibility* model robots may be transparent or opaque. The combination of these two models provides a more difficult setting. Few works have been reported in the *obstructed visibility* model for oblivious robots [1,3–5,10,12,19]. Mainly the *gathering* problem and the *mutual visibility* problem have been studied under this model.

## 5.1 The Gathering Problem for Opaque Robots with No Extent

In the *obstructed visibility* model, solutions to the *gathering* problem exist for the point robots in the 2D plane and 3D space [3,5]. Both the solutions assume that the robots have an agreement in the direction and orientation of one coordinate axis. The solutions work even if initial configuration contains multiplicity points (a point containing multiple robots).

### 5.1.1 In the Euclidean Plane

The work in [3] proposed a solution to the *gathering* problem for a set of asynchronous opaque robots with no extent in the Euclidean plane. The proposed algorithm can tolerate an arbitrary number of crash faults. In crash fault model, a robot may stop working forever. However, the faulty robots remain in the system without performing any further action. A robot can not distinguish between a faulty robot and non-faulty robot. A model which permits at most $f \leq n$ faulty robots among the total $n$ robots is denoted by $(n, f)$ fault model. To describe the algorithm following notations are used:

- Without loss of generality, it is assumed that robots agree on the direction and orientation of $Y$-axis. Let $\mathcal{L}_k(t)$ denote the $k^{th}$ horizontal line from the north t o south direction through the robot positions in $\mathcal{R}(t)$. Robot positions in $\mathcal{V}_i(t)$ are partitioned into following three sub sets,
  - $UP_i(t)$ is the set of all points in $\mathcal{V}_i(t)$ having y-coordinate value greater than that of $r_i(t)$ i.e., $UP_i(t) = \{r_j(t) \in \mathcal{V}_i(t) : r_j(t).y > 0\}$. Note that $UP_i(t)$ is empty for the robots lying on the topmost horizontal line $\mathcal{L}_1(t)$. Let $\mathcal{L}_{u_i}(t)$ be the horizontal line through the points in $UP_i(t)$ having lowest $y$ coordinate.
  - $SM_i(t)$ is the set of all points in $\mathcal{V}_i(t)$ having y-coordinate values equal to $r_i(t)$ i.e., $SM_i(t) = \{r_j(t) \in \mathcal{V}_i(t) : r_j(t).y = 0\}$. Let $\mathcal{L}_{s_i}(t)$ denote the horizontal line through the points in $SM_i(t)$.

- $LW_i(t)$ is the set of all points in $\mathcal{V}_i(t)$ having y-coordinate value less than that of $r_i(t)$ i.e., $LW_i(t) = \{r_j(t) \in \mathcal{V}_i(t) : r_j(t).y < 0\}$. Note that $LW_i(t)$ is empty for the robots lying on the bottommost horizontal line through the points in $\mathcal{R}(t)$. Let $\mathcal{L}_{w_i}(t)$ be the horizontal line through the points in $UP_i(t)$ having highest $y$ coordinate (Fig. 18).

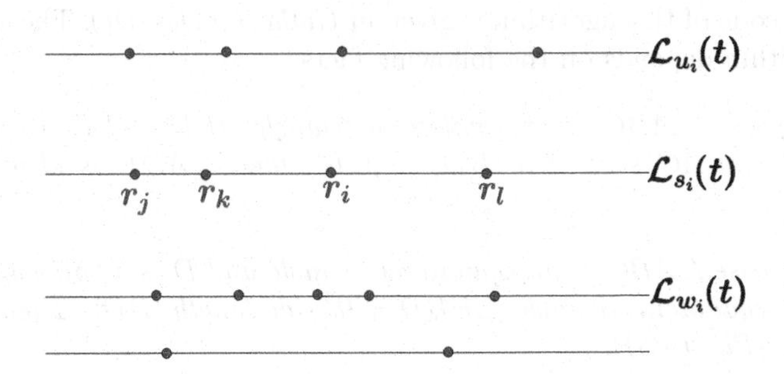


**Fig. 18.** An illustration showing $\mathcal{L}_{u_i}(t)$, $\mathcal{L}_{s_i}(t)$ and $\mathcal{L}_{w_i}(t)$, $\mathcal{M}^i(\mathcal{L}_{s_i}(t)) = \{r_k(t), r_l(t)\}$ and $\mathcal{M}(\mathcal{L}_{s_i}(t)) = \{r_j(t), r_l(t)\}$.

The main idea of the algorithm is as follows: robots try to obtain a unique position in the north direction. If in the initial configuration, there is a unique robot position on $\mathcal{L}_1(t_0)$, then this point serves as the gathering point. Otherwise, robots coordinate their movements to create such a point within finite time. If sets $UP_i(t)$ and $LW_i(t)$ are non-empty, then all the robot positions in these sets are visible to $r_i$. The computation of the destination point for $r_i$ depends on its position. The different scenarios and the corresponding approaches are as follows:

- $\boldsymbol{UP_i(t) = \phi}$:
  In this case, $r_i(t)$ lies on the topmost horizontal line in the north direction i.e. on $\mathcal{L}_1(t)$. If $r_i(t)$ is the only robot position on $\mathcal{L}_1(t)$, robot $r_i$ does not move. Otherwise, there are at least two robot positions on $\mathcal{L}_1(t)$. The computation of the destination point of $r_i$ depends on the number of distinct robot positions on $\mathcal{L}_1(t)$ and its position. Robot $r_i$ computes its destination point as follows:
  - Suppose $r_i(t)$ is a corner point on $\mathcal{L}_1(t)$ and $r_j(t)$ be the point on $\mathcal{L}_1(t))$ visible to $r_i$. Robot $r_i$ computes the equilateral triangle $\Delta r_i(t)T_i(t)r_j(t)$ whose side length is $|\overline{r_i(t)r_j(t)}|$ and $T_i(t)$ is the other vertex which lies in the north of $\overline{r_i(t)r_j(t)}$. Robot $r_i$ moves towards $T_i(t)$ along $\overline{r_i(t)T_i(t)}$.
  - Suppose $r_i(t)$ is not a corner point on $\mathcal{L}_1(t))$. In this case, $r_i$ lies in between two other robots on $\mathcal{L}_1(t)$. Robot $r_i$ computes the point $m_i(t)$ as follows: (i) if $LW_i(t) \neq \emptyset$, the point $m_i(t)$ is the nearest to $r_i(t)$ and is equidistant

from the lines $\mathcal{L}_1(t)$ and $\mathcal{L}_{w_i}(t)$ (ii) otherwise, the point $m_i(t)$ is in the south direction and is at distance $|r_i(t)r_j(t)|$, where $r_j(t)$ is the nearest robot position on $\mathcal{L}_1(t)$. The point $m_i(t)$ is the destination point of $r_i$.

- $\boldsymbol{UP_i(t) \neq \phi}$:
All the robots on $\mathcal{L}_{u_i}(t)$ are visible to $r_i$. Robot $r_i$ moves towards the nearest corner robot position on $\mathcal{L}_{u_i}(t)$ i.e., in $\mathcal{M}(\mathcal{L}_{u_i}(t))$.

The pseudo-code of the algorithm is given in *GatheringObs2D()*. The correctness of the algorithm depends on the following facts.

**Fact 1.** *Suppose $\triangle ABC$ is an equilateral triangle. If $D$ and $E$ are two points on side $AB$ and $AC$ such that $|BD| = |CE|$, then $\triangle ADE$ is also equilateral (Fig. 19(a)).*

**Fact 2.** *Suppose $\triangle ABC$ is an equilateral triangle and $D$ is a point on side $BC$. We draw an equilateral triangle $\triangle EBD$ with side length $|BD|$. Then $E$ lies on the side $AB$ (Fig. 19(b)).*

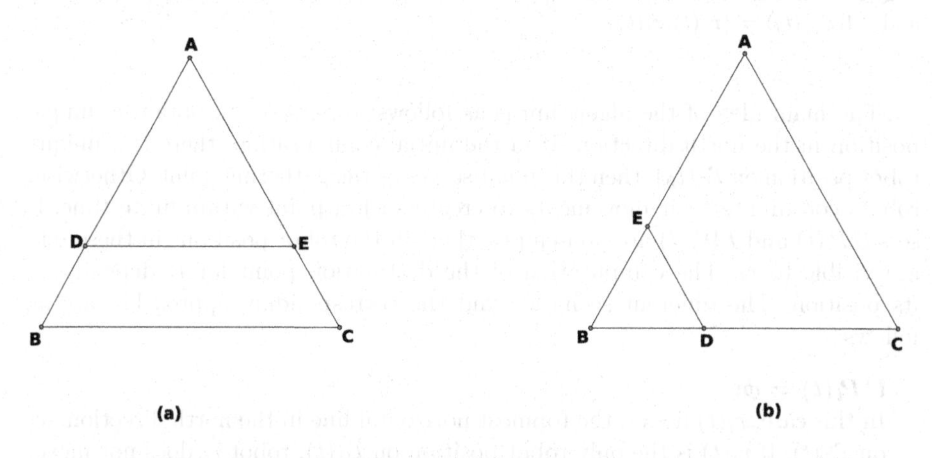


**Fig. 19.** Illustrations showing (a) Fact 1 and (b) Fact 2.

Algorithm *GatheringObs2D()* gathers the robots even if an arbitrary number of robots crash during the execution.

**Theorem 14** [3]. *The gathering problem is solvable for a set of asynchronous robots in $(n, n-1)$ crash fault model when robots have an agreement on one coordinate axis.*

**Algorithm 8.** GatheringObs2D()

**Input**: $r_i(t), \mathcal{V}_i(t)$.
**Output**: A destination point for $r_i$.

1 $UP_i(t) \leftarrow \{r_j(t) \in \mathcal{V}_i(t) | r_j(t).y > 0\}$;
2 $SM_i(t) \leftarrow \{r_j(t) \in \mathcal{V}_i(t) | r_j(t).y = 0\}$;
3 $LW_i(t) \leftarrow \{r_j(t) \in \mathcal{V}_i(t) | r_j(t).y < 0\}$;
4 **if** $UP_i(t) == \emptyset$ **then**
5   **if** $|SM_i(t)| == 1$ **then**
6     $p_i(t) \leftarrow r_i(t)$;
7   **else**
8     $C \leftarrow 1$ if $r_i$ is a corner robot on $\mathcal{L}_{s_i}(t)$, otherwise 0;
9     **if** $C == 1$ **then**
10       $\Delta r_i(t)T_i(t)r_j(t) \leftarrow$ equilateral triangle with side length $|\overline{r_i(t)r_j(t)}|$ and $T_i(t)$ lies in the north direction;
11       $p_i(t) \leftarrow T_i(t)$;
12     **else**
13       **if** $LW_i(t) \neq \emptyset$ **then**
14         $\mathcal{L}_{w_i}(t) \leftarrow$ the horizontal line through robot positions in $LW_i(t)$ having highest $y$ coordinate;
15         $\mathcal{L}^* \leftarrow$ perpendicular line on $\mathcal{L}_{w_i}(t)$ from $r_i(t)$;
16         $l \leftarrow$ perpendicular distance of $\mathcal{L}_{w_i}(t)$ from $r_i(t)$;
17         $p_i(t) \leftarrow$ the point on $\mathcal{L}^*$ in the south direction and at a distance $\frac{l}{2}$ from $r_i(t)$;
18       **else**
19         $\mathcal{L}_{s_i}(t) \leftarrow$ the horizontal line through robot positions in $SM_i(t)$;
20         $\mathcal{L}^* \leftarrow$ perpendicular line on $\mathcal{L}_{s_i}(t)$ at $r_i(t)$;
21         $l \leftarrow$ the distance of the nearest robot on $\mathcal{L}_{s_i}(t)$ from $r_i(t)$;
22         $p_i(t) \leftarrow$ the point on $\mathcal{L}^*$ in the south direction and at a distance $\frac{l}{2}$ from $r_i(t)$;
23 **else**
24   $\mathcal{L}_{u_i}(t) \leftarrow$ the horizontal line through robot positions in $UP_i(t)$ having lowest $y$ coordinate;
25   $p_i(t) \leftarrow$ the nearest corner robot position on $\mathcal{L}_{u_i}(t)$;
26 robot $r_i$ moves towards $p_i(t)$;

### 5.1.2 In the Euclidean 3D Space

The extension of the problems studied in the Euclidean plane to the Euclidean three dimension space is natural. In [5], the *gathering* problem was studied in the three dimension space under *obstructed visibility* model. Two different algorithms were proposed in the $\mathcal{A}$SYNC and $\mathcal{S}$SYNC model. The algorithm in the

$\mathcal{S}$SYNC model can tolerate an arbitrary number of crash faults. The robots are represented as points in the three dimension space. The work considers an agreement in the direction and orientation of only one coordinate axis, namely $Z$-axis. Following notations are used to describe the algorithms.

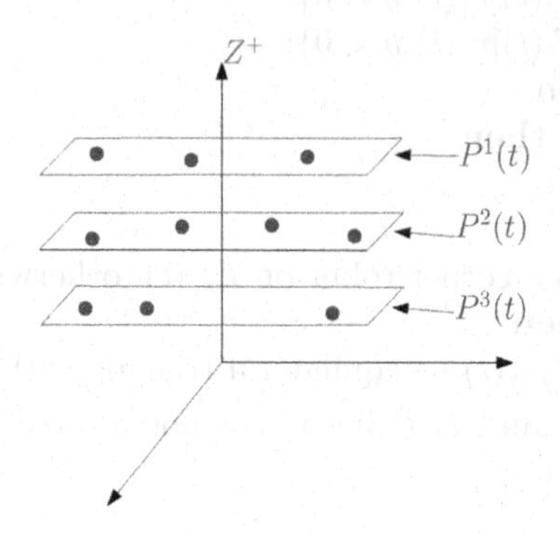


**Fig. 20.** An example of $\mathcal{P}(t)$

- Consider a robot $r_i \in \mathcal{R}$ at time $t$. Let $Z^+$ and $Z^-$ denote the positive and negative directions of $Z$ axis. Planes normal to $Z$ axis are drawn through each point in $\mathcal{R}(t)$. Let $P_i^k(t)$ and $P^k(t)$ denote the $k^{th}$ plane in the direction from $Z^+$ to $Z^-$, containing the points in $\mathcal{V}_i(t)$ and $\mathcal{R}(t)$ respectively. Let $\mathcal{N}_i(t)$ and $\mathcal{N}(t)$ denote the total number of distinct planes for $\mathcal{V}_i(t)$ and $\mathcal{R}(t)$ respectively (Fig. 20).
  Let $R_i^k(t) \subset \mathcal{V}_i(t)$ and $R^m(t) \subset \mathcal{R}(t)$ be the sets of distinct robot positions on the planes $P_i^k(t)$ and $P^k(t)$ respectively.
- Consider $P_i^k(t)$ with $|R_i^k(t)| \geq 2$. Let $S_i^k(t)$ denote the smallest enclosing circle of the robot positions in $R_i^k(t)$ and $\mathcal{O}_i^k(t)$ denote the centre of this circle. Let $SEC(t)$ denote the smallest enclosing circle of the robot positions on $P^1(t)$. Let $W_i(t)$ and $W(t)$ denote the right circular cones with $S_i^1(t)$ and $SEC(t)$ as their bases respectively, axis of the both cones are parallel to $Z$ axis, semi-vertical angles are equal to 45°, vertices $V_i(t)$ and $V(t)$ of the cones respectively are on the upward direction.
- Let $\mathcal{CH}_i^1(t)$ denote the convex hull of the robot positions in $\mathcal{R}_i^1(t)$. A robot $r_i$ lying on a vertex of $\mathcal{CH}_i^1(t)$ is called an *external* robot. All other robots are called *internal* robots. A robot on the plane $P_i^k(t)$ can see all the robot positions on $P_i^{k-1}(t)$ and $P_i^{k+1}(t)$.

- **In the $\mathcal{A}$SYNC model:**
  This section considers a set $n$ asynchronous fault-free robots. When robots are transparent and have one axis agreement, the *gathering* problem has a simple

solution: robots compute the smallest enclosing cone of all the robot positions, having semi-vertical angle 45°, axis of the cone parallel to z-axis and vertex in upward direction. All the robots in the system move straight towards the vertex of the cone. The vertex remains invariant under the straight movements of the robots towards it. Thus, gathering is achieved within finite time. This approach can tolerate arbitrary number of crash faults. However, under obstructed visibility model, this approach does not work.

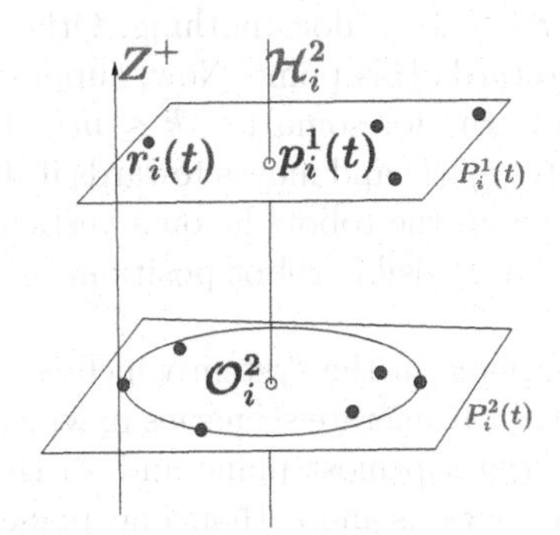


**Fig. 21.** An example of $p_i^1(t)$

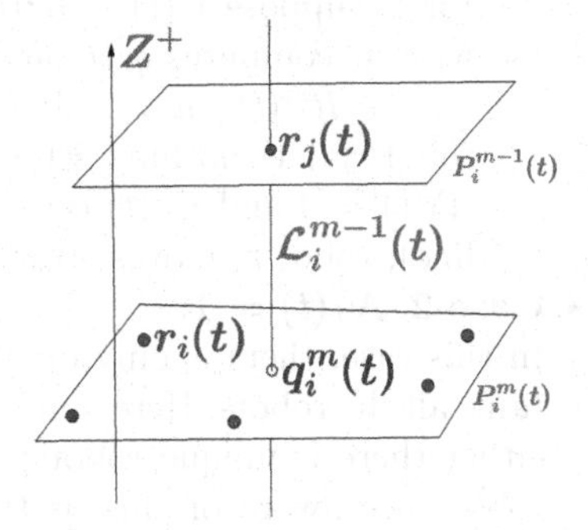


**Fig. 22.** An example of $q_i^m(t)$

Algorithm *GatherAsync3D()* provides a solution to the *gathering* problem under *obstructed visibility* model [5]. The overall idea of the algorithm is as follows: if a unique invariant point is available and all robots can agree on this point, then this point serves as the gathering point. Otherwise, robots movements are coordinated to create such point within finite time. An active robot $r_i \in \mathcal{R}$ at time $t \geq t_0$ acts according to the following strategies.

- **Case-1 $\mathcal{N}_i(t) > 1$:**
  First robots form a vertical line. Then, they start moving in the upward direction to the next visible robot position. The robots move along the vertical line. A robot moves only when it finds no other robot below it. To describe the strategies following notations are used. Let $\mathcal{H}_i^2(t)$ be the perpendicular line from the point $w$ on the plane $P_i^1(t)$, where $w$ is defined as follows: if $|R_i^1(t)| > 1$, $w = \mathcal{O}_i^2(t)$, otherwise $w$ is the unique robot position on $P_i^2(t)$. Let $\mathcal{H}_i^2(t)$ intersect $P_i^1(t)$ at $p_i^1(t)$ (Fig. 21). For a plane $P_i^m(t) \neq P_i^1(t)$, if $|R_i^{m-1}(t)| = 1$, let $\mathcal{L}_i^{m-1}(t)$ be the perpendicular line from $r_j(t) \in R_i^{m-1}(t)$ on the plane $P_i^m(t)$. Let $\mathcal{L}_i^{m-1}(t)$ intersect $P^m(t)$ at $q_i^m(t)$ (Fig. 22). All the robots on $P_i^1(t)$ and $P_i^m(t)$ (including all the robots not visible to $r_i$ but lie on these planes), can compute the points $p_i^1(t)$ and $q_i^m(t)$ respectively. For $m > 1$, we define the following predicate:

$D_i^m(t) : |R_i^{m-1}(t)| = 1 \wedge$ all the robots on $P_i^m(t)$ lie on $\mathcal{L}_i^{m-1}(t)$ i.e., at $q_i^m(t)$.

The robot $r_i$ acts according to the following:

* **Case-1.1 $r_i(t) \in R_i^1(t)$:**
  In this case $r_i$ lies on $P^1(t)$. If $|R_i^1(t)| = 1$, robot $r_i$ does nothing. Otherwise, $r_i$ computes $p_i^1(t)$ and moves straight towards it along $\overline{r_i(t)p_i^1(t)}$.
* **Case-1.2 $r_i(t) \notin R_i^1(t)$:**
  First, suppose $r_i(t) \in R_i^2(t)$. If $|R_i^1(t)| > 1$, $r_i$ does nothing. Otherwise, it computes $q_i^2(t)$ and moves towards this point. Now, suppose $r_i(t) \in R_i^m(t)$, $m > 2$. If $D_i^k(t)$ is not true for some $1 < k \leq m-1$, robot $r_i$ does nothing. Else, it computes $q_i^m(t)$ and moves towards it. If $|\mathcal{V}_i(t)| = 1$ and $D_i^2(t)$ is satisfied (when all the robots lie on a vertical line), robot $r_i$ moves straight towards the visible robot position.

- **Case-2. $\mathcal{N}_i(t) = 1$:**
  In this case, there is only one horizontal plane in the system which contains all the robots. Here, our strategies try to generate scenarios in which either there is unique robot position on the top most plane and all the robots are aware of this or the system contains more than one plane, containing robot positions. The different strategies for robot $r_i$ are as follows:
  * **Case-2.1 $|R_i^1(t)| = 1$:**
    Robot $r_i$ does nothing. In this case, gathering is achieved.
  * **Case-2.2 $|R_i^1(t)| \geq 2$:**
    - **Case-2.2.1 $r_i$ is an internal robot:**
      Let $q_i(t)$ be a point below the plane $P_i^1(t)$ at a distance $\frac{l}{2}$ from $r_i(t)$ and lies on the vertical line through $r_i(t)$, where $l$ is the distance of the nearest robot position from $r_i(t)$. Robot $r_i$ moves straight towards $q_i(t)$.
    - **Case-2.2.2 $r_i$ is an external robot:**
      If $r_i$ does not lie on the boundary of $S_i^1(t)$, it does not move. Otherwise, $r_i$ computes cone $W_i(t)$ and moves straight towards $V_i(t)$.

**Theorem 15** [5]. *The gathering problem for $n \geq 2$ asynchronous robots in three dimensional space is solvable under the obstructed visibility model when robots have an agreement in exactly one coordinate axis and they never become faulty.*

**Algorithm 9.** GatherAsync3D()

**Input:** $r_i(t) \in \mathcal{R}(t)$, $\mathcal{V}_i(t)$.
**Output:** A destination point for robot $r_i$.

1 Compute $\mathcal{P}_i(t) = (P_i^1(t), P_i^2(t), \ldots, P_i^k(t))$ ;
2 $\mathcal{N}_i(t) \leftarrow |\mathcal{P}_i(t)|$;
3 **if** $\mathcal{N}_i(t) > 1$ **then**
4  **if** $r_i(t) \in R_i^1(t)$ **then**
5   **if** $|R_i^1(t)| = 1$ **then**
6    $p_i(t) \leftarrow r_i(t)$;
7   **else**
8    **if** $|R_i^2(t)| > 1$ **then**
9     $S_i^2(t) \leftarrow$ smallest enclosing circle of the points in $R_i^2(t)$;
10     $w \leftarrow$ centre of $S_i^2(t)$;
11    **else**
12     $w \leftarrow r_j(t) \in R_i^2(t)$;
13    $\mathcal{H}_i^2 \leftarrow$ the perpendicular line from $w$ on $P_i^1(t)$;
14    $p_i(t) \leftarrow$ intersection point between $\mathcal{H}_i^2(t)$ and $P_i^1(t)$;
15  **else**
16   **if** $r_i(t) \in R_i^2(t)$ **then**
17    **if** $|R_i^1(t)| > 1$ **then**
18     $p_i(t) \leftarrow r_i(t)$ ;
19    **else**
20     $u \leftarrow r_j(t) \in R_i^1(t)$;
21     $\mathcal{L}_i^1 \leftarrow$ the perpendicular line from $u$ on $P_i^2(t)$;
22     $q_i^2(t) \leftarrow$ intersection point between $\mathcal{L}_i^1(t)$ and $P_i^2(t)$;
23     **if** $|\mathcal{V}_i(t)| == 2 \wedge r_i(t) == q_i^2(t)$ **then**
24      $p_i(t) \leftarrow u$ ;
25     **else**
26      $p_i(t) \leftarrow q_i^2(t)$ ;
27   **else**
28    $r_i(t) \in R_i^m(t)$;
29    $D_i^k(t) \leftarrow |R_i^{k-1}(t)| = 1 \wedge$ all the robots on $P_i^k(t)$ lie at $q_i^k(t)$;
30    **if** $D_i^k(t) == True\ \forall k < m$ **then**
31     $v \leftarrow r_l(t) \in R_i^{m-1}(t)$;
32     $\mathcal{L}_i^{m-1} \leftarrow$ the perpendicular line from $v$ on $P_i^m(t)$;
33     $p_i(t) \leftarrow$ intersection point between $\mathcal{L}_i^{m-1}(t)$ and $P_i^m(t)$;
34    **else**
35     $p_i(t) \leftarrow r_i(t)$ ;
36 **else**
37  **if** $|R_1^i(t)| == 1$ **then**
38   $p_i(t) \leftarrow r_i(t)$;
39  **else**
40   **if** $r_i$ *is internal* **then**
41    $p_i(t) \leftarrow r_i(t)$;
42   **else**
43    $W_i(t) \leftarrow$ right circular cone with $S_i^1(t)$ as base, axis parallel to $Z-$axis, semi-vertical angle 45°, vertex on the upward direction;
44    $p_i(t) \leftarrow$ the vertex of $W_i(t)$;
45 $r_i$ moves towards $p_i(t)$ along the line segment $\overline{r_i(t)p_i(t)}$;

- **In the $\mathcal{S}$SYNC model:**
This section presents a gathering algorithm for a set of $n$ opaque robots with no extent in the 3D space. The algorithm can tolerate an arbitrary number of crash faults. For an active robot $r_i$ at time $t$, the different solution strategies are as follows:

**Algorithm 10.** GatherSsync3D()

```
Input: r_i(t) ∈ R(t), V_i(t).
Output: A destination point for robot r_i.
1  Compute P_i(t) = (P_i^1(t), P_i^2(t), ..., P_i^k(t)) ;
2  N_i(t) ← |P_i(t)|;
3  if r_i(t) ∈ R_i^1(t) then
4      if |R_i^1(t)| == 1 then
5          p_i(t) ← r_i(t);
6      else
7          if r_i is internal then
8              L ← vertical line through r_i(t);
9              if N_i(t) == 1 then
10                 l ← min{dist(r_i(t), r_j(t)) : r_j(t) ∈ V_i(t)};
11                 p_i(t) ← the point at a distance l/2 on L in downward
                   direction;
12             else
13                 l ← distance between R_i^1(t) and R_i^2(t);
14                 p_i(t) ← the point at a distance l/2 on L in downward
                   direction;
15         else
16             W_i(t) ← right circular cone with S_i^1(t) as base, axis parallel to
               Z−axis, semi-vertical angle 45°, vertex on the upward
               direction;
17             p_i(t) ← the vertex of W_i(t);
18 else
19     r_i(t) ∈ R_i^k(t);
20     p_i(t) ← nearest external robot on P_i^{k−1}(t) from r_i(t);
21 r_i moves towards p_i(t);
```

- **Case 1: $r_i(t) \in R_i^1(t)$:** If $|R_i^1(t)| = 1$, robot $r_i$ does not change its position. Otherwise, (i) if $r_i$ is an external robot, it computes the cone $W_i(t)$, marks $V_i(t)$ as its destination point and moves straight towards $V_i(t)$ along the surface of $W_i(t)$. If $r_i$ is an internal robot, it computes a point $q_i(t)$ below the plane $P_i^1(t)$ such that the point $q_i(t)$ is at a distance $\frac{l}{2}$ from $r_i(t)$ and lies on the vertical line through $r_i(t)$, where $l$ is defined as follows: (a) if $\mathcal{N}_i(t) > 1$, $l$ is the distance between the planes $P_i^1(t)$ and $P_i^2(t)$ or (b) if $\mathcal{N}_i(t) = 1$, $l$ is the distance of the nearest robot position from $r_i(t)$.

- **Case 2: $r_i(t) \notin R_i^1(t)$:** Let $r_i$ lie on the plane $P_i^k(t)$, $k \geq 2$. Let $r_j(t)$ be an external robot position on $P_i^{k-1}(t)$ nearest to $r_i(t)$ (if tie, broken arbitrarily). Robot $r_i$ moves towards $r_j(t)$ along the line segment $\overline{r_i(t)r_j(t)}$.

**Theorem 16** [5]. *The gathering problem is solvable for a set of semi-synchronous opaque robots with no extent in the 3D space under (n, n − 1) model, when robots have one-axis agreement.*

### 5.2 The Gathering Problem for Opaque Robots with Extent

For theoretical reasons, robots are represented as points. However, in reality, even a small robot occupy some space. This has motivated the researchers to study different coordination problems for a system of autonomous robots in which each robot has some extent. Robots are represented as discs of unit radius in the Euclidean plane. This restricts two robots to occupy the same position simultaneously. Since robots have some extent they block the movements of the other robots through them. They may obstruct the vision of other robots. This setting is more realistic than point robots.

#### 5.2.1 Gathering Under the $\mathcal{A}$SYNC Model

The *gathering* problem for a swarm of opaque robots with extent is a natural extension of the problem for point robots. The *gathering* problem in this model was first studied in [10]. Since robots have some dimension and they can not share same position, the requirements of the gathering pattern is different from point robots. Instead of gathering a point, robots require to form a connected configuration in which robots touch each other on their boundaries. A *connected configuration* of the robot positions is a configuration in which between any two robot positions, there exists a polygonal line each of whose points belongs to some robot [10]. Following works under $\mathcal{A}$SYNC model assume unlimited visibility range for the robots.

**Definition 4** *(The Gathering Pattern). A set of autonomous opaque robots with extent require to coordinate their movements in such a way that within finite time they form a connected configuration and all of them are mutually visible [10].*

By the position of a robot, we mean the position of its centre and the distance between two robots are measured as the distance between their centres. Following describes two distinct gathering algorithms proposed in [10]. The first algorithm is for a set of three robots and second one is for four robots.

- **Gathering Three Opaque Robots with Extent:**
  Let $\mathcal{R} = \{r_1, r_2, r_3\}$ be the set of three opaque robots with extent. The gathering of these robots is based on the following fact.

**Fact 3.** *For every triangle $ABC$ with all angles smaller than or equal to* $120°$*, there exists a unique point $X$ inside $\triangle ABC$ such that all angles $\angle AXB$, $\angle BXC$ and $\angle CXA$ are exactly* $120°$.

---

**Algorithm 11.** Gathering3Fat()

---

**Input:** $r_i(t) \in R(t)$, $\mathcal{V}_i(t)$.
**Output:** A destination point for robot $r_i$.

```
1  K ← 1 if r_1(t), r_2(t), r_3(t) are non-collinear, otherwise 0;
2  if K == 1 then
3      D(t) ← 1 if all angles of Δ_123(t) are less than equal to 120°, otherwise
       0;
4      if D(t) == True then
5          X ← the point X inside Δ_123(t) such that
           ∠r_1(t)Xr_2(t) = ∠r_2(t)Xr_3(t) = ∠r_3(t)Xr_1(t) = 120°;
6          d_i(t) ← dist(r_i(t), X);
7          if d_i(t) ≤ 2√3/3 then
8              p_i(t) ← r_i(t);
9          else
10             p_i(t) ← the point u on r_i(t)X (segment) at a distance 2√3/3 from X;
11     else
12         B ← 1 if the angle made at r_i(t) by two other robot positions is
           greater than 120°, otherwise 0;
13         if B == 1 then
14             p_i(t) ← r_i(t);
15         else
16             r_j(t) ← the robot position at which the angle made by two
               other robot positions is greater than 120°;
17             p_i(t) ← the point u on r_i(t)r_j(t) (segment) at a distance 2 from r_j(t);
18 else
19     G ← 1 if all the robots are visible to r_i, otherwise 0;
20     if G == 1 then
21         L(t) ← the line on which all robots lie;
22         L*(t) ← the perpendicular line to L(t) at r_i(t);
23         p_i(t) ← the point u on L*(t) at a distance 1 from r_i(t) such that u
           has non-negative local x coordinate;
24     else
25         p_i(t) ← r_i(t);
26 Move to p_i(t) along the line segment r_i(t)p_i(t);
```

---

The robots act according to the following configurations:

- **Robots are non-collinear:** Consider the triangle $\Delta_{123}(t)$ formed by $r_1(t), r_2(t)$ and $r_3(t)$. A robot $r_i \in \mathcal{R}$ performs actions in following ways:

  * If robots form a triangle with all angles smaller than or equal to $120°$, $r_i$ first computes the point $X$ inside $\Delta_{123}(t)$ such that $\angle r_1(t)Xr_2(t) = \angle r_2(t)Xr_3(t) = \angle r_3(t)Xr_1(t) = 120°$. If $r_i$ is at a distance $\frac{2\sqrt{3}}{3}$ from $X$, it does not move. Otherwise, it moves towards the point $p_i(t)$ lying on $\overline{r_i(t)X}$ and at a distance $\frac{2\sqrt{3}}{3}$ from $X$.
  * Suppose the largest angle of $\Delta_{123}(t)$ lies strictly between $120°$ and $180°$. If the largest angle is created at $r_i(t)$, robot $r_i$ does not move. Otherwise, suppose the largest angle is created at $r_j(t)$. Robot $r_i$ moves towards the point $p_i(t)$ lying on $\overline{r_i(t)r_j(t)}$ and at a distance 2 from $r_j(t)$.
- **Robots are collinear:** Let all the robot positions are on the line $\mathcal{L}(t)$. If $r_i$ does not see two other robots, it does not move. Otherwise, $r_i$ moves towards the point $p_i(t)$ lying on the perpendicular line to $\mathcal{L}(t)$ at a distance 1 from $r_i(t)$.

– **Gathering Four Opaque Robots with Extent:** This section considers a set of four asynchronous opaque robots with extent and describes a gathering algorithm for them. Gathering in this case is more complex than the gathering of three robots. Since robots obstruct both movements and visions of the other robots and they work asynchronously, the gathering of four robots is more involved. Instead of giving techniques to compute destination points for the robots depending on the global configurations, nine different situations are identified depending on what they perceive about the configuration. These nine situations are a complete partition of all possible positions of the robots. For each situation, a target point is computed according to its local view. Due to asynchrony, it may happen that two different robots treat two different situations at the same time. Efforts are made to develop strategies to synchronize the behavior of the robots. However, such synchronization is not always possible and two robots may execute different procedures for two different situations at the same time. Such cases are monitored very carefully so that within finite time their movements are coordinated.

  The idea of the algorithm is as follows. In general there are two different situations. In one situation, all the robot positions form a convex *quadrilateral.* In this case, robots move along the diagonals of the quadrilateral. In other situation, three robot positions form a triangle with the fourth one inside it. Here, the robot inside the triangle does not move and rest of the robots move towards the internal robot. While dealing with these two basic situations, there are other possible situations which the system may encounter. These situations are described in the following.

  1. **gathering:** The robots form a connected configuration and they are mutually visible to each other. In this situation gathering is achieved.
  2. **four aligned:** The centres of all four robots are collinear i.e., they lie on same line segment. The robots on the two end points of the segment are external robots and the robots lying in between them are internal robots. In this situation, the external robots do not move. The internal robots move a small distance along the perpendicular directions to their

line of collinearity. This situation is converted to *quadrilateral* or *triangle* situation (described below).

3. **partial visible:** Robots form a convex quadrilateral and vision of two robots are collectively obstructed by the two other robots lying in between. The two external robots do not move. The two internal robots move in such a way that this situation is converted to the *locking* situation.
4. **locking:** Robots are mutually visible. They form a convex quadrilateral whose one diagonal has length $\sqrt{8}$ (Fig. 23(b)). The robots corresponding to the larger diagonal move towards the two other robots until they touch the other robots. This brings to the *gathering* situation.

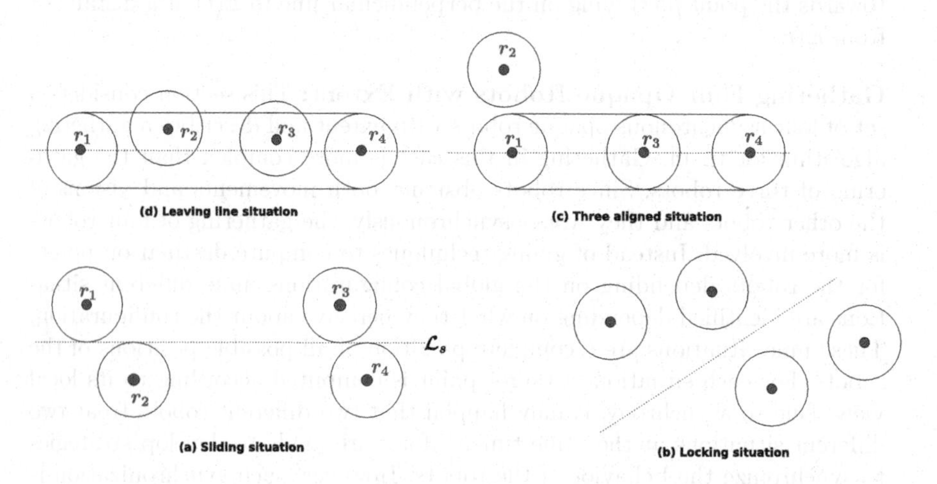


**Fig. 23.** Illustrations of different situations in the gathering of four opaque robots with extent

5. **leaving line:** None of the above four situations holds. Suppose $r_2$ and $r_3$ (internal robots) lie in between $r_1$ and $r_4$ (external robots) such that (i) $r_2$ and $r_3$ intersects the smallest width rectangular annulus containing $r_1$ and $r_4$ in more than one point (ii) the segments $\overline{r_1(t)r_4(t)}$ and $\overline{r_2(t)r_3(t)}$ are not perpendicular to each other (Fig. 23(d)). Without loss of generality suppose that the segment $\overline{r_1(t)r_4(t)}$ is horizontal. The distance between each external robot from the vertical projection of each internal robots on $\overline{r_1(t)r_4(t)}$ is at least 2. The external robots do not move. The internal robots move without obstructing the other robots to reach one of the following situations: *quadrilateral, three aligned* or *locking*.
6. **three aligned:** Exactly three robots are collinear and the robots are not in leaving line situation (Fig. 23(c)). The robot which is not collinear with the three other robots, moves to change this situation into *locking* situation.

7. **sliding:** Robots are mutual visible to each other. There is a common tangent $\mathcal{L}_s$ to the bounding circles of the robots such that (i) there are two pairs of robots and each pair of robots are separated by $\mathcal{L}_s$ and (ii) in each pair, the distance of tangency points with $\mathcal{L}_s$ is at most $\frac{1}{3}$ (Fig. 23(a)). Each pair of robots moves towards the other pair by maintaining the tangency with $\mathcal{L}_s$. This situation is converted to the *locking* situation.
8. **quadrilateral:** None of the above situations holds. Robots form a quadrilateral in which all robots are mutually visible. The robots move along the diagonals of the quadrilateral until they form a rectangle consisting of two symmetric pairs of tangents to the same line $\mathcal{L}$ and tangent to the other robot in the pair. Different actions are taken by the robots depending on the perpendicular and non-perpendicular diagonals. This situation reaches any one of the following: *sliding, locking* or *gathering* situation.
9. **triangle:** Robots are not in living line situation. Three robots form a triangle and the centre of the fourth robot lies inside this triangle. In this situation robots move to gathering around the internal robot. The robots at the vertices of the triangle move towards the internal robot until they touch it. The internal robot does not move.

The correctness proof is quite involved. The basic idea of the correctness proof is that the transition diagram of different situations, shown in Fig. 24, is acyclic and has a unique sink at the gathering situation.

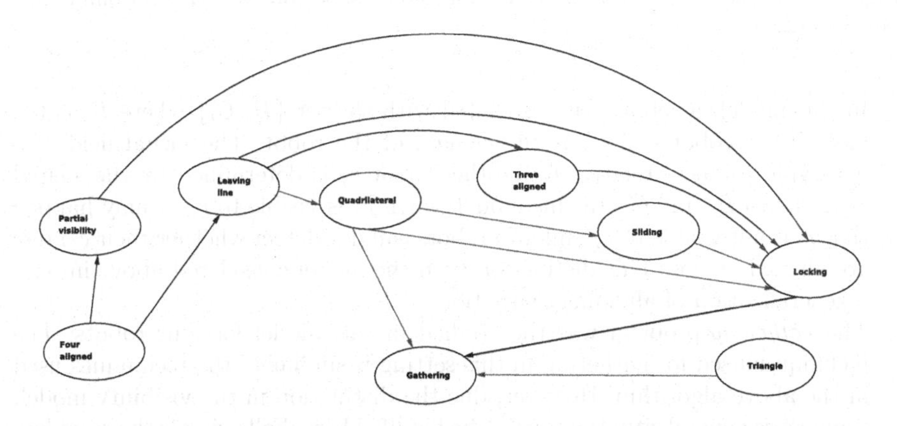


**Fig. 24.** Transition diagram between situations 1–9 and arrows show possible transitions between them in the gathering of four opaque robots with extent

**Theorem 17** [10]. *The gathering of $n \leq 4$ asynchronous opaque robots with extent is possible without any extra assumption.*

– **Gathering with Slim Omni-directional Cameras:** In the above algorithm, it is assumed that two robots are visible to each other if there exists a line segment which intersects both robots at more than one point and this line does not intersect any other robot in between. In practice, to implement this vision sensor model, robots must have sensors around their whole perimeters. This is costly, heavy and computation intensive. One of the ways to overcome this problem is to install a single camera (possibly with several mirrors) inside the robots [19]. This is cost efficient and light weighted. This vision sensor model is significantly less powerful. Each robot is represented by a disc of radius $l$ and each of them have a slim omni-directional camera that is represented as a disc of radius $l_c < l$ (Fig. 25).

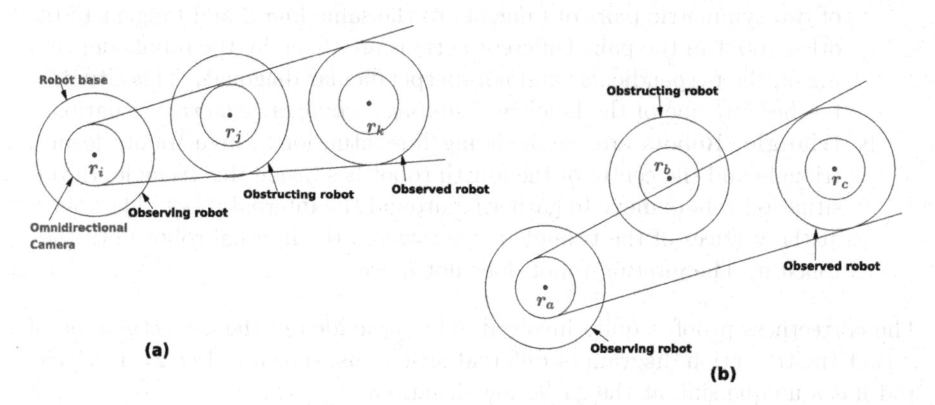


**Fig. 25.** An illustration showing visible and non-visible robots with slim omnidirectional cameras

In this model, a robot $r_i$ is associated with the set $\{B_i, C_i\}$ where $B_i$ is the base of the robot and $C_i$ is the camera of the robot. The visual field of a observing robot $r_i$ relative to another robot $r_j$ is determined by the shared tangents to $B_j$ and $C_i$. In this model, when $r_j$ is visible to $r_i$, it may happen that $r_i$ is not visible to $r_j$ and also robots can not detect whether their centers are aligned or not. This distinction with the model considered above implies a generalization of alignment situations.
The *gathering* problem was the studied in this model for four robots. The techniques used for gathering in this setting is similar to the techniques used in the above algorithm. However, due the distinction in the visibility model, some more general situations are to be handled here. Following is the complete list of different situations of robot configurations: (1) *gathering* (2) *partial visible* (3) *general partial visible* (4) *locking* (5) *leaving line* (6) *sliding* (7) *quadrilateral* and (8) *triangle*. The situations (1) and (4)–(8) are same as defined in the above algorithm and slight variations of the corresponding solutions work for these situation. Situations (2) and (3) are different here and corresponding actions of the robots are as follows:

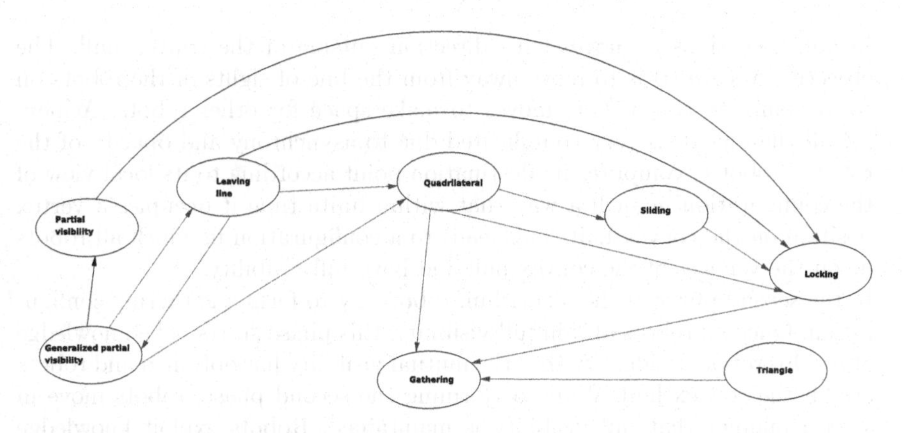


**Fig. 26.** Transition diagram between situations 1–8 and arrows show possible transitions between them

**(2) partial visible:** Robots form a convex quadrilateral and vision of two robots are obstructed by at least one robot lying in between. The two external robots does not move. The two internal robots move in such a way that this situation is converted to the *locking* situation.
**(3) general partial visible:** Robots are not in partial visible situation. At least one robot has partial visibility. In this situation, the two external robots do not move. The two internal robots move to recover the full visibility. The movements of the two internal robots convert this situation to any one of the situations: *leaving line* or *quadrilateral* or *partial visibility* situation.

The transitions between different situations are shown in the Fig. 26.

**Theorem 18** [19]. *The gathering problem is solvable for $n \leq 4$ asynchronous opaque robots with extent when they have slim omnidirectional cameras.*

- **Gathering more than Four Opaque Robots with Extent:** The work in [1], the gathering of $n \geq 5$ asynchronous opaque robots with extent was studied. The robots have unlimited visibility range. The robots have common *chirality* and *non-rigid* movements. Robots know the value of $n$. Following distributed algorithm solves the *gathering* problem under this model.
  The overall basic idea of the gathering algorithm is as follows. The algorithm has two phases. In first phase, robots obtain mutual visibility among themselves. Since the robots are asynchronous and non-transparent, this task is challenging. This challenge is overcome by bringing the robots at the vertices of a convex hull. In order to do so, the robots that are on the convex hull do not move inside and robots that are inside the convex hull move on the convex hull. A robot $r_i \in \mathcal{R}$ on the boundary of the convex hull moves in two situations: (i) $r_i$ finds that it obstructs vision of the other robots on the convex hull (ii) $r_i$ realizes that there is not "enough space" for robots that are inside the convex hull to be placed on the boundary of the convex hull.

In both occasions $r_i$ moves with direction outside of the convex hull. The objective in case (i) is to move away from the line of sights of the robots on convex hull. In case (ii), $r_i$ moves to make space for other robots. Achieving all these goals is very complicated due to asynchrony and opacity of the robots. Robot $r_i$ computes its destination point according to its local view of the configuration in such a way that within finite time it occupies a vertex position on the convex hull. This leads to a configuration in which all robots lie on the vertices of the convex hull and have full visibility.
In the second phase of the algorithm, robots try to form a gathering configuration. Once all robots obtain full visibility, this phase starts. The knowledge of $n$ helps robots to identify that the mutual visibility has obtained and robots are on the convex hull. While performing the second phase, robots move in such a manner that full visibility is maintained. Robots exploit knowledge of $n$ and the common unit distance to maintain mutual visibility. However, asynchrony again makes things complicated.
Performing above two phases, robots eventually form a connected configuration and terminate. This provides a solution to the *gathering* problem. The correctness of the above algorithm is proved using a state-machine representation of the model.

**Theorem 19** [1]. *The gathering problem is solvable for a set of $n \geq 2$ opaque robots with extent under $\mathcal{A}$SYNC model, when they have common chirality.*

### 5.2.2 Gathering Under the $\mathcal{F}$SYNC Model

The work in [6] considered a combination of the *limited* and *obstructed visibility* model. The work has presented strategies to gather a set of $n$ oblivious, synchronous opaque robots with extent. The robots are non-transparent and have limited visibility range $V$. The robots do not have any global coordinate system. Since robots have limited visibility range and $n \geq 2$, it may not be possible to obtain mutual visibility among the robots. Definition of a gathering configuration is weaker than before: the robots are in gathering configuration if they form a connected configuration (as defined in above section). It is not necessary that robots are mutually visible to each other in the gathering configuration. The set of robots are divided into two sub-groups in each round: *perimeter robots* and *inside robots*. A robot $r_i$ is a *perimeter robot* if all robots in $\mathcal{V}_i(t)$ lie in a sector centred at $r_i(t)$ with sector angle at most 120° (Fig. 27). The set of perimeter robots at time $t$ is denoted by $\mathcal{RP}(t)$. The other robots in $\mathcal{R} \backslash \mathcal{RP}(t)$ are *inside robots* and the set of *inside robots* is denoted by $\mathcal{RI}(t)$.

The overall idea of the gathering algorithm is as follows. The perimeter robots move towards the inside robots. However, this can lead to flocking of the robots at the boundary of the visibility graph. To tackle this problem, inside robots try to move towards furthest visible robot positions. In order to execute these ideas, destination points of the robots are to be computed in such a way that the mutually visible robots remain visible in each round and the initial configuration reaches a gathering configuration within finite time. Consider a robot $r_i \in \mathcal{R}$.

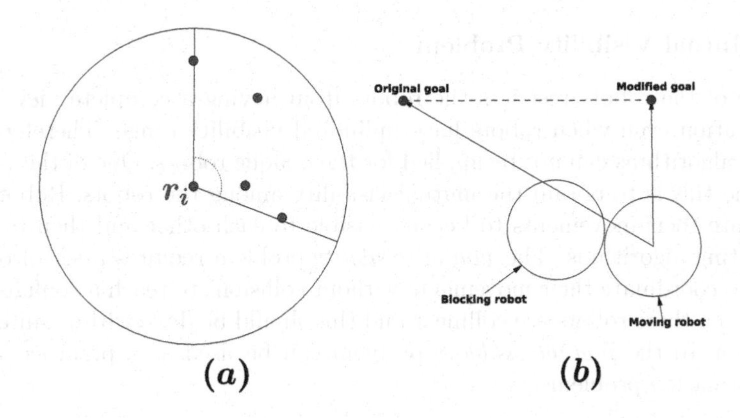


**Fig. 27.** An example of (a) perimeter robot (b) method Slip

Let $r_{d_i}(t)$ be one of the furthest robot positions in $\mathcal{V}_i(t)$. A point $g(t)$ is defined as follows:

$$g(t) = c\frac{r_{d_i}(t) - r_i(t)}{V}(r_{d_i}(t) - r_i(t)),$$

where $c = 1$ if $r_i \in \mathcal{RP}(t)$, otherwise $c = \frac{1}{2}$.

The direction of movement of $r_i$ is along $\overline{g(t)r_i(t)}$. To maintain connectivity with the visible robots, $r_i$ uses same distance limitation strategy as use by Ando et al. in [2], discussed in Sect. 4.1.1. Robot $r_i$ moves towards $g(t)$ by a distance $m = min\{|g(t)|, limit\}$, where $limit$ is defined in the algorithm *CONVERGE-*$\mathcal{S}$SYNC*()* in Sect. 4.1.1.

Since robots have extent, they block the movements of the other robots through them. This may lead to deadlock situations. However, to overcome these situations robots used a method called *Slip* as defined below. The blocked robot modifies its original direction of movements so that the new direction is tangential to the blocker robot (Fig. 27). The new destination point is the projection of the old destination point on the new direction of movement. So, the robot can get closer to its original destination point. The method *Slip* is used only when the number of blocker robot is one. If the number of blockers is zero or more than one, then nothing is to be done with the original destination.

To demonstrate the correctness of the above algorithm, computer simulations in MATLAB were done. The above algorithm is compared with the following two algorithms: (i) the algorithm proposed by Ando et al. [2] for point like robots is applied to the opaque robots with extent and (ii) the modified version of Ando et al.'s algorithm with the method *Slip* to avoid the blocking effect. It was found that in the test cases the algorithm described here performs better than the two other algorithms.

The study in [8] also presented a distributed algorithm for the opaque robots with extent under limited visibility and provided simulation based results to verify the proposed algorithm. In this work, the gathering point is known to the robots a priori.

### 5.3 Mutual Visibility Problem

Opacity of the robots restricts the robots from having a complete view of the configuration even when robots have unlimited visibility range. Therefore, traditional algorithms can not be applied for the opaque robots. One of the ways to overcome this is to obtain the mutual visibility among the robots. Robots first coordinate their movements to become visible to each other and then to apply the existing algorithms. The *mutual visibility* problem requires a set of opaque robots to coordinate their movements without collisions to reach a configuration in which no three robots are collinear and this should be done within finite time. A solution to the *mutual visibility* problem can be used as a preprocessing of other formation problems.

#### 5.3.1 Mutual Visibility Under the $\mathcal{S}$SYNC Model

The work in [12] was the first to address the *mutual visibility* problem for a set of $n$ opaque robots with no extent and unlimited visibility range under the $\mathcal{S}$SYNC model. In the initial configuration robots occupy distinct locations in the Euclidean plane. The robots know the value of $n$. The basic idea of the algorithm is to move the robots in such a way that within finite time they occupy distinct vertices of convex $n$-gone, without colliding with each other. When this is done each robot is visible to the others and they terminate the algorithm ($n$ is known to the robots).

The description of the algorithm is as follows. Consider the convex hull $\mathcal{CH}(t)$ of the robots locations at time $t$. The robots lying on the boundary of $\mathcal{CH}(t)$ are called *external* robots and the robots lying in its interior are the *internal* robots. A robot can identify whether it is an external or an internal robot. The basic strategy of the algorithm is to move the external robots inside $\mathcal{CH}(t)$ to shrink the convex hull. This process makes an internal robot external and it also moves on activation. Continuing this process, within finite time, all the robots become external. Once this situation is achieved, all the robots can see each other. The knowledge of $n$ helps the robots to identify this situation and they can terminate. The destination points of the external robots are computed in the following way. An external robot $r_i$ locates clockwise and counter-clockwise neighbors on $\mathcal{CH}(t)$. Let these two robot positions be $r_c(t)$ and $r_{cw}(t)$. These two positions are visible to $r_i$. The destination point of $r_i$ lies in triangle $\Delta r_c(t) r_i(t) r_{cw}(t)$. This movement shrinks $\mathcal{CH}(t)$ and may make some internal robots external. The destination point of $r_i$ lies in a smaller triangle, shaded in grey in Fig. 28. This helps the robots to avoid collisions with other moving robots and also not to become an internal robot. *MUTUAL-$\mathcal{S}$*SYNC*()* is the pseudo-code of the algorithm.

**Algorithm 12.** MUTUAL-$\mathcal{S}$SYNC()

**Input**: $r_i(t) \in \mathcal{R}(t), \mathcal{V}_i(t), n$

**Output**: A destination point for robot $r_i$.

1 $\mathcal{CH}_i(t) \leftarrow$ the convex hull of $\mathcal{V}_i(t)$;
2 $\mathcal{H}_i^+(t) \leftarrow$ the set of all robot positions on $\mathcal{CH}_i(t)$;
3 $\mathcal{H}_i^*(t) \leftarrow$ the set of non-degenerate vertices of $\mathcal{CH}_i(t)$;
4 **if** $|\mathcal{H}_i(t)| = n$ **then**
5    $p_i(t) \leftarrow r_i(t)$
6 **else**
7    **if** $|\mathcal{V}_i(t)| = 1$ **then**
8      $a \leftarrow r_j(t) \in \mathcal{V}_i(t)$;
9      $\mathcal{L} \leftarrow$ a direction perpendicular to $\overline{r_i(t)a}$;
10      $p_i(t) \leftarrow$ a point on $\mathcal{L}$ at a distance $\frac{1}{2}|\overline{r_i(t)a}|$;
11    **else**
12      **if** $r_i(t) \in \mathcal{H}_i^+(t)$ **then**
13        $r_c(t) \leftarrow$ clockwise neighbor of $r_i(t)$ in $\mathcal{H}_i^+(t)$;
14        $r_{cw}(t) \leftarrow$ counter-clockwise neighbor of $r_i(t)$ in $\mathcal{H}_i^+(t)$;
15        $\gamma \leftarrow \frac{1}{2}$;
16        **if** $r_i(t) \notin \overline{r_c(t)r_{cw}(t)}$ **then**
17          **for** *each* $r_j(t) \in \mathcal{V}_i(t) \setminus \{r_i(t)\}$ **do**
18            Let $\alpha, \beta$ be such that $r_j(t) = \alpha \cdot r_c(t) + \beta \cdot r_{cw}(t)$;
19            **if** $\alpha + \beta < \gamma$ **then**
20              $\gamma \leftarrow \alpha + \beta$
21        $u \leftarrow \gamma \cdot \frac{(2r_c(t)+r_{cw}(t))}{3}$;
22        $v \leftarrow \gamma \cdot \frac{(r_c(t)+2r_{cw}(t))}{3}$;
23        $w \leftarrow$ midpoint of $\overline{uv}$;
24        **if** $w \notin \mathcal{R}(t)$ **then**
25          $p_i(t) \leftarrow w$
26        **else**
27          $r_k(t) \leftarrow$ nearest robot position from $u$ on $\overline{uv}$;
28          $p_i(t) \leftarrow$ midpoint of $\overline{ur_k(t)}$;
29      **else**
30        $e \leftarrow$ nearest edge of $\mathcal{CH}_i(t)$ from $r_i(t)$;
31        $p_i(t) \leftarrow$ midpoint of $e$;
32 $r_i$ moves towards $p_i(t)$;

The correctness of the above algorithm depends on the following invariant: $\mathcal{CH}(t+1) \subset \mathcal{CH}(t)$, $t \geq t_0$. The above algorithm can tolerate a single crash fault. Algorithm *MUTUAL-S*SYNC*()* also provides solutions for the *convex hull formation* problem and the *near gathering* problem under the *obstructed visibility* model.

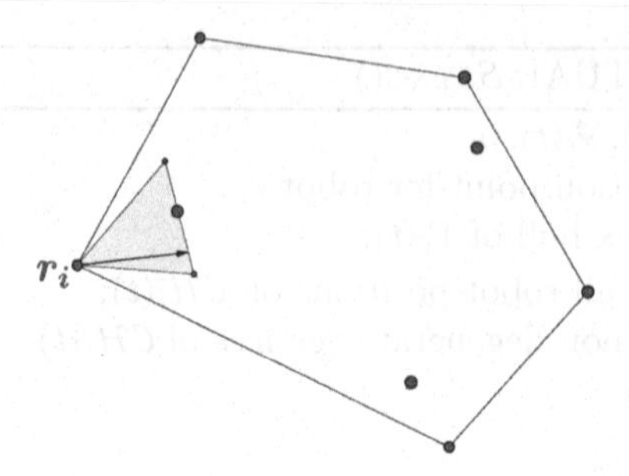


**Fig. 28.** An example of movement of an external robot $r_i$

**Theorem 20** [12]. *The mutual visibility problem is solvable under the* $\mathcal{S}$SYNC *model when robots know n, the total number of robots in the system.*

### 5.3.2 Mutual Visibility Under the $\mathcal{A}$SYNC Model

The *mutual visibility* problem is solvable in $\mathcal{A}$SYNC when robots have one-axis agreement and they know the value of $n$ [4]. It is also assumed that robots have unlimited visibility range and initially no two robots occupy the same position.

Let $\widetilde{C}_L$ denote the set of all robot configurations in which all robots lie on a straight line and $\widetilde{C}_{NL}$ the set of all robot configurations which contain at least three non-collinear robot positions. Let $\widetilde{C}_{GP}$ denote the set of all configurations in which no three robots are collinear. A robot $r_i$ is called an *non-terminal robot* if it lies between two other robot positions on a line of collinearity. Otherwise, $r_i$ is called a *terminal* robot. The straight line joining $r_i(t)$ and $r_j(j)$ is denoted by $\mathcal{L}_{ij}(t)$. The perpendicular distance of the line $\mathcal{L}_{ij}(t)$ from the point $r_k(t)$ is denoted by $d^k_{ij}(t)$. Let $D_i(t)$ be minimum distance of any two robots in $\mathcal{V}_i(t)$.

Consider $\mathcal{R}(t_0)$, the initial robot configuration. If $\mathcal{R}(t_0)$ contains at least three collinear robot positions, the movements of the robots are planned in such a way that after a finite number of movements there are no three collinear robots in the system. If $\mathcal{R}(t_0)$ is in $\widetilde{C}_L$, it is converted into a configuration in $\widetilde{C}_{NL}$. The different strategies are as follows.

(A) **Eligible robots for movements:** A terminal robot $r_i$ is eligible for movement at time $t$ only if it satisfies any one of the following three conditions:
- $|\mathcal{V}_i(t)| < n-1$ and $Y$-axis of $r_i$ does not contain any other robot position on it.
- $|\mathcal{V}_i(t)| < n-1$, $Y$-axis of $r_i$ contains at least one robot position and $r_i(t)$ has the highest $y$-coordinate value among all the robot positions on its $Y$-axis.
- $|\mathcal{V}_i(t)| = n-1$, $Y$-axis of $r_i$ contains at least one robot position and $r_i(t)$ has the highest $y$-coordinate value among all the robot positions on its $Y$-axis. This is essential to avoid deadlocks in the system.

The non-terminal robots do not move.

(B) **Computing Destination Point:** While computing destination point, robot $r_i$ takes care of the followings: (i) its movement does not block the visibility of the other robots (ii) it does not collide with others. The destination point of $r_i$ depends on whether $\mathcal{R}(t)$ is in $\widetilde{C}_L$ or in $\widetilde{C}_{NL}$. To compute destination point, two components are computed: (i) the amount of movement and (ii) the direction of movement.

- **Case-1:** $\mathbf{\mathcal{R}(t) \in \widetilde{C}_{NL}}$. Three non-collinear robots become collinear if the triangle, formed by their respective positions, collapses into a line due to their movements. Thus, destination points of the robots are computed in such a way that this never happens.
  **The direction of movement:** Let $\theta_{ij}(t)$ denote the angle made by $\mathcal{L}_{ij}(t)$ with $Y^+$. Let $\alpha_i(t) = minimum\{\theta_{ij}(t) : r_j \in \mathcal{V}_i(t)$ and $r_j$ does not lie on $Y$-axis of $r_i\}$ (tie, if any, is broken arbitrarily). Let $Bisec_i(t)$ denote the $\frac{1}{n}$th bisector of $\alpha_i(t)$ closest to $Y^+$. It is a ray from $r_i(t)$ and the angle made by it with $Y^+$ is strictly less than $\frac{\pi}{2}$ since $n \geq 3$. Robot $r_i$ moves along $Bisec_i(t)$.
  **The amount of displacement:** The maximum amount of displacement of $r_i$ should be restricted in such a way that it does not create any new collinearity during or after the execution of its movement. If $r_j(t)$ and $r_k(t)$ are in $\mathcal{V}_i(t)$, the destination point of $r_i$ should lie far enough from $\mathcal{L}_{jk}(t)$ so that even if all the three robots $r_i$, $r_j$ and $r_k$ move, they do not become collinear. Let $d_i(t) = minimum\{d^k_{ij}(t), d^j_{ik}(t), d^i_{jk}(t) : \forall r_j(t), r_k(t) \in \mathcal{V}_i(t)\}$. Let

$$\sigma_i(t) = \frac{1}{n^4} minimum\{d_i(t), D_i(t)\}$$

- **Case-2:** $\mathbf{\mathcal{R}(t) \in \widetilde{C}_L}$. In this case, all the robots lie on a straight line, say $\hat{\mathcal{L}}$. There are $n-2$ non-terminal robots. The movement of even one terminal robot converts the present configuration into a configuration in $\widetilde{C}_{NL}$. Suppose $r_i$ is one of the two terminal robots on $\hat{\mathcal{L}}$.
  **The direction of movement:** If there is no other robot position on the $Y$-axis of $r_i$, robot $r_i$ moves along $Y^+$. Otherwise, $r_i$ moves along $X^+$.
  **The amount of displacement:** In this case $|\mathcal{V}_i(t)| = 1$. Let

$$\sigma_i(t) = \frac{1}{2} D_i(t).$$

Let $p_i(t)$ be the point on the direction of movement of $r_i$ (i.e, on $Y^+$ or $Bisec_i(t)$ or $X^+$) at distance $\sigma_i(t)$ from $r_i(t)$. The destination point of $r_i(t)$ is $p_i(t)$.

**Algorithm 13.** ComputeDestination()

```
Input: r_i(t) ∈ R(t) and V_i(t).
Output: A destination point of r_i
1  D_i(t) ← minimum{|r_j(t)r_k(t)| : ∀r_j, r_k ∈ {r_i(t), V_i(t)}};
2  for ∀r_j(t), r_k(t) ∈ V_i(t) do
3      L_ij(t) ← the straight line through r_i(t) and r_j(t);
4      d^k_ij(t) ← perpendicular distance of L_ij(t) from r_k(t);
5  if |V_i(t)| ≥ 2 then
6      d_i(t) ← minimum{d^k_ij(t), d^j_ik(t), d^i_jk(t) : ∀r_j, r_k ∈ V_i(t)};
7      U ← minimum{d_i(t), D_i(t)};
8      σ_i(t) ← (1/n^4) U;
9      for ∀r_j(t) ∈ V_i(t) do
10         θ_ij(t) ← the angle between L_ij(t) and Y+;
11     α_i(t) ← minimum{θ_ij(t) : r_j ∈ V_i(t)};
12     Bisec_i(t) ← Bisector of α_i(t);
13     p_i(t) ← the point on Bisec_i(t) at a distance σ_i(t) from r_i(t);
14 else
15     σ_i(t) ← (1/2) D_i(t);
16     L̂ ← the line containing all the robots in R(t);
17     if L̂ is coincident with Y axis then
18         DIR_i(t) ← X+
19     else
20         DIR_i(t) ← Y+
21     p_i(t) ← the point on DIR_i(t) at a distance σ_i(t) from r_i(t);
22 return p_i(t);
```

**Algorithm 14.** MUTUAL-$\mathcal{A}$SYNC()

```
Input: V_i(t).
Output: Robots in general position.
1  while |V_i(t)| < n − 1 ∨ Y-axis contains other robots do
2      if r_i is non-terminal then
3          p_i(t) ← r_i(t);
4      else
5          if Y-axis contains no other robots then
6              p ← ComputeDestination(r_i(t), V_i(t)) ;
7          else
8              if r_i(t) is on the top of its own Y-axis then
9                  p_i(t) ← ComputeDestination(r_i(t), V_i(t)) ;
10             else
11                 p_i(t) ← r_i(t);
12     r_i(t) moves towards p_i(t);
```

**Theorem 21** [4]. *The mutual visibility problem is solvable under* $\mathcal{A}$SYNC *model when robots have agreement on one local coordinate axis and they know* $n$*, the total number of robots in the system.*

## 6 Conclusions

The restricted visibility model depicts more realistic setting. This chapter presents a study on the existing results for different geometric formation problem under the *restricted visibility* model. The study reveals many open questions which may be addressed in the future.

In all the existing works with limited visibility, it is assumed that robots never develop faults at any stage of the execution. However, in real scenarios, faults may occur during the execution. When robots have limited visibility range, it will be interesting to define appropriate fault models and to extend the existing results in the new model.

Recently, researchers have considered the three dimensional Euclidean space as the deployment area of the robots. Very limited works have been done in this model. The extensions of different formation problems to the 3D space is natural and challenging research task.

The use of visible lights to overcome different setbacks, during the designs of the algorithms, has recently got attention. The *mutual visibility* problem has been extensively studied under this model when robots have unlimited visibility. One possible direction of future research can be the study of different formation problems for the robots with visible lights under limited visibility model. Using visible lights, more efficient algorithms in terms of time and other parameters can be designed.

## References

1. Agathangelou, C., Georgiou, C., Mavronicolas, M.: A distributed algorithm for gathering many fat mobile robots in the plane. In: Proceedings of the ACM Symposium on Principles of Distributed Computing, PODC, pp. 250–259 (2013)
2. Ando, H., Oasa, Y., Suzuki, I., Yamashita, M.: Distributed memoryless point convergence algorithm for mobile robots with limited visibility. IEEE Trans. Robot. Autom. **15**(5), 818–828 (1999)
3. Bhagat, S., Chaudhuri, S.G., Mukhopadhyaya, K.: Fault-tolerant gathering of asynchronous oblivious mobile robots under one-axis agreement. J. Discret. Algorithms **36**, 50–62 (2016)
4. Bhagat, S., Chaudhuri, S.G., Mukhopadhyaya, K.: Formation of general position by asynchronous mobile robots under one-axis agreement. In: Kaykobad, M., Petreschi, R. (eds.) WALCOM 2016. LNCS, vol. 9627, pp. 80–91. Springer, Cham (2016). https://doi.org/10.1007/978-3-319-30139-6_7
5. Bhagat, S., Chaudhuri, S.G., Mukhopadhyaya, K.: Gathering of opaque robots in 3D space, pp. 2:1–2:10 (2018)

6. Bolla, K., Kovacs, T., Fazekas, G.: Gathering of fat robots with limited visibility and without global navigation. In: Rutkowski, L., Korytkowski, M., Scherer, R., Tadeusiewicz, R., Zadeh, L.A., Zurada, J.M. (eds.) EC/SIDE -2012. LNCS, vol. 7269, pp. 30–38. Springer, Heidelberg (2012). https://doi.org/10.1007/978-3-642-29353-5_4
7. Cohen, R., Peleg, D.: Convergence of autonomous mobile robots with inaccurate sensors and movements. SIAM J. Comput. **38**(1), 276–302 (2008)
8. Cord-Landwehr, A., et al.: Collisionless gathering of robots with an extent. In: Černá, I., et al. (eds.) SOFSEM 2011. LNCS, vol. 6543, pp. 178–189. Springer, Heidelberg (2011). https://doi.org/10.1007/978-3-642-18381-2_15
9. Cord-Landwehr, A., Fischer, M., Jung, D., Meyer auf der Heide, F.: Asymptotically optimal gathering on a grid. In: Proceedings of the 28th ACM Symposium on Parallelism in Algorithms and Architectures, SPAA 2016, pp. 301–312 (2016)
10. Czyzowicz, J., Gasieniec, L., Pelc, A.: Gathering few fat mobile robots in the plane. Theoret. Comput. Sci. **410**(6–7), 481–499 (2009)
11. Degener, B., Kempkes, B., Langner, T., Meyer auf der Heide, F., Pietrzyk, P., Wattenhofer, R.: A tight runtime bound for synchronous gathering of autonomous robots with limited visibility. In: Proceedings of the 23rd Annual ACM Symposium on Parallelism in Algorithms and Architectures, SPAA, pp. 139–148 (2011)
12. Di Luna, G.A., Flocchini, P., Poloni, F., Santoro, N., Viglietta, G.: The mutual visibility problem for oblivious robots (2014)
13. Di Stefano, G., Navarra, A.: Optimal gathering on infinite grids. In: Felber, P., Garg, V. (eds.) SSS 2014. LNCS, vol. 8756, pp. 211–225. Springer, Cham (2014). https://doi.org/10.1007/978-3-319-11764-5_15
14. Flocchini, P., Prencipe, G., Santoro, N.: Distributed Computing by Oblivious Mobile Robots. Synthesis Lectures on Distributed Computing Theory. Morgan & Claypool Publishers, San Rafael (2012)
15. Flocchini, P., Prencipe, G., Santoro, N., Widmayer, P.: Gathering of asynchronous robots with limited visibility. Theoret. Comput. Sci. **337**(1–3), 147–168 (2005)
16. Flocchini, P., Prencipe, G., Santoro, N., Widmayer, P.: Arbitrary pattern formation by asynchronous, anonymous, oblivious robots. Theoret. Comput. Sci. **407**(1–3), 412–447 (2008)
17. Gordon, N., Elor, Y., Bruckstein, A.M.: Gathering multiple robotic agents with crude distance sensing capabilities. In: Dorigo, M., Birattari, M., Blum, C., Clerc, M., Stützle, T., Winfield, A.F.T. (eds.) ANTS 2008. LNCS, vol. 5217, pp. 72–83. Springer, Heidelberg (2008). https://doi.org/10.1007/978-3-540-87527-7_7
18. Gordon, N., Wagner, I.A., Bruckstein, A.M.: Gathering multiple robotic a(ge)nts with limited sensing capabilities. In: Dorigo, M., Birattari, M., Blum, C., Gambardella, L.M., Mondada, F., Stützle, T. (eds.) ANTS 2004. LNCS, vol. 3172, pp. 142–153. Springer, Heidelberg (2004). https://doi.org/10.1007/978-3-540-28646-2_13
19. Honorat, A., Potop-Butucaru, M.G., Tixeuil, S.: Gathering fat mobile robots with slim omnidirectional cameras. Theoret. Comput. Sci. **557**, 1–27 (2014)
20. Izumi, T., Kawabata, Y., Kitamura, N.: Toward time-optimal gathering for limited visibility model (2015). https://sites.google.com/site/micromacfrance/abstract-tasuke
21. Izumi, T., Potop-Butucaru, M.G., Tixeuil, S.: Connectivity-preserving scattering of mobile robots with limited visibility. In: Dolev, S., Cobb, J., Fischer, M., Yung, M. (eds.) SSS 2010. LNCS, vol. 6366, pp. 319–331. Springer, Heidelberg (2010). https://doi.org/10.1007/978-3-642-16023-3_27

22. Katreniak, B.: Convergence with limited visibility by asynchronous mobile robots. In: Kosowski, A., Yamashita, M. (eds.) SIROCCO 2011. LNCS, vol. 6796, pp. 125–137. Springer, Heidelberg (2011). https://doi.org/10.1007/978-3-642-22212-2_12
23. Kempkes, B., Kling, P., Meyer auf der Heide, F.: Optimal and competitive runtime bounds for continuous, local gathering of mobile robots. In: Proceedings of the 24th ACM Symposium on Parallelism in Algorithms and Architectures, SPAA, pp. 18–26 (2012)
24. Martínez, S.: Practical multiagent rendezvous through modified circumcenter algorithms. Automatica **45**(9), 2010–2017 (2009)
25. Pagli, L., Prencipe, G., Viglietta, G.: Getting close without touching: near-gathering for autonomous mobile robots. Distrib. Comput. **28**(5), 333–349 (2015)
26. Poudel, P., Sharma, G.: Universally optimal gathering under limited visibility. In: Spirakis, P., Tsigas, P. (eds.) SSS 2017. LNCS, vol. 10616, pp. 323–340. Springer, Cham (2017). https://doi.org/10.1007/978-3-319-69084-1_23
27. Prencipe, G.: Impossibility of gathering by a set of autonomous mobile robots. Theoret. Comput. Sci. **384**(2–3), 222–231 (2007)
28. Souissi, S., Défago, X., Yamashita, M.: Using eventually consistent compasses to gather memory-less mobile robots with limited visibility. ACM Trans. Auton. Adapt. Syst. **4**(1), 9:1–9:27 (2009)
29. Suzuki, I., Yamashita, M.: Distributed anonymous mobile robots: formation of geometric patterns. SIAM J. Comput. **28**, 1347–1363 (1999)
30. Yamauchi, Y., Yamashita, M.: Pattern formation by mobile robots with limited visibility. In: Moscibroda, T., Rescigno, A.A. (eds.) SIROCCO 2013. LNCS, vol. 8179, pp. 201–212. Springer, Cham (2013). https://doi.org/10.1007/978-3-319-03578-9_17

# Asynchronous Robots on Graphs: Gathering

Serafino Cicerone[1], Gabriele Di Stefano[1], and Alfredo Navarra[2(✉)]

[1] Dipartimento di Ingegneria e Scienze dell'Informazione e Matematica, Università degli Studi dell'Aquila, L'Aquila, Italy
{serafino.cicerone,gabriele.distefano}@univaq.it

[2] Dipartimento di Matematica e Informatica, Università degli Studi di Perugia, Perugia, Italy
alfredo.navarra@unipg.it

**Abstract.** Gathering a swarm of robots is one of the basic tasks in distributed computing. Varying of the robots' capabilities as well as on the environments where robots move lead to very different approaches. In general, the problem requires the design of a distributed algorithm that brings all robots to meet at some common location, not known in advance. We consider asynchronous robots subject to the well-established `Look-Compute-Move` model. Each time a robot wakes up, it perceives the current configuration in terms of robots' positions (`Look`), it decides whether and where to move (`Compute`), and makes the computed move (`Move`), if any. Starting from the case of robots moving in the Euclidean plane, we highlight pros and cons for robots moving along the edges of a graph. We survey on the most recent results about robots moving in general graphs and in specific topologies like trees, rings, grids, and cliques. Further, we show how the design of an algorithm for general graphs naturally leads to optimization issues. In particular, we survey on optimal gathering algorithms in terms of total number of edges traversed by robots in order to accomplish the gathering task. Also in this case, results concern general graphs and specific topologies. In doing so, we highlight how the problem and the resolution algorithms change when optimal constraints are included.

**Keywords:** Asynchrony · Mobile robots · Gathering
Discrete environment

## 1 Robots' Model

In this section we recall the main characteristics of the robots' model, independently from the environment in which robots move, typically the Euclidean plane or a graph.

The work has been supported in part by the European project "Geospatial based Environment for Optimisation Systems Addressing Fire Emergencies" (GEO-SAFE), contract no. H2020-691161 and by the Italian National Group for Scientific Computation (GNCS-INdAM).


P. Flocchini et al. (Eds.): Distributed Computing by Mobile Entities, LNCS 11340, pp. 184–217, 2019.
https://doi.org/10.1007/978-3-030-11072-7_8

The gathering problem asks for a distributed algorithm that brings all robots to occupy a common location. Solving the gathering problem depends on the capabilities one assumes for robots. A common approach in distributed computing is to detect the minimal capabilities that are necessary so as robots can perform simple basic tasks, e.g. gathering. The rationale behind this approach is twofold: it is theoretically interesting to answer the minimality question; the weaker is the model assumed to solve a task, the wider is its applicability, including more powerful robots prone to faults. In this chapter robots are considered to be:

- *Anonymous*: no unique identifiers;
- *Autonomous*: no centralized control;
- *Dimensionless*: no occupancy constraints, no volume, modeled as geometric points;
- *Oblivious*: no memory of past events;
- *Homogeneous*: they all execute the same *deterministic* algorithm;
- *Silent*: no means of direct communication;
- *Disoriented*: no common coordinate system, no common left-right orientation;
- *Asynchronous*: no common clock, robots' activities are independent.

**Look-Compute-Move Model.** At any point in time, a robot is either *active* or *inactive*. All robots are initially inactive, i.e. they are all idle. When active, a robot executes a `Look-Compute-Move` (LCM) cycle by performing the following three operations in sequence, each of them associated with a different state:

- `Look`: The robot observes the environment. The result of this phase is a snapshot of the positions of all robots with respect to its own perception.
- `Compute`: The robot executes the designed algorithm, using the data sensed in the Look phase as input. In the Euclidean plane, the result of this phase is a target point along with a trajectory to reach it. If the environment is modeled by a graph, the results of this phase is a vertex among the neighbors of the vertex in which the robot currently resides (at most one edge per cycle can be traversed by a robot).
- `Move`: The robot moves toward the computed target. If the target is the current position, then the robot stays still, i.e. it performs what is called a *null* movement.

The amount of time to complete a full LCM-cycle is assumed to be finite but unpredictable. Moreover, the adversary determining the computational cycles timing is assumed to be *fair*, that is, each robot performs its LCM-cycle within finite time and infinitely often. Without such an assumption the gathering would be unsolvable as the adversary could prevent some robots to ever move.

**Multiplicity Detection.** It is very common (as dictated by impossibility results) that in combination with the LCM-model, robots are endowed with the so-called *multiplicity detection* capability (see e.g. [7,24]). This is a well-studied capability associated to robots within the LCM model, it can be defined in various forms and concerns robots moving both in the Euclidean plane and in graphs.

**Definition 1 (Multiplicity).** *When more than one robot resides on the same point/vertex $x$, then $x$ is said to be occupied by a* multiplicity.

There exist four types of multiplicity detection investigated so far, with the aim to reduce as much as possible the capabilities of the robots.

**Definition 2 (Types of multiplicity detection).** *A robot is said to have the:*

- global weak multiplicity detection *ability when it is able to detect whether a multiplicity exists at any given point/vertex;*
- global strong multiplicity detection *ability when it is able to compute the exact number of robots composing a multiplicity at any given point/vertex;*
- local-weak multiplicity detection *ability when it is able to detect whether a multiplicity exists, at a given point/vertex, only if the robot is part of the multiplicity itself;*
- local-strong multiplicity detection *ability when it is able to compute the exact number of robots composing a multiplicity, at a given point/vertex, only if the robot is part of the multiplicity itself.*

### 1.1 A Look to the Euclidean Plane

Before focusing on robots moving in graphs, it is worth to remark some basic concepts arising in the setting of robots moving in the Euclidean plane.

Apart for the case of just two robots proved to be unsolvable, in [7] the gathering problem has been completely solved when robots move in the Euclidean plane in the very weak asynchronous setting, without any common orientation and with global-weak multiplicity detection. One may wonder whether similar approaches can be applied also in the context of robots moving in graphs.

The resolution strategy proposed in [7] exploits the concept of *Smallest Enclosing Circle* (SEC) of the robots. This is the unique circle that encloses all robots. Being unique, also its center is unique and hence if the strategy makes robots move so as to create a multiplicity on the center of the SEC then the gathering can be easily finalized. The main difficulty is to maintain the same SEC as long as the multiplicity is not created. In fact, movements toward the center of the SEC may modify the SEC itself, and hence its center. The SEC and its center are then used in combination with another interesting point, the so called *Weber point*. Let $d(s,t)$ be the distance of the points $s$ and $t$ in the Euclidean plane.

**Definition 3 (Weber-point).** *Given a set of robots $R$, a point $p$ is said to be the* Weber point *if it minimizes the quantity $\sum_{r \in R} d(p,r)$.*

It is proved that in the Euclidean plane the Weber point is unique (if the robots are not collinear), and that movements of robots toward such a point do not modify the Weber point. Unfortunately, the Weber point is not computable in general [8]. However, in [7] it has been exploited for some specific configurations and in combination with the SEC provide the main ingredients for the resolution strategy.

One further point that has been sometimes exploited in similar contexts is the so-called *Center of Gravity.*

**Definition 4 (Center of Gravity).** *Given a set of robots $R$, let $(r_x, r_y)$ be the coordinates of robot $r$. The* Center of Gravity *is the point of coordinates* $(x = \frac{\sum_{r \in R} r_x}{|R|}, y = \frac{\sum_{r \in R} r_y}{|R|})$.

The Center of Gravity is unique and easily computable, but unfortunately changes as soon as robots move toward it. In fact, it is mainly used for solving the *Convergence* problem [9] rather than the gathering, where robots have to get closer and closer instead of occupying exactly the same point.

**Rigid vs Non-rigid Movements.** Concerning movements of robots, they can be assumed to be *rigid* or not. A rigid movement implies that a robot always reaches its target at the end of an LCM-cycle. A non-rigid movement instead does not provide such a guarantee. However, the distance traveled within a move is neither infinite nor infinitesimally small. More precisely, the adversary has the power to stop a moving robot before it reaches its destination, but there exists an unknown constant $\delta > 0$ such that if the destination point is closer than $\delta$, the robot will reach it, otherwise the robot will be closer to it of at least $\delta$. Note that, without this assumption, an adversary would make impossible for any robot to ever reach its destination.

**Asynchrony.** As robots are assumed asynchronous, all their actions might happen in any moment, within finite time. This implies that while a robot is acquiring its snapshot during a Look phase, other robots might be idle, looking, computing or moving. In particular, if a robot is moving, it can be seen at any position of its performed trajectory. As robots acquire only one snapshot along a whole LCM-cycle, and they are assumed to be oblivious, there is no mean to detect whether other robots are moving or not. More than that, it is not possible even to decide whether other robots are performing or are intended to perform a move. This is an information missing in the snapshot, so robots cannot know the current status of any other robot.

## 1.2 Pros and Cons for Being in Graphs

One main assumption that can be done when robots move in graphs is that moves always assume vertices as target, and they can be considered as *instantaneous*. This results in always perceiving robots on vertices and never on edges during Look phases. The rationale behind this assumption is that the graph may model a communication network, whereas robots model software agents. In such a context, with destinations always represented by vertices, instantaneous moves also imply rigid movements, which is clearly a simplification for designing resolution algorithms as outlined in the previous section. Moreover, differently from the Euclidean case, robots on graphs cannot be seen while moving, but only at the moment they may start moving or when they arrived.

In a graph a *center* is a vertex that minimizes the maximum distance (in terms of hops/number of edges) from any other vertex. When the center of a graph is unique, such a vertex can be found by all robots and the gathering can be easily accomplished. Unfortunately, a graph may admit more than one center.

Another useful vertex where robots may agree to meet is the counter-part of the Weber-point defined in the Euclidean plane. In graphs, a *Weber point* can be defined as a vertex $x$ that minimizes the sum of the shortest paths (in terms of hops/number of edges) from each robot toward $x$. This can be easily computed in graphs and if unique it can be found by all robots. Unfortunately, in graphs the Weber point is not necessarily unique.

Being in graphs may help when the topology implies some special property. For instance, in stars or rooted trees it is very easy for the robots to agree on a specific vertex where to gather, that is the center of the star or the root of the tree, respectively. Contrary, sometimes it might be a disadvantage to be constrained in a graph as trajectories are forced by the topology whereas in the Euclidean plane, robots may perform suitable trajectories in order to avoid undesirable configurations.

As last remark, it is worth mentioning that in graphs, concepts like center of gravity or SEC (and hence its center) are not defined, whereas they are largely exploited in the Euclidean plane.

## 2 Problem Definition and General Impossibility Results

In this section we consider undirected graphs as the environment where robots move. In such an environment, the classical gathering problem is called *Gathering on graph*. We now provide a formal definition.

A simple undirected graph $G = (V, E)$, with vertex set $V$ and edge set $E$, represents the topology where robots are placed on. A function $\mu : V \longrightarrow \mathbb{N}$, represents the number of robots on each vertex of $G$, and we call $(G, \mu)$ a *configuration* whenever $\sum_{v \in V} \mu(v)$ is bounded and greater than zero. A configuration is *initial* if each robot lies on a different vertex (i.e., $\mu(v) \leq 1 \; \forall v \in V$). A configuration is *final* if all the robots are on a single vertex (i.e., $\exists u \in V : \mu(u) > 0$ and $\mu(v) = 0, \; \forall v \in V \setminus \{u\}$). The Gathering on graph problem can be formally defined as the problem of transforming an initial configuration into a final one. A *gathering algorithm* for this problem is a deterministic distributed algorithm that brings the robots in the system to a final configuration in a finite number of LCM-cycles from any given initial configuration, regardless of the adversary. We say that an initial configuration $C = (G, \mu)$ is *ungatherable* if there are no gathering algorithms with respect to $C$. We say that an algorithm *assures* the gathering if it achieves the gathering regardless the adversary.

Before presenting the gathering algorithms defined in the literature, we recall from [19] the notions of configuration automorphisms and symmetries to be applied to general graphs, and accordingly we provide general impossibility results.

## 2.1 Configuration Automorphisms and Symmetries

Two undirected graphs $G = (V_G, E_G)$ and $H = (V_H, E_H)$ are *isomorphic* if there is a bijection $\varphi$ from $V_G$ to $V_H$ such that $\{u, v\} \in E_G$ if and only if $\{\varphi(u), \varphi(v)\} \in E_H$. An *automorphism* on a graph $G$ is an isomorphism from $G$ to itself, that is a permutation of the vertices of $G$ that maps edges to edges and non-edges to non-edges. The set of all automorphisms of $G$ forms a group called *automorphism group* of $G$ and denoted by $\text{Aut}(G)$.

The concept of isomorphism can be extended to configurations in a natural way: two configurations $(G, \mu)$ and $(G', \mu')$ are isomorphic if $G$ and $G'$ are isomorphic via a bijection $\varphi$ and for each vertex $v$ in $G$, $\mu(v) = \mu'(\varphi(v))$. An *automorphism* on a configuration $(G, \mu)$ is an isomorphism from $(G, \mu)$ to itself and the set of all automorphisms of $(G, \mu)$ forms a group that we call *automorphism group* of $(G, \mu)$, denoted by $\text{Aut}((G, \mu))$.

Given an isomorphism $\varphi \in \text{Aut}((G, \mu))$, the *cyclic subgroup* of order $p$ generated by $\varphi$ is given by $\{\varphi^0, \varphi^1 = \varphi, \varphi^2 = \varphi \circ \varphi, \ldots, \varphi^{p-1}\}$ where $\varphi^0$ is the identity isomorphism. If $H$ is a subgroup of $\text{Aut}((G, \mu))$, the *orbit* of a vertex $v$ of $G$ is $Hv = \{\gamma(v) \mid \gamma \in H\}$. If $|\text{Aut}(G)| = 1$, that is $G$ admits only the identity automorphism, then $G$ is said *asymmetric*, otherwise it is said *symmetric*. Analogously, if $|\text{Aut}((G, \mu))| = 1$, we say that $(G, \mu)$ is *asymmetric*, otherwise it is *symmetric*.

The next theorem provides a sufficient condition for a configuration to be ungatherable, but we first need the following definition:

**Definition 5.** *Let $C = ((V, E), \mu)$ be a configuration. An isomorphism $\varphi \in Aut(C)$ is called* partitive *on $V' \subseteq V$ if the cyclic subgroup $H = \{\varphi^0, \varphi^1 = \varphi, \varphi^2 = \varphi \circ \varphi, \ldots, \varphi^{p-1}\}$ generated by $\varphi$ has order $p > 1$ and is such that $|Hu| = p$ for each $u \in V'$.*

Note that, in the above definition, the orbits $Hu$, for each $u \in V'$, form a partition of $V'$. The associated equivalence relation is defined by saying that $u$ and $v$ are equivalent if and only if there exists a $\gamma \in H$ with $\gamma(u) = v$. The orbits are then the equivalence classes under this relation; two elements $u$ and $v$ are equivalent if and only if their orbits are the same; i.e., $Hu = Hv$. Moreover, note that $\mu(u) = \mu(v)$ whenever $u$ and $v$ are equivalent.

The following two theorems from [19] provide general impossibility results for the gathering problem on graphs.

**Theorem 1.** *Let $C = ((V, E), \mu)$ be a non-final configuration. If there exists $\varphi \in Aut(C)$ partitive on $V$ then $C$ is ungatherable.*

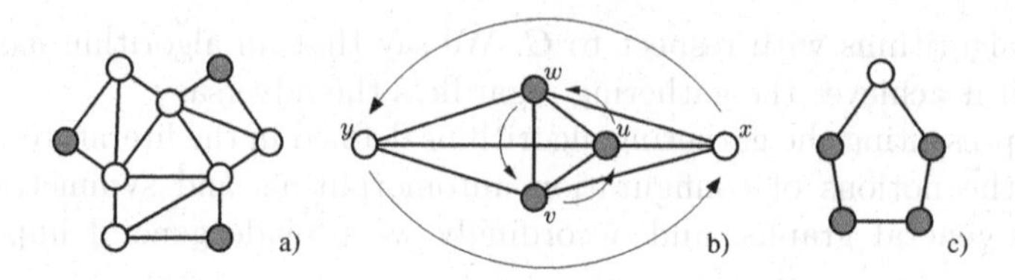


**Fig. 1.** A gray vertex indicates the presence of one robot. (a) Configuration $C_1$ admitting a partitive isomorphism on $V$: the sets of the partition are the three central vertices, the vertices with robots, and the three remaining vertices. (b) Configuration $C_2$ admitting a non-partitive isomorphism that maps $v$ in $u$, $u$ in $w$, $w$ in $v$, $x$ in $y$ and $y$ in $x$. (c) Configuration $C_3$ admitting a non-partitive isomorphism with two sets of the partition of size two, and one of size one.

*Remark 1.* It is worth to remark that the above theorem requires the existence of an automorphism $\varphi$, which in turn is based on the function $\mu$ defining the exact number of robots on each vertex. Hence, Theorem 1 holds when the robots are endowed with the global-strong multiplicity detection. Stating a negative result, it follows that such a theorem holds even when considering weaker robots (i.e., without global-strong multiplicity detection).

In Fig. 1a, it is shown a partitive configuration $C_1$ where each vertex belongs to an orbit of size three. By the above theorem we deduce that $C_1$ is ungatherable, since each move allowed by an algorithm can be executed synchronously by all the three robots due to an adversary. This would always produce a new partitive configuration.

Figure 1b, shows a configuration $C_2$ admitting an isomorphism which is not partitive. In this case the gathering is possible. In fact, moving robots among the three occupied vertices may produce the same configuration if the three robots move concurrently in the same direction. Hence, a gathering algorithm can move the three robots towards the two empty vertices (which can be always recognized as the vertices with minimum degree). Once all the three robots have moved, a multiplicity is created. The multiplicity either contains all the robots or just two. In the first case the gathering has been accomplished. In the second case, the gathering is finalized by letting the single robot move towards the multiplicity. Clearly, such a strategy would require robots endowed with some form of multiplicity detection. Finally, Fig. 1c shows a configuration $C_3$ admitting a non-partitive isomorphism: it is shown in [23] that $C_3$ is ungatherable. This example shows there exists ungatherable configurations even if they do not admit any partitive isomorphism.

The following theorem shows that some configurations can be gathered only at some predetermined vertices. For instance, the only vertex where gathering can be potentially finalized in configuration $C_3$ (cf Fig. 1) is the empty one.

**Theorem 2.** *Given a configuration* $C = ((V, E), \mu)$, *and a subset of vertices* $V' \subset V$, *if there exists an automorphism* $\varphi \in Aut(C)$ *that is partitive on* $V \setminus V'$, *with* $\mu(v) = 0$ *for any* $v \in V'$, *then, any gathering algorithm can not assure the gathering on a vertex in* $V \setminus V'$.

As final remark, when symmetries occur, it is possible that an algorithm elect one single robot to move, but more than one can move concurrently. In such a case there might occur a so called *pending move*. This happens when, due to the asynchrony, one of the robots allowed to move performs its entire LCM-cycle while one of the others does not perform the `Move` phase, i.e. its move is pending. Clearly, all robots performing their cycle are not aware whether there is a pending move, that is they cannot deduce the global status from their view. The presence of pending moves greatly increases the difficulty of designing gathering algorithms for symmetric configurations.

## 3 Gathering Algorithms for Specific Topologies

In this section we describe gathering algorithms for specific topologies, namely complete graphs, trees, rings, and grids.

### 3.1 Complete Graphs

We provide an original result about complete graphs. In fact, apart for all configurations caught by Theorem 1, the result shows that asynchronous robots can never accomplish the gathering task if the underlying graph is a clique.

**Theorem 3.** *Given an initial configuration* $C = (G, \mu)$, *if* $G$ *is a clique then the gathering problem is unsolvable.*

*Proof.* By hypothesis both $G$ and the robots are anonymous. When $G = (V, E)$ is a clique there exists an automorphism $\varphi \in \mathrm{Aut}(C)$ that makes robots pairwise equivalent. In other words, if $\mathcal{A}$ is any gathering algorithm for $C$, then any move planned by $\mathcal{A}$ can be performed by any robot. In particular, each move can only specify whether the robot must move toward an empty vertex or toward a vertex already occupied. If the move is toward an empty vertex, then the adversary can decide to make only one robot move, hence obtaining a new configuration $C'$ isomorphic to $C$. From $C'$, of course the same move will be applied but the adversary can make another robot moving, hence respecting the fairness constraint.

If the move is toward a vertex already occupied, then the adversary can always make all robots move concurrently toward different destinations, in such a way that the robots just exchange their positions on the same set of occupied vertices. Then $\mathcal{A}$ will always produce a configuration isomorphic to $C$. □

### 3.2 Trees

In this section, results about gathering on trees are presented (cf [11]). Let $T$ be a tree and let $C = (T, \mu)$ be an initial configuration. Based on well-known results about the trees, within $T$ there is either one center or there are two neighboring centers [26]. In the former case, no matter the initial distribution of the robots, each of them can move towards the center, concurrently. The gathering will be eventually finalized, even without any multiplicity detection assumption.

In the latter case, some more specific arguments are required. Assume there are two centers in $T$, namely $c_1$ and $c_2$, and let $T_1$ and $T_2$ be the two subtrees rooted at $c_1$ and $c_2$, respectively, when the edge connecting $c_1$ and $c_2$ is removed. In such a scenario we say that $C$ is *balanced* when the sub-configurations $(T_1, \mu)$ and $(T_2, \mu)$ are isomorphic. It is easy to observe that if $C$ is balanced then there exists a partitive automorphism of order two in $C$, and hence, by Theorem 1 configuration $C$ is ungatherable.

If $C$ has two centers $c_1$ and $c_2$ but is not balanced, the following gathering algorithm can be applied. If the number of vertices occupied in $T_1$ is smaller than that in $T_2$, then all robots in $T_1$ are moved towards $c_2$. Once $T_1$ gets empty, all robots in $T_2$ should be moved towards $c_2$ in order to finalize the gathering. If the number of vertices occupied in $T_1$ is equal to that in $T_2$, it is always possible to determine which subtree is less than the other with respect to a natural ordering on labeled trees (see [1,4]). To define the smaller tree as the one with the robots closer to the root, it is possible to associate label 1 to empty vertices, and label 0 to vertices occupied by robots. Then the algorithm would exploit this ordering in order to detect the robots to move from one subtree towards the root of the other one. If a robot moves over a vertex already occupied, the number of occupied vertices in the original subtree decreases. As soon as one robot moves towards the other subtree, the number of robots in the two subtrees is no longer equal and the previous strategy can be applied.

The following theorem summarizes the above arguments.

**Theorem 4 (Gathering on trees).** *Let $T$ be a tree and $C = (T, \mu)$ be an initial configuration. Then, $C$ is gatherable if and only if $C$ is not balanced.*

### 3.3 Rings

In this section, results about gathering on rings are presented (cf [3,6,11,12, 14–17,20–25]). After providing some necessary definitions, impossibility results are summarized when the global-strong multiplicity detection is assumed. Then, differences between the case of global-weak and local-weak multiplicity detection assumptions are presented. In particular, when the global-weak multiplicity detection is assumed, a full characterization of the gatherable configurations is provided [12]. When the local-weak multiplicity detection is assumed, a very few cases are left open [14]. However, the different techniques used to accomplish the gathering task among the approached scenarios are very interesting for further investigations in robot-based computing systems.

A ring is composed of vertices $\{v_0, \cdots, v_{n-1}\}$, $n \geq 3$, where $v_i$ is connected to $v_{i+1 \bmod n}$ for any $0 \leq i < n$. The model assumes that $k < n$ robots are placed over the $n$ vertices of a ring, and in the *initial configurations* each vertex is occupied by at most one robot. Depending on the movements imposed by the running algorithms, multiplicities may occur.

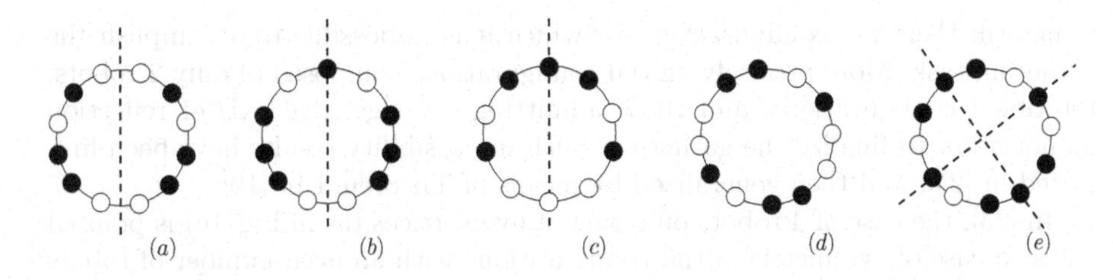


**Fig. 2.** Symmetric and rotational initial configurations on a ring. White vertices are empty while each black vertex is occupied by one robot.

For the purpose of characterizing symmetries is this topology, a ring of $n$ vertices can be intended as a regular polygon with $n$ vertices embedded in the Euclidean plane. Then, it follows that a ring has $n$ axes of reflection and admits $n$ rotations. If $n$ is even then half of these axes pass through two opposite vertices, and the other half through the midpoint of opposite edges. If $n$ is odd then all axes pass through a vertex and the midpoint of the opposite edge. In particular, the general notion of symmetric configuration provided in Sect. 2.1 can be specialized to rings as follows (cf Fig. 2): a symmetric configuration with exactly one axis of reflection has an *edge-edge symmetry* if the axis goes through two edges (Fig. 2a); it has a *vertex-edge symmetry* if the axis goes through one vertex and one edge (Fig. 2b); it has a *vertex-vertex symmetry* if the axis goes through two vertices (Fig. 2c); it has a *robot-on-axis symmetry* if there is at least one vertex on the axis of symmetry occupied by a robot (both Figs. 2b and c). A configuration is called *rotational* if it is invariable under non-trivial (i.e., non-complete) rotations (Figs. 2d and e). Figure 2d admits a rotation of 180°, without axes of reflection, whereas Fig. 2e still admits a rotation of 180° but with two axes of reflection.

It is worth to remark that different assumptions about multiplicity detection may provide different information to robots. For instance, when the global-weak multiplicity is considered, a configuration is said reflective if the ring admits a geometrical axis of symmetry that reflects single robots into single robots, multiplicities into multiplicities, and empty vertices into empty vertices. In this case, a configuration might be considered symmetric even though the two halves of the ring cut by the axis do not contain the same number of robots. This can happen if two symmetric multiplicities at the two halves are composed of a different number of robots. However, symmetric peculiarities of initial configurations are invariant with respect to the assumed multiplicity detection, as multiplicities are not allowed at the beginning.

**Impossibility Results.** In [24], it is proved that the gathering is unsolvable if the multiplicity detection capability is completely removed in either of its forms. When the multiplicity detection is assumed, even in its strong and global form, still there are configurations for which it is impossible to accomplish the gathering task. More precisely, initial configurations composed of only 2 robots, rotational configurations, and those admitting an edge-edge axis of reflection do not allow to finalize the gathering. Such impossibility results have been first stated in [24], and then generalized by means of Theorem 1 in [19].

In [23], the case of 4 robots on a ring of five vertices (as in Fig. 1c) is pointed out as a case of symmetric initial configurations with an even number of robots that does not allow any gathering algorithm. In general, a specific set of configurations with four robots has been defined and studied in the literature as the $SP4$ configurations.

**Definition 6.** *Let $C$ be a reflective configuration with four robots on an odd ring. If the odd interval of vertices cut by the axis is bigger than the even one, then $C$ is said to belong to the set of $SP4$ configurations.*

The case of 4 robots on a five vertices ring belongs to $SP4$. Other configurations in $SP4$ formally proved to be ungatherable can be found in [6,16,17]. Actually, some configurations in $SP4$ could be gatherable (see, e.g. [3]) but they require strategies that are difficult to generalize or to integrate with the algorithms designed for other types of configurations. According to such difficulties and following the conjecture posed in [17], we consider the whole set $SP4$ as formed by ungatherable configurations.

The set of all the ungatherable configurations for rings is denoted by $\mathcal{U}_R$, and is described in Table 1. For all initial configurations not belonging to $\mathcal{U}_R$, various gathering algorithms have been provided, depending also on the assumptions concerning the multiplicity detection capability. Whenever clear by the context, we refer to initial configurations simply as configurations.

**Table 1.** Resume of the known impossibility results about gathering in a ring under the `Look-Compute-Move` model even assuming global-strong multiplicity detection. All the mentioned configurations are initial and form the set $\mathcal{U}_R$. Symbols $n$ and $k$ refer to the number of vertices and number of robots, respectively.

| Configuration type | $n$ | $k$ | Papers |
|---|---|---|---|
| Rotational or with edge-edge axis | - | - | [19,24] |
| - | - | $k = 2$ | [24] |
| SP4 | Odd | $k = 4$ | [12,16,17,23] |

**Local-Weak Multiplicity Detection.** In this section we consider the gathering problem for $k$ robots endowed with local-weak multiplicity detection in a ring of $n$ vertices. In this context, Table 2 provides an overview of all the results obtained so far. In particular, an algorithm starting from initial configurations where the number of robots $k$ is strictly smaller than $\lfloor \frac{n}{2} \rfloor$ was designed in [20].

In [21], the case where $k$ is odd and strictly smaller than $n-3$ was solved. In [22], the authors provide an algorithm for the case where $n$ is odd, $k$ is even, and $10 \leq k \leq n-5$. Recently, the case of asymmetric configurations was solved in [13].

**Table 2.** Resume of the known feasibility results about gathering in a ring under the `Look-Compute-Move` model with the local-weak multiplicity detection. All the mentioned configurations are initial and do not belong to $\mathcal{U}_R$. Symbols $n$ and $k$ refer to the number of vertices and number of robots, respectively.

| Configuration type | $n$ | $k$ | Papers |
|---|---|---|---|
| Asymmetric | - | $k < \lfloor \frac{n}{2} \rfloor$ | [20] |
| - | - | Odd, $k < n-3$ | [21] |
| - | Odd | Even, $10 \leq k \leq n-5$ | [22] |
| Asymmetric | - | - | [13] |
| - | - | $k < n-4$, $k \neq 4$ | [14] |

The most complete result is given in [14], where authors provide a full characterization (except for few pathological cases) of the initial configurations for which the gathering problem can be solved. In particular, for any $k < n-4$ and $k \neq 4$, authors characterize the initial configurations from which the gathering problem is solvable. In particular, they design an algorithm that solves the problem starting from any initial configuration with $k < n-4$, $k \neq 4$, robots empowered by the local-weak multiplicity detection. Similarly to the case of $k = 4$ in [12] and $(n, k) = (7, 6)$ in [15], the cases left out from this characterization ($k = 4$ and $k \geq n-4$), if gatherable, would require specific algorithms difficult to generalize.

The next theorem represents the main result provided in [14].

**Theorem 5 (Gathering on rings, local-weak multiplicity detection).** *Let $R$ be any ring with $n$ vertices and let $(R, \mu)$ be an initial configuration with $k$ robots, $k < n-4$ and $k \neq 4$. If robots are endowed with the local-weak multiplicity detection then $C$ is gatherable.*

In the following we provide some basic ideas of the proposed algorithm LWM-GATHERING [14].

*Notation.* A *configuration* $C$ is defined by the $k$ vertices occupied by robots. In what follows, any configuration is seen as a binary sequences where "0" represents an occupied vertex while "1" stands for an empty vertex. More formally, given a configuration $C$, and for any $i \leq n$, let $S_i = (r_0^i, \cdots, r_{n-1}^i) \in \{0,1\}^n$ be the sequence such that $r_j^i = 0$ if $v_{i+j \bmod n}$ is occupied in $C$ and $r_j^i = 1$ otherwise, $0 \leq j < n$. Intuitively, $S_i$ represents the positions of robots, starting at vertex $v_i$.

A *representation* of $C$ is any sequence in $\mathcal{S}_C = \{\mathcal{S}_i, \overline{(\mathcal{S}_i)}\}_{i<n}$, where $\overline{(\mathcal{S}_i)}$ is $\mathcal{S}_i$ reversed. The view from a vertex/robot $v_i$ is the minimum between $\mathcal{S}_i$ and $\overline{\mathcal{S}_i}$, this also represents the *snapshot* of a configuration acquired by a robot during the Look phase. A *supermin* of $C$ is any representation of $C$ that is minimum in the lexicographical order. The supermin of $C$ is denoted as $C^{\min}$. By $C_i^{\min}$ we denote the representation of the supermin starting from the $i$-th position, that is rotating $C^{\min}$ of $i$ positions. In any supermin $(r_0, \cdots, r_{n-1})$, if $k < n$ then $r_{n-1} = 1$. From their view, all robots can compute the supermin of a configuration.

It is easy to see that each robot has all the information to compute whether it has to move or not according to the acquired configuration during its Look phase (i.e. its snapshot). For instance, suppose that from a given configuration $C$, with supermin $C^{\min} = (r_0, r_1, \ldots, r_{n-1})$, an algorithm makes the robot at $r_i$ move toward $r_{i+1}$. Let $C' = (r'_0, r'_1, \ldots, r'_{n-1})$ be the *local* view of a generic robot $r$. Then, $r$ computes the supermin and checks whether $C' = C_i^{\min}$ or $C' = \overline{(C_i^{\min})}$. If one of such cases occurs, then it deduces it is candidate to move toward $r'_1$ or $r'_{n-1}$, respectively.

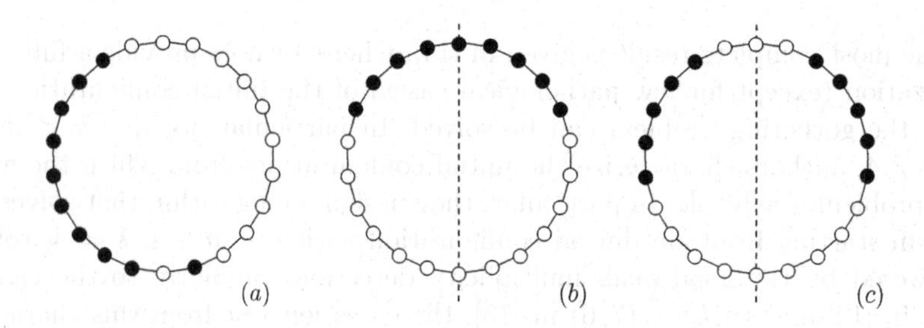


**Fig. 3.** Configurations achieved at the end of Algorithm ALIGN. $(a)$: Configuration $C^a$, $(b)$: Configuration $C^b$, $(c)$: Configuration $C^c$.

*Subroutine* ALIGN. Starting from any initial asymmetric configuration, a subroutine of the main algorithm called ASYM allows to achieve a particular configuration called $C^a = (0^{k-1}, 1, 0, 1^{n-k-1})$ made of $k-1$ consecutive robots, one empty vertex and one robot (see Fig. 3a). Basically, Algorithm ASYM ensures that, from any initial asymmetric configuration, one robot can be uniquely detected and is moved to an unoccupied neighbor by achieving another initial configuration while strictly decreasing the supermin. Algorithm ALIGN generalizes ASYM by handling all initial configurations (not only asymmetric).

In this generalization, several difficulties are overcome. First, in initial symmetric configurations, it is not possible to ensure that a unique robot will move. In such a case, the algorithm may allow a robot $r$ to move, while $r$ is reflected by the axis of symmetry to another robot $r'$. Since $r$ and $r'$ are indistinguishable and execute the same algorithm, then $r'$ should perform the same (symmetric) move. However, due to asynchrony, $r$ may move while the corresponding move of $r'$ is postponed (i.e. $r'$ has performed the `Look` phase but not yet the `Move` phase). The configuration reached after the move of $r$ has a potential pending move (the one of $r'$ that will be executed eventually). To deal with this problem, the algorithm ensures that all the reached configurations that might have a pending move can be always detected as asymmetric configurations with a unique pending move. Therefore, in such a case, the algorithm forces to perform the pending move. That is, contrary to [13] where Algorithm ASYM ensures to only go through asymmetric configurations, the subtlety consists in possibly going from an asymmetric configuration to a symmetric one. To detect such configurations, it is defined the notion of adjacent configurations. Consider an algorithm $\mathcal{A}$ and a procedure M allowed by $\mathcal{A}$, that is algorithm $\mathcal{A}$ performs M for some configuration; possibly, procedure M moves two (symmetric) robots. An asymmetric configuration $C$ is *adjacent* to a symmetric configuration $C'$ with respect to procedure M if $C$ can be obtained from $C'$ by applying M to only one of the robots that can move according to M and the algorithm performs M on $C$. In other words, if $C$ is adjacent to $C'$ with respect to M, there might exist a pending move permitted by M in $C$. Another difficulty is to ensure that all met configurations are allowed for the gathering problem.

Algorithm ALIGN allows to reach one of the configurations $C^a$, $C^b$, or $C^c$ having supermin $(0^{k-1}, 1, 0, 1^{n-k-1})$, $(0^k, 1^{n-k})$, or, $(0^{\frac{k}{2}}, 1^j, 0^{\frac{k}{2}}, 1^{n-k-j})$ for $k$ even and $j < \frac{n-k}{2}$, respectively (see Fig. 3).

*Description of the Gathering Algorithm.* The algorithm makes use of procedure ALIGN to reach one among the following configurations: $C^a = (0^{k-1}, 1, 0, 1^{n-k-1})$ with $k$ even, $C^b = (0^k, 1^{n-k})$, with $k$ or $n$ odd, $C^c = (0^{\frac{k}{2}}, 1^j, 0^{\frac{k}{2}}, 1^{n-k-j})$, with $k$ even and $j$ or $n$ odd.

Since to solve gathering it is necessary to create a multiplicity, configurations containing multiplicities must be handled. According to the assumed local-weak multiplicity detection, each robot perceives a multiplicity only if it is part of it. So, it cannot deduce the actual total number of robots. The algorithm is structured in a way that procedure ALIGN is invoked only at the end, that is once it is sure that the current configuration does not belong to those directly managed for gathering.

There are various moves allowed according to the current configuration. From $C^a$ the gathering will be accomplished by compacting the sequence of consecutive robots from the tail toward the head, the only robot without neighboring robots. Once only two vertices are occupied, only one of them contains a multiplicity. In order to finalize the gathering, the robot not composing the multiplicity (the original head) moves toward the other occupied vertex. The finalization of the gathering from $C^b$ and $C^c$ is accomplished in the middle vertex cut by the axis of symmetry, closest to the robots. In these cases, symmetric moves must be performed. The main difficulties then come from asynchrony.

**Global-Weak Multiplicity Detection.** Concerning the case of global-weak multiplicity detection capability, Table 3 summarizes the progresses made within this context. Entries referring to solved cases must be intended as concerning configurations not belonging to the ungatherable ones.

**Table 3.** Resume of the known feasibility results about gathering in a ring under the Look-Compute-Move model with the global-weak multiplicity detection. All the mentioned configuration concern initial configurations not belonging to $\mathcal{U}_R$. Symbols $n$ and $k$ refer to the number of vertices and number of robots, respectively.

| Configuration type | $n$ | $k$ | Papers |
|---|---|---|---|
| Asymmetric | - | - | [24] |
| Symmetric | - | Odd | [24] |
| Symmetric | - | $k > 18$ | [23] |
| Symmetric | - | k=4 | [25] |
| Symmetric | - | k=6 | [15] |
| - | - | - | [12] |

As shown in the table, increasing the capability of the robots from local-weak to global-weak multiplicity detection does not improve much on feasibility results. In fact, the only difference between the general algorithm provided for the local-weak multiplicity detection is restricted to the $k < n - 4$ and $k \neq 4$, whereas the algorithm designed for the global-weak multiplicity detection works for any configuration not in $\mathcal{U}_R$ (see [12]).

**Theorem 6 (Gathering on rings, global-weak multiplicity detection).** *Let $R$ be a ring and $C = (R, \mu)$ be an initial configuration composed by robots endowed with the global-weak multiplicity detection. Then, $C$ is gatherable if and only if $C \notin \mathcal{U}_R$.*

### 3.4 Grids

In this section, results achieved in [10] are reported. The authors consider the gathering problem on an anonymous and undirected grid of $n \times m$ vertices, with $m \geq n$. As for rings, the grid is intended as embedded in the Euclidean plane, with all edges of the same size. The main assumption that distinguishes these results from those obtained on rings is *the lack of any multiplicity detection capability*: if a vertex is occupied by more than one robot, it is not perceived by the robots, even if they reside on such a vertex. As usual, in initial configurations each vertex is occupied by at most one robot. During a Look operation, a robot perceives the relative locations on the grid of occupied vertices, regardless of the number of robots at a vertex.

The current configuration of the system can be described in terms of the view of a robot $r$ which is performing the Look operation at the current moment. A configuration seen by $r$ is denoted as an $n \times m$ matrix $M$ that has elements belonging to the set $\{0, 1\}$. Value 0 represents an empty vertex, and 1 represents an occupied vertex. Since the grid is anonymous and undirected, each robot can perceive the current configuration with respect to different rotations and reflections leading to any view of the grid satisfying the $n \times m$ dimension. In particular, when $n = m$ each of the 4 rotations and 4 reflections provides a feasible view.

**Definition 7.** *A configuration is:*

1. rotational *if it is invariant with respect to rotations of* $90°$ *or* $180°$*, where the rotation point coincides with the geometric center of the grid;*
2. vertical-symmetric *(*horizontal-symmetric, diagonal-symmetric, *resp.) if it is invariant after a reflection with respect to a vertical (horizontal, diagonal - in case of square grids, resp.) axis passing through the geometric center of the grid.*

**Table 4.** Resume of the known impossibility results about gathering in a grid under the `Look-Compute-Move` model without any multiplicity detection capability. All the mentioned configurations are initial and form the set $\mathcal{U}_G$.

| Parity | Dimension | Symmetry type | Paper |
|---|---|---|---|
| Odd × even | - | Rotational | [10, 19] |
| | - | Vertical-symmetric | |
| Even × even | - | Rotational | |
| | - | Vertical-symmetric | |
| | - | Horizontal-symmetric | |
| | 2 × 2 | - | |

We simply say that a configuration is *symmetric* if any of the cases in the second item of the previous definition applies. The set of all the ungatherable configurations for grids is denoted by $\mathcal{U}_G$, and it is described in Table 4. Except for the case of $2 \times 2$ grids, all configurations in $\mathcal{U}_G$ can be detected by simply applying Theorem 1 as they are all partitive configurations. In [10] it is shown that all the configurations not belonging to $\mathcal{U}_G$ are gatherable; in the following, we describe the gathering algorithms.

**Odd × Odd Grids.** This case is trivially solvable, in fact in odd × odd grids, a robot can always detect, during its `Look` operation, the central vertex of the grid $M[\lceil \frac{n}{2} \rceil, \lceil \frac{m}{2} \rceil]$, regardless of its possible view. This means that all the robots can move toward the center, concurrently.

**Odd × Even Grids.** When the initial configuration does not belong to $\mathcal{U}_G$, it is always possible to devise an algorithm achieving the gathering without any multiplicity detection.

The idea is to distinguish between the two vertices that are the central vertices of the odd borders of the grid. If $m$ ($n$, respectively) is odd, then the two mentioned vertices are given by positions $M[1, \lceil \frac{m}{2} \rceil]$ and $M[n, \lceil \frac{m}{2} \rceil]$ ($M[\lceil \frac{n}{2} \rceil, 1]$ and $M[\lceil \frac{n}{2} \rceil, m]$, respectively). The line connecting those two vertices will be denoted as the NS line. One of the two extreme vertices on the NS line will be the place where the gathering is finalized. In order to select the gathering vertex, a robot considers the line passing through the central edges of the even sides of the grid (denoted as the EW line) dividing the grid into two halves. The idea is to distinguish a north and a south part among the two halves and the gathering vertex will be the one in the north half. The north is the half with more vertices occupied by robots, if any. If the number of occupied vertices in the two halves is the same, then some more computations are required. In both cases, the robots move from the south to the north until all the robots will be in the north part. Note that, during such a stage, if multiplicities are created in the south, then the number of occupied vertices decreases with respect to the north part. If multiplicities are created in the north, it means that at least a robot has moved from the south to the north part, still preserving the required distinction.

Once all the robots belong to one half of the grid, then they are allowed to move, during their Move operation, towards the gathering vertex. In fact, such a vertex is well-defined and cannot change as the robots are not allowed to move to the other half of the grid.

**Even × Even Grids.** For the even×even case, in [10] it is shown that all the initial configurations not in $\mathcal{U}_G$ are gatherable without any multiplicity detection, except for the case of 2 × 2 grids. In order to achieve this result, it is first assumed that at least one vertex on the border of the grid is occupied. Then, the gathering vertex is identified among the eight sequences of distances (number of empty vertices) between occupied vertices obtained by traversing the grid starting from the four corners and proceeding towards the two possible directions (see, e.g. Fig. 4).

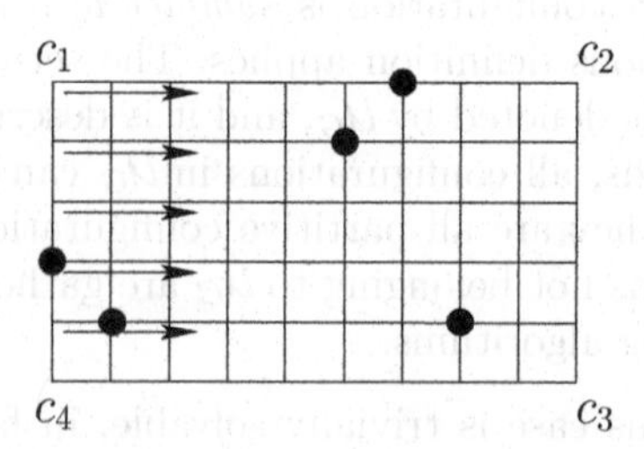


**Fig. 4.** Case of a 6 × 10 grid. The arrows indicate the horizontal direction of the reading from corner $c_1$, it gives (6, 8, 14, 10, 5, 12). The other seven sequences read by the robots are: (3, 6, 20, 4, 9, 13) from $c_1$ vertically, (3, 10, 24, 2, 5, 11) and (16, 1, 6, 26, 4, 2) from $c_2$ horizontally and vertically, respectively, (12, 5, 10, 14, 8, 6) and (13, 9, 4, 20, 6, 3) from $c_3$, (11, 5, 2, 24, 10, 3) and (2, 4, 26, 6, 1, 16) from $c_4$. The *minimal* sequence is (2, 4, 26, 6, 1, 16) and $c = c_4$.

The lexicographically smallest sequence between the two readings from any corner is associated to the corner itself. In rectangular grids, these two sequences can be equal but it is possible to distinguish one of them by assuming the reading in the direction of the smallest side. The *minimal* sequence is defined as follows. If the configuration is symmetric, it is the smallest sequence between the two sequences associated to the two corners through which passes the axis of symmetry, otherwise it is the smallest among the four sequences associated to the four corners. In any case there exists a minimal sequence which identifies either a single corner, of two symmetric corners. In the former case, the identified corner will be the vertex where to finalize the gathering. In the latter case, the middle vertex on the border between the two identified corners will be the vertex where all robots gather. The main difficulties come from designing moves that maintain the same detected gathering vertex during all movements.

The following statement summarizes the results for grids.

**Theorem 7 (Gathering on grids).** *Let $G$ be a grid and $C = (G, \mu)$ be an initial configuration composed by robots endowed with no multiplicity detection capability. Then, $C$ is gatherable if and only if $C \notin \mathcal{U}_G$.*

According to Remark 1, the above result states that the gathering problem on grids is fully characterized.

# 4 From General Algorithms to Optimality

In this section, we consider the gathering problem on graphs when a cost measure is associated with the provided solution, and we ask for optimality. Depending on the chosen cost measure, different strategies can be designed.

## 4.1 Optimization Problems for Robot-Based Computing Systems

In this section, we recall from [5] how the classical notion of optimization problems (cf. [2]) is extended to optimization problems solvable in the context of robot-based computing systems. According to the new framework, we describe how the gathering problem can be reformulated as an optimization problem.

Let $\Pi$ be an optimization problem for robot-based computing systems. For the sake of clarity, we address the minimization case only. The maximization case can be derived analogously. Problem $\Pi$ consists of a triple $(\mathcal{I}, sol, mis)$, where:

- $\mathcal{I}$ is the set of instances (i.e., all possible initial configurations);
- *sol* is a function that maps each initial configuration $C \in \mathcal{I}$ to the set $sol(C)$ of feasible solutions of $C$;
- given an instance $C \in \mathcal{I}$ and a solution $s \in sol(C)$, then $mis(C, s)$ denotes the real positive measure of $s$, and the function *mis* is called *objective function* for $\Pi$.

The goal of $\Pi$ with respect to an instance $C$ is to find an optimal solution, that is, a feasible solution $s \in sol(C)$ such that $mis(C, s)$ equals the value

$$\min\{mis(C, s') : s' \in sol(C)\}.$$

In the following, *opt* will denote the function mapping an instance $C$ to the measure of an optimum solution in $sol(C)$. Given an instance $C$ and a solution $s \in sol(C)$, we define the *performance ratio* of $s$ with respect to $C$ as

$$R(C, s) = \frac{mis(C, s)}{opt(C)}.$$

The performance ratio is always a number greater than or equal to 1 and is as close to 1 as the solution $s$ is close to the optimal solution.

Now, let $\mathcal{A}$ be an algorithm for $\Pi$ and $C$ be an initial configuration. Even though we are dealing with deterministic algorithms, different executions of $\mathcal{A}$ starting from the same initial configuration $C$ can lead to different solutions. In fact, in the described asynchronous setting, an execution depends on the time required by the scheduled activities, and this is implemented by the behavior of the adversary. Then, there exists a set $sol_{\mathcal{A}}(C) \subseteq sol(C)$ of solutions, each corresponding to a possible execution of $\mathcal{A}$ starting from $C$. If $\mathcal{A}(C)$ is a solution $s \in sol_{\mathcal{A}}(C)$ which maximizes $mis(C, s)$, that is

$$\mathcal{A}(C) = \arg \max_{s \in sol_{\mathcal{A}}(C)} mis(C, s),$$

then:

- we say that $\mathcal{A}$ is an *optimal* algorithm for $\Pi$ if $R(C, \mathcal{A}(C)) = 1$ for each instance $C \in \mathcal{I}$;
- given a function $f : \mathbb{N} \rightarrow (1, \infty)$, we say that $\mathcal{A}$ is a $f(n)$-*approximation* for $\Pi$ if $R(C, \mathcal{A}(C)) \leq f(|C|)$ for each instance $C \in \mathcal{I}$. Here $|C|$ denotes the size of a configuration $C$.

**Defining the Gathering as an Optimization Problem.** We can now formalize the gathering problem as an optimization problem $(\mathcal{I}, sol, mis)$, where $\mathcal{I}$ is the set of all the possible initial configurations for the gathering problem. Given an initial configuration $C = (G, \mu)$, an execution of a gathering algorithm $\mathcal{A}$ for $C$ can be seen as a set $\mathcal{P}$ containing paths in the graph $G$. Each path in $\mathcal{P}$ models all the movements performed by a specific robot starting from its initial position and ending on the final gathering vertex. A movement performed by a robot during a computational cycle corresponds to a subpath made of just two adjacent vertices. So, a gathering solution is a set of paths, each one starting from a distinct vertex and ending on the same final gathering vertex. The set of all the gathering solutions for a configuration $C$ defines $sol(C)$. Concerning $mis(C, \mathcal{P})$, different measures can be considered. In the remain of the section we assume:

– $mis(C, \mathcal{P})$ corresponds to $\sum_{P \in \mathcal{P}} length(P)$, where $length(P)$ is the length of the path $P$ expressed as the number of edges in the path.

Other cost measures can be certainly investigated. For instance, an interesting case might be when $mis(C, \mathcal{P})$ corresponds to $\max_{P \in \mathcal{P}} length(P)$. In other words, the goal corresponds to minimize the maximum distance traversed by a single robot.

Going back to the cost measure that will be considered in the next sections concerning the minimization of the lengths of all the paths traversed by all robots, this implies that robots must gather on a Weber point by moving along shortest paths. A Weber point in a graph, in fact, assures the minimization of the considered function *mis*. However, in order to correctly compute the Weber points, robots must be aware of the exact number of robots occupying vertices, that is they need the global-strong multiplicity detection. Consequently, for all the results reported in this section we assume the global-strong multiplicity detection.

### 4.2 Weber Points in Graphs

In this section, general results that allow to define optimal gathering algorithms are recalled from [19]. Such results are based on the notion of Weber points on graphs (cf Sect. 1.2).

Given a configuration $(G, \mu)$, with $G = (V, E)$, the *centrality* of each $v \in V$, is

$$c_{G,\mu}(v) = \sum_{u \in V} d(u, v) \cdot \mu(u).$$

A vertex $v \in V$ is a *Weber point* of $G$ if it has the minimal centrality, that is, $c_{G,\mu}(v) = \min\{c_{G,\mu}(u) \mid u \in V\}$. Whenever clear by the context, the centrality of a vertex $v$ is simply denoted by $c_\mu(v)$, or simply $c(v)$.

By definition, a Weber point is a vertex that has the overall minimal distance from all the robots in the configuration. Then, an algorithm that gathers all the robots on a Weber point via shortest paths is optimum with respect to the total number of moves. More formally, a gathering algorithm must define the sequence of moves for each robot, leading to a final configuration. A move is the change of the position of a single robot from a vertex $u$ to an adjacent vertex $v$. This equals to change the configuration from, say $(G, \mu)$ to $(G, \mu')$, where $\mu(w) = \mu'(w)\ \forall w \in V \setminus \{u, v\}$, $\mu'(u) = \mu(u) - 1$ and $\mu'(v) = \mu(v) + 1$. A robot perceives its position on the graph $G$ if $(G, \mu)$ is asymmetric. Whereas, if $(G, \mu)$ admits a non-identity isomorphism $\varphi$, a robot cannot distinguish its position at $u$ from $\varphi(u)$. As a consequence, two robots (e.g., one on $u$ and one on $\varphi(u)$) can decide to move simultaneously, as any algorithm is unable to distinguish between them. This fact greatly increases the difficulty to devise a gathering algorithm for symmetric configurations.

We start by observing that the gathering problem can be characterized as follows:

**Proposition 1.** *A configuration $((V, E), \mu)$ is final (i.e., all robots have gathered) if and only if there exists a vertex $v \in V$ such that $c(v) = 0$.*

Along the text, we say that a robot on a vertex $u$ *moves towards a vertex $v$* if it moves to a vertex adjacent to $u$ along a shortest path between $u$ and $v$.

**Theorem 8.** *Given any configuration $((V, E), \mu)$ with Weber points in $X \subseteq V$, a move of a robot towards a Weber point $x$ gives rise to a configuration $((V, E), \mu')$ with Weber points in $X' \subseteq V$ such that:*

*1. $c_{\mu'}(v) = c_\mu(v) - 1$ for each $v \in X'$;*
*2. $x \in X'$;*
*3. $X' \subseteq X$.*

When the configuration admits a unique Weber point (or a Weber point can be uniquely determined), the above theorem suggests an optimal gathering algorithm that also exploits concurrency among robots. In fact, regardless other robots, each one can move towards the only Weber point via the shortest path, until finalizing the gathering.

**Corollary 1.** *Let $C = ((V, E), \mu)$ be a configuration. Then:*

- *if $C$ admits only one Weber point then the gathering can be achieved by an optimal algorithm;*
- *if there exists a real function $f : V \longrightarrow \mathbb{R}^+$ such that $f$ admits only one minimum on the set of Weber points, then gathering can be achieved by an optimal algorithm.*

We now provide optimal gathering algorithms for specific topologies, namely trees, rings and infinite grids.

### 4.3 Optimal Gathering on Trees

In this section, we recall from [19] the characterization of optimal gathering on trees. To this aim, a general algorithm that always achieves the optimal gathering starting from configurations not falling into the hypothesis of Theorem 1 is described. This algorithm exploits interesting properties resulting from the tree topology.

Let $(T, \mu)$ be a configuration for a tree $T$, and $a$ and $b$ two of its vertices. We denote by $P_{ab}$ the path between $a$ and $b$ of length $|P_{ab}|$. Tree $T$ can be decomposed into three subtrees, see Fig. 5: the one containing $a$ when removing from $T$ the edge incident to $a$ in $P_{ab}$, and denoted by $T_a$; the one containing $b$ when removing the edge incident to $b$ in $P_{ab}$, and denoted by $T_b$; and the third one obtained from $T$ by removing both $T_a$ and $T_b$, and denoted by $T_{ab}$. Let $L_a = \sum_{v \in T_a} \mu(v)$ and $L_b = \sum_{v \in T_b} \mu(v)$, that is the number of robots in $T_a$ and $T_b$, respectively.

**Theorem 9.** *Let $(T, \mu)$ be a configuration for any tree $T$. Then, the following properties hold:*

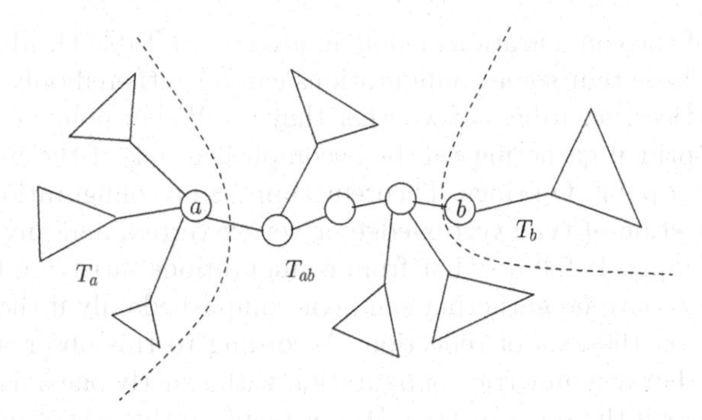


**Fig. 5.** Partitioning of a tree into three subtrees.

- *given two distinct Weber points a and b, $T_{ab}$ does not contain any robots;*
- *given two distinct Weber points a and b, $L_a = L_b$;*
- *the Weber points form a path;*
- *if the number of robots is odd, then there exists only one Weber point.*

The above properties imply the existence of a simple optimal gathering algorithm when the number of robots is odd. A complete characterization about the existence of optimal gathering algorithms on trees is given by the next theorem. It shows that an optimal algorithm exists unless there is an automorphism that maps each vertex to a different one.

**Theorem 10 (Optimal gathering on trees).** *There exists an optimal gathering algorithm for a configuration $C = (T, \mu)$ on a tree $T$ if and only if $C$ is not balanced.*

The algorithm exploits similar properties of that presented in Sect. 3.2 [11] dealing with feasibility only. It is known, in fact, that a tree admits either one or two centers. Then, for optimality purposes the algorithm selects opportunely the Weber point closest to one center. Considering Fig. 6a, it is easy to provide configurations where this algorithm performs the gathering in two moves, while the algorithm in Sect. 3.2 requires $n$ moves.

Before concluding this section, it is interesting to characterize the disposal of Weber points on the degenerate case of paths. This is of practical interest in the next section for characterizing Weber points on rings.

**Lemma 1.** *Given a configuration $(P, \mu)$ where $P$ is any path graph, the set of Weber points is constituted either by one occupied vertex, or by one subpath whose extremes are occupied.*

### 4.4 Optimal Gathering on Rings

In this section, we summarize results from [19] that fully characterize optimal gathering on rings. We start remarking that the gathering cannot be optimally solved in any of the configurations belonging to $\mathcal{U}_R$ (cf Table 1). Moreover, from Theorem 2, we have that some configurations can be gathered only at some predetermined vertices, regardless of whether they are Weber points or not. Hence, in such cases optimal gathering can be accomplished only if the predetermined vertex is a Weber point. On rings, Theorem 2 applies on configurations admitting one axis of reflection of type vertex-edge or vertex-vertex, and any vertex lying on the axis is empty. It follows that from configurations satisfying the hypothesis of Theorem 2, optimal gathering can be accomplished only if there is at least a Weber point on the axis of reflection. According to this observation, in this section we say that a symmetric configuration with exactly one axis of reflection has a *not-wp axis* if there not exists a Weber point on the axis of reflection. The set of all configurations where the gathering cannot be optimally solved in rings is denoted by $\mathcal{U}^*_R$, and it is described in Table 5.

**Table 5.** Resume of the impossibility results about optimal gathering in a ring under the `Look-Compute-Move` model even assuming global-strong multiplicity detection. The first three rows are derived from Table 1. All the mentioned configurations are initial and form the set $\mathcal{U}^*_R$. Symbols $n$ and $k$ refer to the number of vertices and number of robots, respectively.

| Configuration type | $n$ | $k$ | Papers |
|---|---|---|---|
| Rotational or with edge-edge axis | - | - | [19,24] |
| - | - | $k = 2$ | [24] |
| SP4 | Odd | $k = 4$ | [12,16,17,23] |
| With not-wp axis | - | - | [18] |

In [19], authors provided an algorithm able to assure optimal gathering in each configuration not belonging to $\mathcal{U}^*_R$. Before describing this algorithm, we recall some useful properties concerning the disposal of Weber points on rings.

**Lemma 2.** *Given a configuration* $(R, \mu)$ *on any ring* $R$, *if an empty vertex* $u$ *is a Weber point then also its neighbors are Weber points.*

By the above lemma, as for the path case, if there exists a sequence of vertices that are Weber points, then the extremes of such a sequence are Weber points occupied by robots. It is worth noting that on rings there might occur more than one of such sequences.

As first result on rings, the next lemma provides an algorithm that assures optimal gathering from asymmetric configurations. Actually, such an algorithm will be used later as part of the general algorithm for providing optimal gathering in each configuration not belonging to $\mathcal{U}^*_R$.

Let $C = (\mu(v_0), \mu(v_1), \ldots, \mu(v_{n-1}))$ be one of the possible views computed by a robot occupying vertex $v_0$ during its `Look` phase according its clockwise direction. We denote by $\overline{C} = (\mu(v_0), \mu(v_{n-1}), \mu(v_{n-2}), \ldots, \mu(v_1))$,

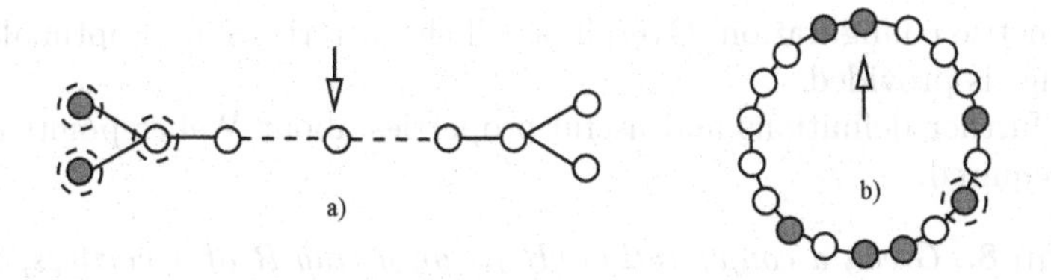


**Fig. 6.** A gray vertex indicates the presence of one robot; Dashed circled vertices are Weber points. Dashed line stands for an undefined sequence of empty vertices. Vertices pointed by an arrow represent the gathering vertices with respect to algorithms in [11] for a), and in [24] for b).

and by $C_i$ the configuration obtained by reading $C$ starting from $v_i$, that is $C_i = (\mu(v_i), \mu(v_{(i+1) \bmod n}), \ldots, \mu(v_{(i+j) \bmod n}))$.

By referring to Fig. 6b, the robot pointed by the arrow has view $C = (1,0,0,0,1,0,1,0,1,1,0,1,0,0,0,0,0,1)$ if it reads in the clockwise direction. Then, $\overline{C} = (1,1,0,0,0,0,0,1,0,1,1,0,1,0,1,0,0,0)$. Its lexicographical maximum view is $\overline{C}$, while the absolute maximum view of the configuration is $\overline{C_9} = (1,1,0,1,0,1,0,0,0,1,1,0,0,0,0,0,1,0)$.

**Lemma 3.** *Given an asymmetric initial configuration* $(R, \mu)$ *on any ring* $R$ *of* $n$ *vertices, it is always possible to assure optimal gathering.*

The main difference of the optimal algorithm with that provided in [24] dealing just with feasibility is in the choice of the vertex where a multiplicity is created. Once this is done, the two algorithms finalize the gathering on the created multiplicity by moving robots along the shortest paths towards it.

The algorithm proposed in [24] considers the longest interval $I$ of empty vertices. Among the two intervals of empty vertices neighboring to $I$, the shortest one was reduced by moving the robot delimiting $I$. Ties were broken by the asymmetry of the configuration. The described move was repeated until creating a multiplicity.

Here [19], the algorithm suitably selects a Weber point (or an interval of Weber points) hence moving robots towards it. By Theorem 8, the selected Weber point remains as such, whereas other Weber points disappear move after move. Moves are performed so as the configuration remains asymmetric, at least until only one Weber point remains. From there on, the finalization becomes easy.

Figure 6b shows a configuration where the optimal algorithm requires 25 moves while the algorithm in [24] takes 35 moves. It is easy to provide worsen instances where $I$ is far apart from Weber points, hence resulting in a much larger difference with our algorithm in terms of computed moves.

For a fair comparison, we remark that the algorithm in [24] deals with the global-weak multiplicity detection while in [19] the global-strong version is assumed.

We now summarize on the results for gathering algorithms dealing with all configurations that allow optimal gathering, exploiting the algorithm designed for asymmetric configuration. Overall, a full characterization of optimal gathering on rings is provided.

Some further definitions and useful properties about Weber points on rings are still required.

**Definition 8.** *Given a configuration $(R, \mu)$ on a ring $R$ of $n$ vertices, two vertices are said* antipodal *if their distance is $\lfloor \frac{n}{2} \rfloor$. Two robots are said* antipodal *if they lie on two antipodal vertices.*

**Lemma 4.** *Given a configuration $C = (R, \mu)$ on any ring $R$ with an even number of vertices, if vertex $u$ is a Weber point then it is a Weber point also in the configuration $C'$ obtained from $C$ by removing all the pairs of antipodal robots.*

**Corollary 2.** *Given a configuration $C = (R, \mu)$ on a ring $R$ with an even number of vertices, if it contains only pairs of antipodal robots then $C$ is rotational.*

It is worth noting that when a configuration satisfies the hypothesis of Corollary 2 then by Lemma 4 all its vertices are Weber points since all the vertices of an empty ring have zero centrality.

**Lemma 5.** *Let $C = (R, \mu)$ be a configuration on any ring $R$ with an even number of vertices, and not admitting rotations. If vertex $v$ is a Weber point in $C$ then its antipodal vertex is not a Weber point.*

**Lemma 6.** *Given a configuration $C$ on any ring $R$ with an odd number of vertices, if two adjacent vertices $u$ and $v$ are two Weber points, and at most one of them is occupied, then vertex $w$ whose antipodal vertices are $u$ and $v$ is not a Weber point.*

We are now ready to define another part of the main algorithm to deal with symmetric configurations with an odd number of robots.

**Lemma 7.** *Given a symmetric configuration $C = (R, \mu)$ on any ring $R$ with an odd number of robots, then optimal gathering can be accomplished.*

The algorithm provided in the proof of Lemma 7 works as follows. If $C$ is symmetric, there must be exactly one robot lying on the axis of symmetry since the number of robots is odd. If there is only one Weber point, then optimal gathering is achieved by exploiting Corollary 1. If there are at least two Weber points, the algorithm then moves the robot on the axis towards one of the two possible directions, indiscriminately. By Theorem 8, all the Weber points contained in the semi-ring where the robot moved are maintained, while all the Weber points in the other semi-ring (that was originally symmetric) disappear. The only exception might be the antipodal vertex (if any) with respect to the original location of the moved robot when the ring is composed of an even number of vertices. Again, if after the move there is only one Weber point, then optimal gathering is accomplished by exploiting Corollary 1. If there are at least two Weber points, the obtained configuration can be symmetric or asymmetric (possibly containing a multiplicity). In the former case, the algorithm moves again the new robot on the axis, and by [24] we are assured that this can happen a finite number of times until reaching an asymmetric configuration. From asymmetric configurations, Lemma 2 can be exploited to finalize the gathering.

For the case of even number of robots on symmetric configurations, we need two more definitions.

**Definition 9.** *Given a symmetric configuration on a ring $R$ of $n$ vertices, a vertex $v$ is called* north *if it lies on the axis and it is a Weber point. The edge whose endpoints are the antipodal vertices (in case of vertex-edge symmetry) of $v$ or its antipodal vertex (in case of vertex-vertex symmetry) is called* south.

It is worth nothing that we use the above definition only for symmetric configurations with single axis of type vertex-vertex or vertex-edge, with a Weber point on the axis. In particular, in case of vertex-vertex symmetry the definition is not ambiguous. In fact, from Lemma 5 the two vertices on the axis cannot be both Weber points. Contrary, if both are not Weber points and are empty, by Theorem 2 optimal gathering cannot be accomplished, hence we do not need such definitions. As we are going to see, if both are not Weber points but are occupied, optimal gathering can be assured but the strategy does not require to define north and south. Whereas, another definition required together with north and south is the following.

**Definition 10.** *Given a symmetric configuration on a ring $R$, the line orthogonal to the axis of symmetry, cutting $R$ on two edges into two subrings whose size differ of at most one vertex in favor of the southern side is called the* horizon.

We are now ready to describe the optimal gathering algorithm.

**Theorem 11 (Optimal gathering on rings).** *There exists an optimal gathering algorithm for a configuration $C = (R, \mu)$ on a ring $R$ of $n$ vertices if and only if $C \notin \mathcal{U}^*_R$.*

The algorithm provided in the proof of Theorem 11 works as follows.

Configurations with an odd number of robots have been solved by Lemma 7, and robots can always recognize they are in such a case by computing $k = \sum_{i=0}^{n-1} \mu(v_i)$.

The asymmetric case has been already solved by Lemma 3. The proposed technique must be slightly modified in order to integrate it with symmetric cases, hence obtaining a unique optimal gathering algorithm characterizing all possible configurations.

If configuration $C$ admits a single axis of reflection passing through two robots, by Lemma 5, the vertices where such robots lie cannot be both Weber points. If one is a Weber point and it is the only one among all vertices, then by Corollary 1 optimal gathering can be accomplished. If there is more than one Weber point in $C$ then the algorithm makes move one of the robots on the axis (towards any direction) as follows.

If there is an odd number of Weber points (the north is a Weber point, necessarily), then the robot occupying the south is moved. In doing so, the number of Weber points remains odd since those initially residing at one side of the axis of symmetry have disappeared but not the one on the north. The obtained configuration can be still reflective (of type robot-robot or vertex-vertex) with a Weber point on the axis and less Weber points than the original one. The case of vertex-vertex reflection will be discussed later, while for the case of robot-robot reflection same arguments can be applied again.

If there is an even number of Weber points (the north is not a Weber point), then the robot occupying the north is moved unless it creates a new axis of symmetry. In such a case, the one on the south is moved and we are sure that the configuration becomes asymmetric.

If $C$ is reflective without robots on the axis, the algorithm allows only moves towards north where the gathering will be accomplished, eventually. The north, which is a Weber point, must exist as otherwise, by Theorem 2, optimal gathering is not possible.

In general, from Lemma 2, the set of Weber points in $R$ is given by a set of paths, and by hypothesis there is at least one path of Weber points (possibly made of just one vertex) containing the north. Moreover, due to symmetry and by Lemmas 5 and 6, the number of such paths is odd and the two adjacent vertices at one side of $R$ divided by the horizon cannot be both Weber points unless they are both occupied.

Similarly to the algorithm proposed in [23], here the symmetry of the configuration is maintained until a single multiplicity in the north is created. Due to asynchrony, either one or two symmetric robots move. In the former case, robots can detect whether the current asymmetric configuration may have been obtained from a symmetric one, and hence they can recognize the unique robot that can (re)-establish the original symmetry. Note that, the algorithm leads to a symmetric configuration even though the initial configuration is asymmetric but obtainable from a symmetric one. This is the only modification required to the algorithm provided in the proof of Lemma 3 for asymmetric configurations.

Moves are accomplished so that the north is maintained as a Weber point, whereas other Weber points disappear move by move. Once a multiplicity is created in the north, then the finalization is easy.

### 4.5 Optimal Gathering on Infinite Grids

In this section, we summarize results from [18] that fully characterize optimal gathering on *infinite grids*. Let an infinite path be the graph $P = (\mathbb{Z}, E)$ with $E = \{\{i, i+1\} : i \in \mathbb{Z}\}$. An infinite grid is defined as the Cartesian product $G^{\infty} = P \times P$. A vertex of the grid is then an ordered pair of integers called *coordinates*. Given $G^{\infty}$, then $C = (G^{\infty}, \mu)$ is a configuration on $G^{\infty}$.

Notice that on infinite grids the topology does not help in detecting a gathering vertex. Nonetheless, the interest in infinite grids also arises from the fact that they represent a natural discretization of the plane. We detect all the specific configurations where gathering cannot be performed. For all other configurations, we describe the basis of a distributed algorithm that exploits the global-strong multiplicity detection and, assures gathering on a Weber point by letting robots move along the shortest paths toward such a vertex, i.e., the algorithm is optimal in terms of moves.

Let $C = (G^{\infty}, \mu)$ be a configuration and $S_C$ be the minimal (finite) sub-grid containing all the occupied vertices of the infinite grid $G^{\infty}$, and $(S_C, \mu)$ be the corresponding configuration. It is worth mentioning that $S_C$ may change while robots move. As a consequence, even though $S_C$ is a finite grid, the approach described in Sect. 3.4 [10] for finite grids cannot be applied (as it is strongly dependent on the dimensions of the grid where robots reside).

During the Look phase, a robot perceives $(S_C, \mu)$ and it is able to recognize its position on $S_C$ if $(G^{\infty}, \mu)$ is asymmetric. Whereas, if $(G^{\infty}, \mu)$ admits an isometry $\varphi$ different from the identity, a robot cannot distinguish its position at $u$ from $\varphi(u)$, unless $u = \varphi(u)$.

In an infinite grid, the center of a rotation can be a vertex, or the center of an edge, or the center of the area surrounded by four vertices, whereas the angle of rotation can be of 90° or 180°. Reflections axis can be horizontal (vertical), passing through vertices or through the middle of edges, or diagonal (45°), passing through vertices. If we assume the infinite grid embedded in a Cartesian plane, it is not difficult to see that other than rotations and reflections it admits also *translations*, that is a shifting of the vertices by applying the same displacement to each vertex. Regarding translations, even if they are possible for infinite grids, they do not belong to any automorphism group of configurations as these are defined for a finite (not null) number of robots. Note that, an infinite number of robots (or no robots at all) is required also when the configuration admits two parallel axes of symmetry, one axis and one center of rotation not lying on the axis, or two distinct centers of rotation. Moreover the automorphism group of a configuration with a finite number of robots is finite.

In order to check whether the current configuration could have been obtained from a symmetric one, we introduce the concept of *previous position* for a robot. Sometimes, an algorithm simulates itself by considering a configuration $C'$ which is identical to the current configuration $C$ but for the position of one robot $r$. If an execution of the algorithm can lead from $C'$ to $C$ then the simulated position of $r$ in $C'$ is called a *previous position* for $r$. This method will be used to detect possible pending moves when $C'$ is symmetric.

According to Corollary 1, in a configuration that admits only one Weber point the gathering can be achieved by an optimal algorithm.

**Impossibility Results.** We have already observed that, in general, a partitive configuration is ungatherable. In infinite grids, this result implies that all initial configurations with an axis of symmetry not passing through vertices or admitting a rotation with a center not coinciding with a vertex, are ungatherable. In fact, all such configurations are partitive with orbits of size at least two, and only those admitting rotations of 90° have orbits of size four. For the sake of convenience, we introduce the following terminology: a configuration $(G^{\infty}, \mu)$ is (1) *edge-symmetric* if it has an axis of symmetry not passing through vertices, (2) *rotational* if it admits a rotation with a center not coinciding with a vertex, (3) *two-robots* if it contains only two robots (or equivalently, two multiplicities of the same size), and (4) *four-corners* if it contains only four robots (or equivalently, four multiplicities of the same size) disposed on the corners of $S_C$.

The set of all the ungatherable configurations for infinite grids is denoted by $\mathcal{U}_{G^\infty}$, and it is described in Table 6.

**Table 6.** Resume of the impossibility results about gathering in an infinite grid under the `Look-Compute-Move` model even assuming global-strong multiplicity detection. All the mentioned configurations are initial and form the set $\mathcal{U}_{G^\infty}$.

| Configuration type | Papers |
|---|---|
| Edge-symmetric | Theorem 1 |
| Rotational | Theorem 1 |
| Two-robots | [18] |
| Four-corners | [18] |

It is worth noting (see [10]) that gathering on finite grids was possible without any multiplicity detection due to the existence of special vertices like corners. Here instead we assume the global-strong multiplicity detection in order to be sure that robots can always compute the correct Weber points, even when multiplicities occur.

In the remaining of this section, we recall from [18] an algorithm which is able to optimally solve the gathering problem in each configuration not belonging to $\mathcal{U}_{G^\infty}$. It is important to point out that in the case of infinite grids each time the problem is solvable, then it can be done in an optimal way.

**One-Dimensional Grids.** We first consider infinite paths as grids with one row and infinitely many columns.

**Lemma 8.** *If the number of robots k is odd, then there exists only one Weber point. If k is even, then all vertices of the subpath delimited by the two central robots (including the vertices where such robots lie) are Weber points.*

**Theorem 12.** *Optimal gathering on one-dimensional grids is always achievable except for configurations with only two robots or admitting partitive automorphisms.*

In the previous theorem, the ungatherable cases simply follow from the first three rows of Table 6. When the number of robots is odd, from Lemma 8 there exists only one Weber point and optimal gathering can be achieved by Corollary 1.

Let us observe how the optimal gathering is achievable when the number of robots is even. In such a case, if the configuration is symmetric, then the subpath of Weber points must be odd as otherwise the configuration is partitive. The idea is then to move the robots delimiting the Weber points toward the central vertex. If both move synchronously, the configuration remains symmetric but the interval of Weber points is reduced until only the Weber point at the central vertex remains. If only one moves, it is possible to recognize the robot that has to move to (re)-establish the symmetry. In fact, considering the two intervals of free vertices neighboring the robots delimiting the Weber points, the algorithm allows to move the robot delimiting the shortest interval.

When the number of robots is even, but the configuration is asymmetric, then either it is at one move from a possible symmetry which is allowed, or one of the two robots delimiting the Weber points can be chosen to move toward the other one without creating a symmetry until only one Weber point remains.

Finally, when there is only one Weber point, from Corollary 1, all robots can move safely toward it. It is worth to notice that such an algorithm also works when the input configuration admits multiplicities.

**Two-Dimensional Grids.** We now describe a general optimal algorithm to solve the gathering problem for each configuration $C = (G^{\infty}, \mu)$ such that $C \notin \mathcal{U}_{G^{\infty}}$. From Corollary 1, if the configuration $C$ admits only one Weber point, then optimal gathering can be accomplished. Another characterization is provided by considering $S_C$, and in particular the projections of the robots to the two generating paths $P_1$ and $P_2$ of $G^{\infty}$. Given a robot on a generic vertex $(i, j)$ of $G^{\infty}$, its projections on $P_1$ and $P_2$ are a robot on vertex $i$ and a robot on vertex $j$, respectively. This gives rise to two configurations $(P_1, \mu_1)$ and $(P_2, \mu_2)$ such that $\mu_1(v) = \sum_j \mu((v, j))$ and $\mu_2(v) = \sum_i \mu((i, v))$. As the movements on a grid are either vertical or horizontal, solving the gathering with respect to the two dimensions separately, solves the general problem.

**Theorem 13.** *Given any configuration $C = (G^{\infty}, \mu)$ with $G^{\infty} = P_1 \times P_2$, if $(P_1, \mu_1)$ and $(P_2, \mu_2)$ are optimally gatherable, then also $C$ is optimally gatherable.*

The optimal gathering considered in the previous theorem is obtained by simply considering $(P_1, \mu_1)$ and $(P_2, \mu_2)$ separately. Each time a robot wakes-up, it can move with respect to any of the two instances indiscriminately, as they are independent to each other. Theorem 12 guarantees optimal gathering on both the instances even though they might contain multiplicities.

Note that there are gatherable configurations that do not satisfy the assumptions of Theorem 13. Hence, a more general strategy must be designed in order to cope with all the gatherable configurations.

Given a configuration $C = (G^{\infty}, \mu)$, let $G_{\mathtt{WP}}(C)$ be the subgraph induced by its Weber points. The next theorem provides a useful characterization about the arrangement of Weber points in a configuration.

**Theorem 14.** *Given any configuration $C = (G^\infty, \mu)$ with $G^\infty = P_1 \times P_2$, $G_{\mathtt{WP}}(C)$ is a finite grid defined by the Cartesian product of the subpaths induced by the Weber points belonging to $(P_1, \mu_1)$ and $(P_2, \mu_2)$.*

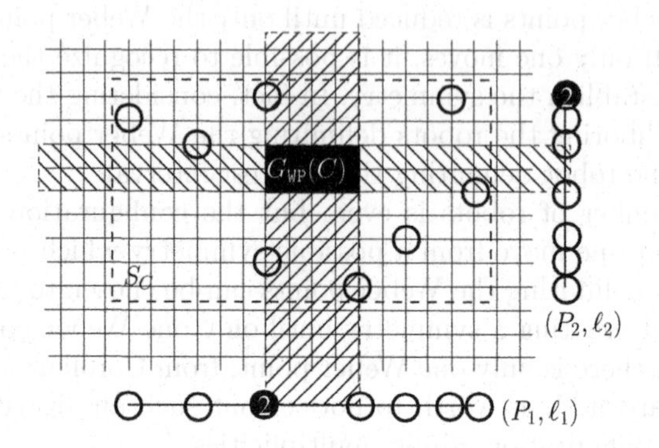


**Fig. 7.** A sample configuration $C$ which induces $S_C$, $G_{\mathtt{WP}}(C)$, and its projections to the sides of $S_C$.

By referring to Fig. 7, it is worth noting that $G_{\mathtt{WP}}(C)$, for some configuration $C$, is in general a finite grid where robots can occupy only the corners. Moreover, all the vertices belonging to the strips from $G_{\mathtt{WP}}(C)$ to the borders of $S_C$ cannot be occupied, but for the ones sharing coordinates with the border of $G_{\mathtt{WP}}(C)$. These robots will be said to *determine* $G_{\mathtt{WP}}(C)$. Note that, given a configuration $C$ with $k$ robots, evaluating the set of Weber points has time complexity $O(|S_C| \times k)$.

**Grids with an Odd Number of Robots.** By Lemma 8, an odd number of robots implies a single Weber point for each instance on the two paths generating $G^\infty$. By Theorem 14, the Cartesian product of those two Weber points constitutes the only Weber point of the configuration, hence by Corollary 1 optimal gathering can be assured. Then, the following results follows.

**Corollary 3.** *If the number of robots in a grid $G^\infty$ is odd, then optimal gathering can be accomplished.*

**Grids with an Even Number of Robots.** When the number of robots is even, the proposed algorithm from [18] considers various cases.

As first distinction, it considers whether $S_C$ has all sides of odd length and its center is a Weber point. If this is the case, then gathering can be accomplished at the center of $S_C$. The idea at the basis of the strategy is to move all the robots not lying on the border of $S_C$ toward the center that becomes the only Weber point of the current configuration. From there on, all the other robots can join the unique Weber point. This can be easily realized if the number of robots is "sufficiently large", while if it is too small, then specific strategies are required.

If $S_C$ has a side of even length or its center is not a Weber point, then a different strategy is designed. In this case, the final gathering vertex will be detected at the end of the process, as robots are moved so as to reduce the number of Weber points in the configuration until one Weber point remains. From there on Corollary 1 can be applied. Moves are performed according to the number of corners of $S_C$ occupied by robots.

The next theorem summarizes the obtained results about infinite grids.

**Theorem 15 (Optimal gathering on infinite grids).** *There exists an optimal gathering algorithm for a configuration $C = (G^\infty, \mu)$ on an infinite grid $G^\infty$ if and only if $C \notin \mathcal{U}_{G^\infty}$.*

## 5 Conclusion

The chapter surveys on the latest results concerning the gathering task for robots moving on graphs. This poses interesting observations with respect to the case of robots moving on the Euclidean plane. As first consequence, there are much more cases where gathering cannot be solved, whereas in the Euclidean plane these concern only the case of 2 robots. This is possibly due to the fact that the movements of the robots are restricted to the edges of the input graph, hence limiting the choice to a resolution algorithm. However, considering different topologies led to completely different resolution strategies, and it seems rather unfeasible to design an algorithm independent on the topology. An exception is given for instance by those graphs admitting one center, or configurations admitting one Weber point. In such cases, gathering can be easily performed regardless the underlying topology.

We remark that the analysis of the problem performed so far should be extended not only to other topologies, but also to graphs that share some properties considered useful for algorithmic purposes (i.e., graphs with bounded treewidth).

This survey on the graph environment also motivates to investigate on different assumptions, as for instance varying on the scheduler. So far, in fact, as gathering on the Euclidean plane has been characterized for asynchronous robots, it has been considered the asynchronous case at the basis of the graph environment. This is wrongly motivated by considering graphs as an easier case with respect to the Euclidean plane where also synchronous and semi-synchronous schedulers have been investigated. It comes out that asynchronous robots can always accomplish the gathering task on the Euclidean plane except for the case of 2 robots that is solvable only when synchronous robots are considered. What happens in graphs? How results are affected by switching to the synchronous or the semi-synchronous cases? On complete graphs, for instance, being synchronous may allow the resolution of some configurations as the case where only one vertex is empty and all other vertices are occupied.

Being on graphs has also permitted to deal with optimization issues when the objective function concerns the total length (in terms of hops) of the trajectories traversed by the robots. In the Euclidean plane, in fact, this was impossible as the Weber-point is not computable, even though it is unique. Also in this case, the investigation of different schedulers or different objective functions might constitute interesting directions for future work. Another interesting problem might be to find an algorithm achieving a best approximation ratio (as defined in Sect. 4.1) for gathering. Then, it would be meaningful to talk about a best solution even for gatherable configurations for which achieving ratio 1 is not possible (for example when only global weak multiplicity detection is available).

As last remark, it is worth mentioning the hybrid environment introduced in [5], where robots move on the Euclidean plane, but the gathering must be accomplished on one of the so-called *meeting points*. These are a finite set of vertices visible to the robots during their Look phase. Such a model allows robots to freely move on the Euclidean plane, but limits robots to meet only at pre-established places as happens on graphs.

## References

1. Aho, A., Hopcroft, J., Ullman, J.: Data Structures and Algorithms. Addison Wesley, Boston (1983)
2. Ausiello, G., Crescenzi, P., Gambosi, G., Kann, V., Marchetti-Spaccamela, A., Protasi, M.: Complexity and Approximation. Combinatorial Optimization Problems and Their Approximability Properties. Springer, Heidelberg (1999). https://doi.org/10.1007/978-3-642-58412-1
3. Bonnet, F., Potop-Butucaru, M., Tixeuil, S.: Asynchronous gathering in rings with 4 robots. In: Mitton, N., Loscri, V., Mouradian, A. (eds.) ADHOC-NOW 2016. LNCS, vol. 9724, pp. 311–324. Springer, Cham (2016). https://doi.org/10.1007/978-3-319-40509-4_22
4. Buss, S.R.: Alogtime algorithms for tree isomorphism, comparison, and canonization. In: Gottlob, G., Leitsch, A., Mundici, D. (eds.) KGC 1997. LNCS, vol. 1289, pp. 18–33. Springer, Heidelberg (1997). https://doi.org/10.1007/3-540-63385-5_30
5. Cicerone, S., Di Stefano, G., Navarra, A.: Gathering of robots on meeting-points: feasibility and optimal resolution algorithms. Distrib. Comput. **31**(1), 1–50 (2018)
6. Cicerone, S., Di Stefano, G., Navarra, A.: "Semi-asynchronous": a new scheduler for robot based computing systems. In: Proceedings of the 38th IEEE International Conference on Distributed Computing Systems, (ICDCS), pp. 176–187. IEEE (2018)
7. Cieliebak, M., Flocchini, P., Prencipe, G., Santoro, N.: Distributed computing by mobile robots: gathering. SIAM J. Comput. **41**(4), 829–879 (2012)
8. Cockayne, E.J., Melzak, Z.A.: Euclidean constructibility in graph-minimization problems. Math. Mag. **42**(4), 206–208 (1969)
9. Cohen, R., Peleg, D.: Convergence properties of the gravitational algorithm in asynchronous robot systems. SIAM J. Comput. **34**(6), 1516–1528 (2005)
10. D'Angelo, G., Di Stefano, G., Klasing, R., Navarra, A.: Gathering of robots on anonymous grids and trees without multiplicity detection. Theoret. Comput. Sci. **610**, 158–168 (2016)

11. D'Angelo, G., Di Stefano, G., Navarra, A.: Gathering asynchronous and oblivious robots on basic graph topologies under the look-compute-move model. In: Alpern, S., Fokkink, R., Gąsieniec, L., Lindelauf, R., Subrahmanian, V. (eds.) Search Theory: A Game Theoretic Perspective, pp. 197–222. Springer, New York (2013). https://doi.org/10.1007/978-1-4614-6825-7_13
12. D'Angelo, G., Di Stefano, G., Navarra, A.: Gathering on rings under the look-compute-move model. Distrib. Comput. **27**(4), 255–285 (2014)
13. D'Angelo, G., Di Stefano, G., Navarra, A., Nisse, N., Suchan, K.: Computing on rings by oblivious robots: a unified approach for different tasks. Algorithmica **72**(4), 1055–1096 (2015)
14. D'Angelo, G., Navarra, A., Nisse, N.: A unified approach for gathering and exclusive searching on rings under weak assumptions. Distrib. Comput. **30**(1), 17–48 (2017)
15. D'Angelo, G., Di Stefano, G., Navarra, A.: Gathering six oblivious robots on anonymous symmetric rings. J. Discret. Algorithms **26**, 16–27 (2014)
16. D'Emidio, M., Frigioni, D., Navarra, A.: Characterizing the computational power of anonymous mobile robots. In: Proceedings of the 36th IEEE International Conference on Distributed Computing Systems, (ICDCS), pp. 293–302. IEEE (2016)
17. Di Stefano, G., Montanari, P., Navarra, A.: About ungatherability of oblivious and asynchronous robots on anonymous rings. In: Lipták, Z., Smyth, W.F. (eds.) IWOCA 2015. LNCS, vol. 9538, pp. 136–147. Springer, Cham (2016). https://doi.org/10.1007/978-3-319-29516-9_12
18. Di Stefano, G., Navarra, A.: Gathering of oblivious robots on infinite grids with minimum traveled distance. Inf. Comput. **254**, 377–391 (2017)
19. Di Stefano, G., Navarra, A.: Optimal gathering of oblivious robots in anonymous graphs and its application on trees and rings. Distrib. Comput. **30**(2), 75–86 (2017)
20. Izumi, T., Izumi, T., Kamei, S., Ooshita, F.: Mobile robots gathering algorithm with local weak multiplicity in rings. In: Patt-Shamir, B., Ekim, T. (eds.) SIROCCO 2010. LNCS, vol. 6058, pp. 101–113. Springer, Heidelberg (2010). https://doi.org/10.1007/978-3-642-13284-1_9
21. Kamei, S., Lamani, A., Ooshita, F., Tixeuil, S.: Asynchronous mobile robot gathering from symmetric configurations without global multiplicity detection. In: Kosowski, A., Yamashita, M. (eds.) SIROCCO 2011. LNCS, vol. 6796, pp. 150–161. Springer, Heidelberg (2011). https://doi.org/10.1007/978-3-642-22212-2_14
22. Kamei, S., Lamani, A., Ooshita, F., Tixeuil, S.: Gathering an even number of robots in an odd ring without global multiplicity detection. In: Rovan, B., Sassone, V., Widmayer, P. (eds.) MFCS 2012. LNCS, vol. 7464, pp. 542–553. Springer, Heidelberg (2012). https://doi.org/10.1007/978-3-642-32589-2_48
23. Klasing, R., Kosowski, A., Navarra, A.: Taking advantage of symmetries: Gathering of many asynchronous oblivious robots on a ring. Theoret. Comput. Sci. **411**, 3235–3246 (2010)
24. Klasing, R., Markou, E., Pelc, A.: Gathering asynchronous oblivious mobile robots in a ring. Theoret. Comput. Sci. **390**, 27–39 (2008)
25. Koren, M.: Gathering small number of mobile asynchronous robots on ring. Zeszyty Naukowe Wydzialu ETI Politechniki Gdanskiej. Technologie Informacyjne **18**, 325–331 (2010)
26. Santoro, N.: Design and Analysis of Distributed Algorithms. Wiley, Hoboken (2007)

# Oblivious Robots on Graphs: Exploration

David Ilcinkas(✉)

CNRS, LaBRI, Univ. Bordeaux, Talence, France
david.ilcinkas@labri.fr
https://www.labri.fr/perso/ilcinkas/

**Abstract.** This chapter focuses on the problem of exploring a graph by a team of mobile robots endowed with vision. More precisely, we consider here mobile robots operating under the Look-Compute-Move paradigm in discrete environments modeled as graphs. The goal for these robots is to explore the graph in which they are, that is to visit all vertices of the graph.

**Keywords:** Oblivious robots · Graph exploration
Terminating exploration · Exclusive perpetual exploration

## 1 Introduction

In this chapter, we consider mobile entities, called robots, operating under the `Look-Compute-Move` paradigm. An activated robot starts first by taking an instantaneous snapshot of its environment (the `Look` phase), then it computes whether and where it wants to move (the `Compute` phase), and finally it moves to the decided new position (the `Move` phase). Robots operating under the `Look-Compute-Move` paradigm are classically considered in continuous environments, usually the plane. The studies on these robots were however recently extended to the case of discrete environments, modeled as graphs (see [6] and [23] for short surveys on the subject). This chapter focuses on these discrete environments. One motivation to consider discrete environments is to get rid of possibly annoying geometric considerations in order to focus on issues directly related to the weaknesses of the robots (anonymity, obliviousness, etc.), to the symmetries of the environment, and to the asynchrony. Another motivation is more practical and comes from the fact that vision sensors do not have an infinite precision. Considering discrete environments thus acknowledges the fact that many vision sensors output digital and thus discrete snapshots of the environment.

We consider in this chapter the graph exploration problem, in which the robots have to visit every vertex. More precisely, two variants of the problem were studied so far in the literature (in the considered model). The first variant is called *terminating exploration* and requires that, first, each vertex is visited by at least one robot, and second, that eventually all robots stop moving. The second variant is called *exclusive perpetual exploration* and requires that, first, each robot visits every vertex of the graph infinitely often, and second, that no


P. Flocchini et al. (Eds.): Distributed Computing by Mobile Entities, LNCS 11340, pp. 218–233, 2019.
https://doi.org/10.1007/978-3-030-11072-7_9

two robots traverse the same edge at the same time nor visit the same vertex at the same time. Exploring a graph is a fundamental task in mobile robot computing and can be used, for example, to search for a specific information, or to discover and list all the services provided by the vertices. Exploration (perpetual in particular) is also interesting for maintenance purposes, where the robots can check forever whether the vertices are properly functioning. Finally, the exclusivity property models the physical constraints that the robots may have if, for example, they operate in environments with limited available space that prevent them from crossing each other or being at the same place.

The chapter is structured as follows. Section 2 defines more precisely the model and the problems, and gives a first simple preliminary result. Sections 3 and 4 respectively consider the terminating exploration and the exclusive perpetual exploration problems. In both cases, known results from the literature are presented, and then the usual tools and techniques used in these works are summarized. Finally Sect. 5 discusses what concerns the correctness of the results, while Sect. 6 concludes the chapter.

## 2 Model and Preliminaries

### 2.1 Model

*The Environment.* We model the environment as a simple undirected connected graph $G = (V, E)$. The number $|V|$ of vertices is usually denoted $n$, while the number $|E|$ of edges is usually denoted $m$. The graph is assumed to be anonymous: neither vertices nor edges are labeled (or, equivalently, such labels cannot be seen by the robots).

*The Robots.* On this graph operate mobile entities called *robots*. They can move from vertex to vertex via the edges of the graph. The robots are all anonymous and identical, i.e. they all execute the same algorithm. They have no direct means of communication. Unless otherwise specified, the robots will be assumed to be oblivious: they do not have persistent memory. When several robots occupy the same vertex, we say that there is a *tower* on this vertex.

*The* `Look-Compute-Move` *Cycle.* The robots operate by repeatedly executing `Look-Compute-Move` cycles. In the `Look` phase, a robot takes an instantaneous egocentric snapshot of its environment. This includes the structure of the graph around it, and the presence of robots on the seen vertices. The structure of the snapshot will be detailed later. Note however that all robots are perceived on vertices, not on edges. In the `Compute` phase, the robot decides whether to move or not, and in the first case to which neighboring vertex. For oblivious robots, this computation is solely based on the last snapshot. Finally, in the `Move` phase, the robot moves to the chosen neighbor, or stays idle if it decided to do so. Moves are considered instantaneous, which is consistent with the fact that robots are seen on vertices in the snapshots. Note nevertheless that this model has been shown equivalent to the model with continuous moves (but still considering that robots are always seen on vertices), and this equivalence is certified using the Coq proof assistant [1].

*Timing Assumptions.* Different levels of asynchrony are classically considered in the literature. In the fully synchronous model $\mathcal{F}$SYNC, all robots execute their Look-Compute-Move cycles simultaneously. Differently speaking, at each round, every robot executes its full Look-Compute-Move cycle. In the semi-synchronous model $\mathcal{S}$SYNC, at each round, a non-empty subset of the robots, chosen by an adversary, execute a full Look-Compute-Move cycle. Finally, in the asynchronous model $\mathcal{A}$SYNC, the timing between the different phases of the Look-Compute-Move cycles performed by the different robots is arbitrary, with the only constraint that, for each robot, the time between two consecutive phases is finite (but possibly unbounded). As a consequence, a move can be performed based on an outdated snapshot in this model.

*The Snapshot.* During the Look phase, a robot perceives the structure of the graph and the presence of robots around it within a *visibility radius* $\rho$ given by the model. More precisely, the snapshot taken by a robot consists of the rooted subgraph induced by the vertices at distance at most $\rho$ from the vertex occupied by the robot and, for each seen vertex, of the perceived number of robots on it. The accuracy of the perceived number of robots is given by another model parameter called the *multiplicity detection.* If weak multiplicity detection is assumed, a robot is only able to distinguish between the presence of "zero", "one", or "more than one" robots on a seen vertex. On the contrary, strong multiplicity detection assumes that a robot knows the exact number of robots that are present on a seen vertex. Orthogonally, multiplicity detection can be either local or global. In the case of local multiplicity detection, a robot only knows the multiplicity of the vertex it occupies (whether in the weak or the strong sense), while in the case of global multiplicity detection, a robot knows the multiplicity of all vertices.

*Configurations, Views, and Symmetries.* The description of the graph, together with the indication of the exact number of robots located on each vertex, constitute a *configuration.* The *view from vertex* $u$ is any rooted graph isomorphic to the subgraph induced by the vertices at distance at most $\rho$ (the visibility radius) from $u$, and the corresponding perceived number of robots on these vertices (depending on the multiplicity detection assumption). In the Look phase, a robot is given a view from the vertex it is located on. Therefore, symmetries of the graph are somehow still present in the snapshot. Indeed, for example in a ring, if a robot lies on an axis of symmetry of the configuration, then it will not be able to differentiate one direction from the other. Therefore, if it decides to move to a neighboring vertex, the actual move will be to a neighbor chosen by the adversary. More generally and more formally, we say that two vertices $v$ and $v'$ are *similar* with respect to $u$ if there exists a view from vertex $u$ such that there exist two vertices $w$ and $w'$ of the view and two isomorphisms $\phi$ and $\phi'$ such that $\phi(w) = \phi'(w) = u$, $\phi(w') = v$, and $\phi'(w') = v'$. Intuitively, $v$ and $v'$ are similar with respect to $u$ if $v$ and $v'$ are indistinguishable for a robot located in $u$ (taking into account the visibility radius and the multiplicity detection). If a robot decides in the Compute phase to move to a vertex $v$, then in the Move phase it will actually move to any vertex $v'$ similar to $v$ with respect to the robot current position, and the choice of $v'$ is made by the adversary.

*Terminating Exploration.* A team of robots solves the problem of *terminating exploration* in a graph family $\mathcal{G}$ if, for any graph $G \in \mathcal{G}$, for any behavior of the adversary controlling the asynchrony and the choices between similar neighbors, and starting from any initial configuration without towers, each vertex of the graph is visited by *at least one* robot and the robots eventually reach a configuration in which no robots ever move.

*Exclusive Perpetual Exploration.* A team of robots solves the problem of *exclusive perpetual exploration* in a graph family $\mathcal{G}$ if, for any graph $G \in \mathcal{G}$, for any behavior of the adversary controlling the asynchrony and the choices between similar neighbors, and starting from any initial configuration without towers, each vertex of the graph is visited by *every* robot *infinitely often* and the exclusivity property is satisfied. This property is satisfied if no two robots are ever located at the same vertex at the same time, and no two robots ever cross the same edge at the same time in opposite directions.

## 2.2 Preliminaries

Let us first start by giving some justifications about the choice of the set of initial configurations, which is defined as the set of configurations without towers for both problem variants. First, note that the exclusivity property of the exclusive perpetual exploration problem is violated in any configuration with at least one tower. The set of configurations without towers is thus the maximal set of initial configurations that is meaningful for this problem. Concerning terminating exploration, note that its definition implies the existence of a configuration in which all robots decide not to move. If not all vertices are occupied by a robot, then this configuration must not be an initial configuration for the problem to be solvable, if the robots are oblivious. Differently speaking, for terminating exploration, the set of initial configurations must not contain all possible configurations when the robots are oblivious and are less than the number of vertices (which is the case in all the papers of the literature considered here). Taking as initial configurations the configurations without towers is thus also rather natural for this problem. Finally, one may also look for *universal* algorithms [22]. Given a number of robots in a specified model, a universal algorithm is an algorithm solving the problem from any initial configuration which is solvable in the considered setting.

Most papers in the literature on the subject concern specific families of graphs. The most commonly studied family is the family of all rings. Other studied graphs are the trees, the grids, the tori and some variations of these graphs. Given such a family, the usual focus is then on the *smallest and/or largest exploring teams*, that is on the numbers $\kappa^-(n)$ and $\kappa^+(n)$ defined as the respectively smallest and largest numbers of robots that can explore any $n$-vertex graph of the given family. In the following, the considered family and exploration type will be clear from the context.

We now present a first simple technical result, inspired from Lemma 2.1 in [20], which allows nevertheless to already draw some conclusions on the value $\kappa^-(n)$ in the case of the rings.

**Lemma 1.** *Let $n \geq 3$ and $k < n/2$ be two positive integers such that $k$ divides $n$. Then both terminating exploration and perpetual exclusive exploration of an $n$-vertex ring are not deterministically solvable by a team of $k$ (possibly non-oblivious) robots, even with full visibility, global strong multiplicity detection, and in the $\mathcal{F}$SYNC model.*

*Proof.* Let us fix any algorithm for a team of $k$ robots. Consider as initial configuration a configuration in which the $k$ robots are regularly scattered around the ring. The configuration being perfectly periodic, all robots have the same view. If a robot decides to move to a neighbor (both neighbors are similar with respect to the robot's current position), then all robots decide to move (they have the same initial state), and the adversary makes them move in the same direction. Therefore, the configuration stays periodic and all robots still have the same state (which may be different from the initial state if the robots are non-oblivious). We make the adversary continue to act that way. More specifically, the adversary makes them move in some fixed direction for each odd round, and in the other direction for each even round (forgetting about rounds in which the robots decide to stay idle). In such an execution, and until they decide to stay idle, each robot keeps going back and forth between its initial location and one specific neighbor. Since $n/k \geq 3$, there exist $k$ vertices which are never visited by the robots. Thus exploration (whatever its type) is not solved by this algorithm. □

As a corollary, a ring of size $n$ equal to three times the least common multiple of $1, 2, \cdots, k-1$ cannot be deterministically explored by less than $k$ robots. Some calculations using the Prime Number Theorem show that $k = \Theta(\log n)$ in that case, see [20]. Therefore, there exists a positive constant $c$ such that, for infinitely many $n$, we have $\kappa^-(n) \geq c \log n$ for deterministic algorithms.

## 3 Exploration with Stop

In this section, we will almost only consider oblivious robots, that is robots using in the `Compute` phase only the snapshot taken in the preceding `Look` phase. In particular, the robots do not have access to time, and thus they do not know whether this is the beginning of the execution or not when they see a configuration without towers.

### 3.1 Known Results

#### In the Rings

As already noted in Sect. 2.2, and already proven by Flocchini et al. in [20], there exists a positive constant $c$ such that, for infinitely many $n$, we have $\kappa^-(n) \geq c \log n$ for deterministic algorithms. This in fact remains true for probabilistic algorithms [15], but only for the asynchronous model $\mathcal{A}$SYNC. Indeed, in the semi-synchronous model $\mathcal{S}$SYNC, a constant number of robots, namely 4 probabilistic robots, are necessary and sufficient to solve the terminating exploration problem in any $n$-vertex ring with $n > 4$, see [15].

Let us now focus on deterministic algorithms. The lower bound 4 on $\kappa^-(n)$ still holds in $\mathcal{S}$SYNC (in $\mathcal{F}$SYNC, only 3 is a clear lower bound for every $n$, in particular when $n$ is odd). When $n$ is even, $\kappa^-(n) \geq 5$ in the $\mathcal{S}$SYNC model [21]. These values are somehow optimal. Indeed, 4 robots can explore the rings of odd size in $\mathcal{S}$SYNC [21], and provided that n is not a multiple of 5, a team of 5 robots can explore the $n$-vertex ring even in the $\mathcal{A}$SYNC model [21]. As pointed out, $\kappa^-(n)$ may be logarithmic in $n$ for infinitely many values of $n$, but this cannot go worse in the sense that $\kappa^-(n)$ is always in $O(\log n)$. Indeed, for any $k \geq 17$ that is co-prime with $n$, a team of $k$ robots can explore the $n$-vertex ring [20].

Note that all the results presented here so far for the case of the rings are strong with respect to multiplicity detection in the following sense. All lower bounds (impossibility results) are valid even with global strong multiplicity detection, while all upper bounds (algorithms) are assuming global weak multiplicity detection. Moreover, note that the results are valid for sufficiently large values of $n$, and may vary for small values of $n$.

Limited visibility has also been considered, for deterministic algorithms, and in the case of global (up to the visibility radius) strong multiplicity detection. When the visibility radius $\rho$ is 1, even a limited amount of asynchrony renders the problem impossible to solve: there are no deterministic algorithms working in the $\mathcal{S}$SYNC model, for any number $k < n$ of robots [11]. In the same paper, the authors show that 5 robots are necessary in the $\mathcal{F}$SYNC model, and they present an algorithm for 5 robots working when all robots are on consecutive vertices in the initial configuration. If robots however have 1 bit of persistent memory which is visible/accessible by any robot within their visibility radius (the $\mathcal{LUMINOUS}$ model), then three (in $\mathcal{F}$SYNC) or four (in $\mathcal{S}$SYNC and $\mathcal{A}$SYNC) robots are necessary and sufficient to solve terminating exploration [22] (for specific initial configurations). These numbers are reduced by one for (non-exclusive) perpetual exploration.

The cases $\rho = 2$ and $\rho = 3$ are considered by Datta et al. in [12]. For $\rho = 2$, there exists an algorithm in the $\mathcal{A}$SYNC model for 7 robots when all robots are on consecutive vertices in the initial configuration. The number of robots can be reduced to 5 when $\rho = 3$. Finally, for $\rho = 3$, there exists an $\mathcal{A}$SYNC algorithm for 7 robots that can handle more general initial configurations: the robots start in a position such that they are "connected by vision" but need not to be on consecutive vertices.

**In the Trees**

The trees [18], and the sub-case of the lines [19], have only been considered in the deterministic setting and assuming global weak multiplicity detection.

In trees, the absence of port numbers (the anonymous graph assumption) makes empty leaves having the same parent indistinguishable (they are similar with respect to the parent). Therefore, in order to explore sibling leaves despite any choice of the adversary concerning similar vertices, at least one robot must be sent to each leaf attached to a given parent. If a vertex has more than two leaves attached to it and all of them are occupied by robots, then at least two of them are similar, having both either a single robot or a tower. The adversary can thus make these robots merge if the algorithm decides to move them. Therefore, one can prove that $\Omega(n)$ robots are necessary in some trees (at least in complete ternary trees) in the $\mathcal{S}$SYNC model. Note that this lower bound heavily relies on the weak multiplicity assumption. For trees of maximum degree 3, less robots may be used: $O(\log n/\log\log n)$ robots are sufficient in such trees, even in the $\mathcal{A}$SYNC model. This number is actually necessary for some trees because $\Omega(\log n/\log\log n)$ robots are necessary to explore complete binary trees, even with global strong multiplicity detection and in the $\mathcal{F}$SYNC model.

In lines, symmetries are much more limited, and the solvable cases are fully characterized. A team of $k < n$ robots can solve terminating exploration in the $n$-vertex line if and only if $k = 3$, or $k \geq 5$, or $k = 4$ and $n$ is odd. The lower bounds are proved in the $\mathcal{S}$SYNC model while the upper bounds hold even in the $\mathcal{A}$SYNC model.

**In the Grids and Tori**

The situation for grids [13] and tori [14] resembles the situation for lines and rings.

In grids, where symmetries are limited, we have $\kappa^-(i,j) = 3$ for all $(i,j)$-grids (except the $(2,2)$-grid and the $(3,3)$-grid for which we have $\kappa^-(2,2) = 4$ and $\kappa^-(3,3) = 5$). The lower bound holds even for probabilistic algorithms, in the $\mathcal{S}$SYNC model, and assuming global strong multiplicity detection, while the algorithm proving the upper bound is deterministic, works in the $\mathcal{A}$SYNC model, and assumes global weak multiplicity detection.

Tori have much more symmetries and thus require more robots. Indeed, $\kappa^-(i,j) \geq 5$ for deterministic algorithms solving the terminating exploration problem in $(i,j)$-tori. Allowing probabilistic algorithms, we have $\kappa^-(i,j) = 4$ (for sufficiently large tori). The lower bounds assume global strong multiplicity detection while the upper bound assumes global weak multiplicity detection. All results for tori are proved in the $\mathcal{S}$SYNC model.

**In the General Graphs**

The case of arbitrary graphs has been considered by Chalopin et al. in [10]. More precisely, the paper considers arbitrary graphs with port numbers, that is graphs for which, at each vertex, the incident edges are distinguished by local port numbers from 1 to the degree of the vertex. The class $\mathcal{H}_k$ is then defined as the class of rigid configurations of $k$ robots, i.e. the class of configurations of $k$ robots (the graphs with port numbers and the positions of the $k$ robots) such that there is no non-trivial automorphism preserving the port numbers and the robots locations. Chalopin, Flocchini, Mans, and Santoro studied the terminating exploration problem in the $\mathcal{A}$SYNC model in these classes, assuming global weak multiplicity detection. They proved that exploration is impossible for $k < 3$ robots, they characterized the graphs that are explorable in $\mathcal{H}_3$, and they show that all graphs are explorable in $\mathcal{H}_4$ and in every $\mathcal{H}_k$ with an odd $k > 3$. The case of even $k > 4$ is left open but can be reduced to the existence of a gathering algorithm for $\mathcal{H}_k$.

## 3.2 Usual Tools and Techniques

**Impossibility Results**
The lower bounds on the number of robots that is necessary to solve terminating exploration are of similar flavor for the different considered topologies, and use the following arguments.

First of all, since the specification of the problem requires termination, there must exist a configuration in which no robot decides to move. When $k < n$, and because of obliviousness, this configuration cannot be an initial configuration, and thus at least one tower must be formed. This already proves that a single robot is never sufficient (if $n > 1$).

Then note that any suffix of a valid execution must remain valid as long as the first configuration of the truncated execution has no towers. Indeed, such a configuration can be an initial configuration, and, because of obliviousness, these two executions (the initial one and the truncated one) both respect the algorithm. Exploration must thus be performed after a tower is formed and keeping at least one tower in each configuration.

The next observation is that towers are difficult to move in asynchronous environments. Indeed, even in the semi-synchronous $\mathcal{S}$SYNC model, if the robots in a tower decide to move, then only one may be activated and the tower may be destroyed. As this is often the case (in particular for a small number of robots), let us assume that it is impossible to move towers. Therefore exploration must be performed by at least another robot and thus 2 robots are not sufficient either.

In order to obtain a larger lower bound, one generally needs a further observation about the suffix of the execution in which all configurations contain at least one tower. This observation is the fact that any two configurations in this suffix must be distinguishable from the point of view of the robots. Indeed, if this is not the case, the adversary can make the execution periodic by repeating the same choices, and the termination requirement is not fulfilled. Besides, at each step of the execution, at most $k$ new vertices are explored, and possibly even less if some robots are blocked in a stationary tower. Therefore, there must exist sufficiently many distinguishable configurations with a tower for the problem to be solvable with $k$ robots. Such a counting argument is generally sufficient to obtain rather good lower bounds.

**Algorithmic Techniques**
The algorithms presented in the different papers also have some similarities. Indeed, as previously seen, exploration must be performed after a tower is formed, and at least a tower must be kept until termination. Therefore, the algorithms for terminating exploration by oblivious robots generally consist of three phases: a set-up phase in which no tower is created but a special configuration is reached, a tower-creation phase in which a tower is created, and finally the exploration phase in which some of the robots explore the graph.

The first phase is usually the most complex one. Indeed, this phase starts from an arbitrary initial configuration while the two other phases start from specific configurations (or classes of configurations). The special configurations that the robots try to reach in the set-up phase are generally configurations without towers where robots are gathered next to each other. In the rings, a special configuration is typically a configuration in which all robots form a block by positioning themselves on consecutive vertices. In the trees, all robots go toward the leaves. In the grids, the robots go towards one of the corners.

Reaching such a special configuration is generally highly non trivial, in particular because robots are oblivious. This constraint prevents the robots from remembering what their plan was at the beginning of the execution. Intuitively, if one constructs a directed graph whose vertices are the possible configurations and in which there is an arc between two configurations if one can reach the second configuration from the first one by applying one step of the algorithm, then this directed graph must be acyclic. This may be not too hard to achieve in graphs for which there are no non-trivial automorphisms, but it can be very tricky in graphs with a lot of symmetries.

The issue with symmetries is that it may be hard, or even impossible, to break them. Therefore, an algorithm may be forced to schedule several robots having the same view of the environment to move in the current configuration. Combined with the asynchrony, several different configurations can be obtained depending on the choices made by the adversary. In the ring for example, it may happen that from a symmetric configuration only one robot is scheduled to move by the adversary and the obtained configuration is again symmetric but with a different axis of symmetry. Even worse, in the asynchronous $\mathcal{A}$SYNC model, one robot may decide to move but the adversary decides to delay this move while the other robots are progressing. The delayed move is said to be *pending* in this case. When this move is finally allowed by the adversary, it does not necessarily correspond to the current situation and may create issues. It is thus often very difficult to design an algorithm that avoids cycling among the configurations.

One can nevertheless express two guidelines that algorithm designers should try to follow. The first one is to minimize the numbers of robots that decide to move in a given configuration. Typically, in an asymmetric configuration, the algorithm should designate a single robot to move. The second guideline could be expressed as follows. In a given configuration, if a robot may have a pending move (for example because the current configuration may come from a symmetric configuration where several robots with the same view were designated to move), then the algorithm should choose this robot to move. Indeed, this would force the adversary to execute the pending move.

Once a special configuration is reached, it is however rather easy to form a tower, since robots are located on contiguous vertices. In rings for example, the robot in the middle of an odd block of robots can simply move in an arbitrary direction to form a tower with its neighbor.

In the exploration phase, usually just a few robots, 1 or 3, are actually exploring the graph. The other robots are used to keep track of the process and to break symmetries. In rings, grids and tori, a tower and a few stationary robots are sufficient to break symmetries. The logarithmic number of robots that may be needed in some rings, see Sect. 2.2, is only due to the fact that smallest numbers of robots allow periodic configurations. On the contrary, the $\Theta(\log n/\log\log n)$ robots used in the trees are really used by the algorithm. Indeed, in very regular trees (typically the complete binary trees), there are many a priori indistinguishable leaves and the robots need a way to distinguish them. For this purpose, the robots maintain a counter that stores the number of explored leaves. This allows the exploration team (two robots forming a tower, and an isolated robot) to know which leaf is the next one to explore. This counter is visually implemented in the tree by locating the $\Theta(\log n/\log\log n)$ robots at carefully chosen positions.

# 4 Exclusive Perpetual Exploration

In this section, we will only consider deterministic algorithms. Recall that contrary to terminating exploration, exclusive perpetual exploration requires that each vertex is visited by *every* robot, and so infinitely often. Moreover, the exclusivity property forbids the robots to cross each other on an edge or to be on the same vertex at the same time.

## 4.1 Known Results

**With Memory**

The exclusive perpetual exploration problem has first been investigated in the case of robots having memory, in the fully synchronous $\mathcal{F}$SYNC model. In [3], given any graph, a labeled mobility tree is defined and a parameter $q$ is associated to it. The authors then prove that $\kappa^+(n) \leq n - q$ for any $n$-vertex graph of associated parameter $q$, even with infinite visibility radius.

It turns out that this bound is tight for the partial grids with sense of direction. These are subgraphs of the grids such that edges are locally labeled by the cardinal points N, S, E, W. More precisely, a graph can be explored if and only if the number $k$ of robots is less than or equal to this bound $n - q$, in the case of an infinite visibility radius. For a null visibility radius, the problem is impossible to solve. For visibility radius 1, the problem is solvable if and only if $k \leq n - q$ except when $q = 0$, in which case the condition is $k \leq n - 1$, see [4].

**In the Rings**

As already noted in Sect. 2.2, exclusive perpetual exploration cannot be solved when the number $k$ of robots divides the number $n$ of vertices. Since towers are not permitted, the same reasoning can be applied to configurations in which vertices without robots are regularly spaced. Therefore, the problem for $k = n - k'$ robots cannot be solved as well when $k'$ divides $n$, see [5]. Because of symmetries and of the exclusivity property, the problem cannot be solved either when the number of robots is even. All these results hold even in the $\mathcal{F}$SYNC model. Finally, in the same paper, the authors claim that for $n$ and $k$ co-prime, in the $\mathcal{A}$SYNC model, $\kappa^-(n) = 3$ if $n \geq 10$ (and is larger otherwise), and $\kappa^+(n) = n-5$ (for $k$ odd). The algorithm in [5] justifying $\kappa^-(n) = 3$ is actually not correct for $n = 10$, but a corrected version is given in [9], see Sect. 5.2 for details.

Focusing on rigid initial configurations, i.e. on initial configurations such that there is no non-trivial automorphism preserving the robots locations, the situation is a bit different. In [17], the authors present an algorithm for $k$ robots solving the exclusive perpetual exploration problem in $n$-vertex rings in the $\mathcal{A}$SYNC model when $n \geq 10$ and $5 \leq k \leq n - 3$, except for $k = 5$ and $n = 10$.

This latter case has been investigated by Bonnet et al. in [7], where a generic method is proposed, and implemented, to list all possible protocols using only rigid configurations of a given number $k$ of robots in a given graph, in the semi-synchronous $\mathcal{S}$SYNC model. The authors used this method to prove that exclusive perpetual exploration is impossible for $k = 5$ robots in a ring of $n = 10$ robots. The proof uses the full specification of the problem in the sense that, if one relaxes the definition of the problem by allowing vertices to be visited by only some robots and not all of them, then the problem becomes solvable for this case (still in the $\mathcal{S}$SYNC model).

**In the Grids**

Similarly as for terminating exploration, exclusive perpetual exploration has also been considered in grids [8]. Contrary to the papers considering the partial grids with sense of direction, this paper considers grids without holes and without any edge labels or port numbers (and thus without sense of direction), focuses on oblivious algorithms, and assumes the $\mathcal{A}$SYNC model. The main result of the paper is a proof that $\kappa^-(n) = 3$ in all $n$-vertex grids having at least two rows and two columns, except for the $(2, 2)$- and $(2, 3)$-grids in which exclusive perpetual exploration is impossible for any $k$.

### 4.2 Usual Tools and Techniques

**The Labeled Mobility Tree and Its Associated Parameter $q$**

We now describe in more details how the labeled mobility tree and its associated parameter $q$ are defined from any graph. Recall that this parameter $q$ is used to bound the maximal number of robots that can solve exclusive perpetual exploration in a graph.

Consider any graph $G$. The first step is to label the vertices with labels from the set $\{0, 1, 2\}$ as follows. Label a vertex with 0 when it is of degree 2 and it does not belong to any non-singleton bridge-less subgraph of $G$ (i.e. it does not belong to any 2-edge-connected component of $G$ of size at least 2). Label a vertex with 1 if it is a leaf of $G$ or if it belongs to a non-singleton bridge-less subgraph of $G$. Label a vertex with 2 in any remaining case.

The second step consists in compressing each maximal non-singleton bridge-less subgraph of $G$ (i.e. each non-singleton 2-edge-connected component of $G$) into a single vertex with label 1. The obtained tree is called the labeled mobility tree associated to $G$.

A mutual exclusion path in this labeled mobility tree is then a path whose extremities have a label different from 0 but whose internal vertices (if they exist) have label 0. The length of such a path is defined as the number of edges of the path plus the number of extremities with label 2. The parameter $q$ associated to the labeled mobility tree, and thus to the graph $G$, is then simply the maximal length of a mutual exclusion path.

Intuitively, a mutual exclusion path is a sequence of bridges of $G$ acting as a long bridge of the graph separating two parts of it. If a robot wants to move from one part to the other (which it has to do infinitely often to solve the problem), then it must traverse this long bridge without crossing or meeting any other robot. This more or less explains why exclusive perpetual exploration cannot be solved by more than $n - q$ robots.

**Algorithmic Techniques for Oblivious Robots in Rings and Grids**
Similarly as for terminating exploration, an algorithm roughly defines a directed graph of configurations. In the case of exclusive perpetual exploration, instead of having a DAG (directed acyclic graph) pointing to a set of terminal configurations (configurations without outgoing edges), the DAG points toward some (generally just one) cycles of configurations achieving exploration. Differently speaking, the robots try to enter a cycle of configurations such that every vertex is explored by every robot when this cycle of configurations is performed forever.

In general grids with three robots for example, the robots gather towards a corner and then two robots become more or less stationary, mainly acting as symmetry breakers, while the third robot explores a large part of the grid (at least half of it). When the explorer terminates its part, the robots are in the same position as before but near the opposite corner and with permuted roles. After six such phases, every robot has explored the whole grid.

In rings with three robots, the exploration is performed simultaneously by all the robots. The purpose of the algorithm is to make the robots form a small asymmetric pattern that moves along the cycle forever like a worm. More precisely, the basic pattern consists of two robots on consecutive vertices and the third robot, marking the tail of the worm, a little bit further (two vertices are left empty between the head and the tail). Let us denote such a pattern by 11001 (1 denotes an occupied vertex while 0 denotes an empty vertex). The worm is then moved by moving repeatedly each robot at a time in one direction from the head to the tail. Hence, from a pattern 11001, the pattern 101001 is formed, then the pattern 110001, and finally the tail moves forming back the pattern 11001, but all robots are now shifted by one vertex in the same direction.

Obviously, as for terminating exploration, the difficult part is the initial phase in which the robots have to deal with asynchrony and symmetries while forming a predefined pattern of the exploration phase.

## 5 Correctness and Related Questions

As it is often the case in distributed computing, proving impossibility results and the correctness of algorithms may be challenging. This section discusses the related techniques used in the literature.

### 5.1 Hand-Written Proofs

Because of asynchrony and of the symmetries, many executions are possible given an initial configuration and an algorithm. Proofs of correctness are thus often based on a case-by-case analysis, usually rather tedious. This can also be the case for impossibility proofs, where a wide variety of possible algorithms may be explored. In both cases, this is especially true for small graphs, for which general arguments may be more difficult to find.

The situation can sometimes be summarized by exhibiting a DAG (directed acyclic graph) of configurations allowing to visualize the different cases and justifying the convergence to a specific configuration (or a class of configurations). This may be convenient for small graphs but, again, may be intractable for arbitrarily large graphs.

As a consequence, potential functions may also be used to prove convergence. It is however not always simple to find such a potential function and one solution consists in combining all these approaches. Typically, one can use a case-by-case analysis to distinguish several classes of configurations, construct a directed graph of accessibility among these classes and prove that this directed graph is almost a DAG, except for few cycles for which a potential function allows to prove that such a cycle is used a finite number of times.

Such a combined approach gives a proof in which one can have some confidence, but even such proofs are prone to human errors. In particular, it is usually very difficult to convince oneself that no tricky sub-cases have been forgotten. This lack of confidence in human-written (and human-checked) proofs gave rise in the recent years to papers using formal verification tools.

### 5.2 Formal Verification

The first use of automated tools for studying the graph exploration problem in the `Look-Compute-Move` paradigm concerns the exclusive perpetual exploration of small rings [7]. As already mentioned in Sect. 4.1, a generic method is proposed, and implemented, to list all possible protocols of a given number $k$ of robots in a given graph, in the semi-synchronous $\mathcal{S}$SYNC model. The authors use this method to prove that exclusive perpetual exploration is impossible for $k = 5$

robots in a ring of $n = 10$ robots, and to list all algorithms solving the weaker version of the problem for the same setting ($k = 5$ and $n = 10$).

A more general formal verification tool was introduced a few years later by Bérard et al. [9]. This one is compatible with all three models of (a)synchrony, and can handle both variants of the exploration problem. First, the terminating exploration algorithm for rings from [20] has been studied. The first and most difficult phase of the algorithm, the set-up phase, is proved correct for small values of $k$ (at most 21), and small values of $n$ (at most 22), and even for some settings not covered by the hand-written proof in [20]. Second, the exclusive perpetual exploration algorithm for rings from [5] has been studied. This time, a counter-example is found, for the case of $k = 3$ robots in a ring of size $n = 10$. This counter-example is a particular execution starting from a symmetric configuration and using the asynchronous model $\mathcal{A}$SYNC to maintain pending moves (i.e. moves that are already computed but not yet executed) even when the symmetry has already been broken. Basically, the worm that we described in Sect. 4.2 has been stretched in such a way that the two robots forming its head do not agree on which direction the worm is moving. This eventually leads to a collision between these two robots. Note that the existence of an ambiguity on the moving direction of the worm heavily relies on the small size of the ring. Besides, the same paper presents a corrected version of the algorithm, which is formally proved correct for $n$ up to 16 and manually proved correct for larger values of $n$.

Finally, Doan et al. [16] also study the exclusive perpetual exploration algorithm for 3 robots in rings from [5], and the same counter-example is found. However, the model checker is different as well as the modeling details, allowing for potentially different performance of the verification.

## 6 Conclusion and Perspectives

The two variants of the graph exploration problem have been well studied in rings, and in some other topologies. Not surprisingly, the amount of asynchrony and of symmetries influences the amount of robots needed to solve the problem, especially for the terminating exploration problem. Probabilistic approaches may also help, but further studies are needed on this aspect. Also, it would be interesting to extend the investigation to larger families of graphs, typically to planar graphs.

An interesting research axis concerns the limited visibility. Indeed, having a limited vision sounds much more realistic, and the first results on this subject seem to indicate that interesting performance can still be achieved despite this limitation. Another direction of research concerning vision could be to consider non-egocentric views, but with sense of direction. This corresponds to the situations in which a camera can see the whole theater of operation, for example when it is attached to the ceiling in a room or embedded in a satellite.

Finally, some effort is required to further formalize and verify the results in the domain. Recent papers showed that hand-written proofs could be flawed, because of the many cases to be considered due to asynchrony and symmetries. Formal verification through model checking seems to also bear some limitations, due to the combinatorial explosion of the problems. An interesting but challenging alternative would consist in using proof assistants like COQ in order to formally prove the results for arbitrary values of the parameters, and not just for small values like in model checking so far. Simple impossibility results in the flavor of Lemma 1 have recently been certified using the Pactole COQ framework [2].

## References


1. Balabonski, T., Courtieu, P., Pelle, R., Rieg, L., Tixeuil, S.,Urbain, X.: Continuous vs. discrete asynchronous moves: a certified approach for mobile robots. Research report, Sorbonne Université, CNRS, Laboratoire d'Informatique de Paris 6, LIP6, F-75005 Paris, France (2018). See also SSS 2018 proceedings. https://hal.sorbonne-universite.fr/hal-01762962
2. Balabonski, T., Pelle, R., Rieg, L., Tixeuil, S.: A foundational framework for certified impossibility results with mobile robots on graphs. In: Proceedings of the 19th International Conference on Distributed Computing and Networking, ICDCN 2018, pp. 5:1–5:10. ACM, New York (2018). https://doi.acm.org/10.1145/3154273.3154321
3. Baldoni, R., Bonnet, F., Milani, A., Raynal, M.: Anonymous graph exploration without collision by mobile robots. Inf. Process. Lett. **109**(2), 98–103 (2008). https://doi.org/10.1016/j.ipl.2008.08.011
4. Baldoni, R., Bonnet, F., Milani, A., Raynal, M.: On the solvability of anonymous partial grids exploration by mobile robots. In: Baker, T.P., Bui, A., Tixeuil, S. (eds.) OPODIS 2008. LNCS, vol. 5401, pp. 428–445. Springer, Heidelberg (2008). https://doi.org/10.1007/978-3-540-92221-6_27
5. Blin, L., Milani, A., Potop-Butucaru, M., Tixeuil, S.: Exclusive perpetual ring exploration without chirality. In: Lynch, N.A., Shvartsman, A.A. (eds.) DISC 2010. LNCS, vol. 6343, pp. 312–327. Springer, Heidelberg (2010). https://doi.org/10.1007/978-3-642-15763-9_29
6. Bonnet, F., Defago, X.: Exploration and surveillance in multi-robots networks. In: 2011 Second International Conference on Networking and Computing, pp. 342–344, November 2011. https://doi.org/10.1109/ICNC.2011.66
7. Bonnet, F., Défago, X., Petit, F., Potop-Butucaru, M., Tixeuil,S.: Discovering and assessing fine-grained metrics in robot networks protocols. In: 2014 IEEE 33rd International Symposium on Reliable Distributed Systems Workshops, pp. 50–59, October 2014. https://doi.org/10.1109/SRDSW.2014.34
8. Bonnet, F., Milani, A., Potop-Butucaru, M., Tixeuil, S.: Asynchronous exclusive perpetual grid exploration without sense of direction. In: Fernàndez Anta, A., Lipari, G., Roy, M. (eds.) OPODIS 2011. LNCS, vol. 7109, pp. 251–265. Springer, Heidelberg (2011). https://doi.org/10.1007/978-3-642-25873-2_18
9. Bérard, B., Lafourcade, P., Millet, L., Potop-Butucaru, M., Thierry-Mieg, Y., Tixeuil, S.: Formal verification of mobile robot protocols. Distrib. Comput. **29**(6), 459–487 (2016). https://doi.org/10.1007/s00446-016-0271-1
10. Chalopin, J., Flocchini, P., Mans, B., Santoro, N.: Network exploration by silent and oblivious robots. In: Thilikos, D.M. (ed.) WG 2010. LNCS, vol. 6410, pp. 208–219. Springer, Heidelberg (2010). https://doi.org/10.1007/978-3-642-16926-7_20

11. Datta, A.K., Lamani, A., Larmore, L.L., Petit, F.: Ring exploration by oblivious agents with local vision. In: 2013 IEEE 33rd International Conference on Distributed Computing Systems, pp. 347–356, July 2013. https://doi.org/10.1109/ICDCS.2013.55
12. Datta, A.K., Lamani, A., Larmore, L.L., Petit, F.: Enabling ring exploration with myopic oblivious robots. In: 2015 IEEE International Parallel and Distributed Processing Symposium Workshop, pp. 490–499, May 2015. https://doi.org/10.1109/IPDPSW.2015.137
13. Devismes, S., Lamani, A., Petit, F., Raymond, P., Tixeuil, S.: Optimal grid exploration by asynchronous oblivious robots. In: Richa, A.W., Scheideler, C. (eds.) SSS 2012. LNCS, vol. 7596, pp. 64–76. Springer, Heidelberg (2012). https://doi.org/10.1007/978-3-642-33536-5_7
14. Devismes, S., Lamani, A., Petit, F., Tixeuil, S.: Optimal torus exploration by oblivious robots. In: Bouajjani, A., Fauconnier, H. (eds.) NETYS 2015. LNCS, vol. 9466, pp. 183–199. Springer, Cham (2015). https://doi.org/10.1007/978-3-319-26850-7_13
15. Devismes, S., Petit, F., Tixeuil, S.: Optimal probabilistic ring exploration by semi-synchronous oblivious robots. Theor. Comput. Sci. **498**, 10–27 (2013). https://doi.org/10.1016/j.tcs.2013.05.031
16. Doan, H.T.T., Bonnet, F., Ogata, K.: Model checking of a mobile robots perpetual exploration algorithm. In: Liu, S., Duan, Z., Tian, C., Nagoya, F. (eds.) SOFL+MSVL 2016. LNCS, vol. 10189, pp. 201–219. Springer, Cham (2017). https://doi.org/10.1007/978-3-319-57708-1_12
17. D'Angelo, G., Di Stefano, G., Navarra, A., Nisse, N., Suchan, K.: Computing on rings by oblivious robots: a unified approach for different tasks. Algorithmica **72**(4), 1055–1096 (2015). https://doi.org/10.1007/s00453-014-9892-6
18. Flocchini, P., Ilcinkas, D., Pelc, A., Santoro, N.: Remembering without memory: tree exploration by asynchronous oblivious robots. Theor. Comput. Sci. **411**(14), 1583–1598 (2010). https://doi.org/10.1016/j.tcs.2010.01.007
19. Flocchini, P., Ilcinkas, D., Pelc, A., Santoro, N.: How many oblivious robot scan explore a line. Inf. Process. Lett. **111**(20), 1027–1031 (2011). https://doi.org/10.1016/j.ipl.2011.07.018
20. Flocchini, P., Ilcinkas, D., Pelc, A., Santoro, N.: Computing without communicating: ring exploration by asynchronous oblivious robots. Algorithmica **65**(3), 562–583 (2013). https://doi.org/10.1007/s00453-011-9611-5
21. Lamani, A., Potop-Butucaru, M.G., Tixeuil, S.: Optimal deterministic ring exploration with oblivious asynchronous robots. In: Patt-Shamir, B., Ekim, T. (eds.) SIROCCO 2010. LNCS, vol. 6058, pp. 183–196. Springer, Heidelberg (2010). https://doi.org/10.1007/978-3-642-13284-1_15. arXiv:0910.0832 [cs]
22. Ooshita, F., Tixeuil, S.: Ring exploration with myopic luminous robots [cs], May 2018. See also SSS 2018 proceedings. arXiv:1805.03965
23. Potop-Butucaru, M., Raynal, M., Tixeuil, S.: Distributed computing with mobile robots: an introductory survey. In: 2011 14th International Conference on Network-Based Information Systems, pp. 318–324, September 2011. https://doi.org/10.1109/NBiS.2011.55

# Fault-Tolerant Mobile Robots

Xavier Défago[1], Maria Potop-Butucaru[2(✉)], and Sébastien Tixeuil[2]

[1] School of Computing, Tokyo Institute of Technology, Tokyo, Japan
[2] Sorbonne Université, CNRS, Laboratoire d'Informatique de Paris 6, LIP6, 75005 Paris, France
maria.potop-butucaru@lip6.fr

**Abstract.** This chapter surveys crash tolerance, self-stabilization, Byzantine fault-tolereance, and resilience to inaccuracies for the main building blocks in mobile robots networks: gathering, convergence, scattering, leader election, and flocking.

**Keywords:** Fault-tolerant · Mobile robots · Distributed algorithms

## 1 Introduction

Swarm Robotics envisions groups of mobile robots self-organizing and cooperating toward the resolution of common objectives. In many cases, such groups of robots are deployed in adverse environments (e.g. space, deep sea). Thus, a group must be able to self-organize in the absence of any prior infrastructure and ensure coordination in spite of the presence of faulty robots as well environmental changes. A faulty robot can stop its execution (crash) or start to behave in an arbitrary way either due to some external factors (e.g. electromagnetic fields, attacks) or to some inaccurate information received transmitted by its own sensors.

Fault-tolerance literature in robot networks addressed some of the main tasks robots execute such as gathering, convergence, scattering, leader election, or flocking.

The *gathering problem*, also known as the *rendezvous* problem when only two robots are involved, is a fundamental coordination problem in cooperative mobile robotics. In short, given a set of robots with arbitrary initial location and no initial agreement on a global coordinate system, gathering requires that all robots, following their algorithm, reach the very same location (which is not initially predefined) within a *finite* number of steps, and remain there.

A relaxed version of this problem, *convergence*, does not impose the actual reaching of the same location in finite time (instead, for any real number $\varepsilon > 0$, robots must be within $\varepsilon$ of one another in finite time). Gathering and convergence can serve as basis for many other protocols, such as constructing a common coordinate system or arranging robots in a specific geometrical pattern. Similar to the Consensus problem in conventional distributed systems, gathering and convergence have a simple definition but the existence of a solution greatly depends


P. Flocchini et al. (Eds.): Distributed Computing by Mobile Entities, LNCS 11340, pp. 234–251, 2019.
https://doi.org/10.1007/978-3-030-11072-7_10

on the synchrony of the system as well as the nature of the faults that may possibly occur.

The dual problem of gathering is the *scattering* problem. Scattering requires that, starting from an arbitrary configuration, eventually no two robots share the same location. While most works about gathering assume that robots start from distinct locations, scattering makes no such hypothesis: robots can start from arbitrary locations.

Other agreement-related problem are leader election and flocking. *Leader election* consists in reaching a configuration such that a unique robot can be distinguished by every robot in the network. As robots are generally uniform and execute the same code, the leader robot usually has a particular position in the geometric shape that is formed by the robots, that prevents confusion with any other robot.

Now, *flocking* denotes the ability for a group of robots to follow a leader (also called a flock-head), often also maintaining a particular geometric shape throughout the movement. Flocking is related to leader election as two variants of the problem were studied: one with an existing leader that is known beforehand, and one with no *a priori* leader (hence, leader election is generally the first step to consider when solving the flocking problem).

## 2 Faults and Schedulers

### 2.1 Faults

A *faulty* robot is a robot whose behavior deviates from its specification. A *correct* robot is one that never fails.

There are two classical classes of faults.

- A *crash* fault occurs when a faulty robot unexpectedly and permanently stops performing any actions.
- A *Byzantine* fault occurs when a faulty robot can behave arbitrarily. This includes omitting to move or moving to arbitrary locations with the deliberate intent to prevent correct robots from solving the desired problem.

Notice that in models without explicit communication using either lights or message passing, a Byzantine robot can only influence the other robots through its movements.

Then, robots may experience less severe faults: *transient* faults that place the network in an arbitrary configuration (more details are given in Sect. 5), and inaccuracies (both for sensing devices and the actuating devices that are embedded on robots, see Sect. 6).

### 2.2 Schedulers

A scheduler decides, for every configuration, which subset of the robots is active (i.e., allowed to perform their actions). A correct robot is *movable* if it there is a possibility that it actually moves when it is activated by the scheduler (*i.e.* it either moves in the case of a deterministic algorithm, or moves with probability $p > 0$ in the case of a probabilistic algorithm). In the following we consider the following schedulers [22]:

- *unfair arbitrary*: At each activation, a non-empty subset of robots is activated. A non-triviality condition ensures that, in any infinite execution, infinitely often, a movable correct robot becomes active.
- *unfair centralized*: The scheduler is unfair (as described above) with the additional restriction that at most one (*i.e.*, exactly one) robot is activated at each activation.
- *fair arbitrary*: At each activation, any non-empty subset of the robots is activated, with the guarantee that every robot becomes active infinitely often in an infinite execution.
- *fair centralized*: The scheduler is fair (see above) with the additional guarantee that no more than one (i.e., exactly one) robot is activated at each activation.
- *fair k-bounded*: The scheduler is fair with the additional guarantee that there exists some bound $k > 0$ such that, between any two consecutive activations of any robot, no other robot is activated more than $k$ times. The bound may be known or unknown to the robots. In the sequel we assume that robots do not know the scheduler bound.
- *round-robin*: The scheduler is fair 1-bounded and centralized. This implies that the robots are activated always in the same sequence.
- *fully synchronized*: Every robot is active at every activation.

It should be noted that given two schedulers $A$ and $B$, $A$ is stronger than $B$ if the set of all possible executions allowed by scheduler $A$ strictly contains the set of all executions allowed by scheduler $B$. As a result, any algorithm that is correct under scheduler $A$ is also correct under scheduler $B$. Likewise, any impossibility proven under scheduler $B$ also holds under scheduler $A$. Défago *et al.* [22] propose a hierarchy of schedulers depicted in Fig. 1.

## 3 Crash Faults

In the crash faults model the problems that have been extensively studied are: gathering, convergence and flocking. In the following we recall the main results.

### 3.1 Gathering and Convergence Under Crash Faults

One of the first steps in the direction on studying gathering feasibility in the context of crash faults is due to Agmon and Peleg [2]. They proved that gathering of correct robots (referred as *weak gathering*) can be achieved in the SSYNC model if at most one robot may crash. Their algorithm makes use of multiplicity detection, and restricts the set of admissible initial configurations to the distinct configurations (that is, the configurations where at most one robot occupies a particular position).

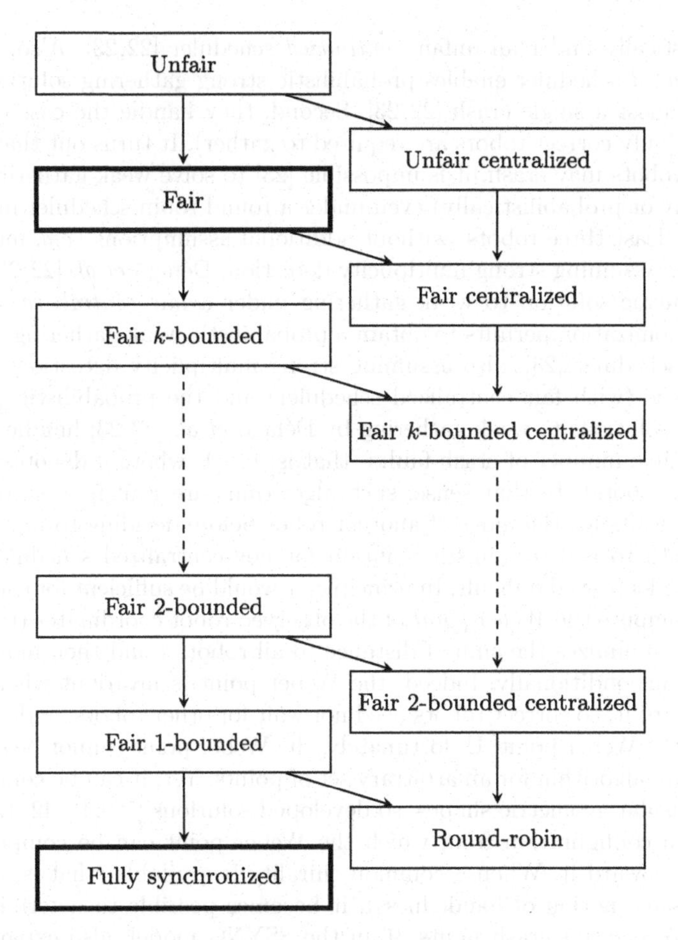


**Fig. 1.** Relationships between scheduler classes. Conventional models are highlighted: SSYNC [49] and ASYNC [43] are fair, and the fully synchronous model FSYNC [49] is its namesake.

Then, Défago *et al.* [22,23] studied the limits of gathering feasibility in both fault-free and fault-prone environments, considering scheduler variants such as the fair $k$-bounded centralized scheduler, for some parameter $k$, for the SSYNC model. The main results obtained by Défago *et al.* [22,23] are as follows. First, they consider the case of "strong" gathering, that is, having *all* robots (both correct and crashing robots) gather at a particular location. Obviously, if two robots are initially crashed at distinct locations, strong gathering is impossible. Now, even when a single robot may crash, it is impossible [22,23] to strongly gather a networks of $n \geq 3$ robots deterministically, even with a round robin scheduler and multiplicity detection. The same impossibility result holds with a probabilistic algorithm [22,23] and a fair centralized scheduler. However, the case of 2 robots is solvable probabilistically under an unfair scheduler and deterministically under an unfair *centralized* scheduler [22,23]. Also, assuming a fair *bounded* scheduler enables probabilistic strong gathering solutions in the case of at most a single crash [22,23]. Second, they handle the case of "weak" gathering (only correct robots are required to gather). It turns out that when at least two robots may crash, it is impossible [23] to solve weak gathering (deterministically or probabilistically) even under a round robin scheduler in any network of at least three robots, without additional assumptions (*e.g.* multiplicity detection). Assuming strong multiplicity detection, Défago *et al.* [22,23] provide a deterministic solution to weak gathering under a fair *centralized* scheduler. Using randomization permits to obtain a probabilistic weak gathering algorithm for a fair scheduler [23], also assuming strong multiplicity detection. Both the deterministic (with fair centralized scheduler) and the probabilistic (with fair scheduler) solutions to weak gathering by Défago *et al.* [22,23] handle the maximum possible number of crash faults (that is, $n-1$, where $n$ denotes the total number of robots). In that sense, such algorithms are wait-free, so no correct robot ever waits for the move of another robot before deciding to move itself.

The path to a deterministic solution for non-centralized schedulers in the SSYNC model proved difficult. In principle, it would be sufficient for each correct robot to compute the *Weber point* of the observed robot coordinates (that is, the point that minimizes the sum of distance to all robots), and then move toward this point unconditionally. Indeed, the Weber point is invariant when a robot moves toward it, so correct robots need not wait for other robots, and eventually gather at the Weber point. Unfortunately, the Weber point cannot be computed by any finite algorithm for an arbitrary set of points. Yet, it can be computed for some particular geometric shapes, so developed solutions [7,9,12,42] try to first construct a configuration from which the Weber point can be computed, and then move toward it. When a common chirality is available (that is, all robots have the same notion of handedness), it becomes possible to deterministically tolerate up to $n-1$ crash faults [9] in the SSYNC model, also expanding the set of initial configurations to those that are not bivalent (so, all feasible initial configurations in a deterministic context are considered).

When the robots agree on a common direction (*e.g.*, North) it becomes possible to solve weak gathering without common agreement on chirality and without any restriction on the set of initial configurations [6]. However, the assumption that a common direction is available trivializes the problem of weak gathering as it also sufficient to solve gathering starting from a bivalent configuration, but not necessary [38] (in that sense, agreeing on a common direction is a stronger problem than weak gathering).

Bramas and Tixeuil [12] proved that neither of these extra assumptions is necessary for deterministic weak gathering under a non-centralized scheduler. They present a deterministic gathering protocol in SSYNC model that can start from any non-bivalent configuration (the largest possible set in the classical model), yet does not require to assume that all robots share a common direction (as in [6]), nor a common chirality (as in [9]). The protocol retains the ability to tolerate up to $n-1$ crash faults, that it, it is wait-free. Their protocol makes use of strong multiplicity detection.

Recently, both Pattanayak *et al.* [42] and Bhagat and Mukhopadhyaya [7] considered the problem of crash tolerant weak gathering when only weak multiplicity detection is available, still assuming the SSYNC model with no agreement on chirality. However, none of those works retains the generality of the initial configuration (the weakest condition in the deterministic case being the non-bivalent configuration): Pattanayak *et al.* [42] assume that initial configurations have at most one multiplicity point, while Bhagat and Mukhopadhyaya [7] assume initial distinct configurations. With respect to the number of crashes tolerated, Pattanayak *et al.* [42] can handle up to $n-1$ crashes (hence, it is wait free), and Bhagat and Mukhopadhyaya [7] can go up to $\lfloor n/2 \rfloor$ crashes.

The problem of convergence in the presence of crash faults was addressed by Cohen and Peleg [17,18], where algorithms based on the move of correct robots to the center of gravity of the observed robots are presented. Those algorithms execute in the ASYNC model and $n$ robots can tolerate up to $f$ crash faults whenever $n > f + 1$.

Overall, positive results for crash-tolerant gathering only exist up to the SSYNC model. The existence of solution in the ASYNC model is open.

### 3.2 Leader Election and Flocking Under Crash Faults

In robot networks, leader election has been studied in the SSYNC model by Suzuki and Yamashita [48]. The authors propose a solution where robots share the same coordinate space and start from distinct positions, so it is easy to elect the robot with *e.g.* the topmost and leftmost coordinates as leader. This approach does not require the robots to move, so it naturally tolerates up to $n-1$ crash faults, where $n$ is the total number of robots. Another way to single out a leader is due to Dieudonné and Petit [25] for robots organized on the same circle (but *not* on a regular polygon); their approach does not require movement (hence works with $n-1$ crashed robots) nor a common coordinate system, but the set of initial admissible configurations is quite specific. A full characterization of geometric patterns that enable leader election without moving (and hence tolerating up to $n-1$ crash faults) is due to Dieudonné *et al.* [24]. Interestingly, their characterization can be concisely expressed using concepts from language theory: Lyndon words.

The flocking problem, although largely discussed from an engineering point of view [39,40,44], was first studied from a distributed algorithms point of view mainly by Gervasi and Prencipe [32,33]. The authors propose non-uniform algorithms where robots have basically two roles: one robot is the leader robot, and all other robots are follower robots. It is furthermore assumed that all follower robots know the leader robot. Obviously, when the leader robot crashes, the flocking is compromised.

To tolerate the crash fault of *any* possible robots, one must design an algorithm that executes correctly without a predefined leader (and is hence uniform).

Canepa and Potop-Butucaru [15] proposed a uniform probabilistic flocking architecture that permits to leader to emerge from an algorithmic computation, in the most general ASYNC model. However, their approach is based on the assumption that the elected leader, and consequently the flock, do not change their direction and trajectory. Also, the algorithm of Canepa and Potop-Butucaru [15] does not tolerate crash failures.

Fault-tolerant (but not self-stabilizing) flocking was addressed by Souissi *et al.* [47] and Yang *et al.* [51] in the SSYNC model. Souissi *et al.* [47] propose a fault tolerant flocking algorithm in the SSYNC model using a leader oracle that provides robots the current head, and a failure detector. A subsequent work by Yang *et al.* [51] implements a failure detector in the SSYNC model assuming a fair $k$-bounded scheduler (for a constant value $k$), persistent memory of the robots (that is, the robots are *not* oblivious), agreement on one coordinate axis, and a common chirality for all robots. Intuitively, having correct robots move as often as possible permits to detect crashed robots if one can remember past actions of other robots and is activated sufficiently often.

Canepa *et al.* [14] presented a uniform probabilistic flocking algorithm in the SSYNC model. Their solution operates without assuming strong hypotheses (robots are oblivious, the initial configuration is arbitrary and may include multiplicity points, robots do not share any coordinate system or chirality) and no pre-existing leader is assumed (mandated by the uniform property). Also, the flocking adjusts to changes of directions and velocity of the elected flock-head. The algorithm of Canepa *et al.* [14] tolerates crash faults if crashed robots disappear from the sensors of other robots, making it suitable for managing flocks of UAVs: When the current flock-head disappears or is damaged and not recognized as a correct robot, the remaining robots agree on a new head and the flock can continue its task.

## 4 Byzantine Faults

To the best of our knowledge only two problems have been specifically addressed in the Byzantine faults model: gathering and convergence.

### 4.1 Gathering Under Byzantine Faults

Agmon and Peleg [2] prove that no deterministic gathering algorithm that tolerates up to *one* Byzantine robot in a three robots system can exist in the SSYNC model. However, the FSYNC model, deterministic gathering is solvable. Moreover, in the FSYNC model, they provide a deterministic algorithm that solves gathering whenever $n > 3f$, where $n$ denotes the total number of robots, and $f$ denotes the maximum number of Byzantine nodes. In small FSYNC systems of three robots, with at most one Byzantine robot, deterministic gathering remains solvable. Their impossibility result in the case of SSYNC makes use on the very low capabilities of the robots and the high power of the environment: the scheduler is fair, the robots are oblivious and uniform, and there is no agreement on the coordinate system.

Défago *et al.* [22,23] improve the impossibility results of Agmon and Peleg [2] by considering weakened forms of schedulers (hence lowering the power of the adversary), that are fair $k$-bounded, for some $k$ that depends on the impossibility result. In more details, Défago *et al.* [22,23] show that if the scheduler is round robin (a weaker form of fair 2-bounded scheduler), then no deterministic nor probabilistic solution exits whenever $n > f+1$ and $f \geq 1$. For small systems of three robots, Défago *et al.* [22,23] prove that no deterministic gathering algorithm can exists, even with only a single Byzantine failure, a fair $k$-bounded centralized scheduler (for some constant $k \geq 2$, and multiplicity detection. Interestingly, the number of Byzantine in the systems yields impossibilities with respect to the bound $k$ of the fair $k$-bounded centralized scheduler, as Défago *et al.* [22,23] demonstrate that when $n$ is even, a scheduler bound $k \geq \left\lceil \frac{n-f}{f} \right\rceil$ implies no deterministic gathering algorithm, while when $n$ is odd, a scheduler bound $k \geq \left\lceil \frac{n-f}{f-1} \right\rceil$ implies a similar result.

Izumi *et al.* [35,36] further extends impossibility results to stronger robots in the SSYNC model with a fair $n$-bounded centralized scheduler: robots may not be uniform (so each may execute a different code), may use persistent memory (so each may keep track of past configurations), may share a common orientation on coordinate systems. Still, Izumi *et al.* [35,36] prove that if even only one robot may be Byzantine, deterministic gathering is impossible in this setting. Interestingly, their proof is based on the distributed BG-simulation proposed by Borowsky *et al.* [8], and extends to other problems than gathering, namely *circle formation* (having all correct robots reach a position such that they are placed on different locations on the boundaries of a common cycle) and *line formation* (having all correct robots reach a position such that they occupy different positions on a common line), which are also proved impossible to solve under the same hypotheses.

Overall, the only positive results so far with respect to deterministic gathering in the presence on Byzantine robots are in the FSYNC model. The lower bound on the number of Byzantine robots that can be tolerated is still open (the upper bound of $n > 3f$ in the general case [2] is not proved tight). In the SSYNC and ASYNC models, the existence of gathering algorithms that can tolerate at least one Byzantine robots is open. An extensive map of possibility/impossibility results and open research problems about Byzantine tolerant gathering was presented by Défago *et al.* [23].

### 4.2 Convergence Under Byzantine Faults

The specification of convergence being less stringent than that of gathering, it is worth investigating whether this leads to better Byzantine tolerance.The feasibility of convergence in Byzantine-prone robot networks was specifically addressed by Bouzid *et al.* [10,11] and Auger *et al.* [3] in the case of uni-dimensional robot networks, where robots evolve on one dimension only (that is, the continuous space the robots evolve in is an infinite line). In more details, Bouzid *et al.* [11] establish a connection between the convergence problem in robot networks, and the distributed *approximate agreement* problem (that requires correct processes to decide, for some constant $\varepsilon > 0$, values that are a most $\varepsilon$ apart and within the range of initial values). However, the lower bounds and the possibility results are different in the two cases, as robots need not *decide* termination, they just go closer to one another forever. Interestingly, the positive results (*a.k.a.* the algorithms) by Bouzid *et al.* use techniques that are similar to the ones by Dolev *et al.* [30] and by Abraham *et al.* [1] for approximate agreement with Byzantine failures. A key ingredient of the algorithm is the *trimming* of extremal values: the robots look at all positions occupied by other robots, and remove the $f$ higher and the $f$ lower of those coordinates (in their own coordinate system, so the notion of higher and lower may vary from robot to robot), then, they approach the set of robots whose positions are *not* extremal. So, if $f$ Byzantine robots are present, they cannot prevent the correct one from moving toward the other correct ones by taking some extremal position. Then, if Byzantine robots are present in the core, correct robots still get closer to other correct robots, and convergence eventually holds.

In the FSYNC model, Bouzid *et al.* [11] show that deterministic convergence is feasible when the number of robots $n$ is strictly greater than $2f$, $f$ denoting the maximum number of Byzantine robots in the network. So, in FSYNC, convergence permits to handle more Byzantine robots than gathering (the gathering algorithm by Agmon and Peleg [2] requires $n > 3f$).

In the SSYNC and ASYNC models, but assuming a fair $k$-bounded scheduler (for some arbitrary constant $k$), Bouzid *et al.* propose an algorithm that solves convergence whenever $n > 3f$. It turns out that the bounds are tight for the considered schedulers [11]: if $n = 2f$, no deterministic convergence algorithm exists in FSYNC, if $n = 3f$, no deterministic convergence algorithm exists in SSYNC (and hence in ASYNC) even with a fair 2-bounded scheduler. Those two lower bounds were formally certified in Coq by Auger *et al.* [3].The case of a fair scheduler in the ASYNC model is more intricate. Bouzid *et al.* [10] provide a deterministic convergence algorithm for mobile robots with a fair scheduler assuming $n > 5f$. This algorithm is *cautious* in the sense that correct robots always move *inside* the range (in the unidirectional space) of other correct robots. Now, for the class of cautious algorithm, Bouzid *et al.* [10] show that $n > 5f$ is actually a necessary condition when the scheduler is only fair (and not fair $k$-bounded) even in the SSYNC model. Overall, in the SSYNC and ASYNC models, convergence with Byzantine robots is still solvable (albeit with restrictions as the system becomes more asynchronous), while gathering is essentially impossible to solve as soon as one Byzantine robot occurs.

## 5 Self-stabilization

Self-stabilization is a property of distributed systems that was introduced by Dijkstra in 1974 [29]. It refers to the ability to recover from arbitrary transient failures. In the context of mobile robot networks, self-stabilization may translate to the following abilities:

- For oblivious robots in the FSYNC or SSYNC models, self-stabilization refers to the ability to start in an arbitrary initial position (including positions with multiplicity points);
- For oblivious robots in the ASYNC model, self-stabilization refers to the ability to start in an arbitrary initial position, with arbitrary pending moves;
- For luminous robots in the FSYNC or SSYNC models, self-stabilization refers to the ability to start in an arbitrary initial position, with arbitrary colors for each robot (in the color domain set);
- For luminous robots in the ASYNC model, self-stabilization refers to the ability to start in an arbitrary initial position, with arbitrary colors for each robot, with arbitrary pending moves.

### 5.1 Gathering and Convergence

A fundamental result of Suzuki and Yamashita [49] shows that in the SSYNC models, no deterministic algorithm can solve gathering for two robots without additional assumptions. In [22,23] the authors prove that the problem of *self-stabilizing* gathering is impossible even under a *round-robin scheduler*. Then they prove the same impossibility for distinct gathering with any algorithm of a class called *rapid* algorithms. Secondly, still without multiplicity detection, they prove that self-stabilizing gathering can be solved probabilistically under a fair bounded scheduler (with an arbitrary but finite bound) when $n \geq 3$ and under an unfair scheduler when $n = 2$, by exhibiting a simple algorithm that solves the problem.

The impossibility results in [49] were also extended using the Coq proof assistant [20] to the bivalent case, that is when an even number of robots is initially evenly split in exactly two locations. So, in the context of self-stabilization, no deterministic solution to gathering can exists in SSYNC, hence in ASYNC.

The FSYNC model yields positive results. When strong multiplicity detection is available, the algorithm that moves robots to their center of gravity by Cohen and Peleg [18] can be used to obtain a self-stabilizing deterministic gathering protocol. When there is no multiplicity detection, using the center of gravity of inhabited locations as suggested by Balabonski *et al.* [5] yields deterministic self-stabilizing gathering in FSYNC.

With the SSYNC model, a way to avoid initial bivalent configuration is to assume an odd number of robots. Dieudonné and Petit [28], assuming global strong multiplicity detection, provided a deterministic and self-stabilizing solution for an odd number of robots to gathering. Alternatively, probabilistic approaches in the SSYNC model allowed self-stabilizing gathering without assumptions on the number of robots [16], and showed the trade-off between the time complexity and the availability of global-weak multiplicity detection or global-strong multiplicity detection (resp., local-weak multiplicity detection or local-strong multiplicity detection). In the discrete space setting (in this case, ring-shaped networks), Ooshita and Tixeuil [41] proved that no solution to gathering may exist if it is either: *(i)* deterministic, *(ii)* operating in ASYNC, *(iii)* using local-strong and global weak multiplicity detection only. As a result, they provide a SSYNC probabilistic algorithm that assumes global-strong multiplicity detection. Another approach to circumvent the impossibility of gathering two robots in SSYNC is to endow each robot with a *light* bulb [21], capable of emitting a fixed number of colors visible to all other robots. In the SSYNC model, Viglietta [50] proved that being able to emit lights of only two colors is sufficient to solve the gathering problem for two robots.

In the ASYNC setting, for the restricted case of two robots gathering, the ability to use lights proved invaluable. Viglietta [50] proved that an algorithm that *only makes use of observed colors* to decide on its next move cannot gather two robots in the ASYNC model using only two colors. In the same paper, Viglietta showed [50] that three colors and the ability to detect null distances is sufficient for gathering two robots in ASYNC. The general approach of Das et al. [21] makes use of four colors per robot in ASYNC. Recently, Heriban *et al.* [34] proved that two colors are sufficient to deterministically gather two robots in ASYNC, starting from any possible initial color and pending move.

The case of self-stabilizing convergence in the ASYNC model is investigated by Cohen and Peleg [18]. In more details, their algorithm that makes every robot move toward the center of gravity of observed robot positions is self-stabilizing in the sense that initial target points of robots may be arbitrary (Cohen and Peleg [18] only requires that the initial set of target points is bounded, so a robot cannot target a position that is infinitely far away) as well as the initial position of the robots. However, the algorithm by Cohen and Peleg makes use of multiplicity detection to compute the center of gravity of the robots (if two robots share the same position, the weight of that position is twice as much as if it were inhabited by a single robot).

### 5.2 Scattering

With oblivious robots, Scattering intrinsically mandates self-stabilizing solutions, as arbitrary many robots may be located at a single location in the initial configuration.

It turns out that neither deterministic gathering [49] nor scattering [27] is possible without additional assumptions. Furthermore, while some extra assumptions allow deterministic gathering (e.g. coordinate system, multiplicity detection), oblivious deterministic scattering remains impossible unless the very specific setting where there are no *clones* [49]: Two robots are considered to be clones of each other if they have the same local coordinate system and the same initial position, and they always become active simultaneously. All other settings mandate probabilistic approaches.

Probabilistic solutions to scattering were proposed in the FSYNC and SSYNC settings. The first probabilistic algorithms to solve mobile robot scattering without multiplicity detection were given by Dieudonné and Petit [26,27]. The algorithms are based on the following simple scheme: after the Look phase, a robot computes the Voronoi diagram [4] of the observed positions, and then tosses a coin ($\frac{1}{4}$ [26] or $\frac{1}{2}$ [27]) to either remain in position, or move toward an arbitrary position in its Voronoi cell. The fact that a robot may only move within its Voronoi cell preserves the fact that initially distinct robots (that is robots occupying distinct positions) remain distinct thereafter. This invariant and the positive probability that two robots on the same point separate implies the eventual scattering of all robots. A later study [16] shows that the scattering algorithm [27] converges in expected $O(\log n \log\log n)$ rounds. In the same paper, a new probabilistic algorithm was presented, with the assumption that robots are aware of the total number of robots. This protocol is optimal in time as it scatters any $n$-robots configuration in expected $O(1)$ rounds. If the total number of robots $n$ is known, then robots are able to choose uniformly at random a position within their Voronoi cell among $2n^2$ possibilities, inducing an expected $O(1)$ rounds scattering time. In the limited visibility setting [37] (the visibility capability of each robot has a constant radius, and visual connectivity has to be maintained throughout scattering), the time lower bound grows to expected $n$ rounds for scattering $n$ robots. None of the aforementioned works investigated the number of random bits used in the scattering process.

Bramas and Tixeuil [13] investigated the amount of randomness (that is, the number of random bits used by the robots) that is necessary to achieve mobile robots scattering. In more details, they first defined a canonical scattering algorithm, that encompasses all previous solutions, and is tantamount to selecting the number of possible locations that are selected uniformly at random by the robots. Then, they proved that $n \log n$ random bits are necessary to scatter $n$ robots in any setting for all scattering algorithms (not only canonical algorithm). Also, they give a sufficient condition for a canonical scattering algorithm to be random bit optimal (namely, the number of possible locations must be polynomial in the number of observed positions). As it turns out that previous solutions for scattering [16,26,27] satisfy the condition of Bramas and Tixeuil [13], they are hence proved random bit optimal for the scattering problem. Finally, they investigate the time complexity of scattering algorithms, when strong multiplicity detection is not available. They prove that such algorithms cannot converge in constant time in the general case and in $o(\log\log n)$ rounds in the case of random bits optimal algorithms (in this last setting, the best known upper bound was $O(\log n \log\log n)$ [26,27]). On the positive side, Bramas and Tixeuil [13] provide a family of scattering algorithms that converge as fast (but not $O(1)$) as needed, without using multiplicity detection. Also, they give a particular protocol among this family that is random bit optimal ($O(n \log n)$ random bits are used) and time optimal ($O(\log\log n)$ rounds are used). This improves the time complexity of previous results in the same setting by an expected $\log n$ factor. All the bounds concerning random bits complexity hold with high probability[1].

To our knowledge, there are currently no published results about scattering with luminous robots. However, the following remarks can be made. First, if the initial configuration may contain two robots whose lights are the same color, deterministic scattering is impossible: simply place those two robots on the same initial position, and assume they have the same coordinate system, always

[1] An event occurs with high probability if it occurs with probability greater than $1 - o(1/n^{\varepsilon})$ with $\varepsilon > 0$.

activate them simultaneously; as they are clones, they never separate, and scattering is never achieved. Second, if the initial configuration only contains robots with unique light colors, and colors can be ordered, then deterministic scattering can be solved in SSYNC. At each multiplicity point, the robot whose light has the smallest color moves arbitrarily in its Voronoi cell.

# 6 Unreliable Sensors and Actuators

Since all algorithms rely on sensors in a way or another, it is natural to question the reliability of such sensors and what happens when they fail. The literature considers two kinds of sensors, namely, sensors giving the position of the robots during the Look phase, and sensors providing shared information such as a common direction, e.g., provided by a compass.

## 6.1 Gathering with Unreliable Compasses

Flocchini *et al.* [31] proposed a gathering algorithm for oblivious robots with *limited visibility* in the ASYNC model, where robots share the knowledge of a common direction as provided by a compass. A natural question that arises is what happens when these compasses are unreliable.

Several authors [38,46] have defined classes and classified unreliable compasses:

- *Perfect compasses* [31] all indicate the same global north direction.
- *Eventually consistent compasses* [46] initially indicate arbitrary directions and may fluctuate arbitrarily but, after some finite (but unknown) global stabilization time, all compasses indicate the same absolute direction.
- *$\phi$-inaccurate compasses* [38] differ at most by an angle $\phi$ from some absolute north direction (unknown to the robots). In other words, a pair of $\phi$-inaccurate compasses can differ by as much as $2\phi$ at any time $t$. The special cases when compasses are invariant and $\phi = 0$ represents perfect compasses, and when $\phi = \frac{\pi}{2}$ the model brings no more information than having no compasses.

*Eventually Consistent Compasses.* Souissi *et al.* [46] first prove that the unmodified gathering algorithm of Flocchini *et al.* [31] cannot tolerate unreliable compasses by exhibiting an execution that leads to a (non-recoverable) partition of the visibility graph. They then propose an algorithm in ASYNC to achieve gathering under limited visibility when the robots are equipped with *eventually consistent* compasses. The difficulty is to avoid any combination of moves that can lead two mutually visible robots from ever breaking their visibility bound.

*Inaccurate Compasses.* Izumi *et al.* [38] consider the problem of gathering two robots (rendezvous problem) when robots are equipped with inaccurate compasses. Under unlimited visibility, rendezvous is known to be impossible without compasses but trivial with perfect compasses. With oblivious robots, the problem is also trivial with eventually consistent compasses. The question is about compasses with bounded errors (i.e., inaccurate compasses).

The paper refines the models of inaccurate compasses by considering the *dynamicity* of the compasses.

- *Fully dynamic compasses* may vary at any time during an execution. In particular, this means that the move operation may be affected by changes. In this case, rendezvous is trivially impossible.
- *Semi-dynamic compasses* may vary only between the end of a cycle and the beginning of the next one, but do not change during a cycle.
- *Fixed compasses* always indicate a constant direction. This corresponds to the assumptions of the $\phi$-inaccurate compasses described earlier.

The possibility/impossibility of gathering heavily depends on the model and the value of bound $\phi$. The results presented by Izumi *et al.* [38] are summed up in Table 1.

**Table 1.** Summary of known results on rendezvous (gathering of two robots) with $\phi$-inaccurate compasses [38].

| Synchrony | Compasses | $\phi = 0$ | $\phi < \frac{\pi}{6}$ | $\phi < \frac{\pi}{5}$ | $\phi < \frac{\pi}{4}$ | $\phi < \frac{\pi}{3}$ | $\phi < \frac{\pi}{2}$ | $\phi = \frac{\pi}{2}$ |
|---|---|---|---|---|---|---|---|---|
| SSYNC | Fixed | Possible | | | | | | Imposs. |
| | Semi-dynamic | Possible | | | | Impossible | | |
| ASYNC | Fixed | Possible | | | | | | Imposs. |
| | Semi-dynamic | Possible | | ?(open)? | | Impossible | | |

*Model Equivalence.* Souissi [45] observes that eventually consistent compasses are weaker or equal to perfect compasses, depending on visibility and obliviousness. With oblivious robots and full visibility, eventually consistent compasses are equivalent to a perfect compass. More generally speaking, any *oblivious and self-stabilizing* algorithm that relies on perfect compasses also works under eventually consistent compasses.

Now, although this was not previously observed to our knowledge, it is possible to define *eventually bounded-error compasses* as a combination of eventually consistent compasses and $\phi$-inaccurate compasses by assuming that a bound on directional errors holds only after some global stabilization time. For the same reasons and under conditions similar to those stated above, eventually bounded-error compasses have the same computational power as $\phi$-inaccurate compasses.

## 6.2 Convergence with Inaccurate Sensors and Actuators

Cohen and Peleg [19] study the problem of convergence under inaccurate sensors and movements. They consider that the values of the positions returned by the look-operation, computations, and movement control may all be subject to inaccuracies.

In this context, they show that the classical approach that consists in going toward the center of gravity of robot position does not work well: while it works in FSYNC, it fails in ASYNC.

Instead, they propose a refined approach, based on calculating the center of gravity of the group of robots, and also estimating the maximum possible error in the center-of-gravity calculation. Then, in the new algorithm, a robot makes no movement if it is within the maximum possible error from the center of gravity. If it is outside the circle of error, it moves towards the center of gravity but only up to the bounds of the circle of error. Cohen and Peleg [19] show that the new approach permits to solve convergence in FSYNC and SSYNC when robots move in bidimensional Euclidean space, and also solve in ASYNC when robots operate in unidimensional Euclidean spaces (*a.k.a*, an infinite line).

## References

1. Abraham, I., Amit, Y., Dolev, D.: Optimal resilience asynchronous approximate agreement. In: Higashino, T. (ed.) OPODIS 2004. LNCS, vol. 3544, pp. 229–239. Springer, Heidelberg (2005). https://doi.org/10.1007/11516798_17
2. Agmon, N., Peleg, D.: Fault-tolerant gathering algorithms for autonomous mobile robots. SIAM J. Comput. **36**(1), 56–82 (2006)
3. Auger, C., Bouzid, Z., Courtieu, P., Tixeuil, S., Urbain, X.: Certified impossibility results for Byzantine-tolerant mobile robots. In: Higashino, T., Katayama, Y., Masuzawa, T., Potop-Butucaru, M., Yamashita, M. (eds.) SSS 2013. LNCS, vol. 8255, pp. 178–190. Springer, Cham (2013). https://doi.org/10.1007/978-3-319-03089-0_13
4. Aurenhammer, F.: Voronoi diagrams: a survey of a fundamental geometric data structure. ACM Comput. Surv. **23**(3), 345–405 (1991)
5. Balabonski, T., Delga, A., Rieg, L., Tixeuil, S., Urbain, X.: Synchronous gathering without multiplicity detection: a certified algorithm. In: Bonakdarpour, B., Petit, F. (eds.) SSS 2016. LNCS, vol. 10083, pp. 7–19. Springer, Cham (2016). https://doi.org/10.1007/978-3-319-49259-9_2
6. Bhagat, S., Gan Chaudhuri, S., Mukhopadhyaya, K.: Fault-tolerant gathering of asynchronous oblivious mobile robots under one-axis agreement. In: Rahman, M.S., Tomita, E. (eds.) WALCOM 2015. LNCS, vol. 8973, pp. 149–160. Springer, Cham (2015). https://doi.org/10.1007/978-3-319-15612-5_14
7. Bhagat, S., Mukhopadhyaya, K.: Fault-tolerant gathering of semi-synchronous robots. In: Proceedings of 18th International Conference on Distributed Computing and Networking (ICDCN), Hyderabad, India, p. 6, January 2017
8. Borowsky, E., Gafni, E., Lynch, N.A., Rajsbaum, S.: The BG distributed simulation algorithm. Distrib. Comput. **14**(3), 127–146 (2001)
9. Bouzid, Z., Das, S., Tixeuil, S.: Gathering of mobile robots tolerating multiple crash faults. In: Proceedings of 33rd IEEE International Conference on Distributed Computing Systems (ICDCS), Philadelphia, PA, USA, pp. 337–346, July 2013
10. Bouzid, Z., Gradinariu Potop-Butucaru, M., Tixeuil, S.: Byzantine convergence in robot networks: the price of asynchrony. In: Abdelzaher, T., Raynal, M., Santoro, N. (eds.) OPODIS 2009. LNCS, vol. 5923, pp. 54–70. Springer, Heidelberg (2009). https://doi.org/10.1007/978-3-642-10877-8_7

11. Bouzid, Z., Potop-Butucaru, M.G., Tixeuil, S.: Optimal Byzantine-resilient convergence in uni-dimensional robot networks. Theoret. Comput. Sci. **411**(34–36), 3154–3168 (2010)
12. Bramas, Q., Tixeuil, S.: Wait-free gathering without chirality. In: Scheideler, C. (ed.) Structural Information and Communication Complexity. LNCS, vol. 9439, pp. 313–327. Springer, Cham (2015). https://doi.org/10.1007/978-3-319-25258-2_22
13. Bramas, Q., Tixeuil, S.: The random bit complexity of mobile robots scattering. Int. J. Found. Comput. Sci. **28**(2), 111–134 (2017)
14. Canepa, D., Defago, X., Izumi, T., Potop-Butucaru, M.: Flocking with oblivious robots. In: Bonakdarpour, B., Petit, F. (eds.) SSS 2016. LNCS, vol. 10083, pp. 94–108. Springer, Cham (2016). https://doi.org/10.1007/978-3-319-49259-9_8
15. Canepa, D., Potop-Butucaru, M.G.: Stabilizing flocking via leader election in robot networks. In: Masuzawa, T., Tixeuil, S. (eds.) SSS 2007. LNCS, vol. 4838, pp. 52–66. Springer, Heidelberg (2007). https://doi.org/10.1007/978-3-540-76627-8_7
16. Clément, J., Défago, X., Potop-Butucaru, M.G., Izumi, T., Messika, S.: The cost of probabilistic agreement in oblivious robot networks. Inf. Process. Lett. **110**(11), 431–438 (2010)
17. Cohen, R., Peleg, D.: Robot convergence via center-of-gravity algorithms. In: Královič, R., Sýkora, O. (eds.) SIROCCO 2004. LNCS, vol. 3104, pp. 79–88. Springer, Heidelberg (2004). https://doi.org/10.1007/978-3-540-27796-5_8
18. Cohen, R., Peleg, D.: Convergence properties of the gravitational algorithm in asynchronous robot systems. SIAM J. Comput. **34**(6), 1516–1528 (2005)
19. Cohen, R., Peleg, D.: Convergence of autonomous mobile robots with inaccurate sensors and movements. SIAM J. Comput. **38**(1), 276–302 (2008)
20. Courtieu, P., Rieg, L., Tixeuil, S., Urbain, X.: Impossibility of gathering, a certification. Inf. Process. Lett. **115**(3), 447–452 (2015)
21. Das, S., Flocchini, P., Prencipe, G., Santoro, N., Yamashita, M.: Autonomous mobile robots with lights. Theoret. Comput. Sci. **609**, 171–184 (2016)
22. Défago, X., Gradinariu, M., Messika, S., Raipin-Parvédy, P.: Fault-tolerant and self-stabilizing mobile robots gathering. In: Dolev, S. (ed.) DISC 2006. LNCS, vol. 4167, pp. 46–60. Springer, Heidelberg (2006). https://doi.org/10.1007/11864219_4
23. Défago, X., Potop-Butucaru, M.G., Clément, J., Messika, S., Raipin Parvédy, P.: Fault and Byzantine tolerant self-stabilizing mobile robots gathering - feasibility study. CoRR abs/1602.05546 (2016)
24. Dieudonné, Y., Levé, F., Petit, F., Villain, V.: Deterministic geoleader election in disoriented anonymous systems. Theoret. Comput. Sci. **506**, 43–54 (2013)
25. Dieudonne, Y., Petit, F.: Circle formation of weak robots and Lyndon words. Inf. Process. Lett. **101**(4), 156–162 (2007)
26. Dieudonné, Y., Petit, F.: Robots and demons (the code of the origins). In: Crescenzi, P., Prencipe, G., Pucci, G. (eds.) FUN 2007. LNCS, vol. 4475, pp. 108–119. Springer, Heidelberg (2007). https://doi.org/10.1007/978-3-540-72914-3_11
27. Dieudonné, Y., Petit, F.: Scatter of robots. Parallel Process. Lett. **19**(1), 175–184 (2009)
28. Dieudonné, Y., Petit, F.: Self-stabilizing gathering with strong multiplicity detection. Theoret. Comput. Sci. **428**, 47–57 (2012)
29. Dijkstra, E.W.: Self-stabilizing systems in spite of distributed control. Commun. ACM **17**(11), 643–644 (1974)
30. Dolev, D., Lynch, N.A., Pinter, S.S., Stark, E.W., Weihl, W.E.: Reaching approximate agreement in the presence of faults. J. ACM (JACM) **33**(3), 499–516 (1986)

31. Flocchini, P., Prencipe, G., Santoro, N., Widmayer, P.: Gathering of asynchronous robots with limited visibility. Theoret. Comput. Sci. **337**(1–3), 147–168 (2005)
32. Gervasi, V., Prencipe, G.: Flocking by a set of autonomous mobile robots. Technical report TR-01-24, Universitat di Pisa (2001)
33. Gervasi, V., Prencipe, G.: Coordination without communication: the case of the flocking problem. Discret. Appl. Math. (2003)
34. Heriban, A., Défago, X., Tixeuil, S.: Optimally gathering two robots. In: Bellavista, P., Garg, V.K. (eds.) Proceedings of the 19th International Conference on Distributed Computing and Networking. ICDCN 2018, 4–7 January 2018, Varanasi, India, pp. 3:1–3:10. ACM, New York (2018)
35. Izumi, T., Bouzid, Z., Tixeuil, S., Wada, K.: The BG-simulation for Byzantine mobile robots. CoRR abs/1106.0113 (2011)
36. Izumi, T., Bouzid, Z., Tixeuil, S., Wada, K.: Brief announcement: the BG-simulation for Byzantine mobile robots. In: Peleg, D. (ed.) DISC 2011. LNCS, vol. 6950, pp. 330–331. Springer, Heidelberg (2011). https://doi.org/10.1007/978-3-642-24100-0_32
37. Izumi, T., Kaino, D., Potop-Butucaru, M.G., Tixeuil, S.: On time complexity for connectivity-preserving scattering of mobile robots. Theoret. Comput. Sci. **738**, 42–52 (2018)
38. Izumi, T., et al.: The gathering problem for two oblivious robots with unreliable compasses. SIAM J. Comput. **41**(1), 26–46 (2012)
39. Qadi, A., Huang, J., Farritor, S., Goddard, S.: Localization and follow-the-leader control of a heterogeneous group of mobile robots. IEEE/ASME Trans. Mechatron. **11**, 205–215 (2006)
40. Lindhe, M.: A flocking and obstacle avoidance algorithm for mobile robots. Ph.D. thesis, KTH Stockholm (2004)
41. Ooshita, F., Tixeuil, S.: On the self-stabilization of mobile oblivious robots in uniform rings. Theoret. Comput. Sci. **568**, 84–96 (2015)
42. Pattanayak, D., Mondal, K., Ramesh, H., Mandal, P.S.: Fault-tolerant gathering of mobile robots with weak multiplicity detection. In: Proceedings of the 18th International Conference on Distributed Computing and Networking, 5–7 January 2017, Hyderabad, India, p. 7. ACM (2017)
43. Prencipe, G.: Corda: distributed coordination of a set of autonomous mobile robots. In: Proceedings of ERSADS, May 2001, pp. 185–190 (2001)
44. Renaud, P., Cervera, E., Martinet, P.: Towards a reliable vision-based mobile robot formation control. In: 2004 IEEE/RSJ International Conference on Intelligent Robots and Systems, Sendai, Japan, 28 September–2 October 2004, pp. 3176–3181. IEEE (2004)
45. Souissi, S.: Fault-resilient cooperation of autonomous mobile robots with unreliable compass sensors. Ph.D. thesis, Graduate School of Information Science, Japan Advanced Institute of Science and Technology (JAIST), September 2007
46. Souissi, S., Défago, X., Yamashita, M.: Using eventually consistent compasses to gather memory-less mobile robots with limited visibility. ACM Trans. Auton. Adapt. Syst. **4**(1), 9:1–9:27 (2009)
47. Souissi, S., Izumi, T., Wada, K.: Oracle-based flocking of mobile robots in crash-recovery model. Theoret. Comput. Sci. **412**(33), 4350–4360 (2011)
48. Suzuki, I., Yamashita, M.: A theory of distributed anonymous mobile robots formation and agreement problems. Technical report, Department of Electrical Engineering and Computer Science, Wisconsin University of Milwaukee, June 1994
49. Suzuki, I., Yamashita, M.: Distributed anonymous mobile robots: formation of geometric patterns. SIAM J. Comput. **28**(4), 1347–1363 (1999)

50. Viglietta, G.: Rendezvous of two robots with visible bits. In: Flocchini, P., Gao, J., Kranakis, E., Meyer auf der Heide, F. (eds.) ALGOSENSORS 2013. LNCS, vol. 8243, pp. 291–306. Springer, Heidelberg (2014). https://doi.org/10.1007/978-3-642-45346-5_21
51. Yang, Y., Souissi, S., Défago, X., Takizawa, M.: Fault-tolerant flocking for a group of autonomous mobile robots. J. Syst. Softw. **84**(1), 29–36 (2011)

# Robots with Lights

Giuseppe Antonio Di Luna[1(✉)] and Giovanni Viglietta[2]

[1] Aix-Marseille Université, LIS, CNRS, Université de Toulon, Toulon, France
g.a.diluna@gmail.com
[2] JAIST, Nomi, Japan
johnny@jaist.ac.jp

**Abstract.** The classic Look-Compute-Move model of oblivious robots has many strengths: algorithms designed for this model are inherently resistant to a large set of failures that can affect the memory of the robots and their communication capabilities.

However, modern technologies allow for cheap and reliable means of communication and memorization. This is especially true if relatively low performances are needed, such as very limited communication bandwidth or constant memory. A theoretical model that expands the classic Look-Compute-Move by adding a minimal ability to communicate and remember is the model of robots with lights. In this model each robot carries a luminous source that it can modify at every cycle. The robot decides the color of its light during its Compute phase, and the light assumes such a color at the beginning of the next Move phase. Other robots can see the color of this light during their Look phases. The light will remain unaltered until the robot that carries it decides to change its color.

Typically, the number of available colors is very limited, i.e., it is constant with respect to the number of robots in the system.

In this chapter we will discuss the hierarchy of $\mathcal{F}$SYNC, $\mathcal{S}$SYNC, and $\mathcal{A}$SYNC models when lights are present, we call this model $\mathcal{LUMINOUS}$. Moreover, we will see how lights are applied to solve classic problems such as rendezvous and forming a sequence of patterns. Finally, we will see how lights have been exploited in models where the visibility of robots is limited by the presence of obstructions.

## 1 Introduction

In the classic Look-Compute-Move model of oblivious robots, the absence of persistent memory and explicit communication ensures that any algorithm for such weak model can be implemented in a wide range of harsh scenarios where communicating is not a reliable option, e.g., a hostile environment where communication jamming is a possibility. Moreover, algorithms designed for this model are inherently resistant to a large set of failures that can affect the memory of the robots and their communication capabilities.

Fortunately, modern technologies allow for cheap and reliable means of communication and storage. This is especially true if relatively low performances are needed, such as very limited communication bandwidth or constant memory.


P. Flocchini et al. (Eds.): Distributed Computing by Mobile Entities, LNCS 11340, pp. 252–277, 2019.
https://doi.org/10.1007/978-3-030-11072-7_11

A theoretical model that expands the classic Look-Compute-Move model by adding a minimal ability to communicate and remember is the model of robots with lights [7,8]. In this model each robot carries a luminous source that it can modify at every cycle. The robot decides the color of its light during its Compute phase, and the light assumes such a color at the beginning of the next Move phase. Other robots can see the color of this light during their Look phases. The light will remain unaltered until the robot that carries it decides to change its color.

Typically, the number of available colors is very limited, i.e., it is constant with respect to the number of robots in the system. Interestingly, using light to communicate is something feasible in the real world, and it has been used to implement real communication channels [19].

Essentially, lights allow robots to perform communication. Moreover, in case an robot can see its own light, the light itself also serves as memory. The impact of using lights with a constant number of colors is drastic, and greatly increases the computational power of mobile robots.

In this chapter we will discuss the hierarchy of $\mathcal{F}$SYNC, $\mathcal{S}$SYNC, and $\mathcal{A}$SYNC models when lights are present, that is when robots are $\mathcal{LUMINOUS}$. Moreover, we will see how lights are applied to solve classic problems such as Rendezvous and forming a sequence of patterns. Finally, we will see how lights have been exploited in models where the visibility of robots is limited by the presence of obstructions.

*Chapter Outline.* The outline of the Chapter is the following. We start with Sect. 2; devoted to formalising the $\mathcal{LUMINOUS}$ model. Specifically, the section describes how the Look-Compute-Move model is modified to incorporate colored lights. This amounts to letting each robot decide the color of its own light at each cycle, and stipulating that the snapshots taken by robots also contain light information. After formalising the model in Sect. 3, we investigate the computational power of lights. We discuss the relationship between $\mathcal{F}$SYNC, $\mathcal{S}$SYNC, and $\mathcal{A}$SYNC when robots are endowed with lights, focusing on robots on plane and mentioning the difference for agents on graph. Section 4 studies the Rendezvous problem with lights. Rendezvous is unsolvable without lights, but it becomes solvable when lights are introduced. In Sect. 5 we discuss how $\mathcal{LUMINOUS}$ robots can form a sequence of patterns. In Sect. 6 we study the problem of making obstructive robots mutually visible, showing how lights can be used to solve problems when the visibility is not unlimited. The Chapter terminates with the conclusive Sect. 7.

## 2 Model

When lights are introduced, the usual model of oblivious robots of Chap. 1, is extended in a natural way. Each robot carries a visible light, which at any time has a color chosen from a palette set $\mathcal{B}$. (The size of $\mathcal{B}$ will often be informally referred to as the "number of lights".) In the basic model with lights, the palette set is the same for all robots, and it contains a special color *Off*, which is the color of every robot's light when the execution starts. Each robot can then freely alter the color of its own light, choosing it from $\mathcal{B}$, at each `Compute` phase.

More specifically, the usual `Look-Compute-Move` phases are modified as follows.

- `Look`: In the `Look` phase, a robot $r$ receives an instant snapshot, which is a multiset of pairs of the form $(p_i, c_i)$, where $p_i$ is a point in the plane and $c_i \in \mathcal{B}$ is a color. The meaning of such a pair is that, at the time the snapshot was taken, the robot $r$ could see a robot $s$ located in $p_i$ carrying a light with color $c_i$. As in the usual model, $p_i$ is the position of $s$ as expressed in the local coordinate system of $r$.
  In the basic model with lights, robots have full visibility, and therefore snapshots always contain information about every robot in the system. When more restrictive visibility models are considered, such as the obstructed visibility model of Sect. 6 of this chapter, a snapshot may contain less information.
- `Compute`: In the `Compute` phase, a robot, using the snapshot obtained in the most recent `Look` phase, chooses a destination point and a new color in $\mathcal{B}$ for its own light.[1]
- `Move`: During the `Move` phase, a robot first changes the color of its light to the color chosen in the most recent `Compute` phase (this action is atomic, i.e., it is instantly executed), and then it proceeds to move to its destination according to the classic `Look-Compute-Move` model.

With $\mathcal{M}\text{ODEL}^m$, we indicate the model $\mathcal{M}\text{ODEL}$ where robots carry lights whose colors are chosen from a palette set of size $|\mathcal{B}| = m$. For example, with $\mathcal{F}\text{SYNC}^{O(1)}$ we indicate the $\mathcal{F}\text{SYNC}$ model with lights of a constant number of colors (constant with respect to the number of robots in the system). When the set of problems solvable in model $\mathcal{M}\text{ODEL}$ is included in the set of problems solvable in $\mathcal{M}\text{ODEL}'$ we write $\mathcal{M}\text{ODEL} \subseteq \mathcal{M}\text{ODEL}'$ (we use $\subset$ to indicate the strict inclusion).

## 3 Computational Power of Lights

The presence of lights have an impact on the relative computational power of the $\mathcal{F}\text{SYNC}$-$\mathcal{A}\text{SYNC}$-$\mathcal{S}\text{SYNC}$ models[2]. When we refer to robots without lights we have that, for $\mathcal{OBLOT}$ there are problems that are solvable in $\mathcal{F}\text{SYNC}$ but not in $\mathcal{S}\text{SYNC}$ (e.g., the *Gathering* problem). However, between $\mathcal{S}\text{SYNC}$ and $\mathcal{A}\text{SYNC}$ the only known result is the trivial $\mathcal{A}sync \subseteq \mathcal{S}sync$.

In this section we investigate what happens when a constant number of lights is available.

[1] Note that setting the light at the end of the compute phase is equivalent to set the light at the beginning of the move phase.

[2] We say $\mathcal{F}\text{SYNC}$ model as shorthand for the model of $\mathcal{OBLOT}$ (or $\mathcal{LUMINOUS}$) robots with $\mathcal{F}\text{SYNC}$ scheduler.

### 3.1 The Relationship Between $\mathcal{A}$SYNC$^{O(1)}$ and $\mathcal{S}$SYNC

[8] shows that $\mathcal{A}$SYNC$^{O(1)}$ is more powerful than $\mathcal{S}$SYNC, that is $\mathcal{S}$SYNC $\subset$ $\mathcal{A}$SYNC$^{O(1)}$. This is done in two steps, first it is shown how to simulate any algorithm for $\mathcal{S}$SYNC in $\mathcal{A}$SYNC$^{O(1)}$ using a constant number of lights. Then it is shown that there exists a problem solvable in $\mathcal{A}$SYNC$^{O(1)}$ but not in $\mathcal{S}$SYNC.

**Simulating $\mathcal{S}$SYNC in $\mathcal{A}$SYNC$^{O(1)}$.** The simulator of [8] uses five colors: Trying ($T$), Waiting ($W$), Moving ($M$), Stopping ($S$), Finished ($F$), and five states, one for each different color.

The idea is to synchronize robots in a simulated cycle, namely the Mega-Cycle. During a Mega-Cycle, each robot executes one step of the simulated algorithm $\mathcal{A}$. A new Mega-Cycle starts only when all robots terminated the execution of the previous Mega-Cycle. Each Mega-Cycle is structured as follows. Initially, all robots have light $T$. A robot with light $T$ once activated tries to simulate one activation of $\mathcal{A}$, in doing so it first checks if all other robots have color in $T$ or $S$. In such case, the robot sets its color to $M$ and it executes $\mathcal{A}$. Otherwise, if there is an robot with color $M$, the robot from color $T$ switches to color $W$ and it does nothing. Intuitively, this check avoid that a robot provides to $\mathcal{A}$ a snapshot containing robots that are moving, this is consistent with $\mathcal{S}$SYNC model.

If a robot has color $M$ and no robot in its snapshot has color $T$, then it goes to color $S$ and it does nothing. An robot with color $W$ goes back to $T$ only if there is no robot with color $M$. A robot with color $S$ goes to color $F$ if all robots have color in $\{S, F\}$. A robot with color $F$ goes to color $T$ if all robots have color in $\{T, F\}$. The previous rules ensure that, in a Mega-Cycle, each robot executes exactly one activation of $\mathcal{A}$. It is also easy to see that all robots execute the Mega-Cycle: until there is some robot in $W$ or $T$, there will be at least one robots that enters in $M$. A new Mega-Cycle starts when some robot goes from $F$ to $T$. The state diagram of the simulator is shown in Fig. 1.

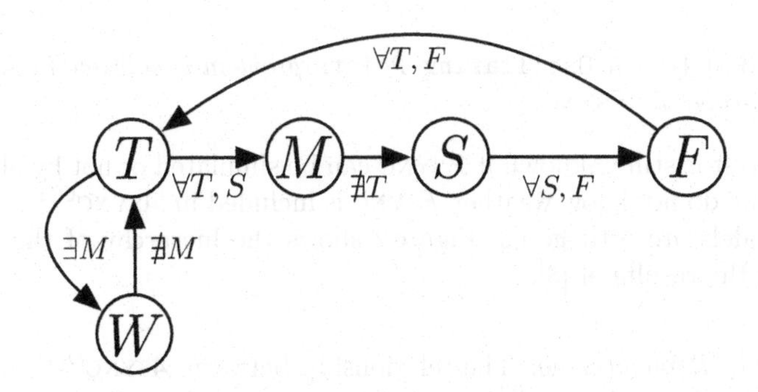


**Fig. 1.** State diagram of the simulator for $\mathcal{S}$SYNC in $\mathcal{A}$SYNC$^{O(1)}$ of [8]. On each arrow there is label specifying a property of the snapshot that has to be verified to do the corresponding state transition. As example, the label between node $T$ and node $S$ is $\forall T, S$, that means: each robot in the snapshot is either in state $T$ or $S$.

This simulator gives the following theorem:

**Theorem 1** ([8]). $\mathcal{S}$SYNC $\subseteq$ $\mathcal{A}$SYNC$^{O(1)}$.

**The Additional Power of $\mathcal{A}$SYNC$^{O(1)}$.** It is well known that Rendezvous of two oblivious anonymous robots cannot be solved in $\mathcal{S}$SYNC, see [26]. However, it is possible to create an algorithm that solves Rendezvous in $\mathcal{A}$SYNC$^{O(1)}$. The first paper showing this has been [8], where Rendezvous is solved using 4 colors. Other papers investigated the Rendezvous improving this first result in many directions. The Rendezvous will be the core of Sect. 4, the interested reader can refer to that Section.

The existence of several Rendezvous algorithms in $\mathcal{A}$SYNC$^{O(1)}$ and the Theorem 1, leads to:

**Theorem 2** ([8]). $\mathcal{S}$SYNC $\subset$ $\mathcal{A}$SYNC$^{O(1)}$.

### 3.2 The Relationship Between $\mathcal{A}$SYNC$^{O(1)}$ and $\mathcal{F}$SYNC

The relationship between $\mathcal{A}$SYNC$^{O(1)}$ and $\mathcal{F}$SYNC is still not clear. It has been shown in [8] that there exist a problem solvable in $\mathcal{A}$SYNC$^{O(1)}$ and not in $\mathcal{F}$SYNC. The problem is the `Oscillating Points`.

In the `Oscillating Points` problem two robots, initially starting in distinct positions, have to move in such a way to alternate configurations in which their relative distance decreases (*near* configuration), and configurations in which their distance increases (*far* configuration). The intuitive reason why such problem is unsolvable in $\mathcal{F}$SYNC is that the problem specification implicitly needs robots to remember weather they are in *near* or *far* configuration. This is not a problem when lights are present, being lights persistent a certain color can be associated with a specific configuration, in [8] 4 colors are used to solve `Oscillating Points`.

**Theorem 3** ([8]). *The* `Oscillating Points` *problem is solvable in* $\mathcal{A}$SYNC$^{O(1)}$, *and is unsolvable in* $\mathcal{F}$SYNC.

However, it is still unknown if $\mathcal{F}$SYNC can be simulated or not by $\mathcal{A}$SYNC$^{O(1)}$. Therefore, we do not know weather $\mathcal{F}$SYNC is included in $\mathcal{A}$SYNC$^{O(1)}$, or weather the two models are orthogonal. Figure 2 shows the hierarchy of the models in the light of the results of [8].

*The Power of Remembering.* The relationship between $\mathcal{A}$SYNC$^{O(1)}$ and $\mathcal{F}$SYNC is completely clear when we allows robots to remember the snapshot of the previous round (model $\mathcal{A}$SYNC$^{O(1)}$ + Snapshot). In this case, it is possible to simulate any $\mathcal{F}$SYNC algorithm, see [8]. The interesting fact is to contrast this with what happens when lights are not available. In [26] it has been shown that $\mathcal{A}$SYNC is weaker than $\mathcal{F}$SYNC even when robots remember an unlimited amount of snapshots.

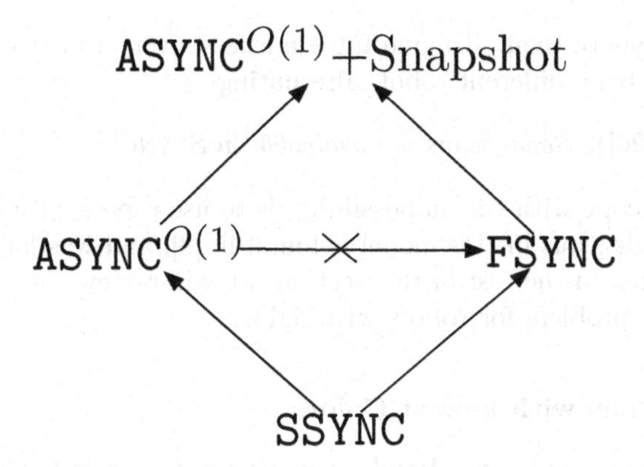


**Fig. 2.** Hiearchy of $\mathcal{A}$SYNC$^{O(1)}$, $\mathcal{F}$SYNC and $\mathcal{S}$SYNC. An arrow indicates that the source model is included in the destination. A strikethrough arrow indicates that the source model is not included in the destination.

*Robots on Graph with Lights.* In [10] is investigated the relationship between $\mathcal{A}$SYNC$^{O(1)}$ and $\mathcal{F}$SYNC when robots move in a discrete environment, that is modelled as a graph. Robots operate following the `Look-Compute-Move` cycle, and the snapshot is the entire graph. Two new problems are introduced to show the separation of $\mathcal{F}$SYNC and $\mathcal{A}$SYNC$^{O(1)}$: the `Pattern Series Chasing` (solvable in $\mathcal{F}$SYNC and not in $\mathcal{A}$SYNC$^{O(1)}$) and the `Forth and Back` (solvable in $\mathcal{A}$SYNC$^{O(1)}$ but not in $\mathcal{F}$SYNC). This implies that, when robots on graph are considered, $\mathcal{A}$SYNC$^{O(1)}$ and $\mathcal{F}$SYNC are orthogonal models.

## 4 Rendezvous

Rendezvous is the special case of Gathering (see Chap. 4, Sect. 2) where the system consists of exactly two robots whose task is to move to the same point, no matter where and when, and then stop forever. This special case is surprisingly hard, due to the lack of "environmental landmarks" that may help the two robots agree on a common rendezvous point. In contrast, when more than two robots are present, it is relatively easy in most cases to implicitly agree on a small subset of robots that should gather first, while the others provide a visible static reference frame that helps circumvent limitations such as asynchrony and non-rigidity[3].

As shown in [26], the Rendezvous problem is unsolvable in $\mathcal{S}$SYNC. Indeed, suppose the local reference frames of the two robots are oriented symmetrically: since they have symmetric views, they always compute symmetric destination points. As long as the destination points they compute are different, the scheduler activates them both and lets them move. Whenever they compute the same destination point (which must be their midpoint, by symmetry), the scheduler

[3] In a rigid model robots always reach the destination when performing the move. In a non-rigid model robots may be stopped before reaching the destination, however they travel of at least a fixed unknown $\delta > 0$.

activates only one of them. To guarantee fairness, every time this happens, the scheduler activates a different robot, alternating.

**Theorem 4** ([26]). *Rendezvous is unsolvable in* $\mathcal{S}$SYNC.

One way to cope with this impossibility is to use robots with lights: the first solution to Rendezvous in this model is found in [8], and was later improved in several directions. In the rest of this section, we will review some approaches to the Rendezvous problem for robots with lights.

### 4.1 Rendezvous with Fewest Colors

The focus of [28] is solving the Rendezvous problem for robots with lights using the minimum number of colors. The problem is solved in a variety of models, which combine different schedulers ($\mathcal{F}$SYNC, $\mathcal{S}$SYNC, $\mathcal{A}$SYNC), rigidity, and self-stabilization. The concept of rigidity is defined in Chap. 1, Sect. 2.4, while self-stabilization (see [14]) is the additional requirement that the two robots solve Rendezvous regardless of their initial colors (as opposed to stipulating that their lights have a specific predefined color when the execution starts).

Figure 3 summarizes the results of [28]: there is a hierarchy of 12 models obtained by combining the aforementioned parameters in all possible ways (an asterisk indicates that the solution has to be self-stabilizing). The number in parentheses after each model indicates that there is an algorithm that solves Rendezvous under that model using that many colors.

Under a certain assumption, all the numbers in Fig. 3 are minimal. The assumption is that the robots cannot use information about their distance to compute their destination points, but can only compute it as a function of their respective colors. More precisely, if a robot is located in $p$ and performs a `Look` when the other robot is in $q$, then the destination point it computes must be of the form $(1-\lambda)p+\lambda q$, where $\lambda \in \mathbb{R}$ is computed as a function of the two robots' lights at the moment the `Look` was performed. All the algorithms presented in [28] are of this kind, as well.

The left side of Fig. 3 is easy to obtain, because in non-rigid $\mathcal{F}$SYNC there is a trivial algorithm that solves Rendezvous even in the basic model (i.e., with only one color): the algorithm makes each robot move to the midpoint of their current locations. Even if movements are non-rigid, the robots either meet or approach each other by at least $2\delta$ at every turn, hence meeting in a finite number of turns.

For non-rigid $\mathcal{S}$SYNC and rigid $\mathcal{A}$SYNC there is an algorithm that uses only two colors, namely $A$ and $B$, shown in Fig. 4. In the case of non-rigid $\mathcal{S}$SYNC, the algorithm is also self-stabilizing.

Labels on arrows indicate the color that is seen on the other robot and the $\lambda$ parameter of the resulting move, i.e., the destination of the next `Move` with respect to the position of the other robot. So, "0" stands for "do not move", "1/2" means "move to the midpoint", and "1" means "move to the other robot". Roughly speaking, the idea of this algorithm is to make the robots approach each

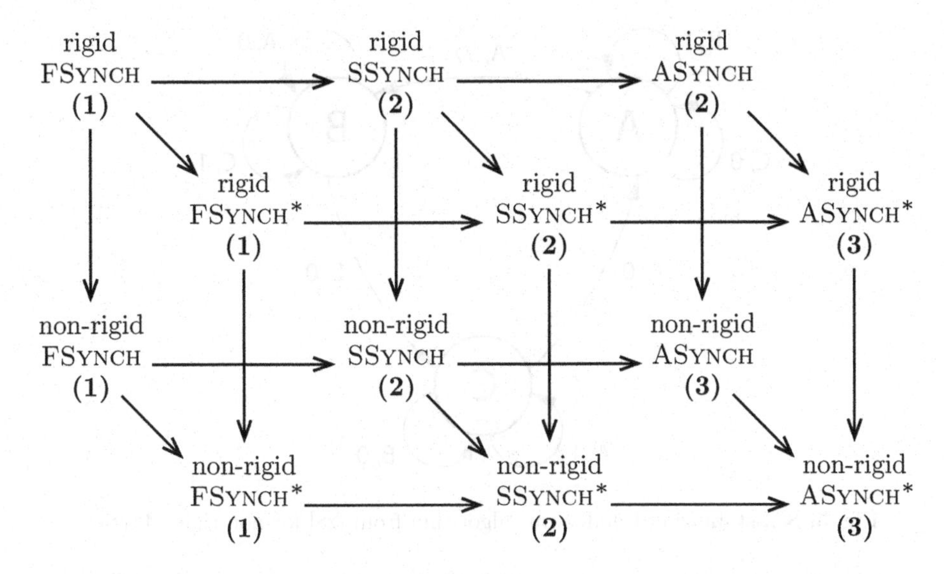


**Fig. 3.** Summary of results of [28]. For each model in the hierarchy, there exists a Rendezvous algorithm using the number of colors in parentheses. An asterisk indicates that the algorithm is self-stabilizing. If robots cannot use distance information in their computations, all these numbers are optimal.

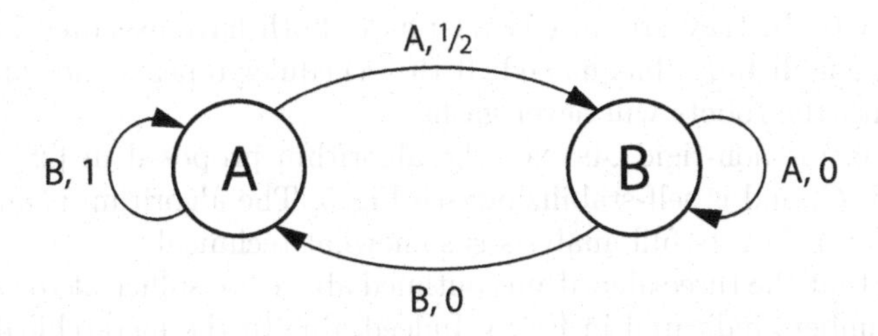


**Fig. 4.** Rendezvous algorithm from [28] for non-rigid $\mathcal{S}$SYNC (self-stabilizing) and rigid $\mathcal{A}$SYNC

other by moving toward the midpoint as long as their lights have the same color: if the scheduler keeps them synchronous, they eventually meet. If, on the other hand, the scheduler does not keep them synchronous, the two robots eventually see each other in different colors. Therefore the $A$-colored robot moves to the other robot's location, while the $B$-colored robot waits.

For non-rigid $\mathcal{A}$SYNC, the above algorithm fails. To see why, let $r$ and $s$ be the two robots, and let them both start with color $A$ at distance greater than $2\delta$. If the scheduler lets them both perform an entire cycle but stops them as soon as they have moved by $\delta$, they end up in color $B$ a positive distance apart. Now, let both robots perform a Look phase, implying that both of them will eventually turn $A$. We let robot $r$ finish the current cycle and perform a new Look, while the other robot $s$ waits, still in color $B$. Hence, $r$ will stay $A$ and move to $s$'s

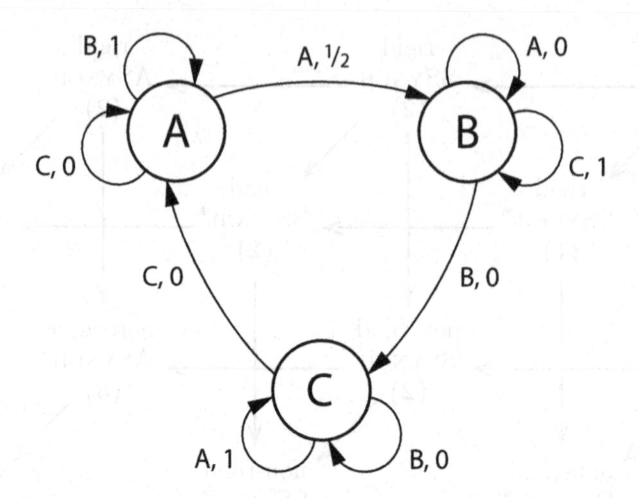


**Fig. 5.** Self-stabilizing Rendezvous algorithm from [28] for non-rigid $\mathcal{A}$SYNC

position. Now we let $s$ finish the current cycle and perform a new Look. So $s$ will turn $B$ and move to the midpoint $m$. We let $r$ finish the current cycle, thus reaching $s$, and perform a whole new cycle, turning $B$. Finally, we let $s$ finish the current cycle, thus turning $B$ and moving to $m$. As a result, both robots are again set to $B$, they are in a WAIT phase, both have executed at least one cycle, and their distance has halved. If the scheduler repeats the same pattern of activations, the robots will never gather.

Therefore, for non-rigid $\mathcal{A}$SYNC, the algorithm proposed in [28] uses three colors, $A$, $B$, $C$, and is self-stabilizing: see Fig. 5. The algorithm is an extension of that of Fig. 4, but its full analysis is somewhat technical.

Observe that the three algorithms outlined above are sufficient to establish all the color numbers indicated in Fig. 3. Indeed, due to the hierarchical structure of the models, if there is an arrow from model $X$ to model $Y$ in Fig. 3, then any algorithm for model $Y$ also works for model $X$. The matching lower-bound proofs can be found in [28]. Summarizing, we have the following theorem.

**Theorem 5** ([28]). *The Rendezvous problem is solvable in each of the 12 models of Fig. 3 using the number of lights indicated in parentheses under each model. If robots are not allowed to use distance information in their computations, these numbers are optimal.*

Finally, [28] shows how to refine the algorithm of Fig. 5 to detect termination: that is, to let the robots acknowledge that they have gathered, in order to turn off or "switch gears" and start performing a new task. Although this is not a requirement of the Rendezvous problem, it is a useful feature to add. The new algorithm still uses only three colors, but it also uses distance information, although robots only need distinguish between zero and non-zero distances.

Later improvements to [28] indicate that, if the robots are allowed to use distance information in their computations, they only require 2 colors to solve

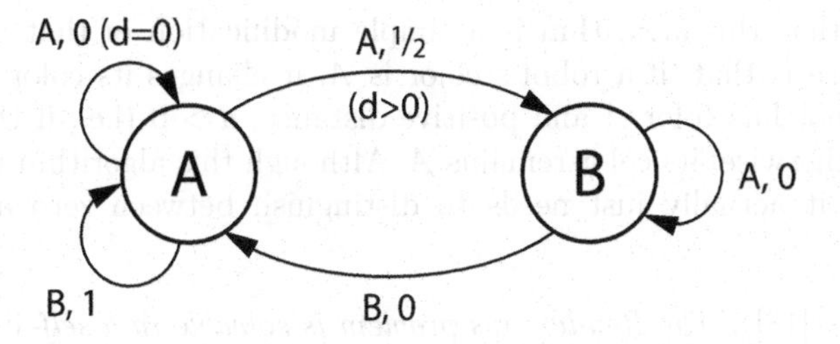


**Fig. 6.** Self-stabilizing Rendezvous algorithm from [18] for non-rigid $\mathcal{A}$SYNC using distance information

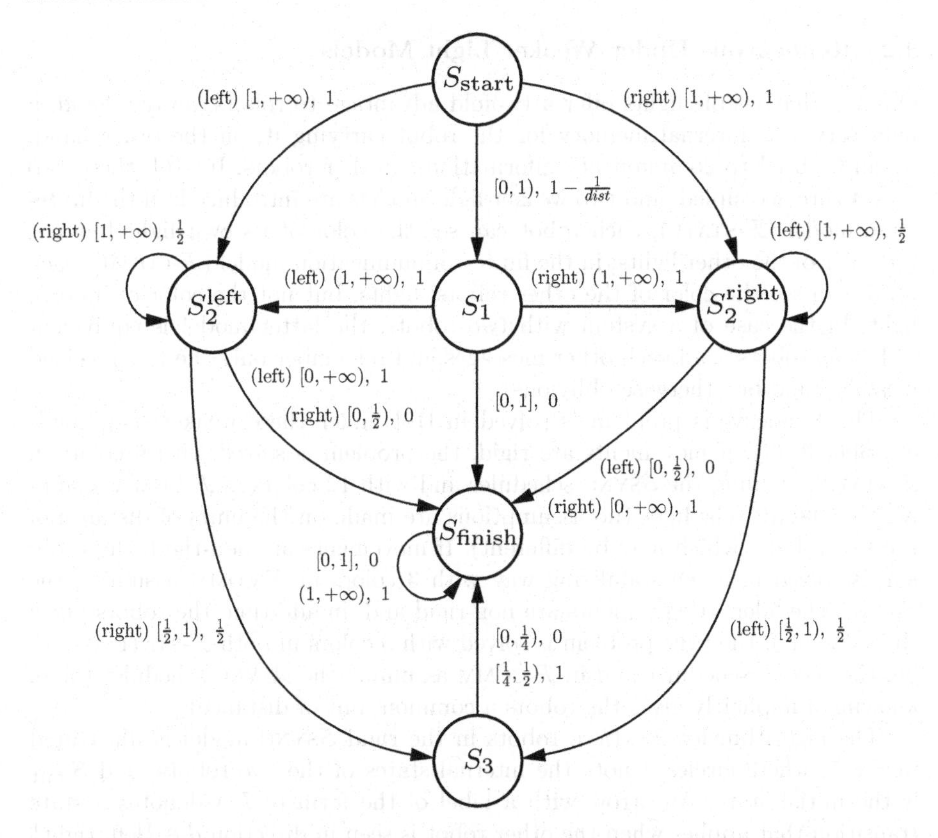


**Fig. 7.** Rendezvous algorithm from [16] for rigid $\mathcal{S}$SYNC robots in $\mathcal{F}$-STATE

Rendezvous, even under the non-rigid $\mathcal{A}$SYNC scheduler, and even in a self-stabilizing way. First, [20] proposed an algorithm that assumes the robots to know the value of the parameter $\delta$ related to non-rigid movements (see Chap. 1, Sect. 2.4). Then, [18] managed to drop even this assumption. The Rendezvous algorithm of [18] is shown in Fig. 6.

Observe that this algorithm is a simple modification of that of Fig. 4: the only difference is that, if a robot's color is $A$, it changes its color to $B$ only if the other robot has color $A$ and positive distance, $d > 0$ (i.e., if they have not gathered); otherwise, its color remains $A$. Although this algorithm uses distance information, it actually just needs to distinguish between zero and non-zero distances.

**Theorem 6** ([18]). *The Rendezvous problem is solvable in a self-stabilizing way with 2 colors under the non-rigid* $\mathcal{A}$SYNC *scheduler, provided that distance information can be used in the computations.*

### 4.2 Rendezvous Under Weaker Light Models

Observe that visible lights offer a twofold advantage to robots: on one hand, a light serves as internal memory for the robot carrying it; on the other hand, it can be used to communicate information to other robots. In [16], these two aspects are decoupled, and two weaker light models are introduced: in the finite-state model ($\mathcal{F}$-STATE), each robot can see the color of its own light but not the color of the other lights; in the finite-communication model ($\mathcal{F}$-COMM), each robot can see the color of the other robots' lights, but not the color of its own light. In the case of a system with two robots, the latter model is equivalent to letting robots send each other messages and remember only the last received message (and be otherwise oblivious).

The Rendezvous problem is solved in [16] under these weaker light models. Specifically, if movements are rigid, the problem is solved with 6 colors in $\mathcal{F}$-STATE assuming the $\mathcal{S}$SYNC scheduler and with 12 colors in $\mathcal{F}$-COMM assuming the $\mathcal{A}$SYNC scheduler (no assumptions are made on the units of distance of the two robots, which may be different). If movements are non-rigid, the problem is solved in a self-stabilizing way with 3 colors in $\mathcal{F}$-COMM assuming the $\mathcal{S}$SYNC scheduler. If movements are non-rigid and, in addition, the robots know the value of $\delta$, then the problem is solved with 3 colors in both $\mathcal{F}$-STATE assuming the $\mathcal{S}$SYNC scheduler and in $\mathcal{F}$-COMM assuming the $\mathcal{A}$SYNC scheduler (here, knowing $\delta$ implicitly gives the robots a common unit of distance).

The algorithm for $\mathcal{F}$-STATE robots in the rigid $\mathcal{S}$SYNC model is illustrated in Fig. 7, where circles denote the internal states of the two robots, and $S_{\text{start}}$ is the initial state. An arrow with a label of the form $(d)I, \lambda$ denotes a state transition that applies when the other robot is seen in direction $d \in \{\text{left}, \text{right}\}$ and its observed distance lies in the interval $I \subset \mathbb{R}$. The parameter $\lambda$ is as in Sect. 4.1, and defines the destination point with respect to the other robot's position. For example, a robot in state $S_{\text{start}}$ perceiving the other at distance $\geqslant 1$ on the right will move to the position of the other robot and will change state to $S_2^{\text{right}}$. Note that a robot can arbitrarily assign a left and a right side to the line that connects it to the other robot, and this assignment does not change as the execution progresses.

According to the algorithm, the robots try to reach a configuration where they both observe each other at distance not smaller than 1 (i.e., their own unit of distance). From this configuration, they attempt to meet in the midpoint. If they never meet because they are never activated in the same turn, eventually one of them notices that its observed distance is lower than 1. This implies a breakdown of symmetry that allows the robots to finally gather.

In order to reach the aforementioned desired configuration where they both observe a distance not smaller than 1, the two robots first move away from each other if they are too close. When they are far enough, they memorize the side on which they see each other (left or right), and try to switch positions. If only one of them is activated, they gather; otherwise they detect a side switch, and they can finally apply the above protocol, which leads to gathering.

This is complicated by the fact that the robots may disagree on the distances they observe, because they have different units of distance. To overcome this disagreement, they use their ability to detect a side switch to understand which distance their partner observed. If the desired configuration is not reached because of a disagreement, a breakdown of symmetry occurs, which is immediately exploited to gather anyway. As soon as the two robots are in the same location at the end of a cycle, they never move again, and Rendezvous is solved.

**Theorem 7** ([16]). *The Rendezvous problem is solvable with 6 colors in $\mathcal{F}$-STATE under the rigid $\mathcal{S}$SYNC scheduler.*

Observe that the above algorithm makes a fundamental use of the fact that the scheduler is rigid and $\mathcal{S}$SYNC. For instance, the correct detection of a side switch by a robot relies on the fact that the other robot is not currently in the middle of a movement while it is observed (hence the scheduler is not $\mathcal{A}$SYNC), or it could be seen on a side and then switch side by the end of the current move. Similarly, the algorithm relies on the fact that the robots can reliably move away from each other and reach a distance not smaller than 1. In a non-rigid setting, they may be stopped too soon, in such a way that both end up in state $S_1$ but still detect a distance smaller than 1. From that point on, they will never move again, because each of them will incorrectly assume that the other robot will measure a distance greater than 1.

Now let us consider the $\mathcal{F}$-COMM model. The Rendezvous algorithm for $\mathcal{F}$-COMM robots in the rigid $\mathcal{A}$SYNC model is shown in Fig. 8, where the initial state of both robots is called "Test". The meaning of an arrow from state $X$ to state $Y$ is that if a robot observes that the light of the other robot has color $X$, then the first robot sets its own light to color $Y$. If an arrow has a label in the form of a predicate on $d$, it means that the transition only happens if the observed distance $d$ between the two robots satisfies the predicate. Moreover, a boldface label $\lambda$ on an arrow has the same meaning as in Sect. 4.1. If such a $\lambda$ is followed by a predicate on the distance $d$ in parentheses, the robot moves only if the predicate is satisfied, and stays still otherwise. For example, if a robot located in $p$ sees the other robot located in $q$ and in state "Both <1", and their distance $d$ is positive, it assumes state "Moving Away". If, in addition, the distance $d$ is less than 1, it also moves to the point $(1/2 + 1/d) \cdot p + (1/2 - 1/d) \cdot q$.

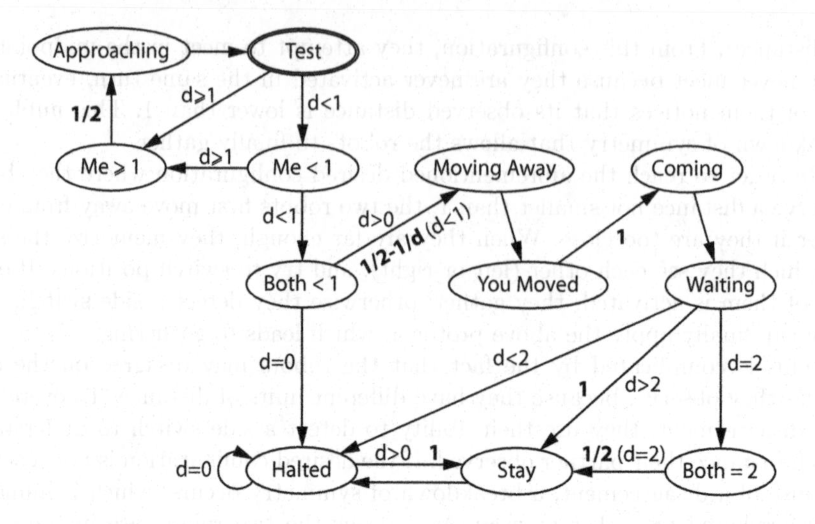


**Fig. 8.** Rendezvous algorithm from [16] for rigid $\mathcal{A}$SYNC robots in $\mathcal{F}$-COMM

According to the algorithm, the two robots try to reach a configuration where they both see each other at distance smaller than 1. To do so, they first communicate to each other whether or not the distance they observe is smaller than 1 (recall that they may disagree, because their units of distance may differ). If one robot acknowledges that its partner has observed a distance not smaller than 1, it reduces the distance by moving to the midpoint.

The process repeats until both robots observe a distance smaller than 1. At this point, if they have not gathered yet, they try to compare their units of distance in order to break symmetry. They move away from each other in such a way that their final distance is the sum of their respective units of distance. Before proceeding, they attempt to switch positions. If, due to asynchrony, they failed to be in the same state at any time before this step, they end up gathering. Instead, if their execution has been synchronous up to this point, they finally switch positions. Now, if the robots have not gathered yet, they know that their distance is actually the sum of their units. Because each robot knows its own unit, they can tell if one of them is larger. If a robot has a smaller unit, it moves toward its partner, which waits.

Otherwise, if their units are equal, they apply a straightforward protocol: as soon as a robot wakes up, it moves toward the midpoint and tells its partner to stay still. If both robots do so, they gather in the midpoint. If one robot is delayed due to asynchrony, it acknowledges the order to stay still and tells the other robot to come.

**Theorem 8** ([16]). *The Rendezvous problem is solvable with 12 colors in* $\mathcal{F}$-COMM *under the rigid* $\mathcal{A}$SYNC *scheduler.*

Once again, the above algorithm crucially uses rigidity, for instance when the robots switch positions and assume that their current distance must be the sum

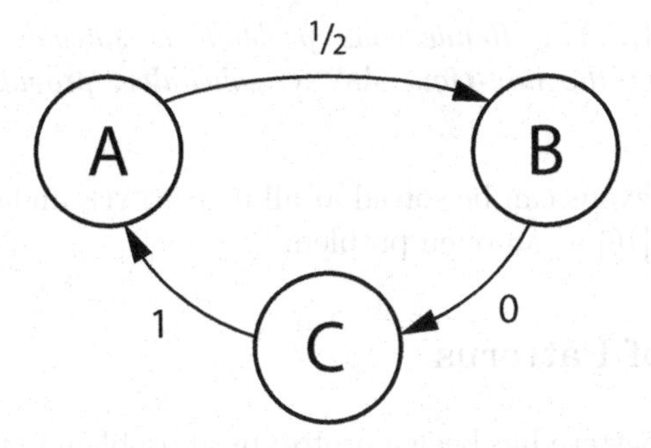


**Fig. 9.** Rendezvous algorithm from [16] for rigid $\mathcal{S}$SYNC robots in $\mathcal{F}$-COMM

of their units. In a non-rigid setting, they could be stopped too soon, and both detect a distance smaller than 2. From that point onward, if they are activated synchronously and rigidly, they keep switching positions without ever gathering.

For $\mathcal{F}$-COMM robots under the non-rigid $\mathcal{S}$SYNC scheduler, there is a simple self-stabilizing algorithm, shown in Fig. 9. The meaning of an arrow from state $X$ to state $Y$ is that if a robot observes that the light of the other robot has color $X$, then the first robot sets its own light to color $Y$. The label $\lambda$ on each arrow has the same meaning as in Sect. 4.1.

Let us analyze this algorithm. Assume first that both robots start in the same state and both are activated at each turn. Then they always have equal states, and they cycle through states $A$, $B$, and $C$ forever. Every time they are both in state $A$, they move toward the midpoint, and their distance reduces by at least $2\delta$. Eventually, it becomes so small that they actually gather.

Otherwise, if at any point the two robots are in different states, they will remain in different states forever. In this case their distance will never increase, and they will periodically be found in states $B$ and $C$, respectively. Whenever this happens, the robot in state $C$ retains its state and waits until the other robot is activated and moves toward it by at least $\delta$. As soon as their distance becomes not greater than $\delta$ and they turn again $B$ and $C$, they finally gather.

**Theorem 9** ([16]). *The Rendezvous problem is solvable in a self-stabilizing way with 3 colors in $\mathcal{F}$-COMM under the non-rigid $\mathcal{S}$SYNC scheduler.*

Finally, if the robots are non-rigid but know the value of $\delta$, they can solve Rendezvous with 3 colors both in $\mathcal{F}$-STATE under the $\mathcal{S}$SYNC scheduler and in $\mathcal{F}$-COMM under the $\mathcal{A}$SYNC scheduler. The two algorithms are relatively simple, because not only do the robots know at what point they can assume that all movements will be rigid (i.e., when their distance is at most $\delta$), but knowing $\delta$ in their respective reference frames also implicitly gives them a common unit of distance. The details of the two algorithms are found in [16].

**Theorem 10** ([16]). *The Rendezvous problem is solvable with 3 colors in $\mathcal{F}$-STATE and under the non-rigid $\mathcal{S}$SYNC scheduler, provided that the robots know the value of $\delta$.*

**Theorem 11** ([16]). *The Rendezvous problem is solvable with 3 colors in $\mathcal{F}$-COMM and under the non-rigid $\mathcal{A}$SYNC scheduler, provided that the robots know the value of $\delta$.*

Whether Rendezvous can be solved at all in $\mathcal{F}$-STATE under the rigid $\mathcal{A}$SYNC scheduler is left in [16] as an open problem.

# 5 Sequence of Patterns

Forming a specific pattern has been a prototypical problem in the oblivious robot model, see [15,17,29] and Chap. 3. The general version of this problem specifies that a set of robot has to form a specific pattern (up to rotation or scaling). In this section we use the concepts of symmetricity and of equivalence class of robots, shortened in class, defined in Chap. 3.

## 5.1 Sequence of Patterns Without Light

A natural extension of the pattern formation is the one in which robots have to form a sequence of patterns, see [9]. Let $\mathcal{S} = \langle S_0, \ldots, S_{m-1} \rangle$ be a sequence of distinct patterns. A set of robots forms $\mathcal{S}$, starting from a configuration $\Gamma$, if it forms the infinite periodic sequence $\mathcal{S}^\infty = \langle S_0, S_2, \ldots, S_{m-1} \rangle^\infty$, obtained by repeating forever the sequence $\mathcal{S}$. It is not hard to see that in the oblivious model there are several restrictions on $\mathcal{S}$, depending on the configuration $\Gamma$. Clearly, the relationship between the symmetricity of $\Gamma$ and the one of any pattern in $\mathcal{S}$ has to be the same of the classic pattern formation (see Chap. 3).

Moreover, the following conditions are necessary: the symmetricity of each pattern in $\mathcal{S}$ is the same; the number of points in each pattern has to be the same. It is not hard to see why the previous conditions are necessary: once two robots share the same position, they could be bonded forever by always activating them at the same time, thus it is not possible for the number of points in the patterns to change; the condition on the symmetricity is also obvious and it comes from similar considerations.

Interestingly, in $\mathcal{S}$SYNC with **rigid** robots the above conditions are also sufficient, see [9].

## 5.2 Sequence of Patterns with Light

Forming a sequence of patterns in the $\mathcal{A}$SYNC model with **non-rigid** luminous robots has been studied in [6]. Note that, in [6] a pattern in $\mathcal{S}$ is only a set of robots positions, without any restriction on lights color, i.e. two patterns are the same even if the colors of the robots in each pattern are completely different.

In the following we will refer to positional class to indicate a set of points that share the same view when colors are not considered, and we will refer to chromatic class to indicate a set of points that share the same view when colors are considered. When lights are available, it is possible to form a sequence $\mathcal{S}$ even if: there are repeating pattern in $\mathcal{S}$; the number of points in each pattern is not the same; and, patterns have different symmetricity. The only constraint on $\mathcal{S}$ is that the symmetricity of any pattern in $\mathcal{S}$ has to be divided by the symmetricity of the initial configuration $\Gamma$. This in contrast with what happens when lights are not available.

In the algorithm of [6] colors are used to synchronize the various phase of the algorithm, and to keep the symmetry broken when robots from distinct positional classes move in such a way to end up in the same positional class, e.g. two robots go to the same point but have two different colors. Moreover, colors are also used to encode information about the sequence. Specifically, if the same pattern $S$ appears in two different positions in $\mathcal{S}$ it will have a different coloring allowing robots to distinguish which instance of $S$ is.

More in details, the algorithm is divided in five phases:

- Leader Identification: In this phase the robots elect one class as leader class. A light with color *Gold* is used to uniquely mark this class in the next phases of the algorithm.
- Pattern Identification: During the pattern identification the leader class moves in such a way to uniquely identify the specific pattern $S_j$ of $\mathcal{S}$ that has to be formed. In this phase colors are used for synchronization.
- Separation: In the separation step robots that are in the same positional class, but in different chromatic classes, separate to form different positional class. Colors are used to symmetry breaking purpose. At the of this phase each chromatic class is on a different circle. All circles are concentric.
- Rotation and composition: Finally, in this two phases the robots dispose themselves to build pattern $S_j$.

**Theorem 12** ([6]). *A set of* **non-rigid** *luminous robots in* $\mathcal{A}$SYNC *starting from an initial configuration* $\Gamma$ *can form a sequence of pattern* $\mathcal{S}$ *if for any* $P \in \mathcal{S}$, *the symmetricity of* $P$ *is divided by the symmetricity of* $\Gamma$.

## 6 Mutual Visibility

In this Section we consider the setting where robots obstruct the visibility of each other, and we focus on the Mutual Visibility problem. Such problem has been the object of several recent papers [11–13,22–25,27], and a variety of solutions have been proposed investigating trade-offs between time complexity and number of lights. For this reason, it is interesting to study the Mutual Visibility problem to understand how lights can be used in the design of sophisticated algorithms.

*Problem Statement.* In the Mutual Visibility problem a set of robots initially positioned in an arbitrary configuration have a to reach a final configuration $\mathcal{C}_f$, where for each pair of distinct robot positions $p_j, p_i$ in $\mathcal{C}_f$ it does not exist any robot $r_s$ on the segment connecting $p_i$ and $p_j$. We say that robots are *mutually visible* in configuration $\mathcal{C}_f$. See Fig. 10.

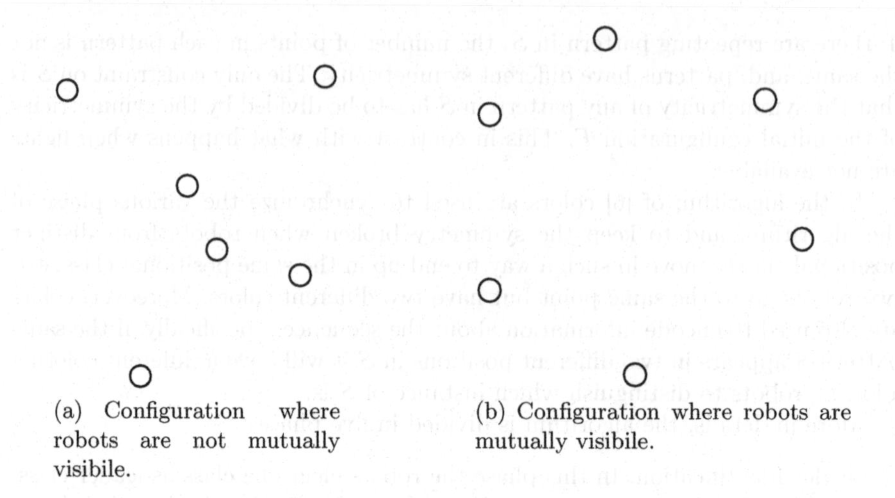

(a) Configuration where robots are not mutually visibile.

(b) Configuration where robots are mutually visibile.

**Fig. 10.** Example of initial and final configuration for Mutual Visibility.

*Preliminary Definitions.* Given a configuration $\mathcal{C}_t$, at time $t$, $\mathcal{H}(t)$ denotes the convex hull of $\{p_1(t), p_2(t), \cdots, p_n(t)\}$ at time $t$. The robots lying on its boundary are the *external robots*, the ones lying in its interior are the *internal robots*. Note that, a robot may not know where the convex hull's vertices are located, because its view may be obstructed by other robots. However, it can easily determine whether it is an external or an internal robot, i.e. a robot $r$ is *external* when there are two robots in its snapshot such that the angle formed by $r$ and them is at least $\pi$ and there is no other robot in that angle.

### 6.1 Main Strategies

Analysing the literature, it is possible to identify three main meta-strategies *Shrink*, *Contain* and *Local*. Roughly, each of these techniques works as follow:

- *Shrink*: In this strategy the external robots shrink towards a single point. In doing so internal robots progressively become external. The strategy terminate when all robots are vertices of the convex hull.
- *Contain*: This strategy is based on two phases *interior depletion* and *vertex adjustment*. In the first phase, the internal robots move towards the convex hull. In the second phase the robots on the convex hull move to become vertices.
- *Local*: In the *Local* strategy each robot does a constant number of steps, sometimes a single step, based on its local view and moving of a small distance from its initial position. Doing this it tries to decrease the number of collinear robots.

These three strategy have been proposed in [11]. In the first two strategies the final configuration does not only ensure Mutual Visibility but it also solves the

Convex Formation problem, arranging the robots as vertices of a convex polygon. The existing algorithms for Mutual Visibility use some variation of these strategies, where the modifications are used to obtain special properties such as fast solutions, optimal number of moves, or resilience to faulty robots.

In the following we will assume that the initial configuration is not a line. In case the initial configuration is a line, it is easy recognizable by each robots, and they can run a simple custom algorithm to move themselves in a configuration where there is a proper convex hull.

**Strategy *Shrink*.** The main idea of strategy *Shrink*, first proposed in [11,13], is to move the external robots on the vertices of the convex hull towards the inside of the convex hull. The final purpose is to shrink the convex hull (see Fig. 11) while not decreasing the number of external robots. The robots use two colors: *Off*, *Vertex*. Initially all robots have a pre-defined color *Off*, the robots that are also vertices become *Vertex* to signal their special positioning. By shrinking the convex hull, internal robots become external, and, eventually, vertices of the convex hull.

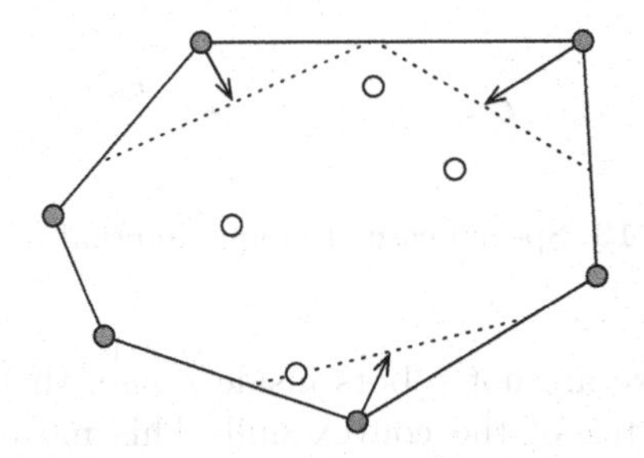

**Fig. 11.** Motion of vertices in *Shrink*.

*Details of* Shrink. We will explain the details of the strategy *Shrink* of [11] designed for **rigid** $\mathcal{S}$SYNC. A vertex robot $r$ in position $p$ moves inside the triangle formed by itself and its own two neighbors on the convex hull's boundary[4] (note that such neighbors are necessarily visible). This triangle is $\triangle pab$ of Fig. 12. The only robots moving are the ones on the vertices of the hull. The move is designed in a careful way, and the robot $r$ moves according to this three rules:

- (Rule 1) To avoid collision with other moving vertices $r$ does not go outside the triangle $\triangle puv$, where $u$, $v$ are the midpoints of edges $pa$ and $pb$.
- (Rule 2) In order to keep being external, $r$ does not cross any line passing through a robot inside $\triangle puv$ and parallel to $ab$. In case there is a robot in $\triangle puv$, robot $r$ moves on the closest of such lines. In this last case the number of external robots is increased, see Fig. 12(a).

[4] The neighbors are the adjacent robots on the convex hull boundary.

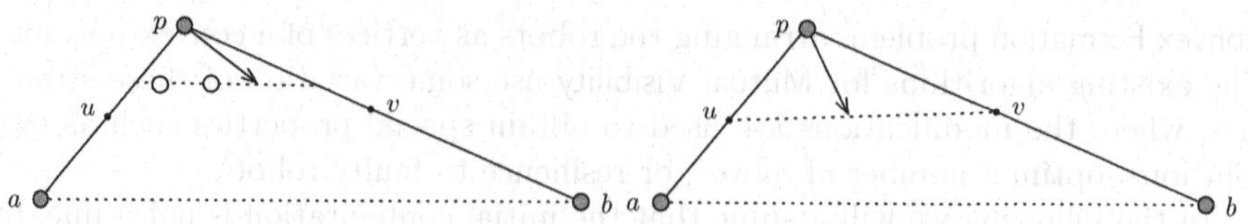


(a) robot $r$ in position $p$ moves carefully to increase external robots and to remain external.

(b) Default move of robot $r$ in position $p$ to shrink the convex hull.

**Fig. 12.** Movements of an external robot according to the presence or not of an internal robot in the area of movement, $\Delta puv$.

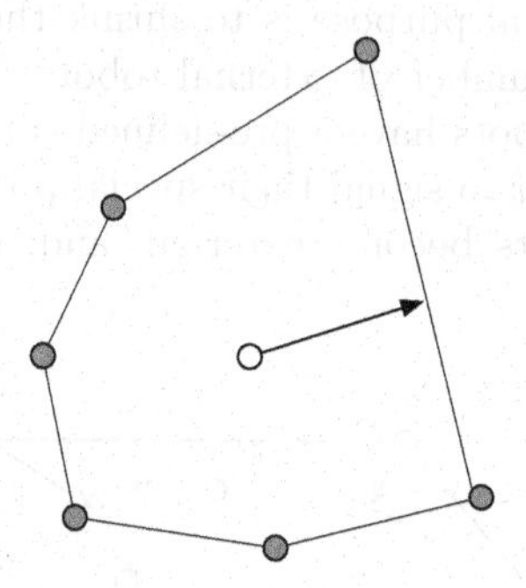

**Fig. 13.** Special case of unique internal robot.

- (Rule 3) When there are not robots inside $\Delta puv$, then $r$ moves on the line $uv$ remaining a vertex of the convex hull. This move allows [11] to prove that the convex hull shrinks in a such a way to converge to a single point. This convergence property is a key point to prove that all internal robots eventually became external.

Notice that, if a vertex robot $r$ moves using Rule 3 and one of its neighbor, let it be $a$, is not a vertex then $a$ becomes a vertex. With this observation it is clear that once all robots are external, they eventually become vertices of the convex hull.

There is only one special case: when there exists an unique internal robots. In this case the robot has to move because it could be positioned in the converge point of the shrinking procedure. Thus, it will be never reached by the others.

A custom move is needed in this case. When an internal robots sees that all the other robots are vertices, then it knows that it is the unique internal robots, and it moves to an edge of the convex hull, see Fig. 13.

Robots terminate when they reach a strictly convex configuration, that is when they all see each other with color *Vertex*. Notice, that in this algorithm lights are used for termination detection.

It has been shown that a protocol based on *Shrink* correctly terminates in **rigid** $\mathcal{S}$SYNC using two colors. Moreover, it is possible to slightly modify it so to solve Mutual Visibility also in the model without lights, but when robots have knowledge of $n$ (the total number of robots in the system). Since lights are only used for termination, without them and with knowledge of $n$ a robots terminate when it sees a strictly convex configuration, and $n$ other robots. It is easy to see that, when $n$ is unknown, *Shrink* uses an optimal number of colors.

**Theorem 13** ([11]). *Protocol* Shrink *solves* Mutual Visibility *by* **rigid** *robots in* $\mathcal{S}$SYNC *with* 2 *colors, or with* no *colors if the robots know* $n$.

The strength of *Shrink* strategy is that it requires a minimal number of colors. However, the convergence shown in Theorem 13 does not work in a **non-rigid** or $\mathcal{A}$SYNC model.

**Strategy *Contain*.** The meta-strategy *Contain*, first presented in [11,12] for **non-rigid** robots in $\mathcal{S}$SYNC, consists of two successive stages: *interior depletion* and *vertex adjustment*. The algorithm uses three colors: *Off*, *External*, *Adjusting*. In the interior depletion stage, the internal robots move towards the boundary of the convex hull. At the end of this stages all robots are external. In the vertex adjustment stage, the external robots make small adjustments to finally reach a strictly convex configuration. Let $\mathcal{H}'(t)$ be the convex hull of the internal robots at time $t \in \mathbb{N}$, see Fig. 14(a).

*Details of Contain.* More precisely, strategy *Contain* works as follows.

- *Interior depletion*: Initially, all robots have lights *Off*. Once an external robot is activated it switches light to *External* and it does nothing. The robots that move are the one on the border $\mathcal{H}'$. Eventually, a robot realizes to be on the border of $\mathcal{H}'$, this can be done locally once enough external robots have been activated. A vertices $r$ of $\mathcal{H}'$ moves on the border of $\mathcal{H}$. There are three possible way for the robot to move:
  1. When $r$ is the only internal robots, it moves towards the midpoint of the closest edge.
  2. When $r$ believes to be a vertex of a non degenerate $\mathcal{H}'$, i.e. the robots in $\mathcal{H}'$ do not form a line, then $r$ try to move to $\mathcal{H}$. It does so, only when it is able to identify correctly identify an edge of $\mathcal{H}$ where it can move without colliding with other moving robots. As example, if $r$ is a vertex forming an acute internal angle of $\mathcal{H}'$, then it moves to $\mathcal{H}$ by remaining inside the zone delimited by the extension of the edges of $\mathcal{H}'$ to which it belongs, see Fig. 14(a).
  3. When $r$ is an extreme of the line $\mathcal{H}'$, it moves towards $\mathcal{H}$ by using a direction that has a right angle away oriented away from $\mathcal{H}'$, see Fig. 14(b).
- *Vertex adjustment*: When a robots $r$, vertex of $\mathcal{H}$, with light *External* sees only robots with light *External*, then it makes an adjustment move, the same of strategy *Shrink* of Fig. 12(b). Before doing the move it sets its light to *Adjusting*. After this adjustment, the neighbors of $r$ on $\mathcal{H}$ will be vertices, if they were not both vertices before the move. This *Adjusting* light is used by $r$ to remember it adjusted itself. A robot with light *Adjusting*, once activated it switches to *External* and it terminates.

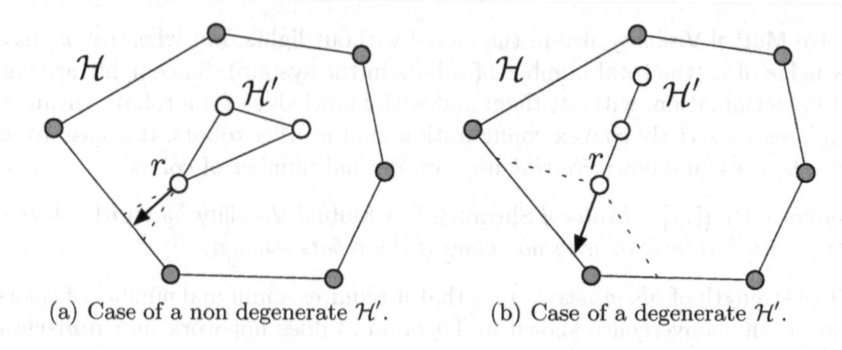


(a) Case of a non degenerate $\mathcal{H}'$. (b) Case of a degenerate $\mathcal{H}'$.

**Fig. 14.** Movements of a robot on the border of $\mathcal{H}'$ in *Contain*.

**Theorem 14** ([11]). *Protocol* Contain *solves* Mutual Visibility *by* **non-rigid** *robots in* $\mathcal{S}$SYNC *with* 3 *colors.*

Note that *Contain* works even when the system is **non-rigid**, this is in contrast with *shrink*. However, it does have a greater number of lights and the algorithm is more complex.

In [11] slight variations of protocol *Contain* are presented to solve the problem under various conditions and knowledge. Let $\delta$ be the minimum distance travelled by a robot.

**Theorem 15** ([11]). Mutual Visibility *can be solved in* $\mathcal{S}$SYNC *by* **non-rigid** *robots with* no *colors, if they know* $\delta$ *and* $n$*; it can be solved with* 2 *colors, if the robots know only* $\delta$. Mutual Visibility *can be solved in* $\mathcal{A}$SYNC *by* **rigid** *robots with* 3 *colors, and in* $\mathcal{A}$SYNC *by* **non-rigid** *robots, if they agree on the direction of one coordinate axis.*

**Strategy *Local.*** The meta-strategy *Local* has been proposed in [11]. *Local* is the only one of the three meta-strategies in which the final solution does not solve Convex Formation. In this strategy, the first time a robot is activated, it makes a small move to a new position avoiding to stop on, or to trespass, any line connecting two visible robots, see Fig. 15. Only two colors are used: *Off*, *Moved* that is turned on before a robot move. This simple idea solves Mutual Visibility using two colors for the sequential scheduler $\mathcal{S}$EQUENTIAL (a particular case of $\mathcal{S}$SYNC where a single robot is activated at each step). Also notice that using this strategy, any robot moves at most once, in the entire execution.

**Theorem 16** ([11]). *Protocol* Local *solves* Mutual Visibility *by* **non-rigid** *robots in* $\mathcal{S}$EQUENTIAL *using 2 colors.*

## 6.2 State of the Art for the Number of Lights

The solutions above have been improved, in number of colors employed and in the model where Mutual Visibility is solvable. In [22] an Algorithm following the

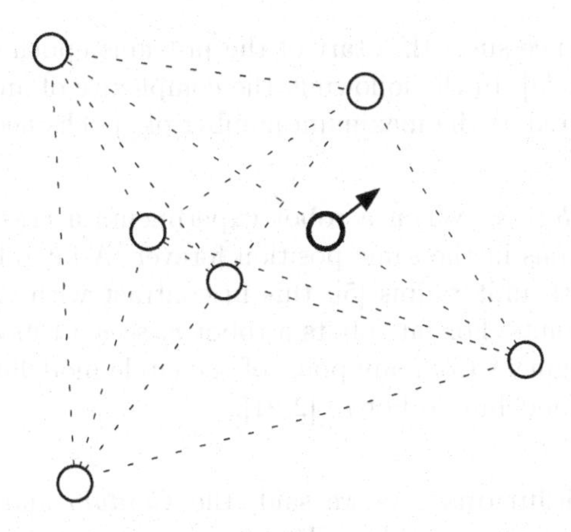

**Fig. 15.** Motion of a robot in *Local*: the robot with a bold circle moves in such a way to never cross or reach a segment connecting two visible robots.

*Contain* meta-strategy has been proposed, which allows to solve the problem in $\mathcal{A}$SYNC **non-rigid** with no colors but agreement on one axis, and in $\mathcal{S}$SYNC **non-rigid** with only 2 colors.

**Theorem 17** ([22]). Mutual Visibility *can be solved in* $\mathcal{S}$SYNC *by* **non-rigid** *robots, in* $\mathcal{A}$SYNC *by* **rigid** *robots, and in* $\mathcal{A}$SYNC *by* **non-rigid** *robots, if they agree on the direction of one coordinate axis, using* 2 *colors.*

Moreover, another solution has been proposed based on the *Local* strategy [3]. This algorithm solves Mutual Visibility (but not Convex Formation) in $\mathcal{A}$SYNC **non-rigid** with 7 colors. The main idea is to enforce an order between the robots movements, by allowing only robots that are not collinear with other robots (the *terminal* robots) to move; in this way a local move monotonically increases the number of terminal robots.

**Theorem 18** ([3]). Mutual Visibility *can be solved in* $\mathcal{A}$SYNC *by* **non-rigid** *robots using* 7 *colors.*

### 6.3 Time Complexity and Fault Tolerance

Apart from minimising the number of colors in Mutual Visibility solutions, other works have focused on designing time efficient solutions [23–25, 27], or solution for environments where the robots may fails [1], or when robots are fat [21].

*Time Complexity.* When we consider $\mathcal{F}$SYNC the time complexity of an algorithm is measured by the maximum number of rounds needed to terminate. In case of $\mathcal{S}$SYNC or $\mathcal{A}$SYNC the definition is not straightforward. [24, 25] proposes the concept of epoch. Each epoch is an interval of time. A new epoch starts when the previous epoch ends, and an epoch ends as soon as all robots have been activated at least once since the start of the previous epoch (or time $t = 0$ for the first epoch), see [4]. In the following, the complexity of an algorithm $\mathcal{S}$SYNC or $\mathcal{A}$SYNC is measured as the maximum number of epochs needed to terminate.

*Failures and Fat Robots.* When a robot experiments a crash failure, it stops moving and it remains in the same position forever. A fat robot is modelled as circular entities with unit radius [5], this in contrast with the classical model where robots are points. For fat robots a robot $r_i$ sees a robot $r_j$ if there exists a not obstructed segment from any point of the circle modelling robot $r_i$ to any point of the circle modelling robot $r_j$ [2,21].

**Time Efficient Solutions.** As we said, the *Contain* and *Shrink* strategies solve the Convex Formation problem. The *Local* strategy solves only the Mutual Visibility. The *Local* solution proposed in [3] solves Mutual Visibility in $\mathcal{A}$SYNC by **non-rigid** robots in $\mathcal{O}(n)$ steps and a constant number of moves for each robot.

*Efficient Solutions for* Convex Formation. The first work proposing an efficient solution for Convex Formation has been [27], an algorithm for $\mathcal{F}$SYNC **rigid** uses the *Contain* meta-strategy to solve the problem in $\mathcal{O}(\log n)$ rounds (the algorithm, however, allows collisions). This runtime is done by proposing an *interior depletion* phase where internal robots goes on the convex hull in $\mathcal{O}(\log n)$ rounds. The *vertex adjustment* also ends in $\mathcal{O}(\log n)$ rounds, and in this case robots on the edge of the $\mathcal{H}$ move to create a strictly convex configuration. This is in contrast with the classic *Contain* where vertex move.

In [24] the authors propose an algorithm for $\mathcal{S}$SYNC **rigid** that follows the *Contain* meta-strategy and solves Mutual Visibility in $\mathcal{O}(1)$ epochs. The algorithm uses an initial step, *corner moving*, in which vertex of $\mathcal{H}$ does one adjustment movement. After this adjustment, the vertex will be visible to all internal robots. The movement is similar to the one used by *Shrink*, with the additional idea of never cross, or reach, locations where a collinearity with internal robots can be created. After the adjustment, the convex hull $\mathcal{H}$ will be detectable by internal robots, that can move on its border in a constant number of epochs. In the last phase, the robots on the edges of $\mathcal{H}$ move to create a strictly convex configuration.

A similar strategy, with some modifications, works in $\mathcal{A}$SYNC **rigid** with time complexity $\mathcal{O}(\log n)$ [25].

Finally, in $\mathcal{F}$SYNC **rigid**, when $n$ is known, a linear time solution exists, without using any light [23]. The current state of the art from a time complexity perspective is summarized in Table 1, where only collision-less solutions are considered.

**Fault-Tolerance and Fat Robots.** In [1] it is investigated how to solve Mutual Visibility in $\mathcal{S}$SYNC **rigid**, with agreement on the axes, when a single robot may experience a crash failure. The algorithm uses 3 colors, and it is based on a variation of the *Shrink* strategy. Notice that, if there is a crashed robot placed

**Table 1.** Collisions-less solutions for Convex Formation, for $\mathcal{F}$SYNC the time complexity is measured using the number or rounds. For $\mathcal{S}$SYNC and $\mathcal{A}$SYNC, is the number of epochs.

| Paper | Scheduler | Time | # Colors |
|---|---|---|---|
| [23] | $\mathcal{F}$SYNC **rigid** | $\mathcal{O}(n)$ | 0 |
| [24] | $\mathcal{S}$SYNC **rigid** | $\mathcal{O}(1)$ | 12 |
| [25] | $\mathcal{A}$SYNC **rigid** | $\mathcal{O}(\log n)$ | 25 |

exactly in the convergence point of the *Shrink* procedure, then the classic *Shrink* strategy could fail. It could be possible for pairs of robots on the convex hull to be collinear with the faulty robot, and to keep this collinearity by moving in a symmetric way. This is solved in [1], by doing a special move once a configuration with a single internal robot is reached.

Finally, Mutual Visibility has been solved, in $\mathcal{F}$SYNC **rigid**, when robots are fat using 10 colors, in time $\mathcal{O}(n)$ [21].

# 7 Conclusions

We have seen that lights greatly enhance the capability of robots. They create a new computational landscape where the relationships between $\mathcal{A}$SYNC, $\mathcal{S}$SYNC and $\mathcal{F}$SYNC are different from the usual oblivious model. Besides this new relationship, a wide set problems can now be solved. A prototypical example is the Rendezvous problem.

Another advantage is the possibility to solve old problems under weaker assumptions, see the reduction of the restrictions needed on the sequences of formable patterns by luminous robots.

Finally, the uses of lights allows the designing of fast and fault tolerant algorithms in models where robots obstruct each other.

The model of luminous robot is still relatively new, and there are many open problems.

# References

1. Aljohani, A., Sharma, G.: Complete visibility for mobile robots with lights tolerating a faulty robot. In: Proceedings of the 32nd IEEE International Parallel and Distributed Processing Symposium Workshops (IPDPS Workshops), pp. 834–843 (2017)
2. Aljohani, A., Poudel, P., Sharma, G.: Fault-tolerant complete visibility for asynchronous robots with lights under one-axis agreement. In: Rahman, M.S., Sung, W.-K., Uehara, R. (eds.) WALCOM 2018. LNCS, vol. 10755, pp. 169–182. Springer, Cham (2018). https://doi.org/10.1007/978-3-319-75172-6_15

3. Bhagat, S., Mukhopadhyaya, K.: Optimum algorithm for mutual visibility among asynchronous robots with lights. In: Spirakis, P., Tsigas, P. (eds.) SSS 2017. LNCS, vol. 10616, pp. 341–355. Springer, Cham (2017). https://doi.org/10.1007/978-3-319-69084-1_24
4. Cord-Landwehr, A., et al.: A new approach for analyzing convergence algorithms for mobile robots. In: Aceto, L., Henzinger, M., Sgall, J. (eds.) ICALP 2011. LNCS, vol. 6756, pp. 650–661. Springer, Heidelberg (2011). https://doi.org/10.1007/978-3-642-22012-8_52
5. Czyzowicz, J., Gasieniec, L., Pelc, A.: Gathering few fat mobile robots in the plane. Theor. Comput. Sci. **410**, 81–499 (2009)
6. Das, S., Flocchini, P., Prencipe, G., Santoro, N.: Forming sequences of geometric patterns with oblivious mobile robots. In: Proceedings of the 7th International Conference on FUN with Algorithms (FUN), pp. 113–124 (2014)
7. Das, S., Flocchini, P., Prencipe, G., Santoro, N., Yamashita, M.: The power of lights: synchronizing asynchronous robots using visible bits. In: 32nd IEEE International Conference on Distributed Computing Systems (ICDCS), pp. 506–515 (2012)
8. Das, S., Flocchini, P., Prencipe, G., Santoro, N., Yamashita, M.: Autonomous mobile robots with lights. Theor. Comput. Sci. **609**, 171–184 (2016)
9. Das, S., Flocchini, P., Santoro, N., Yamashita, M.: Forming sequences of geometric patterns with oblivious mobile robots. Distrib. Comput. **28**, 131–145 (2015)
10. D'Emidio, M., Frigoni, D., Navarra, A.: Synchronous robots vs asynchronous lights-enhanced robots on graphs. Electron Notes Theor. Comput. Sci. **322**, 169–180 (2016)
11. Di Luna, G.A., Flocchini, P., Gan Chaudhuri, S., Poloni, F., Santoro, N., Viglietta, G.: Mutual visibility by luminous robots without collisions. Inf. Comput. **254**, 392–418 (2017)
12. Di Luna, G.A., Flocchini, P., Gan Chaudhuri, S., Santoro, N., Viglietta, G.: Robots with lights: overcoming obstructed visibility without colliding. In: Felber, P., Garg, V. (eds.) SSS 2014. LNCS, vol. 8756, pp. 150–164. Springer, Cham (2014). https://doi.org/10.1007/978-3-319-11764-5_11
13. Di Luna, G.A., Flocchini, P., Poloni, F., Santoro, N., Viglietta, G.: The mutual visibility problem for oblivious robots. In: Proceedings of the 26th Canadian Computational Geometry Conference (CCCG), pp. 348–354 (2014)
14. Dijkstra, E.W.: Self-stabilizing systems in spite of distributed control. Commun. ACM **17**, 643–644 (1974)
15. Flocchini, P., Prencipe, G., Santoro, N., Widmayer, P.: Arbitrary pattern formation by asynchronous oblivious robots. Theor. Comput. Sci. **407**, 412–447 (2008)
16. Flocchini, P., Santoro, N., Viglietta, G., Yamashita, M.: Rendezvous with constant memory. Theor. Comput. Sci. **621**, 57–72 (2016)
17. Fujinaga, N., Yamauchi, Y., Ono, H., Kijima, S., Yamashita, M.: Pattern formation by oblivious asynchronous mobile robots. SIAM J. Comput. **44**, 740–785 (2015)
18. Heriban, A., Defago, X., Tixeuil, S.: Optimally gathering two robots. In: Proceedings of the 19th International Conference on Distributed Computing and Networking (ICDCN), pp. 3:1–3:10 (2018)
19. Khan, L.U.: Visible light communication: applications, architecture, standardization and research challenges. Dig. Commun. Netw. **2**, 78–88 (2017)
20. Okumura, T., Wada, K., Katayama, Y.: Optimal asynchronous rendezvous for mobile robots with lights. Arxiv, CoRR abs/1707.04449 (2017)

21. Sharma, G., Alsaedi, R., Bush, C., Mukhopadyay, S.: The complete visibility problem for fat robots with lights. In: Proceedings of the 19th International Conference on Distributed Computing and Networking (ICDCN), pp. 21:1–21:4 (2018)
22. Sharma, G., Busch, C., Mukhopadhyay, S.: Mutual visibility with an optimal number of colors. In: Bose, P., Gąsieniec, L.A., Römer, K., Wattenhofer, R. (eds.) ALGOSENSORS 2015. LNCS, vol. 9536, pp. 196–210. Springer, Cham (2015). https://doi.org/10.1007/978-3-319-28472-9_15
23. Sharma, G., Bush, C., Mukhopadyay, S.: Brief announcement: complete visibility for oblivious robots in linear time. In: Proceedings of the 29th ACM Symposium on Parallelism in Algorithms and Architectures (SPAA), pp. 325–327 (2017)
24. Sharma, G., Vaidyanathan, R., Trahan, J.L., Busch, C., Rai, S.: Complete visibility for robots with lights in $O(1)$ time. In: Bonakdarpour, B., Petit, F. (eds.) SSS 2016. LNCS, vol. 10083, pp. 327–345. Springer, Cham (2016). https://doi.org/10.1007/978-3-319-49259-9_26
25. Sharma, G., Vaidyanathan, R., Trahan, J.L., Bush, C., Rai, S.: O(log n)-time complete visibility for asynchronous robots with lights. In: Proceedings of the 32nd IEEE International Parallel and Distributed Processing Symposium (IPDPS), pp. 513–522 (2017)
26. Suzuki, I., Yamashita, M.: Distributed anonymous mobile robots: formation of geometric patterns. SIAM J. Comput. **28**, 1347–1363 (1999)
27. Vaidyanathan, R., Bush, C., Trahan, J.L., Sharma, G., Rai, S.: Logarithmic-time complete visibility for robots with lights. In: Proceedings of the 29th IEEE International Parallel and Distributed Processing Symposium (IPDPS), pp. 375–384 (2015)
28. Viglietta, G.: Rendezvous of two robots with visible bits. In: Flocchini, P., Gao, J., Kranakis, E., Meyer auf der Heide, F. (eds.) ALGOSENSORS 2013. LNCS, vol. 8243, pp. 291–306. Springer, Heidelberg (2014). https://doi.org/10.1007/978-3-642-45346-5_21
29. Yamashita, M., Suzuki, I.: Characterizing geometric patterns formable by oblivious anonymous mobile robots. Theor. Comput. Sci. **411**, 2433–2453 (2010)

# Formal Methods for Mobile Robots

Maria Potop-Butucaru[1], Nathalie Sznajder[1], Sébastien Tixeuil[1(✉)], and Xavier Urbain[2]

[1] Sorbonne Université, CNRS, Laboratoire d'Informatique de Paris 6, LIP6, 75005 Paris, France
Sebastien.Tixeuil@lip6.fr

[2] Université Claude Bernard Lyon-1, LIRIS CNRS UMR 5205, Université de Lyon, Lyon, France

**Abstract.** Most existing work in the literature typically ensures the correctness of mobile robot protocols via *ad hoc* handwritten proofs, which are both cumbersome and error-prone.

This paper surveys state-of-the-art results about applying formal methods approaches (namely, model-checking, program synthesis, and proof assistants) to the context of mobile robot networks. Those methods already proved useful for bug-hunting in published literature, designing correct-by-design optimal protocols, and certifying impossibility results and protocols.

**Keywords:** Formal methods · Mobile robots
Distributed algorithms · Model checking · Program synthesis
Proof certification · Proof assistant

## 1 Introduction to Formal Methods

This section reviews the main formal methods techniques that have been used in the context of distributed computing, and autonomous mobile robots in particular.

### 1.1 Model Cheking

Model-checking [9,32] is a technique that was developed for the verification of various models: finite ones but also in some cases infinite, parameterised, or even timed models. It has been successfully used for the verification of distributed systems from classical shared memory (consensus, transactional memory) to population protocols [26,27,33,53,57,59,63,76]. Unfortunately, it was proved in [6] that parameterised model checking is undecidable, and this general result was followed by several stronger ones for specific models, for instance in [46]. In such cases, a classical line of work consisted in combining model-checking with other techniques like abstraction, induction, etc., as first proposed in [64] or [31]. These techniques were largely used since [7,17,24,41].


P. Flocchini et al. (Eds.): Distributed Computing by Mobile Entities, LNCS 11340, pp. 278–313, 2019.
https://doi.org/10.1007/978-3-030-11072-7_12

### 1.2 Games and Protocols Synthesis

In the formal methods community, automatically synthesising programs that would be correct by design is a problem that raised interest early [1,30,65,69]. Actually, this problem goes back to Church [29]. When the program to generate is intended to work in an open system, maintaining an on-going interaction with a (partially) unknown environment, it is known since [22] that seeing the problem as a *game* between the system and the environment is a successful approach. The system and its environment are considered as opposite players that play a game on some graph, the winning condition being the specification the system should fulfill however the environment behaves. Then, the classical problem in game theory of determining winning strategies for the players is equivalent to finding how the system should act in any situation, in order to always satisfy its specification. The case of mobile autonomous robots that we focus on in this paper falls in this category of problems: the robots may evolve (possibly indefinitely) on a ring, making decisions based on some global state of the system at each time instant. The vertices of graph on which the players will play would then be some representation of the different global positions of the robots on the ring. The presence of an opposite player (or environment) is motivated by the absence of chirality of the robots: when a robot is on an axis of symmetry, it is unable to distinguish its two sides one from another, hence to choose exactly *where* it moves; this decision is supposed to be taken by the opposite player.

### 1.3 Certification and Proof Assistants

Mechanical proof assistants are proof management systems in which a user can express data, programs, theorems and proofs. In sharp contrast with automated provers, they are mostly interactive, and thus require some kind of expertise from their users. Sceptical proof assistants provide an additional guarantee by checking mechanically the soundness of a proof *after* it has been interactively developed.

Various proof assistants emerged since the 60's, to name a few: Agda [3], NqThm [21] and its relative ACL2 [2], PVS [70], Mizar [67], Coq [35], Isabelle/HOL [68], etc.

In the context of program verification, Isabelle/HOL and Coq are amongst the most widely used; both are based on type theory. They have been successfully employed for various tasks such as the formalisation of programming language semantics [61], certification of an OS kernel [55], verification of cryptographic protocols [4], certification of RSA keys [75], mathematical developments as involved as the 4-colours theorem [50], the Feit-Thompson theorem [51], or the Kepler Conjecture [74].

During the last twenty years, the use of tool-assisted verification has extended to the validation of distributed processes.

In the context of process algebras, which can be used to describe and verify algorithms built from merge, sequential composition and encapsulation,

Fokkink [48] and Bezem *et al.* [16] use a proof assistant to prove the equality between two processes, one of them being a specification.

TLA/TLAPS [58,60] can enjoy an Isabelle back-end for its provers [40]. Gascard and Pierre [49] focus on interconnection networks that are symmetric: rings, tori, hypercubes. Based on a compositional approach of certified components, their work makes use of Nqthm.

Cansell and Méry's contribution to the RIMEL project [23] addresses the class of local computation (LC) algorithms. A catalogue of case studies like election algorithms, spanning tree construction, and even Mazurkiewicz's enumeration algorithm have been developed in Event-B. The code of these algorithms is obtained by successive refinements starting from an abstract machine that translates directly to a specification. This code is *annotated* with logical formulas—mainly invariants on the state of the system—the proofs of which generate verification conditions through a calculus of weakest preconditions.

Küfner *et al.* [56] propose a methodology to develop (using Isabelle) proofs of properties of fault-tolerant distributed algorithms in an asynchronous message passing style setting. They focus on correctness proofs only.

Chou's methodology [28] is based on the HOL proof assistant. It aims at proving properties of concrete distributed algorithms through simulation with abstract ones. The methodology does not allow to prove impossibility results.

Castéran *et al.* [25] use CoQ to state and prove invariants but also generic results about sub-classes of LC systems, thanks to Castéran and Filou's library Loco [62]. Genericity is worth emphasising here as the approach is not limited to *particular instances* of algorithms. Castéran *et al.* actually propose proofs of negative results in CoQ for some kinds of distributed algorithms in this graph relabelling setting.

Deng and Monin [42] use CoQ to prove the correctness of distributed self-stabilising protocols in the population protocol model. This model permits to describe interactions of an arbitrary large size of mobile entities, however the considered entities lack movement control and geometric awareness that are characteristic of robot networks.

As a matter of fact, surprisingly few works consider using mechanised assistance for networks of *mobile entities.*

# 2 Formal Modelling

## 2.1 Robots on Graphs as Automaton Composition

In this section we describe the model proposed by Bérard *et al.* for the robots, the demons, and the system resulting from their interactions. This model encompasses all three FSYNC, SSYNC, and ASYNC operating modes, but assumes that individual robots can only operate in a discrete setting (that is, a graph).

**Robot Modelling.** All robots execute the same algorithm [47], hence the behaviour of each of them can be described by the finite automaton of Fig. 1. They operate in *Look*, *Compute*, and *Move cycles.*

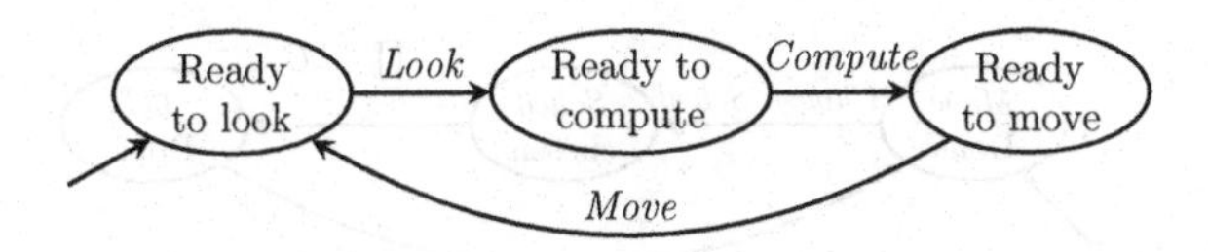


**Fig. 1.** A generic automaton for the robot behaviour.

To start a cycle, a robot takes a snapshot of its environment, which is represented by the *Look* transition. Then, it computes its future location, represented by the *Compute* transition. Finally, the robot moves along an edge of the graph according to its previous computation; this effective movement is represented by the *Move* transition.

The algorithm is implemented in the *Compute* transition, hence the "Ready to move" state is divided into as many parts as there are possible movements according to the protocol under study.

Note that the original model [73] abstracts the precise time constraints (like the computational power or the locomotion speed of robots) and keeps only sequences of instantaneous actions, assuming that each robot completes each cycle in finite time. This model can be reduced by combining the *Look* and *Compute* phases to obtain the *LC* phase. This is simply done by merging the two states "Ready to look" and "Ready to compute" into a single state "Ready to Look-Compute".

**Demon Modelling.** Unlike robots that have the same behaviour regardless of the model, the demon is parameterised by the execution model and by the number of robots. It is also modelled by a finite automaton, one for each variant of the execution model. By synchronising one of these demons with robot automata, we obtain an automaton that represents the global behaviour of robots in the chosen model.

To describe these demon models, we consider a set $Rob = \{r_1, \ldots, r_k\}$ of robots. We denote by $LC_i$ (resp. $Move_i$), the *LC* (resp. *Move*) phase of robot $r_i$. Note that $LC_i$ and $Move_i$ are actually sets of possible actions in the corresponding phases. For a subset $Sched \subseteq Rob$, we denote the synchronisation of all $LC_i$ (resp. $Move_i$) actions of all robots in *Sched* by $\prod_{r_i \in Sched} LC_i$ (resp. $\prod_{r_i \in Sched} Move_i$).

In the SSYNC model, a non-empty subset of robots is scheduled for execution at every phase, and operations are executed synchronously. In this case, the automaton is a cycle, where a set $Sched \subseteq Rob$ is first chosen. In this cycle the *LC* and *Move* phases are synchronised for this set of robots. A generic automaton for SSYNC is described in Fig. 2(a). Actually, the "*Sched* chosen" state has to be divided into $2^k$ states, where $k$ is the number of robots, in order to represent all possible sets *Sched*.

The FSYNC model is a particular instance of the SSYNC model, where all robots are scheduled for execution at every phase, and operate synchronously thereafter. In each global cycle, $Sched = Rob$, hence all global cycles are identical.

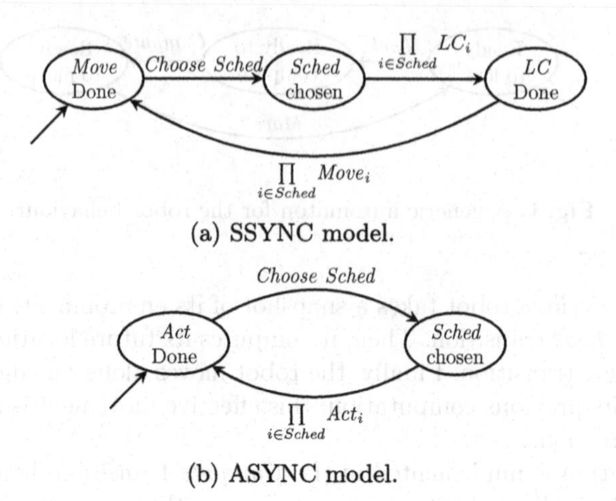


(a) SSYNC model.

(b) ASYNC model.

**Fig. 2.** The demons automata.

The ASYNC model is totally asynchronous: any finite delay may elapse between *LC* and *Move* phases. During each phase a set *Sched* is chosen, and all robots in this set execute an action: the action $Act_i$ is either in $LC_i$ or in $Move_i$ depending on the current state of robot $r_i$. Hence, a robot can move according to an outdated observation. The automaton for this demon is depicted in Fig. 2(b).

**System Modelling.** To describe the global model, we denote by

$$Pos = \{0, \ldots, n-1\} \subseteq \mathbb{N}$$

the set of positions on the graph. A configuration of the system is a mapping $c : Rob \to Pos$ associating with each robot $r$ its position $c(r) \in Pos$. Hence, in a graph of $n$ nodes with $k$ robots, there are $n^k$ possible configurations.

The model of the system is an automaton

$$M = (S, s_0, A, T)$$

obtained by the synchronised product of $k$ robot automata and all the possible configurations, as defined above. The demon is used to define the synchronisation function. The alphabet of actions is $A = \prod_{r_i \in Rob} A_i$, with $A_i = LC_i \cup Move_i$ for each robot $r_i$. In this product, states are of the form $s = (s_1, \ldots, s_k, c)$ where $s_i$ is the local state of robot $r_i$, and $c$ is the configuration. An initial state is of the form $s_0 = (s_{1,0}, \ldots, s_{k,0}, c)$ where $s_{i,0}$ is the initial local state of robot $r_i$, and $c$ is an arbitrary configuration.

A transition of the system is labelled by a tuple

$$a = (a_1, \ldots, a_k)$$

where $a_i \in A_i \cup \{\varepsilon, -\}$ for all $1 \leq i \leq k$ and

$$(s_1, \ldots, s_k, c) \xrightarrow{a} (s'_1, \ldots, s'_k, c')$$

if and only if for all $i$, $s_i \xrightarrow{a_i} s'_i$, and $c'$ is obtained from $c$ by updating the positions of each robot $r_i$ such that $a_i \in Move_i$. To represent the scheduling, we denote by $\prod_{r_i \in Sched} Act_i$ the action $(a_1, \ldots, a_k)$ such that $a_i = -$ if $r_i \notin Sched$, and $a_i \in LC_i \cup Move_i \cup \{\varepsilon\}$ otherwise.

### 2.2 Protocol Synthesis and Reachability Games

To enable robot protocol synthesis (that is, the automatic generation of robot protocols for a given problem in a given setting), the approach of Millet *et al.* [66] is to reuse the modelling presented in Sect. 2.1 for robots, schedulers, and their interactions, in the framework of reacheability games.

We now present classical notions on this subject. If $A$ is a set of symbols, $A^*$ is the set of finite sequences of elements of $A$ (also called *words*), and $A^\omega$ the set of infinite such sequences, with $\varepsilon$ the empty sequence. We note $A^+ = A^* \setminus \{\varepsilon\}$, and $A^\infty = A^* \cup A^\omega$. For a sequence $w \in A^\infty$, we denote its *length* by $|w|$. If $w \in A^*$, $|w|$ is equal to its number of elements. If $w \in A^\omega$, $|w| = \infty$. For all words $w = a_1 \cdots a_k \in A^*$, $w' = a'_1 \cdots \in A^\infty$, we define the *concatenation* of $w$ and $w'$ by the word noted $w \cdot w' = a_1 \cdots a_k a'_1 \cdots$. We sometimes omit the symbol and simply write $ww'$. If $L \subseteq A^*$ and $L' \subseteq A^\infty$, we define $L \cdot L' = \{w \cdot w' \mid w \in L, w' \in L'\}$.

A game is composed of an *arena* and *winning conditions.*

**Arena.** An arena is a (finite in our context) graph $\mathcal{A} = (V, E)$ in which the set of vertices $V = V_p \uplus V_o$ is partitioned into $V_p$, the vertices of the protagonist, and $V_o$ the vertices of the opponent. The set of edges $E \subseteq V \times V$ allows to define the set of successors of some given vertex $v$, noted $vE = \{v' \in V \mid (v, v') \in E\}$.

**Plays.** To play on an arena, a token is positioned on an initial vertex. Then the token is moved by the players from one vertex to one of its successors. Each player can move the token only if it is on one of her own vertices. Formally, a play is a path in the graph, i.e., a finite or infinite sequence of vertices $\pi = v_0 v_1 \cdots \in V^\infty$, where for all $0 < i < |\pi|$, $v_i \in v_{i-1}E$. Moreover, a play is finite only if the token has been taken to a position without any successor (where it is impossible to continue the game): if $\pi$ is finite with $|\pi| = n$, then $v_{n-1}E = \emptyset$.

**Strategies.** A strategy for the protagonist determines where she brings the token whenever it is her turn to play. To do so, the player takes into account the history of the play, and the current vertex. Formally, a strategy for the protagonist is a (partial) function $\sigma : V^* \cdot V_p \to V$ such that, for all sequence (representing the current history) $w \in V^*$, all $v \in V_p$, $\sigma(w \cdot v) \in vE$ (i.e. the move is possible with respect to the arena). A strategy $\sigma$ is *memoryless* if it does not depend on the history. Formally, it means that for all $w, w' \in V^*$, for all $v \in V_p$, $\sigma(w \cdot v) = \sigma(w' \cdot v)$. In that case, we may simply see the strategy as a function $\sigma : V_p \to V$.

Given a strategy $\sigma$ for the protagonist, a play $\pi = v_0 v_1 \cdots \in V^\infty$ is said to be *$\sigma$-consistent* if for all $0 < i < |\pi|$, if $v_{i-1} \in V_p$, then $v_i = \sigma(v_0 \cdots v_{i-1})$. Given an initial vertex $v_0$, the *outcome* of a strategy $\sigma$ is the set of plays starting in $v_0$ that are $\sigma$-consistent. Formally, given an arena $\mathcal{A} = (V, E)$, an initial vertex $v_0$ and a strategy $\sigma : V^* V_p \to V$, we let

$$Outcome(\mathcal{A}, v_0, \sigma) = \{v_0 \pi \in V^\infty \mid v_0 \pi \text{ is a } \sigma\text{-consistent play}\}$$

**Winning Conditions, Winning Plays, and Winning Strategies.** The *winning condition* for the protagonist is defined as a subset of the plays $Win \subseteq V^\infty$. Then, a play $\pi$ is *winning* for the protagonist if $\pi \in Win$. In this work, we focus on the simple case of reachability games: the winning condition is then expressed according to a subset of vertices $T \subseteq V$ by $Reach(T) = \{\pi = v_0 v_1 \cdots \in V^\infty \mid \exists 0 \leq i < |\pi| : v_i \in T\}$. This means that the protagonist wins a play whenever the token is brought on a vertex belonging to the set $T$. Once it has happened, the play is winning, regardless of the following actions of the players.

Given an arena $\mathcal{A} = (V, E)$, an initial vertex $v_0 \in V$ and a winning condition $Win$, a *winning strategy* $\sigma$ for the protagonist is a strategy such that any $\sigma$-consistent play is winning. In other words, a strategy $\sigma$ is winning if $Outcome(\mathcal{A}, v_0, \sigma) \subseteq Win$. The protagonist wins the game $(\mathcal{A}, v_0, Win)$ if she has a winning strategy for $(\mathcal{A}, v_0, Win)$. We say that $\sigma$ is winning on a subset $U \subseteq V$ if it is winning starting from any vertex in $U$: if $Outcome(\mathcal{A}, v_0, \sigma) \subseteq Win$ for all $v_0 \in U$. A subset $U \subseteq V$ of the vertices is *winning* if there exists a strategy $\sigma$ that is winning on $U$.

**Solving a Reachability Game.** Given an arena $\mathcal{A} = (V, E)$ and a subset of the vertices $T \subseteq V$, one wants to determine the set $U \subseteq V$ of winning positions for the protagonist, and a strategy $\sigma : V^* V_p \to V$ for the protagonist, that is winning on $U$ for $Reach(T)$.

Figure 3 represents a reachability 2-player game. We recall now a well-known result [52] on reachability games:

**Theorem 1.** *The set of winning positions for the protagonist in a reachability game can be computed in linear time in the size of the arena. Moreover, from any position, the protagonist has a winning strategy if and only if she has a memoryless winning strategy.*

### 2.3 Parameterised Modelling with Presburger Formulae

Previous approaches (see Sects. 2.1 and 2.2) consider that size of the ring, $n$, is fixed, as well as the number $k$, of robots. In this section, we develop the parameterised model of Sangnier et al. [72] that makes use of Presburger formulae to express systems where the ring size $n$ is arbitrary (and a parameter of the model).

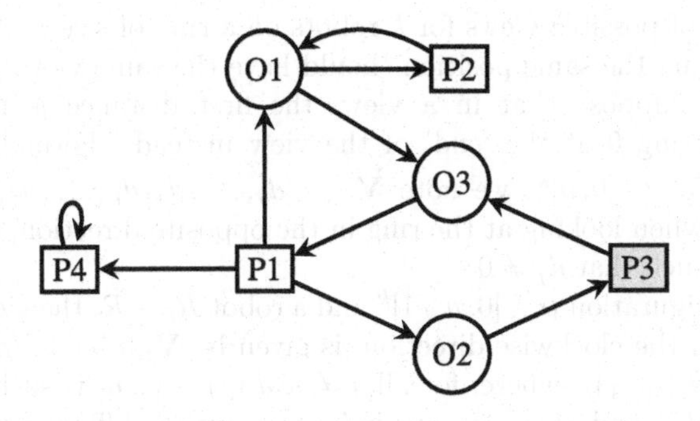


**Fig. 3.** A two-player game. In this figure protagonist vertices are represented by rectangles and antagonist vertices by circles. The winning condition is $Reach(\{P3\})$. Any path in the graph is a play. From P2 the protagonist has no winning strategy. From P1 a (memoryless) winning strategy is to go to O2. Winning positions are $\{P1, P3, O2, O3\}$.

**Preliminaries.** For $a, b \in \mathbb{Z}$ such that $a \leq b$, we denote by $[a, b]$ the set $\{c \in \mathbb{Z} \mid a \leq c \leq b\}$. For $a \in \mathbb{Z}$ and $b \in \mathbb{N}$, $a \odot b$ denotes the natural $d \in [0, (b-1)]$ such that there exists $j \in \mathbb{Z}$ and $a = b.j + d$ (for instance $-1 \odot 3 = 2$).

We recall the definition of Existential Presburger (EP) formulae. Let $Y$ be a countable set of variables. First we define the grammar for terms $\mathtt{t} ::= x \mid \mathtt{t} + \mathtt{t} \mid a \cdot \mathtt{t} \mid \mathtt{t} \mod a$, where $a \in \mathbb{N}$ and $x \in Y$ and then the grammar for formulae is given by $\phi ::= \mathtt{t} \bowtie b \mid \phi \wedge \phi \mid \phi \vee \phi \mid \exists x.\phi$ where $\bowtie \in \{=, \leq, \geq, <, >\}$, $x \in Y$ and $b \in \mathbb{N}$. We sometimes write a formula $\phi$ as $\phi(x_1, \ldots, x_k)$ to underline that $x_1, \ldots, x_k$ are the free variables of $\phi$. The set of Quantifier Free Presburger (QFP) formulae is obtained by the same grammar deleting the elements $\exists x.\phi$. Note that when dealing with QFP formulae, we allow as well negations of formulae.

We say that a vector $V = \langle d_1, \ldots, d_k \rangle$ satisfies an EP formula $\phi(x_1, \ldots, x_k)$, denoted by $V \models \phi$, if the formula obtained by replacing each $x_i$ by $d_i$ holds. Given a formula $\phi$ with free variables $x_1, \ldots, x_k$, we write $\phi(d_1, \ldots, d_k)$ the formula where each $x_i$ is replaced by $d_i$. We let $[\![\phi(x_1, \ldots, x_k)]\!] = \{\langle d_1, \ldots, d_k \rangle \in \mathbb{N}^k \mid \phi(d_1, \ldots, d_k) \models \phi\}$ be the set of models of the formula.

**Configurations and Robot Views.** We consider a fixed number $k > 0$ of robots and, except when stated otherwise, we assume the identities of the robots are $\mathcal{R} = \{R_1, \ldots, R_k\}$. We sometimes identify $\mathcal{R}$ with the set of indices $\{1, \ldots, k\}$. On a ring of size $n \geq k$, a *(k,n)-configuration* of the robots (or simply a configuration if $n$ and $k$ are clear from the context) is given by a vector $\mathbf{p} \in [0, n-1]^k$ associating to each robot $R_i$ its position $\mathbf{p}(i)$ on the ring. We assume w.l.o.g. that positions are numbered in the clockwise direction.

A *view* of a robot on this configuration gives the distances between the robots, starting from its neighbour, i.e. the robot positioned on the next occupied node (a distance equals to 0 meaning that two robots are on the same node). A *view* $\mathbf{V} = \langle d_1, \ldots, d_k \rangle \in [0, n]^k$ is a $k$-tuple such that $\sum_{i=1}^{k} d_i = n$ and $d_1 \neq 0$. We let

$\mathcal{V}_{n,k}$ be the set of possible views for $k$ robots on a ring of size $n$. Notice that all the robots sharing the same position should have the same view.

We as well suppose that in a view, the first distance is not 0 (this is possible by putting 0 at the 'end' of the view instead). Formally, for a view $\mathbf{V} = \langle d_1, \ldots, d_k \rangle \in [0, n]^k$, we note $\overleftarrow{\mathbf{V}} = \langle d_j, \ldots, d_1, d_k, \ldots, d_{j+1} \rangle$ the corresponding view when looking at the ring in the opposite direction, where $j$ is the greatest index such that $d_j \neq 0$.

Given a configuration $\mathbf{p} \in [0, n-1]^k$ and a robot $R_i \in \mathcal{R}$, the view of robot $R_i$ when looking in the clockwise direction, is given by $\mathbf{V_p}[i \rightarrow] = \langle d_i(i_1), d_i(i_2) - d_i(i_1), \ldots, n - d_i(i_{k-1}) \rangle$, where, for all $j \neq i$, $d_i(j) \in [1, n]$ is such that $(\mathbf{p}(i) + d_i(j)) \odot n = \mathbf{p}(j)$ and $i_1, \ldots, i_k$ are indexes pairwise different such that $0 < d_i(i_1) \leq d_i(i_2) \leq \cdots \leq d_i(i_{k-1})$. When robot $R_i$ looks in the opposite direction, its view according to the configuration $\mathbf{p}$ is $\mathbf{V_p}[\leftarrow i] = \overleftarrow{\mathbf{V_p}[i \rightarrow]}$.

**Protocols.** In our context, a protocol for networks of $k$ robots is given by a QFP formula satisfying some specific constraints.

**Definition 1 (Protocol).** *A protocol is a QFP formula $\phi(x_1, \ldots, x_k)$ such that for all views $\mathbf{V}$ the following holds: if $\mathbf{V} \models \phi$ and $\mathbf{V} \neq \overleftarrow{\mathbf{V}}$ then $\overleftarrow{\mathbf{V}} \not\models \phi$*

A robot uses the protocol to know in which direction it should move according to the following rules. As we have already stressed, all the robots that share the same position have the same view of the ring. Given a configuration $\mathbf{p}$ and a robot $R_i \in \mathcal{R}$, if $\mathbf{V_p}[i \rightarrow] \models \phi$, then the robot $R_i$ moves in the clockwise direction, if $\mathbf{V_p}[\leftarrow i] \models \phi$ then it moves in the opposite direction, if none of $\mathbf{V_p}[i \rightarrow]$ and $\mathbf{V_p}[\leftarrow i]$ satisfies $\phi$ then the robot should not move. The conditions expressed in Definition 1 imposes hence a direction when $\mathbf{V_p}[i \rightarrow] \neq \mathbf{V_p}[\leftarrow i]$. In case $\mathbf{V_p}[i \rightarrow] = \mathbf{V_p}[\leftarrow i]$, the robot is disoriented and it can hence move in one direction or the other. Note that such a semantics enforces that the behaviour of a robot is not influenced by its direction. In fact consider two symmetrical configurations $\mathbf{p}$ and $\mathbf{p}'$ such that $\mathbf{V_p}[i \rightarrow] = \overleftarrow{\mathbf{V_{p'}}[i \rightarrow]}$ for each robot $R_i$. If $\mathbf{V_p}[i \rightarrow] \models \phi$ (resp. $\mathbf{V_p}[\leftarrow i] \models \phi$), then necessarily $\mathbf{V_{p'}}[\leftarrow i] \models \phi$ (resp. $\mathbf{V_{p'}}[i \rightarrow] \models \phi$), and the robot in $\mathbf{p}'$ moves in the opposite direction than in $\mathbf{p}$ (and the symmetry of the two configurations is maintained).

We now formalise the way movement is decided. Given a protocol $\phi$ and a view $\mathbf{V}$, the moves of any robot whose clockwise direction view is $\mathbf{V}$ are given by:

$$move(\phi, V) = \begin{cases} \{+1\} & \text{if } \mathbf{V} \models \phi \text{ and } \mathbf{V} \neq \overleftarrow{\mathbf{V}} \\ \{-1\} & \text{if } \overleftarrow{\mathbf{V}} \models \phi \text{ and } \mathbf{V} \neq \overleftarrow{\mathbf{V}} \\ \{-1, +1\} & \text{if } \mathbf{V} \models \phi \text{ and } \mathbf{V} = \overleftarrow{\mathbf{V}} \\ \{0\} & \text{otherwise} \end{cases}$$

Here $+1$ (resp. $-1$) stands for a movement of the robot in the clockwise (resp. anticlockwise) direction.

**Different Possible Semantics.** We now describe different transition relations between configurations. Robots have a two-phase behaviour: (1) look at the ring, and (2) according to their view, compute and perform a movement. In this context, we consider three different modes. In the *semi-synchronous mode*, in one step, some of the robots look at the ring and move. In the *synchronous mode*, in one step, all the robots look at the ring and move. In the *asynchronous mode*, in one step a single robot can either choose to look at the ring, if the last thing it did was a movement, or to move, if the last thing it did was to look at the ring. As a consequence, its movement decision is a consequence of the view of the ring it has in its memory. In the remainder of the paper, we fix a protocol $\phi$ and we consider a set $\mathcal{R}$ of $k$ robots.

*Semi-synchronous Mode.* We begin by providing the semantics in the semi-synchronous case.

For this matter we define the transition relation $\hookrightarrow_\phi \subseteq [0, n-1]^k \times [0, n-1]^k$ (simply noted $\hookrightarrow$ when $\phi$ is clear from the context) between configurations. We have $\mathbf{p} \hookrightarrow \mathbf{p}'$ if there exists a subset $I \subseteq \mathcal{R}$ of robots such that, for all $i \in I$, $\mathbf{p}'(i) = (\mathbf{p}(i) + m) \odot n$, where $m \in \mathit{move}(\phi, \mathbf{V_p}[i \rightarrow])$, and for all $i \in \mathcal{R} \setminus I$, $\mathbf{p}'(i) = \mathbf{p}(i)$.

*Synchronous Mode.* The transition relation $\Rightarrow_\phi \subseteq [0, n-1]^k \times [0, n-1]^k$ (simply noted $\Rightarrow$ when $\phi$ is clear from the context) describing synchronous movements is very similar to the semi-synchronous case, except that all the robots have to move. Then $\mathbf{p} \Rightarrow \mathbf{p}'$ if $\mathbf{p}'(i) = (\mathbf{p}(i) + m) \odot n$ with $m \in \mathit{move}(\phi, \mathbf{V_p}[i \rightarrow])$ for all $i \in \mathcal{R}$.

*Asynchronous Mode.* The definition of transition relation for the asynchronous mode is a bit more involved, for two reasons: first, the move of each robot does not depend on the current configuration, but on the last view of the robot.

As a consequence, an *asynchronous configuration* is a tuple $(\mathbf{p}, \mathbf{s}, \mathcal{V})$ where $\mathbf{p} \in [0, n-1]^k$ gives the current configuration, $\mathbf{s} \in \{\mathbf{L}, \mathbf{M}\}^k$ gives, for each robot, its internal state ($\mathbf{L}$ stands for ready to look and $\mathbf{M}$ stands for compute and move) and $\mathcal{V} \in \mathcal{V}_{n,k}^k$ stores, for each robot, the view (in the clockwise direction) it had the last time it looked at the ring.

The transition relation for asynchronous mode is hence defined by a binary relation $\rightsquigarrow_\phi$ (or simply $\rightsquigarrow$) working on $[0, n-1]^k \times \{\mathbf{L}, \mathbf{M}\}^k \times \mathcal{V}_{n,k}^k$ and defined as follows: $\langle \mathbf{p}, \mathbf{s}, \mathcal{V} \rangle \rightsquigarrow \langle \mathbf{p}', \mathbf{s}', \mathcal{V}' \rangle$ if and only if there exist $R_i \in \mathcal{R}$ such that the following conditions are satisfied:

- for all $R_j \in \mathcal{R}$ such that $j \neq i$, $\mathbf{p}'(j) = \mathbf{p}(j)$, $\mathbf{s}'(j) = \mathbf{s}(j)$ and $\mathcal{V}'(j) = \mathcal{V}(j)$,
- if $\mathbf{s}(i) = \mathbf{L}$ then $\mathbf{s}'(i) = \mathbf{M}$, $\mathcal{V}'(i) = \mathbf{V_p}[i \rightarrow]$ and $\mathbf{p}'(i) = \mathbf{p}(i)$, i.e. if the robot that has been scheduled was about to look, then the configuration of the robots won't change, and this robot updates its view of the ring according to the current configuration and change its internal state,
- if $\mathbf{s}(i) = \mathbf{M}$ then $\mathbf{s}'(i) = \mathbf{L}$, $\mathcal{V}'(i) = \mathcal{V}(i)$ and $\mathbf{p}'(i) = (\mathbf{p}(i) + m) \odot n$, with $m \in \mathit{move}(\phi, \mathcal{V}(i))$, i.e. if the robot was about to move, then it changes its internal state and moves according to the protocol, and *its last view of the ring.*

**Runs.** A semi-synchronous (resp. synchronous) $\phi$-run (or a run according to a protocol $\phi$) is a (finite or infinite) sequence of configurations $\rho = \mathbf{p}_0\mathbf{p}_1\ldots$ where, for all $0 \leq i < |\rho|$, $\mathbf{p}_i \hookrightarrow_\phi \mathbf{p}_{i+1}$ (resp. $\mathbf{p}_i \Rightarrow_\phi \mathbf{p}_{i+1}$). Moreover, if $\rho = \mathbf{p}_0 \cdots \mathbf{p}_n$ is finite, then there is no $\mathbf{p}$ such that $\mathbf{p}_n \hookrightarrow_\phi \mathbf{p}$ (respectively $\mathbf{p}_n \Rightarrow_\phi \mathbf{p}$).

An asynchronous $\phi$-run is a (finite or infinite) sequence of asynchronous configurations $\rho = \langle \mathbf{p}_0, \mathbf{s}_0, \mathcal{V}_0\rangle\langle \mathbf{p}_1, \mathbf{s}_1, \mathcal{V}_1\rangle \cdots$ where, for all $0 \leq i < |\rho|$, $\langle \mathbf{p}_i, \mathbf{s}_i, \mathcal{V}_i\rangle \rightsquigarrow_\phi \langle \mathbf{p}_{i+1}, \mathbf{s}_{i+1}, \mathcal{V}_{i+1}\rangle$ and such that $\mathbf{s}_0(i) = \mathbf{L}$ for all $i \in [1, k]$. Observe that the value of $\mathcal{V}_0$ has no influence on the actual asynchronous run, since any robot starts its computation by a look, hence changing the value of $\mathcal{V}_0$.

We let $\mathbf{Post}_{\mathrm{ss}}(\phi, \mathbf{p}) = \{\mathbf{p}' \mid \mathbf{p} \hookrightarrow_\phi \mathbf{p}'\}$, $\mathbf{Post}_{\mathrm{s}}(\phi, \mathbf{p}) = \{\mathbf{p}' \mid \mathbf{p} \Rightarrow_\phi \mathbf{p}'\}$ and $\mathbf{Post}_{\mathrm{as}}(\phi, \mathbf{p}) = \{\mathbf{p}' \mid \text{ there exist } \mathcal{V}, \mathbf{s}', \mathcal{V}' \text{ s.t. } \langle \mathbf{p}, \mathbf{s}_0, \mathcal{V}\rangle \rightsquigarrow_\phi \langle \mathbf{p}', \mathbf{s}', \mathcal{V}'\rangle\}$, with $\mathbf{s}_0(i) = \mathbf{L}$ for all $i \in [1, k]$. Note that in the asynchronous case we impose all the robots to be ready to look.

We respectively write $\hookrightarrow_\phi^*$, $\Rightarrow_\phi^*$ and $\rightsquigarrow_\phi^*$ for the reflexive and transitive closure of the relations $\hookrightarrow_\phi$, $\Rightarrow_\phi$ and $\rightsquigarrow_\phi$ and we define $\mathbf{Post}^*_{\mathrm{ss}}(\phi, \mathbf{p})$, $\mathbf{Post}^*_{\mathrm{s}}(\phi, \mathbf{p})$ and $\mathbf{Post}^*_{\mathrm{as}}(\phi, \mathbf{p})$ by replacing in the definition $\mathbf{Post}_{\mathrm{ss}}(\phi, \mathbf{p})$,$\mathbf{Post}_{\mathrm{s}}(\phi, \mathbf{p})$ and $\mathbf{Post}_{\mathrm{as}}(\phi, \mathbf{p})$ the relations $\hookrightarrow_\phi$, $\Rightarrow_\phi$ and $\rightsquigarrow_\phi$ by their reflexive and transitive closure accordingly.

**Definition 2.** *A protocol $\phi$ is said to be* uniquely-sequentializable *if, for all configuration $\mathbf{p}$, there is at most one robot $R_i \in \mathcal{R}$ such that*

$$move(\phi, \mathbf{V}_{\mathbf{p}}[i \rightarrow]) \neq \{0\}.$$

When $\phi$ is uniquely-sequentializable at any moment at most one robot moves. Consequently, in that specific case, the three semantics are equivalent.

**Problems Under Study.** In this section, we aim at verifying properties on protocols where we assume that the number of robots is fixed (equals to $k > 0$) but the size of the rings is parameterised and satisfies a given property. Note that when executing a protocol the size of the ring never changes. For our problems, we consider a ring property that is given by a QFP formula `Ring`$(y)$, a set of bad configurations given by a QFP formula `Bad`$(x_1, \ldots, x_k)$ and a set of good configurations given by a QFP formula `Goal`$(x_1, \ldots, x_k)$.

We then define two general problems to address the verification of such algorithms: the $\mathsf{SAFE}_m$ problem, and the $\mathsf{REACH}_m$ problem, with $m \in \{\mathrm{ss}, \mathrm{s}, \mathrm{as}\}$.

The $\mathsf{SAFE}_m$ problem is to decide, given a protocol $\phi$ and two formulae `Ring` and `Bad` whether there exists a size $n \in \mathbb{N}$ with $n \in [\![\texttt{Ring}]\!]$, and a $(k, n)$-configuration $\mathbf{p}$ with $\mathbf{p} \notin [\![\texttt{Bad}]\!]$, such that $\mathbf{Post}^*_m(\phi, \mathbf{p}) \cap [\![\texttt{Bad}]\!] \neq \emptyset$.

The $\mathsf{REACH}_m$ problem is to decide given a protocol $\phi$ and two formulae `Ring` and `Goal` whether there exists a size $n \in \mathbb{N}$ with $n \in [\![\texttt{Ring}]\!]$ and a $(k, n)$-configuration $\mathbf{p}$, such that $\mathbf{Post}^*_m(\phi, \mathbf{p}) \cap [\![\texttt{Goal}]\!] = \emptyset$. Note that the two problems are not dual due to the quantifiers.

### 2.4 A Generic Framework for Formal Proofs

Previous approaches only tackle the case of discrete spaces (and even a specific such discrete space, the ring shaped graph), and are thus unsuitable for establishing formal results in the continuous space. This section describes a generic framework that can be used for both the continuous and the discrete space. Also, the framework allows to get results for an arbitrary number of robots.

**Mechanical Formal Proof.** A formal proof approach to obtain mechanically certified protocols/properties faces two main and separate phases in a verified development.

The first one is the *specification* phase, where all objects, definitions, algorithms, statements and expected properties are expressed *without any ambiguity*, in a higher order type theoretic functional environment. The lack of ambiguity is a key feature to enable the early detection of inconsistencies between the problem specification, the algorithmic proposal, and the execution model. Ideally, there should be no need to be an expert with the proof assistant to use a formal framework in this phase.

The second one is the *proof* phase, where properties are shown to hold for the relevant executions. This phase is, of course, more demanding on the expertise side, and one goal when devising a such a formal framework is to provide useful libraries, and proof techniques, that can be reused in other contexts. This allows for more automation for the protocol designer, and useful assets for reusability. Developing a protocol may amount to modifying its code several times, either to fix a newly discovered bug, or to ease the writeup of the proofs. In such a setting, a correction in the algorithm leads to a modification in the algorithm definition, and to a replay of the proofs certification process after adapting the proof scripts written previously. The mechanised verification of the proofs makes this process fast and trustworthy, compared to a purely handcrafted approach.

Note that most of the formal definitions about a problem under study shall be common to all results: on the one hand it ensures that the very same problem is considered in the various settings, on the other hand it brings formal guaranties on the absence of any goal gap between impossibility results, and protocol that are designed.

All of this calls for a strong *genericity* of the formal development. The use of modules, for example, allows one to achieve a reasonable level of parametricity.

**Pactole and the Coq Proof Assistant.** The Pactole[1] framework enabled the use of high-order logic to certify impossibility results, as well as soundness of protocol, for swarms of autonomous mobile robots.

To certify results and to guarantee the soundness of theorems, it uses Coq, a Curry-Howard-based interactive proof assistant enjoying a trustworthy kernel. Coq is based on *type theory*. Its (functional) language *Gallina* is a very expressive

[1] http://pactole.lri.fr.

λ-calculus: the Calculus of Inductive Constructions (CIC) [36]. In this context, datatypes, objects, algorithms, theorems and proofs can be expressed in a unified way, as terms.

λ-abstraction is denoted **fun** `x:T ⇒t`, and the application of `t` to `u` is denoted by the juxtaposition `t u`. A proof development with Coq consists in trying to build, interactively and using tactics, a λ-term the type of which corresponds to the proven theorem (Curry-Howard style).

The kernel of Coq is a *proof checker* that checks the validity of proofs written as CIC-terms. Indeed, in this framework, a term is a *proof* of its type, and checking a proof consists in typing a term. Roughly speaking, the small kernel of Coq simply type-checks λ-terms to ensure soundness.

*A theorem or a lemma can only be saved/defined in the system if it comes with its type-checked proof.*

Very powerful features of Coq include the ability to define *inductive types* to express inductive data types and inductive properties.

*Coinductive* types are also invaluable to express and to reason about infinite data types, and properties on them, like for instance on the infinite behaviour of robot swarms. Streams are a paradigmatic example of coinductive objects.

Coq enjoys a module system with signatures (called **`Module Type`**) that can be implemented by modules. A signature $\Sigma$ may contain declarations for objects that the modules implementing $\Sigma$ will define. The declaration **`Parameter`** `x : A` will, for example, introduce a parameter `x` of type `A`.

Signatures or modules can contain definitions of objects, giving them a value: **`Definition`** `x := a` defines the object `x` with the value `a`. A module `M` that implements a signature $\Sigma$ can define any parameter `x : A` of $\Sigma$ by providing a value of type `A`. Similarly, a signature can state a property that has to be satisfied by its parameters (that is, an **`Axiom`**), and this statement will be properly defined in a module that provides an actual proof for it (that is, by turning the axiom into a **`Lemma`**).

These language, assets, and a well designed formal framework allow the user to characterise properties, context, and protocols without too cumbersome verbosity or intricacy.

All the results hereafter mentioned, and the relevant Coq scripts and development are available from http://pactole.lri.fr for Coq 8.6.

**Formalising the Model.** The description of the system consists of three parts: space and its topology, capabilities of robots and of movement, and, of course, the computation model itself with its synchrony properties. The first two parts shall define the *configuration* of the system (space and intrinsic robot capabilities), the last two (movement capabilities, and computation) shall define the *evolution* process.

**Space and Topology.** The fundamental operations needed on the space the robots move into are usually not hardwired to a certain type of space. For the sake of generality the space (of locations) is a parameter of the formal model.

It is kept abstract and is encapsulated into a module `Location` that provides a core type `Location.t` denoting the positions in the actual space, together with useful functions and operations (a decidable equality, an origin, a distance, etc.).

Pactole adresses in its development mainly two kinds of spaces: Continuous spaces, most notably the Euclidean plane $\mathbb{R}\times\mathbb{R}$, and Discrete spaces, in particular graphs. Depending on the use-case, the location module may be instantiated with $\mathbb{R}$ if one considers the real line [8,38], or $\mathbb{R} \times \mathbb{R}$ with the relevant arithmetics for the real plane [13,39], etc. Regarding graphs, and so as to allow for their comfortable use, Pactole provides a general template that the user can instantiate with the kind of graphs needed [14]. Though rather lightweight and restricted compared to, for instance, the library LOCO for local computation on graphs [25], this template connects naturally with the main signature for the spaces where robots evolve, which provides simple means of specifying discrete spaces and reasoning about them. The library provides in particular a specialised version for rings.

### Robots and Sensor Capabilities

*Robots, Conformations.* A set of robots of size $N = nB + nG$ consists of the disjoint union of two sets of identifiers: a set of Byzantine robots, of size $nB$, and a set of Good robots, of size $nG$. Both are isomorphic to segments of $\mathbb{N}$.

The *conformation* of a robot describes its state. It may simply consist in its location in space; under certain assumptions it may also include some internal state (as we shall see for ASYNC).

```
Record Info : Type := (* some state description *) .
Record RobotConf := { loc :> Location ; robot_info: Info } .
```

*Configuration and Spectrum.* The *configuration* represents the conformation of any robot designated by its identifier. In other words, it is simply a *function* that maps robots' identifiers to their respective conformation.

```
Definition configuration := identifier → RobotConf.
```

Note that as the whole formal model in COQ is functional, such a representation shall prove comfortable to work with.

The conformation of robots may include some internal states, and thus may contain information that should not be observed by other robots. Indeed, in some cases, assumptions may require that local sensors cannot tell robots apart (anonymity), or detect wether they are correct or Byzantine, or are endorsed with the detection of multiplicity (that is the knowledge of the exact number of robots inhabiting a location in space), etc.

These restrictions of the model can be ensured by the notion of *spectrum*, described below, which characterises what a robot's sensors can perceive of the global system. The forbidden information is pruned from the configuration, using a function `from_config` that returns a spectrum, which is the only input of a robot's function computing its destination.

Spectra form an arbitrary type that is part of the description of the model, and contributes to its genericity. To this goal, Pactole defines a module type `Spect`, parameterised in particular by the type of location, the number of robots, etc. This module type gathers the datatype `Spect.t` of the spectrum, and required or usefull properties and functions: the definition of (a decidable) equality on spectra, a conversion function (`Spect.from_config`) turning a configuration into a spectrum, a formula (`Spect.is_ok`) expressing the relation between a configuration and its spectrum, and the fact that the conversion function satisfies it:

```
∀ config, Spect.is_ok (Spect.from_config config) config
```

More precisely, `Spect.is_ok` ensures the adequacy between the configuration and the information in the spectrum. It may for example ascertain that the locations in a spectrum correspond to actual locations of robots in the relevant configuration.

```
Module Type Spect(Location : DecidableType)(N : Size). (* ... *)
  (* Spectra are abstract decidable types. *)
  Parameter t : Type.
  (* They are equipped with an equality relation *)
  Parameter eq : t → t → Prop.
  (* which is an equivalence relation *)
  Parameter eq_equiv : Equivalence eq.
  (* and which is decidable. *)
  Parameter eq_dec : ∀ x y : t, {eq x y} + {¬ eq x y}.

  (* Turning a configuration into a spectrum
     i.e. erasing information. *)
  Parameter from_config : Config.t → t.
  (* Equal configurations give equal spectra. *)
  Declare Instance from_config_compat :
    Proper (Config.eq ⇒ eq) from_config.

  (* An abstract predicate validating spectra
     for a configuration.  *)
  Parameter is_ok : t → Config.t → Prop.
  (* from_config gives a correct spectrum. *)
  Parameter from_config_spec :
    ∀ config, is_ok (from_config config) config.
End Spectrum.
```

When needed, those abstract properties may be instantiated in accordance to the requirements and assumptions.

*Example 1.* When robots are anonymous, can see the whole space, and can detect multiplicity, a convenient datatype for a spectrum may be a *multiset* of the inhabited locations, the multiplicities of an elements, that is of a place, being the number of robots at this very place.

If, however, the anonymous robots are *not* endorsed with multiplicity detection, then one may choose instead for spectrum the *set* of inhabited locations.

In the case where the considered space is a discrete graph, *without shared vertex naming or origin*, a relevant spectrum may be a multiset (or set depending of multiplicity detection) of locations (vertices), with a vertex marked as being the current location of the robot perceiving that spectrum.

*Robogram.* The protocol together with relevant properties define a *robogram.* It consists of:

- an algorithm pgm that represents the protocol itself, taking a spectrum as its input and returning a destination location, and
- properties stating relevant constraints, for example compatibility statements, range and validity constraints for the computed destination, etc.

*Protocol Function.* Robograms may be naturally defined in a *completely abstract manner*, without any concrete code, in the Coq model.

```
Record robogram := {
  pgm :> Spect.t → Location.t;
  ...(logical properties and constraints)...}.
```

This and the higher order calculus of Coq allow for *quantification* over robograms, and thus for proofs of *impossibility.*

Robograms can be instantiated with a concrete version of the protocol. It suffices to provide the actual code for pgm, and of course prove the relevant properties. For example, Cohen and Peleg [34] propose a protocol asking anonymous and oblivious robots enjoying multiplicity detection to gather towards their barycentre. A suitable spectrum in that context is a multiset of inhabited locations, and the destination is the weighted barycentre of elements in that multiset. This translates simply[2] in Gallina into:

```
Definition CP_protocol_pgm (s : Spect.t) : R2.t :=
  wbarycenter (Spect.M.elements s).
```

*Properties and Constraints.* The second part of the robogram record consists of properties on the pgm function. The basic one is the compatibility statement ensuring that destination locations resulting from pgm when its provided spectra are equivalent (for the equivalence relation on spectra Spect.eq) are the same (for the equality on locations Location.eq).

```
pgm_compat : Proper (Spect.eq ⇒ Location.eq) pgm
```

[2] The translation of integral multiplicities into real weights is omitted in this example.

In addition to the compatibility property, and depending on the context, most notably on space, range and validity constraints for the computed destination may be expressed here. When dealing, for example, with graphs in a discrete setting, one has to ensure that the computed destination vertex `lpost` is actually a neighbour of the current location. That is: one can find an edge `e` linking the current location (pointed in the used spectrum for graphs) to `lpost`, which translates into (`Eeq` being the equality on edge, lifted to options with `opt_eq`):

```
pgm_range :
  ∀ (spect: Spect.t), ∃ lpost e,
      (pgm spect = lpost)
      ∧ (Graph.Eeq
            (Graph.find_edge (Spect.get_current spect) lpost)
            (Some e))
```

**Formalising the Model, Movements and Synchronicity.** Beside the capabilities of robots, there are many variants in Suzuki & Yamashita's model, and thus many results all of which having to be candidate to the certification process.

For instance, movements may be assumed to be *rigid*, robots always reaching the destination they computed. But one may instead suppose that robots' movements are *flexible*, meaning the demon can stop a robot before it has completed its journey to the computed destination.

Different assumptions may also address the synchronicity of executions, allowing the demon to be either FSYNC, SSYNC, or even ASYNC, and enjoying various properties like, for example, fairness.

The formal model must then allow for the expression of such diversity. It is based on a function, named `round`, at the core of the development, and which describes the evolution of the system according to the computation model.

*Core of the Model: Demons, and* `rounds`*.* A *demonic action* characterises the selection of robots that will undergo model dependant changes, and associates them to some choices from the (Maxwell's) demon, which can be a new local frame of reference, a movement ratio, etc. Demonic actions consist of:

- A function `relocate_byz` of relocation for Byzantine robots, i.e. a function mapping Byzantine identifiers to their new conformation;
- A function `step` mapping identifiers to relevant demonic choices (model dependant);
- Properties of compatibility and coherence of the aforementioned choices.

*SSYNC Rigid Demonic Action.* In an SSYNC setting where movements are rigid, robots are oblivious, and no global frame of reference is shared, `step` will simply associate to each robot an option: `None` if it is not activated, or `Some f` if it is activated, in which case `f` is a function computing the change of frame of reference from the global demonic one into the robot's new local one.[3] Compatibility properties will ensure that equivalence in names leads to the same choices, that the zoom factor of the new referential is non-zero, and that the location of the robot is the origin of its new frame of reference.

```
Record demonic_action := {
  relocate_byz : Names.B → Config.RobotConf;
  step : Names.ident → option (Location.t → Sim.t);
  step_compat :
    Proper (eq ⇛ opt_eq (Location.eq ⇛ Sim.eq)) step;
  step_zoom :
    ∀ id sim c,
      step id = Some sim → (sim c).(Sim.zoom) ≠ 0ℝ;
  step_center :
    ∀ id sim c,
      step id = Some sim → Location.eq (sim c).(Sim.center) c}.
```

Note that the result of step is a function returning a transformation (similarity) from a given "point of view" location.

*SSYNC Flexible Demonic Action.* A flexible model allows the demon to interrupt robots before they actually reach their target. It is nevertheless assumed that if the goal is unmet, a minimum (absolute) distance $\delta$ is travelled. Of course, the value of $\delta$ is *unknown* to the robots, as they are just aware that some $\delta$ exists.

If movements are flexible, information about the distance travelled has to be provided somehow, the remainder of the model staying the same. A ratio is hence also given by step, denoting the part of the complete journey to the computed destination that will be actually travelled. This ratio is integrated to the compatibility property. An additional constraint ensures that it lies between 0 and 1.

```
Record demonic_action := {
  relocate_byz : ...
  step :
    Names.ident → option ((Location.t → Sim.t) (* frame change *)
                          * R);                (* travel ratio *)
  step_flexibility :
    ∀ id sim, step id = Some sim → (0 ≤ snd sim ≤ 1)ℝ
  step_compat : ...
  step_zoom :  ...
  step_center : ...}.
```

*Demons, Rounds, and Executions.* A *demon* (or scheduler) is simply an infinite sequence, that is a stream, of demonic actions.

One obtains successive configurations by running the robogram according the current demonic action and configuration. This is done by a function round that computes new conformations in a configuration, *for each robot identifier* r, according to a demonic action da. For the SSYNC flexible case:

[3] Since the robot is oblivious, it has no mean to remember its past frame of reference.

- If the robot is not activated (the demonic action step returns None), its associated conformation remains unchanged,
- If the demonic action step returns a change of frame and a travel ratio,
  - If r is Byzantine, its conformation is given directly by relocate_byz,
  - If r is a good robot, the robogram applies:
    1. The provided local frame of reference is used to convert the configuration according to the relevant local point of view,
    2. The resulting (local) configuration is transformed into a spectrum using from_config,
    3. *The robogram receives the obtained spectrum as a parameter, and returns a target location (in the local frame),*
    4. The chosen destination is obtained, from the target location, using the travel ratio provided by the action,
    5. Depending on the flexibility parameter $\delta$, the actual destination is either the chosen destination or the target one, it is converted from the local frame to the global one, and the conformation is updated accordingly.

Taking the robot identifier as a parameter, one obtains this way a function that maps conformations to identifiers, that is a new configuration.

The fundamental function round for the SSYNC flexible model is thus:

```
Definition round (δ: R) (r: robogram) (da: demonic_action)
  (config: Config.t) : Config.t :=
  (** for a given robot, we compute the new configuration *)
  fun id ⇒
    let c := config id in                         (** c: id's conformation
                                                    as seen by demon *)
    match da.(step) id with          (** Is the robot activated? *)
      | None ⇒ c                    (** not activated, do nothing *)
      | Some (sim, mv_ratio) ⇒ (** activated with                *)
                                 (** similarity [sim (conf g)] *)
                                 (** and move ratio [mv_ratio] *)
        match id with
        | Byz b ⇒ da.(relocate_byz) b
          (* Byzantine robots are relocated by the demon *)
        | Good g ⇒
              (* configuration expressed in the frame of g *)
          let frame_change := sim (config (Good g)) in
          let local_config := Config.map
              (Config.app frame_change) config in
                                        (* apply r on spectrum *)
          let local_target :=
              r (Spect.from_config local_config) in
              (* the demon chooses a point on the line to the
                     target by [mv_ratio] *)
          let chosen_target :=
              Location.mul mv_ratio local_target in
                (* and back to the demon's frame of ref... *)
```

```
        {| Config.loc :=
              frame_change⁻¹
                (if δ ≤ (Location.dist (frame_change⁻¹
                    chosen_target) c.(Config.loc))
                 then chosen_target
                 else local_target);
            Config.robot_info := c.(Config.robot_info) |}
                                            (* irrelevant here *)
      end
  end.
```

The infinite sequence of consecutive configurations following a demon's demonic actions defines an *execution*. To define a full execution, the function `execute rbg d config` iterates `round`, starting from configuration `config`, using robogram `rbg` and demon `d`.

Note that a flexible model is more general than a rigid one. However, so as to keep development as simple as possible, Pactole provides theorems stating the equivalence between the rigid model and the flexible model when the ratio is always 1. That is, in that context: any rigid demon can be simulated with a flexible demon, and any flexible demon that does not interrupt movements can be simulated with a rigid demon.

*The Case of ASYNC.* Balabonski *et al.* [10,11] proposed an extension of Pactole to asynchronous execution schemes. They use it to establish and prove formally an equivalence between a model of discrete graphs where robots can only be at vertices and the more realistic model of continuous moves along the graph's edge but with discrete observation.

For the sake of simplicity, the remainder of this section is limited to rigid movements.

In ASYNC, the formal model must be enriched to reflect the lack of synchronisation and of uniformity of robots' actions [10,11]. In this context, the conformations of robots (`RobotConf`) gather the current location, and information about, at least, movement. We shall consider here that this information consists of source and target locations. The field `robot_info` in a conformation may thus be considered as a record of two locations: `source`, and `target`. This allows for some robots to move while others are looking or computing.

A robot is considered to be *moving* whenever its current and target locations differ. It becomes *idle* when it reaches its target location.

When a robot is idle, it can start a new cycle with a simple Look/Compute action performing the usual Look and Compute phases.[4]

A change of state of the robot is the result of a demonic action.

*ASYNC Demonic Actions.* A demonic action can request a moving robot to pursue its movement towards its designated target, or an idle robot to initiate a

[4] As the computation is based on a snapshot taken during the Look phase only, any event taking place after Look cannot have any impact on its result, and one may thus merge the two phases Look and Compute into a Look/Compute one.

new move, that is a Look/Compute phase. These action cases replace the option in the SSYNC model: either the distance travelled along an ongoing *Move*, or a frame of reference for a robot for a *Look/Compute*.

```
Inductive action {A} :=
  | Move (dist: A)                                    (* moving distance *)
  | LookCompute (Location.t → Sim.t).          (* change ref frame *)
```

The definition of a demonic action is hence as follows:

- relocate_byz: as for the SSYNC case, and when relevant, demonic actions also relocate Byzantine robots in an arbitrary way,
- step now returns the choice Move or LookCompute, that is, an action,
- step_compat is similar to the SSYNC case, the equivalence for options being replaced by the equivalence for actions,
- a new constraint step_LookCompute is added to ensure that only idle robots (that is, robots that are at their target location) may receive an order to look and compute.

*ASYNC Round.* In the case of ASYNC, rounds are not defined by Look-Compute-Move cycles but by changes of conformations.

Note that Byzantine robots are relocated directly on LookCompute actions, and ignore Move ones. Depending on what the demonic action step returns for a robot r:

- Move action: r carries further its ongoing move, its current location is updated to the location it reached during this move (the way this reached location is computed may depend on the underlying space).
  - If r reaches its target location, it is now idle,
  - If r does not reach its target location, it stays moving,
  - If r was already at its target location, it stays idle.
- LookCompute action: a new target location is defined as follows:
  1. The local frame of reference provided by the action is used to compute the configuration according to the relevant local point of view,
  2. The resulting local configuration is transformed into a spectrum using from_config,
  3. *The obtained spectrum is passed as a parameter to the robogram, which returns the target location (in the local frame),*
  4. The target location is converted from the local frame to the global one.

  The robot's conformation is updated with the obtained location as new target, and with the current location as new source.
  - If r's current and target locations differ, the robot switch from idle to moving,
  - If r's current and target location are equal, the robot stays idle.

Executions are defined as usual. Note that a step in an ASYNC execution does not always imply a change in the multiset of inhabited locations, as some robots may undergo a change of state only.

*Reasoning About Demons.* Demons are just streams, and being thus defined as a coinductive construct they allow for a relatively easy specification of their properties. Pactole defines the usual temporal operators $\Diamond$, $\circ$, and $\Box$ to help expressing temporal properties about executions, written respectively `Stream.eventually`, `Stream.next`, and `Stream.forever` in our formalisation. However, the logic of Coq being much more expressive, one can define new temporal operators or new properties directly on an execution.

When for example the Look-Compute-Move cycle is atomic, FSYNC demons have all robots activated in each demonic action, SSYNC demons only have a subset of robots activated in each action, etc. Additional properties like *fairness*-related constraints (fair, unfair, $k$-fair, etc.) are expressed as logical propositions on demons.

*Fairness* is for instance defined as follows in the libraries: a demon `d` is fair to robot `r` if either

- `r` is activated by the current demonic action (that is the head `hd` of the demon), or
- `r` is not activated but will be for a demonic action in the remainder (that is the tail `tl`) of the demon.[5]

The translation into Gallina is almost straightforward.

```
Inductive LocallyFairForOne r (d : demon) : Prop :=
  | NowFair : step (Stream.hd d) r ≠ None
              → LocallyFairForOne r d
  | LaterFair : step (Stream.hd d) r = None
              → LocallyFairForOne r (Stream.tl d)
              → LocallyFairForOne r d.
```

A demon is thus fair if at any point it is fair for every robot.

```
Definition Fair : demon → Prop :=
  Stream.forever (fun d ⇒ ∀ r, LocallyFairForOne r d).
```

This high level modelling allows for the encoding of many types of demons, and for their theoretical formal study. The relevant library includes inclusions and equivalence theorems about demons, for example that a fully-synchronous demon is semi-synchronous, or that a $k$-fair demon is also $(k+1)$-fair, etc.

## 3 A Summary of Obtained Results

### 3.1 Early Result Obtained by Ad Hoc Tools

In the context of mobile robots operating in discrete space, two early attempts, by Devismes *et al.* [43] and by Bonnet *et al.* [18,19], investigate the possibility of automated verification of mobile robots protocols. The first paper uses

[5] Note that this demonic action is accessible, *i.e.* finitely reachable, since the property is defined as *inductive*.

LUSTRE [54] to describe and verify the problem of exploration with stop of a $3 \times 3$ grid by 3 robots in the SSYNC model, and to show by exhaustive searching that no such protocol can exist. The second paper considers the perpetual exclusive exploration by $k$ robots of $n$-sized rings, and generates mechanically all *unambiguous* protocols for $k$ and $n$ in the SSYNC model (that is, all protocols that do *not* have symmetrical configurations). Those two works are restricted to the simpler SSYNC model rather than the more general and more complex ASYNC model. Second, they are either specific to a hard-coded topology (*e.g.*, a $3 \times 3$ grid [43]) that prevents easy reuse in more generic situations, or make additional assumptions about configurations and protocols to be verified (*e.g.* unambiguous protocols [18,19]) that prevent combinatorial explosion but forbid reuse for proof-challenging protocols, which would most benefit from automatic verification.

### 3.2 Generic Results Obtained by Model Checking and Synthesis

In the discrete setting of ring-shaped networks, model-checking proved useful to find bugs in existing literature [15,44,45] and assess formally published algorithms [5,15,43] for some particular instances.

Automatic program synthesis (for the problem of perpetual exclusive exploration in a ring-shaped discrete space) is due to Bonnet *et al.* [19], and can be used to obtain automatically algorithms that are "correct-by-design". The approach was refined by Millet *et al.* [66] for the problem of gathering in a discrete ring network. As all aforementioned approaches are designed for a bounded setting where both the number of locations and the number of robots are known, they cannot permit to establish results that are valid for any number of locations.

Recently, Aminof *et al.* [5] presented a general framework for verifying properties about mobile robots evolving on graphs, where the graphs are a parameter of the problem. While the model of Suzuki and Yamashita could be encoded in their framework, their undecidability proof relies on persistent memory used by the robots, hence is not applicable to the case of oblivious robots we consider here. Also, they obtain decidability in a sub-case that is not relevant for robot protocols like those we consider. Moreover, their decision procedure relies on MSO satisfiability, which does not enjoy good complexity properties and cannot be implemented efficiently for the time being.

Sangnier et al. [72] tackled the more general problem of verifying protocols for swarms of robots for any number of locations. Their formal definition of the problem (presented in Sect. 2.3) permits to describe the protocol as a quantifier free Presburger formula. This logic, weak enough to be decidable, is however powerful enough to express existing algorithms in the literature. Objectives of the robots are also described by Presburger formulae and they consider two problems: when the objective of the robots is a safety objective – robots have to avoid the configurations described by the formula (SAFE), and when it is a reachability objective (REACH). It turns out that REACH is undecidable in any semantics, but SAFE is decidable in FSYNC and SSYNC. Now, when the protocol is *uniquely-sequentializable*, safety properties become decidable even in the asynchronous case.

### 3.3 Generic Results Obtained by Proof Assistants

Putting things into practice amounts to instantiating the model and the environment, to providing a formal definition of the problem, and finally to *stating* the main result/theorem. This is it for the specification job, but of course one has to prove formally the theorem in order to *define* it into the framework.

This way of proceeding is detailed in a few examples below.

**Convergence.** Auger *et al.* [8] proposed the first formal certification that Byzantine-resilient convergence on the real line $\mathbb{R}$ of oblivious robots that enjoy strong global multiplicity was impossible to achieve in a rigid SSYNC setting, when the *ratio* of Byzantine robots over their total number was above a certain bound. Several theorems of this kind have been first stated and pen-and-paper proved in [20].

Note that this is a result of impossibility, i.e. a quantification over *all* protocols, and that there is no bound over the number of robots. Further note that the space is unbounded and continuous.

*The Problem.* Given any initial configuration of robots, the problem of *convergence* requires the robots that are correct to approach asymptotically the same, but unknown beforehand, location. That is, in an SSYNC setting, an execution is said to be convergent when for any $\varepsilon > 0$ there exists a number of rounds $N_\varepsilon \in \mathbb{N}$ and a location $l_\varepsilon$ (in the particular context of [20], $l_\varepsilon \in \mathbb{R}$) such that for all $n > N_\varepsilon$, all correct robots at round $n$ are no further than $\varepsilon$ from $l_\varepsilon$.

In other words, Convergence expresses that all correct robots will eventually be imprisoned forever in a disc of centre $c$ and of radius $\varepsilon$...

The translation into the framework is as follows: `imprisoned` expresses that at each point in an execution `exe`, the conformation of any robot locates it within $\varepsilon$ of a centre `c`.

```
Definition imprisoned (c: Location.t) (ε: R) (exe: execution):
  Prop :=
  Stream.forever
    (Stream.instant
      (fun config ⇒ ∀ g, Location.dist c
      (config (Good g)).(Config.loc) ≤ ε))
    exe.
```

The fact that this situation has to happen eventually is described by `attracted`:

```
Definition attracted (c: Location.t) (ε: R) (exe: execution):
  Prop :=
  Stream.eventually (imprisoned c ε) exe.
```

A robogram is a fair solution to the problem of Convergence for any fair demon if, for all positive $\varepsilon$, one can find a point c to which robots are `attracted` in any execution induced by the demon.

```
Definition FairSolConvergence (r : robogram) : Prop :=
  ∀ (config : Config.t) (d : demon), Fair d
  → ∀ (ε : R), 0 < ε
  → ∃ (c : Location.t), attracted c ε (execute r d config).
```

Auger *et al.* formally proved in [8] the following result from Bouzid *et al.* [20] for rigid movements:

[20] **(Theorem 4.3):** *It is impossible to achieve convergence if* $n \leq 2f$ *in the FSYNC uni-dimensional model, where n denotes the number of robots and f denotes the number of Byzantine robots.*

However, for the sake of simplicity, We consider here the case where Byzantine robots are exactly one third of the total number of robots. This situation suffices to illustrate a proof of impossibility, with bounds neither on space, nor on the number of robots, and with Byzantine failures.

*The Model.* The set of locations is $\mathbb{R}$ and we define `Location.t` as the type of Coq's axiomatic reals. The module of locations is completed accordingly, with the relevant arithmetic and properties.

The set of robots consists of a non zero number $nB$ of Byzantine robots, and a number $nG$ of correct (Good) robots that is twice the number of Byzantine ones.

```
Parameter nB: nat.
Hypothesis nB_non_0 : nB ≠ 0ℕ.
Definition nG := (2 * nB)ℕ.
Definition nB := nB.
```

Conformations of robots are limited to their current location.

As robots can detect multiplicity over the whole universe, an adequate spectrum is the multiset of inhabited locations.

Configuration evolve according to the SSYNC rigid version of `round`. Robots are oblivious, the demonic actions thus select activated robots and provide their change-of-frame function. Byzantine robots are simply relocated, that is, associated to a possibly new location in the configuration.

*The Main Theorem.* With the assumptions stated in the previous paragraph, the theorem stating the impossibility to achieve convergence is simply that any robogram is not a fair solution to Convergence.

```
Theorem noConvergence : ∀ r, ¬(FairSolConvergence r).
```

The proof of this theorem alone is about 30 lines of Coq; it just sets the counter-example in action. The whole file that defines Convergence, the main theorem and the counter-example set-up (demon, intermediate lemmas, etc.) is about 190 lines of specification for 315 lines of proof tactics.

**Gathering.** When robots have to meet actually on the exact same spot, one addresses the problem of *Gathering*. A comprehensive survey on the formalisation of Gathering has been given in [12].

*The Problem.* A configuration `config` in which all robots inhabit the very same position `pt` is characterised by `gathered_at`:

```
Definition gathered_at (pt : Location.t) (config : Config.t) :=
  ∀ g : Names.G, Location.eq (config (Good g)) pt.
```

After a successful gathering along a given execution `e`, the robots stay forever at the same position `pt`. This is characterised with property `Gather`:

```
Definition Gather (pt: Location.t) (e : execution) : Prop :=
  Stream.forever (Streams.instant (gathered_at pt)) e.
```

Finally, robots all have to reach the same position `pt` in finite time, and stay there forever. In other words, property `Gather` will hold eventually:

```
Definition WillGather (pt : Location.t) (e : execution) :
  Prop := Stream.eventually (Gather pt) e.
```

Without any additional initial condition, a robogram achieving Gathering under a demon $d$ fulfils `FullSolGathering`.

```
Definition FullSolGathering (r : robogram) (d : demon) :=
  ∀ config, ∃ pt: Location.t,
    WillGather pt (execute r d config).
```

If conditions on the initial configuration are required, basically if the initial configuration is authorised, that requirement is added in the expression of `FullSolGathering` to obtain `ValidSolGathering`.

```
Definition ValidSolGathering (r : robogram) (d : demon) :=
  ∀ config,
    ¬invalid config
    → ∃ pt : Location.t, WillGather pt (execute r d config).
```

*Impossibility: Gathering of an Even Number.* Courtieu *et al.* [38] first provided a formal specification of Gathering in Pactole, using three properties that exactly reflect the mathematical description of the problem.

Impossibility of Gathering for two oblivious robots in rigid SSYNC was established by Suzuki and Yamashita in their seminal paper [73]. It was slightly extended in [38] to the case where initial configurations may not have all their robots at distinct positions, and for any positive even number of robots.

The environment is instantiated as in the previous example, though without any Byzantine robot, and with the addition that the number of robots `N.nG` is even. So as to keep the statement of the main theorem short, the last constraint is stated as a hypothesis.

```
Parameter N.nG : nat.                    (* number of robots *)
Hypothesis even_nG : Nat.Even N.nG.      (* assumed to be even *)
```

`N.nG` is supposed non-null, hence there are at least two robots.

We stress that the remainder of the model (the definition of rounds, and capabilities/spectra), is exactly the same as before, hence *theorems are stated and proved with the same model assumptions.*

For all robograms `r`, all integers $k \geq 1$, `r` does not solve Gathering against all $k$-fair demons `d`.

```
Theorem noGathering :
  ∀ r k, (1≤k) → ¬(∀ d, kFair k d→ FullSolGathering r d).
```

The file dedicated to the main theorem itself is about 200 lines of specifications for 440 lines of proof scripts.

The proof of the aforementioned theorem relies on the fact that, no matter what the algorithm does, one can always build a demon such that from an invalid position, that is a certain kind of configuration that is not a successful one, the execution resulting from the algorithm and the demon always stays invalid, hence the algorithm fails. In this case, the failing initial configuration is a *bivalent* one: a configuration where robots are equally distributed over two distinct locations only.

*Correctness: Gathering with Initial Conditions.* It turns out that, when such invalid initial configurations (*i.e.* bivalent) are forbidden, one may achieve Gathering, as Courtieu *et al.* proved formally, both in $\mathbb{R}$ [37] and $\mathbb{R}^2$ [39]. As the former can be seen as a particular case of the latter case, we focus our presentation on $\mathbb{R}^2$.

The module of locations is now instantiated by a real metric space with base type `R×R`, and with the usual notions of distance, and vector operations. Note that a library of results and properties over $\mathbb{R}^2$ is provided in the framework, as well as some geometry folklore, namely properties about barycenters, triangles, enclosing circles, etc.

There are at least three robots, none of which is Byzantine.

Further note that since capabilities and synchronicity are still those of previous examples, rounds and spectra stay the same.

Courtieu *et al.* provide a protocol that realises Gathering provided that the initial configuration is not bivalent [39]. The full algorithm is as follows:

**Protocol** gatherR2 (s:Spect.t) **returns** (dest:$\mathbb{R}^2$) :=
**if** max(s) = **{**p**}** **then** dest := p  *(* one max tower *)*
**else begin**
*(* Compute target *)*
  **if** support(s) ∩ SEC(s) = **{**p1,p2,p3**}** **then**  *(* triangle cases *)*
    **if** equilateral(p1,p2,p3) **then** target := barycenter(p1,p2,p3)
      **else if** isosceles(p1,p2,p3) **then** target := opposite of base(p1,p2,p3)
      **else** target := opposite of longest(p1,p2,p3)

```
   else target := center(SEC(s));                                  (* other cases *)
(* only dirty robots move to target, if any, otherwise clean robots can move *)
  if ∀p ∈ s, p ∈ SEC(s) or p = target then dest := target
  else if (0,0)∈ SEC(s) or (0,0) = target then dest := (0,0)
end
```

where *dest* is the target position computed by the protocol. support($s$) and max($s$) denote respectively the support set and set of maximal multiplicity elements of s. SEC($s$) denotes the smallest enclosing circle of positions in $s$.

The protocol relies on a notion of "cleanliness", clean robots being those on the smallest enclosing circle (SEC). A target is computed depending on clean robots only. Unclean robots shall move and gather first as they cannot change the target. Clean robots may then move, possibly changing the SEC and modifying the target.

Target is defined as follows. In the critical situations where exactly three inhabited positions are on the SEC, it depends on the shape of the triangle (here isosceles *excludes* equilateral):

```
Function target_triangle (pt1 pt2 pt3 : R2.t) : R2.t :=
 match classify_triangle pt1 pt2 pt3 with (* Kind of triangle? *)
 | Equilateral ⇒ barycenter3_pts pt1 pt2 pt3  (* To barycenter *)
 | Isosceles p ⇒ p
 | Scalene ⇒ opposite_of_max_side pt1 pt2 pt3
 end.

Function target (s : Spect.t) : R2.t :=
 match on_SEC (Spect.support s) with  (* inhabited loc. on SEC?*)
 | nil ⇒ (0, 0)                                      (* None? *)
 | pt :: nil ⇒ pt         (* Unique loc. on SEC? ⇒ gathered! *)
 | pt1 :: pt2 :: pt3 :: nil ⇒ target_triangle pt1 pt2 pt3
 | _ ⇒ center (SEC l)             (* Gen. case: center of SEC *)
 end.
```

The translation of the algorithm into Gallina is simply:

```
Definition gatherR2_pgm (s : Spect.t) : R2.t :=
 match Spect.support (Spect.max s) with   (* max height towers?*)
 | nil ⇒ (0, 0)              (* None? only happens when no robot *)
 | pt :: nil ⇒ pt              (* Unique highest tower? go there *)
 | _ :: _ :: _ ⇒                                    (* Otherwise *)
   if is_clean s then target s else      (* All on SEC/target ? *)
   if (0, 0) ∈ (SECT s) then (0, 0) else target s
 end.
```

Note that this is almost exactly an actual robot code.

One of main difficulties in the design of the algorithm is to avoid the invalid configuration in any execution. This is done by testing the existence of a unique tower of maximal height and moving towards it if it exists. One can then prove that starting from a non invalid configuration, an invalid one can never appear.

```
Theorem never_invalid :
  ∀ da config, ¬invalid config
    → ¬invalid (round gatherR2 da config).
```

It is interesting to notice that the proof of this theorem in $\mathbb{R}^2$ has been reused from the $\mathbb{R}$ case as it is *exactly* the same.

Robogram gatherR2 achieves Gathering *provided that* the initial configuration is valid (not bivalent), the expression of soundness shall thus use property ValidSolGathering, bivalent configurations being the invalid ones.

The main theorem [39] states that for any fair demon, gatherR2 realises Gathering from any non-bivalent (*i.e.* non-invalid) configuration.

```
Theorem Gathering_in_R2 :
  ∀ d, Fair d → ValidSolGathering gatherR2 d.
```

The proof is led by well-founded induction on a well-founded measure decreasing for each execution round. If all robots are gathered, then it is done. If not, by fairness some robots will have to move, thus a robot will be amongst the first to move. (Formally, this is an induction using fairness.) One concludes by using the induction hypothesis (of the well-founded induction) as this round decreases the measure on configurations. This proof of the main theorem is interestingly small as it is only 20 lines long. The whole file dedicated to specification and certification of the algorithm consists of 480 lines of definitions, specification and intermediate lemmas, and 2750 lines of actual proof.

*Correctness, Weaker Robots, Flexible Movements.* Movements have however not to be always rigid. One may also assume different capabilities for robot sensors, for example denying them the ability to detect the exact number of robots at a location, only knowing if it inhabited or not.

FSYNC Gathering admits a solution in that setting, as shown and formally proved by Balabonski *et al.* [13]. Assuming different capabilities for the robots or for the way they move amounts to modifying parts of the instantiation.

There are at least two robots, none of which is Byzantine.

As there is no detection of multiplicity, the number information cannot be provided as an entry to the algorithm. Instead of a multiset of robots' locations as before, one can now take as a relevant spectrum the *set* of inhabited locations.

In this flexible FSYNC setting, demonic actions provide for each of the oblivious robots both its new self-centred frame of reference, and the ratio of its actual movement over (the distance to) its computed destination. The new conformation of any robot is thus determined in function of this ratio, and the parameter $\delta$.

The specification of space is, though, left untouched.

The algorithm by Balabonski et al. [13] simply brings robots to the barycenter of inhabited locations, that is to the barycenter of the elements in the spectrum.

```
Definition ffgatherR2_pgm (s : Spect.t) : R2.t :=
  barycenter (Spect.M.elements s).
```

Here again the proof of the main theorem is based on a well-founded induction over a measure (well-founded, and decreasing for each round). It is about 20 lines long. The whole file dedicated to the theorem, thus including the definition of that measure, and a proof of its properties, consists of 170 lines of specifications for 1015 lines of actual proofs script.

*Exploration in Graphs.* Problems on graphs, and on some of their particular instances, like rings, have also been investigated using a formal framework, in particular Exploration with Stop (or Terminating Exploration).

Achieving *Exploration with stop* (or Terminating Exploration) consists in:

1. having all vertices to be inhabited at some point during the execution (namely, exploration), and
2. ensuring that robots eventually stop moving once the exploration is complete.

This specification is formalised using, for the exploration part, a predicate `Will_be_visited` such that, if `v` is a vertex and `exc` is an execution, then `Will_be_visited v e` holds if and only if there is a robot inhabiting the vertex `v` in at least one configuration of `exc`:

```
Definition Will_be_visited v exc :=
  Stream.eventually (Stream.instant (is_visited v)) exc.
```

where `is_visited v c` holds for any configuration `c` (in this case, the head of `exc`) that has a robot on the vertex `v`. In a successful exploration, all of the vertices will be visited.

An execution is said to be *terminating* if it contains a configuration identical to all of its successive following ones. In other terms, the execution has a point from which it *stalls* forever. This property is formalised by combining the usual temporal operators:

```
Definition Stall (exc : execution) :=
  Config.eq (Stream.hd exc) (Stream.hd (Stream.tl exc)).

Definition Stopped (exc : execution) :=
  Stream.forever Stall exc.

Definition Will_stop (exc : execution) :=
  Stream.eventually Stopped exc.
```

A full solution to the problem of terminating exploration must then be a robogram the execution of which, for any demon, and from any configuration, satisfies the previous properties: exploration and termination.

```
Definition FullSolExplorationStop  (r : robogram) :=
  ∀ d config, (∀ l, Will_be_visited l (execute r d config))
              ∧ Will_stop (execute r d config).
```

Exploration may not be possible to achieve depending on the parameters of the model, and Balabonski *et al.* [14] prove formally the impossibility for oblivious robots with multiplicity detection to explore a ring, starting from a configuration without tower, in an fair SSYNC setting, when the number of robots divides the number of vertices, *i.e.* the size of the ring.

The module of location is instantiated with the provided Ring module, built from the graph interface implemented with a ring $\mathbb{Z}/n\mathbb{Z}$.

There are $k$ robots, at least two, and none of which is Byzantine.

A suitable spectrum in this context is a multiset of inhabited locations, one of which being highlighted as being the robot's position. The spectrum is extracted up to isomorphism so as to get rid of any vertex name information. This isomorphism is the one provided by the demonic action.

Expressing that starting for a *valid* configuration (that is a configuration without any tower) a protocol can achieve Exploration with Stop for all fair demons is straightforward as it suffices, akin to the Gathering case where bivalent configurations are forbidden, to add the initial conditions to the characterisation of a terminating exploration.

```
Definition ValidSolExplorationStop (r : robogram) :=
  ∀ (c : configuration) (d : demon), Valid_starting_conf c
    → Fair d
    → (∀ l,  Will_be_visited l (execute r d c))
      ∧ Will_stop (execute r d c).
```

where the predicate `Valid_starting_conf` has been here defined to hold for configurations without towers.

The main theorem states that no robogram fulfils `ValidSolExplorationStop` when the number $k$ of robots divides the size $n$ of the ring.

```
Theorem no_exploration_k_divides_n :
  (n mod k) = 0 → ∀ (r : robogram),
    ¬(ValidSolExplorationStop r).
```

The relevant file consists of around 220 lines of specifications for 2200 lines of actual proof scripts.

## 4 Conclusion

A significant amount of recent research effort focused on using formal methods for the purpose of building correct mobile robot protocols.

On the one hand, Model-checking and its derivatives is easy to use but the current (intuitive) model it is based upon has reach its limits (only discrete space is modelled, small fixed set of robots, parametrization hinders the set of specification that can be verified). To use those tools in other settings (such as continuous space, or arbitrary number of robots, etc.), a different (more abstract) model is required. A key difficulty when doing so is to ensure that the initial properties about mobile robots one wants to verify are still relevant using the new modelling.

On the other hand, Proof assistant and the Pactole framework proved modular enough to accommodate all settings seen in the mobile robots literature, prove both impossibilities and algorithms, for an arbitrary number of robots. The downside of the methodology is its relative difficulty in writing actual proofs that permit to define the theorems that have been specified. However, thanks to recent libraries developed in Pactole, specifying models, problems, and properties has become much easier.

## References


1. Abadi, M., Lamport, L., Wolper, P.: Realizable and unrealizable specifications of reactive systems. In: Ausiello, G., Dezani-Ciancaglini, M., Della Rocca, S.R. (eds.) ICALP 1989. LNCS, vol. 372, pp. 1–17. Springer, Heidelberg (1989). https://doi.org/10.1007/BFb0035748
2. ACL2. http://www.cs.utexas.edu/users/moore/acl2/
3. Agda. http://wiki.portal.chalmers.se/agda/pmwiki.php
4. Bacelar Almeida, J., Barbosa, M., Bangerter, E., Barthe, G., Krenn, S., Zanella Béguelin, S.: Full proof cryptography: verifiable compilation of efficient zero-knowledge protocols. In: Yu, T., Danezis, G., Gligor, V.D. (eds.) ACM Conference on Computer and Communications Security, pp. 488–500. ACM (2012)
5. Aminof, B., Murano, A., Rubin, S., Zuleger, F.: Automatic verification of multi-agent systems in parameterised grid-environments. In: Jonker, C.M., Marsella, S., Thangarajah, J., Tuyls, K. (eds.) Proceedings of the 2016 International Conference on Autonomous Agents & Multiagent Systems, Singapore, 9–13 May 2016, pp. 1190–1199. ACM (2016)
6. Apt, K.R., Kozen, D.: Limits for automatic verification of finite-state concurrent systems. Inf. Process. Lett. **22**(6), 307–309 (1986)
7. Arons, T., Pnueli, A., Ruah, S., Xu, Y., Zuck, L.: Parameterized verification with automatically computed inductive assertions? In: Berry, G., Comon, H., Finkel, A. (eds.) CAV 2001. LNCS, vol. 2102, pp. 221–234. Springer, Heidelberg (2001). https://doi.org/10.1007/3-540-44585-4_19
8. Auger, C., Bouzid, Z., Courtieu, P., Tixeuil, S., Urbain, X.: Certified impossibility results for byzantine-tolerant mobile robots. In: Higashino, T., Katayama, Y., Masuzawa, T., Potop-Butucaru, M., Yamashita, M. (eds.) SSS 2013. LNCS, vol. 8255, pp. 178–190. Springer, Cham (2013). https://doi.org/10.1007/978-3-319-03089-0_13
9. Baier, C., Katoen, J.-P.: Principles of Model Checking. MIT Press, Cambridge (2008)
10. Balabonski, T., Courtieu, P., Pelle, R., Rieg, L., Tixeuil, S., Urbain, X.: Brief announcement: continuous vs. discrete asynchronous moves: a certified approach for mobile robots. In: Izumi, T., Kuznetsov, P. (eds.) SSS 2018. LNCS. Springer, Heidelberg (2018). https://doi.org/10.1007/978-3-030-03232-6_29
11. Balabonski, T., Courtieu, P., Pelle, R., Rieg, L., Tixeuil, S., Urbain, X.: Continuous vs. discrete asynchronous moves: a certified approach for mobile robots. Research report, Sorbonne Université, CNRS, Laboratoire d'Informatique de Paris 6, LIP6, F-75005 Paris, France (2018)

12. Balabonski, T., Courtieu, P., Rieg, L., Tixeuil, S., Urbain, X.: Certified gathering of oblivious mobile robots: survey of recent results and open problems. In: Petrucci, L., Seceleanu, C., Cavalcanti, A. (eds.) FMICS/AVoCS -2017. LNCS, vol. 10471, pp. 165–181. Springer, Cham (2017). https://doi.org/10.1007/978-3-319-67113-0_11
13. Balabonski, T., Delga, A., Rieg, L., Tixeuil, S., Urbain, X.: Synchronous gathering without multiplicity detection: a certified algorithm. ACM Trans. Comput. Syst. (2018). https://doi.org/10.1007/s00224-017-9828-z
14. Balabonski, T., Pelle, R., Rieg, L., Tixeuil, S.: A foundational framework for certified impossibility results with mobile robots on graphs. In: Proceedings of International Conference on Distributed Computing and Networking, Varanasi, India, January 2018
15. Bérard, B., Lafourcade, P., Millet, L., Potop-Butucaru, M., Thierry-Mieg, Y., Tixeuil, S.: Formal verification of mobile robot protocols. Distrib. Comput. **29**(6), 459–487 (2016)
16. Bezem, M., Bol, R., Groote, J.F.: Formalizing process algebraic verifications in the calculus of constructions. Form. Asp. Comput. **9**, 1–48 (1997)
17. Bjørner, N., et al.: STeP: deductive-algorithmic verification of reactive and real-time systems. In: Alur, R., Henzinger, T.A. (eds.) CAV 1996. LNCS, vol. 1102, pp. 415–418. Springer, Heidelberg (1996). https://doi.org/10.1007/3-540-61474-5_92
18. Bonnet, F., Défago, X., Petit, F., Potop-Butucaru, M., Tixeuil, S.: Brief announcement: discovering and assessing fine-grained metrics in robot networks protocols. In: Richa and Scheideler [71], pp. 282–284 (2012)
19. Bonnet, F., Défago, X., Petit, F., Potop-Butucaru, M., Tixeuil, S.: Discovering and assessing fine-grained metrics in robot networks protocols. In: 33rd IEEE International Symposium on Reliable Distributed Systems Workshops, SRDS Workshops 2014, Nara, Japan, 6–9 October 2014, pp. 50–59. IEEE Computer Society (2014)
20. Bouzid, Z., Potop-Butucaru, M.G., Tixeuil, S.: Optimal byzantine-resilient convergence in uni-dimensional robot networks. Theor. Comput. Sci. **411**(34–36), 3154–3168 (2010)
21. Boyer, R.S., Moore, J.S.: A Computational Logic Handbook. Academic Press, Cambridge (1988)
22. Büchi, J.R., Landweber, L.H.: Solving sequential conditions by finite-state strategies. Trans. Amer. Math. Soc. **138**, 295–311 (1969)
23. Cansell, D., Méry, D.: The event-B modelling method: concepts and case studies. In: Bjørner, D., Henson, M.C. (eds.) Logics of Specification Languages, pp. 47–152. Springer, Heidelberg (2007). https://doi.org/10.1007/978-3-540-74107-7_3
24. Cansell, D., Méry, D., Merz, S.: Diagram refinements for the design of reactive systems. J. Univ. Comp. Sci. **7**(2), 159–174 (2001)
25. Castéran, P., Filou, V., Mosbah, M.: Certifying distributed algorithms by embedding local computation systems in the Coq proof assistant. In: Bouhoula, A., Ida, T. (eds.) Symbolic Computation in Software Science (SCSS 2009) (2009)
26. Charron-Bost, B., Debrat, H., Merz, S.: Formal verification of consensus algorithms tolerating malicious faults. In: Défago, X., Petit, F., Villain, V. (eds.) SSS 2011. LNCS, vol. 6976, pp. 120–134. Springer, Heidelberg (2011). https://doi.org/10.1007/978-3-642-24550-3_11
27. Chatzigiannakis, I., Michail, O., Spirakis, P.G.: Algorithmic verification of population protocols. In: Dolev, S., Cobb, J., Fischer, M., Yung, M. (eds.) SSS 2010. LNCS, vol. 6366, pp. 221–235. Springer, Heidelberg (2010). https://doi.org/10.1007/978-3-642-16023-3_19

28. Chou, C.-T.: Mechanical verification of distributed algorithms in higher-order logic. Comput. J. **38**, 158–176 (1995)
29. Church, A.: Logic, arithmetics, and automata. In: Proceedings of International Congress of Mathematicians, pp. 23–35 (1963)
30. Clarke, E.M., Emerson, E.A.: Design and synthesis of synchronization skeletons using branching time temporal logic. In: Proceedings of IBM Workshop on Logics of Programs (1981)
31. Clarke, E.M., Grumberg, O., Jha, S.: Verifying parameterized networks using abstraction and regular languages. In: Lee, I., Smolka, S.A. (eds.) CONCUR 1995. LNCS, vol. 962, pp. 395–407. Springer, Heidelberg (1995). https://doi.org/10.1007/3-540-60218-6_30
32. Clarke, E.M., Grumberg, O., Peled, D.A.: Model Checking. MIT Press, Cambridge (1999)
33. Clément, J., Delporte-Gallet, C., Fauconnier, H., Sighireanu, M.: Guidelines for the verification of population protocols. In: Distributed Computing Systems, pp. 215–224. IEEE (2011)
34. Cohen, R., Peleg, D.: Convergence properties of the gravitational algorithm in asynchronous robot systems. SIAM J. Comput. **34**(6), 1516–1528 (2005)
35. Coq. https://coq.inria.fr/
36. Coquand, T., Paulin, C.: Inductively defined types. In: Martin-Löf, P., Mints, G. (eds.) COLOG 1988. LNCS, vol. 417, pp. 50–66. Springer, Heidelberg (1990). https://doi.org/10.1007/3-540-52335-9_47
37. Courtieu, P., Rieg, L., Tixeuil, S., Urbain, X.: A Certified Universal Gathering Algorithm for Oblivious Mobile Robots. CoRR, abs/1506.01603 (2015)
38. Courtieu, P., Rieg, L., Tixeuil, S., Urbain, X.: Impossibility of gathering, a certification. Inf. Process. Lett. **115**, 447–452 (2015)
39. Courtieu, P., Rieg, L., Tixeuil, S., Urbain, X.: Certified universal gathering in $\mathbb{R}^2$ for oblivious mobile robots. In: Gavoille, C., Ilcinkas, D. (eds.) DISC 2016. LNCS, vol. 9888, pp. 187–200. Springer, Heidelberg (2016). https://doi.org/10.1007/978-3-662-53426-7_14
40. Cousineau, D., Doligez, D., Lamport, L., Merz, S., Ricketts, D., Vanzetto, H.: TLA$^+$ proofs. In: Giannakopoulou, D., Méry, D. (eds.) FM 2012. LNCS, vol. 7436, pp. 147–154. Springer, Heidelberg (2012). https://doi.org/10.1007/978-3-642-32759-9_14
41. de Alfaro, L., Manna, Z., Sipma, H.B., Uribe, T.E.: Visual verification of reactive systems. In: Brinksma, E. (ed.) TACAS 1997. LNCS, vol. 1217, pp. 334–350. Springer, Heidelberg (1997). https://doi.org/10.1007/BFb0035398
42. Deng, Y., Monin, J.-F.: Verifying self-stabilizing population protocols with Coq. In: Chin, W.-N., Qin, S. (eds.) Third IEEE International Symposium on Theoretical Aspects of Software Engineering (TASE 2009), Tianjin, China, pp. 201–208. IEEE Computer Society, July 2009
43. Devismes, S., Lamani, A., Petit, F., Raymond, P., Tixeuil, S.: Optimal grid exploration by asynchronous oblivious robots. In: Richa and Scheideler [71], pp. 64–76 (2012)
44. Doan, H.T.T., Bonnet, F., Ogata, K.: Model checking of a mobile robots perpetual exploration algorithm. In: Liu, S., Duan, Z., Tian, C., Nagoya, F. (eds.) SOFL+MSVL 2016. LNCS, vol. 10189, pp. 201–219. Springer, Cham (2017). https://doi.org/10.1007/978-3-319-57708-1_12

45. Doan, H.T.T., Bonnet, F., Ogata, K.: Model checking of robot gathering. In: Aspnes, J., Bessani, A., Felber, P., Leitão, J. (eds.) 21st International Conference on Principles of Distributed Systems, OPODIS 2017, Lisbon, Portugal, 18–20 December 2017, volume 95 of LIPIcs, pp. 12:1–12:16. Schloss Dagstuhl - Leibniz-Zentrum fuer Informatik (2017)
46. Esparza, J., Finkel, A., Mayr, R.: On the verification of broadcast protocols. In: 14th Annual Symposium on Logic in Computer Science, pp. 352–359. IEEE (1999)
47. Flocchini, P., Prencipe, G., Santoro, N.: Distributed Computing by Oblivious Mobile Robots. Morgan & Claypool Publishers, San Rafael (2012)
48. Fokkink, W.: Modelling Distributed Systems. EATCS Texts in Theoretical Computer Science. Springer, Heidelberg (2007). https://doi.org/10.1007/978-3-540-73938-8
49. Gascard, E., Pierre, L.: Formal proof of applications distributed in symmetric interconnexion networks. Parallel Process. Lett. **13**(1), 3–18 (2003)
50. Gonthier, G.: Formal proof—the four-color theorem. In: Notices of the AMS, vol. 55, p. 1370, December 2008
51. Gonthier, G.: Engineering mathematics: the odd order theorem proof. In: Giacobazzi, R., Cousot, R. (eds.) POPL, pp. 1–2. ACM (2013)
52. Grädel, E., Thomas, W., Wilke, T. (eds.): Automata Logics, and Infinite Games. LNCS, vol. 2500. Springer, Heidelberg (2002). https://doi.org/10.1007/3-540-36387-4
53. Guerraoui, R., Henzinger, T.A., Singh, V.: Model checking transactional memories. Distrib. Comput., 129–145 (2010)
54. Halbwachs, N., Caspi, P., Raymond, P., Pilaud, D.: The synchronous dataflow programming language lustre. Proc. IEEE **79**(9), 1305–1320 (1991)
55. Klein, G., et al.: seL4: formal verification of an operating system kernel. Commun. ACM **53**(6), 107–115 (2010)
56. Küfner, P., Nestmann, U., Rickmann, C.: Formal verification of distributed algorithms. In: Baeten, J.C.M., Ball, T., de Boer, F.S. (eds.) TCS 2012. LNCS, vol. 7604, pp. 209–224. Springer, Heidelberg (2012). https://doi.org/10.1007/978-3-642-33475-7_15
57. Kulkarni, S.S., Bonakdarpour, B., Ebnenasir, A.: Mechanical verification of automatic synthesis of fault-tolerant programs. In: Etalle, S. (ed.) LOPSTR 2004. LNCS, vol. 3573, pp. 36–52. Springer, Heidelberg (2005). https://doi.org/10.1007/11506676_3
58. Lamport, L.: Byzantizing paxos by refinement. In: Peleg, D. (ed.) DISC 2011. LNCS, vol. 6950, pp. 211–224. Springer, Heidelberg (2011). https://doi.org/10.1007/978-3-642-24100-0_22
59. Lamport, L., Merz, S.: Specifying and verifying fault-tolerant systems. In: Langmaack, H., de Roever, W.-P., Vytopil, J. (eds.) FTRTFT 1994. LNCS, vol. 863, pp. 41–76. Springer, Heidelberg (1994). https://doi.org/10.1007/3-540-58468-4_159
60. Lamport, L., Shostak, R., Pease, M.: The Byzantine generals problem. ACM Trans. Program. Lang. Syst. **4**(3), 382–401 (1982)
61. Leroy, X.: A formally verified compiler back-end. J. Autom. Reason. **43**(4), 363–446 (2009)
62. Loco. http://www.labri.fr/~casteran/Loco
63. Lu, T., Merz, S., Weidenbach, C.: Towards verification of the pastry protocol using TLA$^+$. In: Bruni, R., Dingel, J. (eds.) FMOODS/FORTE -2011. LNCS, vol. 6722, pp. 244–258. Springer, Heidelberg (2011). https://doi.org/10.1007/978-3-642-21461-5_16

64. Manna, Z., Pnueli, A.: Temporal verification diagrams. In: Hagiya, M., Mitchell, J.C. (eds.) TACS 1994. LNCS, vol. 789, pp. 726–765. Springer, Heidelberg (1994). https://doi.org/10.1007/3-540-57887-0_123
65. Manna, Z., Wolper, P.: Synthesis of communicating processes from temporal logic specifications. ACM Trans. Program. Lang. Syst. **6**(1), 68–93 (1984)
66. Millet, L., Potop-Butucaru, M., Sznajder, N., Tixeuil, S.: On the synthesis of mobile robots algorithms: the case of ring gathering. In: Felber, P., Garg, V. (eds.) SSS 2014. LNCS, vol. 8756, pp. 237–251. Springer, Cham (2014). https://doi.org/10.1007/978-3-319-11764-5_17
67. Mizar. http://mizar.uwb.edu.pl/
68. Nipkow, T., Wenzel, M., Paulson, L.C. (eds.): Isabelle/HOL. LNCS, vol. 2283. Springer, Heidelberg (2002). https://doi.org/10.1007/3-540-45949-9
69. Pnueli, A., Rosner, R.: On the synthesis of a reactive module. In: Proceedings of POPL 1989, pp. 179–190. ACM (1989)
70. PVS. http://pvs.csl.sri.com/
71. Richa, A.W., Scheideler, C. (eds.): SSS 2012. LNCS, vol. 7596. Springer, Heidelberg (2012). https://doi.org/10.1007/978-3-642-33536-5
72. Sangnier, A., Sznajder, N., Potop-Butucaru, M., Tixeuil, S.: Parameterized verification of algorithms for oblivious robots on a ring. In: Stewart, D., Weissenbacher, G. (eds.) 2017 Formal Methods in Computer Aided Design, FMCAD 2017, Vienna, Austria, 2–6 October 2017, pp. 212–219. IEEE (2017)
73. Suzuki, I., Yamashita, M.: Distributed anonymous mobile robots: formation of geometric patterns. SIAM J. Comput. **28**(4), 1347–1363 (1999)
74. Flyspeck Development Team. The Flyspeck Project. https://code.google.com/p/flyspeck/
75. Théry, L., Hanrot, G.: Primality proving with elliptic curves. In: Schneider, K., Brandt, J. (eds.) TPHOLs 2007. LNCS, vol. 4732, pp. 319–333. Springer, Heidelberg (2007). https://doi.org/10.1007/978-3-540-74591-4_24
76. Tsuchiya, T., Schiper, A.: Verification of consensus algorithms using satisfiability solving. Distrib. Comput. **23**, 341–358 (2011)

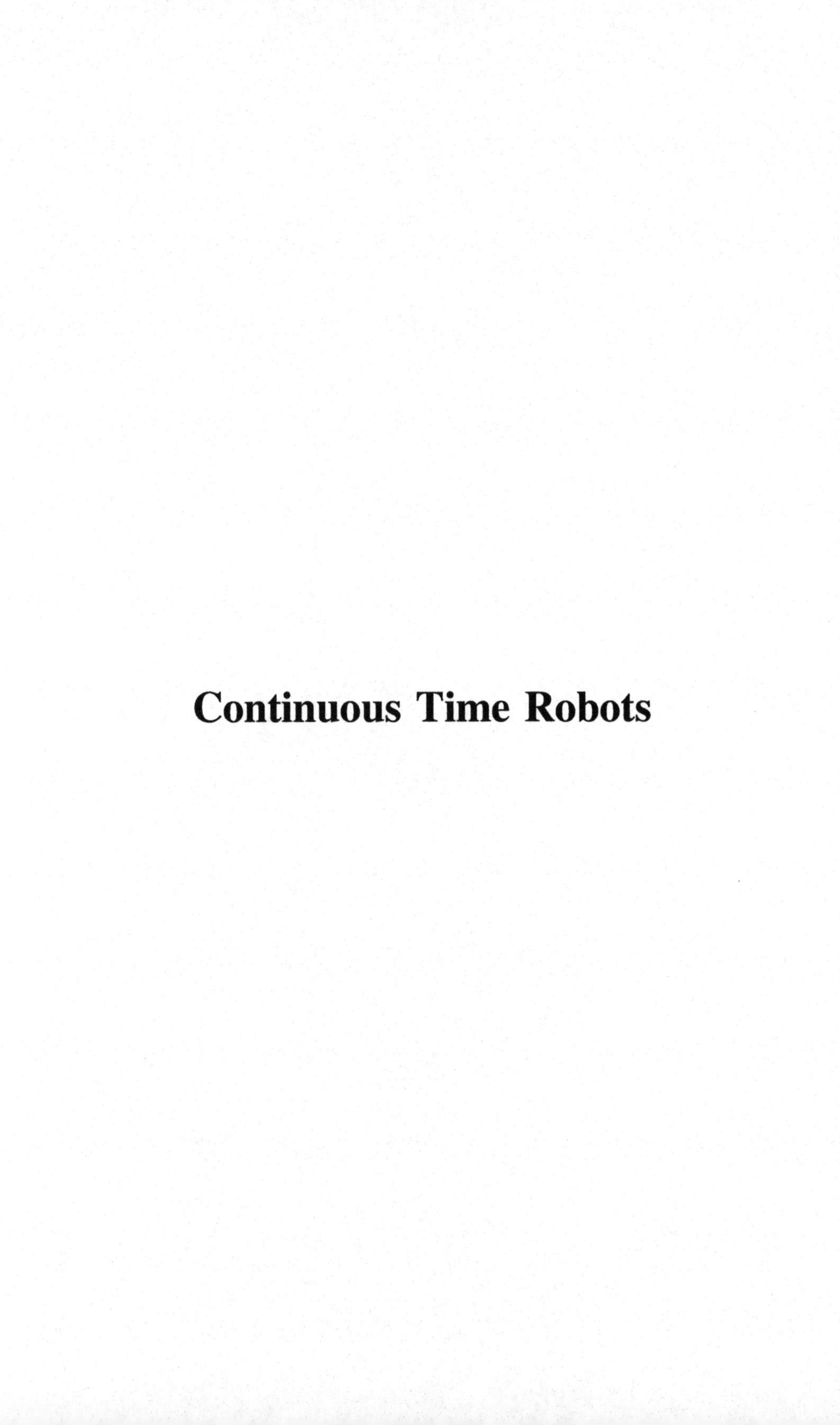

# Continuous Time Robots

# Continuous Protocols for Swarm Robotics

Peter Kling[1(✉)] and Friedhelm Meyer auf der Heide[2]

[1] Universität Hamburg, Hamburg, Germany
peter.kling@uni-hamburg.de

[2] University of Paderborn, Paderborn, Germany
fmadh@uni-paderborn.de

**Abstract.** We consider simple models of swarms of identical, anonymous robots: they are points in the plane and "see" only their neighbors (robots within distance one). We will deal with distributed local protocols of such swarms that result in formations like "gathering at one point". The focus will be on protocols assuming a continuous time model. We present upper and lower bounds on their run time and energy consumption, and compare different protocols both theoretically and experimentally.

**Keywords:** Robots · Continuous · Gathering

## 1 Introduction

Envision a scenario where $n$ mobile robots, each having a limited viewing range, are placed in the Euclidean plane and are supposed to establish a certain formation. To reach this formation, each robot has to plan and perform its movement based solely on the positions of the other robots within its viewing range, which we normalize to 1. In particular, the robots are not provided with global view, communication, or long term memory.

This chapter considers one of the most basic formation problems: the GATHERING problem. Here, the $n$ robots must move such that, eventually, they gather at a single, not predetermined point. While performing their protocols, it is crucial that the visibility graph spanned by the robots stays connected; if a robot loses sight of all other robots, no deterministic, local protocol can be guaranteed to reconnect the lost robot to the remaining formation.

Most protocols for such formation problems are based on some kind of discrete round model. For example, Ando et al. [1] and Degener et al. [4] show that gathering can be achieved with a simple protocol by robots with a limited visibility in a synchronous, discrete time model in $O(n^2)$ rounds. For the same problem with an unlimited viewing range, Cohen and Peleg [3] analyze a simple algorithm in several asynchronous time models. Further publications consider the GATHERING problem in similar time models, see for example [2,6,9–11,16,17].

All of the above models have in common that they are based on the so-called `Look-Compute-Move` *(LCM)* model. That is, the robots act in rounds, where each


P. Flocchini et al. (Eds.): Distributed Computing by Mobile Entities, LNCS 11340, pp. 317–334, 2019.
https://doi.org/10.1007/978-3-030-11072-7_13

round consists of a `Look` operation, a `Compute` operation, and a `Move` operation. During the `Look` operation, a robot determines the positions of all visible robots in its vicinity. During the `Compute` operation, the observed information is used to determine a target point. Finally, during the `Move` operation, the robot moves towards the previously computed target point. The specific models differ in whether these operations are executed synchronously or asynchronously (or something in between).

A different approach is to use a continuous time model. This was first done by Gordon et al. [8] for the Gathering problem. In such a continuous time model, all robots perpetually and at the same time measure and adjust their movement paths. This causes the trajectories of the robots, who are assumed to have some constant maximum moving speed, to become (continuous) curves. While this continuous motion model is somewhat idealized and might seem unrealistic – given that we assume there is no delay between the robots' sensors and actors – it is comparatively close to real applications [14].

This chapter presents some of the more recent results that introduced general techniques to analyze the time required by such distributed robot formation protocols in the continuous time model to reach their goal and discusses other aspects specific to this model.

**Chapter Outline.** We continue with a formal model description and with the introduction of continuous robot formation protocols in Sect. 2. Afterward, we collect some auxiliary results in Sect. 3 that will be used throughout the chapter. In Sect. 4 we introduce a general class of (not necessarily distributed) robot formation protocols, so-called *contracting protocols*, and study their gathering time. Section 5 introduces a specific protocol of this class and sketches the proof that this protocol solves the gathering problem in asymptotical optimal time. We conclude this chapter in Sect. 6, where we discuss the issue of collisions and how to avoid them.

## 2 Model Description and Continuous Protocols

Consider a set of $n$ autonomous, mobile robots in the Euclidean plane $\mathbb{R}^2$. The robots have no spatial dimension and each of them has its own, local coordinate system. In particular, there is no common "origin" notion and no common directional notions like "left" or "right". All robots have a *visual range* of 1, allowing them to perceive other robots that are within this distance and to determine the relative position of such robots. However, neither can robots distinguish other robots from one another nor is there any form of multiplicity detection: a robot can only distinguish whether there is either *no* robot or *at least one* robot at a given position. There is a continuous notion of time and we assume robots move with a *maximum speed* of 1. We assume an idealized sensor-actor mechanism, allowing a robot to act instantly on any data perceived from within its visual range.

In order to convey the basics and characteristics of continuous robot formation problems (in contrast to their more standard discrete time pendants), we concentrate on the probably most basic formation problem: *gathering.*

**Definition 1 (Gathering Problem).** *Consider $n$ autonomous, mobile robots in the Euclidean plane. In the* Gathering *problem, we seek to gather all robots at a single point.*

A *(gathering) protocol* is an algorithmic description that takes the current robot positions and determines how each robot moves. We are mostly interested in *distributed protocols*, where each robot determines its movement by the same algorithm that uses only information available to the respective robot (basically the relative positions of other robots within its visual range). The quality of a protocol is measured by its *gathering time*, the worst-case time (over the set of all possible initial robot positions) required by the protocol to gather all robots.

**General Notation.** Before we dive deeper into the specifics of continuous robot formation protocols, we introduce some general notation. For an integer $m \in \mathbb{N}$ define $[m] := \{1, 2, \ldots, m\}$. For $x \in \mathbb{R}^2$ we use $\|x\|_2$ to denote the Euclidean norm (vector length) of $x$. For $S_1, S_2 \subseteq \mathbb{R}$ and $c_1, c_2 \in \mathbb{R}$ define $c_1 \cdot S_1 + c_2 \cdot S_2 := \{c_1 \cdot s_1 + c_2 \cdot s_2 \mid s_1 \in S_1, s_2 \in S_2\}$. For $x, y \in \mathbb{R}^2$ let $\angle(x, y) \in [-\pi, \pi]$ denote the signed[1] angle formed by the vectors $x$ and $y$.

Fix an arbitrary ordering of the $n$ robots. A *configuration* $c = (c_i)_{i=1}^n$ is a vector whose $i$-th element $c_i \in \mathbb{R}^2$ specifies the positions of the $i$-th robot. Define the polygon $\mathrm{CH}_c \subseteq \mathbb{R}^2$ as the convex hull of all robot positions in configuration $c$. Furthermore, let $\mathrm{Bord}_c := \{i \mid c_i \in \partial\mathrm{CH}_c\}$ be the set of robots that are at the boundary of the configuration's convex hull. Similarly, let $\mathrm{Corn}_c \subseteq \mathrm{Bord}_c$ be the set of robots that are at a vertex of the polygon $\mathrm{CH}_c$. A robot from $\mathrm{Bord}_c$ is called *border robot* and a robot from $\mathrm{Corn}_c$ is called *corner robot.* The *diameter* $\Delta_c := \max\{\|c_i - c_j\|_2 \mid i, j \in [n]\}$ of configuration $c$ is the maximum distance between any two robots.

The *visibility graph* $G_c = (V_c, E_c)$ of configuration $c$ is the Euclidean graph whose vertices are the robots' positions and in which two vertices are connected by an edge if and only if the two corresponding robots are within viewing range of each other. More formally, $V_c := \{c_i \mid i \in [n]\}$ and $E_c := \{\{u, v\} \mid u, v \in V_c, u \neq v, \|u - v\|_2 \leq 1\}$. A configuration is called *connected* if $G_c$ is connected. We use $\mathcal{C}_n$ to denote the set of all connected configurations of $n$ robots.

**Continuous Protocols.** A continuous robot formation protocol $\mathcal{P}$ specifies how, at any time $t \geq 0$, each robot calculates its *velocity vector*, which dictates the robot's current speed and its movement direction. Given a continuous robot formation protocol $\mathcal{P}$, let $c(t) = (c_i(t))_{i=1}^n$ be the configuration at time $t$, such that the function $c_i: \mathbb{R}_{\geq 0} \to \mathbb{R}^2$ is the trajectory of the $i$-th robot. For $i \in [n]$ and $t \geq 0$ let $v_i(t)$ denote the velocity vector of the $i$-th robot at time $t$.

[1] Without loss of generality, we say an angle is positive if it is measured counterclockwise and negative if it is measured clockwise.

That is, at time $t$ robot $i$ moves with speed $\|v_i(t)\|_2$ in direction $v_i(t)$. We use $s_i(t) := \|v_i(t)\|_2$ as a shorthand for the speed of robot $i$ at time $t$.

The trajectories $c_i$ are continuous but not necessarily differentiable. Indeed, robots can change their speed and movement direction non-continuously, resulting in a non-differentiable trajectory. However, natural protocols have right-differentiable trajectories, and we restrict our study to such protocols. In particular, this allows us to see robot $i$'s velocity vector $v_i\colon \mathbb{R}_{\geq 0} \to \mathbb{R}^2$ as the (right) derivative of $c_i$ and we can write $v_i = \dot{c}_i$, where the differentiation is understood to be a right derivative whenever necessary.

It will be useful to consider how robots move relative to each other. For this purpose, we define the angles $\beta_{i,j}(t) := \angle(v_i(t), c_j(t) - c_i(t))$. That is, $\beta_{i,j}(t)$ is the signed angle between the velocity vector of robot $i$ and the line segment connecting robot $i$ and robot $j$.

Figure 1 illustrates the notions introduced above.

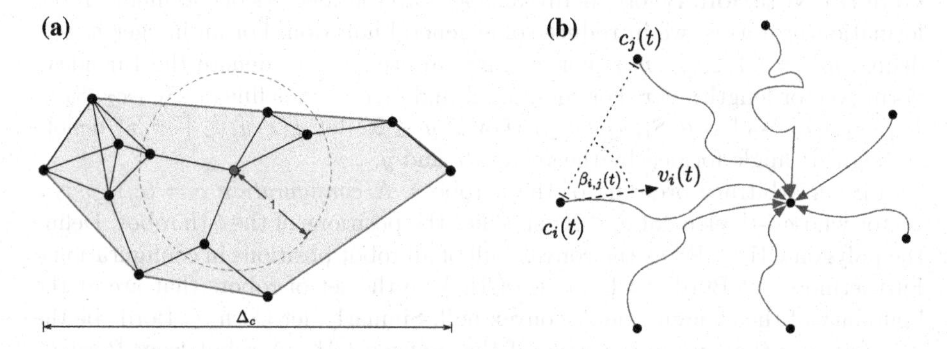


**Fig. 1.** (a) An example configuration $c$ for $n = 13$ robots with the corresponding visibility graph $G_c$ and the convex hull $\mathrm{CH}_c$. The convex hull has 7 border robots, 6 of which are also corner robots. (b) The current movement vector $v_i(t)$ of a robot $i$, its current relative positional angle $\beta_{i,j}(t)$ with respect to robot $j$ and example trajectories (in blue). (Color figure online)

## 3 Auxiliary Results

This section collects some basic results that turn out to be useful in the analysis of continuous robot formation protocols. We start with some simple trigonometric inequalities.

**Lemma 2**

(a) *For* $\alpha \in [0, \pi/2]$ *we have* $\cos\alpha \geq 1 - 2\alpha/\pi$.
(b) *For* $\alpha \geq 0$ *we have* $\cos\alpha \geq 1 - \alpha^2/2$.
(c) *For* $\phi \in [0, 1]$ *and* $\alpha \in [0, \pi]$ *we have* $\cos(\phi\cdot\alpha) + \cos((1-\phi)\cdot\alpha) \geq 2\cdot(1 - \alpha/\pi)^2$.

*Proof*

(a) The statement follows from basic calculus by realizing that the graph of $f(x) = 1 - 2x/\pi$ is a secant line that intersects the graph of $\cos(x)$ at $x = 0$ and $x = \pi/2$.

(b) The statement follows by realizing that $\cos\alpha$ and $1 - \alpha^2/2$ are equal for $\alpha = 0$ and that the first term's derivative $(-\sin\alpha)$ is as least as large as the second term's derivative $(-\alpha)$ for any $\alpha \geq 0$.

(c) For $x, y \in \mathbb{R}$ we have the trigonometric identity $\cos x + \cos y = 2 \cdot \cos((x+y)/2) \cdot \cos((x-y)/2)$ (see [15, Identity 4.21.8]). We apply this to the left-hand side of the desired inequality and calculate

$$\begin{aligned}\cos(\phi \cdot \alpha) + \cos((1-\phi) \cdot \alpha) &= 2 \cdot \cos\Bigl(\frac{\alpha}{2}\Bigr) \cdot \cos\Bigl(\frac{2\phi - 1}{2} \cdot \alpha\Bigr) \\ &\geq 2 \cdot \Bigl(\cos\Bigl(\frac{\alpha}{2}\Bigr)\Bigr)^2 \geq 2 \cdot \Bigl(1 - \frac{\alpha}{\pi}\Bigr)^2.\end{aligned} \tag{1}$$

The first inequality uses $(2\phi - 1)/2 \cdot \alpha \in [-\alpha/2, \alpha/2] \subseteq [-\pi/2, \pi/2]$. Thus, the cosine in the expression is minimized when $(2\phi - 1)/2 \cdot \alpha = \alpha/2$. The second inequality uses the lemma's first statement. □

The next lemma is central for analyzing the gathering time of protocols. Our general progress measure is based on how the robots' convex hull changes over time. We quantify this change by studying how the distance between neighboring corner robots changes. To this end, the following lemma expresses the change in the distance between two robots $i$ and $j$ in terms of their current speeds $s_i(t)$ and $s_j(t)$ and their direction of movement relative to each other (the angles $\beta_{i,j}(t)$ and $\beta_{j,i}(t)$ between their velocity vectors and their connecting line). See Fig. 2 for an illustration.

**Lemma 3.** *Consider a robot formation protocol $\mathcal{P}$ and fix two robots $i$ and $j$. The distance $d(t) := \|c_i(t) - c_j(t)\|_2$ between $i$ and $j$ at time $t$ changes with speed*

$$\dot{d}(t) = -s_i(t) \cdot \cos\beta_{i,j}(t) - s_j(t) \cdot \cos\beta_{j,i}(t). \tag{2}$$

*Proof.* Define $D\colon \mathbb{R}_{\geq 0} \to \mathbb{R}^2, t \mapsto c_j(t) - c_i(t)$, such that $d(t) = \|D(t)\|_2$. We use $D_x(t)$ to refer to the $x$-component of $D(t)$ and, similarly, $D_y(t)$ to refer to the $y$-component of $D(t)$. Fix a time $t \geq 0$. Without loss of generality, we can translate and rotate the coordinate system such that $D(t) = (d(t), 0)$. This immediately yields

$$\dot{d}(t) = \left(\frac{D_x(t)}{d(t)}, \frac{D_y(t)}{d(t)}\right) \cdot \begin{pmatrix}\dot{D}_x(t) \\ \dot{D}_y(t)\end{pmatrix} = \dot{D}_x(t). \tag{3}$$

Note that $\dot{D}_x(t)$ is the $x$-component of $\dot{D}(t)$, which can be written as

$$\begin{aligned}\dot{D}(t) &= \dot{c}_j(t) - \dot{c}_i(t) = v_j(t) - v_i(t) \\ &= s_j(t) \cdot (-\cos\beta_{j,i}(t), \sin\beta_{j,i}(t)) - s_i(t) \cdot (\cos\beta_{i,j}(t), \sin\beta_{i,j}(t)).\end{aligned} \tag{4}$$

Together, Eqs. (3) and (4) imply the lemma's statement. □

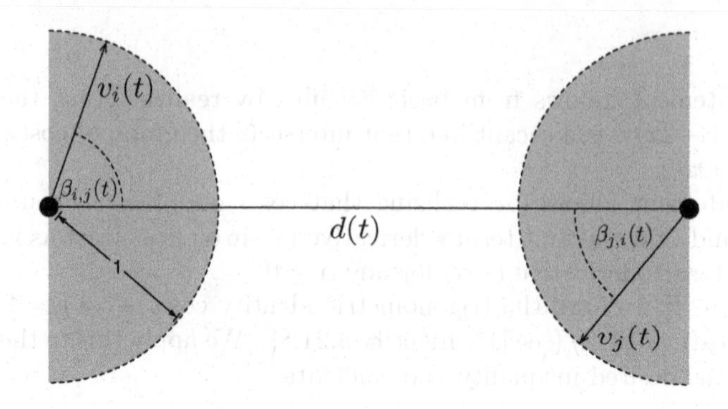


**Fig. 2.** Illustration of Lemma 3

## 4 Contracting Robot Formation Protocols

A natural property of protocols that solve the GATHERING problem is that robots move "towards each other", causing the convex hull of all robot positions to contract over time. In the following we define the class of *contracting* robot formation protocols – which formalizes this intuitive property – and prove general upper and lower bounds for this class.

Define $p(t) := |\{ c_i(t) \mid i \in \mathrm{Corn}_{c(t)} \}|$ as the number of vertices of the polygon $\mathrm{CH}_{c(t)}$. Let $m_1(t), m_2(t), \dots, m_{p(t)}(t)$ denote the vertices ordered counterclockwise along the boundary of $\mathrm{CH}_{c(t)}$ (starting at an arbitrary vertex). For convenience, define $m_0(t) := m_{p(t)}(t)$ and $m_{p(t)+1}(t) := m_1(t)$.

**Definition 4 (Length).** *The* length $\mathfrak{l}(t)$ *of the configuration at time* $t$ *is*

$$\mathfrak{l}(t) := \sum_{\iota=1}^{p(t)} \|m_\iota(t) - m_{\iota-1}(t)\|_2. \tag{5}$$

*We use the shorthand* $\mathfrak{l} := \mathfrak{l}(0)$ *to denote the length of the initial configuration.*

Note that the robots have solved the GATHERING problem at time $t$ if and only if $\mathfrak{l}(t) = 0$. This property (and the fact that most reasonable protocols do not increase a configuration's length) make the length a good way to measure the progress of a gathering protocol. Next we define a quite general class of (not necessarily distributed) robot formation protocols. The definition, which is originally from [13], simply requires that corner robots move with maximum speed in some direction within the robots' convex hull.

**Definition 5 (Contracting [13]).** *Consider a continuous robot formation protocol* $\mathcal{P}$. *We say* $\mathcal{P}$ *is* contracting *if for any time* $t \geq 0$ *with* $\mathfrak{l}(t) > 0$ *each corner robot* $i \in \mathrm{Corn}_{c(t)}$ *moves with speed* $s_i(t) = 1$ *along the velocity vector* $v_i(t) \in (m_{\iota-1}(t) - m_\iota(t)) \cdot \mathbb{R}_{\geq 0} + (m_{\iota+1}(t) - m_\iota(t)) \cdot \mathbb{R}_{\geq 0}$.

See Fig. 3 for an illustration.

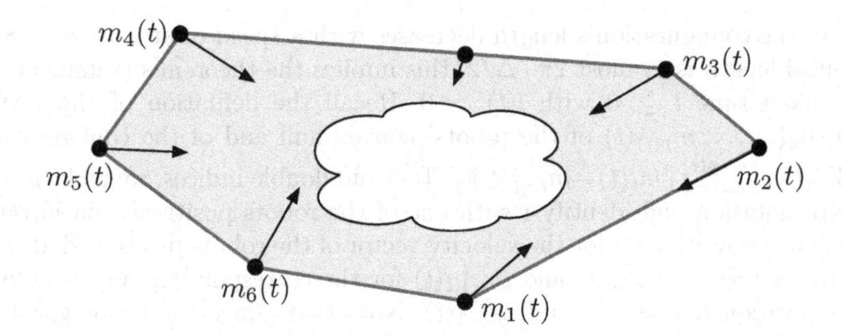


**Fig. 3.** Possible movement vectors for a contracting protocol. Note that the velocity vectors of corner robots have all the same length (1), while the border robot may move at a slower speed. The movement of robots within the convex hull is completely unrestricted.

Even though Definition 5 does not specify the movement of general border robots, it has an interesting implication for their movement. Consider a border robot $j \in \mathrm{Bord}_{c(t)} \setminus \mathrm{Corn}_{c(t)}$ between two corner robots $i_1, i_2 \in \mathrm{Corn}_{c(t)}$ at time $t$ of a contracting robot formation protocol $\mathcal{P}$. Assume that $j$ "falls behind" at time $t$, such that for any small enough $\epsilon > 0$ $j$ is a corner robot (at a position different from $i_1$ and $i_2$) in configuration $c(t + \epsilon)$. Definition 5 requires that $j$ moves with speed 1 towards the line connecting $i_1$ and $i_2$ for any such $\epsilon$. In the worst case this line moves with a speed of at most 1 away from $j$ (if both $i_1$ and $i_2$ move accordingly). Thus, the distance between $j$ and the line connecting $i_1$ and $i_2$ cannot increase at time $t + \epsilon$. But if that holds for any $\epsilon > 0$, $j$ cannot leave the line in the first place. In other words, contracting protocols will not create new corners along the robots' convex hull.

The above phenomenon illustrates an important aspect of continuous strategies and is also known as *Zenoness* (see [8]). The continuous nature of the system – which allows for an instant and continuous course correction – makes it possible that collinear robots (like corner robots and any border robots between them) remain collinear.

### 4.1 Gathering Time of Contracting Protocols

With the above notions we can formulate a general result that bounds the gathering time of *any* contracting robot formation protocol.

**Theorem 6** ([13]). *Consider a continuous robot formation protocol $\mathcal{P}$ started in a configuration of $n$ robots and of diameter $\Delta$. If $\mathcal{P}$ is contracting, then it has gathering time at most $\pi \cdot n \cdot \Delta/8$.*

*Proof.* Fix a time $t$ and consider the convex hull $\mathrm{CH}_{c(t)}$ formed by the robots' positions at time $t$. The robots have gathered at time $t$ if and only if the length $\mathfrak{l}(t)$ of the robots' convex hull equals zero. We show that at any time $t$ with

$\mathfrak{l}(t) > 0$, the configuration's length decreases with a speed of at least $8/n$. Since the initial length is at most $2\pi \cdot \Delta/2$, this implies the theorem's statement.

So fix a time $t \geq 0$ with $\mathfrak{l}(t) > 0$. Recall the definition of the vertices $m_1(t), m_2(t), \ldots, m_{p(t)}(t)$ of the robots' convex hull and of the configuration's length $\mathfrak{l}(t) = \sum_{\iota=1}^{p(t)} \|m_\iota(t) - m_{\iota-1}(t)\|_2$. To avoid double indices, we make a slight abuse of notation and identify $\iota$ with one of the robots positioned on $m_\iota(t)$. In particular, we write $v_\iota(t)$ for the velocity vector of the robots positioned at $m_\iota(t)$. Similarly we write $\beta_{\iota,\iota-1}(t)$ and $\beta_{\iota-1,\iota}(t)$ for the corresponding angles between robots positioned at $m_\iota(t)$ and $m_{\iota-1}(t)$. Note that, since the robot formation protocol $\mathcal{P}$ is contracting, these vertices move with speed $s_\iota(t) = 1$. Define $d_\iota(t) := \|m_\iota(t) - m_{\iota-1}(t)\|_2$ as the distance between the corner robots at $m_\iota(t)$ and $m_{\iota-1}(t)$. By Lemma 3

$$\dot{d}_\iota(t) = -\cos\beta_{\iota,\iota-1}(t) - \cos\beta_{\iota-1,\iota}(t). \tag{6}$$

For $\iota \in [p(t)]$ let $\alpha_\iota(t) \in [0, \pi]$ denote the inner angle of the polygon $\mathrm{CH}_{c(t)}$ at vertex $m_\iota(t)$. Since the robot formation protocol $\mathcal{P}$ is contracting, the velocity vector $v_\iota(t)$ points towards the inside of the robots' convex hull, such that $\alpha_\iota(t) = \beta_{\iota,\iota-1}(t) + \beta_{\iota,\iota+1}(t)$. Since $\mathfrak{l}(t) = \sum_{\iota=1}^{p(t)} d_\iota(t)$, we can take the derivative of $\mathfrak{l}(t)$ and apply Eq. (6) to get

$$\begin{aligned} \dot{\mathfrak{l}}(t) &= \sum_{\iota=1}^{p(t)} (-\cos\beta_{\iota,\iota-1}(t) - \cos\beta_{\iota-1,\iota}(t)) \\ &= -\sum_{\iota=1}^{p(t)} (\cos\beta_{\iota,\iota-1}(t) + \cos\beta_{\iota,\iota+1}(t)). \end{aligned} \tag{7}$$

Applying Lemma 2(c) to the last expression yields

$$\dot{\mathfrak{l}}(t) \leq -2 \cdot \sum_{\iota=1}^{p(t)} \left(1 - \frac{\alpha_\iota(t)}{\pi}\right)^2 = -\frac{2}{\pi^2} \cdot \sum_{\iota=1}^{p(t)} (\pi - \alpha_\iota(t))^2. \tag{8}$$

Now we first use the Cauchy-Schwarz inequality and then that the sum of internal angles of a polygon with $p$ vertices equals $(p-2) \cdot \pi$ to get

$$\begin{aligned} \dot{\mathfrak{l}}(t) &\leq -\frac{2}{p(t) \cdot \pi^2} \cdot \left( \sum_{\iota=1}^{p(t)} (\pi - \alpha_\iota(t)) \right)^2 \\ &= -\frac{2}{p(t) \cdot \pi^2} \cdot (p(t) \cdot \pi - (p(t) - 2) \cdot \pi)^2 \\ &\leq -\frac{8}{p(t)} \leq -\frac{8}{n}. \end{aligned} \tag{9}$$

As argued at the beginning of the proof, this yields the desired statement. □

In Theorem 6 we did not restrict the configuration to be connected. As long as the protocol (which could be executed by an omniscient observer that knows where to move the robots) is contracting, the stated bound holds. If we start in a connected configuration, as required for any truly distributed protocol restricted by the robots' visual range, we see that the initial diameter is at most $n-1$, yielding a bound of $\pi/8 \cdot n^2$. In fact, if instead of bounding the initial configuration's length by $2\pi \cdot \Delta/2$ in the proof of Theorem 6 we use the bound $\mathfrak{l}(0) \leq 2(n-1)$ (which can easily be shown, see [12]), we get the following corollary:

**Corollary 7** **([13]).** *Consider a continuous robot formation protocol $\mathcal{P}$ started in a connected configuration of $n$ robots. If $\mathcal{P}$ is contracting, then it has gathering time at most $n^2/4$.*

### 4.2 Worst-Case Contracting Protocols

It is not difficult to see, that the upper bound of $O(n^2)$ from Corollary 7 is tight in the sense that there are contracting protocols and corresponding initial configurations for which the gathering time is *at least* $\Omega(n^2)$. In fact, if we assume that all robots are positioned at corners of a regular polygon and move towards their counterclockwise neighbor, we get a situation that resembles the so-called *n-bugs problem* [18], which is known to have a quadratic convergence speed. The next lemma provides this result and a simple proof using our terminology.

**Lemma 8.** *There is a contracting robot formation protocol $\mathcal{P}$ and a connected initial configuration $c \in \mathcal{C}_n$ such that the gathering time is at least $n^2/(2\pi^2)$.*

*Proof.* Assume the initial configuration is such that the robots' convex hull is an $n$-sided regular polygon with edge length 1 (the robots' visual range) and assume that each robot moves with speed 1 towards its counterclockwise neighbor. Since all robots move symmetrically, the configuration will stay an $n$-sided regular polygon (which eventually degenerates to a point) at any time $t \geq 0$. Now fix a time $t \geq 0$ with $\mathfrak{l}(t) > 0$. Consider Eq. (7) from the proof of Theorem 6, which gives the speed at which the configuration's length changes. Since the current convex hull is an $n$-sided regular polygon, we have for each $\iota \in \{1, 2, \ldots, p(t)\}$ that $\beta_{\iota,\iota+1} = 0$ and $\beta_{\iota,\iota-1} = (n-2)\cdot\pi/n = \pi - 2\pi/n$. With this, Eq. (7) simplifies to

$$\begin{aligned} \dot{\mathfrak{l}}(t) &= \sum_{\iota=1}^{n} -(\cos(\pi - 2\pi/n) + \cos 0) \\ &= -n \cdot (1 - \cos(2\pi/n)) \\ &\geq -n \cdot 2\pi^2/n^2 = -2\pi^2/n, \end{aligned} \tag{10}$$

where the inequality uses Lemma 2(b). Since the initial configuration has length $n$, this yields the lemma's claim. □

See Fig. 4 for an illustration of Lemma 8. When introducing the MOVE-ON-BISECTOR protocol in the next section, we will also encounter a "best-case" contracting protocol (global MOVE-ON-BISECTOR).

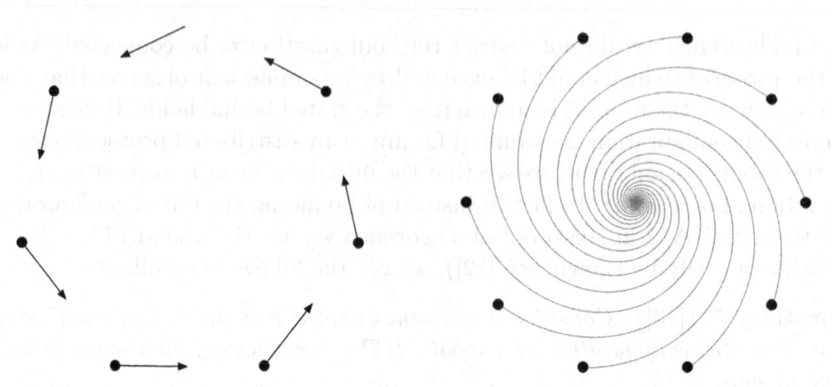

(a) The robots' velocity vectors for $n = 7$. (b) The robots' trajectories for $n = 10$.

**Fig. 4.** Configuration, velocity vectors, and trajectories of the $n$-bug problem from Lemma 8.

# 5 Near-Optimal Continuous Protocols for Gathering

Section 4 showed that any contracting robot formation protocol has at most quadratic gathering time and that this bound is tight in the sense that there are contracting protocols (and corresponding initial configurations) that have at least quadratic gathering time. However, there is still hope that a specific contracting protocol can have a much better – maybe even linear – gathering time. This is indeed true and has been proved for the so-called MOVE-ON-BISECTOR protocol in [5]. This section provides a sketch of this result; see [5] for the full proof (and further related results).

## 5.1 The MOVE-ON-BISECTOR Protocol

At each time $t$, each robot $i$ observes all positions of its neighbors (i.e., other robots within its visual range). It then computes the *local convex hull* $\mathrm{CH}_i(t)$ of its own and any observed positions. Given this local convex hull, robot $i$ performs the following actions:

(a) If $i$ is on a vertex of $\mathrm{CH}_i(t)$, it moves with speed 1 along the angle bisector of this vertex.
(b) If $i$ is not on a vertex but on the boundary $\partial\mathrm{CH}_i(t)$ of the local convex hull, it moves with the corresponding line such that it stays on it and maintains the ratio of distances between its two neighbors on the boundary.
(c) If $i$ is strictly inside $\mathrm{CH}_i(t)$, it does not move.

This protocol was originally suggested by Gordon et al. [8] (using a slightly different description and not under this name). See Fig. 5a for an illustration.

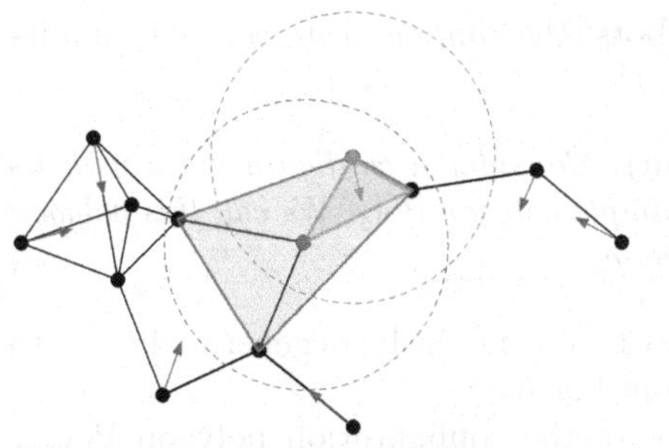

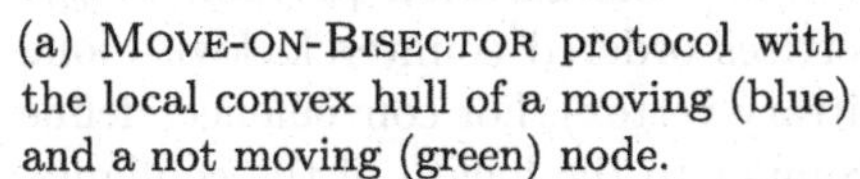

(a) Move-on-Bisector protocol with the local convex hull of a moving (blue) and a not moving (green) node.

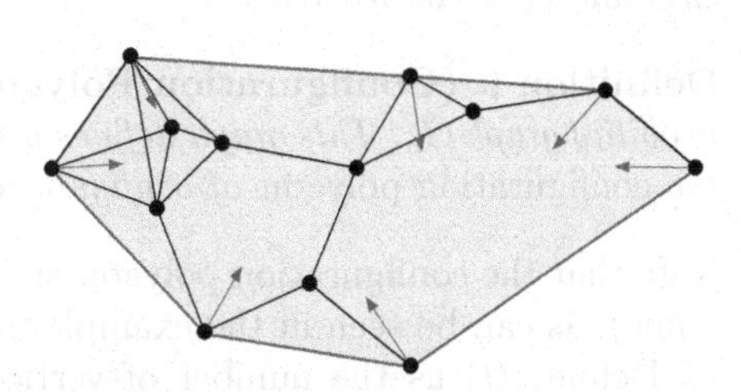

(b) Global Move-on-Bisector protocol for the same configuration.

**Fig. 5.** (Local) Move-on-Bisector vs. global Move-on-Bisector. Note that the lengths of the velocity vectors is not accurate. (Color figure online)

**Global Move-on-Bisector – A Best-Case Contracting Protocol.** Before we analyze the Move-on-Bisector protocol, let us give an intuition why it should be fast. Remember that our definition of contracting robot formation protocols is not restricted to *distributed* protocols. So, consider a global variant of Move-on-Bisector, where the only difference in the protocol description is that robots use the global convex hull $\mathrm{CH}_{c(t)}$ instead of their respective local convex hull $\mathrm{CH}_i(t)$. This is obviously not a distributed protocol, as robots do not have, in general, knowledge of the global convex hull. However, it is a contracting protocol (see also Fig. 5b). In fact, in a sense this is an "optimal" contracting gathering protocol: Using an analysis similar to Sect. 4.1, one can see that, as long as the robots have not yet gathered, the length of the configuration at time $t$ decreases at a constant rate, yielding a gathering time that is linear in the diameter $\Delta$ of the initial configuration. Since robots move at unit speed, this is asymptotically optimal.

Unfortunately, we cannot analyze the actual Move-on-Bisector protocol in the same way as its global counterpart. Since robots move not along the vertices of the global convex hull, the length of the configuration does not necessarily decrease at a constant speed. However, the next section introduces a "local" variant of a configuration's length, which we call *stretch*. With some additional work, we can show (Sect. 5.3) that not only does this stretch decrease at a constant rate but it is also linear in the configuration's diameter. This yields the same asymptotical optimal guarantee as the global Move-on-Bisector protocol.

## 5.2 Preliminaries

Instead of using the robots' convex hull and its circumference as a progress measure, our progress measure for the Move-on-Bisector protocol is based on the simple polygon spanned by the robots (the *configuration polygon*) and its circumference (its *stretch*).

**Definition 9 (Configuration Polygon).** *Consider a configuration $c$ with its visibility graph $G_c$. This graph defines a simple polygon* $\mathrm{Pol}_c$. *We call this polygon the* configuration polygon *of configuration $c$.*

Note that the configuration polygon or parts of it might be degenerated (e.g., to a line), as can be seen in the example from Fig. 6.

Define $q(t)$ as the number of vertices of the configuration polygon $\mathrm{Pol}_{c(t)}$. Let $w_1(t), w_2(t), \dots, w_{q(t)}(t)$ denote the vertices ordered counterclockwise long the boundary of $\mathrm{Pol}_{c(t)}$ (starting at an arbitrary vertex). For convenience, define $w_0(t) := w_{q(t)}(t)$ and $w_{q(t)+1}(t) := w_1(t)$. Similar to the length of the configuration we can define the *stretch* of a configuration as follows:

**Definition 10 (Stretch).** *The* stretch $\mathfrak{s}(t)$ *of the configuration at time $t$ is*

$$\mathfrak{s}(t) := \sum_{\iota=1}^{q(t)} \|w_\iota(t) - w_{\iota-1}(t)\|_2 \tag{11}$$

*We use the shorthand $\mathfrak{s} := \mathfrak{s}(t)$ to denote the stretch of the initial configuration.*

For a vertex $w_\iota(t)$ of the configuration polygon at time $t$ we define $\alpha_\iota(t) \in [0, 2\pi)$ as the *inner* angle of the configuration polygon at that vertex. We use $\mathcal{W}(t) := \{\, \iota \in [q(t)] \mid \alpha_\iota(t) < \pi \,\}$ to characterize the set of all convex angles of the polygon and $\overline{\mathcal{W}}(t) := [q(t)] \setminus \mathcal{W}(t)$ for the remaining (concave) angles. See Fig. 6 for an illustration of the notions defined above.

An important observation is, that a robot at the vertex $w_\iota$ of the configuration polygon at time $t$ is not necessarily on the vertex of its local convex hull. Indeed, if the inner angle at that vertex is concave ($>\pi$), it may be inside its local convex hull and may, thus, not move. However, any robot at a vertex that forms a convex inner angle is guaranteed to move, as they must be on a vertex of their local convex hull. These are exactly the vertices $w_\iota$ with $\iota \in \mathcal{W}$.

Note that while the trajectory of the robots is continuous, the direction in which a robot moves may change in a non-continuous way. This may happen if the visibility graph changes. Moreover, a change in the visibility graph may also influence the length of the current configuration in a non-continuous way. However, it is not hard to see (and it has been proved in [8]) that once two robots are within visual range, they will not lose visibility. Thus, there can be only a finite amount of these discontinuities. Moreover, by the definition of the visibility graph, the number of the configuration polygon's vertices cannot increase at such a discontinuity. This leads to the following observation.

**Observation 11.** *A change in the visibility graph cannot increase the current configuration's stretch.*

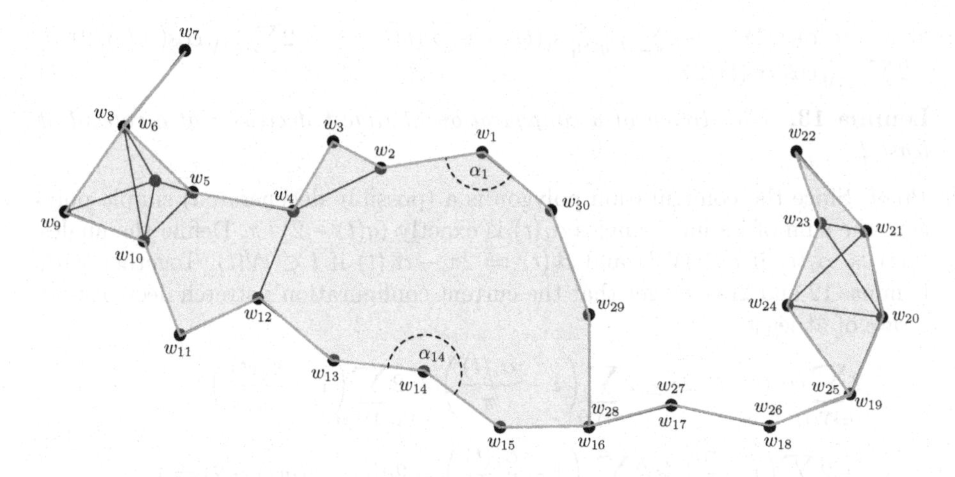


**Fig. 6.** The configuration polygon and its vertices of a configuration $c$ and two examples for the inner angles $\alpha_\iota(t)$ (the image omits the time parameter to improve readability). Note that it is degenerated at several places, causing a single robot position to represent possibly multiple vertices (at most 6 by Lemma 14). The stretch is the length of the red boundary (where degenerated parts are counted twice).

### 5.3 Analysis of the Move-on-Bisector Protocol

We are now ready to sketch the analysis of the Move-on-Bisector protocol. The interested reader will find the full analysis in [5]. First we show, similar to the proof of Theorem 6 a lower bound on the rate at which the current configuration's stretch decreases as long as the robots have not yet gathered (Lemmas 12 and 13). Afterward, we argue that a configuration's stretch is linear in its diameter. Combining these easily yields the desired bound on the gathering time of Move-on-Bisector (Theorem 15).

**Lemma 12.** *The stretch of a configuration at time $t$ changes at a rate of*

$$\dot{\mathfrak{s}}(t) \leq -2 \sum_{\iota \in \mathcal{W}(t)} \cos(\alpha_\iota(t)/2). \tag{12}$$

*Proof.* Fix a time $t \geq 0$ and $\iota \in [q(t)]$. We consider how the distance $d(t) := \|w_\iota(t) - w_\iota(t)\|_2$ changes. Let $s_\iota(t)$ denote the speeds of vertex $\iota$ at time $t$.[2] Set $\beta_\iota(t) := \alpha_\iota(t)$ if $\iota \in \mathcal{W}(t)$ and $\beta_\iota(t) := 2\pi - \alpha_\iota$ if $\iota \notin \mathcal{W}(t)$. By Lemma 3, the distance changes at a rate of

$$\dot{d}(t) = -s_\iota(t) \cdot \cos(\beta_\iota(t)/2) - s_{\iota-1}(t) \cdot \cos(\beta_{\iota-1}(t)/2). \tag{13}$$

Note that the speed of any vertex $w_\iota(t)$ with $\iota \in \mathcal{W}(t)$ equals 1. Summing over all $\iota \in \{1, 2, \ldots, q(t)\}$ we get that the current configuration's stretch changes

[2] As in the proof of Theorem 6, we identify vertices with the robots positioned on them.

at a rate of $\dot{\mathfrak{s}}(t) = -2\sum_{\iota \in [q(t)]} s_\iota(t) \cdot \cos(\beta_\iota(t)/2) \leq -2\sum_{\iota \in \mathcal{W}} \cos(\beta_\iota(t)/2) = -2\sum_{\iota \in \mathcal{W}} \cos(\alpha_\iota(t)/2)$. □

**Lemma 13.** *The stretch of a configuration at time t decreases at a rate of at least 4.*

*Proof.* Since the configuration polygon is a (possibly degenerated) simple polygon, the sum of its inner angles $\alpha_\iota(t)$ is exactly $(q(t) - 2) \cdot \pi$. Define the angles $\beta_\iota(t) := \alpha_\iota(t)$ if $\iota \in \mathcal{W}(t)$ and $\beta_\iota(t) := 2\pi - \alpha_\iota(t)$ if $i \notin \mathcal{W}(t)$. Together with Lemmas 12 and 2(a) we get that the current configuration's stretch decreases at a rate of at least

$$\begin{aligned} 2\sum_{\iota \in \mathcal{W}(t)} \cos(\alpha_\iota(t)/2) &\geq 2\sum_{\iota \in \mathcal{W}(t)} \left(1 - \frac{\alpha_\iota(t)}{\pi}\right) = 2\sum_{\iota \in \mathcal{W}(t)} \left(1 - \frac{\alpha_\iota(t)}{\pi}\right) \\ + 2\sum_{\iota \notin \mathcal{W}(t)} \left(1 - \frac{\pi}{\pi}\right) &\geq 2\sum_{\iota \in [q(t)]} \left(1 - \frac{\alpha_\iota(t)}{\pi}\right) = 2q(t) - 2(q(t) - 2) = 4, \end{aligned} \tag{14}$$

finishing the proof. □

Together, Observation 11 and Lemma 13 yield a bound of $\mathrm{O}(\mathfrak{s})$ on the gathering time. To get our desired bound of $\mathrm{O}(n)$, we use a result from [5] showing that $\mathfrak{s} \leq 6n$. This might seem trivial at first glance: After all, the stretch $\mathfrak{s}$ corresponds to the length of a chain of actual robots that forms the configuration polygon. However, note that while the configuration polygon is simple, it might be degenerated, such that, for example, parts of it may form a line. Thus, the robots along the above-mentioned chain are not necessarily unique. However, using the basic geometry of the underlying visibility graph, one can show that no node appears more than 6 times in this chain.

**Lemma 14** (**[5, Lemma 5.7]**). *For any configuration of n robots and stretch $\mathfrak{s}$, we have $\mathfrak{s} \leq 6n$.*

Combining Observation 11 and Lemmas 13 and 14, we immediately get the following result.

**Theorem 15.** *Consider an initial configuration of n robots and stretch $\mathfrak{s}$. In the worst-case, the* Move-on-Bisector *protocol gathers the robots in time at most $\mathfrak{s}/4 \leq 3n/2$. This is asymptotically optimal for the gathering problem.*

## 6 Avoiding Collisions

The final section of this chapter gives a brief outlook of how one can avoid (early) collisions when gathering multiple point robots. Here, by collision we mean situations where two or more robots share the same position in the Euclidean plane. A configuration is called *collision-free* if no two robots share the same position. When the robots are gathered, we have a collision between all robots. We call this the *final collision*, which is not avoidable if we want to gather. Any collision between two or more robots before all robots are gathered is called an *early collision*. The question we study in this section is:

*Are there continuous gathering protocols that, if started in a collision-free configuration, guarantee that no early collisions happen?*

Such a gathering protocol is called *collision-free.*

Many natural gathering protocols, including the MOVE-ON-BISECTOR protocol, are prone to early collisions. In fact, in the analysis of both discrete and continuous setting, collisions are often seen as a success, as robots that collided behave identical (in synchronous and deterministic robot formation protocols). Thus, there can be at most $n-1$ collisions, allowing us to use the number of collisions as a progress measure. However, from a practical standpoint and, collisions should be avoided as much as possible.

The rest of this section surveys recent results [12,13] that made progress towards answering the above question.

### 6.1 A Candidate for an Almost Collision-Free Gathering Protocol

The GO-TO-THE-CENTER protocol is another simple and natural gathering protocol. Here, each robot moves with speed 1 towards the center of the *minimum enclosing circle* of all robot positions it currently sees.

In the discrete setting, this algorithm was introduced by Ando et al. [1] who showed that it indeed gathers all robots, but the authors provided no bound on the gathering time. A quadratic upper bound in this discrete setting was later shown by Degener et al. [4]. In the continuous setting, we can simply use our framework from Sect. 4: Indeed, it is easy to verify that GO-TO-THE-CENTER is a contracting protocol [13]. This allows us to apply Corollary 7 to get a quadratic bound on its gathering time.

Unfortunately, GO-TO-THE-CENTER is also prone to collisions. In fact, the analysis of its discrete variant [4] is partly based on collisions. However, as noted by Li et al. [12], one can slightly change the definition of GO-TO-THE-CENTER to get a promising candidate for an collision-free, continuous gathering protocol (or almost collision-free; see below).

**GO-TO-THE-GABRIEL-CENTER.** We can rephrase GO-TO-THE-CENTER as follows: each robot moves with speed 1 towards the center of the minimum enclosing circle of its (inclusive) neighborhood in the visibility graph. A variant of GO-TO-THE-CENTER – called GO-TO-THE-GABRIEL-CENTER and due to Li et al. [12] – is defined using the same phrasing but uses the so-called *Gabriel graph* [7] instead of the visibility graph:

**Definition 16 (Gabriel Graph).** *The* Gabriel graph $GG_c = (VG_c, EG_c)$ *of configuration c is a subgraph of the visibility graph. It has the same vertex set* $VG_c = V_c = \{ c_i \mid i \in [n] \}$. *The edge set* $EG_c \subseteq E_c$ *consists of all edges* $\{ u, v \} \in E_c$ *such that the interior of the smallest enclosing circle of u and v does not contain another robot's position.*

See Fig. 7 for an example of how the Gabriel graph differs from the visibility graph.

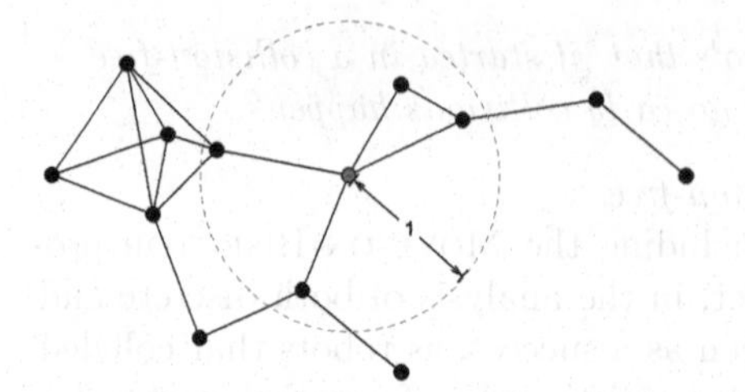


(a) Visibility graph. There is an edge between any two nodes that are within distance 1 of each other.

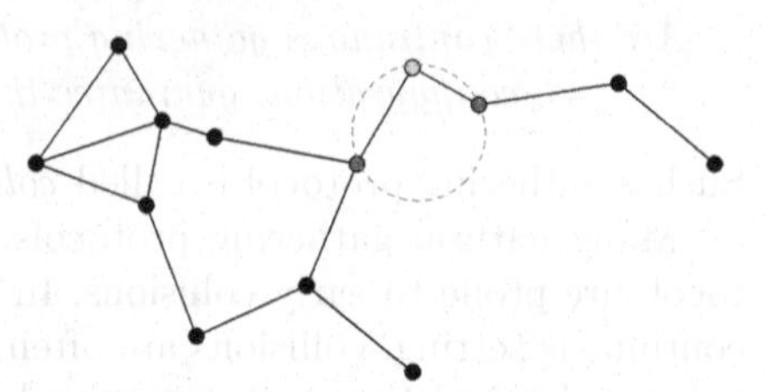

(b) Gabriel graph. The two red nodes are not connected because of the yellow node within their smallest enclosing circle.

**Fig. 7.** Visibility graph vs. Gabriel graph.

With this definition, we can now formally define the GO-TO-THE-GABRIEL-CENTER protocol. Here, at each time $t$ each robot $i$ performs the following actions:

(a) Robot $i$ computes the minimum enclosing circle $\mathfrak{C}_i(t)$ of all robots in its neighborhood (including $i$ itself) of the Gabriel graph $GG_{c(t)}$. Let $T_i(t)$ denote the center of $\mathfrak{C}_i(t)$ (robot $i$' s target point).
(b) If $c_i(t)$ equals $T_i(t)$, robot $i$ moves with $T_i(t)$.
(c) If $c_i(t)$ is different from $T_i(t)$, robot $i$ moves with speed 1 towards $T_i(t)$.

As with the GO-TO-THE-CENTER protocol, it is easy to see (and has been proved in [13]) that the GO-TO-THE-GABRIEL-CENTER protocol is contracting and, thus, gathers in quadratic time.

### 6.2 Collisions in the GO-TO-THE-GABRIEL-CENTER Protocol

Experimental results indicate that for "typical" initial configurations, the GO-TO-THE-GABRIEL-CENTER protocol causes no early collisions. In fact, in the one dimensional case (where all robots start on a line) this can be easily proved [12]. However, in the two dimensional case there are, in fact, some (quite symmetric) situations that exhibit early collisions. See Fig. 8 for an example. One of the central open questions left in [12] is whether the following conjecture is true.

**Conjecture 17.** *The set of initial configurations that lead to early collisions of the* GO-TO-THE-GABRIEL-CENTER *has Lebesgue measure 0.*

A less formal variant of this conjecture states that independent small random perturbations of the robots' initial positions ensure that, with probability 1, there will be no early collisions. Note that even for the special case of $n = 4$ robots, a proof of Conjecture 17 seems challenging.

Even if Conjecture 17 is true, it remains open whether a distributed continuous gathering protocol can achieve truly collisionless gathering. There is currently

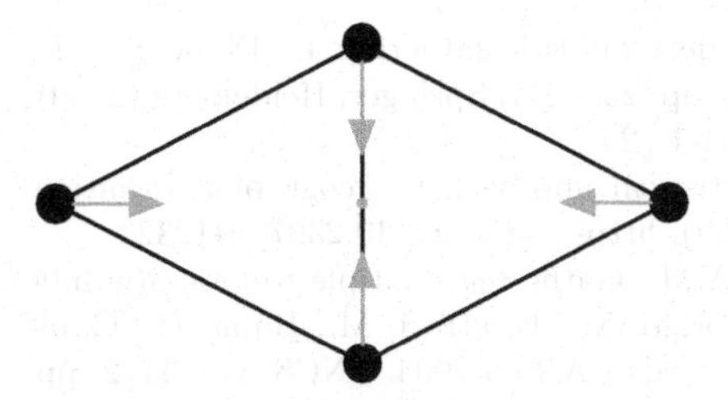

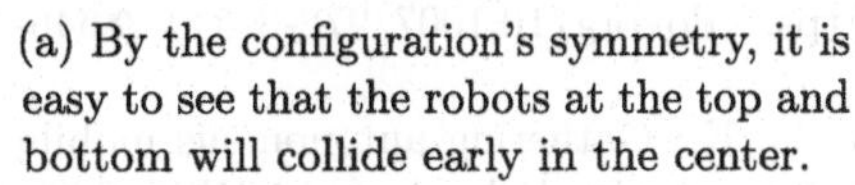

(a) By the configuration's symmetry, it is easy to see that the robots at the top and bottom will collide early in the center.

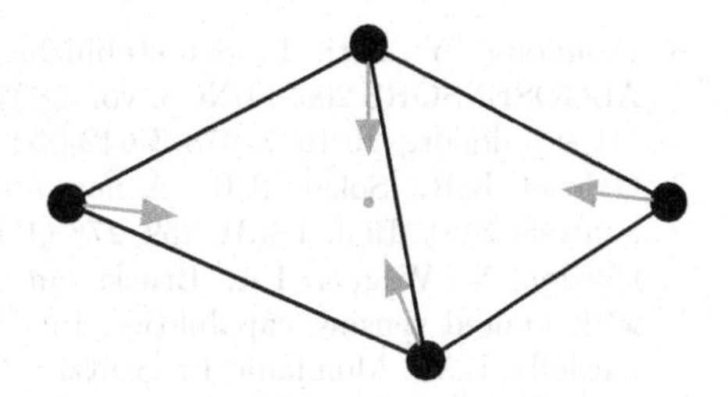

(b) A slight perturbation that destroys the symmetry is enough to ensure that no early collision occurs.

**Fig. 8.** Early collisions in the Go-to-the-Gabriel-Center protocol.

only one continuous gathering protocol [13] that completely avoids early collisions, but it requires a considerably more complex robot model. Namely, robots must be non-oblivious (they have one memory bit), chiral (they share a common left/right orientation), and luminous (the contents of a robot's memory bit is visible to other robots). Additionally, this protocol requires quadratic time to gather all robots, so it is much slower than the Move-on-Bisector protocol from Sect. 5. So even using a more complex robot model like the one above, it would be interesting to find a collision-free continuous gathering protocol that has linear gathering time.

## References


1. Ando, H., Suzuki, I., Yamashita, M.: Formation and agreement problems for synchronous mobile robots with limited visibility. In: Proceedings of 10th International Symposium on Intelligent Control (ISIC), pp. 453–460, August 1995. https://doi.org/10.1109/ISIC.1995.525098
2. Cieliebak, M., Flocchini, P., Prencipe, G., Santoro, N.: Solving the robots gathering problem. In: Baeten, J.C.M., Lenstra, J.K., Parrow, J., Woeginger, G.J. (eds.) ICALP 2003. LNCS, vol. 2719, pp. 1181–1196. Springer, Heidelberg (2003). https://doi.org/10.1007/3-540-45061-0_90
3. Cohen, R., Peleg, D.: Convergence properties of the gravitational algorithm in asynchronous robot systems. SIAM J. Comput. **34**(6), 1516–1528 (2005). https://doi.org/10.1137/S0097539704446475
4. Degener, B., Kempkes, B., Langner, T., Meyer auf der Heide, F., Pietrzyk, P., Wattenhofer, R.: A tight runtime bound for synchronous gathering of autonomous robots with limited visibility. In: Rajaraman R., Meyer auf der Heide, F., (eds.) SPAA 2011: Proceedings of the 23rd Annual ACM Symposium on Parallelism in Algorithms and Architectures, San Jose, CA, USA, 4–6 June 2011 (Co-located with FCRC 2011), pp. 139–148. ACM (2011). https://doi.org/10.1145/1989493.1989515
5. Degener, B., Kempkes, B., Kempkes, P., Meyer auf der Heide, F.: Linear and competitive strategies for continuous robot formation problems. TOPC **2**(1), 2:1–2:18 (2015). https://doi.org/10.1145/2742341

6. Dieudonné, Y., Petit, F.: Self-stabilizing deterministic gathering. In: Dolev, S. (ed.) ALGOSENSORS 2009. LNCS, vol. 5804, pp. 230–241. Springer, Heidelberg (2009). https://doi.org/10.1007/978-3-642-05434-1_23
7. Gabriel, K.R., Sokal, R.R.: A new statistical approach to geographic variation analysis. Syst. Biol. **18**(3), 259–278 (1969). https://doi.org/10.2307/2412323
8. Gordon, N., Wagner, I.A., Bruckstein, A.M.: Gathering multiple robotic a(ge)nts with limited sensing capabilities. In: Dorigo, M., Birattari, M., Blum, C., Gambardella, L.M., Mondada, F., Stützle, T. (eds.) ANTS 2004. LNCS, vol. 3172, pp. 142–153. Springer, Heidelberg (2004). https://doi.org/10.1007/978-3-540-28646-2_13
9. Izumi, T., Katayama, Y., Inuzuka, N., Wada, K.: Gathering autonomous mobile robots with dynamic compasses: an optimal result. In: Pelc, A. (ed.) DISC 2007. LNCS, vol. 4731, pp. 298–312. Springer, Heidelberg (2007). https://doi.org/10.1007/978-3-540-75142-7_24
10. Katayama, Y., Tomida, Y., Imazu, H., Inuzuka, N., Wada, K.: Dynamic compass models and gathering algorithms for autonomous mobile robots. In: Prencipe, G., Zaks, S. (eds.) SIROCCO 2007. LNCS, vol. 4474, pp. 274–288. Springer, Heidelberg (2007). https://doi.org/10.1007/978-3-540-72951-8_22
11. Katreniak, B.: Convergence with limited visibility by asynchronous mobile robots. In: Kosowski, A., Yamashita, M. (eds.) SIROCCO 2011. LNCS, vol. 6796, pp. 125–137. Springer, Heidelberg (2011). https://doi.org/10.1007/978-3-642-22212-2_12
12. Li, S., Meyer auf der Heide, F., Podlipyan, P.: The impact of the gabriel subgraph of the visibility graph on the gathering of mobile autonomous robots. In: Chrobak, M., Fernández Anta, A., Gąsieniec, L., Klasing, R. (eds.) ALGOSENSORS 2016. LNCS, vol. 10050, pp. 62–79. Springer, Cham (2017). https://doi.org/10.1007/978-3-319-53058-1_5
13. Li, S., Markarian, C., Meyer auf der Heide, F., Podlipyan, P.: A continuous strategy for collisionless gathering. In: Fernández Anta, A., Jurdzinski, T., Mosteiro, M.A., Zhang, Y. (eds.) ALGOSENSORS 2017. LNCS, vol. 10718, pp. 182–197. Springer, Cham (2017). https://doi.org/10.1007/978-3-319-72751-6_14
14. Nguyen, H.G., Pezeshkian, N., Raymond, S.M., Gupta, A., Spector, J.M.: Autonomous communication relays for tactical robots. In: Proceedings of the 11th International Conference on Advanced Robotics (ICAR), pp. 35–40 (2003)
15. Olver, F.W., Lozier, D.W., Boisvert, R.F., Clark, C.W.: NIST Handbook of Mathematical Functions, 1st edn. Cambridge University Press, New York (2010)
16. Prencipe, G.: Impossibility of gathering by a set of autonomous mobile robots. Theor. Comput. Sci. **384**(2–3), 222–231 (2007). https://doi.org/10.1016/j.tcs.2007.04.023
17. Souissi, S., Défago, X., Yamashita, M.: Gathering asynchronous mobile robots with inaccurate compasses. In: Shvartsman, M.M.A.A. (ed.) OPODIS 2006. LNCS, vol. 4305, pp. 333–349. Springer, Heidelberg (2006). https://doi.org/10.1007/11945529_24
18. Watton, A., Kydon, D.W.: Analytical aspects of the $n$-bug problem. Am. J. Phys. **37**(2), 220–221 (1969). https://doi.org/10.1119/1.1975458

# Group Search and Evacuation

Jurek Czyzowicz[1], Kostantinos Georgiou[2], and Evangelos Kranakis[3(✉)]

[1] Dépt. d'informatique, Univ. du Québec en Outaouais, Gatineau, QC, Canada
[2] Department of Mathematics, Ryerson University, Toronto, ON, Canada
[3] School of Computer Science, Carleton University, Ottawa, ON, Canada
kranakis@scs.carleton.ca

**Abstract.** Group search and evacuation are fundamental tasks performed by a set of co-operating, autonomous mobile agents. The two tasks are similar in that they both aim to search a given domain so as to locate a target which has been placed at an unknown location in the domain. However they also differ in that the former terminates when the first searcher in the group reaches the target while the latter when the last searcher in the group reaches the target. Variations where termination is determined by some designated agent have also been considered. Depending on the domain being explored we distinguish *linear search* when the target is placed on the infinite line and *circular search* when the target is placed on the perimeter of a disk. The agents move with their own maximum speed, and the goal is to design algorithms that minimize the worst case termination time. Two communication models between the robots are being considered: in the *non-wireless* (or *face-to-face*) *communication* model, robots exchange information only when simultaneously located at the same point, and *wireless communication* in which robots can communicate with one another anywhere at any time. In this paper we survey some of the most interesting recent algorithmic results on search and evacuation concerning mobile agents with and without faults.

**Keywords:** Autonomous agents · Cycle · Evacuation · Exit · Line Search

## 1 Introduction

Search in theoretical computer science is primarily concerned with the algorithmic probing of a well-defined (data-) domain in order to find a stored target object. The main focus of this survey, is on presenting recent algorithmic developments on search performed by a group of collaborating autonomous agents. During the search, the mobile agents are pursuing their own trajectories and are required to locate a target and conclude the task in the minimum amount of time. To begin we introduce some basic concepts and ideas that will be used in later sections.

Research supported in part by NSERC Discovery grants.


P. Flocchini et al. (Eds.): Distributed Computing by Mobile Entities, LNCS 11340, pp. 335–370, 2019.
https://doi.org/10.1007/978-3-030-11072-7_14

## Searchers

Throughout this paper the terms mobile agent, searcher and robot will be identical. We assume that $n$ robots are initially placed at a start position on a geometric domain. The robots can move on a continuous trajectory with a predefined maximum speed (usually one) along a geometric domain (which is either an infinite line or a cycle in the plane).

## Search and Evacuation

We distinguish between search and evacuation. The former succeeds when the first searcher in the group reaches the target and the latter when the last searcher in the group reaches the target. Variations where only distinguished searchers need to reach the target are also considered. In all cases, the target can be identified only by robots that reach its location.

## Linear and Circular Search

The search domain may be either on an infinite line or on a closed curve, like disk. In the first case it is called linear search and was first proposed by Bellman [9] and independently by Beck [8] in a stochastic setting and by [6,7]) in a deterministic setting. In the *linear search model* the environment is an infinite line and the robots start at a given point, called the origin, on this line. An hidden object/target (exit) is placed on the line at a location which is unknown to the robots. In the second case it is sometimes called circular search and the model was first studied in [17]. In the *unit disk search model* the environment is a disk, usually of unit radius; the robots start their movement either at the center of the disk (and they can move anywhere on the plance) or on the disk (and they can move only on the perimeter). The only information robots have is that the target has been placed at an unknown location on the perimeter.

## Communication

Two models of communication between the robots are being considered: in the *non-wireless* (or *face-to-face*) *communication* model, robots exchange information only when they are simultaneously located at the same point, and *wireless communication* in which robots can communicate with one another anywhere at any time.

## Performance Metrics

When operating in geometric environments, performance is measured by the geometric distance traveled, while in discrete settings the number of hops in a trajectory. By default, algorithmic performance is measured with respect to worst case analysis. The competitive ratio of an algorithm is the worst case ratio between the performance of the algorithm and the performance of the best offline algorithm, i.e., an algorithm that knows in advance the location of the "hidden" object. In traditional search the task ends when the first agent finds the object. This is different for evacuation since we are interested in the search *makespan* which refers to the max (or even, total) length of the search strategy when all the agents (or sometimes a specific number of the agents) have finished processing their respective tasks.

**Some Related Literature**

There is a vast literature investigating all aspects of search. Several papers are cited throughout the survey but here we mention only a few books. It is worth cittng the classical book on optimal search [36], the compendium of search problems in [1] and the game theoretic approach in the treatise [5]. Applications of search to foraging and evolution can be found in [30,35]. Further, [33] provides an introduction to the analysis and design of dynamic multiagent networks, and [14] an introduction to the distributed control of robotic networks with a blend of computer science and control theory. Additional specialized monographs are [10,11] as well as the pleasant monograph [34] which provides a different perspective with chases and escapes.

## 2 Linear Search

In this section we focus on linear search on an infinite line and discuss search for robots which may suffer from crash and/or byzantine faults. In the last part of the section we also explore search on linear terrains (a generalization of the infinite line).

### 2.1 Crash Faults

In this section we present the linear search by a collection of robots, some of which may turn out to be faulty [24].

**Model Specifics and Problem Definition**

By $A(n, f)$ we denote a linear search problem using $n$ mobile robots where at most $f$ robots may turn out to be faulty. The robots are placed at the origin of an infinite line. Robots may walk along the line with the same unit speed. At some point on the line, at distance $d$ from the origin, is placed a stationary target that needs to be found by the collection of robots. A robot finds the target when it visits the position of the line where the target is located. A sub-collection of up to $f$ robots may experience crash faults. A faulty robot cannot identify the target despite visiting its location. The search is completed when at least one non-faulty robot finds the target.

The bound $f$ on the number of faulty robots is known to the search algorithm. However, the identities of the faulty robots are unknown to the algorithm. Consequently, the set of the faulty robots is controlled by the adversary, which knows the search algorithm in advance. The distance $d$ to the target position is also unknown to the algorithm, but the performance of the algorithm is measured as a function of $d$. More exactly, the algorithm efficiency is defined by a competitive ratio, which is the worst case ratio of the arrival of the first reliable robot to the target, and the distance $d$ from the source to the target.

For any potential target position, the best adversarial strategy is to choose the first $f$ robots incoming to such position to be faulty. It is clear that, when $n \geq 2f+2$, there exists a simple algorithm with competitive ratio 1 that sends two groups of $f+1$ robots in each direction of the line. Below we focus on efficient search strategies when $f < n < 2f+2$.

**Zig-Zag Strategies**

For $n < 2f+2$, the trajectories of all robots considered are using *zig-zag strategies*, i.e. solutions in which each robot walks alternately in both directions, where its turning points for each direction are more and more distant from the origin. It is useful to illustrate the zig-zag movements using the Cartesian plane in which $x$-axis corresponds to the line of robots' movement and $t$-axis represent time. The trajectory of a robot is represented by a function of time $t$ whose absolute slope is bounded by 1 (as robots move using maximal unit speed). By a turning point $(x_i, t_i)$ we mean that at time $t_i$ the robot is at point $x_i$ of the line and it changes the direction of its movement.

The strategies used are such that the turning points $(x_i, t_i)$ belong to some geometric cone of the Cartesian plane. Let $C_\beta$ denote the cone starting at the origin and extending in the positive direction of $t$-axis such that it is bounded by two semi-lines, each having angle $\beta$ with the $t$-axis. We have

**Definition 1.** *Suppose that at time $a\beta$ a robot visits point $a$ of the line. We say that the robot follows a* zig-zag *movement defined by cone $C_\beta$ and point $(a, a\beta)$ if the robot walks with unit speed inside the cone $C_\beta$ starting at point $(a, a\beta)$ and that it reverses its direction whenever it arrives at the boundary of $C_\beta$.*

Figure 1 illustrates movements of robots defined by the cone $C_\beta$.

It is possible to prove the following lemma.

**Lemma 1** ([24]). *Let $x_0$ be the initial position of a robot on the line at time $t_0$. Consider the zig-zag movement of this robot defined by point $(x_0, t_0)$ and the cone $C_\beta$, where $\beta > 1$. The turning points of the robot are given by the formula*

$$x_i = x_0 \left(\frac{\beta+1}{\beta-1}\right)^i (-1)^i \tag{1}$$

**Proportional Schedules**

When several robots participate in the search, they all move according to zig-zag strategies using the same cone $C_\beta$, but the choice of the parameter $\beta$ depends on the ratio of assumed bound of the faulty robots. Moreover, the most efficient

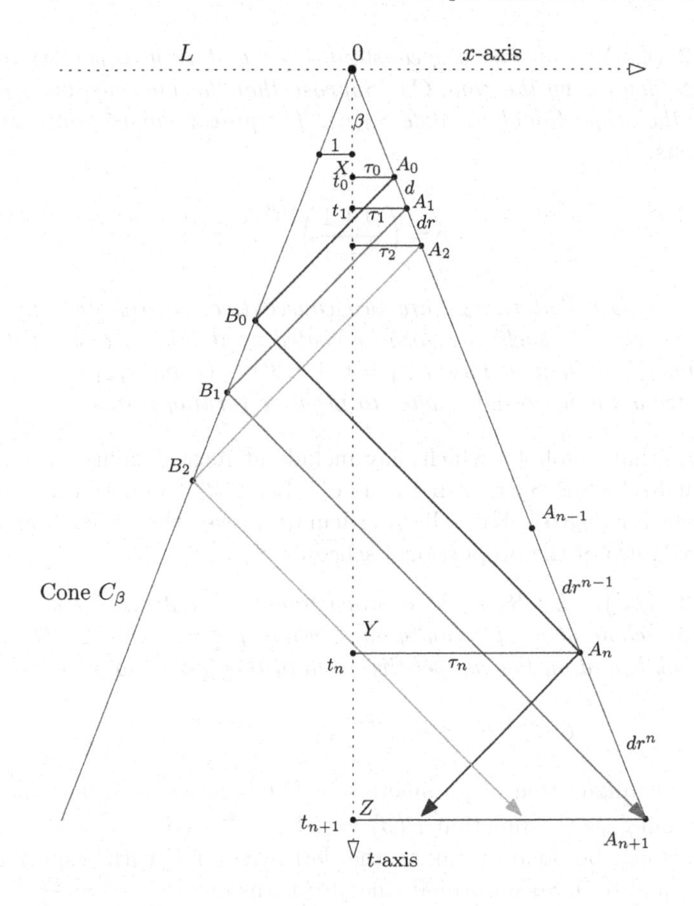


**Fig. 1.** Proportional schedule for $n$ robots $a_0, a_1, \ldots, a_{n-1}$, in the cone $C_\beta$.

search is achieved when the robots' trajectories are forming so-called *proportional schedules* (see Fig. 1). Roughly speaking, for a proportional schedule, the infinite sequence of the consecutive positive $x$-coordinates of the turning points of all robots form a geometric progression. The same is true for the consecutive negative $x$-coordinates of all turning points. More precisely, we give the following definition.

**Definition 2.** *Suppose that a collection of robots performs zig-zag movements defined by the same cone $C_\beta$. Consider the infinite sequence of the consecutive positive turning points $0 < \tau_0 < \tau_1 < \cdots$ obtained from the zig-zag movements of all the robots of the collection. We say that the schedule is* proportional *if for some real value $r$, the ratio $\frac{\tau_{i+1}-\tau_i}{\tau_i-\tau_{i-1}} = r$, for $i = 1, 2, \ldots$. We call $r$ the proportionality ratio of the schedule.*

We have the following lemma.

**Lemma 2** ([24]). *Consider any constant $\beta > 1$ and $n$ robots performing zig-zag movements defined by the cone $C_\beta$. Suppose that the movements of the robots constitute the proportional schedule $S_\beta(n)$. The proportionality ratio of schedule $S_\beta(n)$ equals*

$$r = \left(\frac{\beta+1}{\beta-1}\right)^{2/n}. \tag{2}$$

*Moreover, suppose that $\tau_i, \tau_{i+1}$ are two consecutive, positive turning points of some two robots $a, b$, such that robot $a$ visited $\tau_i$ at time $t_i$ and robot $b$ visited $\tau_{i+1}$ at time $t_{i+1}$. Then we have $t_{i+1} = t_i + \tau_i\beta(r-1)$ and $\tau_{i+1} = r\tau_i$. Note that by symmetry a similar result applies to negative turning points.*

Suppose that $n$ robots, which may include at most $f$ faulty ones execute a proportional schedule $S_\beta(n)$ using cone $C_\beta$. Let $CR_\beta^{n,f}$ denote the competitive ratio of schedule $S_\beta(n)$. The following lemma proves the upper bound on the competitive ratio of the proportional schedule $S_\beta(n)$.

**Lemma 3** ([24]). *Let $S_\beta(n)$ be a proportional schedule executed by $n$ robots, which may include at most $f$ faulty ones, where $f < n < 2f+2$. Then we have the following bound on the competitive ratio of this proportional schedule:*

$$CR_\beta^{n,f} = (\beta+1)^{\frac{2f+2}{n}}(\beta-1)^{1-\frac{2f+2}{n}} + 1. \tag{3}$$

For any configuration of parameters $n, f$ it is possible to find the value of $\beta$ which minimizes the function $F(\beta) := (\beta+1)^{\frac{2f+2}{n}}(\beta-1)^{1-\frac{2f+2}{n}} + 1$, where $\beta > 1$. This can be done by taking the derivative of $F$ with respect to $\beta$ and setting it equal to 0. Such optimal vale of $\beta$ turns out to be $\beta = \frac{4f+4}{n} - 1$.

Consequently, we conclude with the following theorem.

**Theorem 1** ([24]). *Consider a collection of $n$ robots up to $f$ of which are faulty. Then there exists an algorithm $A(n, f)$ performing search on infinite line, whose competitive ratio is at most equal to*

$$\left(\frac{4f+4}{n}\right)^{\frac{2f+2}{n}}\left(\frac{4f+4}{n} - 2\right)^{1-\frac{2f+2}{n}} + 1 \tag{4}$$

Figure 2 illustrates the search algorithm for three robots containing one that may turn out to be faulty. Inside the cone $C_\beta$ may be identified the region $R$ bound by bold polygonal lines (that reminds the Sacrada Familia Barcelonian church). Each point $(x, t)$ inside region $R$ has the property that before time $t$ point $x$ of the line has been visited by at least two robots, hence it is considered successfully searched. If point $(x, t)$ is outside region $R$, that means that only one robot visited $x$ before time $t$ (as that robot may be faulty, $x$ cannot be declared as successfully searched). The competitive ratio of the algorithm is determined by the two lines that pass through the origin, belong to $R$ and have maximal and minimal slopes (thin grey lines $OA$ and $OB$ on Fig. 2). Indeed, the competitive ratio equals the smaller of the absolute values of both slopes. An interested reader may verify that if we perturb slightly the movement of any robot by changing slightly the positions where its trajectory touches the cone $C_\beta$ (i.e. the zig-zag strategy becomes not proportional) the competitive ratio becomes larger. This suggest that the proportional strategies are optimal. Indeed, the optimality of the proportional strategies has been formally proven in the recent work of Kupavskii and Welzl [31].

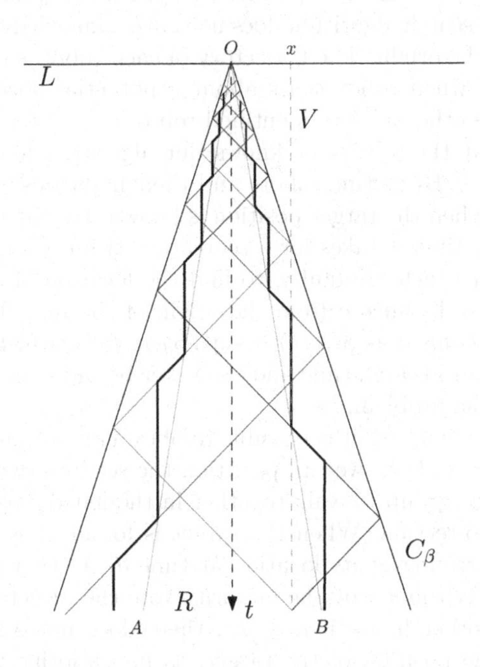


**Fig. 2.** Searching by three robots one of which is faulty.

## 2.2 Byzantine Faults

The section presents linear search when a collection of searchers contains Byzantine robots (cf. [23]). A Byzantine robot may fail to see the target or it may communicate to other robots a position of the target that is not the real one.

### Model Specifics and Problem Definition

A collection of $n$ robots is initially placed at the origin of an infinite line. Each robot can move *left* or *right* along the line with a speed that does not exceed its maximum speed (which is the same for all robots). Robots have a distinct identity and they may communicate wirelessly, so a message sent by any robot is instantaneously heard by all other robots. As the trajectories of all robots are determined in advance, the only possible message that a robot may communicate is that it found the target. Some robots may turn out to experience Byzantine faults and they may fail to communicate a position of the target they find or they may communicate a position of the target which is wrong. Each communicated message is associated with the identity of the robot sending it and it is assumed that a Byzantine robot cannot lie on its identity. There are at most $f$ robots which may behave in Byzantine way.

The search algorithm is designed by a central authority, which knows the bound $f$ on the number of faulty robots, but is unaware which subset of robots is faulty, neither is their behavior predictable (i.e. faulty robots may misreport their findings). The search algorithm does not have a knowledge of the distance $d$ to the target from the origin. The trajectory of each robot is designed so it may be possibly altered when robot hears about a potential position of the target announced by some other, clearly identified robot.

We suppose that the adversary knows our algorithm and may choose the subset of (at most) $f$ Byzantine robots and their malicious actions in order to delay the moment when the target position is known. By $S_d(n, f)$ we denote the *search time*, i.e. the time it takes for a search algorithm using a collection of $n$ robots at most $f$ of which are faulty, to find the location of a target placed at an a priori unknown distance $d$ from the origin of the line. The corresponding *competitive ratio* is defined as $S(n, f) = \sup_d S_d(n, f)/d$, which is the worst case ratio of the algorithm's search time and the lower bound $d$ on the time taken by any algorithm for the problem.

Observe that if $n \geq 4f+2$ it is possible to obtain an optimal algorithm (with competitive ratio 1). Indeed, we can partition the set into two groups of $2f+1$ robots and make one group to walk together in the left direction and the other group in the right direction. When the target is found, it is announced by all non-faulty robots arriving at its location at time $d$. As that group contains at least $f+1$ non-faulty robots, by the majority vote the target position is clearly identified. On the other hand, if $n \leq 2f$, there does not exist any algorithm that may decide the position of the target, as no majority voting is possible. Consequently, the non-trivial solutions are interesting only when $2f < n < 4f+2$.

### Algorithms for Single Byzantine Robot

Following the last observation, and when $f = 1$, it is interesting to consider only the cases of $n = 3, 4$ or 5. The simplest case concerns the collection of $n = 4$ robots.

**Case 1:** $n = 4$. The algorithm instructs two groups of 2 robots to walk together in opposite directions. As each group contains at least one reliable robot, at some point a target position is announced, say at some position at distance $x$ from the origin. If two robots announce (i.e. report that the target is found), then the algorithm finishes in time $d = x$. Otherwise, i.e. when only one robot announces, the groups are instructed to swap their positions and then continue in the opposite directions. If the target is confirmed by the robots arriving at the position being announced, the algorithm finishes in time $3d$. Otherwise, a Byzantine robot is identified and the remaining robots continue, not reacting on any further announcements of the identified Byzantine robot. An announcement by any other robot (at time $2x+d$) finishes the algorithm. As $x < d$, we conclude that

$$S(4,1) \leq 3.$$

**Case 2:** $n = 5$. Two groups of 2 robots continue in opposite directions as in Case 1, and the fifth robot waits, stationary at the origin. Again an announcement must be made at some time $x$ and in case when two robots make the announcement the algorithm finishes at time $d = x$. Suppose then that a single robot makes that announcement. Then the robot which was stationary walks to the announcement point (for simplicity suppose that all other robots wait motionless for time $x$, until the fifth robot reaches the announced position). If this robot may confirm the announcement, then in time $2x = 2d$ the target is found. Otherwise the Byzantine robot is identified, the robots restart walking in the same direction and a further announcement by any other robot concludes the algorithm. The search time equals $x + d$, and as $x < d$ we have

$$S(5,1) \leq 2.$$

**Case 3:** $n = 3$. In this case, [23] conjectured that the best search algorithm was for all three robots to walk together performing a standard cow-path trajectory, that finishes in time $S_d(5,1) < 9d$. However only a lower bound $S(5,1) > 3.93$ was proven in [23]. This lower bound was later improved in [31] to

$$S(3,1) > \frac{8}{3}\sqrt[3]{4} \approx 5.23.$$

### Algorithms for Large Collections of Robots

The proposed upper bounds of the algorithms for large number of robots are usually a function of ratio $r = f/n$. As previously observed, we are interested in the interval $2f + 1 \leq n < 4f + 2$, so we are interested in the case when $1/2 \geq r \geq 1/4$. Observe first, that already for $n = 2f + 1$, the problem is feasible, i.e. there exists an algorithm, which finds the target in finite time. More exactly, for any $f \geq 0$ we have

$$S(2f+1, f) \leq 9.$$

Indeed, we can instruct the robots to walk together along the path and the majority vote before time $9d$ concludes the algorithm.

Asymptotically, for $1/2 < r < 1/4$ we have the competitive ratio

$$1 \leq S(f/r + c, f) \leq 9,$$

where $c$ denotes some constant value. More formally, we can define

$$\hat{S}(r) = \min \ \{q \mid \exists \text{ constant } c_r \text{ such that } \forall f > 0,\ S\left(f/r + c_r, f\right) \leq q\} \qquad (5)$$

Clearly, the larger is the proportion $1 - r$ of the non-faulty robots, the better is the competitive ratio that may be expected from an efficient search. Below is the example of an upper bound for some particular density of Byzantine robots.

**Proposition 1** ([23]). $S\left(\frac{14f+4}{5}, f\right) \leq 3$ *provided* $f \equiv 4 \mod 5$.

The idea of the algorithm providing the bound claimed in Proposition 1 is the following. Two groups of robots, $L$ and $R$, each containing $\frac{7f+2}{5}$ participants, walk in left and right directions, respectively. Suppose that an announcement is made at time $x$. If less than $\frac{2f+2}{5}$ robots vote that target is at $x$ or they vote that the target is not at $x$, then this subgroup is eliminated from consideration (as identified Byzantine robots) and both groups continue walking in their respective directions. So we can assume that the vote is such that each group voting on point $x$ contains at least $\frac{2f+2}{5}$ robots. By symmetry, suppose that the announcement has been made by a sub-collection of group $R$ (i.e. when visiting point $x > 0$). At this point, we send $\frac{3f+3}{5}$ robots belonging to group $L$ to move from their current position at $-x$ to point $x$. At the same time the sub-collections of $\frac{2f+2}{5}$ robots that voted YES and $\frac{2f+2}{5}$ that voted NO are sent from $x$ to $-x$. Once the groups sent swap their positions from $-x$ to $x$ and from $x$ to $-x$, two cases are possible. If the exit is confirmed at $x$ (note that altogether $\frac{7f+2}{5} + \frac{3f+3}{5} = 2f+1$ robots visited $x$ so the state of point $x$ is decided), then the algorithm terminates in time $3x = 3d$. Otherwise we can eliminate from the consideration a set of at least $\frac{2f+2}{5}$ Byzantine robots (present now at point $-x$), so there exist still at most $f' = f - \frac{2f+2}{5} = \frac{3f-2}{5}$ undisclosed Byzantine robots. At that point the number of robots $l'$ present at $-x$ whose state is unknown equals at least

$$l' \geq \frac{7f+2}{5} - \frac{3f+3}{5} + \frac{2f+2}{5} = \frac{6f+1}{5}.$$

At point $x$ we have then $r'$ robots and

$$r' \geq \frac{7f+2}{5} - 2\left(\frac{2f+2}{5}\right) + \frac{3f+3}{5} = \frac{6f+1}{5}.$$

As $l' = r' \geq 2f' + 1$ we have at both points $-x$ and $x$ the sub-collections of robots, each containing more reliable than Byzantine robots. Both groups continue searching away from the origin until a majority vote is present, which concludes the search. The cost of the algorithm is the sum of the search time $d$ and the swap time $2x$. As $x \leq d$ the claim of Proposition 1 is true.

The algorithms for larger values of the ratio $r$ of Byzantine robots may include more than one swap between the groups $L$ and $R$ during the search. The results for these cases are summarized in Table 1.

**Table 1.** Upper bounds on the asymptotic competitive search ratio $\hat{S}(r)$ for various ranges of $r$. Recall that for $r > \frac{1}{2}$ the search problem is infeasible.

| $r$ | $\leq\frac{1}{4}$ | $(\frac{1}{4}, \frac{3}{10}]$ | $(\frac{3}{10}, \frac{1}{3}]$ | $(\frac{1}{3}, \frac{5}{14}]$ | $(\frac{5}{14}, \frac{13}{34}]$ | $(\frac{13}{34}, \frac{19}{46}]$ | $(\frac{19}{46}, \frac{47}{110}]$ | $(\frac{47}{110}, \frac{65}{146}]$ | $(\frac{65}{146}, \frac{157}{396}]$ | $(\frac{157}{396}, \frac{1}{2}]$ |
|---|---|---|---|---|---|---|---|---|---|---|
| UB | 1 | 2 | 3 | 3 | 4 | 5 | 6 | 7 | 8 | 9 |

### 2.3 Linear Terrains

We now explore a setting (first considered in [26]) which generalizes the infinite straight line setting first considered in [6,7] and [8,9], in which the search domain is no longer a straight line but rather a linear terrain with "hills" and valleys. By this we mean that the search is along a curve which is formed by a continuous function (depicted in Fig. 3 whose representation $y = f(x)$ may be known to the robot. As an example, $f(x)$ may be a monotone polygon consisting of $n$ straight-line segments, for some integer $n \geq 1$.

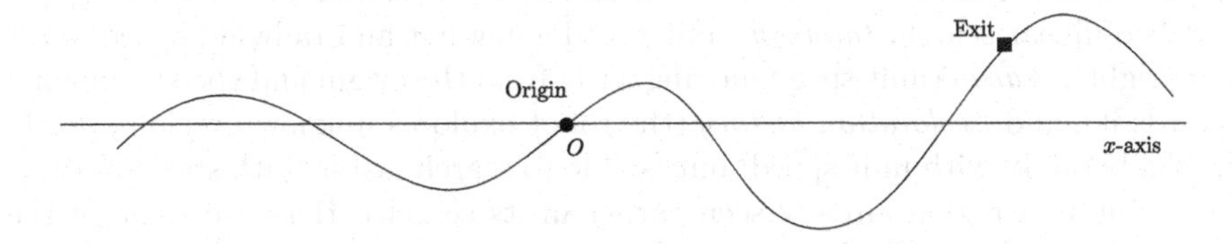


**Fig. 3.** Search in an infinite one-dimensional terrain $y = f(x)$. The robot may move in either direction along the terrain, the point $O$ is the origin (considered as the starting position of the robot) and the exit is located on the terrain at a position unknown to the robot.

#### Model Specifics and Problem Definition

The objective is to design search algorithms that achieve good competitive ratios for the time spent by the robot to complete its search divided by the time spent by an omniscient robot that knows the location of the target. Searching for the exit, could involve variants of the well-known zig-zag, doubling search strategies along the linear terrain. However, the traditional doubling strategy leading to an optimal competitive ratio of 9 for linear search may no longer be adequate and one is required to investigate different approaches and more elaborate strategies for searching that take the shape of the linear terrain $y = f(x)$ into account.

The *canonical* zig-zag search algorithm (or strategy) is parametrized by an infinite sequence of positive distances $X = \{x_k\}_{k \geq 1}$ from the origin $O$ that specifies the turning points of the robot. Obviously, to ensure progress in searching, each trip of the robot away from the origin must cover more distance than the previous trip in the same direction. A natural measure of the efficacy of the zig-zag search strategy $X$, is how well it performs in competition with an omniscient adversary that knows the exact location of the target. If $d$ is the unknown distance of the target from the origin, let $\sigma_X(d)$ be the ratio between the time taken by the robot using the zig-zag strategy $X$ to reach an unknown target divided by the time taken by the adversary to proceed directly to the target (placed at distance $d$ from the origin). In addition, $\sigma_X \triangleq \sup_{d>1} \sigma_X(d)$ denotes the competitive ratio of the strategy $X$. We denote the optimal competitive ratio by $\sigma^*$.

In general, and unlike traditional linear search, the speed of the robot may depend on the physical properties of the terrain. Further, the robot's speed may depend on the direction of travel along the terrain, or on the profile of the terrain, e.g. when the line is inclined the robot may accelerate or decelerate depending on whether it is moving uphill or downhill, respectively. For example, on uphill segments the robot moves with speed 1 while on the $i$-th downhill segment it moves with speed $s_i$, where $s_i = 1 + g \sin \alpha_i \geq 1$, $g$ is the well-known gravitation constant (which is approximately equal to 9.8 $m/s^2$), and $\alpha_i$ is the angle of inclination of the $i$-th (downhill) segment, for $i = 1, 2, \ldots, n$.

**Search Strategies**

The first class of models considered is depicted in Fig. 4 and concerns two-speed models of linear search: *tailwind* (unit speed going left and tailwind speed $s > 1$ going right), *beacon* (unit speed moving away from the origin and speed $s$ moving towards it), and *exploration history* (the robot explores unknown regions slowly and deliberately with unit speed, but is able to search faster–with speed $s$–when it encounters a region already seen earlier in its search). Here are some of the results obtained in [26] (Note that Theorem 3 was independently proved also by [12]).

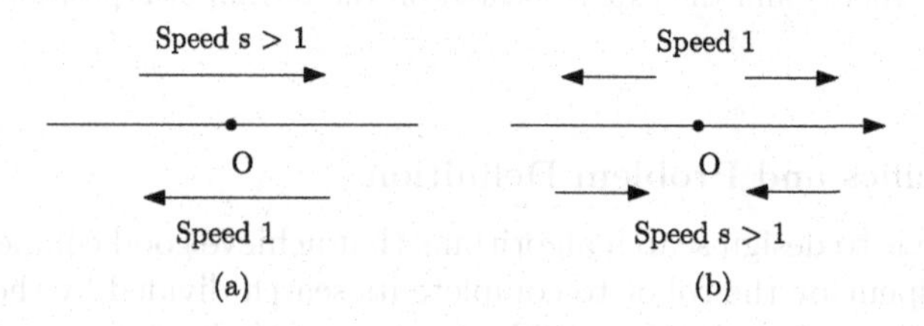


**Fig. 4.** Two-speed models based on (a) absolute direction and (b) direction relative to origin

**Theorem 2 (Tailwind Model, [26]).** *Assume the robot has speed $s \geq 1$ when moving left to right and speed 1 otherwise. For $\alpha, r$ such that $\alpha = (1 - s + \sqrt{(s-1)^2 + 4r^2 s})/(2r)$, $r = \sqrt{2 + (s+1)/\sqrt{s}}$, and $X = \{s, \alpha r, r^2 s, \alpha r^3, \ldots\}$ we have that*

$$2 + 1/s \leq \sigma^* \leq \sigma_X,$$

$$\sigma_X \leq 1 + \frac{s + 2\sqrt{s} + 1}{s + \sqrt{s} + 1} \cdot \frac{s+1}{2s} \cdot \left( s + 1 + \sqrt{(s-1)^2 + 8s + 4\sqrt{s}(s+1)} \right) \quad (6)$$

**Theorem 3 (Beacon Model, [26]).** *The doubling strategy $D$ is optimal for the beacon model, i.e.*

$$\sigma^* = \sigma_D = 5 + \frac{4}{s}. \quad (7)$$

**Theorem 4 (Exploration History Model, [26]).** *Let $r = 1 + \sqrt{2/(s+1)}$, and $X = (r^0, r^1, r^2, \ldots)$ be an expansion strategy. Then, with this strategy, the zig-zag algorithm's competitive ratio satisfies*

$$2 + 1/s \leq \sigma^* \leq \sigma_X = 2 + \frac{1}{s}\left(3 + 2\sqrt{2s+2}\right). \tag{8}$$

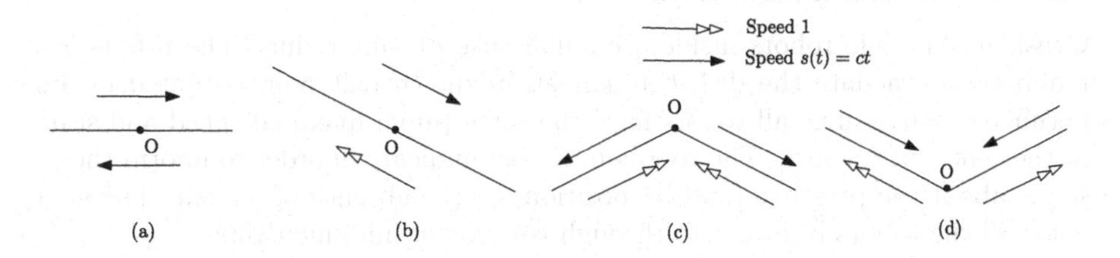


**Fig. 5.** Constant acceleration models: (a) Line (b) Inclined line (c) Hill (d) Valley

The second class of models is depicted in Fig. 5 and concerns constant acceleration models of linear terrain search. In *inclined* linear terrains (the robot can operate in two modes where it is moving with unit speed when moving *uphill* and with constant acceleration when moving *downhill*. The different terrains include an inclined line, a symmetric hill with the hill-top at the origin, or a symmetric valley with the valley-bottom at the origin). Here are some of the results obtained in [26].

**Theorem 5 (Constant acceleration in both directions, [26]).** *Assume the robot is searching with constant acceleration c in either direction, starting from rest initially, as well as at turning points. Then:*

$$3(\sqrt{2} + 1/\sqrt{2}) \leq \sigma^* \leq \sigma_D \leq \frac{2\sqrt{3}}{\sqrt{2}-1} + \sqrt{3} + 1 \tag{9}$$

**Theorem 6 (Moving on an inclined line, [26]).** *Assume the robot moves with acceleration c in the positive direction, and constant speed 1 in the negative direction using the doubling strategy D. Then for any $d \geq 1$,*

$$\sqrt{2c}\sqrt{d} < \sigma_D(d) \leq \sqrt{8c} \cdot \sqrt{d} + O(1). \tag{10}$$

*Furthermore, $\sigma^* \geq \sup_{d>1} \min\{2 + \sqrt{2/(cd)}, \sqrt{2} + \sqrt{cd/2}\}$.*

**Theorem 7 (Starting at the top of a hill, [26]).** *Assume that the robot travels with constant acceleration c away from the origin, and with unit speed towards the origin. Then $\sigma_D(d) = \Theta(\sqrt{d})$ and this is optimal.*

**Theorem 8 (Starting at the bottom of a valley, [26]).** *Assume that the robot travels with constant acceleration c towards the origin, and with unit speed away from the origin. Then for any $d \geq 1$:*

$$\sigma_D(d) \leq 5 + O(d^{-1/2})$$

*Furthermore, $\sigma^* \geq 5$.*

## 3 Evacuation

In this section we discuss search and evacuation which takes place on the perimeter of a closed domain, like circle, triangle or square. We also consider the case of faulty robots.

### 3.1 Evacuating from a Disk

Consider $k$ mobile robots inside a circular disk of unit radius. The robots are required to evacuate the disk through an unknown exit point situated on its perimeter. We assume all robots have the same (unit) maximal speed and start at the centre of the disk. The robots may communicate in order to inform themselves about the presence (and its position) or the absence of an exit. The goal is for all the robots to evacuate through the exit in minimum time.

A single ($k = 1$) robot can find the exit by going to the perimeter and traversing in the clockwise, say, direction. This takes time $1 + 2\pi$ to reach the exit, in the worst case. It is clear that for any $\epsilon > 0$ the robot can cover at most a length $2\pi - \epsilon$ of the perimeter (because its maximum speed is one and the adversary can place the exit in the unvisited portion of the perimeter). Therefore $1+2\pi-\epsilon$ is a lower bound for evacuation, for any $\epsilon > 0$. Hence, $1+2\pi$ is also a tight bound for evacuation of one robot.

**Model Specifics and Problem Definition**

In general, if the $k$ robots are placed in arbitrary initial positions in the interior of the disk then the resulting optimization problem is very difficult and very few (if any) non-trivial evacuation algorithms are known. For this reason in the sequel, the robots are placed at the start at the center of the disk.

An exit is represented as a point on the perimeter of the disk and the robot may locate the exit only if it is colocated with it. Further, two communication models are being considered. *F2F* (Face-to-Face) which requires that the two robots may communicate only if they are colocated at the same time, and *Wireless* in which the robots can communicate regardless of their distance. The more general case where the robots have limited communication range $r > 0$ has never been discussed in the scientific literature.

**Evacuation for Two Robots**

In the sequel we analyze in somewhat detail the evacuation problem for two robots.

Figure 6 depicts two evacuation algorithms which were originally published in [17]: Left-hand-side figure depicts the F2F model and the right-hand-side figure depicts the wireless communication model. There are two robots that need to evacuate from an unknown exit. One robot is represented by the bold arrow and the other by the blank arrow. In both communication models the robots start at the same time at the center $K$ of the disk. In the first part, the two algorithms are identical. The robots move together to the perimeter, say to point $A$, and from there they move in opposite direction along the perimeter. This is where the similarities end.

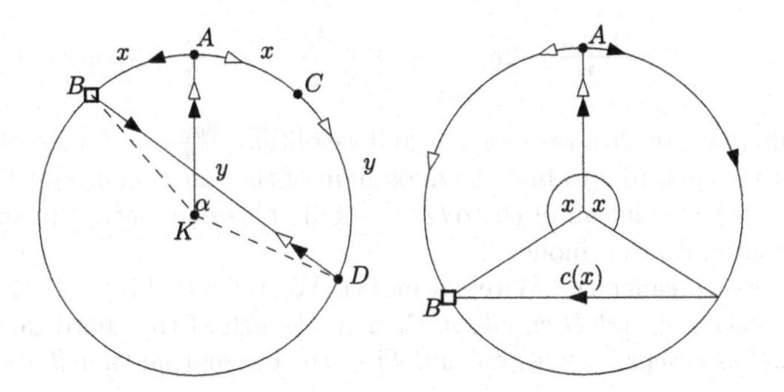


**Fig. 6.** Evacuating two robots from a disk. The robots start at the center $K$ of the disk and the (unknown) exit is located at $B$. Left picture depicts the algorithm in the F2F, while the right picture in the wireless communication model.

1. In the F2F model (see left picture of Fig. 6) when the robot represented by the bold arrow, say, finds the exit at $B$ it makes a cross-cut along the interior of the disk and travels to meet the robot represented by the bold arrow at $D$. From this point on the two robots move together along the interior of the disk to the exit $B$.
2. In the Wireless model (see right picture of Fig. 6) when the robot represented by the blank arrow, say, finds the exit at $B$ it communicates to the robot represented by the bold arrow that it has found the exit and the latter robot moves to the exit $B$ along the interior of the disk.

We can summarize the performance of the two algorithms in the following theorem.

**Theorem 9 (Upper Bounds for 2 Robots,** [17]**).** *Consider two robots starting at the same time from the center of a unit disk.*

1. **(Wireless Model)** *There is an algorithm for evacuating two robots from an unknown exit located on the perimeter of the disk which takes time at most* $1 + \frac{2\pi}{3} + \sqrt{3} \approx 4.826$.
2. **(F2F Model)** *There is an algorithm for evacuating the robots from an unknown exit located on the perimeter of the disk which takes time* $1 + \alpha/2 + 3\sin(\alpha/2)$ *where the angle* $\alpha$ *satisfies the equation* $\cos(\alpha/2) = -1/3$. *It follows that the evacuation algorithm takes time* $\sim$5.74.

*Proof.* For the proof below, we refer to the two pictures in Fig. 6 (left for the F2F and right for wireless model).

First we consider the F2F model. We calculate the time required until both robots reach the exit. Denote $x = |\widehat{BA}| = |\widehat{AC}|$, $y = |BD| = |\widehat{CD}|$ and $\alpha = |\widehat{BD}|$. The total time required is $g(\alpha) = 1 + x + 2y$. Observe that $\alpha = 2x + y$, and $y = 2\sin(\alpha/2)$, because $y$ is a chord of the angle $\alpha$. Substituting $x$ and $y$ in the function $f$ we can express the evacuation time as a function of the angle $\alpha$ as follows

$$g(\alpha) = 1 + \frac{\alpha - y}{2} + 2y = 1 + \frac{\alpha}{2} + \frac{3y}{2} = 1 + \frac{\alpha}{2} + 3\sin(\alpha/2).$$

Now we differentiate with respect to $\alpha$ and we obtain: $\frac{dg(\alpha)}{d\alpha} = \frac{1}{2} + \frac{3}{2}\cos(\alpha/2)$. Set the derivative equal to 0 to find the maximum of the function $g(\alpha)$, which yields as value for $\alpha$ the solution of $\cos(\alpha/2) = -1/3$. This completes the analysis for the F2F communication model.

Second we consider the Wireless model. We refer to Fig. 6. If the angular distance between $A$ and $B$ equals $x$, then the length of the chord taken by the robot $r_2$ equals to $c(x) = 2\sin(\pi - x)$. Thus the evacuation time $T$ satisfies

$$T \leq \max_{0 \leq x \leq \pi} \{1 + x + 2\sin(\pi - x)\} = \max_{0 \leq x \leq \pi} \{1 + x + 2\sin x\}.$$

The function $h(x) = 1 + x + 2\sin x$ in the interval $[0, \pi]$ is maximized at the point $x^* = 2\pi/3$ and $h(x^*) = 1 + 2\pi/3 + \sqrt{3}$. This completes the analysis for the Wireless communication model. □

We also mention the lower bounds for two robots, but the proof is more technical and we refer the reader to [17] for additional details.

**Theorem 10 (Lower Bounds for 2 Robots, [17]).** *Consider two robots starting at the same time from the center of a unit disk.*

1. **(Wireless Model)** *For any algorithm it takes at least* $1 + \frac{2\pi}{3} + \sqrt{3}$ *(*$\approx 4.826$*) time in the worst case for two robots to evacuate from an unknown exit located in the perimeter of the disk.*
2. **(F2F Model)** *It takes at least* $3 + \frac{\pi}{4} + \sqrt{2}$ *(*$\approx 5.199$*) time units for two robots to evacuate from an unknown exit located in the perimeter of the disk.*

Note that the bounds for the F2F communication model are not tight. Table 2 summarizes what is known and indicates the existing gap.

**Table 2.** Upper and lower bounds for the evacuation of 2 robots in the F2F communication model.

| Paper | Upper bound | Lower bound |
|---|---|---|
| [17] | 5.74 | 5.199 |
| [20] | 5.628 | 5.255 |
| [13] | 5.625 | |

## Evacuation for $k \geq 3$ Robots

It is apparent that the evacuation time improves by the collaboration of the robots. This is because a robot can share the search results of its exploration with the other robots in the group (using either F2F or Wireless communication).

Therefore it is not surprising that the evacuation time should improve as the number of robots increases. This is in fact confirmed by the results as listed in Table 3.

**Table 3.** Upper and Lower bounds for $k \geq 3$ robots as proved in [17].

| Model | Bound | $k = 3$ | $k \geq 4$ |
|---|---|---|---|
| F2F | Upper | ~5.09 | $3 + \frac{2\pi}{k} < 4.58$ |
| | Lower | ~4.519 | $3 + \frac{2\pi}{k} - O(k^{-2})$ |
| Wireless | Upper | ~4.22 | $3 + \frac{\pi}{k} + O(k^{-4/3})$ |
| | Lower | ~4.159 | $3 + \frac{\pi}{k} > 3.785$ |

Nevertheless, it is much harder to obtain good bounds for the evacuation of any small number of robots, say three. For example, for three robots [17] gives upper and lower bounds for both the F2F and the wireless communication models, but they are not tight (see Table 3). The interested reader can find additional details for the case of $k = 3$ robots in [17].

Fortunately, it is possible to obtain asymptotically tight bounds for evacuation as the number $k$ of robots tends to infinity. The basic idea is for the robots to explore different parts of the perimeter and share with each other their search results. However the main difficulty is to ensure that when a robot finds (respectively, announces) the exit the rest of the robots are as close to it as possible. An outline of the algorithms are given below.

1. **(F2F Model)** The $k$ robots "spread" at equal angles $2\pi/k$ and upon reaching the perimeter, they all move clockwise (along the perimeter) for $2\pi/k$ time units. In one additional time unit, all robots move to the center of the disk. Since at least one robot has found the exit it can inform the remaining robots and in one additional time unit all robots move to the exit.
2. **(Wireless Model)** The $k$ robots are divided into two groups: Group $G_\alpha$ of size $k_\alpha = \lceil k^{2/3} \rceil$, and Group $G_\beta$ of size $k_\beta = k - k_\alpha$. The robots in group $G_\alpha$ are assigned to "spread" and search a continuos arc $\widehat{AB}$ of length of length $\pi - 2\sqrt{\pi}k^{-1/3}$, while the robots in group $G_\beta$ are assigned to "spread" and search the complement $\widehat{BA}$ of arc $\widehat{AB}$. The robots explore specifically assigned subarcs of the arcs $\widehat{AB}$, $\widehat{BA}$ of length $|\widehat{AB}|/k_\alpha$ $|\widehat{BA}|/k_\beta$, respectively, and upon receiving a notification about the position of the discovered exit they all cut across a chord to the exit.

We can prove that the two algorithms above are asymptotically optimal in their respective communication model (see [17] for additional details).

**Theorem 11 (F2F Model, [17]).** *It is possible to evacuate $k$ robots from an unknown exit located on the perimeter of the disk in time $3 + \frac{2\pi}{k}$. It takes time at least $3 + \frac{2\pi}{k} - O(k^{-2})$ in the worst case to evacuate $k$ robots from an unknown exit located on the perimeter of the disk.*

**Theorem 12 (Wireless Model, [17]).** *If $k \geq 100$ then it is possible to evacuate k robots from an unknown exit located in the perimeter of the disk in time $3 + \frac{\pi}{k} + O(k^{-4/3})$. Moreover, it takes at least $3 + \frac{\pi}{k}$ time in the worst case to evacuate $k \geq 2$ robots from an unknown exit located on the perimeter of the disk.*

### 3.2 Evacuating from Triangles and Squares

Two of the main requirements of the algorithms designed for evacuating robots from a disk were that the robots agree in advance on the search strategy they will follow and also have knowledge of the "shape" of the perimeter on which they need to search for the exit. The former was important so that robots can estimate each other's position at any given time and the latter for traversing the perimeter. Further, any robot that finds the exit can take a "straight-line" short-cut through the interior of the disk so as to either meet another robots or go to the exit. These assumptions can be easily satisfied by any "convex" closed curve (e.g., triangles and squares as depicted in Fig. 7) though this will not necessarily make the optimization problem any simpler regardless of the communication model.

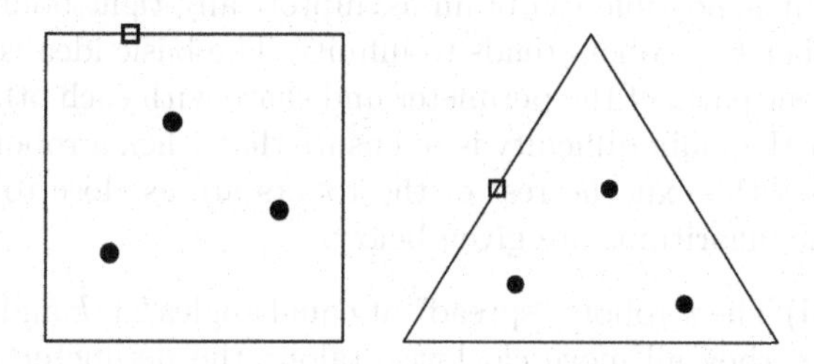

**Fig. 7.** General setting of robot evacuation from equilateral triangles and squares. Robots start in general positions at the interior (or perimeter) and the exit is located on the perimeter of the triangle or square.

#### Model Specifics and Problem Definition

Throughout this section we assume the wireless communication model. Consider an equilateral triangle or square with sides of length 1. A number of robots starting at the same location on the perimeter or in the interior of the triangle or square are required to evacuate from an exit which is located at an unknown location on its perimeter. At any time the robots can move at identical speed equal to 1, and they can cooperate by communicating with each other wirelessly. Thus, if a robot finds the exit it can broadcast "exit found" to the remaining robots which then move in a straight line segment towards the exit to evacuate. Our task is to design robot trajectories that minimize the evacuation time of the robots, i.e., the time the last robot evacuates from the exit. Designing such optimal algorithms turns out to be a very intricate problem and even the case of equilateral triangles turns out to be challenging.

**Evacuation Algorithm**

Consider the case where the robots start at a point $P$ on the perimeter of triangle $ABC$ and at distance $x$ from the midpoint of edge $BC$ (see Fig. 8). Now consider the following algorithm. From the midpoint the robots move in opposite directions along the perimeter, i.e. Robot 1 towards vertex $A$ via vertex $B$, and the other Robot 2 towards vertex $A$ via vertex $C$. When a robot finds the exit it broadcasts "Exit found" to the other robot which immediately goes in a straight line segment to the exit. A similar approach is required when the two robots start together at an interior point of the triangle.

These algorithms are not difficult to analyze and we can prove the following theorems (details can be found in [25]).

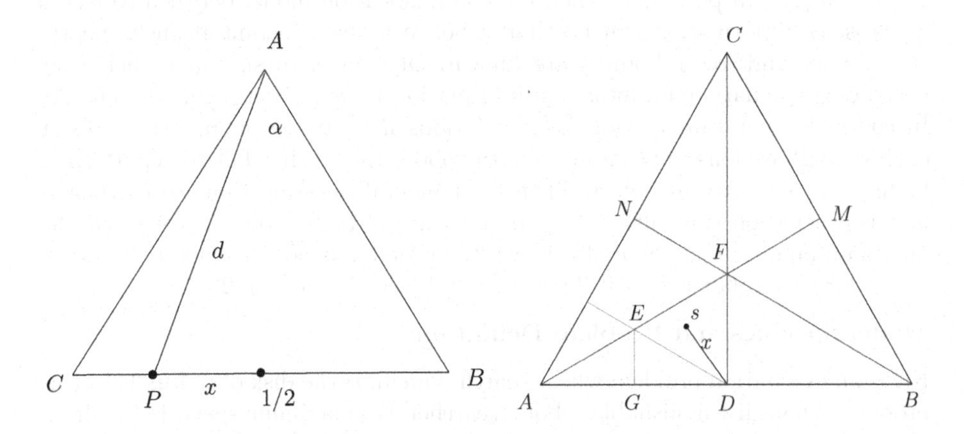


**Fig. 8.** Evacuating from an equilateral triangle. Left: the two robots start at a point $P$ at distance $x$ from the midpoint of edge $BC$. Right: the starting position of the two robots is in $\triangle AFD$. $N, M, D$ are the midpoints of the corresponding sides.

**Theorem 13 (Robots starting on the perimeter, [25]).** *Assume that two robots are initially located on the perimeter of an equilateral triangle at distance $x$ from the closest midpoint of an edge of the triangle. Then $x + \frac{3}{2}$ is a tight bound for evacuating these two robots.*

**Theorem 14 (Robots start in the interior, [25]).** *Assume that two robots are initially located at point $s$ inside the equilateral triangle, and let $x = \min\{d(s, m) \mid m$ is a mid point of an edge$\}$. Then $x + \frac{3}{2}$ is a tight bound for evacuating these two robots.*

Similar techniques can be used to analyze evacuation in unit squares. To sum up the results obtained in [25] include the following.

1. **Equilateral Triangle.** Optimal evacuation trajectories (algorithms) for 2 robots and for any (same) starting position. 3 or more robots starting on the perimeter cannot achieve better evacuation time than two robots.

2. **Square.** Optimal evacuation trajectories (algorithms) for 2 robots for starting positions on the perimeter. 3 or more robots starting at one of the corners cannot achieve better evacuation time than 2 robots.

Additional results for more robots and more generally for regular polygons can be found in [25]. For evacuation from an equilaterla triangle in the F2F model see [16]. An interesting problem concerns evacuation on an ellipse or an arbitrary convex polygon.

### 3.3 Evacuation with Faulty Agents

Evacuating robots from the disk is a well studied problem, first introduced and studied in [17]. In particular when the communication model between robots is wireless, i.e. information can be shared between them instantaneously, nearly tight upper and lower bounds are known. Of course, in such a model what becomes particularly relevant is that information shared among agents is reliable. In contrast, operational algorithmic solutions need to be, in practice, robust against malfunctions. This translates to robots that either fail to report their findings, or even to misreports. The fundamental question that arises then is how is evacuation time affected in the presence of faulty robots. Note that the minimum number $n$ of robots for which the problem is non degenerate is $n = 3$, out of which 1 robot is faulty. This is the subject of study in [22].

#### Model Specifics and Problem Definition

FE is an evacuation problem whose search domain is the disk of radius 1. In this problem, 3 non-distinguishable robots (searchers) of maximum speed 1 start from the center of the disk, and they can communicate with each other their findings instantaneously, i.e. they operate under the wireless model. Somewhere on the perimeter of the disk lies a hidden object (exit) that can be identified only if a robot goes over it. Among the searchers there is distinct robot, thereafter referred as faulty. An evacuation algorithm is a search trajectory for all three robots, in which eventually all robots reach the hidden object. Given a placement of the hidden object, the cost of the search algorithm is the time till the last non-faulty robot reaches the hidden item. The evacuation time of the algorithm is defined as the worst case cost of the algorithm.

Clearly, given a search algorithm, and in the spirit of worst case analysis, the adversary controls not only where the hidden object is placed, but also which of the searchers is faulty. In the same direction, the adversary also controls the actions of the faulty robot, and therefore one needs to determine the limitations of such adversarial choices. In the crash-faulty model, the faulty robot may only fail to report that the hidden object is found, whereas in the byzantine-faulty model, the faulty robot may misreport at any moment its findings. We denote the two evacuation problems in the corresponding faulty models as $FE_c$ and $FE_b$, respectively.

### Disk Evacuation Against Crash Faults

The advantage of trying to evacuate 3 robots in the crash-fault model is that once the location of the hidden object is reported, the remaining non-faulty robot may abandon searching and move toward the exit along a shortest (line segment) path. Maybe the simplest family of algorithms one may consider first is a symmetric-type, in which robots partition the circle in three contiguous arcs. The robots deploy to the endpoints of these arcs, and start searching in the same direction, each of them its own contiguous arc till the exit is found (reported). It is not hard to prove that the best algorithm of this family deploys the searchers in equidistant points, i.e. the three arcs are of length $2\pi/3$ each, and the induced evacuation time is $1 + 4\pi/3 + \sqrt{3}$.

One of the main contributions of [22] is to show how a non-symmetric algorithm can evacuate the two non-faulty robots efficiently. Practically, one may again define a family of, non-symmetric evacuation trajectories this time, as follows. Fix parameter $\beta$. Two robots deploy to an arbitrary point of the disk, with the intension to explore in opposite directions. The third robot is deployed to a point of the disk at arc distance $\beta$ from the deployment point of the other robots, with the intention to explore toward the closest robot (and hence in opposite direction than that robot), see Fig. 9.

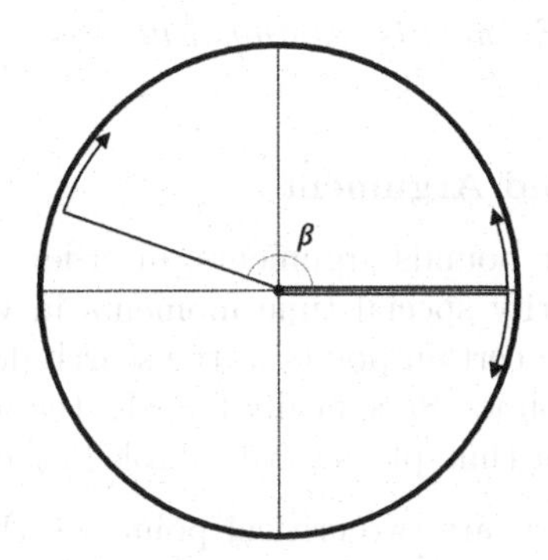


**Fig. 9.** The deployment of the three robots, and their initial direction of movements for the algorithm the shows the upper bound of Theorem 15

Consider now the following two adversarial choices. First, the two non-fault robots are those at arc distance $\beta$ moving toward each other. Assuming that $\beta \leq 4\pi/3$ one can show that the worst placement of the exit is at time $2\pi/3$, after they search their in-between arc segment (this is because the maximizer of function $x+2\sin(x/2)$ is attained at $x = 2\pi/3$) inducing cost $1+\beta/2+2\pi/3+\sqrt{3}$. Second, assume that the non-faulty robots are those moving in the same direction. It is easy to see that the worst placement of the exit makes the non-faulty robots search the perimeter for $2\pi - \beta$, which maintain throughout the exploration arc distance $\beta$ (hence the last robot needs additional time $2\sin(\beta/2)$ to reach the exit). Overall, this second case induces worst evacuation time $1+2\pi-\beta+2\sin(\beta)$.

The best algorithm known for $FE_c$ is exactly the one above that uses as $\beta$ the value that equates the evacuation costs in the aforementioned adversarial inputs.

**Theorem 15** ([17]). *Let $\beta_0$ be the solution to equation $3\beta_0/2 - 2\sin(\beta_0/2) = 4\pi/3 - \sqrt{3}$, where $\beta_0 \approx 2.966$. Then $FE_c$ admits solution with evacuation time at most $1 + \beta_0/2 + 2\pi/3 + \sqrt{3} \approx 6.309$.*

### Disk Evacuation Against Byzantine Faults

The best performance achieved for $FE_c$ with symmetric algorithms turns out to be the best performance known for $FE_b$, but remarkably with a different type of algorithm. The inherent difference in the byzantine-fault model is that once the exit is reported, it firsts needs to be confirmed as a reliable message before the non faulty robot attempts to reach it, as otherwise performance would be suboptimal. Indeed, evacuating in the presence of a byzantine faulty robot who may misreport her findings, all robots are asked to first explore, in the same direction, a contiguous arc segment of length $2\pi/3$. Depending on the report(s) that have been circulated, robots have information to either go to the exit or continue searching the circle for additional time $\frac{2\pi}{3}$. The fact that each point is searched twice, allows them to resolve any conflicts and deduce the real location of the exit. This idea gives rise to the following upper bound.

**Theorem 16** ([17]). $FE_b$ *admits solution with evacuation time at most* $1 + 4\pi/3 + \sqrt{3} \approx 6.92$.

### A Unified Lower Bound Argument

As it is common in lower bounds arguments, in order to show negative results for FE, one has to identify special time moments in which, independently of the algorithm considered, certain points in the search domain have not all been explored by non-faulty robots. Specifically for FE, the following predicate $P(\cdot)$, given any evacuation algorithm, plays a crucial role in the lower bound argument.

$P(T)$: At time $T$, there are two critical points on the circle at arc distance $2\pi/3$, and none of them is visited more than once by any of the three searchers.

One of the main technical contributions in [22] was to prove that $P\left(1 + \frac{13}{7}\sqrt{3}\right)$ is true. Now in $FE_b$, there must be a robot that has visited none of the two critical points. The adversary can chose that robot to be non-faulty, and clearly that robot requires extra at least $\sqrt{3}/2$ to reach any of them. For $FE_b$, the adversary has more power to mislead non-faulty robots, and in fact can call a potentially misreport of the exit. Independently of the performance of the evacuation algorithm, it is shown that the non-faulty robot must be at least $2\pi/3$ away from the true location of the exit, inducing extra cost $2\sin(2\pi/3) = \sqrt{3}$. Overall, the results of the lower bound arguments are summarized below.

**Theorem 17** ([17]). *Let $T_0 = 1 + \frac{13}{7}\sqrt{3}$. Then no algorithm can solve* $FE_c$ *with evacuation time less than $T_0 + \sqrt{3}/2 \approx 5.082$. Also, no algorithm can solve* $FE_b$ *with evacuation time less than $T_0 + \sqrt{3} \approx 5.948$.*

# 4 Multiple Targets on a Circle

Evacuation type problems emerged by requiring collections of robots to identify the location of a hidden item, and to reach it as fast as possible. It is natural to also assume variations in which there may be more than one objects hidden in the domain, and search termination to be defined as the latest time in which the last robot reaches any of the hidden objects. From a practical perspective, the hidden objects can be thought as available exits that are placed in the underlying domain in which searchers collectively try to reach any of them (and evacuate) efficiently, possibly using as partial information the relative distance of the exits, but not their locations. The main reference work for this model with multiple exits described above is [21].

### Model Specifics and Problem Definition

$FE_k$ is an evacuation problem whose search domain is the circle of perimeter 1 (i.e. of radius $1/2\pi$). In this problem, $k$ hidden and identical (non distinuishable) objects (exits) are located on the circle. The relative distance between the hidden items is known, thereafter referred as the map, but not their locations. Two identical robots are placed at arc distance $L$ on the circle, they can move at speed 1, and they can see each other. They can see any of the hidden object only if they are collocated with the object, and they can communicate wirelessly and instantaneously their findings. Their goal is that each of them reaches any of the hidden objects (exits). The evacuation time is defined as the worst case time of the last robot to reach an(y) exit.

There are two variations of the problem, where the initial placement (relative distance) $L$ of the robots is either part of the input, or is the subject of an algorithmic choice based on the provided map. In both variations, the goal is to design trajectories for the two searchers on the circle so as to minimize their evacuation time.

### Multiple Exit Evacuation with Given Robot Placement

Maybe the simplest case of all is the one when the map contains only one exit. The two robots start at known arc distance $L$, and they try to minimize the time that the last robot reaches the exit. It is convenient for the moment to think that the two robots are co-located. Very naturally the robots should start exploring in opposite directions till the exits is found, say at time $x \leq \frac{1}{2}$. Then, the other robot is notified, and can evacuate choosing the shortest route (along the circle) of length $\min\{2x, 1-2x\}$. Given that the adversary controls the location of the exit, that would induce worst case performance

$$\max_{x\in[0,1/2]} \{x + \min\{2x, 1-2x\}\} = 3/4.$$

If on the other hand the robots are at distance $L$, then it is natural to first have them explore the arc between them, taking time $L/2$, and then running the previous algorithm of extra cost 3/4. It turns out that this simple idea is also optimal.

**Theorem 18** ([21]). *When the initial arc distance $L \in [0, 1/2]$ of the two robots is part of the input,* $\text{EME}_1$ *can be solved with evacuation time $\frac{3}{4} + \frac{1}{2}L$. Moreover this is optimal.*

$\text{EME}_k$ becomes more interesting for $k \geq 2$. Since exits are not distinguishable, one needs to identify the critical information that can be deduced from every map. Given an instance of $\text{EME}_k$, e.g. a map for the placement of $k$ exits, the critical parameter that can be utilized algorithmically turns out to be longest arc length of the circle that does not strictly contain any exit. This can be also thought as a pessimistic estimation for the distance of a hidden item from the location of an exit currently found (and reported). A map in which the longest arc not strictly containing an exit has length $D$ will be called a map with critical value $D$, see also Fig. 10 for an example of an $\text{EME}_k$ instance. Note that by definition, any map of $\text{EME}_k$ has critical value $D \in [1/k, 1]$.

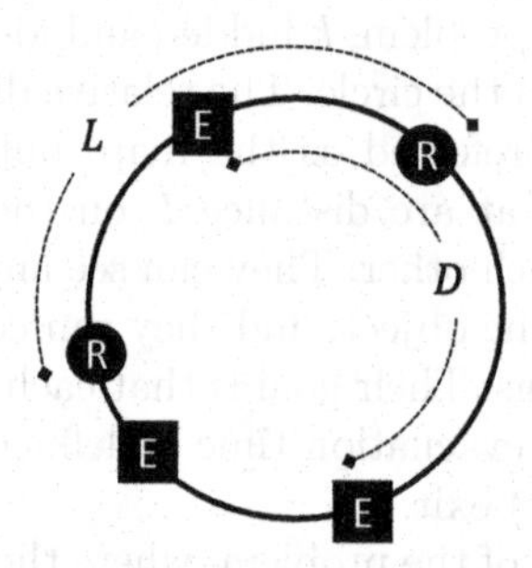


**Fig. 10.** An instance of $\text{EME}_3$. Robots placements are depicted with circles $R$, and exit placements are depicted by squares $E$.

**Theorem 19** ([21]). *When the initial arc distance $L \in [0, 1/2]$ of the two robots is part of the input, a map of* $\text{EME}_k$ *with critical distance $D$ can be evacuated in time*

$$\frac{3}{4}D + \frac{1}{2}L.$$

*Moreover no algorithm can have evacuation time better than $\frac{3}{4}D - \frac{1}{2}L$.*

The upper bound for Theorem 19 is due to an algorithm similar to the optimal algorithm for $\text{EME}_1$. Indeed, when there is only 1 exit, the critical value of the map is $D = 1$, whereas for general critical value $D \in [1/k, 1]$, each of the two robots performs, in the worst case, within an arc of length $D$, which is similar to performing on a circle of perimeter $D$. Nevertheless this argument is not strong enough to derive a matching upper and lower bound. In contrast, when all consecutive exits are equidistant, we do know the best possible evacuation time.

**Theorem 20** ([21]). *When the initial arc distance $L \in [0, 1/2]$ of the two robots is part of the input, a map of* $\mathrm{EME}_k$ *with critical distance $D = 1/k$ can be evacuated it time $\frac{3}{4}D + \frac{1}{2}\lambda$, where*

$$\lambda = \begin{cases} L \bmod \frac{1}{k}, & \text{if } L \bmod \frac{1}{k} \leq \frac{1}{2k} \\ \frac{1}{k} - L \bmod \frac{1}{k}, & \text{if } L \bmod \frac{1}{k} > \frac{1}{2k} \end{cases}$$

*Moreover this performance is optimal.*

The heart of the argument for proving Theorem 20 relies on the idea of reducing $\mathrm{EME}_k$ on the circle of perimeter 1 to $\mathrm{EME}_1$ on a circle of perimeter $1/k$ (which is possible due to that all consecutive exits are equispaced). The fact that evacuation performance is non-monotonic has to do with the initial distance of the robots. When the circle is partitioned in arcs of length $1/k$, their original distance $L$ evaluated $\bmod 1/k$ can be either larger or smaller than $1/2k$. If $\lambda_0 = L \bmod 1/k \leq 1/2k$, then the problem is equivalent to $\mathrm{EME}_1$ on the circle of perimeter $1/k$, with initial robots distance $\lambda_0$. Otherwise, the distance of the two robots in the new instance is $1/k - \lambda_0$.

**Multiple Exit Evacuation with Chosen Robot Placement**

Evacuation problem $\mathrm{EME}_k$, when the initial relative distance $L$ of the robots can be chosen as a function of the given map of critical value $D$, i.e. when we allow $L = L(D)$, requires a more elaborate analysis. At a high level, given initial distance $L$ of the robots, optimal, or near-optimal evacuation trajectories are known. For these trajectories, an adversary can choose when the first exit is found, say at time $x \leq \frac{1}{2} - \frac{1}{2}L$. If the induced evacuation cost is denoted by $g(D, L, x)$, then the evacuation cost of the algorithm is given by $\max_{x \in [0, \frac{1}{2} - \frac{1}{2}L]} \{g(D, L, x)\}$. If in addition we allow the algorithm to pick $L = L(D)$, then one has to solve optimization problem

$$\min_{L \in [0,1/2]} \max_{x \in [0, \frac{1}{2} - \frac{1}{2}L]} \{g(D, L, x)\}$$

in order to determine the best possible evacuation algorithm. This is exactly what the next theorem describes.

**Theorem 21** ([21]). *When the initial arc distance $L \in [0, 1/2]$ of the two robots can be chosen algorithmically based on a map of* $\mathrm{EME}_k$ *with critical distance $D \in [1/k, 1]$, there is an evacuation algorithm with cost at most*

$$\begin{cases} \frac{3}{4}D, & \text{if } D \in [1/k, 4/5), \text{ achieved for } L = 0 \\ 1 - \frac{1}{2}D, & \text{if } D \in [4/5, 6/7) \text{ achieved for } L = \frac{5}{2}D - 2 \\ \frac{5}{4}D - \frac{1}{2}, & \text{if } D \in [6/7, 1], \text{ achieved for } L = 1 - D. \end{cases}$$

## 5 Search and Fetch

Search and Fetch type problems are common in search-and-rescue operations where a hidden object (a victim) not only has to be located, but also has to be carried (fetched) to a designated spot. From a theoretical perspective, which is the focus of the current work, such problems fall under the generic evacuation type problems, where evacuation has to be accomplished only by a subset of the involved objects/mobile agents, hence delivering a combinatorial flavor to the problem. An attempt to model and study such search and fetch problems on the plane with 1 and 2 mobile agents appeared first in [28] and [27], respectively.

**Model Specifics and Problem Definition**

$\text{TE}_\alpha^n$ is an evacuation-type problem whose domain is the unit disk. In this problem, $n \in \{1, 2\}$ mobile agents start from the center of the disk and can move at maximum speed 1 anywhere on the plane. Two hidden objects, a treasure $T$ and an exit $E$, reside on the perimeter of the disk at *known* arc distance $\alpha$. $E$ is immobile, while $T$ can be moved by any mobile agent. Any of the hidden objects can be located only by a robot that walks over it.

An evacuation algorithm for $\text{TE}_\alpha^n$ is composed by the trajectory for each of the $n$ robots which guarantees that $T$ reaches $E$ in a finite amount of time. For any placement of the objects (at known arc distance $\alpha$), the completion time is defined as the time it takes $T$ to reach $E$, i.e. the time to evacuate $T$. The completion (evacuation) time for an evacuation algorithm for $\text{TE}_\alpha^n$ is the worst case evacuation time over all placements of hidden objects $T, E$. Finally, the evacuation time of $\text{TE}_\alpha^n$ is the minimum completion time over all evacuation algorithms.

**Relevant Literature**

$\text{TE}_\alpha^1$ is closely related to search-type problems that involve a searcher and a hider in numerous variants for which the literature is vast [3,4]. Evacuation-type problems similar to $\text{TE}_\alpha^2$ in which the model of computation between the robots becomes relevant but without the fetching/combinatorial component, have been studied is a series of papers [13,17,20,21,25,32]. From a practical perspective, search-and-rescue problems have been studied since the late 90's by the robotics community, e.g. see [29], and extensively by the operations research community, e.g., see [15]. Maybe the most similar problem to $\text{TE}_\alpha^n$ studied before is the one introduced by Alpern in [2], where the domain was discrete (a tree), and the goal of the rescuer was to fetch a treasure hidden in a leaf back to the root of the tree.

**Model Motivation**

The underlying model of $\text{TE}_\alpha^n$ is admittedly simplistic, yet its' importance is fivefold.

*First*, as it will be clear in the remaining of the section, solving optimally either $\text{TE}_\alpha^1$, $\text{TE}_\alpha^2$ seems a particularly challenging task. Moreover any upper or lower bounds for $\text{TE}_\alpha^1$, $\text{TE}_\alpha^2$ require non-trivial and sometimes technical arguments. Together with the fact the known upper and lower bounds are far from being matched, solving $\text{TE}_\alpha^n$ qualifies as a challenging mathematical puzzle.

*Second*, $\text{TE}_\alpha^n$ is by definition an *online problem* where an algorithm is asked to perform well against an unknown input (here the position of the hidden objects).

A fundamental subject of theoretical computer science is exactly to study the boundaries of computational capabilities given limited resources, and for $\mathrm{TE}_\alpha^n$ that would be the knowledge of the input. In that direction, $\mathrm{TE}_\alpha^n$ proposes a compromise between no information and full information about the input. From an algorithmic perspective, $\mathrm{TE}_\alpha^n$ asks the fascinating question of utilizing the partial information available about the input (the distance between the two hidden objects) in order to improve upon an algorithm with no information about the input.

*Third*, $\mathrm{TE}_\alpha^n$ is maybe the first attempt to inject a combinatorial flavour to evacuation type problems. This far, evacuation problems treated robots equivalently, meaning that completion time was oblivious to robot's identities. In contrast, $\mathrm{TE}_\alpha^n$ requires a specific immobile robot, the treasure, to be fetched to a hidden exit, and specifically it poses no constraints to the facilitators (other robots) which try to expedite the evacuation of the treasure. Only very recently, two more papers studied evacuation problems with similar combinatorial-type requirements [18,19].

*Fourth*, and when $n \geq 2$, $\mathrm{TE}_\alpha^n$ emphasizes the relevance of the communication protocol between robots in search-and-rescue operations. Indeed, when access to information is overall restricted in online problems, i.e. no information about hidden objects is available, search protocols where robots are allowed to communicate wirelessly are expected to outperform search protocols where robots can exchange information only by meeting (face-to-face model). In contrast, when partial information becomes available, in our case the distance between the hidden objects, a face-to-face algorithm is better equipped against the uncertainty that even wireless algorithms face, allowing possibly for solutions whose costs do not differ by much in the two communication models.

*Fifth*, the domain of $\mathrm{TE}_\alpha^n$ as well as robots specifications follow closely model specifications of fundamental problems in search theory and rendezvous, and as such $\mathrm{TE}_\alpha^n$ proposes a natural extension of them. Since all of these problems intend to introduce fundamental search/algorithmic techniques with applicability to real life search-and-rescue operations, $\mathrm{TE}_\alpha^n$ in particular becomes relevant when rescuers performance is quantified not by their evacuation time, rather by the evacuation time of the victim they are trying to save.

### 5.1 Searching with One Robot

Search and fetch with one robot, i.e. $\mathrm{TE}_\alpha^1$, has been studied in two variations depending on the precise knowledge regarding the location of the two hidden objects. In one of them, a bound $\alpha$ is known for the exact distance of the objects, while in the other, $\alpha$ is guaranteed to be the distance of the objects. Clearly, the first variation gives rise to an optimal completion time which at least as costly than the one of the second variation. However, somehow surprisingly, providing an optimal algorithm for the first variation is a relatively easy task, compared to the other variation where a similar result is still eluding.

**Knowledge of a Bound of the Critical Distance**

Consider the variation of $\mathrm{TE}^1_\alpha$ where the online algorithm has access to a lower bound $\alpha$ of the actual distance $\ell$ of the two hidden objects, i.e. $\ell \geq \alpha$. In other words, one tries to minimize the worst case completion time of fetching the treasure $T$ to the exit $E$, assuming that the arc-distance between $T, E$ is at least $\alpha$. We denote this variant as $\mathrm{TE}^1_{\geq\alpha}$.

**Theorem 22** ([28]). *An instance of* $\mathrm{TE}^1_{\geq\alpha}$ *where* $T, E$ *are at arc distance* $\ell \geq \alpha$ *can be solved with worst case completion time* $1+2\pi-\alpha+2\sin(\alpha/2)+2\sin(\ell/2)$, *and this is optimal.*

Notice that the description of a treasure evacuation algorithm concerns only the part of the execution till the locations of both objects are identified. Then, robot(s) may fetch the treasure to the exit in an optimal way, given objects' and robot's locations. The algorithm that proves the upper bound of Theorem 22 utilizes nicely the key component to all evacuation protocols for all $\mathrm{TE}^n_\alpha$, and concerns an "arc-avoidance" step during the exploration phase of the algorithm. Indeed, assume that a robot explores the perimeter of the disk, searching still for the first hidden object, and assume that already an arc of length at least $\alpha$ has been explored. Once the first hidden item is located, the other has to be arc-distance at least $\alpha$ away. Therefore cross-cutting along the corresponding chord of length $2\sin(\alpha/2)$ induces total savings of $\alpha - 2\sin(\alpha/2)$. Note that such a move is beneficial independently of whether the first item found is the treasure or the exit. From an adversarial perspective, the first item to be found has to be the exit, so that an extra $2\sin(\ell/2)$ is required for the fetching phase, as well as the two hidden items are to be located (in the worst case) as late in the time horizon as possible. Overall, this explains why the worst case evacuation time of the following algorithm is indeed $1 + 2\pi - \alpha + 2\sin(\alpha/2) + 2\sin(\ell/2)$.

---

**Algorithm 1.** Arc Avoidance Evacuation Algorithm for $\mathrm{TE}^1_{\geq\alpha}$

---

**Step 1:** Starting from an arbitrary point on the disk, start searching cw till the first hidden object $I$ is found.
**Step 2:** Move along the chord connecting $I$ to point at cw arc-distance $\alpha$.
**Step 3.** Continue exploring cw till the second hidden object is found.

---

The lower bound of Theorem 22 is based on a simple observation regarding any evacuation algorithm. If a robot has explored less than $2\pi - \alpha$ of the perimeter of the disk, then there is a chord of length at least $2\sin(\alpha/2)$ (say of corresponding arc length $\ell$) neither of whose endpoints has been visited by the robot. Therefore, any adversary could let any algorithm run for $1 + 2\pi - \alpha - \epsilon$, before she fixes the locations of the two hidden points, and she can do that by making the first explored point to be the exit.

### Knowledge of the Exact Critical Distance

Now we turn our attention to the variant of $\mathrm{TE}^1_\alpha$, denoted as $\mathrm{TE}^1_{=\alpha}$, in which the two hidden items are exactly at arc distance $\alpha$, and $\alpha$ is known. Notably, Algorithm 1 is still applicable, and it is no surprise that Theorem 22 predicts its worst case performance to be equal to $1 + 2\pi - \alpha + 4\sin(\alpha/2)$. With some technical work, one can improve upon the previous upper bound.

The main idea behind the best algorithm known for $\mathrm{TE}^1_{=\alpha}$ relies on the partition of the search space, i.e. the perimeter of the disk, into contiguous segments that will be searched, and others that can be "skipped". Indeed, fix some $\alpha$, and consider a robot having searched an arc of length $a_1 \geq \alpha$. If any of the hidden items lies within $b_1 \leq \alpha$ arc distance from the robot, then the location of the other hidden item can be deduced. Hence, the robot has an incentive to move along the chord of length $2\sin(b_1)$ (without ever missing both hidden items), and continue searching a new arc. If the latter has length $b_2 \geq \alpha$, then the previous argument applies again.

Consider now a partition of the disk into arc of length $a_1, b_1, \ldots, a_n, b_n, a_{n+1}$, such that $b_i \leq \alpha \leq a_i$. A robot can search each of the arcs of length $a_i$, following a "jump" along the chords of length $2\sin(b_i)$, till one of the hidden items is found. Call the search strategy associated with such a partition greedy. If the first hidden item is found by the greedy algorithm while searching arc $a_i, i \leq n$, then the robot may have to check two locations for the second hidden object. However, any jump saves locally time $b_i - 2\sin(b_i/2) > 0$, as long as $b_i > 0$. Moreover, $b_i - 2\sin(b_i/2)$ is monotone in $b_i$, hence it is intuitive that one should choose a partition of the disk that induces as much savings as possible. In that direction, it is natural to pack as many maximal jumps as possible of length $b_i = \alpha$, making sure that if a hidden object is found in the last arc to be searched, then the other object location is deduced from the searched space. Omitting several technicalities, one can show that the number $n_\alpha$ of such jumps must be equal to

$$n_\alpha := \left\lfloor \frac{2\pi - 3\alpha - 2\sin(\alpha)}{2\alpha} \right\rfloor$$

giving rise to the following positive result.

**Theorem 23** ([28]). *For an instance of* $\mathrm{TE}^1_{=\alpha}$*, consider a greedy search algorithm using disk partition* $a_1, b_1, a_2, b_2, \ldots, a_{n_\alpha}, b_{n_\alpha}, a_{n_\alpha+1}$. *If* $n_\alpha < \frac{\pi - \sin(\alpha)}{\alpha} - 2$, *then the worst evacuation time of the algorithm is*

$$2\pi - (n_\alpha + 2)\alpha + 2(n_\alpha + 3)\sin(\alpha/2)$$

*and otherwise, the worst evacuation time is*

$$(n_\alpha + 2)\alpha + 2(n_\alpha + 2)\sin(\alpha/2) + 2\sin\left(\frac{2n_\alpha + 3}{2}\alpha + \sin\alpha\right) + 2\sin\alpha$$

*Moreover the analysis it tight.*

Although it is conjectured that the above upper bound is the best possible, a tight lower bound is still eluding. At the same time, a non trivial lower bound can be obtained by consider a "weaker" adversary that is restricted to choose the locations of the two hidden items after a search algorithm has left only one contiguous unexplored arc of length $\alpha$. Given such a restricted adversary, it is not difficult to show that the most efficient algorithms must be greedy, hence, their trajectories must be determined by a sequence of non-negative reals $a_1, b_1, a_2, b_2, \ldots, a_{l-1}, b_{l-1}, a_l$ for some integer $l \geq 1$, with $b_i \leq \alpha$. Placing the two hidden objects at the very end of the search (where the first item found is the exit) induces cost $1 + \sum_{i=1}^{l} a_i + \sum_{i=1}^{l-1} 2\sin(b_i/2) + 4\sin(\alpha/2)$. Hence, the values of the following family of Non-Linear Programs gives a lower bound for the best evacuation time for $\mathrm{TE}^1_{=\alpha}$

$$\textbf{minimize} \quad 1 + \sum_{i=1}^{l} a_i + \sum_{i=1}^{l-1} 2\sin(b_i/2) + 4\sin(\alpha/2)$$

$$\textbf{subject to} \quad \sum_{i=1}^{l} a_i + \sum_{i=1}^{l} b_i = 2\pi$$

$$b_i \leq \alpha \text{ for } i = 1, 2, \ldots, l$$

$$a_i, b_i \geq 0 \text{ for } i = 1, 2, \ldots, l$$

Some technical work is required to obtain the maximum value of the optimization problems above (over all intefers $l \geq 1$), resulting in the following theorem.

**Theorem 24** ([28]). *No algorithm for* $\mathrm{TE}^1_{=\alpha}$ *has evacuation time better than*

$$1 + \pi + \min \left\{ \begin{array}{l} 4\sin\frac{\alpha}{2} + 2\left(\left\lceil\frac{\pi}{\alpha}\right\rceil - 1\right)\sin\frac{\pi-\alpha}{2\left(\left\lceil\frac{\pi}{\alpha}\right\rceil - 1\right)}, \\ \pi - \alpha\left\lfloor\frac{\pi}{\alpha}\right\rfloor + 2\left(\left\lfloor\frac{\pi}{\alpha}\right\rfloor + 1\right)\sin\frac{\alpha}{2} \end{array} \right\}$$

### 5.2 Searching with Two Robots

Known results for $\mathrm{TE}^1_\alpha$ indicate that optimal trajectories for $\mathrm{TE}^2_\alpha$ must employ complicated trajectories for the two robots using alternating arcs of the disk that are to be searched and skipped. Such complications seem unnecessary, given that efficient algorithms for $\mathrm{TE}^2_\alpha$ are far from being plain-vanilla type, especially in the face-to-face model. Indeed, the work in [27] indicates that even if one considers a special family of search algorithms in which robots do not abandon searching before at least one hidden item is found, efficient search trajectories do require ingenuity.

#### Upper Bound in the Wireless Model

It is worthwhile investigating a simple-minded algorithm for $\mathrm{TE}^2_\alpha$, demonstrating the need for algorithms that are adaptive in $\alpha$, especially given that in the wireless model, robots share information instantaneously.

Indeed, consider an algorithm, thereafter referred as $\mathcal{A}^w_1$, that deploys two robots in an arbitrary point on the disk, and robots start searching in opposite directions till the first hidden object is found, after time, say, $x$. Due to the communication model, robots can coordinate in order to explore all possible locations of the second hidden object, fetching the treasure to the exit in an optimal way (from the time that both objects have been found).

It turns out that $\mathcal{A}_1^w$ performs well when $\alpha$ is not too big, and indeed, one can show that the worst case evacuation time is at most $1 + \pi - \alpha + 4\sin(\alpha/2)$, as long as $\alpha \leq 2\pi/3$. Performance analysis is based on case analysis as to where the hidden objects are placed. It is not difficult to see that one of the worst input configurations occur when the first object found is the exit, say after time $x$, see Fig. 11. If $2x \leq \alpha$, there is still uncertainty as to where the treasure is, even though it must be at distance $2\sin(\alpha/2)$ from the found object. The two robots can search independently the two candidate locations, and as long as $x$ is not too big, the exit founder reaches the actual location of the treasure after the other robot, overall inducing cost at least

$$1 + \alpha/2 - \arcsin(\sin(\alpha/2) - \sin(\alpha)) + 4\sin(\alpha/2),$$

which provably exceeds the bound of $1 + \pi - \alpha + 4\sin(\alpha/2)$ for large enough $\alpha$.

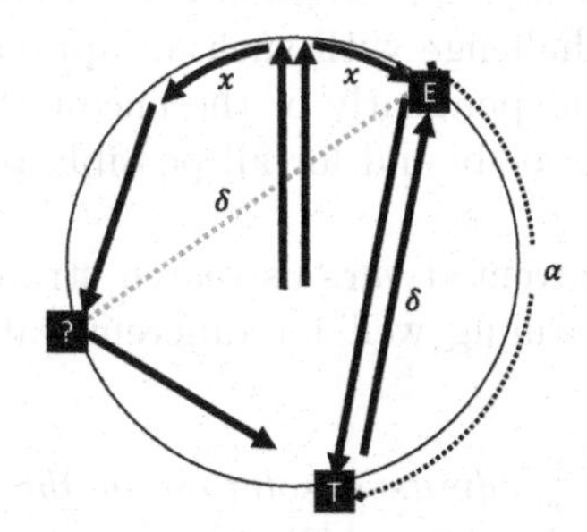


**Fig. 11.** An example of robots' trajectories in which the first interesting point found is the exit, depicted with the square $E$. At distance $\delta = 2\sin(\alpha/2)$ is the other interesting point. The actual location of the treasure is depicted with the square $T$ and it's other possible location by the square?

To circumvent the poor behavior of $\mathcal{A}_1^w$, we employ another search strategy, that we call $\mathcal{A}_2^w$. In this algorithm, robots deploy in two antipodal points of the disk, and search the perimeter in the same direction till the first hidden item is found, say at time $x$. In this case, uncertainty about the location of the second item occurs when $x \leq \alpha$ and when $\pi - x \leq \alpha$, and hence the algorithm can deduce the location of both items in more cases, when $\alpha$ is large enough. In fact, one can show that if $\alpha \geq 2\pi/3$, then the worst evacuation time of $\mathcal{A}_2^w$ is at most $1 + \pi - \alpha + 4\sin(\alpha/2)$. Overall, we have algorithms $\mathcal{A}_1^w, \mathcal{A}_2^w$, each achieving the same upper bound (as a function of $\alpha$) for complementary values of $\alpha$ concluding the following theorem.

**Theorem 25** ([27]). $\mathrm{TE}_\alpha^2$ *admits a solution in the wireless model with evacuation time* $1 + \pi - \alpha + 4\sin(\alpha/2)$.

### Upper Bound in the Face-to-Face Model

The optimal strategy known in the wireless model for $\mathrm{TE}_\alpha^2$ indicates that strategies adaptive to the value of $\alpha$ are necessary in order to achieve efficient evacuation times. Such strategies become more relevant in the face-to-face model where information cannot be shared by the searchers from distance.

In order to circumvent the lack of communication, it is no surprise that efficient algorithms for $\mathrm{TE}_\alpha^2$ in the face-to-face model need to adapt not only with $\alpha$, but as well as with the timing of certain findings. For example, it is natural to aim for search strategies that change behavior as a function of when the first hidden object is found, and of the type of the hidden object. Especially challenging seems that task of having the robots coordinate their trajectories from distance, given that any findings cannot be shared unless the searchers meet. Nevertheless, robots may be able to indirectly exchange information from distance given that they have agreed in advance to meet in predetermined locations assuming certain configurations have been encountered by the robots. If the meeting is realized, then clearly robots exchange information regarding their findings. Similarly, if the meeting is not realized, robots may preclude certain configurations, hence deducing information about the topology of hidden objects indirectly. Of course the challenge with such an approach is (a) to coordinate the robots properly and independently of the encountered configurations, and (b) to achieve the same upper bound for all possible placements of the hidden objects.

In this direction, [27] invented various search strategies in the face-to-face model, each of them performing well for different values of $\alpha$, achieving the following result.

**Theorem 26** ([27]). $\mathrm{TE}_\alpha^2$ *admits a solution in the face-to-face model with evacuation time* $1+\pi-\alpha/2+3\sin(\alpha/2)$.

There are three different search algorithms $\mathcal{A}_1^f, \mathcal{A}_2^f, \mathcal{A}_3^f$, that need to be employed in order to achieve the bound of Theorem 26, which we briefly outline next. All of them, deploy both robots to an arbitrary point of the disk, and have them start exploring in opposite directions till the first hidden object is found, and the algorithms are distinguished with respect to what happens next. Figure 12 depicts possible executions of the three algorithms.

In $\mathcal{A}_1^f$, each robot greedily tries to find locations of the hidden items, and the treasure founder fetches the treasure to the exit without attempting coordination (or any exchange of information via a coordinated meeting) as if there was only one searcher. Some technical analysis shows that this algorithm works well (see Theorem 26) as long as $\alpha$ is not large enough.

$\mathcal{A}_2^f$ is employed either when $\alpha$ is between two critical values, and the treasure is found within a carefully predetermined time window. In this case, the treasure founder goes to the center of the disk, and waits for as long as it would take the other robot to find the exit in a candidate location and come to notify the treasure holder. No matter whether the meeting is realized or not, the treasure holder

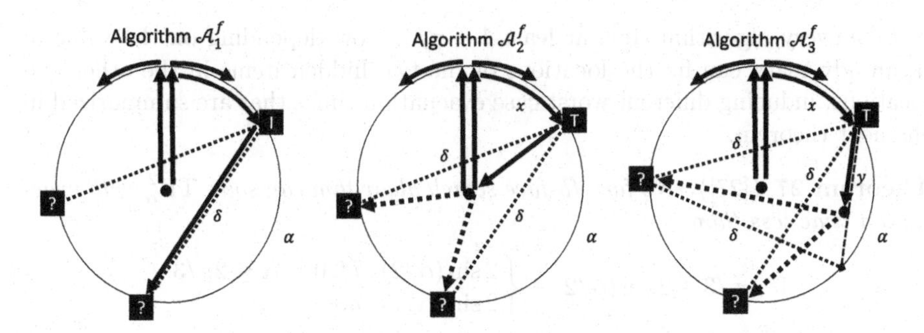


**Fig. 12.** Possible executions of algorithms $\mathcal{A}_1^f, \mathcal{A}_2^f, \mathcal{A}_3^f$. On the left, the first robot that finds an interesting point, depicted with square $I$, tries one of the possible locations of the other interesting point. In the middle, the founder of the first interesting point which is the treasure, goes to the center of the circle, and waits till it collects enough information as to where the exit is (possible trajectories depicted with dotted arrows). On the right, the founder of the treasure follows a special chord specified by points $A, B, C, D$ (described in the definition of $\mathcal{A}_3^f$ below) only up to a total length $y$. Thereafter, there are two possibilities of movements depicted with the dotted arrows, and that depend on the information the robot will have collected about the location of the exit.

deduces the location of the exit, and fetches it inducing worst case evacuation time equal to the one promised by Theorem 26.

Finally, $\mathcal{A}_3^f$ is employed in all other cases and utilizes the most involved trajectories and is based, at a high level, on the following idea. Suppose a hidden object is located at point $A$, and that object is the treasure. If there is uncertainty about the location of the exit, then this must be at arc distance $\alpha$ either cw or ccw, forming an imaginary triangle $ABC$. What would induce a bad performance is the treasure holder to check points $B, C$ alone. Instead, the treasure founder can utilize her knowledge that her peer robot is searching one of the arcs $AB$ or $AC$, and the length of these arcs is known since robots started exploring the disk from the same point. So say that arc is $AB$. Had the other robot found the exit at point $B$, that would induce for her another triangle $ABD$ with the possible locations of the second hidden object. In fact, $\mathcal{A}_3^f$ instructs such a robot to check point $D$ for the object, and if failed to start moving toward point $A$ to notify the treasure founder. Hence, the treasure founder has an incentive, instead of checking her candidate points $B, C$ to first start moving toward point $D$, just in case the other robot has already encountered the other hidden object. No matter whether the meeting is realized, the treasure holder would be able to deduce the location of the exit, fetching to it the treasure efficiently.

**Lower Bounds for $\mathbf{TE}_\alpha^2$**

The only lower bound known for $\mathrm{TE}_\alpha^2$ is for the face-to-face model and relies on an adversary that waits for the two robots to explore the disk until there are three points $A, B, C$ on the disk, where arcs $AB, BC$ are both of length $\alpha$ and at most one of the points $A, B, C$ has been visited. One can argue that this has to take every algorithm time at least $1 + \pi/3$. Now depending on the value of $\alpha$, an adversary can fix the locations of the two hidden items in the other two locations, inducing different worst case evacuation times that are summarized in the next theorem.

**Theorem 27** ([27]). *No face-to-face search algorithm can solve* $\mathrm{TE}^2_\alpha$ *with evacuation time less than*

$$1 + \pi/3 + 2\sin(\alpha/2) + \begin{cases} 2\sin(\alpha/2), & \text{if } 0 \le \alpha \le 2\pi/3 \\ 2\sin(\alpha), & o.w. \end{cases}$$

## 6 Conclusion

Search has always been an inexhaustible source of challenging mathematical optimization problems. In this paper a brief survey has been provided of recent developments in group search and evacuation in linear and circular search domains. The paper is no doubt biased in favor of recent work by the authors and their collaborators. Also note that the survey is not meant to be exhaustive but rather provide the reader with the flavor of some of the recent, challenging and exciting questions on this topic.

**Acknowledgements.** We would like to express our deepest appreciation to our colleagues Stefan Dobrev, Leszek Gasieniec, Maxime Godon, Danny Krizanc, Fraser MacQuarrie, Russell Martin, Dominik Pajak, Oscar Morales Ponce, Lata Narayanan, and Jarda Opatrny for numerous interesting conversations that excited our interests on all aspects of search and evacuation.

## References

1. Ahlswede, R., Wegener, I.: Search Problems. Wiley, Hoboken (1987)
2. Alpern, S.: Find-and-fetch search on a tree. Oper. Res. **59**(5), 1258–1268 (2011)
3. Alpern, S., Gal, S.: The Theory of Search Games and Rendezvous. Springer, Heidelberg (2003). https://doi.org/10.1007/b100809
4. Alpern, S., Fokkink, R., Gasieniec, L., Lindelauf, R., Subrahmanian, V.S.: Search Theory: A Game Theoretic Perspective. Springer, Heidelberg (2013). https://doi.org/10.1007/978-1-4614-6825-7
5. Alpern, S., Gal, S.: The Theory of Search Games and Rendezvous, vol. 55. Springer, Heidelberg (2006). https://doi.org/10.1007/b100809
6. Baeza-Yates, R., Culberson, J., Rawlins, G.: Searching in the plane. Inf. Comput. **106**(2), 234–252 (1993)
7. Baeza-Yates, R., Schott, R.: Parallel searching in the plane. Comput. Geom. **5**(3), 143–154 (1995)
8. Beck, A.: On the linear search problem. Israel J. Math **2**(4), 221–228 (1964)
9. Bellman, R.: An optimal search. SIAM Rev. **5**(3), 274–274 (1963)
10. Bonato, A.: The Game of Cops and Robbers on Graphs. American Mathematical Society, Providence (2011)
11. Bonato, A., Pralat, P.: Graph Searching Games and Probabilistic Methods. CRC Press, Boca Raton (2017)

12. Bose, P., De Carufel, J.-L.: A general framework for searching on a line. Theor. Comput. Sci. **703**, 1–17 (2017)
13. Brandt, S., Laufenberg, F., Lv, Y., Stolz, D., Wattenhofer, R.: Collaboration without communication: evacuating two robots from a disk. In: Fotakis, D., Pagourtzis, A., Paschos, V.T. (eds.) CIAC 2017. LNCS, vol. 10236, pp. 104–115. Springer, Cham (2017). https://doi.org/10.1007/978-3-319-57586-5_10
14. Bullo, F., Cortes, J., Martinez, S.: Distributed Control of Robotic Networks: A Mathematical Approach to Motion Coordination Algorithms. Princeton University Press, Princeton (2009)
15. Cares, J.R.: Operations Research for Unmanned Systems. Wiley, Hoboken (2016)
16. Chuangpishit, H., Mehrabi, S., Narayanan, L., Opatrny, J.: Evacuating an equilateral triangle in the face-to-face model. In: Aspnes, J., Bessani, A., Felber, P., Leitão, J. (eds.) 21st International Conference on Principles of Distributed Systems (OPODIS 2017). Leibniz International Proceedings in Informatics (LIPIcs), vol. 95, pp. 11:1–11:16 (2018)
17. Czyzowicz, J., Gąsieniec, L., Gorry, T., Kranakis, E., Martin, R., Pajak, D.: Evacuating robots via unknown exit in a disk. In: Kuhn, F. (ed.) DISC 2014. LNCS, vol. 8784, pp. 122–136. Springer, Heidelberg (2014). https://doi.org/10.1007/978-3-662-45174-8_9
18. Czyzowicz, J., et al.: Priority evacuation from a disk using mobile robots. In: Lotker, Z., Patt-Shamir, B. (eds.) SIROCCO 2018. LNCS, vol. 11085, pp. 392–407. Springer, Cham (2018). https://doi.org/10.1007/978-3-030-01325-7_32
19. Czyzowicz, J., et al.: God save the queen. In: 9th International Conference on Fun With Algorithms (FUN 2018) (2018)
20. Czyzowicz, J., Georgiou, K., Kranakis, E., Narayanan, L., Opatrny, J., Vogtenhuber, B.: Evacuating robots from a disk using face-to-face communication (extended abstract). In: Paschos, V.T., Widmayer, P. (eds.) CIAC 2015. LNCS, vol. 9079, pp. 140–152. Springer, Cham (2015). https://doi.org/10.1007/978-3-319-18173-8_10
21. Czyzowicz, J., Dobrev, S., Georgiou, K., Kranakis, E., MacQuarrie, F.: Evacuating two robots from multiple unknown exits in a circle. Theor. Comput. Sci. **709**, 20–30 (2018)
22. Czyzowicz, J., et al.: Evacuation from a disc in the presence of a faulty robot. In: Das, S., Tixeuil, S. (eds.) SIROCCO 2017. LNCS, vol. 10641, pp. 158–173. Springer, Cham (2017). https://doi.org/10.1007/978-3-319-72050-0_10
23. Czyzowicz, J., et al.: Search on a line by byzantine robots. In: 27th International Symposium on Algorithms and Computation, ISAAC 2016, Sydney, Australia, 12–14 December 2016, pp. 27:1–27:12 (2016)
24. Czyzowicz, J., Kranakis, E., Krizanc, D., Narayanan, L., Opatrny, J.: Search on a line with faulty robots. In: Proceedings of the 2016 ACM Symposium on Principles of Distributed Computing, PODC 2016, Chicago, IL, USA, 25–28 July 2016, pp. 405–414 (2016)
25. Czyzowicz, J., Kranakis, E., Krizanc, D., Narayanan, L., Opatrny, J., Shende, S.M.: Wireless autonomous robot evacuation from equilateral triangles and squares. In: Papavassiliou, S., Ruehrup, S. (eds.) ADHOC-NOW 2015. LNCS, vol. 9143, pp. 181–194. Springer, Cham (2015). https://doi.org/10.1007/978-3-319-19662-6_13
26. Czyzowicz, J., Kranakis, E., Krizanc, D., Narayanan, L., Opatrny, J., Shende, S.M.: Linear search with terrain-dependent speeds. In: Fotakis, D., Pagourtzis, A., Paschos, V.T. (eds.) CIAC 2017. LNCS, vol. 10236, pp. 430–441. Springer, Cham (2017). https://doi.org/10.1007/978-3-319-57586-5_36

27. Georgiou, K., Karakostas, G., Kranakis, E.: Search-and-fetch with 2 robots on a disk: wireless and face-to-face communication models. CoRR, abs/1611.10208 (2016)
28. Georgiou, K., Karakostas, G., Kranakis, E.: Search-and-fetch with one robot on a disk - (track: wireless and geometry). In: Chrobak, M., Fernández Anta, A., Gąsieniec, L., Klasing, R. (eds.) ALGOSENSORS 2016. LNCS, vol. 10050, pp. 80–94. Springer, Cham (2017). https://doi.org/10.1007/978-3-319-53058-1_6
29. Jennings, J.S., Whelan, G., Evans, W.F.: Cooperative search and rescue with a team of mobile robots. In: ICAR, pp. 193–200. IEEE (1997)
30. Kagan, E., Ben-Gal, I.: Search and Foraging: Individual Motion and Swarm Dynamics. CRC Press, Boca Raton (2015)
31. Kupavskii, A., Welzl, E.: Lower bounds for searching robots, some faulty. CoRR, abs/1707.05077 (2017)
32. Lamprou, I., Martin, R., Schewe, S.: Fast two-robot disk evacuation with wireless communication. In: Gavoille, C., Ilcinkas, D. (eds.) DISC 2016. LNCS, vol. 9888, pp. 1–15. Springer, Heidelberg (2016). https://doi.org/10.1007/978-3-662-53426-7_1
33. Mesbahi, M., Egerstedt, M.: Graph Theoretic Methods in Multiagent Networks. Princeton University Press, Princeton (2010)
34. Nahin, P.J.: Chases and Escapes: The Mathematics of Pursuit and Evasion. Princeton University Press, Princeton (2012)
35. Schwefel, H.-P.P.: Evolution and Optimum Seeking: The Sixth Generation. Wiley, Hoboken (1993)
36. Stone, L.D.: Theory of Optimal Search, vol. 118. Elsevier, Amsterdam (1976)

# Patrolling

Jurek Czyzowicz[1], Kostantinos Georgiou[2], and Evangelos Kranakis[3](✉)

[1] Dépt. d'informatique, Univ. du Québec en Outaouais, Gatineau, QC, Canada
[2] Department of Mathematics, Ryerson University, Toronto, ON, Canada
[3] School of Computer Science, Carleton University, Ottawa, ON, Canada
kranakis@scs.carleton.ca

**Abstract.** Patrolling is concerned with the design of continuous trajectories which specify robots perpetual movements along a curve so that the time between any two consecutive visits to any point of the curve is minimized. In this paper we survey recent rigorous results on patrolling by various number of robots and robots' specifications (e.g., speed), and for various types of curves. We discuss efficient patrolling strategies for mobile agents with various capabilities and behaviors acting on a variety of geometric graph domains.

**Keywords:** Agents · Faulty · Graphs · Patrolling · Speeds
Strategies · Trees · Visibility

## 1 Introduction

Patrolling is defined as the act of perpetual surveillance of either an area or the perimeter of an area by mobile agents in order to monitor, protect or supervise it. It has been a theme of extensive investigation in robotics [2,6,15–17,20,22]. It can be useful in order to detect whether intruders may penetrate an area through its perimeter or even monitor an area itself so as to determine objects or humans that need to be rescued from a disaster environment. For example, network administrators may use mobile agent patrols to detect network failures or to discover web pages which need to be indexed by search engines [20].

Patrolling involves a perpetual movement of the robots that is performed in a static or in a dynamically changing environment. Motivated from Artificial Intelligence studies, patrolling has given rise to beautiful optimization problems [19] and has provided numerous applications in Computer Science, including Infrastructure Security, Computer Games, perpetual domain-surveying, and monitoring in 1D and 2D geometric domains.

The present article is concerned with deterministic patrolling algorithms describing the movement of the autonomous robots so as to patrol a given underlying geometric domain.

Research supported in part by NSERC Discovery grants.


P. Flocchini et al. (Eds.): Distributed Computing by Mobile Entities, LNCS 11340, pp. 371–400, 2019.
https://doi.org/10.1007/978-3-030-11072-7_15

## Roadmap

A road map of the present survey is as follows. In the remainder of the current section we introduce the basic definitions, preliminary and essential concepts pertaining to the autonomous robots, their movement and underlying domain to be patrolled. The main characteristic of the results described in the following four Sects. (2, 3, 4, and 5) is that the initial positions and movements of the robots are specified by a scheduler. The specific topics discussed in the respective sections include robot capabilities, fault tolerance, patrolling fragmented boundaries, and optimal patrolling on trees, Non-centralized algorithms turn out to be more complex than centralized and in the next Sect. 6 we outline rotor based patrolling in which the robots make use of a specific labeling (the rotor) of the vertices and edges of the underlying graph. Finally, in Sect. 7 we provide decentralized strategies in which the robots' trajectories are based on distributed online algorithms.

## Underlying Domain

In a general setting, we are given a domain which is a connected geometric graph $G = (V, E)$, where $V$ is its set of vertices and $E$ its set of edges.

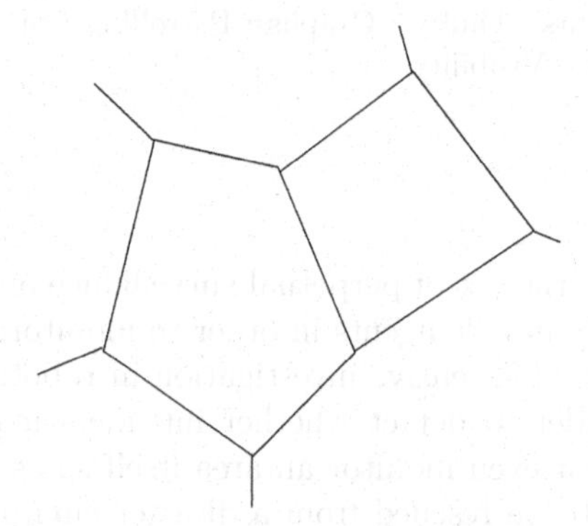

**Fig. 1.** Portion of a connected geometric graph.

Each edge $e \in E$ of the graph $G = (V, E)$ is modelled as a smooth continuous and rectifiable (with continuous derivative) curve of arbitrary positive length represented by its edge weight $w(e)$. By $|E|$ we denote the sum of the lengths of the edges of $G$. At any time a robot may occupy any point belonging to an edge $e$ (this may include interior and endpoints). We denote by $\mathcal{D}_G$ the domain (the union of the edges) along which the robots walk.

## Robots and Movement

Assume that $k$ mobile robots are deployed on such a graph. The robots move perpetually along edges of the domain without exceeding their maximum speed.

The robots patrol the graph by regularly and perpetually visiting all points of the domain in well-defined trajectories.

In a later section we will also consider patrolling with faulty robots. In this case, we will consider a team of $k$ robots (patrolmen), at most $f$, where $f < k$, of which may be unreliable in that they fail to comply with their patrolling duties, e.g., they do not follow their trajectories because either of reduced or failing capabilities.

We assume a continuous traversal model, whereby the movement of the $i$-th robot within $\mathcal{D}_G$ follows a continuous function of time

$$t \rightarrow \pi_i(t) : \pi_i : [0, \infty) \rightarrow \mathcal{D}_G, \text{ for each } i = 1, 2, \ldots, k,$$

and $\pi_i(t)$ denotes the position in $\mathcal{D}_G$ of the $i$-th robot at time $t$. Each robot may move in any direction along an edge not exceeding its maximum speed specifications. If the maximum speed is one then within time interval $[t_1, t_2]$ each robot may travel a distance of at most $t_2 - t_1$; we also assume that when walking at maximum speed, a robot travels the unit distance in unit time, so that time and distance travelled are commensurable. In the sequel we will also consider robots moving with different maximum speeds (sometimes dependent on the task they perform, i.e. patrolling or just walking).

### Patrolling Strategies

By *patrolling strategy* we understand the set $\mathcal{P} = \{\pi_1, \pi_2, \ldots, \pi_k\}$ of infinite trajectories of $k$ robots in $\mathcal{D}_G$, where $\pi_i(t)$ is the point of $\mathcal{D}_G$ occupied by the $i$-th robot at time $t$.

Boundary and area patrolling has been studied in several papers including [1,15,16,21]. It has usually been dealt with using an ad-hoc approach which emphasizes experimental results. Following [6], two basic patrolling strategies are considered: *partition-based* (when the environment is partitioned into parts monitored by individual agents) and *cyclic-based* (when all agents patrol the environment walking in the same direction along some cycle).

For example, Fig. 2 depicts the possible trajectories of four robots in the graph of Fig. 1. The left picture depicts a cyclic strategy in which the four robots are patrolling along the same cycle. The right picture depicts a partition of the graph into two cycles each of which is patrolled by a given subset of the four robots.

### Idleness

Patrolling strategies are evaluated by using the *idleness*, a measure widely used in robotics literature. Consider a patrolling strategy $\mathcal{P}$ for $k$ robots moving along a geometric graph $G$. Then the idleness of strategy $\mathcal{P}$ for graph $G$ (or its *idle time*), denoted by $\Im_k^f(G, \mathcal{P})$ is the supremum of the lengths of time intervals between two consecutive visits to the same point of $\mathcal{D}_G$ (supremum taken over time and over all points of $\mathcal{D}_G$): when up to $f$ robots may be faulty, the adversary, knowing our strategy, may choose a set $F$ of $f$ faulty robots, a point $p$ of the

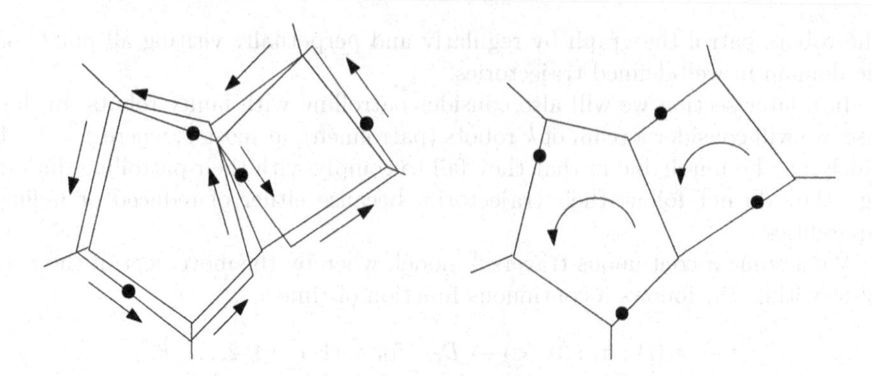

**Fig. 2.** A graph and two possible patrolling strategies. 1. Left figure: a cyclic strategy, and 2. Right figure: a partition strategy with two different cycles.

domain and a time moment $t \geq 0$. The idleness of the strategy is the supremum (taken over all such adversarial choices) of time intervals $T$ such that point $p$ is not visited during the time interval $[t, t+T]$ by any reliable robot. Finally, the idleness of a graph $G$ for $k$ robots, at most $f$ of which may be faulty, is denoted by

$$\Im_k^f(G) := \inf_{\mathcal{P}} \Im_k^f(G, \mathcal{P}).$$

Hence $\Im_k^f(G)$ is the lower bound of idleness over all possible patrolling strategies. When there are no faulty robots (i.e., $f = 0$) we use the notation $\Im_k(G, \mathcal{P}) := \Im_k^0(G, \mathcal{P})$. In the sequel, mention of $G$ and $\mathcal{P}$ will be omitted when easily understood from the context.

## 2 Agent Capabilities

In this section we discuss how speeds and visibility may affect patrolling of a domain by mobile agents.

### 2.1 Agents with Possibly Distinct Maximal Speeds

The patrolling problem is perhaps the most challenging for agents whose maximal speeds are not necessarily the same. In [9] the authors address the patrolling problem for mobile agents with distinct maximal speeds placed in the line and ring environments. The question asked was whether the solutions used in robotics, where the strategy used are either partition-based or cycle-based, are optimal. The results of this paper and the works which followed show that this is the case only for very small collections of agents.

**Model Specifics and Problem Definition**

The set of $k$ agents may be placed at arbitrary positions of the environment and move at velocities not exceeding their respective maximal speeds. We suppose that the agents are numbered according to their decreasing maximal speeds i.e. $v_1 \geq v_2 \geq \ldots \geq v_k > 0$. The studied environments are a segment of a given finite length and a ring of a given size. We assume that the positive speed of an agent corresponds to the counterclockwise traversal of the ring and its left-to-right movement from on the segment. Using a scaling argument we suppose that the time and length are commensurable. In other words, we assume that the length of the ring is equal to 1 (one unit of length), and that an agent using constant speed 1 (one unit of speed) in one unit of time exactly traverses once the entire ring.

**Fence Patrolling**

By *fence patrolling* the robotics community considers perpetual monitoring of the set of points homeomorphic with a segment of finite length. We denote by $S = [0, 1]$ the unit length interval. [9] proposes the following *proportional* algorithm.

1. Let $v_1, v_2, \ldots, v_k > 0$ be the set of agents speeds and $V = \sum_{i=1}^{k} v_i$.
2. Partition $S$ into sub-segments $S_1, S_2, \ldots, S_k$ of length $|S_i| = \frac{v_i}{V}$.
3. Agent $a_i$ walks at its full speed back and forth between the endpoints of $S_i$.

It is easy to show that the idle time of this algorithm equals $2/V$, i.e. it is independent of the number of mobile agents, but it is dependent on the summation of their speeds. Indeed, it may be observed that replacing any subset $S' \subset S$ of agents by a single agent having a maximal speed $v$, such that $v = \sum_{i \in S'} v_i$ results in the proportional algorithm producing the same idle time.

The proposed algorithm is a typical partition-based solution. The authors of [9] proved that the above proportional algorithm results in the optimal idle time for two agents and they asked whether the same algorithm works for any number of agents.

This question has been addressed by Kawamura and Kobayashi in [18] where the authors prove that the proportional algorithm results in the optimal idle time for $k \leq 3$ agents. However, they prove that the algorithm is not optimal in general. The authors of [18] produce a counterexample by giving a set $A$ of six agents. Four agents of set $A$ have maximal speed 1 while the remaining two agents have maximal speeds 1/2 and 7/3. The example given by Kawamura and Kobayashi arranges the movements of the agents of $A$ so that the obtained idle time equals 41/42 of the idle time of the proportional algorithm.

**Patrolling a Unidirectional Boundary**

In this section we suggest an algorithm generating the schedule for the ring patrolling, where all agents are walking in the same, say counterclockwise, direction of the ring. It is easy to show that, having two agents, one of the following two solutions will be optimal:

1. The slower agent is not taken into consideration (it stays immobile) and the idle time is obtained from the faster agent walking at full speed.
2. The faster agent slows down to use the speed of the slower agent. Both agents walk at speed $v_2$ remaining continuously antipodal on the ring.

The idle time obtained from the first schedule above equals $1/v_1$, while the second schedule produces the idle time $1/2v_2$. Therefore the first schedule may be chosen when $v_1 \geq 2v_2$, otherwise the second schedule is preferred.

The idea used for two agents might be extended to the larger collections of agents. The authors of [9] propose the following algorithm.

1. Let $r$ be such that for all $i = 1, 2, \ldots, k$ we have $iv_i \leq rv_r$
2. All agents $a_{r+1}, a_{r+2} \ldots a_k$ are not used by the algorithm
3. Place the agents $a_1, a_2 \ldots a_r$ at equal distances around the circle
4. For each $i = 1, \ldots, r$ agent $a_i$ moves counterclockwise around the circle at speed $v_r$.

The above algorithm is a typical cycle-based solution. This algorithm has been shown in [9] to produce the optimal idle time for any collection of $k \leq 4$ agents and any configuration of their speeds. It was left open in [9] whether the solution produced is optimal for collections of $k$ agents for arbitrary value of $k$ and arbitrary configuration of their speeds. However the authors of [14] have shown that this is not true. The example of [14] contains 32 agents having harmonic maximal speeds $v_i = 1/i$ for $i = 1, 2, \ldots, 32$. The idle time of such a collection of agents on the unidirectional ring equals 1. The example proposes the original arrangement of the 32 agents around the unidirectional ring and their walks that never exceed the respective maximal speeds, resulting in the idle time strictly smaller than 1.

**Patrolling a Boundary with Movements in Both Directions**
In the original paper [9] using agents of distinct speeds, the algorithms proposed for fence patrolling generated partition-based schedules, in which the agents always use their maximal speeds. In the case of unidirectional ring some optimal solutions involve agents that need to use speeds that are not maximal. However the velocities used there are uniform during the entire schedule and the generated schedules are cycle-based. An example was given in [9], that already for 3 agents on the bidirectional ring neither the partition-based, nor the cycle-based strategy achieves the best idle time.

Consider an example of 3 agents with harmonic speeds $v_1 = 1, v_2 = 1/2, v_3 = 1/3$. The proportional schedule would partition the ring into three segments of sizes 2/11, 3/11 and 6/11. Each agent moving back and forth inside its assigned segment would result in the idle time of 12/11. Any construction of the cycle-based strategy results in idle time of 1. As $iv_i = 1$, for each $i = 1, 2, 3$, such solution may use one, two or all three agents. Figure 3 illustrates the movements of the three agents having harmonic maximal speeds, which results in a better idle time that both standard partition-based and cycle-based solutions.

The perpetual movements of the three agents in Fig. 3 repeat in the cycles, each one taking 3 units of time. During such cycle the agent $a_3$ walks counterclockwise at its maximal speed of 1/3 exactly traversing the entire ring. Agent $a_2$ walks counterclockwise for time 5/8 then clockwise for time 1/8, repeating the same procedure four times during an interval of 3 units of time. Finally agent $a_1$ walks counterclockwise for time 1/2 then clockwise for time 1/4, again repeating the same procedure four times during an interval of 3 units of time. The cycle of 3 units of time starts at the same time moment for all three agents. The starting positions of the three agents are chosen so that the worst-case waiting time between the visits of agents $a_3$ and $a_2$, visits of agents $a_2$ and $a_1$ as well as visits of agents $a_1$ and $a_3$ is made equal. This time, represented by interval $\tau$ on Fig. 3 equals 35/36 and it is the idle time of the schedule produced by the algorithm.

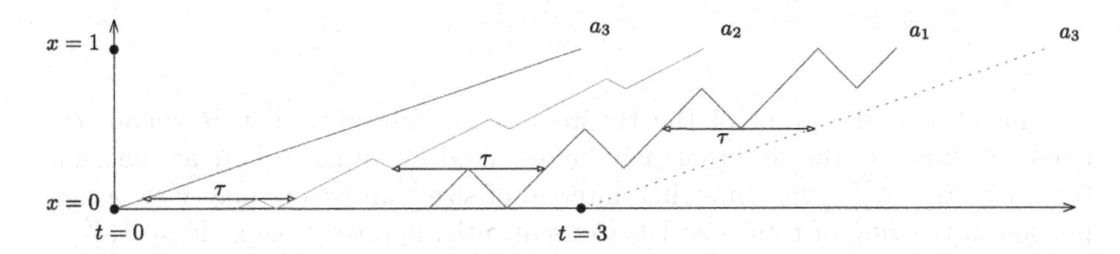


**Fig. 3.** An example of a schedule for three agents having harmonic maximal speeds $v_1 = 1, v_2 = 1/2, v_3 = 1/3$

The solution represented on Fig. 3 is not proven to be optimal, but it is proven to be better than standard partition-based and cycle-based strategies. It seems that in the case of the ring, for no configuration of agents speeds the proportional, partition-based strategy is optimal. For some configuration of speeds (e.g. speeds which are relatively "uniform") the cycle-based strategy would be optimal. However, as shown by the example from Fig. 3, for many speed configurations the optimal solution might involve the movements of agents that are quite particular and difficult to find.

## 2.2 Agents with Distinct Visibility Ranges

The patrolling problem for collections of agents having possibly distinct radii of visibility has been studied in [13]. The agents may walk not exceeding their maximal speed. The paper considers the case when the maximal speed is the same for all agents, as well as the case when the maximal speeds are distinct. Each agent $a_i$, for $i = 1, 2, \ldots, k$, has the radius of visibility $r_i$, which may be possibly distinct for different agents. Therefore, when agent $a_i$ visits point $x$ of the environment all points at distance at most $r_i$ from $x$ are being patrolled at the same time. The environments considered are the open boundary (i.e. unit length segment) as well the closed boundary (unit length ring). The patrolling problem for other graphs is also briefly considered.

**Boundary Patrolling**

Similarly to the case of the previous section, by using the scaling argument, it is assumed that the segment that needs to be patrolled has unit length. Denote by $V$ the sum of all agents' speeds

$$V = \sum_{i=1}^{k} v_i.$$

The known lower bound for the problem is stated in the following theorem.

**Theorem 1** ([13]). *Consider the patrolling problem using $k$ robots having maximal speeds $v_1, \ldots, v_k$ and radii of visibility $r_1, \ldots r_k$, respectively. Any algorithm patrolling the unit ring must achieve the idle time $\Im$ such that*

$$\Im \geq \frac{1 - \sum_{i=1}^{k} 2r_i}{V}.$$

The idea of the proof of the theorem is the following. Let $R$ denote the maximal subset of the ring that may be observed by all the robots at the same time, i.e., $R = \sum_{i=1}^{k} 2r_i$. In a time interval of size $t$ all robots may view a new portion of the ring of total size $Vt$. Consequently, it takes time at least $\frac{1-R}{V}$ to explore the ring. This is the lower bound on the idle time.

In the case of robots having the same unit maximal speed we can design an optimal patrolling algorithm:

1. Place the agents in such positions around the ring that no two agents see the same point of the ring. Moreover let the placement be such that each of $k$ ring segments containing the points, which are not seen by any agent are of the same length.
2. Each agent moves in the counterclockwise direction of the ring with maximal speed.

At their assigned initial positions the agents see the portion of the ring of size $R$. Hence each of the $k$ regions, unseen by any agent, is of the size $(1-R)/k$. As all agents go in the same direction with unit speed, $(1-R)/k$ is the maximal time interval during which a point of the ring remains unseen by a robot. As we have $V = k$ we can conclude the following Corollary.

**Corollary 1** ([13]). *When the collection of robots have the same maximal speed the proportional algorithm achieves the optimal idle time.*

**Fence Patrolling**

Consider a unit length segment (fence) and $k$ agents with possibly distinct visibility ranges and maximal speeds. There exists a natural proportional algorithm for the fence patrolling:

1. Partition the portion $1 - R$ of the fence proportionally to agents' speeds, i.e. let $s_i = (1-R)v_i/V$ for $i = 1, 2 \ldots, k$.
2. Assign to agent $a_i$ a sub-segment of the fence of size $s_i + 2r_i$. Denote by $S_i = [b_i, b_i + v_i + 2r_i]$ the sub-segment of the fence assigned to agent $a_i$.
3. Each agent $a_i$ for $i = 1, 2 \ldots, k$ walks at its maximal speed inside the assigned sub-segment $S_i$, back and forth between the two points $b_i + r_i$ and $b_i + v_i + r_i$.

Observe that the segments $S_i$ assigned to the agents cover the entire fence. Note that, thanks to its visibility range, during its back and forth movement each agent sees all points inside the assigned sub-segment. Because of the proportionality of the assignment, each agent revisits the left (or right) endpoint of the segment assigned to it after the same amount of time $I$. This amount of time equals

$$I = \frac{2s_i}{v_i} = \frac{2(1-R)v_i/V}{v_i} = 2\frac{1-\sum_{i=1}^{k} 2r_i}{V}.$$

The value $I$ is the idle time of the proposed algorithm.

It may be proved that, when the agents maximal speeds are the same, the above algorithm is optimal.

**Theorem 2** ([13]). *Consider a collection of $k$ agents having unit maximal speed and possibly different radii of visibility $r_1, r_2, \ldots, r_k$. The proportional algorithm for the unit length fence achieves the optimal idle time of*

$$2\frac{1-\sum_{i=1}^{k} 2r_i}{k}.$$

When the agents maximal speeds are distinct, the proportional algorithm is not optimal in general. Indeed, the result of [18] shows that the proportional algorithm is suboptimal even with a point visibility range of every agent. However, it is shown in [13] that the proportional algorithm is optimal if the collection contains only two agents. More exactly, the authors of [13] prove the following Theorem.

**Theorem 3** ([13]). *Consider 2 agents $a_1, a_2$ having possibly different maximal speeds $v_1, v_2$ and different radii of visibility $r_1, r_2$. The proportional algorithm using agents $a_1, a_2$ for the unit length fence achieves optimal idle time*

$$\Im = 2\frac{1-2(r_1+r_2)}{v_1+v_2}.$$

**Hardness Results**

In the previous sections it has been shown that in the case of equal-speed agents with possibly different visibility radii the fence patrolling and the boundary patrolling problems may be solved in linear time. It is interesting to know whether the patrolling problem may be also easily solved for other environments. Section 5 shows that for a tree environment there exists a linear-time algorithm for equal-speed agents with point-visibility (i.e. $r_i = 0$ for $i = 1, 2, \ldots, k$). However, the situation is quite different when radii of visibility are allowed to be different. It is possible to prove that the decision problem whether there exists a patrolling algorithm whose idle time is not larger than a given constant is NP-hard for trees. We use a reduction to the PARTITION problem that we recall below.

**Instance:** Finite set $D = \{d_1, d_2, \ldots, d_n\}$ of integers summing up to an even number $S$.

**Question:** Is it possible to partition set $D$ into two subsets each of whose elements sum up to $S/2$?

The PARTITION problem is known to be NP-complete. We have the following Theorem.

**Theorem 4** ([13]). *There exists a tree $T$ and a configuration of agents' visibility radii for which the problem of finding the optimal idle time using equal-speed agents is NP-hard.*

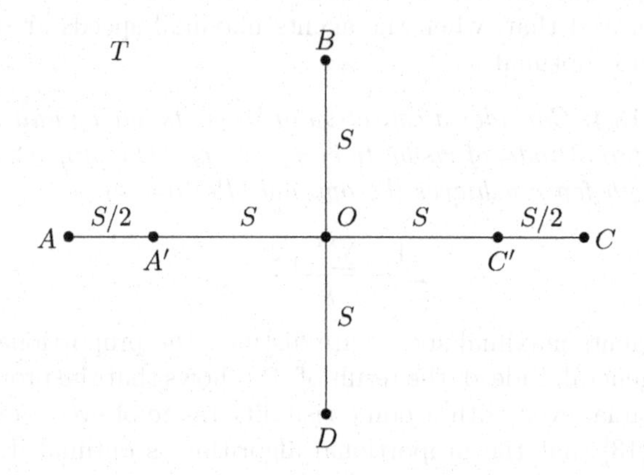


**Fig. 4.** For the proof of Theorem 4

From the above instance of the PARTITION problem we construct the following instance of the patrolling problem. The environment is the tree $T$ illustrated at Fig. 4. $T$ is a star formed by four segments $OA, OB, OC, OD$. The length of each segment $OA$ and $OC$ equals $3S/2$ and the length of segments $OB$ and $OD$ equals $S$. The collection contains $n+1$ agents $A = \{a_0, a_1, \ldots, a_n\}$ with radii of visibility $r_0, r_1, \ldots, r_n$, respectively. We set $r_0 = S$ and $r_i = d_i/2$ for $i = 1, 2, \ldots, n$. Is it possible to design a patrolling algorithm with idle time equal to 0? In other words, is it possible to place the (immobile) agents in the tree $T$ so that each point of $T$ is visible by some agent? Observe that if the agent $a_0$ is not placed at point $O$ of the tree $T$ the entire tree $T$ cannot be made visible by the set of immobile agents $A$. Suppose then that agent $a_0$ is placed at point $O$, covering by its visibility range segments $OA', OB, OC'$ and $OD$. The remaining agents need to cover segments $AA'$ and $CC'$, each one of length $S/2$. By construction, this is possible only if there is a partition of set $D$ into two subsets of equal sum $S/2$.

A more complicated construction is possible in order to prove a more general Theorem.

**Theorem 5** ([13]). *For any fixed real value $I$ there exists a graph $G$ and a configuration of visibilities $r_1, r_2, \ldots, r_k$ of equal-speed agents, so that the decision problem whether the agents can patrol $G$ with idle time at most $I$, is NP-hard.*

## 3 Constraints and Agent Behavior

In this section we survey patrolling under faulty behavior of the robots and discuss what competitive ratio on idle time can be attained when the robots are required to obey certain time constraints.

### 3.1 Patrolling with Unreliable Mobile Agents

The authors of [10] study the patrolling problem when some agents may turn out to be unreliable, i.e. they experience faults during their work. We denote by $A = \{a_1, a_2 \ldots, a_k\}$ the set of agents and by $F \subset A$ the set of agents that turn out to be faulty. The cardinality of subset $F$ is bounded by some constant $f < k$. The algorithm does not have the knowledge of the set $F$, but it knows the constant $f$.

We can suppose that the patrolmen need to report some events that may happen to the points of the network (some sort of failures, presence of intruders, etc.) while visiting such points. A reliable agent walks according to the assigned schedule, perceives an event incoming to a visited point and reports it instantaneously. A faulty agent may deviate from the assigned route, fail to perceive an event, fail to report it or report an inexistent event. Minimization of the idle time means that we want to report a real event by a reliable agent in the shortest possible time after its occurrence. In this sense, since the agents do not communicate, it is irrelevant whether we say that such faults are of crash or Byzantine nature.

It is assumed that an omnipotent adversary knows the patrolling algorithm. As a function of this knowledge, and aiming for a worst case analysis, the adversary may choose:

1. a point $P^*$ of the environment which needs to be patrolled,
2. a time moment $t^*$ after which an event at point $P^*$ starts to occur,
3. a subset $F$ of up to $f$ mobile agents which turn up to be faulty.

Consequently, the idle time $\Im$ of the proposed algorithm is such that at time $t^*$ a robot $a_i \in A \setminus F$ visits point $P^*$, at time $t^* + \Im$ a robot $a_j \in A \setminus F$ visits point $P^*$ and during the time interval $[t^*, t^* + \Im]$ all $f$ robots of set $F$ visit point $P^*$.

**Fence Patrolling**
The environment to be patrolled is a unit length segment $I = [0, 1]$. It is assumed that most robots of the collection $A$ are reliable, more precisely that $f < \frac{k-2}{2}$. The main algorithm for fence patrolling is proposed in [10]. Its idea is to partition the collection $A$ of agents into three groups. Each group is instructed to make an Eulerian tour (i.e., back and forth movement) of some sub-segment of $I$. By an Eulerian tour we mean to place the agents equally spaced around the subsegment and walk in the counterclockwise direction (in the resulting Eulerian cycle) with maximal speed. Hence each agent makes back and forth movements between the endpoints of the subsegment. Each point of the segment, except its extremities,

is visited by the stream of agents traversing it in the left-to-right direction as well as the stream of agents walking over it in the right-to-left direction. The three subsegments are the left subsegment $I_L$ (including point 0 - the left endpoint of $I$), the right subsegment $I_R$ (including point 1) and the entire segment $I$. The lengths of segments $I_L, I_R$ as well as the distribution of the agents among the three groups is a function of parameters $k$ and $f$. More precisely we have the following algorithm.

**Algorithm $A_I$**

1. Consider three segments:

$$I_L = \left[0, \frac{\lceil f/2 \rceil}{k - 2\lceil f/2 \rceil}\right],$$
$$I_M = \left[\frac{\lceil f/2 \rceil}{k - 2\lceil f/2 \rceil}, 1 - \frac{\lceil f/2 \rceil}{k - 2\lceil f/2 \rceil}\right]$$
$$I = [0, 1]$$

2. Assign $\lceil f/2 \rceil$ agents to each segment $I_L, I_R$.
3. Assign the remaining $k - 2\lceil f/2 \rceil$ agents to segment $I$.
4. Each of the three groups of agents perform an Eulerian tour around the assigned segment.

**Theorem 6** ([10]). *The strategy executed by algorithm $A_I$ results in patrolling of segment $I$ with idle time*

$$\Im_k^f(I, A_I) \leq \frac{2\lfloor f/2 \rfloor + 2}{k - 2\lceil f/2 \rceil}.$$

The rough idea of the proof of Theorem 6 is the following. The distance $d$ between two consecutive agents of each Eulerian tour equals $d = \frac{2}{k-2\lceil f/2 \rceil}$.

We bound first the idle time of a point $p$ belonging to an extremal segment $I_L$ (or, by symmetry to $I_R$). The most vulnerable point is the endpoint, say point 0 (as other points are visited twice more often). If some agent assigned to subsegment $I_L$ is non-faulty then this agent visits point 0 every $2\frac{\lceil f/2 \rceil}{k-2\lceil f/2 \rceil}$ time, which is not greater than the time $\frac{2\lfloor f/2 \rfloor+2}{k-2\lceil f/2 \rceil}$ claimed in the Theorem. Otherwise, $\lceil f/2 \rceil$ agents assigned to $I_L$ are faulty and the idle time at point 0 is assured by the agents assigned to segment $I$. As there are at most $f - \lfloor f/2 \rfloor$ remaining faulty agents, the distance between the non-faulty agents in the Eulerian cycle of $I$ is at most $d(f - \lfloor f/2 \rfloor + 1)$ and we have

$$\Im_k^f(I, \mathcal{P}) \leq d(f - \lceil f/2 \rceil + 1) = d(\lfloor f/2 \rfloor + 1) = \frac{2(\lfloor f/2 \rfloor + 1)}{k - 2\lceil f/2 \rceil}. \quad (1)$$

The idle time of a point $p$ outside the extremal segments $I_L$ and $I_R$ is assured by the agents assigned to segment $I$. There are two streams of agents traversing $p$ in both directions. The worst case arises when the two streams contain long sequences $f_1, f_2$ of faulty agents. It may be shown that when $f_1 = f_2$ or $f_1 \neq f_2$

the time between two visits of non-faulty agents is also bound by the value from Eq. (1).

It may be proven that our algorithm is optimal for odd values of $f$ and almost optimal for even $f$. More exactly we have the following Theorem.

**Theorem 7** ([10]). *Consider patrolling of a segment with $k$ robots. Suppose that up to $f$ robots of the collection, such that $f < k/2 - 1$, may turn out to be faulty. For any patrolling strategy $\mathcal{A}$ its idle time is lower-bounded by*

$$\Im_k^f(I, \mathcal{A}) \geq \frac{f+1}{k-f-1}.$$

To sketch the proof of Theorem 7 we observe first that, in order to get a better lower bound, at all times there must be at least $f+1$ agents inside the segment $\left[0, \frac{f+1}{2(k-f-1)}\right]$. By symmetry there must be also at least $f+1$ agents inside the subsegment of $I$ of the same length that contains point 1. We can then use the following Lemma.

**Lemma 1** ([10]). *Consider a graph $G$ and its patrolling strategy $\mathcal{P}$. Let $E'$ be some subset of portions of edges of $G$ and let $|E'|$ denote the sum of their lengths. Suppose that starting from some time moment of the strategy, $E'$ contains at most $r$ robots. Then*

$$\Im_k^f(\mathcal{G}, \mathcal{P}) \geq \frac{(f+1)|E'|}{r}.$$

We then show that if the idle time is smaller than $\frac{f+1}{k-f-1}$, having only the remaining $k-2(f+1)$ agents in the middle part of $I$ would contradict Lemma 1.

**Idle Time of Arbitrary Graphs**

For the general graphs, the lower and upper bounds on the idle time is given by the following Theorem.

**Theorem 8** ([10]). *Consider a connected graph $G$ that needs to be patrolled by $k \geq 2$ robots, at most $f$ of which are faulty. Then*

$$\frac{(f+1)|E|}{k} \leq \Im_k^f(G) \leq \frac{(f+1)CPT(G)}{k}.$$

*where $CPT(G)$ denotes the length of the Chinese Postman Tour of $G$.*

To get the upper bound $\frac{(f+1)CPT(G)}{k}$ it is sufficient to patrol the graph $G$ by $k$ agents placed equidistant around the Chinese Postman Tour $CPT(G)$. The lower bound $\frac{(f+1)E(G)}{k}$ follows from Lemma 1 applied to $E(G)$.

As for any Eulerian graph $G$ we have $E(G) = CPT(G)$, using Theorem 8 we obtain the following Corollary.

**Corollary 2** ([10]). *For any connected Eulerian graph $G$, there exist an optimal patrolling strategy $\mathcal{P}$ using $k \geq 2$ agents that may contain at most $f$ faulty ones and*

$$\Im_k^f(G, \mathcal{P}) = \frac{(f+1)E(G)}{k}.$$

We conclude with the following theorem, which states that the problem of computing the optimal idle time for collections of agents including some faulty ones is NP-hard.

**Theorem 9** ([10]). *Consider a collection of 3 robots, such that one of them, a priori unknown, may turn out to be faulty. There exist graphs for which computing the idle time for such collection of robots of the optimal patrolling strategy is NP-hard.*

The reader might be interested in the following observation. Note that in the solution of the fence patrolling problem, the larger the number $f$ of faulty robots, the longer the idle time of the proposed algorithm. Hence knowledge of the bound $f$, allows us to obtain the best idle time for that bound. Similarly, if we know the time bound $t$ during which an event that is occurring at a point of the network, may remain unreported, we may design an algorithm producing a patrolling schedule whose idle time is at most $t$ while at the same time tolerating the largest possible number of faults. However, it is the knowledge of $f$ in the former case and the knowledge of $t$ in the latter one which is essential.

### 3.2 Patrolling with Time Constraints

The notion of the idle time of a patrolling schedule, and hence the objective of traditional patrolling, assumes that points in the domain to be patrolled are indistinguishable with respect to visitation demands. Even more, one could restate the patrolling problem of a domain, consisting of a collection of points $N$ to be patrolled, as follows; associate each point in $N$ with $\tau$, and request a patrolling schedule so that the time between any two visitations of any point is no more than $\tau$. Clearly, even when $\tau$ is as low as the optimal idle time of patrolling the domain, the problem admits a feasible solution, even though such a solution might be difficult to find. In a more interesting setting, the points of the domain are each associated with different visitation requirements. This is the subject of [7].

**The Model of Patrolling Points with Visitation Requirements**
In the *Path Patrolling Problem with Visitation Requirements* (PPVR), two robots of maximum speed 1 patrol $n$ points $0 = y_1 < y_2 < \ldots < y_n = 1$ placed on a unit interval. Each point $y_i$ is associated with its *visitation frequency requirement* $I(y_i) \in \mathbb{R}_+$. The idle time in PPVR is defined only with respect to points $y_i$, even though the movement of the robots take place in the contiguous unit interval. In that direction, and for a patrolling schedule $\mathcal{P}$, we denote by $w_{\mathcal{P}}(y_i)$ the idle time of point $i$ (and we drop subscript $\mathcal{P}$, when it is clear from the context). A patrolling schedule $\mathcal{P}$ is called *c-feasible* if $w_{\mathcal{P}}(y_i)/I(y_i) \leq c$, for each $i = 1, \ldots, n$. Thus a feasible patrolling schedule is also 1-feasible, and an instance that admits such a schedule will also be called *feasible*. The objective of PPVR is to design patrolling schedules that minimize the worst normalized violation of the idleness times for feasible instances, i.e. $c$-feasible schedules where $c \geq 1$ is as small as possible.

**Structural Properties of Feasible Instances**

Deciding whether a PPVR instance is feasible, is not known to be in $P$, neither $NP-hard$. At the same time, finding $c$-feasible schedules for feasible instances has proved to be a challenging task, and the best result known, due to [7], is the following.

**Theorem 10** ([7]). *Feasible instances to* PPVR *admit* $\sqrt{3}$*-feasible schedules.*

The main challenge behind designing effective patrolling schedules is imposed by the visitation requirements of points $y_i$. Indeed, small values of $I(y_i)$ require that at least one robot is always in good proximity of point $y_i$. We are therefore motivated to introduce notation for the ball

$$R(y_i) := \left[\max\left\{0, y_i - \frac{I(y_i)}{2}\right\}, \min\left\{1, i + \frac{I(y_i)}{2}\right\}\right]$$

around point $i$, i.e. the range within which a robot can be located introducing no violation to the visitation frequency requirement of point $i$. Since also at least one of the extreme points of the unit interval can be assumed to be patrolled only by one robot, we are motivated to group points $y_i$ with respect to whether the extreme points fall within their range:

$$\begin{aligned} S_{00} &:= \{y_i \in S:\ 0, 1 \notin R(y_i)\}, \\ S_{01} &:= \{y_i \in S:\ 0 \notin R(y_i), 1 \in R(y_i)\}, \\ S_{01} &:= \{y_i \in S:\ 0 \in R(y_i), 1 \notin R(y_i)\}, \\ S_{11} &:= \{y_i \in S:\ 0, 1 \in R(y_i)\}. \end{aligned}$$

Note that $S_{00}$ contains points that cannot be patrolled just by one robot, and hence require their coordination. For that reason, instances for which $S_{00}$ is empty are considered easy.

**Theorem 11** ([7]). *Feasible instances of* PPVR *with* $S_{00} = \emptyset$ *admit a partition based 1-feasible patrolling schedule.*

The proof of Theorem 11 relies on a characterization of feasible instances of PPVR with $S_{00} = \emptyset$, according to which those instances are exactly those for which

- $S_{10} \subset \bigcap_{x \in S_{10}} R(x) =: X_{10}$, and $0 \in X_{10}$.
- $S_{01} \subset \bigcap_{x \in S_{01}} R(x) =: X_{01}$, and $1 \in X_{01}$.
- $S \subset [\bigcap_{x \in S_{10}} R(x)] \cup \bigcap_{x \in S_{01}} R(x)] = X_{10} \cup X_{01}$

As a result, a robot may search $X_{10} \setminus X_{01}$ and the other $X_{01}$, and it is easy to see that the schedule is indeed 1-feasible.

On the other hand, feasible instances for which $S_{00} \neq \emptyset$ admit no known characterization, and as a result they admit solutions that only approximate feasibility. Nevertheless, known algorithms are based on a necessary condition that is known for all such instances, partially revealing their structural properties.

**Lemma 2** ([7]). *$S_{00} \subset \bigcap_{x \in S} R(x)$, for all feasible instances of* PPVR.

**Finding *c*-Feasible Patrolling Schedules**
Equipped with Lemma 2, one can easily devise a 4-feasible patrolling schedule for any feasible PPVR instance. The patrolling schedule with such a guarantee involves nested traversals. For this, define $i_0 = \text{argmin}_i I(y_i)$. Then the nested traversal has one robot perpetually going back and forth between the endpoints of the unit interval, and the other going back and forth between the endpoints of $R(y_{i_0})$. The advantage of this strategy is that only one of the robots needs to remember the endpoints of $R(y_{i_0})$.

Improving the feasibility violation to $\sqrt{3}$ as suggested by Theorem 10 requires a much more elaborate algorithm which in particular assumes coordination between the robots. The structural component that one has to utilize in order to efficiently patrol PPVR instances is related to the span of points in $S_{00}$. In particular, define $x_1, x_4$ be such that

$$\bigcap_{x \in S_{00}} R(x) = [x_1, x_4].$$

We call an instance to PPVR $\alpha$-expanding if $x_1 = \frac{\alpha}{1+\alpha} x_4$. This measure quantifies how close $x_1, x_4$ are to the endpoints of the domain of the patrolling instance (and note that $\frac{\alpha}{1+\alpha}$ is monotone in $\alpha$). Different algorithms need to be employed for different ranges of $\alpha$, and the crux of the proof of Theorem 10 is the design of two algorithms that are $(1 + 2\alpha)$ and $\frac{2+\alpha}{1+\alpha}$ feasible. Since one can always run the more efficient one, the worst performance one can get is when both violate the visitation requirement by the same factor, i.e. $\sqrt{3}$, and this happens for $\frac{\sqrt{3}-1}{2}$-expanding instances.

In order to present these algorithms, one has to relate not only $x_1, x_2$ but also the leftmost and rightmost points in $S_{00}$, call them $x_2, x_3$, respectively. By definition of the $x_i$'s, it is not difficult to see that $x_1 \le x_2 < x_4$ and that $x_1 < x_3 \le x_4$, while by breaking symmetry we may always assume that $x_1 \le 1 - x_4$. Now specific to feasible instances is that $x_4 - x_3 \le x_3 - x_1$ and $x_2 - x_1 \le x_4 - x_1$, a claim that can be shown as a corollary of Lemma 2. These conditions are enough to derive that the idle time of the points that have to be patrolled satisfy inequalities

$$I(x) \ge \begin{cases} \max\{2x, 2(1 - x - x_4 + x_1), x_4 - x_1\}, & x \in [0, x_1) \\ 2\max\{x_4 - x, x - x_1\}, & x \in [x_1, x_4] \\ \max\{2(1 - x), 2(x - x_4 + x_1), x_4 - x_1\}, & x \in (x_4, 1] \end{cases}$$

It follows then that one needs patrolling schedules whose worst point visitation frequencies are not too large, as a function of $x_1, \ldots, x_4$.

The algorithm that performs well for small expansion factors $\alpha$ is a simple partitioned based patrolling schedule (requiring no coordination), in which the two robots zig-zag between intervals $[0, x_3]$ and $[x_3, 1]$. The induced visitation waiting times $w(x)$ for each point $x$ can be shown to satisfy $w(x) \le 2\max\{x, x_3 - x\}$. Some exhaustive analysis then can show that $w(x)/I(x) \le 1 + 2\alpha$.

For large values of expansion factor $\alpha$, one has to be more creative, and robots do need to coordinate their movements throughout the execution of the algorithm. In particular, robots need to always maintain distance at least $d$ (to be determined shortly), while they take turns in zig-zaging within interval $[x_1, x_4]$. While moving within $[x_1, x_4]$, if the other robot approaches closer than $d$, then the first robot abandons the zig-zaging, and heads toward one of the endpoints. The process then is repeated, with the next robot to abandon the zig-zaging having to visit the other endpoint. Some technical details require also a special initial placement of the robots, so as to guarantee that the induced visitation waiting times satisfy

$$w(x) \begin{cases} = 2\max\{x, 1-x-d\}, & x \in [0, x_1) \\ \leq 2\max\{x - x_1, x_4 - x\} + d, & x \in [x_1, x_4] \\ = 2\max\{1-x, x-d\}, & x \in (x_4, 1] \end{cases}$$

The performance of the algorithm then is determined by finding

$$\min_d \max_x \frac{w(x)}{I(x)}$$

and some technical analysis is required to show that the best option is to choose

$$d = \frac{1}{1+\alpha} \min\{x_1, x_4 - x_1\},$$

inducing visitation violation $\frac{2+\alpha}{1+\alpha}$.

## 4 Patrolling Fragmented Boundaries

The problem studied in [8] concerns patrolling a boundary such that only some portions of it need to be monitored. These portions are called *vital intervals.* The remaining parts of the boundary (that we call *neutral intervals*) may be traversed by agents, but the idle time of the patrolling strategy relates only to maximal revisitation time of points from vital regions. More exactly we have the following.

**Definition 1.** *The boundary* $I = [0, 1]$ *contains* $n$ *disjoint* vital regions *represented by intervals* $V_1, V_2, \ldots, V_n$*, where* $V_i = [b_i, e_i]$ *and* $b_i \leq e_i < b_{i+1}$ *for* $1 \leq i \leq n-1$*,* $b_1 = 0$ *and* $e_n \leq 1$*. We call*

$$V = \bigcup_{i=1}^{n} V_i$$

*the* vital part *of the boundary* $B$*.*

We consider first the case of an *open curve* $I$, where it is represented by a unit segment called also a *fence* in the literature. Then we consider the *closed curve* $I$, which represents a unit ring, with points 0 and 1 identified. This case is most often referred to as being a *boundary*, i.e. a simple curve bounding some planar region.

Recall that in the robotics literature the typical strategies applied for patrolling were *partition-based*, where each agents perform a back and forth movement inside a region assigned to it, and *cycle-based*, where agents perpetually walk around a cycle inside the environment. We show that, optimal patrolling of a fragmented fence is always partition-based and optimal patrolling of a fragmented boundary is sometimes partition-based and sometimes cycle-based.

### 4.1 Patrolling a Fragmented Fence

In order to describe the region patrolled by a single robot in the partition-based strategy we refer to the concept of a *lid*.

**Definition 2.** *A* $d$-lid *is a closed interval of segment* $I = [0, 1]$ *of length* $d$*. For a vital part* $V \subseteq I$ *we say that it has a* $(d, k)$-lid cover *if there exists a set* $L_d$ *of* $k$ *(not necessarily disjoint)* $d$*-lids such that every point* $p \in V$ *belongs to some* $d$*-lid of* $L_d$*.*

The patrolling algorithm of a fragmented fence needs to find the smallest possible real value $d$ such the set $V$ has a $(d, k)$-lid cover. We make the following Observation.

**Observation:** Consider a vital set $V = \{V_1, V_2, \ldots, V_n\}$. For any $d$-lid and any $k$ it is possible to decide in $O(n + k)$ time whether $V$ has a $(d, k)$-lid cover.

Indeed we can place lids one by one making them adjacent whenever the end of the previous lid falls inside a vital interval. If the previous lid ends in the neutral region, the next lid starts at a beginning of the subsequent vital interval. This way, a single constant-time step permits to remove one element of a lid set or advance in the list $V$. The complexity of process corresponds to that of merging two ordered lists.

Assume w.l.o.g. that the vital intervals are arranged from left to right as $V_i = [b_i, e_i]$, for $i = 1, 2, \ldots, n$ where $b_1 = 0, b_i \leq e_i < b_{i+1}$. To find the size of optimal lid we observe that, the optimal $(d, k)$-lid cover has a property that some sequence of adjacent lids starts at the beginning of some vital interval and it ends at some (perhaps not the same) vital interval. More exactly, we have the following Lemma.

**Lemma 3** ([8]). *Suppose that* $L$ *is the minimal size of lid such that* $V$ *admits* $(d, k)$*-lid cover. Then there exists three integers* $x, y, z$*, such that* $1 \leq x \leq y \leq n$*,* $1 \leq z \leq k$ *and* $L = \frac{e_y - b_x}{z}$*.*

Consequently, we have the following.

**Corollary 3** ([8]). *The optimal lid size may be found in* $O(kn^2 \log n)$*.*

Indeed, using Lemma 3, we can sort in $O(kn^2 \log n)$ time at most $O(kn^2)$ candidate values for the size of optimal lid and then perform binary search to determine the right value.

The authors of [8], using a more involved procedure, propose an algorithm with an improved complexity of $O(kn \log n)$.

**Theorem 12** ([8]). *Consider patrolling a fragmented fence with vital part $V$ by a collection of $k$ agents. The idle time of the optimal patrolling strategy equals $2L$, where $L$ is the smallest value such that $V$ admits a $(L,k)$-lid cover.*

### 4.2 Patrolling a Fragmented Boundary

In the case of closed curve, the strategy patrolling the vital regions may be either partition-based or cycle-based. Consider first the case of a single vital interval $V = \{V_1\}$ in a unit-length ring. It is easy to see that, if the length of $|V_1| > 1/2$ the cycle-based strategy is better than the partition-based one, independently on the number $k$ of agents used. For other configurations of vital intervals it may be not clear, which type of strategy, the partition-based or the cycle-based produces a better idle time. Clearly it depends on the value of $k$.

The optimal partition-based strategy may be obtained by the following approach, which uses the algorithm from the previous section.

**Algorithm** PB

1. For each vital interval $V_i$ cut the boundary at $b_i$ - the starting point of $V_i$, obtaining this way a unit-length segment $S_i$.
2. For each segment $S_i$ find the smallest lid size $L_i$ such that $S_i$ admits $(L_i, k)$-lid cover.
3. $L = \max_{1 \leq i \leq n} L_i$.
4. Optimal idle time for partition-based strategies $= 2L$.

The cycle-based patrolling strategy of a fragmented boundary is independent on the configuration of vital regions and it implies the idle time $1/k$. To find the best patrolling strategy it is then sufficient to compare the value of $1/k$ with the best idle time for the partition-based strategy computed by Algorithm PB. The following Theorem states that there is not any better patrolling strategy than the one discussed above.

**Theorem 13** ([8]). *The idle time $\Im_k(\mathcal{P})$ of any traversal strategy $\mathcal{P}$ in a unit-size fragmented ring satisfies*

$$\Im_k(\mathcal{P}) \geq \min\{1/k, 2L\},$$

*where $L$ is the minimum possible lid size which permits a $(L,k)$-lid cover of the ring.*

The proof of this Theorem is quite involved but details can be found in [8].

## 5 Optimal Idle Time for Trees

Throughout this section we use the term tree to refer to an acyclic geometric graph drawn in the plane so that its edges may only cross at vertices. This section provides an *off-line* schedule placing the $k$ robots at specific initial positions on a tree $T$ such that if the robots move perpetually at speed 1 it will achieve the optimal idle time $2|T|/k$, where $|T|$ denotes the sum of lengths of the edges of $T$. Note that this is well known in the special case when $T$ is a line segment (see [9]).

The main theorem of the present section concerns patrolling on a single type of underlying topology, namely the tree, and is the following.

**Theorem 14** ([12]). *For any tree $T$ and any number $k$ of robots,*

$$\Im_k(T) = \frac{2|T|}{k}. \tag{2}$$

Note that the idle time above is attained when the $k$ robots traverse the tree at their maximum speed 1 along an Eulerian cycle of the tree, while at the same time ensuring that during the traversal consecutive robots remain equidistant on this cycle.

Throughout the proof it is assumed that during their traversal the robots may change direction anywhere on vertices as well as in the interior of edges of the tree. In order to make the Eulerian tour of the tree, every edge is replaced by two anti-parallel edges (of total length $2|T|$).

To prove the lower bound, first we define the following useful concepts of cumulative idle time and caterpillar trees (Fig. 5).

**Definition 3.** *The cumulative idle time on a tree $T$ is defined as $F_T(k) := k\Im_k(T)$, where $\Im_k(T)$ is the optimal idle time for $k$ robots on the tree $T$.*

**Definition 4.** *A d-caterpillar is a tree in which all edges adjacent to leaves have Euclidean distance at most $d$, where $d \geq 1$.*

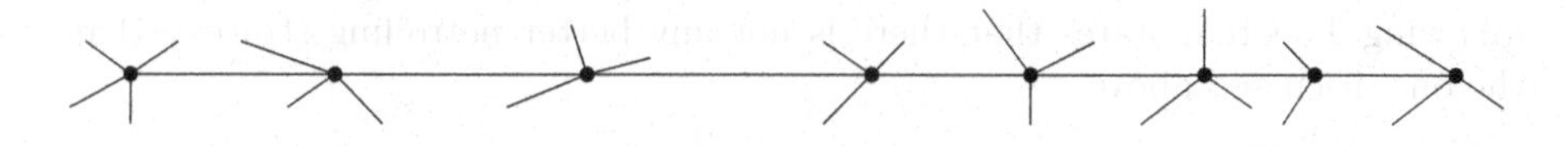

**Fig. 5.** A caterpillar.

The main idea for the proof of the lower bound $\Im_k(T) \geq \frac{2|T|}{k}$ is based on proving the following two properties:

1. Monotonicity of the *cumulative idle time* with respect to doubling the number of robots (see Inequality (3) and Lemma 4), and

2. Validity of the lower bound on caterpillar trees for a sufficiently large number $k$ of robots (see Lemma 6).

First we consider the monotonicity of the cumulative idle time.

**Lemma 4** ([12]). *For any number of robots $k$,*

$$F_T(2k) \leq F_T(k). \tag{3}$$

*Proof.* Consider a patrolling algorithm $\mathcal{A}$ attaining the optimal idle time $\Im_k(T)$. Let $r$ be a given robot. For any real number $t \geq 0$, observe a trajectory traversed by robot $r$ according to algorithm $\mathcal{A}$ during the time interval $[t, t + \Im_k(T)]$ (of duration $\Im_k(T)$). Since during the patrolling the robot moves with speed at most 1, the positions of the robot at times $t$ and $t+\Im_k(T)$ cannot be at distance bigger than $\Im_k(T)$, where the distance is measured along the edges of the tree.

Consider the $k$ original robots as they are moving according to an optimal algorithm $\mathcal{A}$ patrolling the tree with idle time $\Im_k(T)$. We now double the original number $k$ of robots by inserting $k$ additional robots patrolling the tree. The insertion procedure is as follows. Let the original robots be $r_1, r_2, \ldots, r_k$. For the current discussion lets use the notation $I := \Im_k(T)$. Take a snapshot of the robots at some time, say $t$. Look at the robots at time $t+I/2$ as the robots move to new positions. Take $k$ new robots $r'_1, r'_2, \ldots, r'_k$ and place them in the positions previously occupied by robots $r_1, r_2, \ldots, r_k$, respectively, at time $t$. Call robot $r'_i$ the *follower* of robot $r_i$, and similarly robot $r_i$ the *master* of robot $r'_i$. The $k$ new robots $r'_1, r'_2, \ldots, r'_k$ are given the following trajectory:

- from that time on, have each robot $r'_i$ copy faithfully the trajectory of its master, for $i = 1, 2, \ldots, k$.

It is clear that if an arbitrary point $p$ is visited by a robot $r$ at any time $t$ it will be visited again by its follower at time $t+I/2$. Therefore the idle time $\Im_{2k}(T)$ for $2k$ robots is at most half the idle time $\Im_k(T)$ for $k$ robots, i.e., $\Im_{2k}(T) \leq \Im_k(T)/2$. After multiplying both sides by $2k$, this implies Inequality (3). ∎

The following lemma provides an interesting property that will be useful for the proof of the main theorem.

**Lemma 5** ([12]). *If $T$ is a $d$-caterpillar with $d \geq 2$ then there is a $(d-1)$-caterpillar $T'$ which is subtree of $T$ and such that all the leaves of $T$ are within distance 1 of a leaf of $T'$.*

*Proof.* The subtree $T'$ is easily obtained from $T$ by cutting edges of $T$ which are adjacent to leaves. ∎

**Lemma 6** ([12]). *For any caterpillar tree $T$, and for any real number $\epsilon > 0$ there is a sufficiently large integer $k_0$ such that*

$$\Im_k(T) \geq \frac{2|T|}{k} - \frac{2\epsilon}{k},$$

*for all $k \geq k_0$.*

*Proof.* First observe that when two robots meet they can exchange their patrolling roles. Therefore without loss of generality we may assume that no two robots ever cross each other at any point on the tree when coming from opposite directions. Using this, it follows that for any leaf node of the tree whose distance from its parent is more than $\Im_k(T)/2$ a robot always has to stay within distance at most $\Im_k(T)/2$ from this leaf node so as to maintain idle time $\Im_k(T,)$. Thus a robot must be dedicated to patrolling each such leaf; this allows us to chop a segment of length $\Im_k(T)/2$ from each leaf whose adjacent edge length is at least $\Im_k(T)/2$ without affecting the idle time of the tree. By chopping such segments of length $\Im_k(T)/2$ recursively we are left with star graphs (having as centers internal nodes of the tree) and respective edges of lengths at most $\Im_k(T)/2$. Further, since the upper bound $\Im_k(T) \leq \frac{2|T|}{k}$ holds, the quantity $\Im_k(T)/2$ (which is an upper bound on the lengths of the edges of the stars) can be made as small as we wish by making $k$ sufficiently large.

Let $P$ be the central path of the caterpillar tree. The graph resulting from the procedure above consists of $P$ together with "stars of small weight". Consider the sum, say $\Delta$, of all the degrees of all the stars in the tree. Clearly, $\Delta$ depends only on the given tree $T$ and is otherwise independent of the number $k$ of robots. In addition, the sum of the lengths of the edges of all the stars is at most $\Delta\frac{\Im_k(T)}{2} \leq \Delta\frac{|T|}{k}$. Therefore for any $\epsilon > 0$ we can select $k_0$ sufficiently large so that the sum of the lengths of all the edges of all the stars is at most $\Delta\frac{|T|}{k_0} \leq \epsilon$.

It is clear from the above discussion that

$$|P| + \epsilon \geq |T| \geq |P|,$$

for all $k \geq k_0$. However $\Im_k(P) = \frac{2|P|}{k}$, for any $k$. Thus, it is obvious that for all $\epsilon > 0$ there is an integer $k_0$ such that for all $k \geq k_0$ we have that

$$\Im_k(T) \geq I(P,k) = \frac{2|P|}{k} \geq \frac{2|T|}{k} - \frac{2\epsilon}{k}.$$

This proves Lemma 6. ■

Now we are ready to prove the lower bound for caterpillars. We prove the following.

**Lemma 7** ([12]). *For any caterpillar tree $T$ and any number $k$ of robots,*

$$\Im_k(T) = \frac{2|T|}{k}.$$

*Proof.* The upper bound follows from [9]. We now concentrate on the lower bound. Indeed, by Lemma 6 the lower bound

$$\Im_k(T) \geq \frac{2|T|}{k} - \frac{2\epsilon}{k},$$

is valid for any $k \geq k_0$, where $\epsilon$, $k_0$ are selected as specified in Lemma 6. So assume that $k \leq k_0$. Choose an integer $i$ sufficiently large such that $k \leq k_0 \leq 2^i k$. Now observe that the following inequalities are valid

$$
\begin{aligned}
\Im_k(T) &= \frac{F_T(k)}{k} \text{ (by definition)} \\
&\geq \frac{F_T(2^i k)}{k} \text{ (by Lemma 4)} \\
&= \frac{F_T(2^i k)}{2^i k} \cdot \frac{2^i k}{k} \\
&= \Im_{2^i k}(T) \cdot \frac{2^i k}{k} \text{ (by definition)} \\
&\geq \left( \frac{2|T|}{2^i k} - \frac{2\epsilon}{2^i k} \right) \cdot \frac{2^i k}{k} \text{ (by Lemma 6)} \\
&= \frac{2|T|}{k} - \frac{2\epsilon}{k}.
\end{aligned}
$$

The last inequality is valid for any integer $k$ and any real number $\epsilon > 0$. By letting $\epsilon \to 0$ the proof of the lemma is complete. ■

We are now in a position to prove the main theorem which was given at the beginning of the paper.

*Proof* (Theorem 14). Without loss of generality we may assume that the tree is a $d$-caterpillar, for some $d \geq 2$ (in fact, every tree is a $d$-caterpillar, for some $d \geq 2$, provided $d$ is sufficiently large). Now the proof of Identity (2) proceeds by induction on $d$. Recall that Lemma 7 is precisely the base case $d = 1$. Suppose the identity in the theorem is valid for $d - 1$. By Lemma 5, the subtree $T'$ obtained from $T$ by removing all its leaves is a $(d-1)$-caterpillar. Clearly, Identity (2) is valid for $|T'|$, namely $\Im_k(T') = \frac{2|T'|}{k}$. Therefore by repeating the proof of Lemma 6 we can show that for any real number $\epsilon > 0$ there is a sufficiently large integer $k_0$ such that $\Im_k(T) \geq \frac{2|T|}{k} - \frac{2\epsilon}{k}$, for all $k \geq k_0$. In turn, using this last statement we repeat the proof of Lemma 7 to prove the desired identity. This completes the proof of Theorem 14. ■

## 6 Decentralized Strategies

There is a lack of distributed and/or decentralized patrolling algorithms. This may be due not only to the lack of knowledge of the underlying topology but also to the difficulty of incorporating a suitable communication model which will potentially relay information about the geographic location and local patrolling strategy among participating robots. For this reason most of the patrolling strategies proposed so far are centralized algorithms in that the robots receive instructions from a central controller.

In the sequel we describe the rotor which provides a labeling of the underlying graph which may be used to design a patrolling strategy that the robots can use to traverse the graph.

### 6.1 Rotors

We now describe a mechanism, so-called *rotor-router*, which has been extensively studied in the literature as a deterministic alternative to the random walk in undirected graphs. The basic idea of rotor-router is to set locally shared memories at the vertices of the underlying graph. Subsequently, the robots will be updating these shared memories as they visit the vertices of the graph. Thus, in the rotor model, a set of $k$ identical robots is deployed; they start from a selected subset of nodes of the underlying graph, and move around in parallel in synchronous steps. Each node maintains a cyclic ordering of its outgoing edges, and successively propagates visiting robots along its outgoing edges in a round-robin manner, according to a fixed ordering. Therefore, besides being distributed, the rotor-router also provides an on-line algorithm in that the environment to be patrolled is unknown, which also makes it a practical tool for patrolling an unknown environment as well.

For example, the rotor-router algorithm described in [22] works as follows. Let $u$ be a vertex of the graph. Denote by $d(u)$ degree of $u$. Label the edges adjacent to $u$ with numbers $1, \ldots, d(u)$ (called *ports pointers* at $u$). At each vertex $u$ there is a pointer (called *exit port* at $u$), indicating the next (adjacent) edge to be traversed by a robot. Further, a departing robot also updates (increments) the exit port.

Independently from the *initial configuration*, which is defined by an initial placement of the robots, port pointers and initial exit ports at the nodes, after some transient time steps, the system reaches a *stable state*, in which the placement of the robots within the graph and the state of the exit ports (at all the nodes) repeats periodically (this is called *periodic behavior* of rotor-router). Interestingly, it has been proven (see [22]) that for a single robot such *periodicity* is only $2m\Delta$ steps, where $m$ is the number of edges, and $\Delta$ is the diameter of the network. More generally, [22] also show that for $k$ robots, after at most $2(1+1/k)m\Delta$ steps the numbers of edge visits in the network are balanced up to a factor of two.

Thus when viewed as random walk, a single walk achieves a cover time of exactly $\Theta(m\Delta)$ on any connected $n$-node graph with $m$ edges and diameter $\Delta$, and that the robot eventually stabilizes to a traversal of an Eulerian circuit on the set of all directed edges of the graph (see also [3]). Also noted is additional research in [4,5] which show that this cover time is at most $\Theta(m\Delta/\log k)$ and at least $\Theta(m\Delta/k)$ for any graph. This corresponds to a speedup of between $\Theta(\log k)$ and $\Theta(k)$ with respect to the cover time of a single walk. Both of these extremal values of speedup are achieved for some graph classes. Their results are valid for up to a polynomially large number of walks, $k = O(poly(n))$.

## 7 Dynamic Patrolling with Primitive Agents

Specifying agents' perpetual movements (patrolling strategies) so as to minimize the idle time is a particularly challenging task in a distributed environment. The difficulty of this problem touches first, on the invention of efficient strategies/trajectories, and second on their theoretical analysis which can be technical. Patrolling in a distributed environment becomes particularly interesting when agents demonstrate only primitive computation power, i.e. they have limited (or no) memory and they can perform only primitive (or no) computations. In this section we outline a thorough investigation of such a problem due to [11], indicating that *optimal idle times* may be obtained by natural trajectory-inducing processes that converge to efficient, and occasionally optimal, patrolling strategies.

**The Model of One-Way Patrolling**
$n$ mobile agents $R = \{r_1, \ldots, r_n\}$ are each associated with some *patrolling speed* $p_i$ and some *walking speed* $w_i$, where $p_i < w_j$ for $i, j = 1, \ldots, n$. Robots perpetually move along the unit segment $[0, 1]$ (an open curve) in both directions. Agent $r_i$ can be either in *walking state* moving at speed $w_i$ or in *patrolling state* moving at speed $p_i$. Each robot may walk in both directions but its patrolling is always done in the same direction.

In a centralized environment with "intelligent" agents, robots may be centrally coordinated, placed on the unit interval, and associated with walking-patrolling directions and with appropriate turning points so as to minimize the idle time (note that the idleness is defined over all points, uncountably many, of the contiguous unit interval, and a point is visited by a robot only if the latter is in patrolling state). In the decentralized and distributed environment, robots have only primitive capabilities. In particular, agents are *oblivious* and *silent*. Besides their two-speed mobility sates, they perceive the environment by recognizing obstacles (i.e., endpoints or other robots), allowing them to change states and schedules accordingly.

**The Optimal Patrolling Strategy**
Solving the problem of patrolling the unit interval in the one-way patrolling model optimally requires special coordination between the agents. Indeed, suppose a centralized algorithm has control over where the agents will be placed and what their turning points will be for a partitioned based strategy. Each robot will be assigned a subinterval, and will patrol in one direction, moving at speed $s_i$, and walk back to it's initial starting point, moving at speed $w_i$, before it repeats the process. The time needed to execute one iteration of the process is proportional to the length of the assigned subinterval, and to quantity $\frac{1}{p_i} + \frac{1}{w_i}$ which we refer to as the time cycle $\tau_i$ of robot $r_i$. A condition inducing optimal idleness would require that the idle times of points in the unit interval, except from those jointly patrolled by two robots, to be all equal. As a result, in an optimal partition based patrolling schedule of the unit interval, each robot $r_i$ will be associated with a subinterval proportional to $\tau_i$, giving rise to following theorem.

**Theorem 15** ([11]). *The optimal patrolling strategy of a unit interval in the one-way patrolling model is partition based. Each robot $r_i$ is associated with a subinterval of length $\tau_i / \sum_{j=1}^{n} \tau_j$, and the induced idle time becomes*

$$\left(\sum_{j=1}^{n} \tau_j^{-1}\right)^{-1}.$$

The role of a centralized algorithm then would be to instruct each of the robots to move within its assigned subintervals, and to turn direction and moving state, each time the endpoint of the interval is reached. Moreover, under some extra conditions concerning the relative order of the patrolling and walking speeds, the induced *dynamical* system would look perfectly synchronized. Indeed, suppose that all robots are initially placed at the same-side endpoints of their intervals and they all start patrolling in the same direction simultaneously. Since in the optimal patrolling schedule all but a few points have the same idle time, all robots will return to their initial configuration after time equal to the optimal idle time, before all of them repeat the same process. The only fundamental assumption here is that each robot remembers the endpoints of it's interval, and is aware of it's location so as to be able to change direction and state when the endpoint of the interval is reached. A fundamental question then is whether the same solution can be achieved in a decentralized environment with memoryless agents.

**Optimal Patrolling with 2 Primitive Agents**

The optimal schedule of Theorem 15 can be simulated by primitive agents, at least in the restricted case of two robots. In that direction, consider the following dynamical system that can be executed by robots that have no memory, and their only allowed computation is to switch moving state and direction when an obstacle is found. The two robots $r_1 = (p_1, w_1), r_2 = (p_2, w_2)$ are placed at the two endpoints of the unit interval, and they start patrolling toward each other till they collide, say at point $x_1$, when they bounce by changing directions and moving state. Each of the robots will reach their corresponding endpoint of the interval, when they bounce again, by changing directions and returning to the patrolling state, before they collide again, say at point $x_2$. Clearly, this dynamical process induces an infinite sequence of meeting points $x_1, x_2, x_3, \ldots$. It is not difficult to establish a recurrence relation for the $x_i$'s, according to which $x_i = -ax_{i-1} + b$, where

$$a := \frac{\frac{1}{w_1} + \frac{1}{w_2}}{\frac{1}{p_1} + \frac{1}{p_2}}, \text{ and } b := \frac{\frac{1}{p_2} + \frac{1}{w_2}}{\frac{1}{p_1} + \frac{1}{p_2}},$$

and hence

$$x_i = (-a)^i x_0 + b\frac{1-(-a)^i}{1+a}.$$

Since $w_i > p_i$, we see that $a < 1$, and hence the sequence of the meeting points converges to

$$\lim_{i\to\infty} x_i = \frac{b}{1+a} = \frac{\tau_2}{\tau_1 + \tau_2}.$$

We conclude that the primitive dynamical system of two robots converges to the partitioned-based solution of Theorem 15, and hence eventually the patrolling strategy is optimal.

**Patrolling with $n$ Primitive Agents**
The collision dynamics of a primitive system of $n$ agents patrolling a unit interval in the one-way patrolling model are much more involved. For uniformity reasons it is natural to have all robots start from the same endpoint of the unit interval, as well as in the same moving state toward the same direction. Assuming that all robots speeds are distinct, robots' speed will induce a permutation of the robots, starting with the slowest. Since robots will bounce (change direction and moving state) whenever an obstacle is found, the permutation of the robots will be preserved indefinitely. In particular, collisions will occur only between consecutive agents $i, i+1$. We denote the collision-locations of these agents for the $t$'th time by $x_t^i$. As in the case of two agents our goal is to derive a recurrence relation for the sequence of vectors

$$X_t = \left(x_t^1, x_t^2, \ldots, x_t^{n-1}\right)^T .$$

The difficulty in establishing a recurrence for $X_t$ is due to the fact that collisions may involve consecutive agents that are moving in the same direction. Therefore, we are motivated to restrict ourselves to *regular* dynamical systems where collisions take place only between robots in different movement states, and hence moving toward each other. For such systems it can be shown that

$$AX_{t+1} + BX_t = c,$$

where

$$A = \begin{pmatrix} 1/p_2 + 1/w_1 & 0 & 0 & \cdots & 0 & 0 \\ -1/p_2 - 1/w_2 & 1/p_3 + 1/w_2 & 0 & \cdots & 0 & 0 \\ 0 & -1/p_3 - 1/w_3 & 1/p_4 + 1/w_3 & \cdots & 0 & 0 \\ \vdots & \vdots & \vdots & \ddots & \vdots & \vdots \\ 0 & 0 & 0 & \cdots & 1/p_{n-1} + 1/w_{n-2} & 0 \\ 0 & 0 & 0 & \cdots & -1/p_{n-1} - 1/w_{n-1} & 1/p_n + 1/w_{n-1} \end{pmatrix}$$

$$B = \begin{pmatrix} 1/p_1 + 1/w_2 & -1/p_2 - 1/w_2 & 0 & \cdots & 0 & 0 \\ 0 & 1/p_2 + 1/w_1 & -1/p_3 - 1/w_3 & \cdots & 0 & 0 \\ 0 & 0 & 1/p_3 + 1/w_4 & \cdots & 0 & 0 \\ \vdots & \vdots & \vdots & \ddots & \vdots & \vdots \\ 0 & 0 & 0 & \cdots & 1/p_{n-2} + 1/w_{n-1} & -1/p_{n-1} - 1/w_{n-1} \\ 0 & 0 & 0 & \cdots & 0 & 1/p_{n-1} + 1/w_n \end{pmatrix}$$

$$c^T = (0\,0 \ldots 0\; 1/p_n + 1/w_n) .$$

and hence, observing that $A$ is non-singular, we obtain that

$$X_t = (-1)^t \left(A^{-1}B\right)^t + \left(I + A^{-1}B\right)^{-1} \left(I - (-1)^{-1} \left(A^{-1}B\right)^t\right) A^{-1}c.$$

The convergence of vectors $X_t$ is determined now by the convergence of $\left(A^{-1}B\right)^t$, as $t$ goes to infinity, and consequently by the moduli of the eigenvalues of $A^{-1}B$. In particular, if the norms of all eigenvalues of $A^{-1}B$ are at most 1, then $\lim_{t\to\infty} X_t$ exists and the limit vector gives rise to a partition-based patrolling strategy which coincides with the one of Theorem 15 and hence it is optimal.

A natural question that arises then pertains to a characterization of robots $r_i = (p_i, w_i)$ that guarantee that the moduli of the eigenvalues of $A^{-1}B$ are at most 1. In that direction, only sufficient conditions are known, and only for a limited number of robots with special specifications. Indeed, [11] introduces a special family of $n$-robots, called monotone, for which

$$w_n > \ldots > w_1 > p_1 > \ldots > w_n.$$

For monotone collection of $n \leq 3$ robots, which provably are regular as well, it can be shown that all eigenvalues of $A^{-1}B$ lie within the unit disk, and hence the dynamical system induces an optimal partitioned based patrolling strategy. The result can also be extended to $n = 4$ robots by introducing a refinement of monotonicity.

Interestingly, for the general case with $n$ robots, very little is known. The key difficulty lies in the analysis of the eigenvalues of the $(n-1) \times (n-1)$ matrix $A^{-1}B$. Notably, some technical work can derive the characteristic polynomial of the latter matrix, and then algorithmic results can be used to decide whether the roots of the polynomial fall within the unit disk without finding them (see [11] for additional details). Hence, deciding whether a set of robots $r_i = (p_i, w_i)$ induces a dynamic system with convergent collision points, and hence inducing optimal partition based patrolling strategies, can be answered efficiently.

## 8 Conclusion

In this paper we provided a brief survey of recent developments on patrolling in a variety of geometric graph domains. There are interesting open problems everywhere in this research area from designing deterministic centralized algorithms to decentralized distributed and dynamic algorithms. Interestingly enough no probabilistic results are known on patrolling though there is no reason why they could not be forthcoming depending on the interest of the research community. Inventing patrolling algorithms, proving their correctness and analyzing their performance has proved to be quite challenging.

It should be noted that the current survey paper cannot but be somewhat biased in favour of recent work by the authors and their collaborators. Moreover, this survey is not meant to be exhaustive but rather provide the reader with possible research directions and the distinctive flavour of some of the recent algorithmic results on this topic.

**Acknowledgements.** We would like to express our deepest appreciation to our colleagues Huda Chuangpishit, Leszek Gasieniec, Tomasz Jurdzinsk, Adrian Kosowski, Danny Krizanc, Fraser MacQuarrie, Russell Martin, Dominik Pajak, Oscar Morales

Ponce, Lata Narayanan, Jarda Opatrny, and Najmeh Taleb for numerous interesting conversations that excited our interests on all aspects of patrolling.

## References


1. Agmon, N., Kraus, S., Kaminka, G.A.: Multi-robot perimeter patrol in adversarial settings. In: ICRA, pp. 2339–2345 (2008)
2. Almeida, A., et al.: Recent advances on multi-agent patrolling. In: Bazzan, A.L.C., Labidi, S. (eds.) SBIA 2004. LNCS, vol. 3171, pp. 474–483. Springer, Heidelberg (2004). https://doi.org/10.1007/978-3-540-28645-5_48
3. Bampas, E., Gąsieniec, L., Hanusse, N., Ilcinkas, D., Klasing, R., Kosowski, A.: Euler tour lock-in problem in the rotor-router model. In: Keidar, I. (ed.) DISC 2009. LNCS, vol. 5805, pp. 423–435. Springer, Heidelberg (2009). https://doi.org/10.1007/978-3-642-04355-0_44
4. Chalopin, J., Das, S., Gawrychowski, P., Kosowski, A., Labourel, A., Uznański, P.: Lock-in problem for parallel rotor-router walks. arXiv preprint arXiv:1407.3200 (2014)
5. Chalopin, J., Das, S., Gawrychowski, P., Kosowski, A., Labourel, A., Uznański, P.: Limit behavior of the multi-agent rotor-router system. In: Moses, Y. (ed.) DISC 2015. LNCS, vol. 9363, pp. 123–139. Springer, Heidelberg (2015). https://doi.org/10.1007/978-3-662-48653-5_9
6. Chevaleyre, Y.: Theoretical analysis of the multi-agent patrolling problem. In: IAT, pp. 302–308 (2004)
7. Chuangpishit, H., Czyzowicz, J., Gasieniec, L., Georgiou, K., Jurdzinski, T., Kranakis, E.: Patrolling a path connecting a set of points with unbalanced frequencies of visits. CoRR, abs/1710.00466 (2017)
8. Collins, A., et al.: Optimal patrolling of fragmented boundaries. In: 25th ACM Symposium on Parallelism in Algorithms and Architectures, SPAA 2013, Montreal, QC, Canada, 23–25 July 2013, pp. 241–250 (2013)
9. Czyzowicz, J., Gąsieniec, L., Kosowski, A., Kranakis, E.: Boundary patrolling by mobile agents with distinct maximal speeds. In: Demetrescu, C., Halldórsson, M.M. (eds.) ESA 2011. LNCS, vol. 6942, pp. 701–712. Springer, Heidelberg (2011). https://doi.org/10.1007/978-3-642-23719-5_59
10. Czyzowicz, J., Gasieniec, L., Kosowski, A., Kranakis, E., Krizanc, D., Taleb, N.: When patrolmen become corrupted: monitoring a graph using faulty mobile robots. Algorithmica **79**(3), 925–940 (2017)
11. Czyzowicz, J., Georgiou, K., Kranakis, E., MacQuarrie, F., Pajak, D.: Fence patrolling with two-speed robots. In: Proceedings of 5th the International Conference on Operations Research and Enterprise Systems, ICORES 2016, Rome, Italy, 23–25 February 2016, pp. 229–241 (2016)
12. Czyzowicz, J., Kosowski, A., Kranakis, E., Taleb, N.: Patrolling trees with mobile robots. In: Cuppens, F., Wang, L., Cuppens-Boulahia, N., Tawbi, N., Garcia-Alfaro, J. (eds.) FPS 2016. LNCS, vol. 10128, pp. 331–344. Springer, Cham (2017). https://doi.org/10.1007/978-3-319-51966-1_22
13. Czyzowicz, J., Kranakis, E., Pajak, D., Taleb, N.: Patrolling by robots equipped with visibility. In: Halldórsson, M.M. (ed.) SIROCCO 2014. LNCS, vol. 8576, pp. 224–234. Springer, Cham (2014). https://doi.org/10.1007/978-3-319-09620-9_18
14. Dumitrescu, A., Ghosh, A., Tóth, C.D.: On fence patrolling by mobile agents. Electr. J. Comb. **21**(3), P3.4 (2014)

15. Elmaliach, Y., Agmon, N., Kaminka, G.A.: Multi-robot area patrol under frequency constraints. Ann. Math. Artif. Intell. **57**(3–4), 293–320 (2009)
16. Elmaliach, Y., Shiloni, A., Kaminka, G.A.: A realistic model of frequency-based multi-robot polyline patrolling. In: AAMAS, no. 1, pp. 63–70 (2008)
17. Hazon, N., Kaminka, G.A.: On redundancy, efficiency, and robustness in coverage for multiple robots. Robot. Auton. Syst. **56**(12), 1102–1114 (2008)
18. Kawamura, A., Kobayashi, Y.: Fence patrolling by mobile agents with distinct speeds. Distrib. Comput. **28**(2), 147–154 (2015)
19. Kranakis, E., Krizanc, D.: Optimization problems in infrastructure security. In: Garcia-Alfaro, J., Kranakis, E., Bonfante, G. (eds.) FPS 2015. LNCS, vol. 9482, pp. 3–13. Springer, Cham (2016). https://doi.org/10.1007/978-3-319-30303-1_1
20. Machado, A., Ramalho, G., Zucker, J.-D., Drogoul, A.: Multi-agent patrolling: an empirical analysis of alternative architectures. In: Simão Sichman, J., Bousquet, F., Davidsson, P. (eds.) MABS 2002. LNCS, vol. 2581, pp. 155–170. Springer, Heidelberg (2003). https://doi.org/10.1007/3-540-36483-8_11
21. Pasqualetti, F., Franchi, A., Bullo, F.: On optimal cooperative patrolling. In: CDC, pp. 7153–7158 (2010)
22. Yanovski, V., Wagner, I.A., Bruckstein, A.M.: A distributed ant algorithm for efficiently patrolling a network. Algorithmica **37**(3), 165–186 (2003)

# Agents

# Graph Explorations with Mobile Agents

Shantanu Das[(✉)]

Aix-Marseille University, CNRS, LIS, Marseille, France
shantanu.das@lis-lab.fr

**Abstract.** The basic primitive for a mobile agent is the ability to visit all the nodes of the graph in a systematic manner. This chapter considers the exploration of unknown graphs in full detail, for the specific mobile agent model considered in this book. The graph is considered to be finite, undirected and connected. Other than this fact, no prior knowledge of the graph is assumed. Several exploration techniques are introduced and explained for either a single agent, or multiple agents, exploring either labelled or unlabelled graphs. We focus on the efficiency of exploration and consider three different complexity measures, the time taken, the amount of memory used by the agents and the storage needed at each node of the graph. For exploration by multiple agents, we consider collaborative exploration by a team of colocated agents as well as distributed exploration by agents scattered in a graph. The concluding section presents some brief ideas and references on more advanced topics on graph exploration that are not covered in this chapter.

**Keywords:** Mobile agents · Graph exploration · Undirected graph
Deterministic · Anonymous

## 1 Introduction

Most tasks for mobile agents require them to navigate in a graph, visiting all the nodes in a systematic manner. We call the task of visiting all the nodes of an unknown graph as the *Exploration* problem. We will only consider finite and connected graphs in this chapter. Consider a single mobile agent located in one of the nodes of such a graph. The agent can move only along the edges of the graph. The task of Exploration requires the agent to visit each node of the graph at least once. Depending on the objective, further requirements for the agent may be to terminate after visiting all nodes, or to return to the starting node. The former is called "Exploration with stop" while the latter is referred to as "Exploration with return". As an example, if the agent is required to search for some resource (or information) among the nodes of a graph, it must be able to determine whether the target resource is actually present in the graph and if not return back to the starting node and report failure. In such case, termination of the exploration is important. On the hand, if the agent is required to monitor the nodes of the graph to prevent intruders (e.g. guards patrolling an art gallery),


P. Flocchini et al. (Eds.): Distributed Computing by Mobile Entities, LNCS 11340, pp. 403–422, 2019.
https://doi.org/10.1007/978-3-030-11072-7_16

in such cases, we do not require the agent to terminate the exploration but rather continue visiting all nodes repeatedly. Such an exploration where each node needs to be visited infinitely often is called "Perpetual Exploration" of the graph.

Throughout this chapter we assume that the graph to be explored is initially unknown to the mobile agent. Further an agent has only local visibility restricted to the current node where it is located. So, the agent starting at a node $v$ sees only the node $v$ and the incident edges to it; the agent can not even see the neighboring nodes at the other end of these edges. As the agent visits more nodes, it may store in its memory the history of what it has explored so far. The problem of reconstructing the graph from this information obtained in course of the exploration, is called the *Map Construction* problem.

We distinguish the mobile agent exploration problem from the graph traversal problem in the context of the graph data structures (as in [38]). Graph traversal in that context does not need to be continuous as it is possible to return to any previously visited vertex in the data-structure. However, in the context of mobile agent exploration, the agent must always follow a path of consecutive edges in the graphs (i.e. jumps are not allowed). Another important difference is that the agent may not always be able to distinguish between the nodes of the graph during the exploration. We will study both exploration of both unlabelled graphs (where the nodes are anonymous) and labelled graphs (where each node is assigned a unique identifier that is visible to the agent visiting that node). Even for labelled graphs, the ability of the agent to recognize previously visited nodes depends on its capabilities, e.g. the size of its memory.

**The Model:** We assume the usual mobile agent model, with an undirected connected graph $G$ where the edges of the graph are labelled with port numbering $\lambda$ so that an agent at any node $v$ can deterministically choose to traverse one of the adjacent edges, $e$, by selecting the port number $\lambda_e$ of that edge. In particular we assume that a *proper port numbering* which assigns the labels $0, 1, 2, \dots d-1$ to the edges incident to a node of degree $d$. For any such edge-labelled graph $G$, we define the corresponding digraph $\hat{G}$ by replacing each edge $e = (u, v)$ by two directed arcs - one from $u$ to $v$ and the other from $v$ to $u$, marked with the port labels $(\lambda_u(e), \lambda_v(e))$ and $(\lambda_u(e), \lambda_v(e))$ respectively. The nodes of the graph $G$ may be labelled or unlabelled. Throughout this chapter, we will use $n$, $m$, $D$ and $\Delta$ to denote respectively, the number of nodes, the number of edges, the diameter, and the maximum degree of any node, of $G$.

We consider different models for communication and interaction of the agent with the environment, including the whiteboard model, the token model and the face-to-face model. The graph is assumed to be static and does not change during the execution of the algorithm (in particular, there are no failures of any kind). Exploration of dynamic graphs is fully investigated in Chap. 20. This chapter considers only *deterministic* algorithm for exploration of *static* graphs.

**Complexity:** We are interested in efficient algorithms for exploration. We measure the efficiency of an algorithm using three different criteria.

- **[Moves] or [Time]:** The *moves complexity* of an algorithm is the total number of edge traversals performed by all agents, where any single edge traversed by a single agent counts as one move. For exploration by only one agent, the moves complexity is same as the *time* complexity, assuming each edge traversal takes one unit of time (we assume that the graph is unweighted and all edges are similar). When there are multiple agents, we may be interested in the total energy consumption (for exploration by physical robots) or the bandwidth consumption (for software agents exploring communication networks); the moves complexity captures these notions.
- **[Storage]:** The maximum amount of information stored at any node of the graph during the course of the exploration, is called the *storage* complexity of the algorithm. Under the whiteboard model of communication, storage is counted as the minimum size (in bits) of the whiteboard needed at each node of the graph. Under the token model, we count storage as the maximum number of tokens at any node. When the agents do not have the ability to mark the nodes of the graph, we say that the algorithm is a *zero-storage* algorithm.
- **[Memory]:** The *memory* complexity of an exploration algorithm is the maximum size of persistent memory needed by any agent during the exploration. Here *persistent memory* refers to the size of the information carried by the agent when moving from one node to another; usually this does not include the working memory used by the agent while performing computations at any node. If $S$ is the set of states of the agents, then the memory complexity is equal to $\log(|S|)$ bits. If the cardinality of $S$ is a constant, independent of the size of the graph or any other parameters, then the agents are called *finite state automata* and we say that the algorithm is a *constant-memory* algorithm. If the algorithm can be executed by agents having no persistent memory then we say that the algorithm is a *zero-memory* algorithm and that the agents are *oblivious*.

## Bibliographical Notes

The problem of exploring an unknown graph started with the study of mazes, labyrinths and caves, and the need for devising algorithmic strategies to traverse such environments. An early as 1951, Shannon [37] studied exploration by a finite state automaton (a mechanical mouse) moving in a two dimensional maze. Later Budach [8] performed a more rigorous study showing that no automata can explore all mazes. Blum and Kozen [6] showed that either one automaton with just 2 pebbles, or a team of two automata can explore all mazes. These results strongly utilize the orientation information for exploring mazes, thus the results do not hold for arbitrary graphs where it is not possible to use a compass to orientate. Graphs are indeed more difficult to explore than mazes as it was shown by Rollik [36] that any finite team of finite state automata cannot explore all graphs. A tight lower bound of $\Omega(log\, n)$ on the memory complexity of graph exploration was given by Fraigniaud et al. [30], while a matching upper bound was provided by the result of Reingold [35] who gave a log-space exploration algorithm for all graphs. The space complexity of exploration can be reduced to $O(\log \log n)$, when the agent is provided with $O(\log \log n)$ distinct tokens; such an algorithm was provided recently by Disser et al. [24] whose idea was to use a sequence of explored nodes as a tape of the Turing machine and store information by writing on the tape using the tokens.

In terms of time complexity of exploration, the fastest exploration for arbitrary labelled graphs was given by Panaite and Pelc [34] which makes $m + O(n)$ moves thus having only a linear penalty with respect to any optimal traversal algorithm which knows the graph in advance. There has been a lot of interest on achieving fast exploration of graphs using a zero-memory agent, by assigning specific port numbering to the edges of the graph, in order to guide the memoryless agent. This is known as *label-guided exploration*. The fastest such exploration makes $4n - 2$ moves [31], while a lower bound of $2.8n - 2$ has been proved [13]. For a constant memory agent, [13] provided a faster exploration with $3.5n$ steps while only a trivial lower bound $2n - 2$ is known for this case. There are schemes for labelling the nodes of a graph for guiding the exploration by a $O(1)$ memory agent. The most storage efficient scheme, given by Cohen et al. [11] provides one bit labels to nodes (i.e. the nodes are colored in black or white) in such a way that an agent with constant memory can explore the graph in $O(m)$ time. When no preprocessing of the graph is allowed, the same article provided a 2 bit labelling scheme such that the exploring agent can itself assign the labels during the exploration, while still achieving an exploration time linear in the number of edges.

There has been some studies on exploration of specific families of graphs, including trees [23], unoriented grids and tori [4], edge-labelled hypercubes [28] and other interconnection graphs [25]. Exploration of graphs with *sense of direction* labelling has been investigated by Barrière et al. [3].

Exploring directed graphs (digraphs) is more difficult than exploring undirected graphs due to the impossibility of backtracking. For labelled directed graphs that are strongly connected, Deng et al. [20] showed that the time complexity of exploration depends on the so-called *deficiency* of the graph which is the number of edges that need to be added to make the graph Eulerian. The exploration of unlabelled directed graphs was studied by Bender et al. [5] and the best known algorithm for mapping a digraph using a single pebble has a time cost of $O(n^8 \Delta^2)$, when the agent knows the size of the graph (or an upper bound). When the agent does not know any upper bound on the size of the graph, at least $\Omega(\log \log n)$ pebbles are necessary, as shown in the above paper.

## Outline of this Chapter

We first look at some basic techniques for exploration of unknown graphs. We then consider algorithms for single agent exploration based on the optimization criteria: time efficiency, moves efficiency, storage efficiency and memory efficiency. We then consider the problem of map construction when exploring unlabelled graphs. Finally we investigate multi-agent algorithms for exploration as well as map construction and show how the latter is related to problems of gathering and leader election. We conclude with some observations concerning more advanced topics on exploration that have not been considered in this chapter.

## 2 Basic Techniques

*Depth-First-Search (DFS)*: The most standard technique for exploration, called depth-first search, is based on a simple rule: At each node $v$, an agent chooses a unexplored edge (if any) and traverses it; if the agent reaches an already visited node, the agent returns back to node $v$. If there are no unexplored edges at $v$ then the agent backtracks to the node visited prior to arriving at $v$ for the first time. The pseudo-code for this algorithm is given in Algorithm 1.

**Algorithm 1.** Algorithm DFS (Depth-First Search)

```
if current node v has unexplored edges then
    Traverse the unexplored edge with lowest port number;
    if reached node is already visited then
        Return to previous node v;
else
    if current node v is not the starting node then
        Traverse the edge used to reach v for first time;
```

Some properties of DFS exploration:

- The agent needs to distinguish nodes visited for the first time from nodes previously visited during the exploration.
- For each visited node $v$, the agent needs to know (1) which incident edges at $v$ are still unexplored, and (2) which incident edge was used to enter $v$ for the first time.
- The exploration makes exactly $2m$ moves for graphs with $m$ edges. Thus, this is an asymptotically time optimal algorithm.

*Right-Hand-on-the-Wall (RHW)*: The RHW algorithm is a technique used by explorers lost in dark caves, where the idea is to follow the boundary of the cave by feeling it with one hand (assuming the cave is too dark to see anything). A similar strategy can be used for exploring certain port-labelled graphs; the exploration starts at a node $v$ by taking the edge with lowest port number (i.e. port number 0) and at each subsequent step, the agent arriving at a node $u$ through port number $x$, leaves that node by the next port number (either port $x + 1$ or port 0, if $x$ equals the degree of $u$). The algorithm is shown below (Algorithm 2).

**Algorithm 2.** Algorithm RHW (Right Hand on the Wall)

```
p := port number of arrival to current node (initially 0);
d := degree of current node;
Traverse the edge with port number (p + 1) mod d;
```

Some properties of RHW exploration:

- The algorithm RHW performs perpetual exploration on any tree starting from any vertex. This is a zero memory and zero storage algorithm.
- The exploration on the tree can be terminated on reaching the starting vertex from the port number $d(v)$. Thus, exploration with stop can be performed if the agent has the capability to recognize the starting vertex (e.g. by marking with a pebble).
- The exploration with stop makes exactly $2m$ moves for a tree with $m$ edges $(m = n - 1)$.
- On cyclic graphs having nodes of degree $\geq 3$, labelled with arbitrary port numbering, the algorithm may not visit all nodes.

One way of exploring arbitrary connected graphs using the above technique is to perform a preprocessing on the graph $G$ to assign specific port numbering to the edges, that allows an agent performing RHW to explore all nodes of this edge-labelled graph. Consider any spanning tree of the graph $G$. For each node $v$ of $G$, we assign port numbers greater than deg$(v)$ to the non-tree edges and assign port numbers $0, 1, 2, \ldots d - 1$ to the tree edges adjacent to $v$, where $d$ is the degree of node $v$ in the spanning tree. An execution of Algorithm RHW on this graph[1] would visit all edges of the spanning tree and thus visit all nodes of $G$. A more elegant scheme for assigning a proper port numbering to the edges of any graph to allow the RHW algorithm to explore the graph, was provided in [13]. However, note that the preprocessing step of assigning port-numbers requires the intervention of some central authority having prior knowledge of the graph $G$ or at least of a spanning tree of $G$.

*Universal Exploration Sequences (UXS)*: The idea of RHW exploration can be adapted to arbitrary graphs with arbitrary port numbering using the concept of *universal exploration sequences* which is defined below. Given any node $u \in G$, we define the $i$th successor of $u$, denoted by $succ(u, i)$ as the node $v$ reached by taking port number $i$ from node $u$ (where $0 \leq i < d(u)$). Let $(a_1, a_2, \ldots, a_t)$ be a sequence of integers. An *application* of this sequence to a graph $G$ at node $u$ is the sequence of nodes $(u_0, \ldots, u_{t+1})$ obtained as follows: $u_0 = u$, $u_1 = succ(u_0, 0)$; for any $1 \leq i \leq t$, $u_{i+1} = succ(u_i, (p + a_i) \mod d(u_i))$, where $p$ is the port-number at $u_i$ corresponding to the edge $\{u_{i-1}, u_i\}$. A sequence $(a_1, a_2, \ldots, a_t)$ whose application to a graph $G$ at any node $u$ contains all nodes of this graph

[1] Note that since this is not a proper port labelling, the algorithm should have a default action of traversing port number 0 whenever there is no edge with the required port number.

is called a universal exploration sequence (UXS) for this graph. A UXS for a family of graphs is a UXS for all graphs in this family. The following result on the existence of UXS is important for us.

*Property 1.* For any positive integers $n$, there exists a UXS of length $Poly(n)$ for the family of all connected graphs with at most $n$ nodes. Further such a sequence can be computed in polynomial time using a deterministic algorithm.

The above result implies an algorithm for exploration of unlabelled graphs, without the need for any storage at the nodes, provided that the size of the graph is known. The construction for the UXS of polynomial size was given by Reingold [35]. We will call the algorithm that applies the Reingold's UXS for exploration as RUXS algorithm. This algorithm has the following properties:

- This is a zero storage algorithm for exploration of unlabelled graphs.
- The exploration requires knowledge of the size of the graph, $n$, or at least an upper bound on $n$.
- The algorithm performs exploration in polynomial time using $O(\log n)$ bits of agent memory.

*Rotor-Router (RR)*: Another technique for exploration by a memoryless agent uses a pointer saved in each node of the graph, which points to the next incident edge to be explored. The agent arriving at a node $v$ simply leaves node $v$ by the edge pointed to by the pointer and at that point the pointer is incremented (modulo the degree) to point to the next incident edge. Such a system is called *Rotor router* (also sometime called *Propp machine*). Given any connected graph $G$ and some port numbering $\lambda$ on $G$, if all pointer values are properly initialized (in a preprocessing step), then the rotor-router process moves the agent on an Eulerian tour of the corresponding digraph $\hat{G}$ obtained from $G$ by replacing each edge with two directed arcs; thus the agent periodically explores all edges of the graph with a period of at most $2m$ moves. On the other hand, if the pointer values have arbitrary initial values, the process takes some time to stabilize and after this stabilization time, the agent follows an Euler tour of the digraph $\hat{G}$ as before. It was shown by Yanovski et al. [40] that the stabilization time is no more than $2mD$ for any graph of diameter $D$. Thus the rotor-router can perform an exploration of the graph in $O(m \cdot D)$ moves, using zero agent memory and $O(\log \Delta)$ storage space per node. This algorithm is *self stabilizing* as both the agent and the nodes of the graph may be in arbitrary states at the start of the algorithm (no initialization is required).

**Algorithm 3.** Algorithm RRA (Rotor Router Agent)

Let $p$ be the pointer at current node and $e$ be the edge pointed to by $p$
$d :=$ degree of current node;
Set pointer $p := (p + 1) \mod d$;
Traverse the edge $e$;

Some properties of the rotor router algorithm:

- Algorithm RRA is a zero memory algorithm for unlabelled graphs.
- The algorithm performs perpetual exploration visiting each node with a period of at most $2m$, after stabilization.
- The algorithm requires $O(\log \Delta)$ bits of storage space per node and it is a self stabilizing algorithm.

## 3 Single Agent Explorations

### 3.1 Time Efficient Explorations

Any algorithm for exploring an unknown graph must visit all edges of the graph. Thus the time complexity or move complexity of exploration is at least $m$. The algorithm DFS makes $2m$ moves in total for exploring graphs of $m$ edges. Thus algorithm DFS is asymptotically optimal. There has been attempts at reducing further the exploration time for unknown graphs. Panaite et al. [34] gave an exploration algorithm that takes $m + O(n)$ steps in the worst case for arbitrary undirected graphs, using a modified version of DFS that reduces the number of times the agent backtracks.

Although the DFS algorithm and its variants are asymptotically optimal in time and moves, the algorithm may not be optimal in terms of agent memory and storage, depending on the implementation. Note that the algorithm requires the agent to distinguish the visited nodes from unvisited ones. In labelled graphs, each node has a unique identifier and the agent simply needs to memorize the identifiers of all visited nodes, in order to recognize them. Thus the agent requires $O(n \log n)$ memory but no storage is required. On the other hand if the graph is unlabelled, the agent needs to mark each node that it visits (by writing on the whiteboard, or placing a token) to recognize it as a visited node; this requires $O(1)$ bits of storage per node. In both cases, however, the agent still needs memory to remember the paths already visited to allow it to backtrack; In particular the algorithm remembers a spanning tree of the visited subgraph constructed during the exploration (this is often called the DFS tree). So, the algorithm requires $O(n \log n)$ bits of agent memory. We state the following result based on the standard DFS algorithm [38].

**Theorem 1.** *There is an optimal time algorithm for exploration with stop in unlabelled graphs, that requires $O(1)$ bits of storage per node and $O(n \log n)$ bits of agent memory.*

### 3.2 Storage Efficient Explorations

When the nodes of the graph $G$ are labelled with unique identifiers and these are visible to the exploring agent, it is possible to explore the graph using the DFS algorithm using zero storage. This is optimal in terms of storage complexity. However, when the graph is unlabelled, the DFS algorithm requires $O(1)$ bits or 1 pebble per node, and thus $n$ pebbles in total. So, the question is whether it is possible to explore unlabelled graphs without marking the nodes. The following result (from folklore) gives a negative answer.

**Theorem 2.** *There is no zero storage algorithm for exploration with stop of unlabelled graphs irrespective of the agent memory.*

*Proof.* Consider two unlabelled graphs: Let $G_1$ be a simple ring of $n_1 = 3$ nodes and let $G_2$ be a line of $n_2$ nodes. Suppose each edge of the line is labelled with port numbers $(1, 2)$ in a consistent manner (e.g. from left to right) and each edge of the ring is labelled with port numbers $(1, 2)$ in the clockwise direction. With this port numbering, all nodes of $G_2$, except the end-points, look exactly like the nodes of $G_1$ to any exploring agent. If there was an algorithm for exploration with stop, consider the execution of this algorithm on $G_1$; the execution must terminates after a finite number of steps $t$. Now if we take $n_2 = 2t + 2$ and we place an agent at the mid-point of the line $G_2$, then the execution of the same algorithm for $t$ steps could visit only nodes at distance at most $t$ from the starting node, thus the agent would never reach either end-point of the line during this time, thus visiting only the nodes which look identical to the nodes of the ring. Thus the algorithm would terminate without visiting all nodes of $G_2$—a contradiction to the correctness of the algorithm.

Note that the above impossibility is due to the fact that the agent has no prior knowledge of the graph. If the agent knows the size of the graph (either $n$ or $D$, or some upper bound) then it is always possible to perform exploration with stop, without the need to mark the nodes. For example, if the agent known the diameter $D$ of the graph $G$, then an agent starting at node $v$ could systematically traverse all paths of length $D$ starting at node $v$, thus visiting every node of $G$. This is equivalent to traversing the *view* of node $v$ in $G$ to a depth of $D$. When the agent does not know the diameter, the value of $n$ could be used as an upper bound on $D$ and the same procedure would perform an exploration with stop. The time or moves complexity of such an algorithm would be $O(\Delta^n)$ in the worst case, thus this algorithm is exponential in terms of moves/time. It is possible to have a polynomial time exploration with stop using the `RUXS` algorithm as discussed before. The following result follows from the properties of the `RUXS` algorithm.

**Theorem 3** [35]**.** *There is a polynomial time, zero storage algorithm for exploration with stop for all graphs when the value of $n$ is known a priori.*

*Exploration with No Knowledge*: In general, we assume that the agent has no information about the graph that is exploring. When the value of $n$ is not known, or can not be determined, then the agent can repetitively execute the above algorithm, with increasing values of $\hat{n} = 2, 4, 8, \ldots$, where $\hat{n}$ is a guessed value for $n$; when $\hat{n} > n$ all the nodes of the graph would have been visited (without the agent having the knowledge of this fact). This is gives an algorithm for exploration *without termination.*

We now consider storage efficient algorithms for exploration with stop in unlabelled graphs when no prior information about the graph is available. Due to the impossibility result from Theorem 2 we know that that the agent needs to store some information on the nodes. In fact, an agent with a single pebble is capable of performing exploration with stop.

**Theorem 4.** *There is an $O(m \cdot n)$ time and $O(nlogn)$ memory algorithm, using one pebble, for exploration with stop in unknown graphs.*

The algorithm that achieves the above result is a modified version of the DFS algorithm. Since the agent has only one pebble, the pebble needs to be reused for marking each new node that is visited. As before, the agent stores in its memory the DFS-tree $T$ which is a spanning tree of the subgraph already explored by the agent. Whenever the agent explores any unexplored edge to reach some node $v$, it places the token on $v$ and performs a full traversal of $T$ - if the token is encountered during the traversal then node $v$ already belongs to $T$ (so, it's not a new node); otherwise $v$ must be a new node. Now, the agent can return to node $v$, recover the token, and continue the exploration. Thus, the agent makes an additional $O(n)$ moves for each edge of $G$, which gives a complexity of $O(m \cdot n)$ moves.

The above result is tight with respect the storage complexity, so we know that zero storage algorithms are possible only for labelled graphs and impossible for unlabelled graphs, while 1 bit of storage (or one pebble) in total suffices to explore unlabelled graphs. Surprisingly, it is possible to perform zero storage exploration, even if at least one of the nodes of the graph is uniquely labelled (and the rest of the nodes are unlabelled). We say that the unique node $v$ is a *landmark*, any agent arriving at $v$ can recognize it immediately as the landmark node. We now present an algorithm for zero-storage exploration when the agent starts at the landmark node and explores the whole graph.

*Exploration with a Landmark*: The algorithm is based on the fact that each node $w$ can be uniquely identified using an edge-label sequence $P(v, w)$ corresponding to a path from the landmark $v$ to node $w$. On reaching any node $u$ the agent can detect whether or not the node $u$ is distinct from node $w$ by applying the sequence $P(v, w)$ at node $u$ and checking if it leads to the landmark. Thus, the algorithm maintains a set of so-called *Root-paths*, one for each new node discovered during the algorithm. For each edge explored by the algorithm, the agent needs to detect if the node reached has been already visited - this requires performing the checking procedure mentioned before for each path in the set of stored Root-paths. Thus the agent makes at most $O(n^2)$ moves for each new edge explored by the agent. The exploration proceeds in a breadth-first manner, visiting all nodes at depth $h$ from the landmark, before visiting any node at depth $h + 1$. The algorithm requires $O(n \log n)$ bits of agent memory to store the tree containing all the Root-paths (See [9] for more details).

**Theorem 5.** *There is an algorithm taking $O(m \cdot n^2)$ time and using $O(n \log n)$ bits of agent memory, for exploration with stop in unlabelled graphs containing one landmark node.*

Note that the above algorithm can be implemented using one *unmovable* token (i.e. a one-time use token) that can be placed on the starting node to create the landmark. This requires storage facility at only the starting node of $G$. In contrast, the previous algorithm uses one moveable (i.e. reusable) token which requires some storage at each node of the graph at some point during the algorithm.

### 3.3 Memory Efficient Explorations

The optimal algorithm for exploration in terms of memory is a zero-memory algorithm, for example, the RRA algorithm presented before. The algorithm RRA can be used to perform exploration with stop, by adding $O(1)$ bits of the storage at the starting node to distinguish this node and by initializing the pointers at each node to zero on the first visit to the node. The agent can terminate the algorithm when it returns to the starting node and finds that the pointer value is zero. This algorithm requires $O(\log \Delta)$ storage at each node and a single bit of agent memory (to distinguish the starting state). However, if termination is not required then it is possible to have a zero-memory algorithm taking $O(m \cdot D)$ moves in the worst case. Zero-memory agents are like tokens that are moved around by the system and such systems corresponding to many natural physical system (e.g. chip firing games). Thus, there has been a lot of investigations on the properties of such systems. The following result shows the optimality of the RRA algorithm in terms of storage and moves.

**Theorem 6** [32]. *Any zero memory algorithm for exploration requires $\Omega(\log \Delta)$ bits of storage per node and makes $\Omega(n^3)$ moves.*

There has been a lot of interest on the minimal memory required to explore unlabelled graphs without any storage. In particular there have been several studies on exploration of unlabelled graphs by constant memory agents (i.e. agents with $O(1)$ bits memory, sometime called finite automata). Whether such agents can explore graphs of arbitrary size (possibly without termination) was an open question for the long time, until the question was answered negatively in [36]. The following result by Fraigniaud et al. [30] gives an exact lower bound for the memory requirement of a single agent exploring unknown graphs without marking.

**Theorem 7** [30]. *Any zero storage algorithm for exploration requires $\Omega(\log n)$ bits of memory for exploration of all graphs of size $n$, even for constant degree graphs.*

The proof of this theorem is based on exploration of regular graphs of degree $\Delta = 3$. Note that in such a graph every node look the same to a visiting agent, thus the action taken by the agent is only a function of its current state. Any agent having $\log k$ bits of memory can be in $k$ distinct states. If such an agent is exploring a graph of $n > k$ nodes then it must enter two distinct nodes of $G$ in the same state, and thus it must exit the node by the same port number in both cases. Based on this fact, it is possible to construct a regular graph $G$ of degree 3 and $k + 1$ nodes where the agent forever moves in a cycle of size less than $k$, thus never visiting the rest of the nodes of $G$.

The lower bound from the above theorem is matched by the algorithm RUXS which can explore unlabelled graphs using $O(\log n)$ bits of memory and zero storage. The algorithm requires the knowledge of $n$ or an upper bound, which is necessary in unlabelled graphs due to the impossibility result from Theorem 2. However if the nodes of the graph are labelled with unique identifiers, then the knowledge of $n$ is no longer necessary and the algorithm can be adapted to work without any prior knowledge of the graph, while still using $O(\log n)$ bits of memory.

The only remaining question at this stage, is what is the memory complexity of exploration, when the storage space per node is a small constant. Recall that for storage space of $\Omega(\log \Delta)$, it is already possible to have a zero memory algorithm. So, it is natural to ask if there are algorithms using both constant memory and constant storage per node. If fact, it was shown that having only 2 bits of storage per node is sufficient to circumvent the lower bound of Theorem 6 and explore all graphs using constant memory agents.

**Theorem 8** [11]. *There is a polynomial time algorithm for exploring all graphs using $O(1)$ bits of agent memory and only 2 bits of storage per node, without any prior knowledge of the graph.*

The idea of the algorithm is to preprocess the graph assigning labels from a set of three colors to all the nodes of the graph, such that nodes that are at the same depth from the root (i.e. the starting node) are assigned the same color, which is distinct from the color assigned to nodes one level below and one level above. This coloring of the nodes enables the agent to determine after each move whether it moved closer or further, or remained at the same distance from the starting node. The algorithm enables the agent to traverse a spanning tree of the graph in a depth first search manner, with some additional edge traversals at each node, thus having an overall time complexity of $O(m)$ steps. The labelling of the nodes can be done by the agent during the exploration, provided that each node is initialized (i.e. uncolored) at the beginning. This is the only known algorithm that uses both constant memory and constant storage for exploring unknown graphs.

## 4 Map Construction While Exploration

The problem of *Map construction* requires the agent to output a copy of the graph including all port labels, at the end of the exploration. This immediately implies that: (i) The exploration must terminate, and (ii) the agent must have enough memory to store a copy of the graph (i.e. $\Omega(m \log n)$ bits of memory).

When the nodes of the graph have unique labels, exploration is equivalent to map construction; if the agent can remember a full history of all edges traversed by it, this information is sufficient to reconstruct a map of the graph. This is because each node (and each edge) can be uniquely identified on each visit. However when the graph is unlabelled, this is not always the case. For unlabelled graphs, the algorithm *Exploration-with-a-landmark* from Sect. 3.2, can be used to build a map of the graph, since each node can be uniquely identified using its root-path. Thus if the agent is able to mark its hombase with a token, it is possible to solve the problem of map construction. However when marking of nodes is not allowed (i.e. in the Face-to-Face model), it is not always possible to construct a map of the graph, even though it is always possible to perform exploration with stop. In other words, there exists graphs which are not recognizable by an agent even after traversing every edge of the graph and even if the agent has an unbounded amount of memory allowing it to remember everything that it has seen during the exploration.

**Theorem 9** [39]. *There is no zero-storage algorithm for Map-construction in unlabelled graphs, even if the agent has unlimited memory and knows the exact size $n$ of the graph.*

First, consider an agent in a ring of size $n$ where each edge is consistently labelled with port numbers $(0, 1)$ in the clockwise direction. An agent moving in such ring networks cannot distinguish a ring of size $n = 3$ from a ring of size $n = 4$. Thus, clearly map construction is not solvable with only the knowledge of an upper bound on the size of the graph, although this knowledge suffices for exploration as we saw in Sect. 3.2. Even when the agent knows the exact size $n$, there exists graphs of same size that are indistinguishable. This can be explained using the concept of *graph coverings*. We say that a graph $G$ covers a graph $H$ if there is a homomorphism $\varphi$ mapping nodes and edges of $G$ to nodes and edges of $H$ such that for any edge $e$ between adjacent vertices $u, v \in G$, there is an edge connecting $\varphi(u)$ to $\varphi(v)$ in $H$ ($H$ could potentially be a multi-graph). The quotient graph of any graph $G$ is the smallest multi-graph $B$ such that $G$ covers $B$, under an edge-label preserving graph homomorphism (c.f. Chap. 2). If graphs $G_1$ and $G_2$ cover the same quotient graph $B$ then $G_1$ and $G_2$ are indistinguishable to any mobile agent, as all the information that the agent can gather can be represented by the quotient graph $B$. For example, the Fig. 1 below show two graphs of size $n = 16$ which are non-isomorphic, but have the same quotient graph $B$.

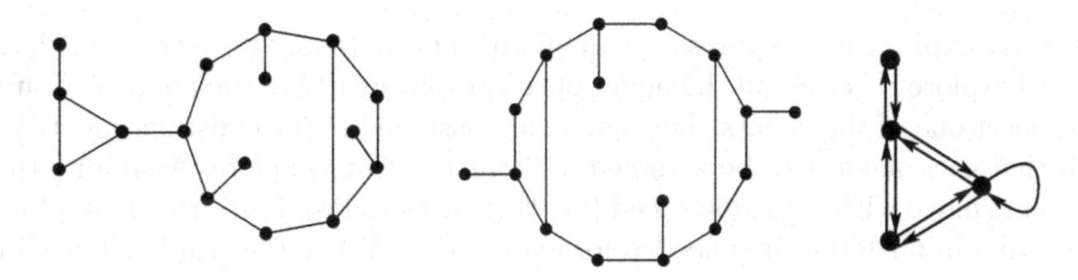

**Fig. 1.** (i) Graphs $G_1$ and $G_2$ that are indistinguishable to a mobile agent (ii) Graph $B$ that is the quotient graph

It is always possible to construct the quotient graph of any graph, given the knowledge of an upper bound $\hat{n}$ on the size of the graph. For example, an agent can apply the algorithm for *view* construction from Sect. 3.2 and collapse the view into the quotient graph by merging all 'similar' vertices (i.e. vertices that have the same view up to a depth of $\hat{n}$). Whenever the graph $G$ is identical to its quotient graph then it is possible to construct a map of the graph. Such graphs are said to be *covering minimal.*

**Theorem 10.** *There exists a zero-storage algorithm for Map-construction of any graph $G$ with port-labelling $\lambda$ if (i) the graph $(G, \lambda)$ is covering minimal, and (ii) the agent knows an upper bound on the size $n$ of the graph.*

There exists a more exact characterization of the class of graphs which allow map construction without marking, provided in [39]. We also remark here that it is possible to have a more efficient algorithm for map construction of covering minimal graphs, using the concept of *signatures*, introduced in [14], to identify the nodes. Each node $v$ in such graphs, can be uniquely identified by a sequence of edge-labels encountered by performing by following a UXS path of sufficient length on the graph starting from node $v$. Thus, it is possible to perform DFS type exploration with a check procedure for each visited node that computes its signature, and compares it with the signatures of the previously visited vertices.

# 5 Multi-agent Explorations

When there are multiple agents available, they can together explore the graph to reduce the time taken for exploration. However this requires coordination and communication between the agents, making the task more difficult than single agent exploration. On the positive side, multi-agent exploration can be robust against failures of some agents.

We consider explorations with either colocated agents, or, with agents initially dispersed in the graph. We denote by $k$ the number of agents present in the graph.

## 5.1 Collective Exploration

Collective exploration requires a team of $k$ agents that start from the same location, to explore together all the nodes of the graph, such that each node is visited by at least one of the agents. The agents are assumed to have distinct identifiers such that each agent can be assigned a distinct path to explore. Assuming that all agents move with the same speed (i.e. they are synchronized), the main objective is to minimize the time needed for exploration. When the graph is known in advance, it is possible to devise a strategy to divide the task among the agents such that each agent travels on a distinct tour and they together span the nodes of the graph. We call this an *offline* strategy for exploration; finding the optimal offline strategy that minimizes the maximum tour length of any agent for a given graph $G$ and team size $k$ is known to be an NP-hard problem even for trees [29]. However, we consider the graph to be a priori unknown and the agents need to design and adapt their strategy in an online fashion as they discover new parts of the graph.

Any optimal exploration algorithm using $k$ agent for exploring a graph of diameter $D$ must take at least $O(D + n/k)$ time. When $G$ is tree, Fraigniaud et al. [29] provided a collective algorithm for exploration in $O(D+n/\log k)$ time. The algorithm has a simple strategy, at each node $v$, the available agents are distributed in a round robin manner among the unexplored edges; whenever a subtree has been explored completely all agents in that subtree move to the parent node. The algorithm uses the whiteboard model for communication, thus any agent arriving at a node $v$ can obtain knowledge about the current distribution of agents in the subtree rooted at $v$. The algorithm achieves a competitive ratio of $k/\log k$ over any optimal offline exploration strategy. The best known lower bound for the competitive ratio of any collective exploration algorithm using $k < \sqrt{n}$ agents, is $\Omega(\log k/\log\log k)$, even with global communication between the agents [27].

For graphs of small diameter, fast exploration by small teams of agents can be achieved by a DFS based algorithm presented in [7] which has a time complexity of $O(n/k + D^{k-1})$. On the other hand, for large teams of agents, there exists an optimal algorithm for exploring general graphs in $O(D)$ time [21], when the number of agents $k$ is at least $D \cdot n^{1+\epsilon}$ for any $\epsilon > 0$. This algorithm does not require whiteboards for communication and works even in the face-to-face model. The basic strategy is to deploy a fixed number of agents from root at each time step. Exploration of the special class of grid graphs with rectangular holes was studied in [33], which presented a collective exploration algorithm with competitive ratio of $O(log^2 D)$ for such graphs.

## 5.2 Distributed Exploration

When multiple mobile agents start from distinct nodes of the graph, coordination among the agents is more difficult. The task of exploration starting from dispersed locations of the graph is called *Distributed Exploration.* Since the agents do not have a common reference point, direct communication is inutile in this situation. Instead we assume that the agents communicate by writing on the nodes (i.e. the whiteboard model of communication). Notice that if the agents have distinct identities, each agent can individually explore the complete graph (e.g. using the DFS algorithm) while marking each visited node with its identifier.

On the other hand if the agents are identical then the marks left on a node by an agent would not be distinguishable from those of another agent. In this case some cooperation between the agents seems necessary. A simple strategy is to use a distributed version of the DFS algorithm which we call the distributed depth-first search (DDFS) [17].

*Distributed Depth-First Search*: Each agent $a$ performs DFS algorithm marking the nodes that it visits (unless they are already marked) and labelling them with a counter that it increments. Each node marked by the agent and the edge used to reach it are added to DFS-tree stored in the memory of the agent. Note that the agent treat nodes marked by any agent as visited nodes. Thus, whenever the agent reaches an already marked node, it backtracks to the previous node, as in the original algorithm. The tree obtained at the end of the traversal is called the territory $T_a$ of the agent $a$. It was shown in [17] that when all agents have completed the algorithm, the territories obtained by the agents in the above process, forms a spanning forest of the graph $G$. Thus, each node of the agent is visited by some agent and the agents together have explored all nodes of the graph. The exploration requires $O(m)$ moves in total (instead of $O(m)$ moves per agent if the agents individually explored the graph).

### 5.3 Collision Free Exploration

When each node of the graph can host at most one agent at any time, then any multi-agent exploration algorithm must prevent collisions (i.e. two agents moving to the same node at the same time). The problem of exploring every node by every agent while ensuring that no node is occupied by more than one agent at any time, is called *Collision-free exploration.* [12] provides such an exploration strategy in labelled graphs for $k$ mobile agents when the mobile agents have 1-hop visibility. The algorithm uses the concept of universal exploration sequences and thus the time complexity is proportional to the length of the UXS. For trees, the paper provides a faster exploration taking $O(n^2)$ time. The algorithms require the agents to start at the same time and always move synchronously so that agents on neighboring nodes swap places without collision.

### 5.4 Map Construction and Leader Election

In Sect. 4 we saw that Map Construction is possible by a single agent that is allowed to mark the nodes of the graph during exploration. However when there are multiple agents dispersed in the graph, then Map Construction is not always possible, since multiple agents mark several distinct node simultaneously, making it difficult to uniquely identify the nodes. In this case, the possibility of map construction depends on the presence of symmetries in the graph as well as the initial location of the agents. We denote by $b : V \to \{0, 1\}$ a bi-coloring of the graph $G(V, E)$ such $b(v) = 1$ if node $v$ is the homebase of an agent and $b(v) = 0$ otherwise.

**Theorem 11.** *It is possible to solve Map-construction in any graph $G$ with $k$ dispersed agents if $(G, \lambda, b)$ is covering minimal with respect to label preserving and color preserving coverings.*

We now briefly describe an algorithm for map construction by multiple agents in graphs where the above condition is satisfied (see [18] for more details). The map construction algorithm proceeds in two phases. In the first phase, each agent performs the distributed depth-first search (`DDFS`) algorithm discussed previously. At the end of this phase the graph is partitioned into a spanning forest and each agent $a$ has a map of its DFS-tree $T_a$. The agent obtains its *territory* by adding to this tree, all outgoing edges that are incident to any node of $T_a$. The territory of an agent (including all edge labels) is encoded as an integer $l_a$ that is used by the agent in the subsequent part of the algorithm. The second phase of the algorithm is a competition between neighboring agents, by comparison of the encoded territories. Each losing agent merges its territory with the corresponding winning agent and terminates the algorithm, while each winning agent updates its territory and the same process is repeated with only the winner agents. If the conditions of Theorem 11 hold then there would eventually be a single winner – *the leader* and the territory of this agent would be a spanning tree of the graph. As a final step, nodes of this spanning tree are assigned unique labels (based on the unique path from the root) and thus all non-tree edges can be identified and added to the map. The main complication in this algorithm is the process of synchronizing the agents during each round of the competition phase. There can be at most $O(\log k)$ such rounds in any successful execution of the algorithm and the overall complexity of the algorithm $O(m \log k)$ moves in total. The algorithm fails to construct a map only if the conditions of Theorem 11 do not hold and in those cases, the agent can detect failure after at most $k - 1$ rounds of the competition phase.

The above algorithm also elects a leader among the agents, which is another fundamental task in distributed computing with agents. In fact the problems of the leader election, gathering and map construction in a distributed setting are almost equivalent, with the only exception in the case of *symmetric* trees where leader election may be impossible but map construction is still possible.

## 6 Ongoing Research and Future Directions

This chapter surveyed the main techniques and algorithms for exploration by one or more agents when the agents are deterministic, fault-free and have no constraints on their movements. Similarly, we also assumed that the environment explored is stable and failure-free, allowing any agent to move in any direction on every edge of the graph. This situation is idealistic although in reality several of these assumptions may not hold. In such cases, the task of exploration may become more difficult. We present some directions for further research on the exploration problem, which have been partially investigated.

### 6.1 Constrained Explorations

When mobile agents have constraints on their movements they may not be able to complete the task of exploration in a single attempt. One typical constraint is the budget constraint (i.e. having limited energy for movement) allowing any mobile agent to traverse at most $B$ edges. The problem of *piece-meal exploration* requires an agent to return to its homebase after at most every $B$ edge-traversals, in order to refuel and continue again. In this case, we need to assume that no nodes are at a distance larger than $B$ from the starting location of the agent. Even with this assumption, the usual techniques for exploration cannot explore the graph efficiently. An algorithm for piecemeal exploration was provided in [1] based on the idea of exploring strips of increasing depths from the starting node, and performing a depth-first search in each strip. This algorithm was later improved upon in [26], achieving an optimal time complexity of $O(m)$ moves in total. However, both these algorithms explore the graph to a depth of $r < B$ and the time complexity increases drastically as $r$ approaches $B$. For piecemeal exploration of graphs of depth $r \leq B$, no efficient algorithm are known. For the family of tree, a piecemeal version of DFS algorithm was presented in [16], and shown to be constant competitive (i.e. the algorithm is at most 10 times worse than any optimal offline piecemeal exploration of the same tree).

When the agents are not able to refuel, it is possible to perform constrained exploration using many agents each having a budget of $B$ edge traversals. For any fixed $k$, the best online algorithm for tree exploration using $k$ agents [27] achieves a competitive ratio of $4-2/k$ on the value of $B$ required for exploration. On the other hand, given any fixed $B$, exploration of all trees of depth at most $B$ can be achieved with a competitive ratio of $O(logB)$ for the value of $k$, and this was shown to be asymptotically optimal, at least in the case of local (face-to-face) communication between agents [15]. When both $k$ and $B$ are fixed, it is not possible to completely explore an unknown tree; in this case the problem of *maximal exploration*, which maximizes the number of nodes visited, has been studied [2]. The algorithm in [2] has a competitive ratio of 3, while a lower bound of 2.17 on the competitive ratio was shown in the same paper.

### 6.2 Fault-Tolerant Explorations

While most of the known results are for exploration in fault-free environments, there have been some preliminary investigations on fault tolerant algorithms for exploration. Exploration with faulty tokens that disappear, have been studied in the context of the gathering problem for many agents where each agent uses one token to mark its homebase [19]. Single agent exploration with Byzantine tokens has also been recently studied in [22]. Here, the tokens are unmovable but may sometime be invisible to the agent. If at least one token is fault-free and the agent knows the total number of tokens, there is an exploration algorithm for any unknown and unlabelled graph. Exploration and map construction of *dangerous* graphs containing black holes and black links, can be performed when there are sufficiently many agents [10] using an extended version of the distributed DFS based algorithm discussed previously. Under the assumption that the faults do not disconnect the graph, the algorithm builds a map of the fault-free part of the graph whenever it is theoretically possible. Other techniques for exploration of dangerous graphs are covered in Chap. 18.

## References

1. Awerbuch, B., Betke, M., Singh, M.: Piecemeal graph learning by a mobile robot. Inf. Comput. **152**, 155–172 (1999)
2. Bampas, E., Chalopin, J., Das, S., Hackfeld, J., Karousatou, C.: Maximal exploration of trees with energy-constrained agents. CoRR, abs/1802.06636 (2018)
3. Barrière, L., Flocchini, P., Fraigniaud, P., Santoro, N.: Rendezvous and election of mobile agents: impact of sense of direction. Theory Comput. Syst. **40**(2), 143–162 (2007)
4. Becha, H., Flocchini, P.: Optimal construction of sense of direction in a torus by a mobile agent. Int. J. Found. Comput. Sci. **18**(3), 529–546 (2007)
5. Bender, M., Fernandez, A., Ron, D., Sahai, A., Vadhan, S.: The power of a pebble: exploring and mapping directed graphs. In: Proceedings of the 30th ACM Symposium on Theory of Computing (STOC 1998), pp. 269–287 (1998)
6. Blum, M., Kozen, D.: On the power of the compass (or, why mazes are easier to search than graphs). In: 19th Symposium on Foundations of Computer Science (FOCS 1978), pp. 132–142 (1978)
7. Brass, P., Cabrera-Mora, F., Gasparri, A., Xiao, J.: Multirobot tree and graph exploration. IEEE Trans. Robot. **27**(4), 707–717 (2011)
8. Budach, L.: Automata and labyrinths. Math. Nachrichten **86**, 195–282 (1978)
9. Chalopin, J., Das, S., Kosowski, A.: Constructing a map of an anonymous graph: applications of universal sequences. In: Lu, C., Masuzawa, T., Mosbah, M. (eds.) OPODIS 2010. LNCS, vol. 6490, pp. 119–134. Springer, Heidelberg (2010). https://doi.org/10.1007/978-3-642-17653-1_10
10. Chalopin, J., Das, S., Santoro, N.: Rendezvous of mobile agents in unknown graphs with faulty links. In: Proceedings of 21st International Symposium on Distributed Computing (DISC), pp. 108–122 (2007)
11. Cohen, R., Fraigniaud, P., Ilcinkas, D., Korman, A., Peleg, D.: Label-guided graph exploration by a finite automaton. ACM Trans. Algorithms **4**(4), 42:1–42:18 (2008)
12. Czyzowicz, J., Dereniowski, D., Gasieniec, L., Klasing, R., Kosowski, A., Pajak, D.: Collision-free network exploration. J. Comput. Syst. Sci. **86**, 70–81 (2017)
13. Czyzowicz, J., et al.: More efficient periodic traversal in anonymous undirected graphs. Theor. Comput. Sci. **444**, 60–76 (2012)
14. Czyzowicz, J., Kosowski, A., Pelc, A.: How to meet when you forget: log-space rendezvous in arbitrary graphs. Distrib. Comput. **25**(2), 165–178 (2012)
15. Das, S., Dereniowski, D., Karousatou, C.: Collaborative exploration of trees by energy-constrained mobile robots. Theory Comput. Syst. **62**(5), 1223–1240 (2018)
16. Das, S., Dereniowski, D., Uznanski, P.: Energy constrained depth first search. CoRR, abs/1709.10146 (2017)
17. Das, S., Flocchini, P., Kutten, S., Nayak, A., Santoro, N.: Map construction of unknown graphs by multiple agents. Theor. Comput. Sci. **385**(1–3), 34–48 (2007)
18. Das, S., Flocchini, P., Nayak, A., Santoro, N.: Effective elections for anonymous mobile agents. In: Asano, T. (ed.) ISAAC 2006. LNCS, vol. 4288, pp. 732–743. Springer, Heidelberg (2006). https://doi.org/10.1007/11940128_73
19. Das, S., Mihalák, M., Šrámek, R., Vicari, E., Widmayer, P.: Rendezvous of mobile agents when tokens fail anytime. In: Baker, T.P., Bui, A., Tixeuil, S. (eds.) OPODIS 2008. LNCS, vol. 5401, pp. 463–480. Springer, Heidelberg (2008). https://doi.org/10.1007/978-3-540-92221-6_29
20. Deng, X., Papadimitriou, C.H.: Exploring an unknown graph. J. Graph Theory **32**(3), 265–297 (1999)

21. Dereniowski, D., Disser, Y., Kosowski, A., Pajak, D., Uznański, P.: Fast collaborative graph exploration. Inf. Comput. **243**, 37–49 (2015)
22. Dieudonne, Y., Pelc, A., Peleg, D.: Gathering despite mischief. ACM Trans. Algorithms **11**(1), 1 (2014)
23. Diks, K., Fraigniaud, P., Kranakis, E., Pelc, A.: Tree exploration with little memory. J. Algorithms **51**, 38–63 (2004)
24. Disser, Y., Hackfeld, J., Klimm, M.: Undirected graph exploration with $\Theta(\log \log n)$ pebbles. In: Proceedings of the ACM-SIAM Symposium on Discrete Algorithms (SODA), pp. 25–39 (2016)
25. Dobrev, S., Flocchini, P., Kralovic, R., Ruzicka, P., Prencipe, G., Santoro, N.: Black hole search in common interconnection networks. Networks **47**(2), 61–71 (2006)
26. Duncan, C.A., Kobourov, S.G., Kumar, V.S.A.: Optimal constrained graph exploration. In: 12th ACM-SIAM Symposium on Discrete Algorithms (SODA), pp. 807–814 (2001)
27. Dynia, M., Łopuszański, J., Schindelhauer, C.: Why robots need maps. In: Prencipe, G., Zaks, S. (eds.) SIROCCO 2007. LNCS, vol. 4474, pp. 41–50. Springer, Heidelberg (2007). https://doi.org/10.1007/978-3-540-72951-8_5
28. Flocchini, P., Huang, M.J., Luccio, F.L.: Decontamination of hypercubes by mobile agents. Networks **52**(3), 167–178 (2008)
29. Fraigniaud, P., Gasieniec, L., Kowalski, D., Pelc, A.: Collective tree exploration. Networks **48**(3), 166–177 (2006)
30. Fraigniaud, P., Ilcinkas, D., Peer, G., Pelc, A., Peleg, D.: Graph exploration by a finite automaton. Theor. Comput. Sci. **345**(2–3), 331–344 (2005)
31. Kosowski, A., Navarra, A.: Graph decomposition for memoryless periodic exploration. Algorithmica **63**(1–2), 26–38 (2012)
32. Menc, A., Pajak, D., Uznanski, P.: Time and space optimality of rotor-router graph exploration. Inf. Process. Lett. **127**, 17–20 (2017)
33. Ortolf, C., Schindelhauer, C.: Online multi-robot exploration of grid graphs with rectangular obstacles. In: 24th ACM Symposium on Parallelism in Algorithms and Architectures (SPAA), pp. 27–36 (2012)
34. Panaite, P., Pelc, A.: Exploring unknown undirected graphs. J. Algorithms **33**, 281–295 (1999)
35. Reingold, O.: Undirected connectivity in log-space. J. ACM **55**(4), 17:1–17:24 (2008)
36. Rollik, H.A.: Automaten in planaren graphen. Acta Informatica **13**(3), 287–298 (1980)
37. Shannon, C.E.: Presentation of a maze-solving machine. In: 8th Conference of the Josiah Macy Jr. Found. (Cybernetics), pp. 173–180 (1951)
38. Tarjan, R.: Depth first search and linear graph algorithms. SIAM J. Comput. **1**(2), 146–160 (1972)
39. Yamashita, M., Kameda, T.: Computing on anonymous networks: part i-characterizing the solvable cases. IEEE Trans. Parallel Distrib. Syst. **7**(1), 69–89 (1996)
40. Yanovski, V., Wagner, I.A., Bruckstein, A.M.: A distributed ant algorithm for efficiently patrolling a network. Algorithmica **37**(3), 165–186 (2003)

# Deterministic Rendezvous Algorithms

Andrzej Pelc(✉)

Département d'informatique, Université du Québec en Outaouais,
Gatineau, QC J8X 3X7, Canada
pelc@uqo.ca

**Abstract.** The task of rendezvous (also called *gathering*) calls for a meeting of two or more mobile entities, starting from different positions in some environment. Those entities are called mobile agents or robots, and the environment can be a network modeled as a graph or a terrain in the plane, possibly with obstacles. The rendezvous problem has been studied in many different scenarios. Two among many adopted assumptions particularly influence the methodology to be used to accomplish rendezvous. One of the assumptions specifies whether the agents in their navigation can see something apart from parts of the environment itself, for example other agents or marks left by them. The other assumption concerns the way in which the entities move: it can be either deterministic or randomized. In this paper we survey results on deterministic rendezvous of agents that cannot see the other agents prior to meeting them, and cannot leave any marks.

**Keywords:** Mobile agent · Rendezvous · Deterministic · Network
Graph · Terrain · Plane

## 1 Introduction

How to meet in an unknown environment? This question has to be answered in many applications. The most obvious and commonly encountered are those where the entities that have to meet are part of the natural world: they are humans or animals. One of the examples cited in [2] is the Astronaut Problem, in which two astronauts land in distant places on a planet, without any orientation, and have to minimize the expected time of getting together. More common examples of situations when humans have to meet is the task of finding a lost hiker in the mountains by rescuers, or meeting guests at the airport by their host. Schelling [39] studied issues related to rendezvous problems: two players want to meet in an unknown town and have only one attempt to make. Schelling emphasized the need of finding "focal points" (such as the main station, or the central square of the town) that are likely to be chosen by both players (without

A. Pelc—Research supported in part by NSERC Discovery Grant 8136 – 2013 and by the Research Chair in Distributed Computing of the Université du Québec en Outaouais.


P. Flocchini et al. (Eds.): Distributed Computing by Mobile Entities, LNCS 11340, pp. 423–454, 2019.
https://doi.org/10.1007/978-3-030-11072-7_17

previous agreement), due to their common cultural background. However, in algorithmic rendezvous problems, focal points often do not exist, when agents have to meet in the empty plane or in a highly symmetric network. Rendezvous tasks are also frequent in the animal world, such as gathering of migratory birds or undersea animals, or penguin parents finding their offspring when they come back with food.

In computer science applications, the most interesting cases concern human-made agents. The first example of such mobile agents are autonomous mobile robots that start in different locations of a planar terrain or a labyrinth, and have to meet. The reason of meeting can be to exchange samples of the ground previously collected by the robots, or exchange information obtained when exploring different parts of the terrain. The second example is that of software agents, i.e., mobile pieces of software that navigate in a computer network in order to perform maintenance of its components or to collect data distributed in nodes of the network. Periodic meeting of software agents is necessary to exchange collected data and plan further actions, possibly depending on those data.

Since rendezvous algorithms do not depend on the physical nature of the mobile entities executing them, but only on their perception capabilities, memory size, mobility characteristics and on the structure of the environment, we will not distinguish between natural and artificial agents, and among the latter between mobile robots and software agents, and we will use the generic name of *agents* regardless of whether the algorithm is to be applied to people, animals, mobile robots, or software agents. In the case of more than two agents, the rendezvous problem is sometimes called *gathering*. For the sake of uniformity, we will call it *rendezvous* also in this case, using the term *gathering* as a synonym.

Since rendezvous problems usually have to be solved without the help of any central monitor coordinating the actions of agents, these problems belong naturally to the domain of distributed computing. There are, however, many scenarios and models under which rendezvous has been studied. Two among many adopted assumptions particularly influence the methodology to be used to accomplish rendezvous.

One of the assumptions specifies whether the agents in their navigation can see something apart from the underlying environment itself, for example other agents or marks left by them. Needless to say, such a capability significantly facilitates the task of rendezvous: for example, two agents seeing each other in the plane may meet approaching each other along the line joining them. The other assumption concerns the way in which the entities move: it can be either deterministic or randomized. In a deterministic scenario, the initial positions of the agents are chosen by an adversary which models the worst-case situation, and each move of the agent is determined only by its current history that may include the identity of the agent (if any), and the part of the environment that the agent has seen to date. By contrast, in a randomized scenario, initial positions of the agents are often chosen at random and their moves may also involve coin tosses. Randomized rendezvous algorithms in networks often use random walks in the underlying graph. The cost of rendezvous is also different in both scenarios: while in deterministic rendezvous the worst-case cost is usually considered (cost being defined as the time or the length of the agents' trajectories until rendezvous), in the randomized scenario it is the expected value of this quantity. In both cases the problem is often to minimize the worst case (resp. expected) cost. Deterministic rendezvous problems usually require combinatorial tools, while randomized rendezvous often calls for analytic methods.

In this chapter we consider the task of rendezvous under the weaker variant of each of the above two assumptions. First we assume that the navigating agents do not see other agents prior to their meeting and cannot leave any marks, and second, we consider only deterministic rendezvous algorithms. The decision of carving out this particular subdomain of the domain of rendezvous algorithms has two reasons. The first and the most obvious of them is the space limitation: the realm of rendezvous algorithms is very large and covering all of it would require a large book rather than a chapter. The second reason is to avoid duplication of information contained in previous surveys and in other chapters of this book.

There are six main previous surveys concerning rendezvous. Chronologically the first of them is [1], almost entirely contained in the second part of the excellent book [2]. Both [1] and [2] concern randomized rendezvous, viewed from the operations research point of view. The third survey is [32]. While its scope is large, the authors concentrate mainly on presenting rendezvous models and compare their underlying assumptions. The book [31] deals mostly with rendezvous problems on the ring, only briefly mentioning other network topologies in this context. The survey [36] is the closest to the present chapter but it covers only deterministic rendezvous in networks. Finally, Chap. 4 of this book covers the rich domain of gathering agents in the plane, under a scenario where agents have contrasting capabilities: they have no memory but enjoy very strong perception capabilities – they can periodically make snapshots in which they see other agents.

The existence of the book [2] is the main reason of our restriction to deterministic rendezvous algorithms, and Chap. 4 of this book is the main reason of our restriction to rendezvous under the scenario where navigating agents cannot see much.

The present chapter differs from [1] and [2] by concentrating on deterministic rather than on randomized scenarios; it differs from [32] by the level of details in treating the rendezvous problem: besides presenting various models under which deterministic rendezvous is studied, we want to report precisely the results obtained under each of them, discussing how varying assumptions influence feasibility and complexity of rendezvous under various settings. This chapter differs from [31] by discussing many different topologies, mostly arbitrary, even unknown graphs, rather than concentrating on a particular type of networks. As mentioned above, this chapter is the closest to [36]: the identity of the author may be a reason. What are the differences with respect to [36]? The survey [36] was published in 2012, reporting results until 2011. During these 7 years many new results concerning deterministic rendezvous appeared in the literature. Since many of them concerned the scenario where agents cannot leave any marks, and the scenario allowing marking of the nodes was considered in [36], we decided to exclude this scenario from the present chapter, and add instead issues concerning rendezvous in the plane, as they often use rendezvous in a specific graph, the infinite grid, as a methodological tool. There is some overlap of this chapter with [36] but we tried to discuss the older results covered there only in a cursory manner, concentrating on the new developments.

A remark is in order concerning the chronology of the reported results. In each case when a journal paper is available, we refer to it, as to the most definitive version. However, the journal version is sometimes published much later than the conference version in which a given result first appeared. This may scramble the precedence relations of results. In such cases we tried to mention the correct order of discoveries.

The rest of the chapter is organized as follows. In Sect. 2 we discuss various scenarios resulting from alternative assumptions adopted for the rendezvous problem, and mention methodological differences of the solutions in different models. The main dividing line in the entire body of research surveyed in this chapter is between the two types of environment in which rendezvous has to take place: one type are networks modeled as undirected graphs, the other is the plane or parts of it, with possible obstacles obstructing moves of the agents. Consequently, Sect. 3 covers rendezvous in networks and Sect. 4 is devoted to rendezvous in the plane. Finally, Sect. 5 contains conclusions and open problems.

## 2 Discussion of Assumptions, Models and Methodology

As announced in the introduction, we will consider the rendezvous problem in two different environments: in networks modeled as undirected graphs and in the plane or its parts. Among many alternative pairs of assumptions adopted for the study of rendezvous, this one is arguably the most basic, as it influences even the precise definition of rendezvous. In both cases, agents are modeled as points. In the first case they navigate in the graph, traversing its edges and visiting its nodes, and in the second case they move freely in the plane, possibly avoiding obstacles. In the case of rendezvous in networks, meeting of the agents is defined as being at the same node in the same time or as being in the same point of an edge at the same time. (We will further discuss submodels in which one or the other of these definitions is applied). In the case when agents move in the plane, they are also modeled as points. However, in particular where rendezvous has to take place in the empty plane, we cannot require the agents to get to the same point of the plane at the same time. To see this, consider an easier problem, when one of the agents is inert and the other one has to find it. (Rendezvous can always be reduced to this case, if the adversary decides to delay the start of one of the agents sufficiently long). Since the walking agent cannot see the inert agent prior to meeting, a correct algorithm would require to construct a trajectory, which is a curve in the plane passing through each given point after walking a finite distance. Such a curve does not exist. Hence, in the case of meeting in the plane, rendezvous is defined more loosely as *approach*: agents have to get at distance 1 of each other. This can be motivated by assuming that the agents have visibility permitting to see at distance 1, and rendezvous is accomplished when agents are inside one another's visibility regions. Hence in the case of the plane we will concentrate on the task of approach.

There is one exception to this change of definition. When agents are in a bounded part of the plane, possibly with "holes", i.e., impenetrable obstacles, then rendezvous defined as getting to the same point at the same time can be required, because in this case agents can meet in a boundary point of the terrain or of one of the obstacles. We will see such a situation in [14].

We will now review the common assumptions used in the literature to consider rendezvous in each of the above scenarios. We first consider the network scenario. The first common assumption is modeling the network as a simple undirected connected graph, whose nodes represent processors, computers or stations of a communication network, or crossings of corridors of a labyrinth, depending on the application, and links represent communication channels in a communication network, or corridors in a labyrinth. Modeling the network as an undirected graph captures the ability of the agents to move in both directions along each link. The assumption that the graph is simple (no self-loops or multiple edges) is motivated by most of the realistic applications, and connectivity of the graph is a necessary condition on feasibility of rendezvous when starting from any initial positions: agents starting in different connected components could not meet. Throughout the chapter, we use the term *graph* to mean a simple undirected connected graph.

The second common assumption is the anonymity of the underlying network: the absence of distinct names of nodes that can be perceived by the navigating agents. There are two reasons for seeking rendezvous algorithms that do not assume knowledge of node identities. The first one is practical: while nodes may indeed have different labels, they may refrain from informing the agents about them, e.g., for privacy or security reasons, or limited sensory capabilities of agents may prevent them from perceiving these names. The latter restriction is mostly applicable to mobile robots whose sensing device may be too weak to read such labels. The other reason for assuming anonymity of the network is methodological. If distinct names of nodes can be perceived by the agents, they can follow an algorithm which guides each of them to the node with the smallest label and stop. Thus the rendezvous problem becomes reducible to graph exploration, which has been well studied.

The last common assumption concerns port numbers at each node. It is assumed that a node of degree $d$ has ports $0, 1, \dots, d-1$ corresponding to the incident edges. Ports at each node can be perceived by an agent visiting this node, but there is no coherence assumed between port labelings at different nodes. (In the case when such a coherence is assumed, for example in the case of an oriented grid, it will be explicitly mentioned). When an agent enters a node, it learns its degree and the port of entry. The reason for assuming the existence of port labelings accessible to agents is the following. If an agent is unable to locally distinguish ports at a node, it may even be unable to visit all neighbors of a node of degree at least 3. Indeed, after visiting the second neighbor, the agent cannot distinguish the port leading to the first visited neighbor from the port leading to the unvisited one. Thus an adversary may always force an agent to avoid all but two edges incident to such a node. Consequently, agents initially located at two nodes of degree at least 3 might never be able to meet. From the practical point of view, assuming the existence of port numbers legible by the agents is a much less problematic assumption than assuming labels of nodes. First, the privacy and security reasons for not divulging node labels do not apply to port numbers. Second, the sensory capabilities of agents required to read port numbers are much smaller than those for reading node labels. For example, one port at a node can be marked by a "red dot" and then consecutive ports can have a pointer *next* from the preceding port. Reading this type of information requires minimal sensory capabilities.

We will now review the common assumptions concerning rendezvous (i.e., approach) in the plane. These assumptions permit an agent to navigate in the plane in the absence of any visual information. It is assumed that the agent has a compass indicating North and that it has a measure of distance. These two features permit to establish a system of orthogonal coordinates and permit to trace arbitrary angles. Consequently, the agent can execute instructions such as "go at distance $x$ in direction *dir*", where *dir* is expressed as an angle from direction North. Notice that the compasses and the measures of length of different agents are not necessarily the same. If they are, this will be explicitly mentioned.

We proceed to the overview of various alternative assumptions yielding different scenarios under which the rendezvous problem is usually considered, both in the network environment and in the plane. There are two such main pairs of assumptions. The first concerns the possibility to distinguish the agents: they can be either anonymous (i.e., identical), or each agent may have a distinct integer label that it knows and can use as a parameter in the executed rendezvous algorithm which is common to all agents. The second pair of alternative assumptions concerns time: agents may move either in a synchronous or in an asynchronous way. We will give the precise definitions later but, roughly speaking, synchronous movement in graphs means that clocks of the agents tick at the same rate, one tick per round, and in each round an agent can either stay in the current node, or move to a neighbor. In the plane, synchronous movement means that the speed of agents is the same. In the asynchronous scenario, the speed of the agents may vary adversarially. (We will see that there is also a semi-synchronous scenario, where speeds of agents are constant but possibly different).

The above possible scenarios imply different methodological approach to rendezvous in each case. The main problem that has to be solved in order to make deterministic rendezvous possible, both in networks and in the plane, is breaking symmetry. To see why this is necessary, consider an oriented ring (i.e., a ring in which ports at all nodes are labeled as follows: 0 the clockwise port and 1 the counterclockwise port) or consider the infinite plane without obstacles. Consider two identical agents starting at distinct nodes of the ring or at any two points in the plane, and running the same deterministic algorithm. It is easy to see that if they start simultaneously and move synchronously, they will never meet. In the ring, at all times they will use the port (at their respective current nodes) having the same label (as their history is the same and the algorithm is deterministic), and hence the distance between them will be always the same. Likewise, in the plane, if the agents start simultaneously, have the same compass, the same measure of length and the same speed, they will traverse parallel trajectories and remain at the same distance at all times.

In the deterministic scenario there are two ways of breaking symmetry. The first is by distinguishing the agents: each of them has a label and the labels are different. Each agent knows its label, but we do not need to assume that it knows the label of the other agent. (If it does, then the solution is the well-known algorithm Wait For Mommy: the agent with the smaller label stays idle, while the other one explores the graph or the plane in order to find it.) Both agents use the same *parametrized* algorithm with the agent's label as the parameter. To see how this can help, consider two agents that have to meet in an oriented ring of known size $n$. As mentioned above, if agents are anonymous (and marking nodes is disallowed), rendezvous is impossible. Now assume that agents have distinct labels $L_1$ and $L_2$. A simple (although inefficient) rendezvous algorithm is: Make $L$ tours of the ring, where $L$ is your label, and stop. Then the agent with larger label will make at least one full tour of the ring while the other one is already inert, thus guaranteeing rendezvous.

The second way of breaking symmetry, available even when agents are anonymous, is by exploiting either non-symmetries of the network itself, or the differences of the initial positions of the agents, even in a symmetric network. This method is impossible to use in the plane without obstacles, and in the network environment it is usable only for some classes of networks, as either the network must have distinguishable nodes that play the role of "focal points" or the initial positions of agents have to be "non-symmetric" (the precise meaning of this condition will be defined later). As a simple example of the application of this method, consider a $n$-node line with two identical agents. If $n$ is odd, then the line contains a central node that both agents can identify and meet at this node. If $n$ is even, (and even when the port labelings are symmetric with respect to the axis of symmetry of the line) but the initial positions of the agents have different distances from their closest extremity, then the following algorithm works: Compute your distance $d$ from the closest extremity of the line, then traverse the line $d$ times and stop. For the same reasons as before, this algorithm guarantees rendezvous, whenever the initial positions of the agents are not symmetrically situated. On the other hand, if they are symmetric (and port labelings are symmetric as well), then it is easy to see that rendezvous is impossible if agents have to meet at a node and do not realize crossings on an edge.

We now discuss methodological implications of the distinction between the second pair of alternative assumptions, yielding the synchronous and asynchronous scenarios. The discussion is for agents with distinct labels. In the first case, the ability of the agents to exploit time, and more precisely, to vary between carefully measured periods of activity, when the agent explores parts of the network or of the plane, and of passivity, when it stays idle, is a powerful tool in the solution of the rendezvous problem. Indeed, agents may exploit differences in their labels to schedule these activity and passivity periods in such a way that at some point the active agent must visit the position of the agent that is currently passive, and thus accomplish rendezvous. No such possibility is available in the asynchronous scenario. In this case, the main methodological tool is constructing trajectories of the agents, again exploiting the differences of their labels, in such a way that parts of these trajectories coincide, and the agents are forced, regardless of their speed, to traverse a common segment of the trajectory at approximately the same time, implying rendezvous. It should be mentioned that, in the asynchronous scenario applied to networks, it may be impossible to meet at a node, and thus the requirement is relaxed to that of meeting at a node or inside an edge.

In order to make the statement of a rendezvous problem precise, we have to point out what exactly is being sought, apart from meeting. The most general question is that of feasibility: for what classes of networks and what initial positions is rendezvous possible under a particular scenario, and when is it possible in the plane? Here a complete solution would be to prove that for some classes of networks and some initial configurations of agents rendezvous is impossible, and to provide a rendezvous algorithm for all other situations. In the case of the plane, the question is whether approach starting from arbitrary unknown positions is always possible under a given scenario. More specific questions concern the amount of resources needed for rendezvous. These are usually of two types. The first is rendezvous *cost*: the maximum number of steps made by an agent until rendezvous, or the maximum time used by the agents to meet. Algorithms minimizing the cost (or its order of magnitude) are sought in this context. The other important resource is *memory*: what is the minimum memory with which agents have to be equipped in order to solve the rendezvous problem in a given class of networks. When only cost optimization is sought, memory of the agents is often assumed to be unbounded and they are modeled as Turing machines. In problems seeking memory minimization (or tradeoffs between memory and time), the model of input/output automata (finite state machines) is usually used. As usual in optimization tasks, a complete solution calls for an algorithm with given cost or memory and for an accompanying lower bound showing that this cost or amount of memory is optimal.

As mentioned before, in Sect. 3 we present results concerning rendezvous in networks, while in Sect. 4 we study the problem of approach in the plane. The study is further subdivided by considering the synchronous and asynchronous scenarios. Further assumptions are added when presenting a particular model. In each case, we first give a precise description of the model, state the problem to be solved, then present the results and often give a high-level description of methods and algorithms used to obtain them.

# 3 Rendezvous in Networks

In this section we survey results on deterministic rendezvous in networks, dividing our considerations into two major scenarios: synchronous and asynchronous.

## 3.1 Synchronous Rendezvous

Agents move in synchronous rounds. In every round, an agent may either remain at the same node or move to an adjacent node. Rendezvous means that all agents are at the same node in the same round. Agents that cross each other when moving along the same edge, do not notice this fact. Two subscenarios are considered: *simultaneous startup*, when both agents start executing the algorithm in the same round, and *arbitrary delay*, when starting rounds are arbitrarily decided by an adversary. In the former case, agents know that starting rounds are the same, while in the latter case, they are not aware of the difference between starting rounds, and each of them starts executing the rendezvous algorithm and counting steps in the round of its own startup.

We will discuss separately the sub scenario of labeled agents, where agents have distinct integer labels that they can use as a parameter in the common deterministic algorithm, and that of anonymous agents, where agents do not have any labels and thus are identical.

**Labeled Agents.** In [18] (whose journal version was published in 2006, but which is based on two earlier conference papers published in 2003 and 2004), rendezvous of two agents is considered and it is indicated that all results can be generalized to an arbitrary number of agents. It is assumed that agents have different positive integer labels, and each agent knows its own label (which is a parameter of the common deterministic algorithm that they use), but is unaware of the label of the other agent. In general, agents do not know the topology of the graph in which they have to meet. It is assumed that the agents have unlimited memory (they are modeled as Turing machines) and the authors aim at optimizing the cost of rendezvous. This cost is defined as the worst-case number of rounds since the startup of the later agent until rendezvous is achieved, where the worst case is taken over all graphs in the considered class, all initial positions of the agents and all possible startup times (decided by an adversary), in the case of the arbitrary delay scenario.

The following notation is used. The labels of the agents are $L_1$ and $L_2$. The smaller of the two labels is denoted by $l$. The delay (the difference between startup times of the agents) is denoted by $\tau$, $n$ denotes the number of nodes in the graph, and $D$ – the distance between initial positions of agents.

The authors introduce the problem in the relatively simple case of trees. They show that rendezvous can be completed at cost $O(n + \log l)$ on any $n$-node tree, even with arbitrary delay. They also show that for some trees this complexity cannot be improved, even with simultaneous startup. Rendezvous in trees is relatively easy for two reasons. First, a tree can be explored with termination and a map of it can be constructed by a single agent, using the *basic walk* which consists in leaving every node by the next port with respect to the entry port at this node (modulo the degree). The second reason is the existence of the central node or the central edge in any tree. Once each agent locates this object independently, the central node plays the role of the "focal point" and rendezvous can be accomplished at linear cost. If there exists a central edge, rendezvous is slightly more complicated and, after identifying this edge, it reduces to rendezvous in the two-node graph. It is this case that is responsible for the $O(\log l)$ additive term in the cost complexity.

As soon as the graph contains cycles, another technique has to be applied. The authors continue the study by concentrating on the simplest class of such graphs, i.e., rings. They prove that, with simultaneous startup, the optimal cost of rendezvous on any ring is $\Theta(D \log l)$. They construct an algorithm achieving rendezvous with this complexity and show that, for any distance $D$, it cannot be improved.

The lower bound $\Omega(D \log l)$ relies on the following idea. The (oriented) ring is partitioned into pieces of equal size $\Theta(D)$ and time is partitioned into segments of the same length. It is observed that at the end of a segment the agent can be either in the same piece as in the beginning of it, or in one of the neighboring pieces. This permits to code the behavior of an agent by a ternary sequence corresponding to its position at the end of each time segment. It is argued that if two agents have the same behavior code, then they cannot meet. Moreover, if the time of rendezvous were $o(D \log l)$, then behavior codes would have to be short, and thus for some two different labels of agents they would have to be the same, as the behavior of an agent before the meeting depends only on its label. Assigning these labels to the agents would preclude rendezvous.

With an arbitrary delay, $\Omega(n + D \log l)$ is a lower bound on the cost required for rendezvous in a $n$-node ring. Under this scenario, two rendezvous algorithms for the ring are presented in [18]: an algorithm with cost $O(n \log l)$, for known $n$, and an algorithm with cost $O(l\tau + ln^2)$, if $n$ is unknown. In the latter case, the cost was later improved to $O(n \log l)$ in [30]. In view of the above lower bound, this cannot be improved in general.

For arbitrary connected graphs, the main contribution of [18] is the first deterministic rendezvous algorithm with cost polynomial in $n$, $\tau$ and $\log l$. More precisely, the authors present an algorithm that solves the rendezvous problem for any $n$-node graph $G$, for any labels $L_1 > L_2 = l$ of agents and for any delay $\tau$ between startup times, in cost $\mathcal{O}(n^5 \sqrt{\tau \log l} \log n + n^{10} \log^2 n \log l)$. The algorithm contains a non-constructive ingredient: agents use combinatorial objects whose existence is proved by the probabilistic method. Nevertheless the algorithm is indeed deterministic. Both agents can find separately the same combinatorial object with the desired properties (which is then used in the rendezvous algorithm). This can be done using brute force exhaustive search which may be quite complex but in the adopted model only moves of the agents are counted and local computation time of the agents does not contribute to cost. Finally, the authors prove a lower bound $\Omega(n^2)$ on the cost of rendezvous in some family of graphs.

The paper is concluded by an open problem concerning the dependence of rendezvous cost on the delay $\tau$. The dependence on the other parameters follows from the results cited above. Indeed, a lower bound $\Omega(n^2)$ on rendezvous cost has been shown in some graphs. The authors also showed that cost $\Omega(\log l)$ is required even for the two-node graph. On the other hand, for agents starting at distance $\Omega(n)$ in a ring, cost $\Omega(n \log l)$ is required, even for $\tau = 0$. However, since the complexity of their algorithm contains a factor $\sqrt{\tau}$, the authors state the following problem:

Does there exist a deterministic rendezvous algorithm for arbitrary connected graphs with cost polynomial in $n$ and $l$ (or even in $n$ and $\log l$) but independent of $\tau$?

A positive answer to this problem was given in [30] (whose conference version was published in 2006). The authors present a rendezvous algorithm for two agents, working in arbitrary connected graphs for an arbitrary delay $\tau$, whose complexity is $O(\log^3 l + n^{15} \log^{12} n)$, i.e., is independent of $\tau$ and polynomial in $n$ and $\log l$. As before, the algorithm contains a non-constructive ingredient, but is deterministic.

The rendezvous algorithms from [18,30], working for arbitrary connected graphs, yield an intriguing question, stated in [18]. While both of them have polynomial cost (the one from [18] depending on $\tau$, and the one from [30] independent of $\tau$), they both use a non-constructive ingredient, i.e., a combinatorial object whose existence is proved using the probabilistic method. As mentioned above, each of the agents can deterministically find such an object by exhaustive search, and then use it in the execution of its algorithm, which keeps the algorithm deterministic, but may significantly increase the time of local computations. In the described model the time of these local computations does not contribute to cost which is measured by the number of steps, regardless of the time taken to compute each step. Nevertheless, it is interesting if there exists a rendezvous algorithm for which both the cost and the time of local computations are polynomial in $n$ and $\log l$. Such an algorithm would have to eliminate any non-constructive ingredients.

This problem was solved in [41] (whose conference version appeared in 2007), using the important notion of a *Universal Exploration Sequence* (UXS) [29]. Let $(a_1, a_2, \ldots, a_k)$ be a sequence of integers. An *application* of this sequence to a graph $G$ at node $u$ is the sequence of nodes $(u_0, \ldots, u_{k+1})$ obtained as follows: $u_0 = u$, $u_1$ is the node joined to $u$ by the edge corresponding to port 0 at $u$; for any $1 \le i \le k$, $u_{i+1}$ is the node joined to $u_i$ by the edge corresponding (at $u_i$) to port $(p + a_i) \mod d(u_i)$, where $p$ is the port number at $u_i$ corresponding to the edge $\{u_i, u_{i-1}\}$ and $d(u_i)$ denotes the degree of node $u_i$. (Informally, an application of $(a_1, a_2, \ldots, a_k)$ corresponds to a walk in the graph in which the current exit port is computed by adding $a_i$ to the current entry port.) A sequence $(a_1, a_2, \ldots, a_k)$ whose application to a graph $G$ at any node $u$ contains all nodes of this graph is called a UXS for this graph. A UXS for a class $\mathcal{G}$ of graphs is a

UXS for all graphs in this class. The solution from [41] uses a result following from [38] that a UXS for the class of all connected graphs with at most $n$ nodes can be computed in time polynomial in $n$. Moreover, the authors propose a rendezvous algorithm working in $O(n^5 \log l)$ rounds (up to factors logarithmic in $n$ and $\log l$). This complexity beats those from [18,30] and makes the algorithm from [41] the currently most efficient rendezvous algorithm working in arbitrary connected graphs.

The paper [22] started the investigation of gathering of a team of labeled agents, some of which can be Byzantine. The size of the team is unknown to the agents. Agents can exchange all currently held information when they meet at a node of the graph. Up to $f$ of the agents are Byzantine. The authors define two levels of Byzantine behavior. A strongly Byzantine agent can choose an arbitrary port when it moves and it can transmit arbitrary information to other agents, while a weakly Byzantine agent can do the same, except changing its label. The main problem investigated in the paper is what is the minimum number of good agents that guarantees deterministic gathering of all of them, with termination. (Of course, Byzantine agents cannot be forced to gather.) The authors solve exactly this Byzantine gathering problem in arbitrary networks for weakly Byzantine agents, and give approximate solutions for strongly Byzantine agents, both when the size of the network is known and when it is unknown. They show that both the strength versus weakness of Byzantine behavior and the knowledge of network size have an important influence on the results.

For weakly Byzantine agents it is shown that any number of good agents permit to solve the problem for networks of known size. If the size is unknown, then this minimum number is $f+2$. More precisely, the authors design a deterministic polynomial algorithm that gathers all good agents in an arbitrary network, provided that there are at least $f+2$ of them. They also prove a matching lower bound showing that if the number of good agents is at most $f+1$, then they are not able to gather deterministically with termination in some networks.

For strongly Byzantine agents the authors give a lower bound of $f+1$, even when the graph is known: they show that $f$ good agents cannot gather deterministically in the presence of $f$ Byzantine agents even in a ring of known size. In order to establish upper bounds, they propose deterministic gathering algorithms for at least $2f+1$ good agents when the size of the network is known, and for at least $4f+2$ good agents when it is unknown. These upper bounds were subsequently improved in [5] to $f+1$ when the size of the network is known and to $f+2$ when it is unknown. Together with the lower bounds from [22], both these upper bounds are tight.

As the authors of [5] point out, the above results show an interesting difference between the scenarios of known vs. unknown size of the network. While for known size, the gap between the number of good agents permitting gathering with weakly and with strongly Byzantine agents is very significant (1 vs. $f+1$) this gap completely disappears for the scenario of unknown size: the minimum number of good agents is then $f+2$, regardless of whether the bad agents are weakly or strongly Byzantine.

In the above papers, gathering algorithms are very inefficient: their time complexity is exponential in the size $n$ of the graph and in the largest label $L$ of a good agent. In [7], the authors make a concession on the proportion of the Byzantine agents within the team, but they significantly improve the efficiency of gathering. Assuming that the agents are in a *strong team*, which is defined as having at least $5f^2+6f+2$ good agents, they design a gathering algorithm that works in the presence of at most $f$ Byzantine agents in time polynomial in $n$ and in the length of the smallest label of a good agent. Moreover, they show how to achieve this gathering using global knowledge of size $O(\log\log\log n)$, and they prove that this size of knowledge is optimal for this task.

Gathering of agents in the presence of a more benign type of faults is considered in [37]. In this paper it is assumed that some agents are subject to *crash faults* which can occur at any time. Two fault scenarios are considered. A *motion fault* immobilizes the agent at a node or inside an edge but leaves intact its memory at the time when the fault occurred. A more severe *total fault* immobilizes the agent as well, but also erases its entire memory. As before, we cannot require faulty agents to gather. Thus the gathering problem for fault prone agents calls for all fault-free agents to gather at a single node, and terminate.

It is observed that when agents move completely asynchronously, gathering with crash faults of any type is impossible. Hence the author considers a restricted version of asynchrony, where each agent is assigned by the adversary a fixed speed, possibly different for each agent. Agents have clocks ticking at the same rate. Each agent can wait for a time of its choice at any node, or decide to traverse an edge but then it moves at constant speed assigned to it. It is appropriate to discuss this model in the section devoted to synchronous algorithms, as methodologically the two scenarios are similar: it is still possible to wait at a node for a prescribed amount of time, and this capability significantly influences algorithm design. When two or more agents are at the same node or in the same point of an edge in the same time, they can see the memory content of other agents at this node or at this point of an edge, except for memory of faulty agents in the case of total faults.

The main results of the paper are a gathering algorithm working for any team of at least two agents in the scenario of motion faults, and a gathering algorithm working in the presence of total faults, provided that at least two agents are fault free all the time. If only one agent is fault free, the task of gathering with total faults is sometimes impossible (recall that termination is required). This shows that in the case of crash faults more faulty agents can be tolerated for gathering than when faults are Byzantine. Both algorithms from [37] work in time polynomial in the size of the graph, in the logarithm of the largest label, in the inverse of the smallest speed, and in the ratio between the largest and the smallest speed.

Rendezvous of two agents subject to transient faults is considered in [9]. Agents do not know the topology of the network or any bound on its size. In each round an agent decides if it remains idle or if it wants to move to one of the adjacent nodes. Agents are subject to *delay faults*: if an agent incurs a fault in a given round, it remains in the current node, regardless of its decision. If it planned to move and the fault happened, the agent is aware of it. The authors consider three scenarios of fault distribution: random (delay faults occur independently in each round and for each agent with constant probability $0 < p < 1$), unbounded adversarial (the adversary can delay an agent for an arbitrary finite number of consecutive rounds) and bounded adversarial (the adversary can delay an agent for at most $c$ consecutive rounds, where $c$ is unknown to the agents). The quality measure of a rendezvous algorithm is its cost, which is the total number of edge traversals.

For random faults, the authors design an algorithm with cost polynomial in the size $n$ of the network and polylogarithmic in the larger label $L$, which achieves rendezvous with probability at least $1 - 1/n$ in arbitrary networks. By contrast, for unbounded adversarial faults they prove that rendezvous is not feasible, even in the class of rings. Under this scenario, the authors design a rendezvous algorithm with cost $O(n\ell)$, where $\ell$ is the smaller label, working in arbitrary trees, and they show that $\Omega(\ell)$ is the lower bound on rendezvous cost, even for the two-node tree. For bounded adversarial faults, the authors construct a rendezvous algorithm working for arbitrary networks, with cost polynomial in $n$, and logarithmic in the bound $c$ and in the larger label $L$.

In [35], the authors use a different approach to counter transient faults. They propose a *self-stabilizing* rendezvous algorithm for two agents navigating in an arbitrary network. A self-stabilizing algorithm has the property that if the agents start from any two nodes with arbitrary memory states, then eventually they will get to the same node in the same round. Thus, even if the agents incur a transient fault of any kind that corrupts their memory, they can meet in finite time after the fault disappears. The authors design a self-stabilizing rendezvous algorithm for arbitrary graphs, without any time guarantees, and construct polynomial-time self-stabilizing rendezvous algorithms for trees and rings. More precisely, the algorithm for trees is polynomial in the size of the tree and in the logarithm of the smaller label, and the algorithm for rings is polynomial in the size of the ring and in both labels.

In [33], the authors consider two main efficiency measures of rendezvous: its time, i.e., the number of rounds until the meeting, and its cost, i.e., the total number of edge traversals. They investigate tradeoffs between these two measures. A natural benchmark for both time and cost of rendezvous in a network is the number of edge traversals needed for visiting all nodes of the network, called the exploration time. The authors express the time and cost of rendezvous as functions of an upper bound $E$ on the time of exploration (where an exploration procedure is known to both agents) and of the size $L$ of the label space. They design two rendezvous algorithms: algorithm `Cheap` has cost $O(E)$ and time $O(EL)$, and algorithm `Fast` has both time and cost $O(E \log L)$. The main contribution of the paper are lower bounds showing that these two algorithms capture the tradeoffs between time and cost of rendezvous almost tightly. They show that any deterministic rendezvous algorithm of cost asymptotically $E$ (i.e., of cost $E + o(E)$) must have time $\Omega(EL)$. On the other hand, they prove that any deterministic rendezvous algorithm with time complexity $O(E \log L)$ must have cost $\Omega(E \log L)$.

In [34], the problem of rendezvous is studied in the framework of *advice*, which is a popular paradigm permitting to measure the amount of information that agents need in order to perform some task in networks. If $D$ is the distance between the initial positions of the agents, then $\Omega(D)$ is a lower bound on the time of rendezvous. However, in the absence of any knowledge about the network, agents usually cannot meet in time $O(D)$. Thus the authors study the minimum amount of information that has to be available *a priori* to the agents in order to achieve rendezvous in optimal time $\Theta(D)$. Following the advice paradigm, this information is provided to the agents at the start by an oracle knowing the entire instance of the problem, i.e., the network, the starting positions of the agents, their wake-up rounds, and both of their labels. The oracle provides the agents with the same binary string called *advice*, which can be used by the agents during their rendezvous algorithm. The length of this string is called the *size of advice*. The goal of the paper is to find the smallest size of advice which enables the agents to meet in time $\Theta(D)$. The authors solve this problem completely by showing that this optimal size of advice is $\Theta(D \log(n/D) + \log \log L)$, where $n$ is the size of the graph, $D$ is the initial distance between the agents, and $L$ is the size of the label space. The upper bound is proved by constructing an advice string of this size, and providing a rendezvous algorithm using this advice that works in time $\Theta(D)$ for all networks. The matching lower bound, which is the most difficult and interesting part of the paper, is proved by constructing classes of networks for which it is impossible to achieve rendezvous in time $\Theta(D)$ with smaller advice.

The authors of [16] investigate the rendezvous problem in graphs under the scenario where during navigation each agent gets some restricted feedback about the position of the other agent. More precisely, they consider distance-aware agents that, in every round, are informed of the distance between them. The authors show that such agents can meet in time $O(\Delta(D + \log l))$, where $D$ is the initial distance between the two agents, $l$ is the smaller label and $\Delta$ is the maximum degree of the graph. Thus, even in a very large graph, distance-aware agents can meet in time polynomial in local parameters of the instance of the rendezvous problem. Moreover, the authors show an almost matching lower bound $\Omega(\Delta(D + \log l / \log \Delta))$ on the time of rendezvous in their scenario.

In most formulations of the synchronous rendezvous problem, meeting is accomplished when the agents get to the same node in the same round. In [24], the authors consider a more demanding task, called *rendezvous with detection*: agents must become aware that the meeting is accomplished, simultaneously declare this and stop. It is clear that in order to signal to the other agent the presence at a given node, agents must communicate, and the awareness of the meeting depends on ways of communication between the agents. The authors study two variations of a very weak model of communication, called the *beeping model*, introduced in [11]. In each round an agent can either listen or beep. In the *local beeping model*, an agent hears a beep in a round if it listens in this round and if the other agent is at the same node and beeps. In the *global beeping model*, an agent hears a *loud* beep in a round if it listens in this round and if the other agent is at the same node and beeps, and it hears a *soft* beep in a round if it listens in this round and if the other agent is at some other node and beeps.

The authors propose a deterministic algorithm of rendezvous with detection working, even for the weaker local beeping model, in an arbitrary unknown network in time polynomial in the size of the network and in the logarithm of the smaller label. However, this algorithm is highly energy consuming: the number of moves that an agent must make, is proportional to the time of rendezvous. Hence the authors ask if *bounded-energy agents*, i.e., agents that can make at most $c$ moves, for some integer $c$, can always achieve rendezvous with detection as well. They observe that this is impossible for some networks of unbounded size. Hence they rephrase the question as follows. Can bounded-energy agents always achieve rendezvous with detection in bounded-size networks? The authors prove that the answer to this question is positive, even in the local beeping model but this ability comes at a steep price of time: the meeting time of bounded-energy agents is exponentially larger than that of unrestricted agents. By contrast, the authors propose an algorithm for rendezvous with detection in the global beeping model that works for bounded-energy agents (in bounded-size networks) as fast as for unrestricted agents.

We conclude this section by discussing rendezvous of agents that have fixed but possibly different speeds. The meeting must be at a node, which precludes the fully asynchronous scenario. Hence, the authors of [8] consider a scenario of agents with restricted asynchrony: agents have the same measure of time but the adversary can arbitrarily impose the speed of traversing each edge by each of the agents. They construct a rendezvous algorithm for such agents, working in time polynomial in the size of the graph, in the length of the smaller label, and in the largest edge traversal time.

**Anonymous Agents.** One of the first papers on synchronous rendezvous of two anonymous agents was [26] where the authors compare the time of deterministic and of randomized rendezvous in trees. For deterministic rendezvous they propose an algorithm working in time linear in the size of the tree, for every initial configuration for which rendezvous is possible, and they show that this time cannot be improved in general, even when agents start at distance 1 in bounded degree trees.

In the case of anonymous agents, rendezvous may be impossible for some initial configurations, in some networks, as witnessed by the example of two agents in the oriented ring, mentioned before. This yields an important feasibility problem, which is to characterize those initial configurations of an arbitrary number of anonymous agents for which rendezvous (gathering) is feasible, and to provide a gathering algorithm working for all such configurations. This problem was attacked and completely solved in [19].

At least two agents start from different nodes of the graph. The adversary wakes up some of the agents at possibly different times. A dormant agent, not woken up by the adversary, is woken up by the first agent that visits its starting node, if such an agent exists. Agents do not know the topology of the graph or the size of the team. The authors considered two scenarios: one when agents know a polynomial upper bound on the size of the graph and another when no bound is known. When several agents are at the same node in the same round, they can exchange all information they currently have. The authors assume that the memory of the agents is unlimited.

An initial configuration of agents, i.e., their placement at some distinct nodes of the graph, is called *gatherable* if there exists a deterministic algorithm (even only dedicated to this particular configuration) that achieves meeting of all agents in one node, regardless of the times at which some of the agents are woken up by the adversary. The authors study the problem of which initial configurations are gatherable and how to gather all of them deterministically by the same algorithm. The problem calls for deciding which initial configurations are possible to gather, even by an algorithm specifically designed for this particular configuration, and for finding a *universal* gathering algorithm that gathers all such configurations. The authors restrict attention only to *terminating* algorithms, in which every agent eventually stops forever.

The main result of the paper is a complete solution of the above gathering problem in arbitrary graphs. The authors characterize all gatherable configurations and give two *universal* deterministic gathering algorithms, i.e., algorithms that gather all gatherable configurations. The first algorithm works under the assumption that an upper bound $n$ on the size of the network is known. In this case their algorithm guarantees *gathering with detection*, i.e., the existence of a round in which, for any gatherable configuration, all agents are at the same node and all declare that gathering is accomplished. If no upper bound on the size of the network is known, the authors show that a universal algorithm for gathering with detection does not exist. Hence, for this harder scenario, they construct a second universal gathering algorithm, which guarantees that, for any gatherable configuration, all agents eventually get to one node and stop, although they cannot tell if gathering is over. The time of the first algorithm is polynomial in the upper bound $n$ on the size of the network, and the time of the second algorithm is polynomial in the (unknown) size of the network itself.

As pointed out by the authors of [19], the problem of gathering an unknown team of anonymous agents in an arbitrary network presents the following major difficulty. The asymmetry of the initial configuration because of which gathering is feasible, may be caused not only by non-symmetric locations of the agents with respect to the structure of the graph, but by their different situation with respect to other agents. Hence the authors had to come up with a new algorithmic idea: in order to gather, agents that were initially identical, must make decisions based on the memories of other agents met to date, in order to make their future behavior different and break symmetry in this way. In the beginning the memory of each agent is empty and in the execution of the algorithm it records what the agent has seen in previous steps of the navigation and what it heard from other agents that it met. Even small differences occurring in a remote part of the graph must eventually influence the behavior of initially distant agents. Agents in different initial situations may be unaware of this difference in early meetings, as the difference may depend on their location with respect to remote agents and thus be revealed only later on. Hence an agent may mistakenly consider that two different agents that it met in different stages of the algorithm execution, are the same agent. Confusions due to this possibility are a significant challenge in the algorithm design, that occurs neither in gathering two (even anonymous) agents nor in gathering many labeled agents.

Rendezvous of two anonymous agents was considered in [12,27]. As mentioned before, in this case rendezvous is not possible for arbitrary networks and arbitrary initial positions of the agents. In order to describe initial positions of the agents for which rendezvous is possible, we need the notion of a *view* from a node of a graph, introduced in [42]. Let $G$ be a graph and $v$ a node of $G$. The *view* from $v$ is the infinite tree $\mathcal{V}(v)$ rooted at $v$ with labeled ports, whose branches are infinite paths in $G$ starting at $v$, coded as sequences of ports. A pair $(u, v)$ of distinct nodes is called *symmetric*, if $\mathcal{V}(u) = \mathcal{V}(v)$. Notice that if there exists a port-preserving automorphism of the graph carrying node $u$ to node $v$ then $(u, v)$ is a symmetric pair but the converse is not always true (cf. Fig. 1). However, this equivalence is true for the particular case of the class of trees.

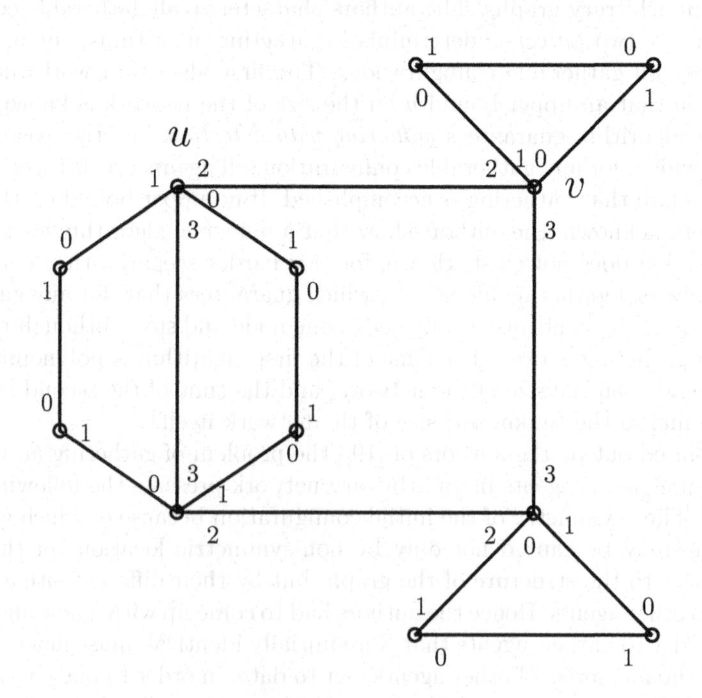


**Fig. 1.** $(u, v)$ is a symmetric pair

Initial positions forming a symmetric pair of nodes are crucial when considering the feasibility of rendezvous in arbitrary graphs. Indeed, it is proved in [12] that rendezvous is feasible, if and only if the initial positions of the agents are not a symmetric pair.

The aim of [12,27] was optimizing the memory size of the agents that seek rendezvous. In order to model agents with bounded memory, the formalism of input/output automata is used. An agent is an abstract state machine $\mathcal{A} = (S, \pi, \lambda, s_0)$, where $S$ is a set of states among which there is a specified state $s_0$ called the *initial* state, $\pi : S \times \mathbb{Z}^2 \to S$, and $\lambda : S \to \mathbb{Z}$. Initially the agent is at some node $u_0$, called its *initial position*, in the initial state $s_0 \in S$. The agent performs an action in each step. Each action can be either a move to an adjacent node or a null move resulting in remaining in the currently occupied node. State $s_0$ determines an integer number $\lambda(s_0)$. If $\lambda(s_0) = -1$, then the agent makes a null move (i.e., remains at $u_0$). If $\lambda(s_0) \geq 0$ then the agent leaves $u_0$ by port $\lambda(s_0) \bmod d(u_0)$, where $d(u_0)$ is the degree of $u_0$. When incoming to a node $v$ in state $s \in S$, the behavior of the agent is as follows. It reads the number $i$ of the port through which it entered $v$ and the degree $d$ of $v$. The pair $(i, d) \in \mathbb{Z}^2$ is an input symbol that causes the transition from state $s$ to state $s' = \pi(s, (i, d))$. If the previous move of the agent was null, (i.e., the agent stayed at node $v$ in state $s$), then the pair $(-1, d) \in \mathbb{Z}^2$ is the input symbol read by the agent, that causes the transition from state $s$ to state $s' = \pi(s, (-1, d))$. In both cases $s'$ determines an integer $\lambda(s')$, which is either $-1$, in which case the agent makes a null move, or a non negative integer indicating the port number by which the agent leaves $v$. This port number is $\lambda(s') \bmod d(v)$, where $d(v)$ is the degree of $v$. The agent continues moving in this way, possibly infinitely.

In order to grasp the assumption that agents are identical, it is assumed that agents are copies $A$ and $A'$ of the same abstract state machine $\mathcal{A}$, starting at two distinct nodes $v_A$ and $v_{A'}$. We will refer to such identical machines as a *pair of agents.* A pair of agents is said to solve the rendezvous problem *with delay* $\tau$ in a class $\mathcal{C}$ of graphs, if, for any graph in the class $\mathcal{C}$ and for any initial positions that are not symmetric, both agents are eventually in the same node of the graph in the same round, provided that they start with delay $\tau$. The memory of an agent is measured by the number of states of the corresponding automaton, or equivalently by the number of bits on which these states are encoded. An automaton with $K$ states requires $\Theta(\log K)$ bits of memory.

In [27] (based on conference papers published by the same authors in DISC 2008 and SPAA 2010) the authors focus attention on optimizing memory size of identical agents that permits them meeting in trees. They assume that the port labeling is decided by an adversary aiming at preventing two agents from meeting, or at allowing the agents to meet only after having consumed a lot of resources, e.g., memory space. This yields the following definition. A pair of agents initially placed at nodes $u$ and $v$ of a tree $T$ solves the rendezvous problem if, for any port labeling of $T$, both agents are eventually in the same node of the tree in the same round.

Nodes $u$ and $v$ of a tree $T = (V, E)$ are *perfectly symmetrizable* if there exists a port labeling $\mu$ of $T$ and an automorphism of the tree preserving $\mu$ that carries one node on the other. According to the above definition, the condition on feasibility of rendezvous can be reformulated as follows: a pair of agents can solve the rendezvous problem in a tree, if and only if their initial positions are not perfectly symmetrizable. Consequently, throughout [27], the authors consider only non perfectly symmetrizable initial positions of the agents.

It is first shown that the minimum size of memory of the agents that can solve the rendezvous problem in the class of trees with at most $n$ nodes is $\Theta(\log n)$. A rendezvous algorithm for arbitrary delay $\tau$, that uses only a logarithmic number of memory bits follows, e.g., from [12]. It is observed in [27] that $\Omega(\log n)$ is also a lower bound on the number of bits of memory that permit the agents to meet in all trees of size linear in $n$. Due to this lower bound, a *universal* pair of finite agents achieving rendezvous in the class of all trees cannot exist. However, the lower bound uses a counterexample of a tree with maximum degree linear in the size of the tree. Hence, it is natural to ask if there exists a pair of finite agents solving the rendezvous problem in all trees of *bounded* degree. The authors give a negative answer to this question. In fact they show that, for any pair of identical finite agents, there is a line for which these agents cannot solve the rendezvous problem, even with simultaneous start. As a function of the size of the trees, this impossibility result indicates a lower bound $\Omega(\log \log n)$ bits on the memory size for rendezvous in bounded degree trees of at most $n$ nodes.

The main topic of [27] is the impact of the delay between startup times of agents on the minimum size of memory permitting rendezvous. The authors show that if this delay is arbitrary, then the lower bound on memory required for rendezvous is $\Omega(\log n)$ bits, even for the line of length $n$. This lower bound matches the upper bound from [12], which shows that the minimum size of memory of the agents that can solve the rendezvous problem in the class of *bounded degree* trees with at most $n$ nodes is $\Theta(\log n)$. By contrast, for simultaneous start, they show that the amount of memory needed for rendezvous depends on two parameters of the tree: the number $n$ of nodes and the number $\ell$ of leaves. Indeed, they construct two identical agents with $O(\log \ell + \log \log n)$ bits of memory that solve the rendezvous problem in all trees with $n$ nodes and $\ell$ leaves. For the class of trees with $O(\log n)$ leaves, this proves an exponential gap in minimum memory size of the agents permitting them to meet, between the scenario with arbitrary delay and with delay zero.

Moreover, it is shown in [27] that the size $\Theta(\log \ell + \log \log n)$ of memory used to solve the rendezvous problem with simultaneous start in trees with at most $n$ nodes and at most $\ell$ leaves is optimal, even in the class of trees with degrees bounded by 3. More precisely, for infinitely many integers $\ell$, the authors show a class of arbitrarily large trees with maximum degree 3 and with $\ell$ leaves, for which rendezvous with simultaneous start requires $\Omega(\log \ell)$ bits of memory. This lower bound, together with the previously mentioned lower bound $\Omega(\log \log n)$ on the number of bits of memory required to meet with simultaneous start in the line of length $n$, implies that the upper bound $O(\log \ell + \log \log n)$ cannot be improved even for trees with maximum degree at most 3.

Trade-offs between the size of memory and the time of rendezvous in trees by identical agents are investigated in [13]. The authors consider trees with a given port labeling and assume that there is no port-preserving automorphism of the tree that carries the initial position of one agent to that of the other (otherwise rendezvous with simultaneous start is impossible). The main result of the paper is a tight trade-off between optimal time of rendezvous and the size of memory of the agents. For agents with $k$ memory bits, it is shown that optimal rendezvous time is $\Theta(n + n^2/k)$ in $n$-node trees. More precisely, if $k \geq c \log n$, for some constant $c$, the authors construct agents accomplishing rendezvous in arbitrary trees of unknown size $n$ in time $O(n+n^2/k)$, starting with arbitrary delay. They also show that no pair of agents can accomplish rendezvous in time $o(n+n^2/k)$, even in the class of lines and even with simultaneous start.

Trade-offs between the size of memory of the agents and the time of rendezvous in trees are investigated in [3] in a slightly different scenario: the difference is in the definition of rendezvous. The authors consider the rendezvous problem of any number of anonymous agents. To handle the case of symmetric trees they weaken the rendezvous requirements: agents have to meet at one node if the tree is not symmetric, and at two neighboring nodes otherwise. In this latter case, some of the agents may finish the algorithm in one of the two nodes and other agents in the other node. The authors observe that $\Omega(n)$ is a lower bound on the time of rendezvous in the class of $n$-node trees and show that any algorithm achieving rendezvous in optimal (i.e., $O(n)$) time must use $\Omega(n)$ bits of memory for each agent. Then they show a rendezvous algorithm that uses $O(n)$ time and $O(n)$ bits of memory per agent. Finally they design a polynomial time algorithm using $O(\log n)$ bits of memory per agent. An additional feature of the algorithms from [3] is that they can also work in an asynchronous setting: each agent independently identifies the node or one of the two nodes where meeting should occur, it reaches this node and stops.

While [27] solves the problem of minimum memory size needed for rendezvous in trees, the same problem for the class of arbitrary graphs is solved in [12]. The authors establish the minimum size of the memory of agents that guarantees deterministic rendezvous when it is feasible. They show that this minimum size is $\Theta(\log n)$, where $n$ is the size of the graph, regardless of the delay between the startup rounds of the agents. More precisely, the authors construct identical agents equipped with $\Theta(\log n)$ memory bits that solve the rendezvous problem in all graphs with at most $n$ nodes, when starting with any delay, and they prove a matching lower bound $\Omega(\log n)$ on the number of memory bits needed to achieve rendezvous, even for simultaneous start. In fact, this lower bound is valid already on the class of rings.

The positive result from [12] is based on a result from [38] which implies that a (usually non-simple) path traversing all nodes can be computed (node by node) in memory $O(\log n)$, for any graph with at most $n$ nodes. Moreover, logarithmic memory suffices to walk back and forth on this path. More precisely the result from [38] states that, for any positive integer $n$, there exists a UXS $Y(n) = (a_1, a_2, \ldots, a_M)$ for the class $\mathcal{G}_n$ of all graphs with at most $n$ nodes, such that $M$ is polynomial in $n$, and for any $i \leq M$, the integer $a_i$ can be constructed using $O(\log n)$ bits of memory.

At a high level, the idea of the algorithm from [12] is the following. The authors introduce the notion of the *signature* $S(u)$ of node $u$ corresponding to a given UXS. This is the sequence of entry and exit ports which are traversed by an application of the UXS at $u$. They show that if $Y$ is a UXS for the class $\mathcal{G}_{n^2+n}$ of all graphs with at most $n^2+n$ nodes, and $S(u)$ denotes the signature of $u$ in a $n$-node graph $G$, corresponding to the UXS $Y$, then for any nodes $v$, $w$ of $G$, $\mathcal{V}(v) \neq \mathcal{V}(w)$ is equivalent to $S(v) \neq S(w)$. Thus, the sequence $S(u)$ can be treated as a compact representation of the view $\mathcal{V}(u)$. Then the authors show that, using logarithmic memory, it is possible to further compress $S(u)$ to a positive integer value of at most $n$ in such a way that different signatures correspond to different values. This numerical value is then used as the label of the agent. Hence agents starting from non-symmetric initial positions get different labels. Once the agents' anonymity is broken, the rest of the meeting procedure is performed in the usual way, by dividing time into segments corresponding to activity/passivity phases. Time segments are long enough to perform a complete exploration of $G$ (using a UXS that can be computed in logarithmic memory). An agent explores $G$ in a single phase allowed for its label and waits in the remaining phases. Hence an agent performing its exploration phase must meet any agent of different label which is inert during this phase.

### 3.2 Asynchronous Rendezvous

In the asynchronous scenario agents no longer perform their moves in synchronized steps. While the agent chooses the adjacent node to which it wants to go, the time at which this move is executed and a possibly varying speed are chosen by an adversary, which considerably complicates rendezvous. It is easy to see that, even in the two-node graph, meeting at a node cannot be guaranteed under this scenario, hence the rendezvous requirement is relaxed by demanding only that meeting occur either at a node or inside an edge. Since meetings inside an edge are allowed, unwanted crossings of edges have to be avoided. Thus, for the asynchronous scenario, an embedding of the underlying graph in the three-dimensional Euclidean space is considered, with nodes of the graph being points of the space and edges being pairwise disjoint line segments joining them. For any graph such an embedding exists. Agents are modeled as points moving inside this embedding.

At any currently visited node, an agent executing a rendezvous algorithm chooses a port number at this node, corresponding to the edge that the agent wants to traverse. However, the way of traversing this edge is decided by the adversary, capturing the asynchronous characteristics of the rendezvous process. When the agent, situated at a node $v$ at time $t_0$, has to traverse an edge modeled as a segment $[v, w]$, the adversary performs the following choice. It selects a time point $t_1 > t_0$ and any continuous function $f : [t_0, t_1] \longrightarrow [v, w]$, with $f(t_0) = v$ and $f(t_1) = w$. This function models the actual movement of the agent inside the line segment $[v, w]$ in the time period $[t_0, t_1]$. Hence this movement can be at arbitrary speed, the agent may be even forced by the adversary to go back and forth, as long as it does not leave the segment and the movement is continuous.

We say that at time $t \in [t_0, t_1]$ the agent is in point $f(t) \in [v, w]$. Moreover, the adversary chooses the starting time of the agent. Hence an agent's trajectory is represented by the concatenation of the functions chosen by the adversary for consecutive edges that the agent traverses. Recall that the choice of the edge incident to a current node is determined by the choice of the port number, belonging to the agent.

For a given algorithm, given starting nodes of agents and a given sequence of adversarial decisions in an embedding of a graph $G$, a rendezvous occurs, if both agents are at the same point of the embedding at the same time. Rendezvous is *feasible* in a given graph, if there exists an algorithm for agents such that for any embedding of the graph, any (adversarial) choice of two distinct labels of agents, any starting nodes and any sequences of adversarial decisions, the rendezvous does occur. The *cost* of rendezvous is defined as the worst-case number of edge traversals by both agents (the last partial traversal counted as a complete one for both agents), where the worst case is taken over all decisions of the adversary.

**Labeled Agents.** The above asynchronous model was introduced in [17] where the authors consider labeled agents. They study asynchronous rendezvous in the infinite line, in the ring and in arbitrary connected graphs, both in the case when the initial instance $D$ between the agents is known, and when it is unknown. In the first two cases they propose several algorithms and analyze their cost. In one situation, for rendezvous in a ring of known size $n$ (but unknown $D$) they propose an algorithm of cost $O(nl)$, where $l$ is the logarithm of the smaller label. This cost is optimal. The cost of their rendezvous algorithms in the infinite line has been subsequently improved in [40]. The rendezvous algorithms from [17] for the infinite line and for the ring are based on the following idea: first transform the label $L$ of the agent in an appropriate way, and then *execute* the transformed label by making some prescribed moves, if the current bit of it is 0 and making symmetric moves otherwise.

However, from the hindsight, the most influential part of [17] was that concerning rendezvous in arbitrary graphs. Here the authors tackle the question of feasibility. They prove that rendezvous is feasible, if an upper bound on the size of the graph is known. As an open problem, the authors state the question if asynchronous deterministic rendezvous is feasible in arbitrary graphs of unknown size. The solution from [17] heavily uses the knowledge of the upper bound on the size.

The general problem of feasibility of asynchronous rendezvous for arbitrary graphs is solved in [15]. The authors propose an algorithm that accomplishes asynchronous rendezvous in any connected countable (finite or infinite) graph, for arbitrary starting nodes. A consequence of this result is a strong positive answer to the above mentioned open problem from [17]: not only is rendezvous always possible, without the knowledge of any upper bound on the size of a finite (connected) graph, but it is also possible for all infinite (countable and connected) graphs.

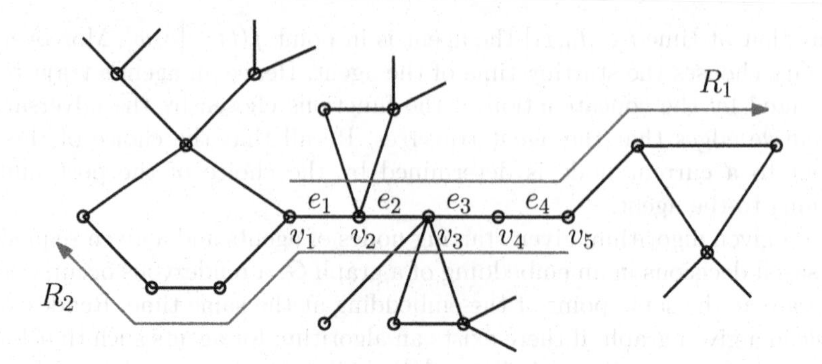


**Fig. 2.** Routes $R_1$ and $R_2$ form a tunnel

The rendezvous algorithm from [15] is based on the notion of a *tunnel.* Consider any graph $G$ and two routes $R_1$ and $R_2$ starting at nodes $v$ and $w$, respectively. (These are sequences of edges, such that consecutive edges in the sequence are incident.) We say that these routes form a tunnel (cf. Fig. 2), if there exists a prefix $[e_1, e_2, \ldots, e_n]$ of route $R_1$ and a prefix $[e_n, e_{n-1}, \ldots, e_1]$ of route $R_2$, for some edges $e_i$ in the graph, such that $e_i = \{v_i, v_{i+1}\}$, where $v_1 = v$ and $v_{n+1} = w$. Intuitively, the route $R_1$ has a prefix $P$ ending at $w$ and the route $R_2$ has a prefix which is the reverse of $P$, ending at $v$. It is proved in [15] that if routes $R_1$ and $R_2$ form a tunnel, then rendezvous is guaranteed, regardless of the decisions of the adversary.

A high-level idea of the algorithm from [15] is to force the routes of any two agents with different labels to form a tunnel, for every possible combination of starting nodes and (distinct) labels of the two agents. This is done by enumerating all quadruples $(i, j, s', s'')$, where $i$ and $j$ are different positive integers and $s'$, $s''$ are finite sequences of natural numbers, and arranging them in one countably infinite sequence. This enumeration is part of the algorithm and is the same for all agents. Then each agent processes quadruples $(i, j, s', s'')$ in the order of their enumeration. Any starting configuration of agent with label $i$ placed at node $v$ and of agent with label $j$ placed at node $w$ by the adversary corresponds to a quadruple $(i, j, s', s'')$, where $s'$ is a sequence of port numbers coding a path from $v$ to $w$ and $s''$ is a sequence of port numbers coding the reverse path from $w$ to $v$. During the processing of a quadruple $(i, j, s', s'')$, a suffix is added to the already constructed initial segment of the route of agents with label $i$ or $j$, in such a way that if labels and initial positions of agents correspond to this quadruple, then the routes of the agents form a tunnel. Since for some quadruple this condition must hold, arbitrary agents placed at arbitrary initial positions in the graph must eventually meet.

The cost of the algorithm depends on the enumeration of the quadruples, and more precisely on the position (in this enumeration) of the quadruple corresponding to the initial configuration of the agents. During each phase of the algorithm, the length of the routes of the two agents corresponding to the currently processed quadruple, is at least doubled. Hence, the complexity of the algorithm is at least exponential in terms of the number of quadruples with the same labels as those of the two agents, that are before the quadruple corresponding to the initial configuration of the agents in the enumeration. Thus the authors conclude their paper with the following natural question:

Does there exist a deterministic asynchronous rendezvous algorithm, working for all unknown connected finite graphs, with complexity polynomial in the labels of the agents and in the size of the graph?

This question is answered in [23] where the authors propose a deterministic asynchronous rendezvous algorithm, working for all unknown connected finite graphs, with cost polynomial in the size of the graph and *in the logarithm* of the smaller label.

The high-level idea of the algorithm from [23] is based on the following observation. If one agent follows a trajectory traversing all edges of the graph during some time segment, then it must either meet the other agent or this other agent must perform at least one complete edge traversal during this time segment, i.e., it must make *progress*. A naive use of this observation leads to the following simple algorithm (which is similar to that from [17]). Let $R(n,v)$ be a trajectory starting at $v$ and traversing all edges of the graph of size at most $n$, and let $\overline{R(n,v)}$ be the reverse trajectory. $R(n,v)$ can be, e.g., based on Reingold's [38] exploration. An agent with label $L$ starting at node $v$ of a graph of size $n$ follows the trajectory $(R(n,v)\overline{R(n,v)})^{(2P(n)+1)^L}$, where $P(n)$ is an upper bound on the length of $R(n,v)$, and stops. Indeed, in this case the number of trajectories $R(n,v)\overline{R(n,v)}$ (that traverse all edges of the graph) performed by the larger agent (i.e., the agent with the larger label) is larger than the number of edge traversals by the smaller agent and consequently, if they have not met before, the larger agent must meet the smaller one after the smaller agent stops, because the larger agent will still perform at least one entire trajectory afterwards. However, this simple algorithm has two major drawbacks. First, it requires knowledge of $n$ (or of an upper bound on it) and second, it is exponential in $L$, while the goal is an algorithm *polylogarithmic* in $L$. Hence the above observation has to be used in a much more subtle way. As the authors of [23] say, their algorithm "constructs a trajectory for each agent, polynomial in the size of the graph and polylogarithmic in the shorter label, i.e., polynomial in its length, which has the following *synchronization* property that holds in a graph of arbitrary unknown size. When one of the agents has already followed some part of its trajectory, it has either met the other agent, or this other agent must have completed some other related part of its trajectory. (In a way, if the meeting has not yet occurred, the other agent has been "pushed" to execute some part of its route.) The trajectories are designed in such a way that, unless a meeting has already occurred, the agents are forced to follow in the same time interval such parts of their trajectories that meeting is inevitable. A design satisfying this synchronization property is difficult due to the arbitrary behavior of the adversary and is the main technical challenge of the paper."

The aim of [21] is to investigate the difference of cost between the synchronous and asynchronous versions of a task executed by mobile agents. The authors show that for some natural task executed by mobile agents in a network, the optimal cost of its deterministic solution in the asynchronous setting has higher order of magnitude than that in the synchronous scenario. The task for which this difference is proved is rendezvous of two agents in an infinite oriented grid. They consider two agents starting at a known distance $D$ in the infinite oriented grid. Agents do not have any global system of coordinates. They have to meet in a node or inside an edge of the grid, and the cost of a rendezvous algorithm is the number of edge traversals by both agents until the meeting. It is proved that in the synchronous scenario rendezvous can be performed at cost $O(D\ell)$, where $\ell$ is the length of the binary representation of the smaller label, while cost $\Omega(D^2 + D\ell)$ is needed for asynchronous completion of rendezvous. Hence, for instances with $\ell = o(D)$, the optimal cost of asynchronous rendezvous is asymptotically larger than that of synchronous rendezvous.

**Anonymous Agents.** The papers [4,10] were among the first to consider asynchronous rendezvous of anonymous agents in graphs. The authors concentrate on particular graphs and use strong additional assumptions. They consider rendezvous in an infinite two-dimensional grid, where ports are consistently labeled $N, E, S, W$, and agents know their initial coordinates in the grid, with respect to a common system of coordinates. Hence, it is even problematic if such agents can be called anonymous (identical), as they are right away differentiated by their different initial coordinates. It is proved in [10] that under these assumptions asynchronous rendezvous can be accomplished at cost $O(d^{2+\epsilon})$, where $d$ is the initial distance between the agents in the grid, and $\epsilon$ is an arbitrary positive real. This result has been generalized and strengthened in [4], under the same assumptions. The authors show an asynchronous rendezvous algorithm working for $\delta$-dimensional infinite grids with cost $O(d^{\delta} polylog(d))$. This complexity is close to optimal, as $\Omega(d^{\delta})$ is a lower bound on the cost of any asynchronous rendezvous algorithm in this setting.

The problem of feasibility of asynchronous rendezvous of anonymous agents in arbitrary graphs is solved in [28]. The authors show that rendezvous is possible if and only if the views from the initial positions of the agents are different, or these positions are connected by a path whose corresponding sequence of port numbers is a palindrome. The authors provide an algorithm guaranteeing deterministic asynchronous rendezvous from all such initial positions in an arbitrary connected graph that is either finite of arbitrary unknown size, or (countably) infinite.

The algorithm is based on the idea of creating a tunnel, similarly as in [15]. However, the main difficulty in designing the algorithm in the present setting is that, as opposed to [15], agents do not have distinct labels allowing them to break symmetry. Hence symmetry can be broken only by inspecting the views of the agents, if these views are different. Even when they are different, the agents cannot know how deeply their views have to be explored to find the first difference. Thus the algorithm proceeds in epochs: in each consecutive epoch each agent explores its view more deeply, and creates a code of this truncated view, subsequently treating it as its temporary label and applying the procedure from [15] to a restricted list of quadruples. If views are different, a tunnel will be eventually created after an epoch with sufficiently high index because in this epoch the truncated views serving as temporary labels of the agents will be different, and the argument from [15] will work. The algorithm in [28] has an additional feature permitting the creation of a tunnel when views of the agents are the same but their initial positions are joined by a path which is a palindrome. The simplest example of such a situation is the two-node graph with agents starting at extremities of the single edge. Views of the agents are the same and the code of the unique simple path joining the initial positions of the agents is the palindrome (00).

## 4 Approach in the Plane

In this section we study the problem of approach in the plane or in terrains that are subsets of the plane: agents modeled as points moving in the terrain and starting at distinct points of it have to get at distance at most 1 of each other. Throughout the section we assume that agents have labels that are different positive integers. Each agent is equipped with a compass and a unit of length.

The problem of approach in the plane of agents that have coherent compasses and the same unit of length can be reduced to the problem of rendezvous in an infinite oriented grid, where rendezvous is defined as getting at the same time to the same node or the same point of some edge of the grid. This means that if rendezvous in the infinite oriented grid can be solved under some set of assumptions about the agents, then the problem of approach in the empty plane can be solved under analogous assumptions.

The problem of feasibility of asynchronous rendezvous in the plane or terrains is solved in [15]. Consider any terrain (bounded or unbounded) that is a (topologically) closed subset of the plane with path-connected interior. (The latter means that for any two interior points of the terrain there exists a path, all of whose points are interior points of the terrain, connecting them). Agents start at arbitrary interior points of the terrain and their trajectories can be any polygonal lines. The authors propose an algorithm that accomplishes approach of any such agents in the terrain in finite time. Compasses and units of length of the agents may even be different.

While the algorithm from [15] works in a very general setting, its drawback is the complexity: similarly as for the rendezvous algorithm for graphs proposed in this paper, the cost of the algorithm for terrains cannot be controlled. In order to find a more efficient algorithm, the authors of [20] consider a scenario of restricted asynchrony. Agents have coherent compasses and the same measure of length and of time, but they are assigned arbitrary, possibly different velocities by an adversary. An agent can stay inert for a chosen amount of time, or it can move in a chosen direction and distance at its assigned speed. Under these restrictions the authors propose an algorithm accomplishing approach in the plane in time polynomial in the unknown initial distance between the agents, in the logarithm of the smaller label and in the inverse of the larger speed. The distance travelled by each agent until approach is polynomial in the first two parameters and does not depend on the third.

The problem of tractable approach in the plane under full asynchrony has been finally solved in [6]. The authors propose an algorithm accomplishing approach in the plane for agents whose speed may vary adversarially. The cost of the algorithm is polynomial in the initial distance between the agents and in the logarithm of the smaller label.

It should be mentioned that the result from [4] concerning asynchronous rendezvous in the infinite oriented two dimensional grid, accomplished at cost $O(d^2 polylog(d))$, where $d$ is the initial distance between the agents in the grid, carries over to the task of approach in the plane under similar strong assumptions as in [4]: agents have coherent compasses and the same unit of length and they know their initial coordinates in the plane, with respect to a common system of coordinates. Similarly as for the grid, this complexity is close to optimal, due to the lower bound $\Omega(d^2)$.

In [25], the authors consider the problem of approach in the plane under the synchronous model. Agents are equipped with coherent compasses and the same unit of length, and have synchronized clocks. They make a series of moves. Each move specifies the direction and the duration of moving. In a null move an agent stays inert for some time, or forever. In a non-null move agents travel at the same constant speed, normalized to 1.

The twist of the model in this paper is restricted feedback that the agents get after each move, that is similar in spirit to the model of distance-aware agents from [16], but weaker. It is assumed that agents have sensors enabling them to estimate the distance from the other agent, but not the direction towards it. The authors consider two models of estimation. In both models an agent reads its sensor at the moment of its appearance in the plane and then at the end of each move. This reading (together with the previous ones) determines the decision concerning the next move. In both models the reading of the sensor tells the agent if the other agent is already present. Moreover, in the *monotone model*, each agent can determine, for any two readings in moments $t_1$ and $t_2$, whether the distance from the other agent at time $t_1$ was smaller, equal or larger than at time $t_2$. It does not, however, get the value of this distance. In the weaker *binary model*, each agent can find out, at any reading, whether it is at distance less than $\rho$ or at distance at least $\rho$ from the other agent, for some real $\rho > 1$ unknown to them. To motivate their model, the authors mention that such distance estimation can be implemented, e.g., using chemical sensors. Each agent emits some chemical substance (scent), and the sensor of the other agent detects it, i.e., the other agent *sniffs*. The intensity of the scent decreases with the distance. In the monotone model it is assumed that the sensors of the agents are very precise and can measure any change of intensity. In the binary model it is only assumed that the sensors can detect the scent below some distance (without being able to measure intensity or its changes) above which the density of the chemical is too weak to be detected.

The authors investigate how the two ways of sensing influence the cost of meeting, defined as the total distance travelled by both agents until the meeting. For the monotone model they present an algorithm achieving meeting in time $O(D)$, where $D$ is the initial distance between the agents. This complexity is of course optimal. For the binary model they show that, if agents start at a distance smaller than $\rho$ (i.e., when they can sense each other initially) then meeting can be guaranteed at cost $O(\rho \log \lambda)$, where $\lambda$ is the larger label, and that this cost cannot be improved in general. It is also observed that, if agents start at distance $\alpha\rho$, for some constant $\alpha > 1$ in the binary model, then sniffing does not help, i.e., the worst-case optimal meeting cost is of the same order of magnitude as without any sniffing ability.

In [14], the authors consider both the task of approach and that of (exact) rendezvous of two agents in a terrain. Exact rendezvous (getting to the same point of the terrain at the same time) is possible because the terrain is a polygon with polygonal holes, and hence exact meeting can take place at the boundary of the terrain or of a hole. Movements of the agents are asynchronous and agents have bounded memory: they are modeled as finite automata. The authors compare the feasibility of the task of rendezvous to that of approach for anonymous and for labeled agents. This gives rise to four scenarios, and the authors show classes of polygonal terrains which distinguish all pairs of them from the point of view of feasibility of rendezvous. The characteristics of the terrain that influence the feasibility of rendezvous and of approach include symmetries of the terrain, boundedness of its diameter, and the total number of vertices of polygons in the terrain.

## 5 Conclusion

In this chapter we surveyed algorithmic results concerning deterministic rendezvous in networks and deterministic approach in terrains of the plane. The appearance of many new results in the last few years is an indication of how vibrant is this domain of distributed computing. On the other hand, hopefully we managed to show that our understanding of rendezvous problems is still very incomplete, and a lot remains to be done. We would like to conclude the chapter by pointing out several avenues of research that this author finds promising. This choice of the open problems is very subjective and reflects the personal taste of the author, rather than their importance on some hypothetical objective scale.

It seems reasonable to classify possible open problems into two categories. The first category is strengthening of the existing results without changing the model under which they were originally obtained. Here the most interesting problems seem those aiming at improving the efficiency of existing algorithms and ultimately finding an algorithm of optimal complexity. In this category we would put forward the problem of finding:

- an optimal-time synchronous rendezvous algorithm in arbitrary graphs
- an optimal-cost asynchronous rendezvous algorithm in arbitrary graphs
- an optimal-cost asynchronous approach algorithm in the plane.

All these problems are formulated for labeled agents. The third problem is for agents with coherent compasses and the same unit of length. As we know, for all these problems polynomial algorithms are known, but, especially in the case of problems 2 and 3, the exponents of the polynomials are very large. Finding algorithms of optimal complexity for any of these scenarios seems to be very challenging. We think that even a significant improvement of the existing algorithms would be a big step forward.

The second category concerns investigating rendezvous under new models. It seems that the interplay between efficiency of rendezvous or approach and the communication capabilities of agents is still poorly understood. This problem has been "touched" in papers [16,24,25] but a more complete analysis of tradeoffs between communication of agents and efficiency of rendezvous under various scenarios is badly needed. A realistic type of communication, especially for agents in the terrains, seems to be wireless. This can be challenging, especially for large teams of agents, as usual problems of wireless communication concerning collisions would have to be tackled.

Another interesting issue are trade-offs between memory of the agents and their sensory capabilities. In this chapter we assumed that agents cannot see other agents prior to meeting but usually they have significant or even unbounded memory. By contrast, in Chap. 4, it is usually assumed that agents cannot remember any information from the previous Look-Compute-Move cycles, but they can take a snapshot of the entire network (or a large part of it) during the Look action. Most applications are probably in between those two extremes: agents have some memory of past events (for example of constant size), but their sensory capabilities are more limited, e.g., they can only perceive the part of the configuration at a given radius from their current position, they cannot see "through" other agents, etc. Studying feasibility of rendezvous under such more balanced scenarios could involve characterizing initial configurations for which rendezvous (or approach in the plane) is possible, and trying to optimize the cost of meeting, which, under very small memory of the agents, is often still unknown.

## References


1. Alpern, S.: Rendezvous search: a personal perspective. Oper. Res. **50**, 772–795 (2002)
2. Alpern, S., Gal, S.: The Theory of Search Games and Rendezvous. International Series in Operations Research and Management Science. Kluwer Academic Publishers, Norwell (2003)
3. Baba, D., Izumi, T., Ooshita, F., Kakugawa, H., Masuzawa, T.: Linear time and space gathering of anonymous mobile agents in asynchronous trees. Theor. Comput. Sci. **478**, 118–126 (2013)
4. Bampas, E., Czyzowicz, J., Gąsieniec, L., Ilcinkas, D., Labourel, A.: Almost optimal asynchronous rendezvous in infinite multidimensional grids. In: Lynch, N.A., Shvartsman, A.A. (eds.) DISC 2010. LNCS, vol. 6343, pp. 297–311. Springer, Heidelberg (2010). https://doi.org/10.1007/978-3-642-15763-9_28

5. Bouchard, S., Dieudonné, Y., Ducourthial, B.: Byzantine gathering in networks. Distrib. Comput. **29**, 435–457 (2016)
6. Bouchard, S., Bournat, M., Dieudonné, Y., Dubois, S., Petit, F.: Asynchronous approach in the plane: a deterministic polynomial algorithm. In: Proceedings of 31st International Symposium on Distributed Computing (DISC 2017), pp. 8:1–8:16 (2017)
7. Bouchard, S., Dieudonné, Y., Lamani, A.: Byzantine gathering in polynomial time. In: Proceedings of 45th International Colloquium on Automata, Languages, and Programming (ICALP 2018), pp. 147:1–147:15 (2018)
8. Bouchard, S., Dieudonné, Y., Pelc, A., Petit, F.: Deterministic rendezvous at a node of agents with arbitrary velocities. Inf. Process. Lett. **133**, 39–43 (2018)
9. Chalopin, J., Dieudonné, Y., Labourel, A., Pelc, A.: Rendezvous in networks in spite of delay faults. Distrib. Comput. **29**, 187–205 (2016)
10. Collins, A., Czyzowicz, J., Gąsieniec, L., Labourel, A.: Tell me where i am so i can meet you sooner. In: Abramsky, S., Gavoille, C., Kirchner, C., Meyer auf der Heide, F., Spirakis, P.G. (eds.) ICALP 2010. LNCS, vol. 6199, pp. 502–514. Springer, Heidelberg (2010). https://doi.org/10.1007/978-3-642-14162-1_42
11. Cornejo, A., Kuhn, F.: Deploying wireless networks with beeps. In: Lynch, N.A., Shvartsman, A.A. (eds.) DISC 2010. LNCS, vol. 6343, pp. 148–162. Springer, Heidelberg (2010). https://doi.org/10.1007/978-3-642-15763-9_15
12. Czyzowicz, J., Kosowski, A., Pelc, A.: How to meet when you forget: log-space rendezvous in arbitrary graphs. Distrib. Comput. **25**, 165–178 (2012)
13. Czyzowicz, J., Kosowski, A., Pelc, A.: Time vs. space trade-offs for rendezvous in trees. Distrib. Comput. **27**, 95–109 (2014)
14. Czyzowicz, J., Kosowski, A., Pelc, A.: Deterministic rendezvous of asynchronous bounded-memory agents in polygonal terrains. Theory Comput. Syst. **52**, 179–199 (2013)
15. Czyzowicz, J., Labourel, A., Pelc, A.: How to meet asynchronously (almost) everywhere. ACM Trans. Algorithms **8**, 37:1–37:14 (2012)
16. Das, S., Dereniowski, D., Kosowski, A., Uznański, P.: Rendezvous of distance-aware mobile agents in unknown graphs. In: Halldórsson, M.M. (ed.) SIROCCO 2014. LNCS, vol. 8576, pp. 295–310. Springer, Cham (2014). https://doi.org/10.1007/978-3-319-09620-9_23
17. De Marco, G., Gargano, L., Kranakis, E., Krizanc, D., Pelc, A., Vaccaro, U.: Asynchronous deterministic rendezvous in graphs. Theoret. Comput. Sci. **355**, 315–326 (2006)
18. Dessmark, A., Fraigniaud, P., Kowalski, D., Pelc, A.: Deterministic rendezvous in graphs. Algorithmica **46**, 69–96 (2006)
19. Dieudonné, Y., Pelc, A.: Anonymous meeting in networks. Algorithmica **74**, 908–946 (2016)
20. Dieudonné, Y., Pelc, A.: Deterministic polynomial approach in the plane. Distrib. Comput. **28**, 111–129 (2015)
21. Dieudonné, Y., Pelc, A.: Price of asynchrony in mobile agents computing. Theoret. Comput. Sci. **524**, 59–67 (2014)
22. Dieudonné, Y., Pelc, A., Peleg, D.: Gathering despite mischief. ACM Trans. Algorithms **11**, 1:1–1:28 (2014)
23. Dieudonné, Y., Pelc, A., Villain, V.: How to meet asynchronously at polynomial cost. SIAM J. Comput. **44**, 844–867 (2015)
24. Elouasbi, S., Pelc, A.: Deterministic rendezvous with detection using beeps. Int. J. Found. Comput. Sci. **28**, 77–97 (2017)

25. Elouasbi, S., Pelc, A.: Deterministic meeting of sniffing agents in the plane. Fundam. Informaticae **160**, 281–301 (2018)
26. Elouasbi, S., Pelc, A.: Time of anonymous rendezvous in trees: determinism vs. randomization. In: Even, G., Halldórsson, M.M. (eds.) SIROCCO 2012. LNCS, vol. 7355, pp. 291–302. Springer, Heidelberg (2012). https://doi.org/10.1007/978-3-642-31104-8_25
27. Fraigniaud, P., Pelc, A.: Delays induce an exponential memory gap for rendezvous in trees. ACM Trans. Algorithms **9**, 17:1–17:24 (2013)
28. Guilbault, S., Pelc, A.: Asynchronous rendezvous of anonymous agents in arbitrary graphs. In: Fernàndez Anta, A., Lipari, G., Roy, M. (eds.) OPODIS 2011. LNCS, vol. 7109, pp. 421–434. Springer, Heidelberg (2011). https://doi.org/10.1007/978-3-642-25873-2_29
29. Koucký, M.: Universal traversal sequences with backtracking. J. Comput. Syst. Sci. **65**, 717–726 (2002)
30. Kowalski, D., Malinowski, A.: How to meet in anonymous network. Theoret. Comput. Sci. **399**, 141–156 (2008)
31. Kranakis, E., Krizanc, D., Markou, E.: The Mobile Agent Rendezvous Problem in the Ring. Morgan and Claypool Publishers, San Rafael (2010)
32. Kranakis, E., Krizanc, D., Rajsbaum, S.: Mobile agent rendezvous: a survey. In: Flocchini, P., Gąsieniec, L. (eds.) SIROCCO 2006. LNCS, vol. 4056, pp. 1–9. Springer, Heidelberg (2006). https://doi.org/10.1007/11780823_1
33. Miller, A., Pelc, A.: Time versus cost tradeoffs for deterministic rendezvous in networks. Distrib. Comput. **29**, 51–64 (2016)
34. Miller, A., Pelc, A.: Fast rendezvous with advice. Theoret. Comput. Sci. **608**, 190–198 (2015)
35. Ooshita, F., Datta, A.K., Masuzawa, T.: Self-stabilizing rendezvous of synchronous mobile agents in graphs. In: Spirakis, P., Tsigas, P. (eds.) SSS 2017. LNCS, vol. 10616, pp. 18–32. Springer, Cham (2017). https://doi.org/10.1007/978-3-319-69084-1_2
36. Pelc, A.: Deterministic rendezvous in networks: a comprehensive survey. Networks **59**, 331–347 (2012)
37. Pelc, A.: Deterministic gathering with crash faults, CoRR abs/1704.08880 (2017)
38. Reingold, O.: Undirected connectivity in log-space. J. ACM **55**, 1–24 (2008)
39. Schelling, T.: The Strategy of Conflict. Oxford University Press, Oxford (1960)
40. Stachowiak, G.: Asynchronous deterministic rendezvous on the line. In: Nielsen, M., Kučera, A., Miltersen, P.B., Palamidessi, C., Tůma, P., Valencia, F. (eds.) SOFSEM 2009. LNCS, vol. 5404, pp. 497–508. Springer, Heidelberg (2009). https://doi.org/10.1007/978-3-540-95891-8_45
41. Ta-Shma, A., Zwick, U.: Deterministic rendezvous, treasure hunts and strongly universal exploration sequences. ACM Trans. Algorithms **10**, 12:1–12:15 (2014)
42. Yamashita, M., Kameda, T.: Computing on anonymous networks: part i-characterizing the solvable cases. IEEE Trans. Parallel Distrib. Syst. **7**, 69–89 (1996)

# Dangerous Graphs

Euripides Markou[1(✉)] and Wei Shi[2]

[1] University of Thessaly, Lamia, Greece
emarkou@dib.uth.gr
[2] Carleton University, Ottawa, Canada
wei.shi@carleton.ca

**Abstract.** Anomalies and faults are inevitable in computer networks, today more than ever before. This is due to the large scale and dynamic nature of the networks used to process big data and to the ever-increasing number of ad-hoc devices. Beyond natural faults and anomalies occurring in a network, threats proceeding from attacks conducted by malicious intruders must be considered. Consequently, there is often a need to quickly isolate and even repair a fault in a network when it appears. Furthermore, despite the presence in a network of faults stemming from malicious entities, we need to identify the latter and their behaviours, and develop protocols resilient to their attacks. Thus, defining models to capture the dangers inherent to various faults, anomalies and threats in a network and studying such threats, has become increasingly important and popular.

Threats in networks can be of two kinds: either mobile or stationary. A malicious mobile process can move along the network, whereas a stationary harmful process resides in a host. One of the most studied models for stationary harmful processes is the *black hole*, which was introduced by Dobrev, Flocchini, Prencipe and Santoro in 2001. A *black hole* models a network node in which a destructive process deletes any visiting agent or incoming data upon arrival, without leaving any observable trace. Conversely, a network may face one or more malicious mobile processes infecting one or more nodes. Given both kinds of threats, a first crucial task consists in searching for and reporting as quickly as possible the location all faulty nodes while using a minimum number of mobile agents. In general, the main issue is to identify the minimal hypotheses under which faulty nodes can be found. This problem has been investigated in both *asynchronous* and *synchronous* networks. A corollary task is to make sure that the protocols designed for solving problems such as gathering and transferring data still work despite the presence of one or more faulty nodes.

In this chapter, we review the state-of-the-art of research pertaining to the presence of faulty nodes in a network. We discuss different models in synchronous and asynchronous networks and for different communication and computation capabilities of the agents. We also address relevant computational issues and present algorithmic techniques and impossibility results.


P. Flocchini et al. (Eds.): Distributed Computing by Mobile Entities, LNCS 11340, pp. 455–515, 2019.
https://doi.org/10.1007/978-3-030-11072-7_18

**Keywords:** Distributed algorithms · Fault tolerance
Black hole search · Networks · Graph exploration · Mobile agents
Computational complexity · Deterministic finite automata

## 1 Introduction

Over the past few decades, as network-based services have become prevalent, so has the need for effective diagnosis of all-too-frequent network anomalies and faults. Among these, a *black hole* is a severe and pervasive problem. A black hole models a computer that is accidentally off-line or a network site in which a resident process (e.g., an unknowingly-installed virus) deletes any visiting agents or incoming data upon their arrival without leaving any observable trace [43]. For example, in a cloud, a node that causes loss of essential data (for the system and/or its users) constitutes a black hole and *de facto* compromises the quality of any service in this cloud. Similarly, any undetectable crash failure of a site in a network transforms that site into a black hole.

In this chapter we study the *Black Hole Search* problem. We present and discuss a number of models for which the problem has been studied. We start by describing some basic ingredients of most of the models. Then we discuss the problem in *synchronous* networks. We chose to present some algorithms and impossibility results for the problem in tree topologies (since this was the first topology studied for synchronous networks) as well as arbitrary graphs. We also discuss some algorithmic techniques for *scattered finite automata* (since there are only few results for agents with limited memory and lots of open questions) on ring and torus topologies. Then we move to *asynchronous* networks. After overviewing the state of the art for black hole search problem in various asynchronous networks, we present algorithms for the problem in ring topology. Finally we briefly mention other problems in networks with black holes as well as other types of malicious behaviour and conclude with interesting open questions and future directions in the area.

### 1.1 Mobile Agents

In distributed computing, the research focus is on the computational and complexity issues of systems composed of autonomous computational entities interacting with each other to solve a problem or to perform a task. While traditionally these autonomous computational entities have been assumed to be static, recent advances in a variety of fields, ranging from robotics to networking, have motivated the community to address the situation of *mobile entities*.

A *mobile agent* is an abstract and autonomous software entity. As such, agents are versatile and robust in changing environments, and can be programmed to work in cooperative teams. Members of such teams may have different complementary specialties, or be duplicates of one another [76]. These agents usually have limited computing capabilities and bounded storage. They all obey an identical set of behavioural rules (referred to as the "protocol") and can move from a node to a neighbouring one. Also, these agents are usually anonymous (i.e., do not have distinct identifiers) and autonomous (i.e., each does its own computing and uses its own memory).

Using such agents offers several potential advantages: they can reduce network load, overcome network latency, encapsulate protocols, execute asynchronously and autonomously, and even adapt dynamically [88]. For example, black hole search may instead rely on the use of a central controller. In this case, the latter must constantly send Ping messages to nodes or, alternatively, require that each node send it periodically a message confirming this node's activity. Both of these strategies lead to heavy network traffic that can be avoided when using mobile agents for such a search.

Recently, an increasing number of investigations are being carried out on the computational and complexity issues arising in systems of mobile entities that can move in a spatial universe. Depending on the nature of the spatial universe, there are two basic settings in which mobile entities are being investigated. The first setting, called sometimes continuous, is when the universe is a region of the $2D$ (or $3D$) space. This is for example the case of robotic swarms, mobile sensor networks, mobile robotic sensors, etc. (as in, e.g., [1,37,110]). The second setting, sometimes called graph world or discrete universe, is when the universe is a simple graph; this is for example the case of mobile agents in communication networks (as in, e.g., [32,72]). In both settings, the research concern is on determining what tasks can be performed by such entities, under what conditions, and at what cost. In particular, a central question is to determine what minimal hypotheses allow a given problem to be solved.

Mobile agents in networks can be thought of as autonomous, goal-oriented software entities that can transport their state from one computational environment to the next and resume their execution in the new environment, thus remaining active as they migrate between computers. This makes them a powerful tool for implementing distributed applications in computer networks. The agents are generally modeled as automata that move on a network modeled as a graph. The first known algorithm designed for graph exploration by a mobile agent, modeled as a finite automaton, was introduced by Shannon [104] in 1951. Since then, several papers have been dedicated to the feasibility of graph exploration by one or more agents. Important properties that have been considered by researchers are as follows:

1. Whether or not the agents are distinguishable, i.e., if they have distinct identities. Anonymous agents are limited to executing precisely the same algorithm, while agents with distinct identities have the potential to execute different algorithms.
2. The size of memory an agent has (e.g., a function of the size of the network or constant memory) and the knowledge an agent has about the network it is on (e.g., a map or the size of the network) and about the other agents.
3. The method through which the agents communicate. For example they may have the ability to read the state of other agents residing at the same node. Or they can communicate via a shared memory space provided at each node

(usually called *whiteboard*), or via message passing, or by leaving indistinguishable markers at nodes or edges (often called *tokens* or *pebbles*).
4. Whether the nodes and/or edges of the network are distinguishable by an agent (as in, e.g., [54,95]) or not (as in, e.g., [14]). The outgoing edges of a node are usually thought of as distinguishable but an important distinction is made between a *globally consistent* edge-labeling (thus giving the agents the ability to navigate) versus a *locally independent* edge-labeling (in that case, even in restricted topologies like a torus, the agents cannot navigate based only on the edge labeling). If the labeling satisfies certain coding properties it is called a *sense of direction* [67].
5. How networks deal with time. In a synchronous network there exists a global clock available to all nodes. This global clock is inherited by the agents. In particular it is usually assumed that in a single step an agent arrives at a node, performs some calculation, and exits the node and that all agents are performing these tasks 'in sync'. In an asynchronous network such a global clock is not available; the speed with which an agent computes or moves between nodes, while guaranteed to be finite, is not a priori determined (as in, e.g., [98]).
6. Whether networks and/or agents may be faulty or not. For example crash or omission failures, faulty edges, Byzantine failures (where a faulty agent and/or node behaves arbitrarily and potentially maliciously) have been considered.

All the above properties have turned out to greatly effect the solvability and efficiency of solution of a number of problems in distributed computing and have been shown to be important for the study of mobile agents as well.

For a given choice of agent plus network model there are a number of important resources for which one can define a complexity measure. Measures that reflect the time and bandwidth efficiency of a given algorithm are of paramount concern. The total bandwidth consumed by an agent depends upon its size (memory) as well as the number of moves it makes during an execution of its algorithm. Generally the size (memory) of an agent is identified with the number of bits required to encode its states, i.e., it is proportional to the log base two of the number of possible states (as in, e.g., [38,73]). Depending on its memory, an agent may be able to make and/or store a map of the network, store the network size, count the number of nodes visited, etc. If the agent sends messages then the size and number of these messages must also count towards any measure of its bandwidth. Other complexity measurements of interest include the size of shared memory required at each node assuming the agents communicate via shared memory, the number of random bits used by a randomized agent and the number and kind of faults an algorithm can successfully deal with.

### 1.2 Exploration in Unsafe Networks

One of the main concerns in distributed mobile computing has to do with the security of both agents and hosts [16,24,76,79,94,102]. The case of harmful mobile agents (representing mobile viruses that infect any visited network site) has been considered in the literature. A crucial task is to decontaminate the infected network; this task is to be carried out by a team of system agents (the cleaners), able to decontaminate visited sites, preventing any reinfection of cleaned areas. This problem is equivalent to the one of capturing an intruder moving in the network. Results on this and related problems have appeared in [5,12,56,57,71,89,108].

Various methods of protecting mobile agents against malicious hosts have been discussed (as in, e.g., [77,78,93,101,102,111]). In [70] results for both harmful nodes (especially for asynchronous networks) and harmful agents are surveyed.

The exploration problem has been studied in unsafe networks which contain malicious hosts of a highly harmful nature, called *black holes*. A black hole is a node which contains a stationary process destroying all mobile agents visiting this node, without leaving any trace. In the *Black Hole Search* problem the goal for the agents is to locate the black hole within finite time. In particular, at least one agent has to survive knowing all edges leading to the black hole. The problem has been introduced by Dobrev, Flocchini, Prencipe and Santoro in 2001 [43]. The only way of locating a black hole is to have at least one agent visiting it. However, since any agent visiting a black hole is vanished without leaving any trace, the location of the black hole must be deducted by some communication mechanism employed by the agents.

In 2006, Flocchini *et al.* [69] scrutinized the black hole search problem for both asynchronous and synchronous networks. That survey also introduced the black hole search problem as a special case of exploring and mapping an unknown environment. While there exists a large body of literature on unknown graph exploration problems, it generally assumes that the underlying network graph does not contain any type of malicious entities [4]. Conversely, work on *dangerous graph search* (e.g., [23]) does address the detection and localization of malicious hosts (such as black holes), malicious agents, and faulty links. In particular, in their 2012 survey [90], Markou *et al.* discussed previous research on identifying hostile nodes. They mainly focused on synchronous special trees, arbitrary trees and arbitrary graphs, with a brief mention of asynchronous rings. More recently, Zarrad *et al.* [112] briefly discuss solutions for black hole search in synchronous and asynchronous networks, however without analyzing the underlying assumptions of these solutions. Results on the black hole search problem have also appeared in surveys [90,97].

### 1.3 Common Models

The Black Hole Search problem has been studied on a number of models. We present here the common assumptions of those models. A list of the common assumptions of those models is provided in Table 1. We will now provide a detailed explanation of each of these assumptions.

**Table 1.** Models and assumptions frequently used for black hole search

| Network synchronization | Communication model | Agent starting location | Knowledge of network |
|---|---|---|---|
| Synchronous network | Pure token | Co-located | No knowledge (e.g., unknown) |
| | Enhanced token | | Edge-labelled (e.g., sense of direction) |
| Asynchronous network | Whiteboard | Dispersed | Network topology (e.g., ring) |
| | Face-to-face | | Complete knowledge (e.g., map) |

#### 1.3.1 Network Synchronization

**Synchronous Network.** A *synchronous network* is a network in which all agents initially wake up at the same time and where it takes a quantum amount of time (called a *time unit*) for an agent to traverse a link or explore a node: All agents are thus synchronized with respect to a global clock. By the end of each time unit, an agent must decide whether to move to a neighbouring node, or stay at its current node, or terminate the algorithm. As such, the complexity of the agent's algorithm in synchronous networks can be measured in terms of the number of time units.

In synchronous networks, a *time-out mechanism* is available to enforce the time synchronization [22,29–31,83]. Such a mechanism allows us to easily identify which agents vanished in the black hole(s). Suppose a team of agents should meet at a node $u$ after $m$ time units, after this time-out, all other agents know that those that do not show up in node $u$ perished in the black hole(s).

Using such a time-out mechanism, the black hole can be located using only 2 agents in any network that has only one black hole present when a network map is available for every agent. In this case the network size is not required to guarantee a solution. For example, let 2 agents, $a$ and $b$, be at a safe node $u$. Assume agent $a$ moves to the neighbouring node $v$ and is expected to return to $u$ while agent $b$ waits at node $u$. As each move takes 1 time unit, if agent $a$ does not come back to node $u$ after 2 units, then agent $b$ knows that agent $a$ is lost and that node $v$ is the black hole. Once agent $b$ knows the location of the black hole, the algorithm can terminate immediately even if there are remaining unexplored nodes in the network.

Furthermore, with this mechanism, it is also possible to know whether or not a black hole exists. More specifically, if all $n$ nodes of the network have been explored by the end of a predefined time-out, we can conclude that there is no black hole in this network, provided that $n$ be known *a priori*. In this case, Klasing *et al.* [84] and Czyzowicz *et al.* [29] solve the black hole search problem under the assumption that there is one or no black holes in the network.

**Asynchronous Network.** Unlike for synchronous networks, there is no global clock mechanism in asynchronous networks. Thus, the agents could initially wake up at different times. Also, the time that an agent takes for every action (sleep or transit) is finite but unpredictable [69]. Therefore, it is impossible to distinguish whether an agent vanished in a black hole or is stuck in a slow link/node of the network since the latter possibility takes an unpredictable amount of time [106]. It follows that the only way to locate a black hole in an asynchronous network is to explore the *entire* network [69]. Consequently, the network size $n$ and the number of black holes $b$ must be known *a priori* in order to count the total number of explored nodes (whether for single or multiple black hole search): only when at least $n - b$ nodes are explored may the algorithm terminate.

While knowing network size $n$ is not required for solving black hole search in synchronous networks, generally it is for asynchronous ones. Also, a network may be disconnected due to the presence of a black hole. In the context of an asynchronous network, this makes it is impossible for an agent to finish exploring the entire network and terminate the algorithm. In order to bypass this roadblock, research papers that study the single black hole search problem in asynchronous networks assume that the network is bi-connected or at least that the network remains connected after removing the black hole node. In contrast, synchronous networks need not be bi-connected for single black hole search. For example, Czyzowicz *et al.* [31] study this problem in tree networks. Finally, we remark that this issue is far more complex when considering multiple black hole search.

### 1.3.2 Communication Models

Given the location of the black hole is initially unknown, regardless of network synchronization, an agent may vanish at any time while it explores. As previously mentioned, in order to systematically identify a black hole, a team of agents is used to locate the black hole. Collaboration between agents is not only necessary; it is essential. To this end, the agents are usually assumed to communicate with each other using one of the four communication models: the *pure token model*, the *enhanced token model*, the *whiteboard model*, and the *face-to-face model*. In the first three of these models, agents have no means for direct communication between themselves.

Before discussing each of these models, we remark that a crucial goal of agent communication is to minimize the number of agents that vanish in a black hole. To this end, it is assumed that at most one agent should be allowed to enter the same node at the same time via the same link. More specifically, when a port is explored for the first time, this initial exploration must involve only one agent that, before entering this port, must somehow indicate to other exploring agents that the node to which this port leads is currently under exploration and thus is to be considered dangerous until proven otherwise. Such a strategy, called *Cautious Walk*, is commonly used in black hole search algorithms for it prevents other agents from entering a node under exploration via the same link. It was first introduced by Dobrev *et al.* [43] to minimize the number of agents that vanish in the black hole. It typically requires that a node be conceptualized as having *ports.* A port can be classified as (a) *unexplored* - no agent has ever passed through this port, or (b) *dangerous* - an agent left via this port but no agent has returned through it, or (c) *safe* - an agent has left and returned through this port. How the status of a port is captured differs between communication models. Regardless, Cautious Walk guarantees that no agent leave a node via a dangerous port. Consequently, if node $v$ is a black hole and can be accessed from port $p$ of a neighbour of $v$, then at most one agent will vanish via $p$.

**Whiteboard Model.** The most powerful inter-agent communication mechanism is having whiteboards at all nodes. Since access to a whiteboard is provided in mutual exclusion, this model could also provide the agents a breaking symmetry mechanism: If the agents start at the same node, they can get distinct identities and then the distinct agents can assign different labels to all nodes. Hence in this model, if the agents are initially co-located, both the agents and the nodes can be assumed to be non-anonymous without any loss of generality.

In the *whiteboard model* introduced by Dobrev *et al.* [43], each node has a bounded amount of storage where information can be written and read by agents. All incoming agents can access the whiteboard of a node in a fair mutual exclusion way and communicate with each other via reading/writing on such whiteboards.

When executing the cautious walk, an agent leaves from a node $u$ to a neighbouring node $v$ via an unexplored port $p$. It marks port $p$ as dangerous by writing on the whiteboard of node $u$. After visiting node $v$, this agent immediately returns to node $u$ in order to update its whiteboard so that the status of $p$ is changed from dangerous to safe.

**Pure Token Model.** In the *pure token model*, each agent has a limited number of tokens that can be placed on or picked up at a node in the course of searching. An agent places one or more tokens at its current node $u$ to indicate that the 'next' node it visits is dangerous. (More precisely, each node has a single location, referred to as its 'center', where to place tokens.) Mutiple tokens may be required in order to capture which of the neighbouring nodes of $u$ is this 'next' node visited by the agent at hand.

The pure token model can be considered as a special whiteboard model with $O(1)$-bit memory on each node. Tokens that can be picked up from a node and placed on another are called *movable tokens.* In contrast, Chalopin *et al.* [21] define *unmovable tokens* as those that cannot be picked up once placed on a node. Usually all tokens are identical (that is, they cannot be distinguished one from the other).

**Enhanced Token Model.** Clearly, the pure token model has strong limitations, in particular with respect to the limited number of messages that can be expressed using a constant number of tokens. In light of such constraints, many researchers (e.g., [41,50,52]) enhance the pure token model in order to increase the information that can be expressed via tokens. More specifically, in the *enhanced token model*, the tokens can be left not only at the 'center' of a node, but also on the ports of a node. But as the number of locations to hold the tokens increases at each node, so does the memory cost of each node. Typically, the memory cost is set to $O(\log n)$ bits in the whiteboard model, $O(\log \Delta)$ in the enhanced token model, and $O(1)$ in the pure token model, where $n$ is the network size and $\Delta$ is the maximum node degree in the network graph [22].

When executing cautious walk under this model, an agent marks a port as dangerous by placing a token at this port before moving to the next node. (As for the whiteboard model, no agent will leave via a dangerous port, that is, in this model, a port at which a token is present [50].) Upon its return, this agent will pick up this token to show that this port is not dangerous. Reusing movable tokens in several different nodes helps minimizing the overall number of tokens used, a challenge not faced in the whiteboard model (since, once written to a whiteboard, a message may be repeatedly accessed by agents over a long period of time, and even be modified). However, the use of movable tokens results in a significantly more complex communication model than (a) one with only unmovable tokens and (b) a whiteboard model (in which messages written to a whiteboard are far easier to use than tokens).

**Face-to-Face Model.** In the *face-to-face model*, agents move through the network in synchronous steps and communicate with each other only when they meet at a node [25]; no other communication method (e.g., whiteboard or tokens) is available. In contrast to the three communication models mentioned above, face-to-face communication does not require that nodes have memory. Finally, clearly, face-to-face communication only applies to synchronous networks: the unpredictability of wake up times and of time required to move and/or compute in asynchronous networks entails agents may never meet.

### 1.3.3 Agent Starting Location

The starting location of an agent is another factor that significantly affects black hole search. Since at least 2 agents are necessary to locate the black hole, the agents could start at the same node or different nodes. More generally, with respect to starting locations, agents may be:

1. Co-located: all the agents initially wake up at the same node, and this node is referred to as *homebase*;
2. Dispersed: the agents wake up at different nodes. The node in which an agent wakes up is its *homebase*. Dispersed agents are also occasionally referred to as *scattered agents*.

In both cases, all homebases are assumed to be safe. Moreover, each dispersed agent only knows its own homebase and, upon waking up, there is no communication between the dispersed agents. In contrast, upon waking up, co-located agents can communicate, which can lead to guaranteed coordination [106].

For synchronous networks, if the face-to-face model is adopted, then only co-located agents must be used: should agents be dispersed, there is a possibility that all will die in the black hole before they ever meet. That is, only co-location guarantees face-to-face communication.

### 1.3.4 Network Knowledge

What agents know about the network considerably affects both the design and complexity of a solution to black hole search. This knowledge includes some of the following: network size, network topology, network direction, edge-labelling and sense of direction.

**Network Size.** *Network size* refers to the total number of nodes in the network. As mentioned before, if the agents do not know the number of nodes nor the number of edges in the network, then the black hole search problem is unsolvable in an asynchronous network. In addition, the problem is also unsolvable in the asynchronous network if the number of black holes is not known *a priori*.

**Network Topology.** *Network topology* refers to the topological structure of the network abstracted as a graph (e.g., a ring, a torus, etc.). Many algorithms are specifically designed for certain network topologies. For example, in [51,53], Dobrev *et al.* provide a protocol called *shadow check* that only works on ring networks. A ring is a fundamental network topology in the context of black hole search for it is the basis for more complex topologies (e.g., torus and hypercube).

In synchronous networks, when the agents have no knowledge of network size nor possess a network map, the black hole search problem can still be solved with only 2 agents if the network topology is known. For example, such solutions exist for rings [22] and tori [20] networks.

In both synchronous and asynchronous networks, when the agents have no topological knowledge, at least $\Delta + 1$ agents are needed in any generic solution, even if the agents are given $n$ and $\Delta$ [69]. If the black hole is a node with degree $\Delta$, then there are $\Delta$ ports leading to the black hole that have to be marked as dangerous. Since one agent vanishes for each dangerous port to mark, and given at least one agent has to survive to eventually report the black hole location, it follows that at least $\Delta + 1$ agents are necessary.

**Network Direction.** *Network direction* refers to whether a graph is directed or undirected (e.g. bi-directional). Most importantly, we remark that most commonly used techniques for black hole search (e.g., the previously mentioned Cautious Walk) can only be used in undirected graphs. Although results for the exploration of directed graphs have appeared since the mid-1990s (e.g. [13,14,73]), the first study dealing with black hole search in directed graphs was published by Czyzowicz *et al.* [28] only in 2010. In [28] it was proved that in directed asynchronous graphs with whiteboards there is an exponential gap on the number of agents needed in order to solve the Black Hole Search problem: $2^{\Delta}$ agents are needed in the worst case (where $\Delta + 1$ agents without a map are sufficient in undirected graphs), where $\Delta$ is the in-degree of the black hole. This is a consequence of the fact that in directed graphs the Cautious-Walk technique cannot be applied. This lower bound holds even in the case of synchronous graphs. However in planar directed graphs with a planar embedding known to the agents, $2\Delta$ agents are needed and $2\Delta + 1$ agents are sufficient. In synchronous directed graphs with whiteboards it was shown in [85] that $O(\Delta \cdot 2^{\Delta})$ agents are sufficient to solve the problem. Additional research on black hole search in directed graphs can be found in [28,85,86].

**Edge-Labelling and Sense of Direction.** An *edge-labelled graph* is one where at each node $x$, there is a distinct label associated with each one of its ports and the incident link of each port. Let $\lambda_x(x,z)$ denote the label associated at $x$ with the link $(x,z) \in E$, and $\lambda_x$ denote the overall injective mapping at $x$. The set $\lambda = \{\lambda_x | x \in V\}$ of those mappings is called a *labelling* and we shall denote by $(G,\lambda)$ the resulting edge-labelled graph. The nodes of $G$ can be anonymous (e.g., without unique names) [48]. When visiting a node in an edge-labelled network, an agent can distinguish the ports of this node, whereas this is not possible in an edge-unlabelled network.

*Sense of direction* occurs in an edge-labelled undirected graph if, from any given node $u$, it is possible to determine whether or not different paths from node $u$ will end in the same node. More precisely, in order to obtain a sense of direction, a *consistent coding function* and a *consistent decoding function* must be defined [68].

For example, in a ring network, if each port is labeled as $A$ or $B$ and such labeling is *consistent*, we say this ring has a sense of direction. Such a labeling is consistent if starting from some specific port and following a specific convention for traversal (e.g., 'A-B-A-B-...-A-B-A' or 'A-A-B-B-...A-A-B-B-A-A'), an agent can traverse the ring of $n$ nodes and return to its starting port. A ring with a consistent labeling (e.g., all ports going in the clockwise direction are labelled *Right*) is commonly referred to as an *oriented* ring. Otherwise a ring is referred to as an *unoriented* one.

We further clarify the relationships between the network direction and the sense of direction in Table 2.

**Table 2.** Relationships between network direction and sense of direction

<table>
<tr><td rowspan="5">Directed graph</td><td colspan="4">Undirected graph</td></tr>
<tr><td rowspan="4">Edge unlabelled</td><td colspan="3">Edge-labelled</td></tr>
<tr><td>Arbitrarily labelled</td><td colspan="2">Consistently labelled</td></tr>
<tr><td rowspan="2">Un-oriented: no sense of direction</td><td>Oriented</td><td>Un-oriented</td></tr>
<tr><td>Sense of direction</td><td>No sense of direction</td></tr>
</table>

**Complete Knowledge.** *Complete knowledge* refers to the case where the agents know the size, topology and sense of direction (e.g., torus with consistent and systematic "N-S-E-W" labelling) of the network. Sometimes, agents are equipped with a network *map* that holds all this knowledge and can also be used to mark the explored nodes during a black hole search [40]. In this model, the black hole search problem becomes much less complex.

### 1.4 Cost Analysis Metrics

Complexity analysis is generally used to compare different solutions to black hole search with respect to specific costs. The most frequently measured costs are:

- Number of agents: the minimal number of agents used to solve the black hole search problem.
- Number of agent moves: the total number of moves performed by all agents from the first agent waking up until the black hole has been located.
- Number of tokens: the minimal number of tokens used by each agent (or by the entire agent team) in order to locate the black hole.
- Memory footprint of agents: the memory overhead of agents. Usually, in models relying on tokens, the agents are designed with a small memory footprint (e.g., an agent can only carry a constant number of tokens at any point in time [20,21]). In other types of models, agents may have a very large memory footprint (e.g., agents carrying a network map [59,83]).
- Memory footprint of nodes: the memory overhead of each node in the network. For example, a $O(\log n)$-bits whiteboard is sufficient for all the algorithms proposed in [8,39]. Recall that the pure token model can be viewed as a whiteboard model with $O(1)$-bit memory on each node when assuming that only a constant number of tokens can be placed at a node [21]. However, in practice, memory overhead is considered mostly for whiteboard models. Instead, not surprisingly, in token models, the number of tokens is taken to be much more relevant.
- Time cost: In synchronous networks, this metric is computed as total number of time units used from when algorithm starts until the black hole is found. Given that, in an asynchronous network, a move of an agent costs finite but unpredictable time, generally time cost is not measured. However, some research [7,8,43] assumes a unitary time delay for each move, which enables the calculation of time complexity. Such a measure is referred to as *ideal time*. Under this assumption, time cost is almost the same as the number of agent moves.

Beyond measures of complexity, evaluations of correctness are also a commonly presented in black hole search work. Most papers in this area use mathematical proofs (e.g., [8,20,29,41,60,63,84]), while only a few researchers conduct simulations and use the results of such experiments to demonstrate correctness [35,106]. For example, Shi *et al.* [106] present their simulation results for three proposed algorithms in addition to providing theoretical proofs. Similarly, D'Emidio *et al.* [35] simulate and compare their own algorithms before further analysis is used to decide which one performs better.

As in many cases that deal with the analysis of algorithms for certain problems, an approach in proving correctness of an algorithm for the Black Hole Search problem, showing upper bounds on time needed for discovering the black hole or proving infeasibility of the problem under a certain model, often uses the notion of an *adversary*. The analysis of the Black Hole Search problem under a model can then be considered as a game between a proposed algorithm and an *adversary* who uses its power to make the algorithm fail. The weakest the model is, the more powerful the adversary is. The adversary can choose all the parameters of the instance of the problem for which the algorithm has no knowledge. For example, the adversary can decide for the initial positions of the agents, the location of the black hole and, depending on the model, the topology or the size of the network, the number of the agents, the ports' labeling, and (in asynchronous networks) the time that an agent needs to cross an edge. A correct algorithm should work of course for any options of the adversary. The problem is infeasible when the adversary has an option to make any algorithm fail.

# 2 Search in Synchronous Dangerous Networks

In this section, we overview solutions for black hole search in synchronous networks. Given no existing research has used the enhanced token or the whiteboard models in synchronous networks, our presentation will follow the different possibilities given in Fig. 1.

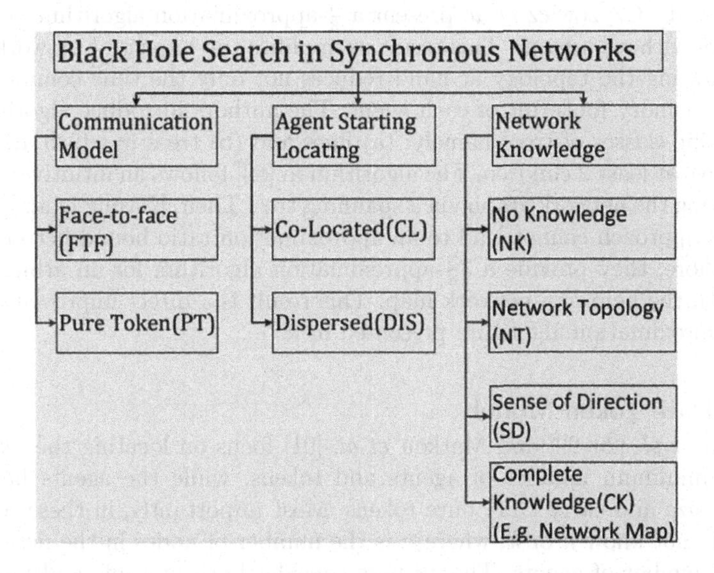


**Fig. 1.** Different variants for black hole search in synchronous networks

## 2.1 Solutions Under Different Communication Models

### 2.1.1 Face-to-Face Model

Recall the face-to-face model is only possible in synchronous networks. According to this model, agents simultaneously present in the same node can communicate with each other using an unlimited number of messages.

Czyzowicz *et al.* [29–31] and Klasing *et al.* [82–84] consider the problem of finding the most efficient solution (in terms of time cost) for the black hole search under the same assumption: 2 co-located agents with maps searching for a black hole in an edge-labelled undirected synchronous network. Instead, Chalopin *et al.* [21] study the problem using a hybrid communication model: agents can carry and place a bounded number of pure tokens *and* can communicate with each other when they meet on a node. Since that work focuses more on the impact of the tokens, we discuss it in the section on the pure token model (Sect. 2.1.2).

Under the assumptions they make, Czyzowicz *et al.* [30] show that the optimal black hole search problem is NP-hard, and propose a 9.3-approximation algorithm for it. Additionally, Klasing *et al.* [84] prove that this problem cannot be approximated in polynomial-time using a constant factor less than $\frac{389}{388}$ (unless $P = NP$), and give a 6-approximation algorithm. In both [84] and [30], each agent carries a network map and starts from the same node. But whereas the algorithm proposed by [30] can solve the problem when there is one and only one black hole in the network, the solution in [84] can first detect whether there is a black hole and then locate this black hole if present. (Recall, as previously mentioned, that such detection is only possible in a synchronous network.)

In [29,31], Czyzowicz *et al.* present a $\frac{5}{3}$-approximation algorithm in an arbitrary tree without a map. This result exemplifies the impact of network knowledge: knowing the topology at hand reduces not only the time complexity but also the memory footprint of each agent. The authors introduce algorithms for two specific classes of trees namely: (a) lines and (b) trees in which all internal nodes have at least 2 children. The algorithm in [83] follows an intuitive approach of exploring the network graph via a spanning tree. Then, Klasing *et al.* [83] prove that this approach cannot lead to an approximation ratio bound better than $\frac{3}{2}$. Furthermore, they provide a $3\frac{3}{8}$-approximation algorithm for an arbitrary network with the help of a network map. This result is a direct improvement from the $\frac{7}{2}$-approximation algorithm presented in [82].

### 2.1.2 Pure Token Model

Chalopin *et al.* [20–22] and Markou *et al.* [91] focus on locating the black hole using a minimum number of agents and tokens, while the agents have $O(1)$ memory size and carry $O(1)$ pure tokens. Most importantly, in these solutions, agents do not know $n$ or $k$, where $n$ is the number of nodes in the network and $k$ is the number of agents. The authors consider both movable and unmovable tokens in rings [21] and tori [20,91] respectively.

As previously mentioned, in [21], Chalopin *et al.* consider the black hole search problem with agents that have hybrid communication capabilities: they can communicate with each other face-to-face when they are in the same node and they can also carry either movable or unmovable tokens. When using movable tokens, 3 agents, each of which carrying only 1 token, are necessary and sufficient for both oriented and un-oriented rings. In contrast, using unmovable tokens, 4 agents are required, each with 2 tokens, for oriented rings and 5 agents, each with 2 tokens, when exploring un-oriented rings. Expressing messages using unmovable tokens is equivalent to writing messages on whiteboards with limited memory. Given this observation, one might expect the use of unmovable tokens to be more 'powerful' than that of movable ones. Interestingly, the results show that using unmovable tokens is more costly than that using movable one with respect to number of agents used. Furthermore, results show that more agents are necessary for un-oriented rings than for oriented rings.

In addition to rings, Chalopin *et al.* [20] also study the oriented torus under the same assumptions: dispersed agents, pure token model and face-to-face communication. They prove that the black hole search problem is unsolvable in synchronous oriented torus in three scenarios: (1) when the number of agents is constant and tokens are unmovable; (2) when using 2 dispersed agents, even if the tokens are movable and the agents have unlimited memory; (3) when using 3 agents with constant memory and 1 movable token each. Ultimately, they show that at least 3 agents, each with 2 movable tokens, are necessary and sufficient to solve the problem in any oriented torus.

In [91], Markou *et al.* study the black hole search problem under the same assumptions as [20] but in an un-oriented torus. The authors discuss four cases of un-oriented tori: from totally un-oriented to semi-oriented (i.e., without an agreement on the orientation in the horizontal or vertical axis, as explained shortly). The authors prove that the black hole search problem cannot be solved in an un-oriented torus using a constant number of agents and tokens if these tokens are unmovable. The authors then consider the use of movable tokens. They prove that the problem is also unsolvable when using any constant number of dispersed agents with 1 movable token each. The authors provide algorithms, each using 5 agents and 3 tokens, for any semi-oriented torus. Finally, they conjecture that at least 5 scattered agents with constant memory, equipped with at least 2 movable tokens, would be able to locate the black hole in a totally un-oriented torus.

## 2.2 Solutions Under Different Agent Starting Locations

Some researchers [29–31,82–84] choose to study the black hole search using co-located agents, others with dispersed ones [20–22,91]. When all agents wake up in the same node, coordination and communication are guaranteed for these co-located agents. This greatly simplifies graph exploration.

### 2.2.1 Co-located Agents

As previously mentioned, adopting face-to-face communication entails using co-located agents (since dispersed agents may all vanish in the black hole before ever meeting.) Given it was shown early that, with complete knowledge of the network, 2 co-located synchronous agents are sufficient to locate the black hole, subsequent work [29–31,82–84] has focused on finding solutions that improve the time cost (as reported in Sect. 2.1.1). Finally, we remark that when using only 2 co-located agents, whether the agents are anonymous or not is irrelevant since an agent can definitely distinguish itself from the other when they meet.

### 2.2.2 Dispersed Agents

In order to extend the results obtained for 2 co-located synchronous agents, Chalopin *et al.* [20–22] and Markou *et al.* [91] consider using dispersed agents under the pure token model (as discussed in Sect. 2.1.2). In these contributions, in contrast to work on co-located agents, the focus is not on time complexity and agent moves but rather on the minimal number of dispersed agents and tokens being used. It is assumed that network size is unknown *a priori* and that agents are restricted to using pure tokens. Thus, given tokens can be placed only at a node, not its ports, coordination between dispersed agents becomes significantly more complex. For example, in a torus, even when an agent sees a token at a node, it still cannot know from which port the previous agent left.

## 2.3 Solutions Under Different Network Knowledge

As previously hinted, network topology may significantly impact on solutions for black hole search. For example, the above-mentioned work on rings [21,22] and tori [20,91] clearly shows that, under the same assumptions, more agents and tokens are needed for a torus than a ring. That is, it appears network topology not only affects the complexity of the network, but also the number of agents and the number of tokens necessary and sufficient to solve the black hole search problem. Furthermore we notice that, with a map of an arbitrary network, [83] offers a $3\frac{3}{8}$-approximation algorithm, whereas [31] presents a $\frac{5}{3}$-approximation algorithm that does not use a map but does know the topology is a tree. This strongly suggests that, even without a map, a solution that relies on knowledge of topology may have better time cost than a solution designed for an arbitrary network (i.e.,, *without* knowledge of topology), even with the help of a map.

Sense of direction is another important consideration. It offers not only consistent edge-labelling, but also a guaranteed method of systematic exploration of the entire graph. (In contrast, without edge-labelling, an agent may not be able to distinguish the edges incident to a node, and thus a whole part of the graph may not be considered during exploration.) Its importance is clearly demonstrated in the results obtained for oriented rings and tori [20–22] that, under the same assumptions, improve on those for un-oriented rings and tori [21,22,91].

Similarly, in [91], Markou *et al.* discuss four levels of network knowledge in a torus, namely: (1) the agents have no agreement on anything regarding the orientation; (2) the agents perceive orthogonal links but they do not agree on which link is horizontal and which is vertical (3) the agents agree on which link is horizontal and which is vertical, but there is no consensus on the orientation of each link; and (4) the agents agree on which link is horizontal and which is vertical and they also agree on the orientation in one of the links. The three latter are called *semi-oriented*. Their results (reported in Sect. 2.1.2) demonstrate solutions for oriented tori are less costly than those for semi-oriented tori, thus emphasizing again the importance of orientation.

### 2.4 Black Hole Search in a Synchronous Tree

The first results on the Black-Hole Search (BHS) problem in *synchronous* networks appeared for two agents operating in tree topologies in [29] (see also the full version of the paper in [31]). We present here the model which was used, give lower bounds on BHS schemes in synchronous trees and describe a fastest algorithm for the problem in some special tree topologies.

#### 2.4.1 Model and Basic Lower Bounds and Tools

We consider a labeled tree $T$ rooted at node $s$ which is the starting node of two agents, and is assumed to be safe ($s$ is not a black hole). Agents are synchronous, have memory, a map of the tree and distinct labels. They can communicate (i.e., reading each-other's states) only when they meet at a node (and not, e.g., by leaving messages at nodes). We assume that there is at most one black hole in the network. The goal is to design an algorithm for the agents which produces a black hole search scheme (*BHS-scheme*) for the input $(T, s)$ (i.e., a pair of sequences of edge traversals (moves) of each of the two agents). Upon completion of the BHS-scheme there should be at least one surviving agent which either knows the exact location of the black hole or knows that there is no black hole in the tree. The surviving agents must return back to $s$ to report this information. An agent can move along the edges of the tree (each move takes one time unit) or can wait at a node.

The time of a black hole search scheme is the number of time units until the completion of the scheme, assuming the worst-case location of the black hole (or its absence, whichever is worse). It is easy to see that the worst case for a given scheme occurs when there is no black hole in the tree or when the black hole is the last unvisited node, both cases yielding the same time. A scheme is called *fastest* for a given input if its time is the shortest possible for this input. We emphasize here that the time of a black hole search scheme should not be confused with the time complexity of an algorithm producing such a scheme.

Since there is at most one black hole in the tree, in any BHS-scheme all nodes of the tree should be eventually visited in the worst case (e.g., when there is no black hole). Hence all edges of the tree should be traversed (*explored*).

When a meeting occurs the agents exchange information about the explored territory (i.e., the set of explored edges). The sequence of steps of a BHS-scheme between two consecutive meetings is called a *phase*. Since an unexplored edge could be incident to a black hole we have:

**Lemma 1** ([31]). *In a BHS-scheme, an unexplored edge cannot be traversed by both agents.*

If one of the agents $a$ attempts to traverse more than one unexplored edges before meeting with the other agent $b$ then $a$ may vanish in the black hole (placed by an adversary somewhere in $a$'s path) and $b$ does not have enough information to decide for the correct location of the black hole. Thus:

**Lemma 2** ([31]). *During a phase of a BHS-scheme an agent can traverse at most* one *unexplored edge.*

Hence in a BHS-scheme, an edge can be explored only in the following way: an agent traverses this edge and then a meeting is scheduled (somewhere within the previously known explored territory) where the agents exchange information about the explored territory. Whether the meeting occurs or not (in the latter case the agent vanished in the black hole) the edge becomes explored. Therefore an unexplored edge could be explored in the next phase only if it is adjacent to the explored territory (which stays connected).

**Lemma 3** ([31]). *At the end of each phase, the explored territory is increased by one or two edges, or the black hole is found.*

We define a *1-phase* to be a phase in which exactly one edge is explored. Similarly, we define a *2-phase* to be a phase in which exactly two edges are explored. In view of Lemma 3, every phase is either a 1-phase or a 2-phase.

Let $(u, v)$ be an unexplored edge with node $u$ being incident to a previously explored edge (or $u \equiv s$). Suppose that the two agents are at $u$ at time $t$. A way of exploring exactly one edge in a phase is the following:

**Procedure** `Probe(`$v$`)`**:** One agent traverses edge $(u, v)$ and returns to node $u$ to meet the other agent who waits. If they do not meet at step $t+2$ then the black hole is at $v$.

We also define a procedure that the two agents could follow to explore two new edges in a phase. Let $(u_1, v_1), (u_2, v_2)$ be two unexplored edges with nodes $u_1, u_2$ being incident to previously explored edges (or $u_1 \equiv u_2 \equiv s$). Suppose that one of the agents is at $u_1$ and the other agent is at $u_2$ at time $t$. The definition of the procedure is the following:

**Procedure** `Split(`$v_1, v_2$`)`**:** One of the agents traverses edge $(u_1, v_1)$ and then returns to $u_1$ while the other one traverses edge $(u_2, v_2)$ and then returns to $u_2$. Then both agents traverse the path $<u_1, u_2>$ (which should be totally within the explored territory since the explored territory is always connected) from different directions until they meet at a node as soon as possible. Let $dist(u_1, u_2)$ denote the number of edges in the path from node $u_1$ to node $u_2$. If the agents do not meet at step $t + \lceil \frac{dist(u_1,u_2)}{2} \rceil + 2$ then the black hole has been found.

Both above procedures resemble the `Cautious-Walk` procedure in the sense that in both procedures the agents explore at most one edge each and then they return to give a message to each-other. Two agents with a map can easily discover the exact location of the black hole (or decide that there is no black hole) in any tree within finite time (e.g., exploring the tree in a Depth First Search traversal using Procedure `Probe`). In fact this simple algorithm guarantees a 4 approximation ratio on the fastest BHS scheme for any arbitrary tree (see at the end of the section for details). The authors [31] give an algorithm which uses both procedures `Probe` and `Split` and solves the BHS problem for a special family of trees in the fastest possible time. Before describing this algorithm we give a general lower bound on the time needed by any algorithm to solve the problem on any tree.

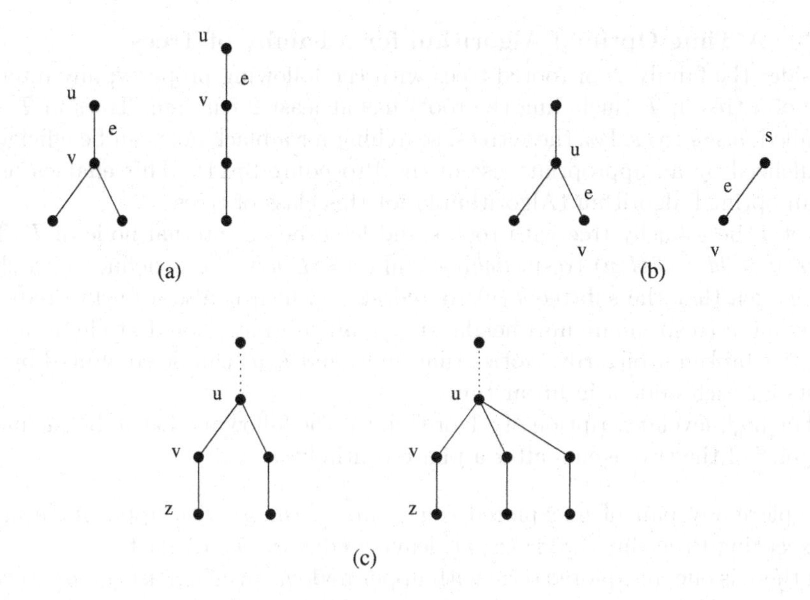


**Fig. 2.** In (a) $e$ is a red edge. In (b) $e$ is a green edge. In (c) all solid edges are blue.

### 2.4.2 A Lower Bound on the Time Needed in Any Tree

Let $T$ be a tree rooted at the starting node $s$ and $e = (u, v)$ be an edge of $T$ (with $v$ being a child of $u$). Consider the following coloring which creates a partition of the edges of the tree.

- assign red color to edge $e$ if node $v$ has at least two descendants,
- assign green color to edge $e$ if $v$ is a leaf and exactly one of the following holds: $u = s$ or the edge $(w, u)$ is a red edge (where $w$ is the parent of $u$),
- assign blue color to edge $e$ if it has none of the above properties.

Red, green and blue edges are shown in Fig. 2.

Let $e = (u, v)$ and $e' = (v, z)$ be two blue edges as shown in Fig. 2c, i.e., $v$ is a child of $u$ and $z$ is a leaf and the unique child of $v$. We call the set of these two edges a *branch*. The set of all branches of blue edges with upper node $u$ is called a *block*. In view of Lemmas 1 and 2 it can be easily proved that:

**Lemma 4** ([31]). *In any BHS-scheme, the following holds: a green edge has to be traversed by the agents at least* 2 *times, a red edge has to be traversed at least* 6 *times and a branch of blue edges requires a total of at least* 6 *traversals.*

Since in any BHS-scheme, each of the two agents can do at most one traversal in every time unit, the following lemma holds:

**Lemma 5** ([31]). *Any BHS-scheme requires at least* 3, 1 *and* $3r$ *time units for the traversals of a red edge, a green edge and a block of* $r$ *branches of blue edges, respectively.*

### 2.4.3 A Time Optimal Algorithm for a Family of Trees

Consider the family $\mathcal{T}$ of rooted trees with the following property: any internal node of a tree in $\mathcal{T}$ (including the root) has at least 2 children. Trees in $\mathcal{T}$ will be called *bushy trees*. For these trees, searching for a black hole can be efficiently parallelized by an appropriate use of the Procedure `Split`. This enables us to get an optimal algorithm (Algorithm 1) for this class of trees.

Let $T$ be a bushy tree with root $s$ and let $u$ be an internal node of $T$. The *heaviest child* $v = H(u)$ (resp. *lightest* child $v = L(u)$) of $u$ is defined as a child $v$ of $u$ such that the subtree $T(v)$ rooted at $v$ (which is also a bushy tree) has a maximum (resp. minimum) height among all subtrees rooted at children of $u$ with ties broken arbitrarily. Notice that $H(u)$ and $L(u)$ can be computed by the agents for each node $u$ in linear time.

The high-level description of Algorithm 1 is the following. Let $m$ be the meeting point of the two agents after a phase (initially $m \equiv s$).

- Explore any pair of unexplored edges $(m, x)$, $(m, y)$ with upper node $m$ by executing Procedure `Split(x,y)`, leaving edge $(m, L(m))$ last.
- If there is one unexplored edge with upper node $m$ (which must be $(m, L(m))$) then one of the agents explores this edge while the other one explores another 'close by' unexplored edge (if any) again using `Split`. If edge $(m, L(m))$ is the last unexplored edge in the tree, explore it by executing `Probe(L(m))`.
- If all edges with upper node $m$ are explored, explore similarly as before any unexplored edges incident to the children of $m$ and to ancestors of $m$.

---

**Algorithm 1.** (2 *agents with memory and a map in a synchronous 'bushy' tree*)

1: $next := s$;
2: **repeat**
3: $v := next$;
4: **for** every pair of unexplored edges $(v, x), (v, y)$ with upper node $v$ **do**
5: Split$(x, y)$, so that edge $(v, L(v))$ is explored last;
6: **end for**
7: **if** there are still unexplored edges in the tree **then**
8: **case 1**: every edge incident to $v$ has been explored:
9: **case 1.1**: there is an unexplored edge incident to a child $w$ of $v$:
10: $next := w$;
11: **case 1.2**: every edge incident to any child of $v$ is explored:
12: let $t$ be the parent of $v$;
13: $next := t$;
14: walk to node $next$;
15: **case 2**: there is an unexplored edge $(v, z)$ incident to $v$:
16: *(* must be $z = L(v)$ *)*
17: $next :=$ Explore-only-child$(v)$;
18: **end if**
19: **until** every edge has been explored
20: walk to node $s$ and report the location of the black hole;

---

Function `Explore-only-child(`$v$`)` takes as input the current node $v$ where both agents reside and returns the new meeting point after the exploration of edge $(v, L(v))$. The description of the procedure is the following:

- If there is an unexplored edge incident to a child $w$ of $v$, $w \neq L(v)$, then the agents explore edge $(w, H(w))$ together with edge $(v, L(v))$ calling Procedure `Split(H(w),L(v))`. The new meeting point is $w$.
- If every edge incident to any child $w$ of $v$, different from $L(v)$, is explored and edge $(v, L(v))$ is not the last unexplored edge in the tree, then find the deepest ancestor $a$ of $v$ having a descendant incident to an unexplored edge (excluding $L(v)$); the agents explore edge $(D(a), H(D(a)))$ (where $D(a)$ is the closest descendant of $a$ with incident unexplored edges), together with edge $(v, L(v))$, by `Split(H(D(a)),L(v))`; the new meeting point is $D(a)$.
- If edge $(v, L(v))$ is the last unexplored edge in the tree then explore it by calling `Probe(L(v))`; the new meeting point is $v$.

Notice that all edges of the tree (except possibly the last one if the number of edges is odd) are explored by calling Procedure `Split`. Observe that in any bushy tree, there are only *red* and *green* edges.

It is proved that the BHS scheme produced by Algorithm 1 traverses any red edge 6 times and any green edge 2 times. Moreover every phase is a 2-phase (i.e. the two agents traverse edges in parallel), except possibly the last phase, and no agent waits in any 2-phase. Hence, in view of Lemma 5 the following theorem holds:

**Theorem 1** ([31]). *Algorithm 1 produces a fastest BHS-scheme for any bushy tree.*

#### 2.4.4 The Case of Arbitrary Trees

For arbitrary trees there is a simple algorithm which guarantees an approximation ratio strictly less than 4, i.e., an algorithm which produces a BHS-scheme whose time is less than 4 times the cost of the fastest BHS scheme for any arbitrary tree and starting node. The algorithm is the following: both agents traverse the tree together in depth first search order and explore each new node with a *probe* phase. This BHS scheme explores a $n$-node tree within $4(n-1) - 2l$ steps, where $l$ is the number of leafs in the tree. In any BHS scheme of an $n$-node tree, each edge has to be traversed at least twice by an agent which gives a total of $2(n-1)$ traversals. Hence at least $n-1$ steps are required which implies a strictly less than 4 approximation factor for the above algorithm.

A better (and still simple) approximation algorithm for arbitrary trees was also proposed in [31] achieving a ratio of $\frac{5}{3}$. The high-level description of the algorithm is the following. Let $v$ be the meeting point of the two agents after a phase (initially $v \equiv s$); the edges with upper node $v$ are explored by calling Procedure `Split` until either all such edges are explored or there is at most one remaining unexplored edge incident to $v$, which is explored by calling Procedure `Probe`; this is repeated for any child of $v$.

The question of whether it is possible to produce a fastest BHS-scheme for any arbitrary tree is open until now. The conjecture is that it is possible but such an algorithm would probably need to take care of many special cases that could occur in an arbitrary tree.

## 2.5 Black-Hole Search in Synchronous Graphs

Producing a fastest Black Hole Search (BHS) scheme for arbitrary graphs was proved to be NP-hard in [82,83]. This result extended in some way a reduction which showed the NP-hardness of finding fastest schemes for a more general version of the Black Hole Search problem in which a set of safe nodes is given (instead of just the starting node) and appeared in [30]. Later, in [84], APX-hardness for both versions in arbitrary graphs was proved. In this section we first describe the (less complicated) reduction that shows the NP-hardness of the general BHS problem and later we discuss the original BHS problem. We also discuss the APX-hardness results and approximation algorithms for both versions.

### 2.5.1 The General Black Hole Search Problem

In [30] the authors studied the BHS problem in synchronous arbitrary graphs adopting a more general scenario in which initially a subset of nodes of the network (instead of just the starting node), containing the starting node, is safe, and the black hole can be located in one of the remaining nodes. Let us call this version of the problem *general Black Hole Search (gBHS)*. It was shown that the problem of finding the fastest possible black hole search scheme by two agents in an arbitrary graph is NP-hard, and a 9.3-approximation algorithm was given for this problem, i.e., a polynomial time algorithm which, given a graph with a subset of safe nodes and a starting node as input, produces, in polynomial time, a black hole search scheme whose time is at most 9.3 times larger than the time of the fastest scheme for this input.

**Model and Terminology.** The model which was used was the same as before with the following differences. Given are:

- a graph $G$ with node $s$ which is the starting node of both agents, and
- a subset $S$ of *safe* nodes containing $s$ ($S$ cannot contain a black hole).

It was again assumed that there is at most one black hole in the network and the goal is to find a gBHS-scheme for the input $(G, S, s)$.

**NP-Hardness of the General Black Hole Search Problem.** To prove NP-hardness of the gBHS problem, the authors presented a reduction from the (NP-complete) Hamiltonian Cycle Problem (HC problem) to the decision version of the gBHS problem (called dgBHS).

**The HC Problem:**

*Instance:* A Graph $G$.
*Question:* Does $G$ contain a Hamiltonian cycle?

**The dgBHS Problem:**

*Instance:* A graph $G'$ with a subset $S$ of safe nodes, starting node $s \in S$, positive integer $X$.
*Question:* Does there exist a gBHS scheme for the input $(G', S, s)$, with time at most $X$?

**Construction.** Let a graph $G$ with $n$ nodes and $e$ edges be an instance of the HC problem. We construct a new graph $G'$ as follows. Call the nodes of graph $G$ old nodes. In each edge of $G$ we add 2 new unsafe nodes adjacent to endpoints of this edge and $M = 4e + 5n - 1$ new safe nodes between them, as in Fig. 3. Let $s$ be any node of the old $n$ nodes. All old nodes except $s$ are considered unsafe. Hence the set of unsafe nodes consists of all old nodes except $s$ and all nodes adjacent to old nodes.

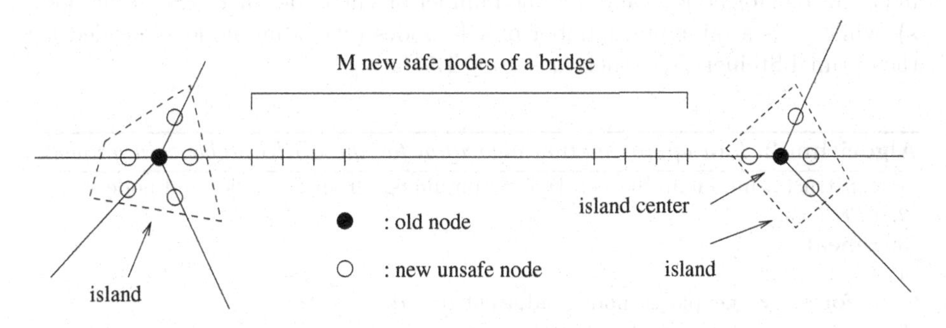


**Fig. 3.** Construction of a dgBHS problem instance

The instance of the dgBHS problem is the graph $G'$ with $n' = n + (M + 2)e$ nodes, the set $S$ of $Me + 1$ safe nodes, node $s$ as a starting node, and the integer $X = M(n + 1) - 1$.

The construction of this instance from the graph $G$ can be clearly done in polynomial time. If there is a Hamilton Cycle $C$ in the old graph $G$ then the two agents can follow this cycle starting at node $s$ and identify the black hole (or discover that there is no black hole) in the new graph $G'$ (constructed as above) in at most $M(n + 1) - 1$ time units as follows:

They explore by probing (one agent goes to an adjacent node while the other agent waits) the nodes which are adjacent to the center of an island (see Fig. 3) except the two nodes which are on $C$ and they return to the center of the island.

Then they explore by probing the unsafe node of the bridge on $C$ in the chosen direction and they walk along the bridge, till they get to the last safe node of it. Subsequently, they explore the adjacent unsafe node $v$ (in the next island) by probing, walk to it, explore the center of this island by probing and walk to it. They repeat the above procedure in every island on $C$ until they reach again node $s$.

In the absence of a Hamiltonian Cycle in graph $G$ the agents need at least $M(n+1)$ time units to identify the black hole (or discover that there is no black hole) in $G'$ (see [30] for more details).

**Theorem 2** ([30]). *Producing a fastest BHS scheme for the gBHS problem in arbitrary graphs is NP-hard.*

**An Approximation Algorithm for the gBHS Problem.** An approximation algorithm (Algorithm 2) for the gBHS problem in arbitrary graphs also appeared in [30]. The algorithm is based on the construction of a *Steiner Tree* of the input graph $G$, where the unsafe nodes of $G$ along with the starting node $s$ are the *required* nodes. Recall that a Steiner Tree for a graph $G = (V, E)$ with the set $R \subseteq V$ of required nodes is any subtree of $G$ containing $R$. We can construct such a Steiner Tree $T$ in polynomial time with approximation ratio $\alpha$, where $\alpha = 1 + \frac{\ln 3}{2} < 1.55$ [80,99]. More specifically, if $x$ is the number of unsafe nodes in $G$ plus one for node $s$, and $y$ is the number of safe nodes in $T$ (excluding node $s$), while $y^*$ is a minimum number of safe nodes (excluding node $s$) needed for the optimal Steiner Tree, then $(x + y) \leq 1.55(x + y^*)$.

**Algorithm 2.** (*An approximation algorithm for the gBHS problem in graphs*)

1: construct a minimum Steiner Tree $T$ containing all unsafe nodes and node $s$;
2: *next* := $s$;
3: **repeat**
4: $v$ := *next*;
5: **for** every unexplored node $z$ adjacent to $v$ **do**
6: probe($z$);
7: **end for**
8: **if** there are still unexplored nodes **then**
9: **case 1**: there is an unexplored node adjacent to a child $w$ of $v$:
10: *next* := $w$;
11: **case 2**: every node adjacent to any child of $v$ is explored:
12: let $t$ be the parent of $v$;
13: *next* := $t$;
14: walk to node *next*;
15: **end if**
16: **until** every node is explored
17: walk to node $s$ and report the location of the black hole;

The high-level description of Algorithm 2 is the following. First a Steiner Tree is constructed as described above. Let $v$ be the meeting point of the two agents after a phase (initially $v \equiv s$); the unexplored children of $v$ are explored by calling procedure probe; if there is an unexplored node adjacent to a child of $v$ then the agents go to that child and repeat. Otherwise the two agents go to the parent of $v$ and repeat.

The time-complexity of Algorithm 2 is polynomial on the size of $G$ and is dominated by the time of constructing the Steiner Tree. Algorithm 2 is in fact a Depth First Search type algorithm with the only difference that any unsafe node is visited using a cautious way (probing). The time spent on traversals of any edge $(u, v)$ ($v$ is a child of $u$) of the tree $T$ is at most 4 units: the worst case is when edge $(u, v)$ leads to an unexplored node $v$ which is not a leaf in $T$, therefore the agents spend 2 time units for probing $v$, 1 time unit to walk to $v$ and another time unit to return to node $u$ - after the exploration of the descendants of $v$. The total time needed by the gBHS-scheme produced by Algorithm 2 is less than $4(x+y)$.

**Lemma 6** ([30]). *Any gBHS scheme for the graph $G$ requires at least $\frac{4}{3}(x+y^*)$ traversals of edges.*

The lower bound of Lemma 6 together with the upper bound achieved by Algorithm 2 and the approximation factor $\alpha$ (<1.55) of the Minimum Steiner Tree lead to the following theorem:

**Theorem 3** ([30]). *Algorithm 2 is an approximation algorithm for the gBHS problem with ratio $\alpha < 9.3$.*

The approximation ratio of the gBHS problem was improved in [84] where a 6-approximation algorithm was given for the problem, using again the approach of minimum spanning trees.

### 2.5.2 The BHS Problem in Arbitrary Graphs

In [82] (see also the full version in [83]) it was proved that the problem of constructing a time optimal Black Hole Search scheme for two agents under the restricted scenario of one safe node, the starting node (i.e., the original BHS problem) is NP-hard even on planar graphs.

**NP-Hardness of Black Hole Search in Planar Graphs.** The NP-hardness of the BHS problem in planar graphs (BHSp problem) was shown by providing a reduction from a particular version of the Hamiltonian Cycle problem (Hamiltonian Cycle on cubic planar graphs (cpHC problem)) to the decision version of the BHSp problem.

Here is the formalization of the two problems:

**The cpHC Problem**

*Instance:* a cubic planar 2-connected graph $G = (V, E)$, and an edge $(x, y) \in E$;
*Question:* does $G$ contain a Hamiltonian cycle that includes edge $(x, y)$?

**The dBHSp Problem**

*Instance*: a planar graph $G' = (V', E')$, with a starting node $s \in V'$, and a positive integer $X$;
*Question*: does there exist a BHS scheme for $G'$ starting from $s$ with time at most $X$?

The cpHC problem above is proved to be NP-complete in [83] by a simple reduction from the Hamilton Cycle problem in cubic planar graphs without the extra requirement that the Hamiltonian cycle passes through a given edge (which was proven NP-complete in [74]). Then with a more technical construction than the one in the previous section the authors reduce the cpHC problem to the dBHSp problem proving that the dBHSp problem is NP-complete:

**Theorem 4** ([83]). *A graph $G$ with $n$ nodes has a Hamiltonian Cycle passing through an edge $(x, y)$ if and only if there is a BHS scheme for a graph $G'$ and a starting node $s$ (constructed from $G$ in polynomial time) with time at most $5n + 2$ units.*

**An Approximation Algorithm for the BHS Problem.** For an approximation algorithm for the BHS problem in an arbitrary graph $G$, a natural approach is the following: First select a spanning tree in $G$ and then explore the graph by traversing the tree edges. Since any BHS scheme of a $n$-node graph requires at least $n - 1$ steps, this approach together with the simple algorithm described at the end of Sect. 2.4 guarantees an approximation ratio of 4 for any arbitrary graph. To follow this spanning-tree approach more effectively we need an algorithm for constructing "good" BHS schemes for trees and an algorithm for computing spanning trees which are "good" for those schemes. A linear-time algorithm which extends the construction of the optimal BHS scheme for *bushy* trees (which has been presented in Sect. 2.4) to the general rooted trees is proposed in [83]. This algorithm still does not guarantee optimality of computed exploration schemes for trees other than bushy trees: the question of computing in polynomial time optimal exploration schemes for general trees remains open. The cost of the exploration scheme computed by this extended algorithm for an arbitrary tree $T$ is given as a function of the number of nodes of different types in tree $T$. Then a heuristic algorithm is presented for the problem of computing a rooted spanning tree $T$ of graph $G$ which gives a relatively small value of that formula. This approach guarantees an approximation ratio of at most $3\frac{3}{8}$.

### 2.5.3 APX-Hardness of the BHS Problem in Arbitrary Graphs

Finally in [84] both versions of the black hole problem discussed in this section were proved to be APX-hard. It was shown that a fastest BHS scheme for the general Black Hole Search problem is not approximable in polynomial time within a $1+\varepsilon$ factor for any $\varepsilon < \frac{1}{388}$, unless $P = NP$. It was also proved that the original BHS problem (in which only the starting node is initially known to be safe) is also APX-hard.

The authors provide an explicit lower bound on the approximability of the General Black Hole Search problem by showing an approximation-preserving reduction from a particular subcase of the Traveling Salesman Problem (TSP), presented in [55]. In this subcase of TSP, distances between nodes are symmetric and satisfy the triangle inequality and between integers 1 and $M$, while the maximum distance $M$ is a constant.

**Lemma 7** ([55]). *It is NP-hard to approximate* TSP(1,M) *within* $1+\varepsilon$ *for any* $\varepsilon < \frac{1}{388}$.

The authors' approach to prove the APX-hardness of the gBHS problem is the following. They first provide a reduction from instances $(G, d)$ of TSP to instances $(G', S, s)$ of the gBHS problem and then they show that if the optimal solution of the gBHS, constructed by the given reduction from an instance of TSP, can be approximated within a $(1+\varepsilon)$ factor, then the optimal solution of the corresponding instance of TSP can be approximated within the same factor.

**Theorem 5** ([84]). *The gBHS problem is not approximable in polynomial time within a factor of* $1+\varepsilon$ *for any* $\varepsilon < \frac{1}{388}$, *unless* $P = NP$.

They also provide a reduction from the TSP problem where the distances between the nodes are either 1 or 2 to the original version of the BHS problem (in which only the starting node is known to be safe) and show that:

**Theorem 6** ([84]). *It is NP-hard to approximate the BHS problem within a factor* $1+\varepsilon$ *for any* $\varepsilon < \frac{1}{2258}$.

## 2.6 Black Hole Search with Scattered Finite Automata

The Black Hole Search (BHS) problem has been extensively studied for co-located agents with memory.

The memory is a critical capability that allows the agents to store (or make) a map of the network, to keep information about (or count) the number of nodes of the graph and generally helps them decisively to explore the network.

In asynchronous networks the co-location of the agents in the whiteboard model gave them the ability to assign different identities to themselves and different labels at nodes while in synchronous networks the co-location gave the agents the ability to *probe* the network in a *cautious way* and discover the black hole.

The question of whether even more weak models allow the solution of the black hole search problem has been also raised. Although there are results for scattered agents (not initially starting at the same node) using tokens (instead of whiteboards) and even not having a map of the graph for asynchronous networks, the memory capability cannot be dropped since in such networks the agents need to know (and store) the number of nodes (or at least an upper bound). However in synchronous networks no knowledge of such bound is generally needed so an interesting question is whether *deterministic finite automata (DFAs)* (i.e., with only a constant memory), initially scattered in the network, without a map and having only a constant number of 'pure' tokens (which can leave only at nodes) can still solve the black hole search problem.

We notice here that Rollik [100] has proved that no finite set of finite automata can cooperatively perform exploration of all cubic planar graphs. Since a finite automaton is more powerful than a token, it means that no finite set of finite automata using any constant number of tokens can explore all cubic planar graphs even when the agents start at the same node and can exchange information when they meet at nodes. It is clear that if the agents cannot even explore the graph (i.e., visiting all nodes) they cannot solve the BHS problem. Hence a challenging question is whether there are synchronous network topologies for which the agents can solve this problem in such a weak model. Even when the agents can communicate when they meet at a node, since they are initially scattered they need to solve the *rendezvous* problem which is far from trivial for identical, anonymous DFAs carrying only indistinguishable tokens.

The BHS problem in this model (initially *scattered anonymous agents* with *constant memory*, carrying only a constant number of *pure* indistinguishable tokens and having the ability to communicate only when they meet at the same node) has been investigated for ring [21] and torus [20] topologies.

### 2.6.1 The Model

The model consists of an anonymous, synchronous network with $k \geq 2$ identical mobile agents that are initially located at distinct nodes called *homebases*. Each mobile agent owns a constant number $t$ of identical tokens which can be placed at any node visited by the agent. The tokens are indistinguishable. Any token or agent at a given node is visible to all agents on the same node, but not visible to agents on other nodes. The agents follow the same deterministic algorithm and begin execution at the same time and being in the same initial state.

A token is called *movable* if it can be put on a node and picked up later by any mobile agent visiting the node. Otherwise the token is called *unmovable* in the sense that, once released, it can occupy only the node where it was released.

Formally a mobile agent was considered as a finite Moore automaton.

All computations by the agents are independent (the agents have no knowledge) of the size of the network. The algorithms for the ring topology work even without knowledge of the number of the agents. There is exactly one black hole in the network. An agent can start from any node other than the black hole and no two agents are initially co-located. Once an agent detects a link to the black hole, it marks the link permanently as dangerous. The goal is that at the end of a black hole search scheme, all links incident to the black hole (and only those links) are marked dangerous and that there is at least one surviving agent. Note that this definition of a successful BHS scheme is slightly different from the original definition. Indeed, in the original definition, it is required that there is at least one surviving agent, and this agent knows the location of all edges incident to the black hole. However, this is impossible in this model since the agents have only constant memory.

### 2.6.2 The Ring Topology

For the ring topology four different scenarios were considered in [21] depending on whether the tokens are movable or not, and whether the agents agree on a common orientation. Surprisingly, the agreement on the ring orientation does not influence the number of agents needed in the case of movable tokens but is important in the case of unmovable tokens.

The lower bounds presented in [21] are very strong in the sense that they do not allow any trade-off between the number of agents and the number of tokens for solving the BHS problem. In particular it was shown that:

- Any constant number of agents, even having unlimited memory, cannot solve the BHS problem with less tokens than depicted in all cases of Table 3.
- Any number of agents less than that depicted in all cases of Table 3 cannot solve the BHS problem even if the agents are equipped with any constant number of tokens and they have unlimited memory.

Meanwhile the algorithms match the lower bounds on the number of tokens, are asymptotically time-optimal and since they do not require any knowledge of the size of the ring or the number of agents, they work in any anonymous synchronous ring, for any number of anonymous identical agents (respecting the minimal requirements of Table 3).

**Table 3.** Results for BHS with DFAs and tokens in synchronous rings

| Tokens are | Ring is | Resources necessary and sufficient | |
|---|---|---|---|
| | | # agents | # tokens |
| Movable | Oriented | 3 | 1 |
| | Unoriented | | |
| Unmovable | Oriented | 4 | 2 |
| | Unoriented | 5 | 2 |

### Impossibility Results

**Oriented Rings.** Consider that the agents agree on the orientation of the ring (i.e., the ports of the edges are consistently labeled as 'left', 'right'). When the tokens are unmovable, a team of any constant number of agents needs at least two tokens per agent to solve the BHS problem. This is due to the fact that with only one unmovable token the agents are forced to always take the same actions including marking (incorrectly in different sized rings) links as dangerous.

**Theorem 7** ([21]). *For any constant k, there exists no algorithm that solves BHS in all oriented rings containing one black hole and k or more scattered agents, when each agent is provided with only one unmovable token. The result holds even if the agents have unlimited memory.*

To solve the BHS problem in a ring, both links leading to the black hole need to be marked as dangerous. Thus, we immediately arrive at the following result.

**Theorem 8** ([21]). *Two mobile agents carrying any number of movable (or unmovable) tokens each, cannot solve the BHS problem in an oriented ring, even if the agents have unlimited memory.*

When the tokens are unmovable, even three agents are not sufficient to solve BHS as shown below. This is due to the following facts: (a) because of the constant number of unmovable tokens the agents cannot guarantee to leave a token at a node which is adjacent to the black hole, and (b) after one of them vanishes in the black hole, the remaining two agents could possibly meet (by exploiting the asymmetry left by the vanished agent) but they cannot discover both incident links to the black hole.

**Theorem 9** ([21]). *Three mobile agents carrying a constant number of unmovable tokens each, cannot solve the BHS problem in an oriented ring, even if agents have unlimited memory.*

**Unoriented Rings.** In an unoriented ring (i.e., the clockwise direction perceived by an agent is handled by an adversary), even four agents do not suffice to solve the BHS problem with unmovable tokens, since two of them can be forced by the adversary to vanish in the black hole at the same time leaving their tokens more than a constant distance away from the black hole and the remaining two agents cannot correctly mark both links incident to the black hole.

**Theorem 10** ([21]). *In an unoriented ring, four agents carrying any constant number of unmovable tokens each, cannot correctly mark any link incident to the black hole, even when the agents have unlimited memory.*

**A BHS Scheme with Movable Tokens**

If each agent has a movable token it can perform a cautious walk type movement using its token. The `Cautious-Walk` procedure consists of the following actions: Put the token at the current node, move one step in the specified direction, return to pick up the token, and again move one step in the specified direction (carrying the token).

We show that only three agents are sufficient to solve BHS, when they have one movable token each. Algorithm 3 achieves this, both for oriented and unoriented rings.

**Theorem 11** ([21]). *Algorithm 3 solves the BHS problem in an unoriented ring with $k \geq 3$ agents having constant memory and one movable token each.*

**BHS Schemes with Unmovable Tokens**

For agents having only unmovable tokens, we use the *probing* technique (as discussed in previous sections) for exploring new nodes. Now in order to use this technique, we need to gather two agents at the same node and break the

**Algorithm 3.** (*Three DFAs with one movable token in synchronous rings*)

1: **repeat**
2: CautiousWalk(Left)
3: **until** you see a token and no agent OR next link is marked Dangerous
4: MarkLink(Left)
5: **repeat**
6: CautiousWalk(Right)
7: **until** you see a token and no agent OR next link is marked Dangerous
8: MarkLink(Right)

symmetry between them, so that distinct roles can be assigned to each of them. This is the main difficulty that has been taken care by the algorithms. The basic idea of our algorithms is the following. We first identify the two homebases that are closest to the black hole (one on each side). These homebases are called *gates*. The gates divide the ring into two segments: one segment contains the black hole (thus, is dangerous); the other segment contains all other homebases (and is safe). Initially all agents are in the safe part and an agent can move to the dangerous part only when it passes through the gate node. We ensure that any agent reaching a gate node, waits for a partner agent in order to perform the probing procedure. We now present two BHS algorithms, one for oriented rings and the other for unoriented rings.

**Oriented Rings.** In this section, we describe an algorithm (`Algorithm BHS-Ring-2`) using at least four agents with two unmovable tokens.

**Description of `Algorithm BHS-Ring-2`:** During the first phase of the algorithm each agent places a token on its homebase, moves left until the next homebase (i.e., next node with a token) and then returns to its homebase to put down the second token. During this phase one agent will fall into the black hole and there will be a unique homebase with a single token (we call this node the *gate* node) and all the other homebases will eventually contain exactly two tokens each. However, the agents may not complete this phase of the algorithm at the same time. Thus during the algorithm, there may be multiple homebases that contain a single token. Whenever an agent reaches any node containing a single token, it waits for a partner agent and then they perform probing in the left direction. One of the agents of a pair eventually falls into the black hole and the other agent marks the edge leading to the black hole and returns to the gate node, waiting for another partner. When another agent arrives at this node, these two agents perform probing in the opposite direction to find the other incident link to the black hole. ■

**Theorem 12** ([21]). *Algorithm BHS-Ring-2 correctly solves the black hole search problem in any oriented ring with 4 or more agents having constant memory and carrying two unmovable tokens each.*

**Unoriented Rings.** For unoriented rings, we need at least 5 agents with two unmovable tokens each. The `Algorithm BHS-Ring-3` for unoriented rings using five agents with two unmovable tokens is similar to the one for oriented rings, except that each agent chooses an orientation. When two agents meet they can agree on the orientation and assign different roles to themselves.

**Description of Algorithm BHS-Ring-3:** Each agent puts one token on its homebase, goes on *its left* until it sees another token and then returns to its homebase. Now the agent goes on its right until it sees a token and then returns again to the homebase. The agent now puts its second token on its homebase. During this operation exactly two agents will fall into the black hole. Each surviving agent walks to its left until it sees a node $u$ with a single token. At this point the agent has to wait, since either there is a black hole ahead, or $u$ is the homebase of an agent $b$ that has not yet returned to put its second token. Since we assume there are at least five agents, at least two of them will meet at a gate. They identify one of the links incident to the black hole and then the remaining agent can join the other agent who waits and identify together the other link incident to the black hole. ∎

**Theorem 13** ([21]). *Algorithm BHS-Ring-3 correctly solves the black hole search problem in an unoriented ring with 5 or more agents having constant memory and carrying two unmovable tokens each.*

### 2.6.3 The Torus Topology

While in ring topologies the exploration of a safe (even unoriented) ring is feasible by one DFA with one unmovable token, in torus topologies the exploration of a (safe) torus is not always possible by a DFA using tokens.

**Impossibility Results in an Oriented Torus**

**Agents with Unmovable Tokens**

**Theorem 14** ([20]). *For any constant numbers $k, t$, there exists no algorithm that solves BHS in all oriented tori containing one black hole and $k$ scattered agents, where each agent has a constant memory and $t$ unmovable tokens.*

The idea of the proof of Theorem 14 is the following: It is shown that an adversary (by looking at the transition function of an agent) can always select a big enough torus and initially place the agents so that no agent visits nodes which contain tokens left by another agent, or meets with another agent. Moreover there are nodes on the torus never visited by any agent. Hence the adversary may place the black hole at a node not visited by any of the agents to make any algorithm fail.

**Agents with Movable Tokens.** The situation with movable tokens is quite different than with unmovable ones since now any oriented (safe) torus can be eventually explored even by one DFA carrying one movable token (exploration without stop, also known as *perpetual* exploration). Hence any lower bound using movable tokens can not be based on impossibility of exploration (visiting all nodes of the torus).

A fairly easy observation is that two agents carrying any number of movable tokens cannot solve the BHS problem in an oriented torus even if the agents have unlimited memory due to the fact that the adversary can always place the agents and the black hole in such a way that one of the agents vanishes while changing vertical rings and the other one while changing horizontal rings. Additionally the following lower bound is shown:

**Lemma 8** ([20]). *There exists no algorithm that could solve the BHS problem in all oriented tori using three agents with constant memory and one movable token each.*

The idea of the proof is that at least two of the three agents are forced to leave their tokens more than a constant number of nodes away (otherwise they are not able to explore the torus) from the black hole before they vanish and then the third agent cannot decide for the correct location of the black hole.

Hence the following theorem holds:

**Theorem 15** ([20]). *At least three agents are necessary to solve the BHS problem in an oriented torus of arbitrary size. Any algorithm solving this problem using three agents requires at least two movable tokens per agent.*

**BHS Schemes in Oriented Tori Using Movable Tokens**

Due to the impossibility result from the previous section, any algorithm for the BHS problem by DFAs should use movable tokens. We describe below an algorithm which leads three agents carrying three tokens each to locate the black hole.

**Algorithm** `BHS-torus-33`. An agent explores one horizontal ring at a time and then moves one step South to the next horizontal ring and so on. When exploring a horizontal ring, the agent leaves one token on the starting node. This node is called the *homebase* of the agent and the token left (called homebase token) will be used by the agent to decide when to proceed to the next horizontal ring. The agent uses the two remaining tokens to repeat Cautious-Walk in the East direction until it has seen twice a node containing one token. Any node containing one token is a homebase either of this agent or of another agent. The agent moves to the next horizontal ring below (using again Cautious-Walk with three tokens) after encountering two homebases. It then repeats the same exploration process for this new ring leaving one token at its new homebase. Whenever the agent sees two or three tokens at the end of a cautious-walk, the agent has detected the location of the black hole: If there are two (resp. three) tokens at the current node, the black hole is the neighboring node $w$ to the East (resp. South). In this case, the agent stops its normal execution and then traverses a cycle around node $w$, visiting all neighbors of $w$ and marking all the links leading to $w$ as dangerous. ∎

**Theorem 16** ([20]). *Algorithm BHS-torus-33 correctly solves the BHS problem using* 3 *or more agents with three tokens each.*

Using a similar technique, the authors present an algorithm that solves the problem using four agents and two tokens. Finally they give a more involved algorithm (using a technique which makes the agents meet when they are 'close' enough) that meets the lower bound, i.e., solves the BHS problem using three agents with two tokens each.

**Theorem 17** ([20]). *The BHS problem can be solved in any oriented torus with exactly three agents carrying two tokens each.*

**The Case of an Unoriented Torus**
The situation in an unoriented torus is quite different than in an oriented one. As new results in [91] show, any constant number of DFAs with one movable token each cannot explore all (safe) unoriented tori and thus the BHS problem cannot be solved. However, surprisingly enough[1], it is shown in [91] that one agent with two movable tokens can explore any totally unoriented (safe) torus which gives a hope that even in a totally unoriented torus the BHS problem can be solved by a small number of agents with movable tokens. In a partially unoriented torus the BHS problem can be solved using five agents with constant memory and three movable tokens [91].

## 3 Search in Asynchronous Dangerous Networks

In the following section, we overview the state of the art for black hole search in various asynchronous networks focusing on the ring topology. Our presentation will follow the different variants given in Fig. 4.

### 3.1 Solutions Under Different Communication Models

In asynchronous networks, agents may wake up at different times. Agents may also never meet each other regardless whether they vanish in the black hole or not. Therefore, face-to-face communication is not of great use to solve the black hole search problem. We will therefore focus on solutions that use a whiteboard or tokens.

#### 3.1.1 Pure Token Model
Flocchini *et al.* [59] first prove that the pure token model is as powerful as the whiteboard model and that, in an arbitrary network, its complexity is the same as that of the whiteboard model if each of the co-located agents carries a map. They also show that 2 co-located agents, each with 1 token, can locate the black hole in a ring topology using a technique called *ping-pong*. In this specific case, when the network topology is known, the agents can achieve this goal without using a map. They further demonstrate that this ping-pong technique can also

[1] Given the old result of Rollik in [100] (discussed in the beginning of the section) which proved that no finite set of finite automata can cooperatively perform exploration of all cubic planar graphs.

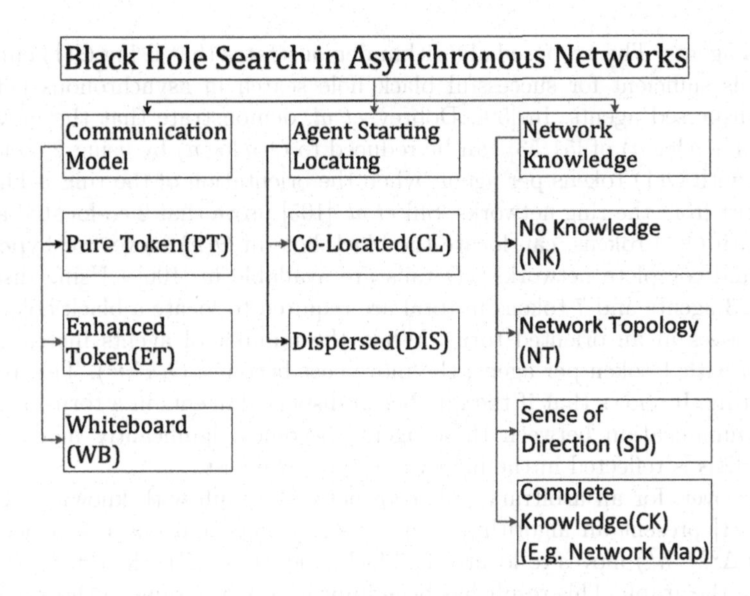


**Fig. 4.** Different variants for black hole search in asynchronous networks

be applied to an arbitrary network if a corresponding network map is available to each agent. In the latter case, it costs $\Theta(n \log n)$ moves to locate the black hole. (Additional details are given in [60].)

It is known that $\Delta + 1$ (Recall that $\Delta$ is the maximum node degree in the network graph.) agents are necessary to locate the black hole when the topology of an asynchronous network is unknown, regardless of the number of tokens used [69]. With the same number of agents and $O(1)$ tokens in total, it is possible to locate the black hole if each agent has a network map available. Balamohan *et al.* study whether $\Delta + 1$ agents, each with $O(1)$ tokens, can still locate the black hole in an unknown network in [6]. They prove that in order to keep the total number of tokens used to $O(1)$, $\Delta + 1$ agents are not sufficient. They then present a protocol that uses $\Delta + 2$ agents, each carrying 3 tokens, to locate a black in an unknown network.

Finally, we remark that, when using the pure token model for black hole search in an asynchronous network, researchers have exclusively considered co-located agents.

### 3.1.2 Enhanced Token Model

Due to the limitations of the pure token model, Dobrev *et al.* [41,50,52,53] and Shi *et al.* [105] use the enhanced token model to further improve the move and agent costs. In all these studies, each agent can carry and most importantly can place in the same node more than 1 token at any time. Given these characteristics, Dobrev *et al.* [51,53] introduce an algorithm to locate the black hole in an un-oriented ring network with dispersed agents. Same as in synchronous networks, coordinating dispersed agents is significantly more complex than using co-located agents. The proposed algorithm demonstrates that using $O(1)$ enhanced tokens is sufficient for successful black hole search in asynchronous networks using dispersed agents. In [50], Dobrev *et al.* demonstrate that the move cost of $O(kn + n \log n)$ of [51,53] can be reduced to $O(n \log n)$ by using 2 co-located agents with $O(1)$ tokens per agent, when the orientation of the ring is known.

Apart from the ring networks, Shi *et al.* [105] prove that 2 co-located agents, each with $O(1)$ tokens, can locate the black hole in $\Theta(n)$ moves for hypercube, torus and complete networks. (Details are available in [106].) Using dispersed agents, 3 agents and 7 tokens in total are required to locate a black hole within $\Theta(n)$ moves in an oriented torus. When the number of agents increases to $k$ ($k > 3$) with 1 token per agent, the move cost becomes $O(k^2n^2)$. This result is interesting. It shows that if the number of dispersed agents in a torus increases, the communication between these agents becomes significantly more complicated. This is reflected in the increase of the move cost.

Moreover, for an arbitrary unknown network graph with known $n$, Dobrev *et al.* [41] present an algorithm using $\Delta + 1$ agents and one token per agent and $O(\Delta^2 M^2 n^7)$ moves to locate the black hole. Here $M$ is the total number of edges of the graph. This result has been improved by the same authors in [42] to $O(\Delta^2 M^2 n^5)$ moves. In contrast, under the same assumption in the whiteboard model, the cost of the algorithm is $\Delta + 1$ agents and $\Theta(n^2)$ moves. For arbitrary unknown network graphs, the costs of the enhanced token model are significantly greater than those of the whiteboard model [41]. However, when a network map is available to the agents, the costs of the enhanced token model can be reduced to the same as those for the whiteboard model [50].

### 3.1.3 Whiteboard Model

In both types of token models, agents can only express very limited messages. This is why the whiteboard model is still the most popular agent communication model and has been studied by many (e.g., [7,8,28,39,40,43–45,47–49,75]).

In addition to presenting solutions to black hole search in asynchronous arbitrary networks [44,48,49], Dobrev *et al.* [45] solve a multiple agents rendezvous problem in a ring network that contains a black hole. In their paper, the final goal of the agents is not only to locate the black hole but also to collect all survived dispersed agents in one node. The authors offer a protocol that can rendezvous $k$ agents in $\Theta(n)$ time units. They claim that when $k$ is unknown, this protocol is also a solution to the black hole search problem. In terms of the time complexity in rings, Dobrev *et al.* [43,47] show that at least $2n - 4$ time units are needed in the worst case and give an algorithm, achieving it using $n - 1$ co-located agents. (Here movement and exploration are assumed to consume one time unit.) Apart from time complexity, the authors also prove that 2 agents are necessary and sufficient and present an algorithm to locate the black hole in $O(n \log n)$ moves, regardless whether the agents are co-located or dispersed, provided the orientation of the ring is known *a priori*. If the ring is un-oriented, 3 dispersed agents are necessary and sufficient. Apart from rings, Dobrev *et al.* [39] (with additional details in [40]) also present a general strategy to locate the black hole in $O(n)$

moves by using 2 co-located agents for some other common interconnected networks, such as *cube-connected cycles, wrapped butterflies, star graphs, chordal rings, hypercubes, tori of restricted diameter*, and in *multidimensional meshes.*

Based on Dobrev's work, Balamohan *et al.* [7] prove that $3n \log_3 n - O(n)$ moves are necessary in an asynchronous ring when 2 co-located agents are used. As for time complexity, Balamohan *et al.* [8] improve the algorithm of [43] to solve the problem in an average of $\frac{7}{4}n - O(1)$ time units when $n-1$ agents are used (with 2 extra time units required in the worst case). The authors also propose another algorithm to locate the black hole in $\frac{3}{2}n - O(1)$ time units on average, using $2(n-1)$ agents without increasing the time complexity in the worst case.

While all the above studies only consider the case of undirected graphs, Czyzowicz *et al.* [28] study the black hole search in directed graphs. They show that at least $2^{\Delta}$ agents are necessary in the worst case, where $\Delta$ is the in-degree of the black hole. If a planar graph with a planar embedding is known to the agents, $2\Delta$ agents are needed, and $2\Delta + 1$ agents are sufficient.

## 3.2 Solutions Under Different Agent Starting Locations

As discussed for the synchronous networks, when the homebases of the agents are dispersed, black hole search is more complex than if all agents wake up in the same node. This is even more so for an asynchronous network: given agents may wake up at different times, coordinating them to locate the black hole with minimal resource cost is a challenge. For example, 2 co-located agents suffice to solve the problem in a complete network in $\Theta(n)$ moves in [106]; while using dispersed agents costs $O(n^2)$ moves.

### 3.2.1 Co-located Agents

The co-located agent model is frequently used in the literature. Many whiteboard based studies adopt this model (e.g., [7,8,28,40,43,44,48,49,75]). Similarly, in token-based research, many choose to solve the problem under this model (e.g., [6,41,42,50,59,60,105,106]). Among these papers, [7,8,43,50,59,60] specifically consider ring networks, while [7,8,43] instead study time complexity. In particular, [8] offers an algorithm that improves the average time from [43]. Moreover, [50,59,60] only use 2 agents, and [50] studies the enhanced token model, while [59,60] investigate the pure token model.

As previously suggested, when the agents are initially co-located, they can easily establish agreements before any exploration. This can greatly help coordinating agents and eventually reducing the resource costs. For example, in a ring network, when the agents are co-located, the orientation is no longer important. This is because when there are only two directions, the agents can certainly make an agreement at the beginning of the exploration on what direction to take. Furthermore, solving the problem using co-located agents in a ring with $n$ nodes is the same as having each agent carry a network map in asynchronous networks. The situation is different when using dispersed agents. That is, unless the orientation of the ring is known, having a map or not leads to different solutions.

The following example (described in [59,60]) illustrates how a pair of co-located agents can locate the black hole using such an 'agreement': 2 agents each with one token start to explore the ring using cautious walk; one going right and the other going left. However, only one agent at a time is allowed to explore. To ensure this, one agent must first 'steal' the token from the other before its start its exploration. Stealing is possible because, during cautious walk, an agent has to leave a token before going to the next node. After such a theft, the agent without a token cannot continue exploration and has to go 'back' to look for a token. This is repeated until one agent vanishes. For example, suppose the right agent goes first. Before the left agent starts, it must first go right and steal the token of the right agent, and then it goes left for exploring. Once the right agent finds its token has gone, it goes left and steals a token from the left agent, and then goes right again. Repeating this process can ensure that only one agent vanishes in the black hole and that the surviving one knows the location of the lost agent.

### 3.2.2 Dispersed Agents

Dispersed agents have been adopted by the research based either on the whiteboard model [43,45,47,61,75] or on the enhanced token model [52,53,105,106]. Furthermore, no one has yet offered solutions to black hole search in asynchronous networks that use dispersed agents carrying pure tokens. The reason for this might be that such a solution is likely to use more pure tokens than one that relies on enhanced tokens.

Both Shi *et al.* [105] and Dobrev *et al.* [43] consider the use co-located agents *and* the use of dispersed ones. More specifically, in [105] authors focus on agent moves in *hypercube, torus*, and *complete networks*, whereas in [43], they measure agent moves and time complexity in ring networks.

Finally, in [52], Dobrev *et al.* solve the black hole search problem using an algorithm called *Pair Elimination* in oriented ring networks. The agents are initially dispersed in the ring and each endowed with $O(1)$ enhanced tokens. This algorithm consists in letting all the agents try to form pairs as soon as they wake up. All paired agents eliminate all the single agents they meet. Each pair has a level. A pair increases its level each time it eliminates another agent. When two pairs meet, the higher level pair always eliminates the lower level pair. Between pairs of the same level, the right pair eliminates the left pair. Eventually only one pair will survive, and one of the two agents forming that pair will locate the black hole. In contrast to the co-located case (for which each agent carries only 1 pure token), pair elimination requires 4 tokens for each agent even when they use the enhanced token model in the dispersed case. (This stems from the fact that communication/coordination among dispersed agents is significantly more complex than the co-located case.)

### 3.3 Solutions Under Different Network Knowledge

Most existing work on black hole search in asynchronous networks (e.g., [8,50, 52,59,60]) assumes agents have *knowledge of incoming links*, which means that when an agent enters a node, it is 'told' which port it used to do so. In turn, this enables this agent to possibly 'go back' to its previous node. Conversely, Glaus *et al.* [75] study arbitrary, unknown distributed systems without knowledge of incoming links. They present a lower bound on the size of the optimal solution, showing that at least $\frac{d^2+d}{2}+1$ co-located agents are necessary and sufficient to locate the black hole. Here $d$ denotes the number of links leading into the black hole (i.e., the node degree of the black hole).

In an un-oriented network, all ports that lead to a black hole should be marked as dangerous, hence $\Delta+1$ agents are necessary. However, in an oriented network, the number of agents that die in the black hole can be reduced by forcing agents to only enter a node from certain directions. For example, given a torus whose nodes have their ports labelled as *north, south, east*, and *west*, Shi *et al.* [106] assume an agent can only enter a node from the west and come out from the east, or enter from the north and come out from the south. With this assumption, only 3 agents are necessary. In contrast, when agents are allowed to enter a node from all four directions, at least 5 agents are necessary.

Dobrev *et al.* [46] prove that without any knowledge, $\Delta+1$ agents are needed and the cost is $\Theta(n^2)$. However, with a sense of direction but lack of information of the network topology, only 2 agents are required to achieve the same cost. The main idea of that algorithm is as follows: (a) the two agents start from the homebase $hb$ and construct at $hb$ a spanning tree of explored nodes (i.e., those visited by one agent); (b) an agent searches this tree and if there is a node with unexplored ports, that agent goes to explore that node in order to make all its ports explored using cautious walk; (c) after each such exploration, the agent comes back to $hb$ and adds that node to the tree as an explored node. The algorithm depends on a agent leaving navigation instructions (i.e., where it is going) to the other agent, each time the former leaves the homebase. The algorithm terminates when the number of explored nodes reaches $n-1$.

Again we observe that the knowledge of the network topology (e.g., ring, hypercube, torus, complete, tree and arbitrary networks) has great impact on results for black hole search. For example, Balamohan *et al.* [7,8], Chalopin *et al.* [21] and Dobrev *et al.* [43,47,50,52] propose algorithms based on ring networks. In the same vein, Shi *et al.* [106] design algorithms for hypercube and torus networks with co-located agents, and for torus and complete networks with dispersed agents. Also, Dobrev *et al.* [39,40] present a general strategy that allows 2 agents to locate the black hole with $O(n)$ moves in some common interconnected networks. In contrast, [6,39,40,44,46,49,59,60] search the black hole in arbitrary networks.

As just mentioned, for an arbitrary network, Dobrev *et al.* [44,46] prove that using the whiteboard model, the black hole search problem can be solved with $\Delta+1$ agents in $\Theta(n^2)$ moves without network maps. Also, recall this result (pertaining to move complexity) can be achieved using only 2 agents provided there is a sense of direction. With complete knowledge of the network, 2 agents are sufficient and the cost can be reduced to $\Theta(n \log n)$. In another paper [48], Dobrev *et al.* present a universal protocol that locates the black hole using at most $O(n + d \log d)$ moves with 2 agents each carrying a network map. Here $d$ is the diameter of the network. Still using 2 agents, the same authors [49] present a strategy that can locate the black hole in $O(\Sigma_{i=1}^{k}|C_i| \log |C_i|)$ moves, here $C = C_1, C_2, ..., C_i..., C_k$ is an open vertex cover by cycles of a 2-connected graph[2].

The point to be grasped is that these results show that having a network map or a sense of direction can significantly reduce the cost complexity in asynchronous networks.

### 3.4 Black-Hole Search in an Asynchronous Ring

The Black-Hole Search (BHS) problem has been introduced in [43] (see also the full version of the paper in [47]) where it was studied for asynchronous ring topologies. We present here some results from this seminal paper. We first discuss the model used and give lower bounds on the number of agents and time-complexity. We then give two algorithms that solve the problem using two or more co-located mobile agents with memory using whiteboards.

#### 3.4.1 Model and Basic Lower Bounds

We consider two anonymous co-located agents with memory. The ring is anonymous, asynchronous and consists of $n$ nodes. On each node there is a *whiteboard* where the agents can leave messages. The access on a whiteboard is done using mutual exclusion and hence the agents can acquire distinct identities by the order in which they access the whiteboard (e.g., the first agent accessing the whiteboard creates a counter on it, initializes it to 1 and gets this identity, while the next agent increases the counter and take its value as its identity). Although the ring is anonymous, the distinct identities assigned to agents, allow them to agree on the clockwise direction. The goal is that after a finite time at least one agent should report back to the starting node the exact location of the black hole.

It is trivial to see that, if there is only one agent, the BHS problem is unsolvable since the only agent would necessarily vanish into the black hole. Hence at least two agents are needed to locate the black hole.

Due to network asynchrony, it is impossible to distinguish between a 'slow' link and a link leading to a black hole. This observation gives us the following two lemmas.

**Lemma 9** ([47]). *The problem of whether there exists or not a black hole in an asynchronous network is unsolvable.*

[2] An *open vertex cover by cycles* (C) is defined as a set of simple cycles such that (a) each vertex of $G$ is covered by a cycle from $C$ and (b) the connectivity graph of these cycles (where each cycle is represented by a vertex, and 2 vertices are connected if the corresponding cycles share an edge) is connected.

**Lemma 10** ([47]). *It is impossible to locate the black hole if the size of the network is unknown.*

The following theorem gives us a lower bound on the number of steps the agents need to locate the black hole in a ring.

**Theorem 18** ([47]). *Any algorithm needs at least $2n-4$ moves to find the black hole in a ring regardless of the number of agents available.*

*Proof.* In order to report the position of the black hole back to the starting node $h$, the agents need to receive information from any other node apart from the node $\mathcal{B}$ containing the black hole. This means that every node apart from $\mathcal{B}$ has to be visited by at least one agent. Suppose that the black hole resides at a distance $n-1$ clockwise. Any agents traveling in counter-clockwise direction would vanish in the black hole but (due to the asynchronous network) the remaining agents cannot decide whether the black hole is at $\mathcal{B}$ or the link to node $\mathcal{B}$ is 'slow'. Hence an agent must travel clockwise to node $n-2$ and then an agent must report back to $h$. Therefore a total number of $2n-4$ steps must be taken.

### 3.4.2 A Protocol for Two Agents

We denote with $U$ and $E$ the unexplored and explored area respectively. We also denote with $U^L$ and $U^R$ the continuous unexplored area adjacent counter-clockwise and clockwise respectively from the explored area. A basic tool which is used in the algorithm is the *Cautious Walk* (introduced in [43,47]):

Consider an agent situated at a node $v_0$ adjacent to an unexplored node $v_1$. The agent explores a previously unexplored area $U_k = <v_1, v_2, ..., v_k>$ in the following way:

*Cautious Walk:*

- before leaving a node $v_i$ going to a node $v_{i+1}$, the agent marks the port leading from $v_i$ to $v_{i+1}$ as *active*,
- immediately after visiting $v_{i+1}$, the agent returns to $v_i$, and marks the port leading from $v_i$ to $v_{i+1}$ as *safe*,
- the agent checks for messages at $v_i$ and (if such a message exists) re-assigns to itself an unexplored area $U_k'$ which has to discover and repeats from the start.

The agents know the size $n$ of the ring. They follow Algorithm 4. A high level description of the algorithm is the following:

Suppose the agents start at node $h$. Using mutual exclusion they write at the whiteboard of node $h$ and get distinct identities as described in the beginning of the section. They also agree on the clockwise orientation. Then they divide the unexplored area $U$ with $|U| = n-1$ into two disjoint paths $U^L$ and $U^R$ with $|U^L| = \lceil \frac{n-1}{2} \rceil$ and $|U^R| = \lfloor \frac{n-1}{2} \rfloor$. Agent 1 explores the area $U^L$ and agent 2 explores the area $U^R$ using *Cautious Walk.* Since there is exactly one black hole, and the two sets are continuous and disjoint, after a finite time exactly one of the agents will finish with the exploration. Suppose without loss of generality that agent 1 finishes. Then agent 1 traverses back through the explored area until it meets a node $u$ whose port leading to a node $v$ has not been marked as safe. At that point agent 1 updates explored area and divides the updated unexplored area into new continuous and disjoint paths $U^L$ and $U^R$ having $U^L$ starting at a node situated counter-clockwise from $h$ and $U^R$ starting at node $v$ situated clockwise from $h$. Assigns $U^L$ to itself and $U^R$ to agent 2 and leaves this message at node $u$. Now agent 1 traverses the explored area and explores its new assigned path $U^L$. Agent 2 either travels through a slow link (not having explored its assigned area) or vanished in the black-hole. In the first case agent 2 will first return at node $u$, will check for messages and update its assigned area for exploration. The agents repeat this procedure until exactly one of them vanishes into the black-hole. Then the other agent which repeats the procedure, will eventually come up with an explored area of size $n-1$. At that point it knows the exact location of the black-hole.

**Algorithm 4.** (*2 agents with memory in an asynchronous ring with whiteboards*)

1: The agents get distinct identities and agree on the clockwise orientation
2: Let $X := h$
3: **while** $|E| < n-1$ **do**
4: Divide $U$ into two continuous disjoint parts $U^L$ (starting counter-clockwise of node $X$) and $U^R$ (starting clockwise of node $X$) of almost equal sizes
5: Agent 1(2) leaves a message at $X$ saying that she will explore $U^L(U^R)$
6: Agent 1(2) explores $U^L(U^R)$ using *Cautious Walk*
7: Agent 1(2) traverses clockwise (counter-clockwise) the explored area until it reaches a node $u$ with a not safe port
8: Update $E$ and $U$
9: Let $X := u$
10: **end while**
11: Report the black-hole location

**Theorem 19** ([47]). *Algorithm 4 locates the black-hole in an asynchronous ring of $n$ nodes with two co-located agents within* $2n \log n + O(n)$ *moves.*

A natural question is the following: Can we decrease the time needed for Black Hole Search if we have $k \geq 2$ available co-located agents?

### 3.4.3 The Case of $n-1$ Agents

Algorithm 5 shows that $n-1$ co-located agents can locate the black hole in an asynchronous ring of $n$ nodes within $2n-4$ moves.

In Algorithm 5, it should be clear that among the $n-1$ agents, exactly one will finish the algorithm, while all the other $n-2$ agents will vanish into the black hole. Also notice that the $n-1$ agents use the whiteboard only at the starting node in order to assign distinct identities to themselves.

**Algorithm 5.** ($n-1$ *agents with memory in an asynchronous ring with whiteboards*)

1: Agents get distinct identities from the set $\{1, 2, ..., n-1\}$ and agree on the clockwise orientation
2: Agent $i$ travels $i-1$ edges in clockwise direction
3: Agent $i$ travels $n-2$ edges in counter-clockwise direction
4: Agent $i$ returns to homebase traveling in clockwise direction
5: Agent $i$ reports that the black hole resides at a distance $i$ clockwise

**Theorem 20** ([47]). *Algorithm 5 lets* $n-1$ *co-located agents find the black hole in an asynchronous ring of* $n$ *nodes within* $2n-4$ *moves.*

*Proof.* Suppose that the black hole resides at a node $\mathcal{B}$ which is $\mathcal{B}$ steps clockwise from node 0 (where the agents start). Then agent with label $\mathcal{B}$ travels $\mathcal{B}-1+n-2+n-(\mathcal{B}+1)=2n-4$.

# 4 Solving Problems in Dangerous Graphs

## 4.1 Multiple Black Hole Search

As previously mentioned, the only way to locate a black hole in an asynchronous network is to have at least one agent visit all the nodes except the black hole. Therefore, the network minus the black hole has to be connected. Otherwise, the presence of the black hole may partition the network into several disconnected subgraphs, making it impossible to visit all nodes. Also recall that, in synchronous networks, with the help of a time-out mechanism, the single black hole search problem can still be solved even if the network is disconnected by the black hole (as is the case for tree networks).

Clearly, the problem becomes more complex when the network contains multiple black holes. For example, Fig. 5 illustrates a ring network containing 3 black holes that disconnect the ring into 3 sub-graphs. Unless there are enough agents starting at specific nodes, locating these multiple black holes cannot be guaranteed even on synchronous networks.

Strategies for finding multiple black holes can be intuitively grouped into three categories, each with different assumptions and results, which are discussed next.

### 4.1.1 Best Effort Without Modifying the Black Hole Search Problem

This strategy tries to find as many black holes as possible without modifying the traditional black hole search problem. In a synchronous network, as discussed in Sect. 1.3.1, finding out whether there is a single black hole is fairly easy given a time-out mechanism. Should the network be disconnected due to the presence of several black holes, some nodes may never be explored. In this case, finding all the black holes is impossible. Otherwise, Cooper *et al.* [25] offer a solution to

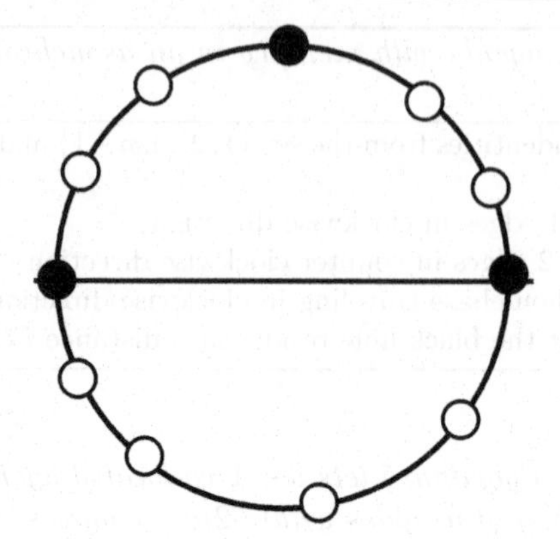

**Fig. 5.** A ring network that is disconnected by multiple black holes, represented as solid circles.

finding all possible black holes. First, they study the multiple black hole search problem in synchronous networks using the face-to-face model. They assume that $k$ co-located agents know the topology of the whole network including the size $n$ and number of black holes $b$. They conclude that any exploration algorithm needs $\Omega(n/k+D_b)$ steps in the worst case to solve a multiple black hole search problem, while $D_b$ is the diameter of the network with at most $b$ nodes deleted. They then provide a general algorithm that performs the exploration in $O(\frac{\frac{n}{k}\log n}{\log\log n}+bD_b)$ steps in an arbitrary network with network maps available to the agents, where $b \leqslant k/2$. In the case where $b \leqslant k/2$, $bD_b = O(\sqrt{n})$ and $k = O(\sqrt{n})$, they give a refined algorithm that performs the exploration in asymptotically optimal $O(n/k)$ steps. Ultimately a node can be identified as a black hole or as a safe node (if and only if it can be reached following a path of safe nodes).

### 4.1.2 Variants of the Black Hole Search Problem

In the traditional black hole search problem, the existence of a black hole is persistent. That is, a black hole is not affected by the arrival of any incoming agent. Cooper *et al.* [26] solve a variant of the multiple black hole search problem in synchronous networks by changing this model. They introduce the notion of a *faulty node*, which is a weak form of a black hole. A faulty node is repaired when first visited by an agent (which, however, vanishes repairing it). And once repaired, this node will permanently behave as a normal one. Hence, when a network contains more than one faulty node, the agents are still able to explore the whole graph. Also, if more than one agent enters the same faulty node at the same time, only one will repair the faulty node and vanish while the others can continue their explorations.

The agents used in [26] know the topology of the whole network, move synchronously, use the face-to-face model, and are initially co-located at the same node. Given a network map, the whole network is first divided into equal partitions of size $O(D)$, where $D$ is the diameter of the network. Consequently, an agent should spend $O(D)$ time to explore one such partition. All the agents start from the same homebase and each agent explores a partition. After $O(D)$ time, if an agent returns to the homebase, it is inferred that the partition it explored contains no faulty node. Should an agent not show up on time, it is assumed to be dead in a faulty node (which it repaired) and the partition to which it was assigned is still marked as unsafe and thus in need of further exploration. After one such iteration of exploration, once all surviving agents come back to the homebase, they will start a new iteration of exploration on the remaining unsafe partitions of the network. This process will be repeated until there are no more unsafe partitions. Eventually, this 'faulty node repair' problem can be solved within $O(\frac{n}{k} + \frac{D \log f}{\log \log f})$ time steps, where $f = min(\frac{n}{k}, \frac{n}{D})$, assuming that the number of faulty nodes is at most $k/2$. It must be emphasized that, in [26], because the face-to-face model leaves no mark on the nodes, once an agent vanishes repairing a node, the other agents cannot know where it vanished. Therefore, ultimately, all faulty nodes are repaired but their locations remain unknown.

D'Emidio *et al.* [35,36] study the same problem under the same conditions as [26] with a slight change to one assumption: if more than one agent enters the same faulty node at the same time, all agents vanish. Trying to make the problem more realistic, the authors however introduce a new behavior: if one agent enters a faulty node $u$, all agents within distance $r$ from $u$ disappear along with the faulty node. D'Emidio *et al.* first prove that the faulty node repair problem is NP-hard even when $b = k = 1$, where $b$ is the number of faulty nodes and $k$ is the number of agents. Second, when $r = 0$ (which means the agents die only when they physically enter a faulty node), using a simple variation of the algorithm of Cooper *et al.*, the faulty node repair problem can be solved in $\Theta(\frac{n}{k-b} + \frac{D \log f}{\log \log f})$ with $k > b$ always true. Otherwise, all agents will vanish. Third, for any $r > 0$, the faulty node repair problem requires $\Omega(n)$ time steps in the worst case. Fourth, when $r = 1$, the faulty node repair problem can be solved in $\Theta(n)$ time steps, and the authors provide two strategies to achieve this bound. Finally, the authors report their experimental results to demonstrate correctness.

Shi *et al.* [96,107] proposed a new attack mode that involves multiple *faulty nodes* that are repairable by software agents. Once repaired, a faulty node behaves like a normal one. However, a *gray virus* can infect again a repaired faulty node due to this node's inherent vulnerability. A gray virus is a piece of malicious software that can infect a repaired node by residing in it and turning it into a black hole. This gray virus has no effect on a normal node or link. Under this attack model, the authors propose solutions to solve the *Faulty Node Repair and Dynamically Spawned Black Hole Search* (*FNR-DSBHS*) problem in an asynchronous ring. In order to study the worst (i.e., the most expensive in terms of number of agents used) case for cost of a repair, the authors assume an agent "dies" after having repaired a faulty node. Moreover, should several agents simultaneously enter a faulty node, one agent will die after repairing that node, whereas all other agents die immediately. Clearly, fewer agents are needed to solve the problem in the case where at least one other agent in the same faulty node survives after the repair.

Contrary to the traditional black hole search in which all agents start in a network with one and only one black hole whose existence is known a priori, in the proposed new attack model, a repaired faulty node can be infected and turned into a black hole at any point in time while the agents traverse the network to try to repair the faulty nodes. This detail drastically changes the nature of the problem at hand in asynchronous networks: the possible scenarios in this case are significantly more complex than for the traditional black hole search, especially with the presence of multiple faulty nodes in need of repair, each of which eliminating all agents that simultaneously enter it. Furthermore, the co-existence of a black hole and multiple faulty nodes in the proposed model significantly increases the difficulty of both repairing faulty nodes and locating this black hole because no agent knows the difference between the two types of nodes a priori.

In an asynchronous network, if a gray virus can move faster than the agents, and/or if multiple gray viruses reside in the network, it is possible that, from an agent's viewpoint, all the repaired nodes appear to be black holes. Consequently, in the presence of a multi-stop gray virus or multiple gray viruses, the *FNR-DSBHS* problem becomes unsolvable in an asynchronous network. Thus both [96,107] study the *FNR-DSBHS* problem under the condition that a gray virus present in the asynchronous ring stops moving once it infects a repaired node.

Shi *et al.* [107] proposed an algorithm to solve this problem in an asynchronous ring network with only one whiteboard (which resides in a node called the *homebase*). The authors conclude that, using this proposed algorithm, $b + 4$ agents can repair all faulty nodes and locate the black hole infected by a virus $v$ within finite time. The algorithm works even when the number of faulty nodes $b$ is unknown a priori.

Peng *et al.* [96] demonstrate that in such a ring, $b + 9$ agents can repair all faulty nodes as well as locate *the* black hole that is infected by this single one-stop gray virus. They also show that in the worst case, within $O(kn^2)$ moves, $b + 9$ agents suffice to repair $b$ faulty nodes and report the location of the black hole that is infected, at any arbitrary point in time, by the one-stop gray virus.

In a different vein, Flocchini *et al.* [62,63] solve the multiple black hole search problem via a *subway model* using co-located agents with the whiteboard model, the number $b$ of black holes being known to the agents. The authors use carriers (the subway trains) to transport agents (the passengers) from node to node (the subway stops), and the carriers move asynchronously in a directed graph. When a carrier enters a node, the agents can either get off from the carrier and explore the node, or stay on the carrier to go to another node. In a traditional black hole search, any incoming data will be deleted, including the carrier. However, in this subway model, the black holes no longer affect the carriers and can only eliminate the agents. At the homebase, there is a whiteboard that is used to record all explored, unexplored and dangerous nodes. Initially, all nodes are recorded as unexplored except the homebase. Once an agent chooses to explore a node, the node will be marked as dangerous until the agent comes back and marks it as explored. Eventually, the algorithm terminates when $n - b$ nodes have been explored, the remaining $b$ dangerous nodes being the black holes. In [62,63], when $k = r + 1$ agents are used (where $r$ is the number of carrier stops at black holes), the number of carrier moves is $O(k \cdot n_C^2 \cdot l_R + n_C \cdot l_R^2)$. Here $n_C$ is the number of subway trains, and $l_R$ is the length of the subway route with the most stops.

Under the same assumption and keeping the same carrier moves as [62,63], Flocchini *et al.* [65] solve the same problem with dispersed agents. Instead of having a whiteboard at the homebase, these authors put the whiteboard on the carriers. Thus, an agent only has to come back to a carrier to update its exploration information.

#### 4.1.3 A Simplifying Assumption

As Fig. 5 suggests, if the presence of multiple black holes results in the network being effectively partitioned by the latter into several disconnected partitions, then it is impossible to visit all nodes without going through a black hole. In order to alleviate this difficulty, some researchers make the assumption that the network minus the black holes is connected.

For example, Flocchini *et al.* explicitly state this simplifying assumption in [64]: "after deleting all the black holes, the network still remains interconnected". (Clearly, without this assumption, it is impossible to locate all the black holes in any given network.) However, these authors also complicate multiple black hole search by adding link failure. That is, in their model, an link failure is locally detectable at an adjacent node. More specifically: (1) an edge is identified by its port number in its incident node and (2) if information about an edge is written on a whiteboard, an agent can notice the absence of an edge with such a port number. If no information about an edge is written (i.e., this edge has disappeared before any agent has visited), it is treated likely it has never existed. It is assumed that any such failure occurs only when no agent is traversing that link, and that the failures do not disconnect the safe part of the network (otherwise dangerous graph exploration is clearly unsolvable). Under this assumption, the authors present an algorithm to solve dangerous graph exploration with link deletions in an arbitrary unknown graph with asynchronous dispersed agents using the whiteboard model. The algorithm can correctly solve the link deletion problem within finite time by marking all safe edges as such, and marking as dangerous every port that is on a safe node leading to a black hole or to a faulty edge (i.e., an edge that has failed). The total number of moves performed by the agents is at most $O(k^2 \cdot n_s + n_s \cdot m + k \cdot n_s \cdot D)$, where $k$ is the number of agents, $n_s$ is the number of safe nodes, and $m$ is the number of edges or links.

Kosowski *et al.* [85,86] also assume that the graph is strongly connected if all black holes are removed. They find out that $O(d \cdot 2^d)$ co-located agents are sufficient to solve the black hole search problem in a directed graph with an arbitrarily large $n$, where the network is synchronous and $d$ is the number of edges leading to the black holes. Furthermore, the authors show that when $d = 2$, 4 agents are always sufficient in synchronous networks. However, in asynchronous networks, at least 5 agents are required when $d = 2$. Finally, when $d = 1$, 2 agents are always sufficient and sometimes required in both synchronous and asynchronous networks.

### 4.2 Other Types of Malicious Behaviour

Beyond studying the traditional black hole and its variants (e.g., (a) the repairable black holes introduced in [26] by Cooper *et al.* and (b) the new subway model presented by Flocchini *et al.* in [62,63]), work on other types of malicious hosts exists and is briefly discussed here.

Dobrev *et al.* [43,45] study the rendezvous problem of mobile agents in asynchronous rings in spite of a black hole. More specifically they study how $k$ dispersed agents can rendezvous in an anonymous, asynchronous ring in spite of a black hole. They first prove that if the ring is unoriented, then it is impossible for $k-1$ agents to rendezvous. They also show that if $k$ is unknown, then rendezvous requires locating the black hole and hence either $k$ or $n$ must be known for rendezvous. Rendezvous of dispersed agents can be solved easily in the "whiteboard" model in an oriented anonymous, asynchronous ring in spite of a black hole when $k$ is known. In this case one agent vanishes in the black hole while the rest will rendezvous within at most $3n-6$ traversals. In an unoriented ring and when $k$ is a known odd number $k-2$ agents can gather within at most $5(n-2)$ traversals. We refer the interested reader to articles [43,45] for more results concerning cases when $k$ is an even number, or unknown in oriented and unoriented rings, etc.

Chalopin *et al.* [23] study the rendezvous problem in a network with *faulty links*. In that model, some of the edges in the graph are dangerous for the agents: any agent that attempts to traverse such an edge (from either direction) simply disappears, without leaving any trace. Notice that if all the edges incident to a node $u$ are faulty, then node $u$ can never be reached by any agent and thus is essentially equivalent to a black hole.

Das *et al.* [33,34] consider the rendezvous problem in spite of a *malicious agent*. The agents that need to gather are called 'honest agents' and their communication model is similar to the face-to-face one. In the network there is a mobile fault which they call *malicious agent*. This agent can physically block the movement of a honest agent to the node it occupies, preventing the honest agents from gathering in some cases. The malicious agent can move arbitrarily fast along the edges of the graph, it has full information about the graph and the location of the agents, and it may even have full knowledge of the actions that will be taken by the honest agents. On the other hand the malicious agent can be detected by a honest agent, when it blocks the movement of this honest agent. They show that even one malicious agent can prevent gathering of honest agents in many cases [34]. For a single malicious agent, the graph must be at least bi-connected otherwise the problem is unsolvable. In [33] they investigate the feasibility of gathering for $k \geq 2$ honest agents in oriented and unoriented ring networks in the presence of the malicious adversary, that blocks other agents from having access to parts of the network. The honest agents are identical (thus anonymous), and they only have local communication capabilities allowing two agents to talk only when they are in the same node. Furthermore since the honest agents have only constant-sized memory, they do not know the size of the ring and the number of agents present in the ring. Since gathering of identical agents in a ring is impossible due to symmetry, even if the agents have infinite memory and know the size of the ring and the number of agents they use the following symmetry breaking mechanism: they assume the existence of a specially marked node in the ring which can be used as *landmark*. Although the use of a landmark node is the simplest mechanism for symmetry breaking that guarantees an immediate solution in fault free networks, in the presence of a malicious agent, the gathering problem still remains challenging as the malicious agent may prevent the honest agents from ever reaching the landmark node, if the malicious agent occupies this node. They provide a characterization of the feasible instances for gathering of $k$ agents in a ring of size $n$, in presence of a malicious agent. In an oriented ring, they show that gathering is always possible and they provide a gathering algorithm for any $k \geq 2$. In an unoriented ring network of $n$ nodes they show that gathering of asynchronous agents is not solvable if $k$ is even while, gathering of synchronous agents is unsolvable when both $n$ is odd and $k$ is even. They show that all the other cases in an unoriented ring are solvable for $k > 2$ agents and provide algorithms for all those cases. In particular, they give (i) an algorithm for asynchronous agents that works for any $n$ whenever $k$ is odd, and (ii) an algorithm for synchronous agents which works for all solvable cases, i.e., when either $k$ is odd or $n$ is even (assuming $k > 2$). For the special case of $k = 2$ agents in an unoriented ring, they prove that rendezvous is impossible for constant-sized memory agents; however they present a protocol for rendezvous of 2 synchronous agents when they have enough memory to count up to $n$. In [34] they study oriented mesh topologies and they prove that the problem can be solved when the honest agents initially form a connected configuration without holes if and only if they can see which are the occupied nodes within a two-hops distance.

Královič and Miklík [87,92] study how the various capabilities of a malicious host affect the solvability of exploration problems in asynchronous networks with whiteboards. They first consider networks with a malicious host (called *gray hole*) which can at any time choose whether to behave as a black-hole or as a safe node. Since the malicious behavior may never appear, the agents might not be able, in certain cases, to decide the location of the malicious host. Hence, they introduce and study the so called *Periodic Data Retrieval* problem in which, on each safe node of the network, an infinite sequence of data is generated over time and these data have to be gathered in the homebase. The goal is to design a protocol for a team of initially co-located agents so that data from every safe node are reported to the homebase, infinitely often, minimizing the total number of agents used. One agent can solve the problem in networks without malicious hosts, where the problem reduces to the *Periodic Exploration* problem (e.g., see [27] and references therein) in which the goal is to minimize the number of moves between two consecutive visits of a node. When the malicious host is a black hole, the Periodic Data Retrieval and the Periodic Exploration problem are solved by the same number of agents. As observed in [87], $n-1$ agents are sufficient for solving the Periodic Data Retrieval problem in any 2-connected network of $n$ nodes with one malicious host when the topology is known to the agents: each of the $n-1$ agents selects a different node of the network and periodically visits all other nodes. The authors show that two agents are not sufficient to solve the problem in a ring with a gray hole and they present a protocol which solves the problem using 9 agents. They also consider a second type of malicious host which behaves as a gray hole and, in addition, can alter the contents of its whiteboard; they show that 27 agents are sufficient to solve the Periodic Data Retrieval problem in a ring, under this type of malicious host. Later Bampas *et al.* [9] refine the model of [87] and improve the solution. They show that at least 4 agents are needed when the malicious host is a gray hole, and at least 5 agents are needed when the malicious host whiteboard is unreliable. On the positive side, they propose an optimal protocol for Periodic Data Retrieval in asynchronous rings with a gray hole, which solves the problem with only 4 agents. Finally, they propose a protocol with 7 agents when the whiteboard of the malicious host is unreliable.

Cai *et al.* [17–19] consider the problem of a *black virus* that, like a black hole, deletes any incoming agent. But, unlike a black hole (which is defined as a static host), a black virus moves from node to node, thus potentially increasing the number of dangerous nodes. Furthermore, unlike a black hole (which can only be located but not removed), a black virus is destroyed if it enters a node that contains an anti-viral system agent. Thus, the only way to remove a black virus is to surround it by anti-viral system agents and force it to move to a neighbouring node that already contains at least one anti-viral system agent. In the same vein, some theoretical work has focused on the *intruder capture* problem (also known as *graph decontamination*): an intruder (a harmful agent) moves through the network infecting nodes and the goal is to remove the intruder from the network using mobile agents. Unlike a black virus, an intruder can only harm nodes, not agents. This problem has been extensively studied as for example in [12,15,66].

Finally, *black hole attack* [3,11,109] is also a research topic remotely related to black hole search. Most importantly, the networks considered for black hole attack are different from those of black hole search: In the latter, the networks are static, while in the former, the networks can be dynamic (e.g., MANET (Mobile Ad-Hoc networks), wireless networks, mobile networks). For example, in MANET, the network topology is only formed once one node needs to send a data package. Khari *et al.* [81] survey security attacks, as well as secured routing protocols in MANET, and offer a definition for black hole attack. Moreover, their survey mentions a variation of black hole attack called *grey hole attack* [2,10,103]: whereas a black hole will delete any incoming data packages, a grey hole only deletes part of the packages.

# 5 Discussion and Future Work

In this section, we highlight some future work in the following three categories:

1. single black hole search in both asynchronous networks and synchronous networks,
2. multiple black hole search,
3. other models of agents and malicious behaviour.

## 5.1 Single Black Hole Search in Both Asynchronous and Synchronous Networks

We first list all possible combinations of different assumptions then organize the single black hole search studies under each such combination. Our findings are presented in Tables 4 and 5. We then point out the remaining combinations that have not yet studied, then identify interesting open problems for future research.

### 5.1.1 Asynchronous Networks

A list of papers studying the Black Hole Search problem in asynchronous networks is shown in Table 4.

We do not include edge-labelling in our discussion because it is widely adopted in the field. Also, since network size must be known *a priori* in asynchronous networks, we do not further mention $n$ in this subsection. Also recall that, when using co-located agents to explore a ring, whether or not the ring is oriented does not affect the move cost of the algorithm. Therefore, we do not discuss separately each of these two possibilities below. Finally, as ring is a special topology, namely the sparsest bi-connected graph, we list it separately in Table 4.

Glaus *et al.* solve the black hole search problem without the knowledge of incoming links in an unknown un-oriented arbitrary network when both the agents and the network nodes have distinct IDs. Whether, under the same assumptions, the black hole search problem is still solvable in an anonymous network when anonymous agents are used, remains an open problem. Solving the black hole search problem without the knowledge of incoming links in an unknown un-oriented arbitrary network also remains an open problem if (a) it is in a synchronous network and/or (b) tokens and/or (c) dispersed agents are used (in lieu of the asynchronous network with whiteboard and co-located agents of [75]).

Balamohan *et al.* identify another open problem in [8]: Is there an algorithm that locates the black hole in $\frac{3}{2}n - O(1)$ time (average case) and $2(n-1)$ time (the worst case) using $n-1$ co-located agents and the whiteboard model?

In [42], Dobrev *et al.* study the black hole search problem assuming no topology knowledge. They prove that the black hole can be located in any graph $G$ within $O(\Delta^2 M^2 n^5)$ moves, using the enhanced token model with $\Delta + 1$ co-located agents, where $\Delta$ is the maximum degree, $M$ is the number of edges and

$n$ is the number of nodes of $G$. Under the same conditions, solving the problem in the whiteboard model only costs $\Theta(n^2)$ moves. Whether there is a solution at a lower move cost for the problem using the enhanced token model when the topology knowledge is unknown remains another open problem. Our intuition is that relaxing the assumption regarding knowledge of topology possibly leads to further reductions of the move cost.

While only ring, hypercube, torus, and complete network have been studied under the enhanced token model, consideration of other topologies under the same assumptions may be another research direction. With respect to the use of the pure token model, whether or not the number of moves can be further reduced by increasing the number of agents and/or by knowing the topology at hand is an open question. Other open problems pertaining to this model include: (a) whether black hole search is solvable in an arbitrary unknown graph and (b) whether dispersed agents can still solve the problem (over co-located agents, which have been the only ones used with this model so far).

**Table 4.** Existing work on black hole search in asynchronous networks.

| # | Communication model | Agent starting location | Network knowledge | Paper |
|---|---|---|---|---|
| 1 | Whiteboard | Co-located | No knowledge | [28,44,46,75] |
| 2 | Whiteboard | Co-located | Network topology | [39,40] |
| 3 | Whiteboard | Co-located | Sense of direction and no network topology | [44,46] |
| 4 | Whiteboard | Co-located | Ring | [7,8,43,47] |
| 5 | Whiteboard | Co-located | Complete knowledge | [39,40,44,46,48,49] |
| 6 | Whiteboard | Dispersed | Un-oriented ring | [43,45,47] |
| 7 | Whiteboard | Dispersed | Oriented ring | [43,45,47] |
| 8 | Enhanced token | Co-located | No knowledge | [41,42] |
| 9 | Enhanced token | Co-located | Oriented ring | [50] |
| 10 | Enhanced token | Co-located | Sense of direction and network topology | [105,106] |
| 11 | Enhanced token | Dispersed | Sense of direction and network topology | [105,106] |
| 12 | Enhanced token | Dispersed | Oriented ring | [52] |
| 13 | Enhanced token | Dispersed | Un-oriented ring | [51,53] |
| 14 | Pure token | Co-located | Ring | [59,60] |
| 15 | Pure token | Co-located | Complete knowledge | [59,60] |
| 16 | Pure token | Co-located | No knowledge | [6] |

### 5.1.2 Synchronous Networks

Single black hole search in synchronous networks is not studied as often as in the more realistic asynchronous networks. Additionally, we have observed that, for black hole search in asynchronous networks, using co-located agents always costs fewer moves than using dispersed agents. As shown in Table 5, the pure token model is only used with dispersed agents. These 4 papers study the black hole search problem in this model on ring and torus topologies leaving it open for other topologies (such as hypercube or mesh): studying the use of co-located agents with the pure token model in a synchronous network could further support this observation. Moreover, it has been proven that, in asynchronous arbitrary networks, the pure token model can offer the same complexity as the whiteboard model provided a network map is available. Whether this is also true for synchronous networks also needs to be studied.

**Table 5.** Existing work on black hole search in synchronous networks.

| # | Communication model | Agent starting location | Network knowledge | Paper |
|---|---|---|---|---|
| 1 | Pure token | Dispersed | Unknown $n$, oriented torus | [20] |
| 2 | Pure token | Dispersed | Unknown $n$, un-oriented torus | [91] |
| 3 | Face-to-face + Pure token | Dispersed | Unknown $n$, oriented or un-oriented ring | [21,22] |
| 4 | Face-to-face | Co-located | Tree | [29,31] |
| 5 | Face-to-face | Co-located | Complete knowledge | [30,82–84] |

Finally, the possibility of using face-to-face communication is one of the several advantages of solving the black hole search problem in synchronous versus asynchronous networks. A hybrid that combines this model with the use of whiteboards or tokens might lead to further reduction on both time costs and agent moves.

## 5.2 Multiple Black Hole Search

As previously mentioned, Flocchini *et al.* [62] solve the multiple black hole search problem with a *subway model* in an asynchronous network using the whiteboard model and co-located agents. These initial results suggest several questions to research such as (a) whether a different communication model can be used, (b) whether a solution for a synchronous network can be obtained (especially given that it is assumed that a carrier takes the same amount of time to travel between two stations) and (c) how such new solutions would compare in costs with the solution put forth by these authors. In particular, with respect to choice of communication model, it may be interesting to assume that no whiteboard is available and that a cell phone network does not work in subways. Consequently, a new model to study would have agents only communicate with each other when they meet in the same subway station or in the same carrier. Alternatively, agents could use 'walkie-talkies' that support two-way communication over short distances. Both of these models-to-study are essentially variants of the face-to-face model. Finally, recall that Flocchini *et al.* (a) assume the network is a directed graph and (b) distinguish between carrier moves and agent moves. This prompts asking whether or not the use of an undirected graph may help reducing (a) the carrier move complexity and/or (b) the total move cost of the agents.

As emphasized by Peng *et al.*, without any additional assumptions, the general *Faulty Node Repair and Dynamically Spawned Black Hole Search* problem [96,107] becomes a multiple black hole search problem and remains unsolvable in an asynchronous network. Therefore, it is worth finding the weakest additional assumptions to make this problem become solvable in the presence of one multiple one-stop *GV*s and in the case of one or more multi-stop *GV*s in an arbitrary unknown network topology. Furthermore, whether the general problem or any of its specializations is solvable in a synchronous network should also be considered.

### 5.3 Other Models of Agents and Malicious Behaviour

Many solutions presented in this chapter rest on the use of agents endowed with unlimited memory so that they can carry a network map or build such a map during the network exploration. In reality, however, the memory of a mobile agent is constrained. Flocchini *et al.* [58] introduce agents with *very* limited memory. More generally, an agent is *oblivious* if all the information it holds is cleared at the end of each computing cycle. In essence, such agents are memoryless, that is, have no memory of any past actions and computations: the decision on the next action can only be based on what has been determined in the current computing cycle. The consequences of using such agents (or agents with similar constrained behavior) remain to be explored. In particular, can the 'absence' of memory in such agents be compensated by the use of a whiteboard or tokens?

Black hole is a particular type of a malicious host with a simple behavior: eliminating every agent instantly without leaving any trace. In fact, a host has many ways to harm the agents: it may not only eliminate any agent residing in it at any point in time, it may also alter an agent's behavior (e.g., alter it to disobey communication protocols, such as following FIFO order), duplicate agents, introduce fake agents or tamper the runtime environment (e.g. changing the contents of the whiteboard). Moreover the malicious threat can be mobile instead of static and the malicious behavior of a host may be periodic instead of constant. Those are all very interesting and challenging future directions on protecting a network from hostile nodes or malicious agents.

## References


1. Agmon, N., Peleg, D.: Fault-tolerant gathering algorithms for autonomous mobile robots. SIAM J. Comput. **36**(1), 56–82 (2006)
2. Agrawal, P., Ghosh, R.K., Das, S.K.: Cooperative black and gray hole attacks in mobile ad hoc networks. In: Proceedings of the 2nd International Conference on Ubiquitous Information Management and Communication, ICUIMC 2008, pp. 310–314. ACM, New York (2008)
3. Al-Shurman, M., Yoo, S.-M., Park, S.: Black hole attack in mobile ad hoc networks. In: Proceedings of the 42nd Annual Southeast Regional Conference, pp. 96–97. ACM (2004)
4. Albers, S., Henzinger, M.R.: Exploring unknown environments. SIAM J. Comput. **29**(4), 1164–1188 (2000)
5. Asaka, M., Okazawa, S., Taguchi, A., Goto, S.: A method of tracing intruders by use of mobile agents. In: Proceedings of of 9th Annual Conference of the Internet Society (1999). http://www.isoc.org/
6. Balamohan, B., Dobrev, S., Flocchini, P., Santoro, N.: Asynchronous exploration of an unknown anonymous dangerous graph with $O(1)$ pebbles. In: Even, G., Halldórsson, M.M. (eds.) SIROCCO 2012. LNCS, vol. 7355, pp. 279–290. Springer, Heidelberg (2012). https://doi.org/10.1007/978-3-642-31104-8_24
7. Balamohan, B., Flocchini, P., Miri, A., Santoro, N.: Improving the optimal bounds for black hole search in rings. In: Kosowski, A., Yamashita, M. (eds.) SIROCCO 2011. LNCS, vol. 6796, pp. 198–209. Springer, Heidelberg (2011). https://doi.org/10.1007/978-3-642-22212-2_18
8. Balamohan, B., Flocchini, P., Miri, A., Santoro, N.: Time optimal algorithms for black hole search in rings. Discret. Math. Algorithms Appl. **3**(4), 457–471 (2011)
9. Bampas, E., Leonardos, N., Markou, E., Pagourtzis, A., Petrolia, M.: Improved periodic data retrieval in asynchronous rings with a faulty host. Theor. Comput. Sci. **608**, 231–254 (2015)
10. Banerjee, S.: Detection/removal of cooperative black and gray hole attack in mobile ad-hoc networks. In: Proceedings of the World Congress on Engineering and Computer Science, pp. 22–24 (2008)
11. Banerjee, S., Sardar, M., Majumder, K.: AODV based black-hole attack mitigation in MANET. In: Satapathy, S., Udgata, S., Biswal, B. (eds.) FICTA 2013. AISC, vol. 247, pp. 345–352. Springer, Cham (2014). https://doi.org/10.1007/978-3-319-02931-3_39
12. Barriere, L., Flocchini, P., Fraigniaud, P., Santoro, N.: Capture of an intruder by mobile agents. In: Proceedings of 14th ACM Symposium on Parallel Algorithms and Architectures, pp. 200–209 (2002)
13. Bender, M.A., Fernández, A., Ron, D., Sahai, A., Vadhan, S.: The power of a pebble: exploring and mapping directed graphs. In: Proceedings of the Thirtieth Annual ACM Symposium on Theory of Computing, pp. 269–278. ACM (1998)
14. Bender, M.A., Slonim, D.: The power of team exploration: two robots can learn unlabeled directed graphs. In: Proceedings of 35th Annual Symposium on Foundations of Computer Science, pp. 75–85 (1994)
15. Blin, L., Fraigniaud, P., Nisse, N., Vial, S.: Distributed chasing of network intruders. In: Flocchini, P., Gąsieniec, L. (eds.) SIROCCO 2006. LNCS, vol. 4056, pp. 70–84. Springer, Heidelberg (2006). https://doi.org/10.1007/11780823_7
16. Borselius, N.: Mobile agent security. Electron. Commun. Eng. J. **14**(5), 211–218 (2002)

17. Cai, J., Flocchini, P., Santoro, N.: Network decontamination from a black virus. In: 2013 IEEE 27th International Parallel and Distributed Processing Symposium Workshops & Ph.D. Forum (IPDPSW), pp. 696–705. IEEE (2013)
18. Cai, J., Flocchini, P., Santoro, N.: Decontaminating a network from a black virus. Int. J. Netw. Comput. **4**(1), 151–173 (2014)
19. Cai, J., Flocchini, P., Santoro, N.: Distributed black virus decontamination and rooted acyclic orientations. In: 2015 IEEE International Conference on Computer and Information Technology; Ubiquitous Computing and Communications; Dependable, Autonomic and Secure Computing; Pervasive Intelligence and Computing (CIT/IUCC/DASC/PICOM), pp. 1681–1688. IEEE (2015)
20. Chalopin, J., Das, S., Labourel, A., Markou, E.: Black hole search with finite automata scattered in a synchronous torus. In: Peleg, D. (ed.) DISC 2011. LNCS, vol. 6950, pp. 432–446. Springer, Heidelberg (2011). https://doi.org/10.1007/978-3-642-24100-0_41
21. Chalopin, J., Das, S., Labourel, A., Markou, E.: Tight bounds for scattered black hole search in a ring. In: Kosowski, A., Yamashita, M. (eds.) SIROCCO 2011. LNCS, vol. 6796, pp. 186–197. Springer, Heidelberg (2011). https://doi.org/10.1007/978-3-642-22212-2_17
22. Chalopin, J., Das, S., Labourel, A., Markou, E.: Tight bounds for black hole search with scattered agents in synchronous rings. Theor. Comput. Sci. (TCS) **509**, 70–85 (2013)
23. Chalopin, J., Das, S., Santoro, N.: Rendezvous of mobile agents in unknown graphs with faulty links. In: Proceedings of 21st International Conference on Distributed Computing, pp. 108–122 (2007)
24. Chess, D.M.: Security issues in mobile code systems. In: Vigna, G. (ed.) Mobile Agents and Security. LNCS, vol. 1419, pp. 1–14. Springer, Heidelberg (1998). https://doi.org/10.1007/3-540-68671-1_1
25. Cooper, C., Klasing, R., Radzik, T.: Searching for black-hole faults in a network using multiple agents. In: Shvartsman, M.M.A.A. (ed.) OPODIS 2006. LNCS, vol. 4305, pp. 320–332. Springer, Heidelberg (2006). https://doi.org/10.1007/11945529_23
26. Cooper, C., Klasing, R., Radzik, T.: Locating and repairing faults in a network with mobile agents. Theor. Comput. Sci. **411**(14–15), 1638–1647 (2010)
27. Czyzowicz, J., et al.: More efficient periodic traversal in anonymous undirected graphs. Theor. Comput. Sci. **444**, 60–76 (2012)
28. Czyzowicz, J., Dobrev, S., Kralovic, R., Miklik, S., Pardubska, D.: Black hole search in directed graphs. In: Proceedings of 16th International Colloquium on Structural Information and Communication Complexity, pp. 182–194 (2009)
29. Czyzowicz, J., Kowalski, D., Markou, E., Pelc, A.: Searching for a black hole in tree networks. In: Higashino, T. (ed.) OPODIS 2004. LNCS, vol. 3544, pp. 67–80. Springer, Heidelberg (2005). https://doi.org/10.1007/11516798_5
30. Czyzowicz, J., Kowalski, D., Markou, E., Pelc, A.: Complexity of searching for a black hole. Fundamenta Informaticae **71**(2,3), 229–242 (2006)
31. Czyzowicz, J., Kowalski, D., Markou, E., Pelc, A.: Searching for a black hole in synchronous tree networks. Comb. Probab. Comput. **16**(4), 595–619 (2007)
32. Das, S., Flocchini, P., Kutten, S., Nayak, A., Santoro, N.: Map construction of unknown graphs by multiple agents. Theor. Comput. Sci. **385**(1–3), 34–48 (2007)
33. Das, S., Focardi, R., Luccio, F., Markou, E., Squarcina, M.: Gathering of robots in a ring with mobile faults. Theor. Comput. Sci. (2018, to appear)

34. Das, S., Luccio, F.L., Markou, E.: Mobile agents rendezvous in spite of a malicious agent. In: Bose, P., Gąsieniec, L.A., Römer, K., Wattenhofer, R. (eds.) ALGOSENSORS 2015. LNCS, vol. 9536, pp. 211–224. Springer, Cham (2015). https://doi.org/10.1007/978-3-319-28472-9_16
35. D'Emidio, M., Frigioni, D., Navarra, A.: Exploring and making safe dangerous networks using mobile entities. In: Cichoń, J., Gębala, M., Klonowski, M. (eds.) ADHOC-NOW 2013. LNCS, vol. 7960, pp. 136–147. Springer, Heidelberg (2013). https://doi.org/10.1007/978-3-642-39247-4_12
36. D'Emidio, M., Frigioni, D., Navarra, A.: Explore and repair graphs with black holes using mobile entities. Theor. Comput. Sci. **605**, 129–145 (2015)
37. Deng, X., Kameda, T., Papadimitriou, C.H.: How to learn an unknown environment I: the rectilinear case. J. ACM **45**, 215–245 (1998)
38. Diks, K., Fraigniaud, P., Kranakis, E., Pelc, A.: Tree exploration with little memory. J. Algorithms **51**, 38–63 (2004)
39. Dobrev, S., Flocchini, P., Královič, R., Prencipe, G., Ruzicka, P., Santoro, N.: Black hole search by mobile agents in hypercubes and related networks. In: OPODIS, vol. 3, pp. 169–180 (2002)
40. Dobrev, S., Flocchini, P., Královič, R., Ružička, P., Prencipe, G., Santoro, N.: Black hole search in common interconnection networks. Networks **47**(2), 61–71 (2006)
41. Dobrev, S., Flocchini, P., Královič, R., Santoro, N.: Exploring an unknown graph to locate a black hole using tokens. In: Navarro, G., Bertossi, L., Kohayakawa, Y. (eds.) TCS 2006. IIFIP, vol. 209, pp. 131–150. Springer, Boston, MA (2006). https://doi.org/10.1007/978-0-387-34735-6_14
42. Dobrev, S., Flocchini, P., Královič, R., Santoro, N.: Exploring an unknown dangerous graph using tokens. Theor. Comput. Sci. **472**, 28–45 (2013)
43. Dobrev, S., Flocchini, P., Prencipe, G., Santoro, N.: Mobile search for a black hole in an anonymous ring. In: Welch, J. (ed.) DISC 2001. LNCS, vol. 2180, pp. 166–179. Springer, Heidelberg (2001). https://doi.org/10.1007/3-540-45414-4_12
44. Dobrev, S., Flocchini, P., Prencipe, G., Santoro, N.: Searching for a black hole in arbitrary networks: optimal mobile agent protocols. In: Proceedings of the Twenty-first Annual Symposium on Principles of Distributed Computing, PODC 2002, pp. 153–162. ACM, New York (2002)
45. Dobrev, S., Flocchini, P., Prencipe, G., Santoro, N.: Multiple agents rendezvous in a ring in spite of a black hole. In: Papatriantafilou, M., Hunel, P. (eds.) OPODIS 2003. LNCS, vol. 3144, pp. 34–46. Springer, Heidelberg (2004). https://doi.org/10.1007/978-3-540-27860-3_6
46. Dobrev, S., Flocchini, P., Prencipe, G., Santoro, N.: Searching for a black hole in arbitrary networks: optimal mobile agents protocols. Distrib. Comput. **19**(1), 1–18 (2006)
47. Dobrev, S., Flocchini, P., Prencipe, G., Santoro, N.: Mobile search for a black hole in an anonymous ring. Algorithmica **48**, 67–90 (2007)
48. Dobrev, S., Flocchini, P., Santoro, N.: Improved bounds for optimal black hole search with a network map. In: Královič, R., Sýkora, O. (eds.) SIROCCO 2004. LNCS, vol. 3104, pp. 111–122. Springer, Heidelberg (2004). https://doi.org/10.1007/978-3-540-27796-5_11
49. Dobrev, S., Flocchini, P., Santoro, N.: Cycling through a dangerous network: a simple efficient strategy for black hole search. In: Proceedings of the 26th IEEE International Conference on Distributed Computing Systems, ICDCS 2006, p. 57. IEEE Computer Society, Washington, DC (2006)

50. Dobrev, S., Královič, R., Santoro, N., Shi, W.: Black hole search in asynchronous rings using tokens. In: Calamoneri, T., Finocchi, I., Italiano, G.F. (eds.) CIAC 2006. LNCS, vol. 3998, pp. 139–150. Springer, Heidelberg (2006). https://doi.org/10.1007/11758471_16
51. Dobrev, S., Santoro, N., Shi, W.: Locating a black hole in an un-oriented ring using tokens: the case of scattered agents. In: Kermarrec, A.-M., Bougé, L., Priol, T. (eds.) Euro-Par 2007. LNCS, vol. 4641, pp. 608–617. Springer, Heidelberg (2007). https://doi.org/10.1007/978-3-540-74466-5_64
52. Dobrev, S., Santoro, N., Shi, W.: Scattered black hole search in an oriented ring using tokens. In: Proceedings of IEEE International Parallel and Distributed Processing Symposium, pp. 1–8 (2007)
53. Dobrev, S., Santoro, N., Shi, W.: Using scattered mobile agents to locate a black hole in an un-oriented ring with tokens. Int. J. Found. Comput. Sci. **19**(6), 1355–1372 (2008)
54. Duncan, C.A., Kobourov, S.G., Kumar, V.S.: Optimal constrained graph exploration. In: Proceedings of 12th Annual ACM Symposium on Discrete Algorithms, pp. 807–814 (2001)
55. Engebretsen, L., Karpinski, M.: TSP with bounded metrics. J. Comput. Syst. Sci. **72**(4), 509–546 (2006)
56. Flocchini, P., Huang, M.J., Luccio, F.L.: Contiguous search in the hypercube for capturing an intruder. In: Proceedings of 18th IEEE International Parallel and Distributed Processing Symposium (2005)
57. Flocchini, P., Huang, M.J., Luccio, F.L.: Decontamination of chordal rings and tori. In: Proceedings of 8th Workshop on Advances in Parallel and Distributed Computational Models (2006)
58. Flocchini, P., Ilcinkas, D., Pelc, A., Santoro, N.: Computing without communicating: ring exploration by asynchronous oblivious robots. In: Tovar, E., Tsigas, P., Fouchal, H. (eds.) OPODIS 2007. LNCS, vol. 4878, pp. 105–118. Springer, Heidelberg (2007). https://doi.org/10.1007/978-3-540-77096-1_8
59. Flocchini, P., Ilcinkas, D., Santoro, N.: Ping pong in dangerous graphs: optimal black hole search with pure tokens. In: Taubenfeld, G. (ed.) DISC 2008. LNCS, vol. 5218, pp. 227–241. Springer, Heidelberg (2008). https://doi.org/10.1007/978-3-540-87779-0_16
60. Flocchini, P., Ilcinkas, D., Santoro, N.: Ping pong in dangerous graphs: optimal black hole search with pebbles. Algorithmica **62**(3–4), 1006–1033 (2012)
61. Flocchini, P., Kellett, M., Mason, P., Santoro, N.: Map construction and exploration by mobile agents scattered in a dangerous network. In: Proceedings of IEEE International Symposium on Parallel & Distributed Processing, pp. 1–10 (2009)
62. Flocchini, P., Kellett, M., Mason, P., Santoro, N.: Searching for black holes in subways. Theory Comput. Syst. **50**(1), 158–184 (2012)
63. Flocchini, P., Kellett, M., Mason, P.C., Santoro, N.: Mapping an unfriendly subway system. In: Boldi, P., Gargano, L. (eds.) FUN 2010. LNCS, vol. 6099, pp. 190–201. Springer, Heidelberg (2010). https://doi.org/10.1007/978-3-642-13122-6_20
64. Flocchini, P., Kellett, M., Mason, P.C., Santoro, N.: Fault-tolerant exploration of an unknown dangerous graph by scattered agents. In: Richa, A.W., Scheideler, C. (eds.) SSS 2012. LNCS, vol. 7596, pp. 299–313. Springer, Heidelberg (2012). https://doi.org/10.1007/978-3-642-33536-5_30

65. Flocchini, P., Kellett, M., Mason, P.C., Santoro, N.: Finding good coffee in Paris. In: Kranakis, E., Krizanc, D., Luccio, F. (eds.) FUN 2012. LNCS, vol. 7288, pp. 154–165. Springer, Heidelberg (2012). https://doi.org/10.1007/978-3-642-30347-0_17
66. Flocchini, P., Luccio, F.L., Song, L.X.: Size optimal strategies for capturing an intruder in mesh networks. In: Communications in Computing, pp. 200–206 (2005)
67. Flocchini, P., Mans, B., Santoro, N.: Sense of direction: definitions, properties, and classes. Networks **32**(3), 165–180 (1998)
68. Flocchini, P., Mans, B., Santoro, N.: Sense of direction in distributed computing. Theor. Comput. Sci. **291**, 29–53 (2003)
69. Flocchini, P., Santoro, N.: Distributed security algorithms by mobile agents. In: Chaudhuri, S., Das, S.R., Paul, H.S., Tirthapura, S. (eds.) ICDCN 2006. LNCS, vol. 4308, pp. 1–14. Springer, Heidelberg (2006). https://doi.org/10.1007/11947950_1
70. Flocchini, P., Santoro, N.: Distributed security algorithms for mobile agents. In: Sao, J., Das, S. (eds.) Mobile Agents in Networking and Distributed Computing. Wiley, Hoboken (2012)
71. Foukia, N., Hulaas, J.G., Harms, J.: Intrusion detection with mobile agents. In: Proceedings of 11th Annual Conference of the Internet Society (2001)
72. Fraigniaud, P., Gasieniec, L., Kowalski, D., Pelc, A.: Collective tree exploration. Networks **48**, 166–177 (2006)
73. Fraigniaud, P., Ilcinkas, D.: Digraphs exploration with little memory. In: Diekert, V., Habib, M. (eds.) STACS 2004. LNCS, vol. 2996, pp. 246–257. Springer, Heidelberg (2004). https://doi.org/10.1007/978-3-540-24749-4_22
74. Garey, M.R., Johnson, D.S., Tarjan, R.E.: The planar hamiltonian circuit problem is NP-complete. SIAM J. Comput. **5**(4), 704–714 (1976)
75. Glaus, P.: Locating a black hole without the knowledge of incoming link. In: Dolev, S. (ed.) ALGOSENSORS 2009. LNCS, vol. 5804, pp. 128–138. Springer, Heidelberg (2009). https://doi.org/10.1007/978-3-642-05434-1_13
76. Greenberg, M.S., Byington, J.C., Harper, D.G.: Mobile agents and security. IEEE Commun. Mag. **36**(7), 76–85 (1998)
77. Hohl, F.: Time limited blackbox security: protecting mobile agents from malicious hosts. In: Vigna, G. (ed.) Mobile Agents and Security. LNCS, vol. 1419, pp. 92–113. Springer, Heidelberg (1998). https://doi.org/10.1007/3-540-68671-1_6
78. Hohl, F.: A framework to protect mobile agents by using reference states. In: Proceedings of 20th International Conference on Distributed Computing Systems, pp. 410–417 (2000)
79. Jansen, W.: Countermeasures for mobile agent security. Comput. Commun. **23**(17), 1667–1676 (2000)
80. Kann, V.: Minimum steiner tree. http://www.nada.kth.se/~viggo/wwwcompendium/node78.html
81. Supriya, K.M.: Mobile ad hoc netwoks security attacks and secured routing protocols: a survey. In: Meghanathan, N., Chaki, N., Nagamalai, D. (eds.) CCSIT 2012. LNICST, vol. 84, pp. 119–124. Springer, Heidelberg (2012). https://doi.org/10.1007/978-3-642-27299-8_14
82. Klasing, R., Markou, E., Radzik, T., Sarracco, F.: Hardness and approximation results for black hole search in arbitrary graphs. In: Pelc, A., Raynal, M. (eds.) SIROCCO 2005. LNCS, vol. 3499, pp. 200–215. Springer, Heidelberg (2005). https://doi.org/10.1007/11429647_17
83. Klasing, R., Markou, E., Radzik, T., Sarracco, F.: Hardness and approximation results for black hole search in arbitrary graphs. Theor. Comput. Sci. **384**(2–3), 201–221 (2007)

84. Klasing, R., Markou, E., Radzik, T., Sarracco, F.: Approximation bounds for black hole search problems. Networks **52**(4), 216–226 (2008)
85. Kosowski, A., Navarra, A., Pinotti, C.M.: Synchronization helps robots to detect black holes in directed graphs. In: Abdelzaher, T., Raynal, M., Santoro, N. (eds.) OPODIS 2009. LNCS, vol. 5923, pp. 86–98. Springer, Heidelberg (2009). https://doi.org/10.1007/978-3-642-10877-8_9
86. Kosowski, A., Navarra, A., Pinotti, C.M.: Synchronous black hole search in directed graphs. Theor. Comput. Sci. **412**(41), 5752–5759 (2011)
87. Královič, R., Miklík, S.: Periodic data retrieval problem in rings containing a malicious host. In: Patt-Shamir, B., Ekim, T. (eds.) SIROCCO 2010. LNCS, vol. 6058, pp. 157–167. Springer, Heidelberg (2010). https://doi.org/10.1007/978-3-642-13284-1_13
88. Lange, D., Oshima, M.: Seven good reasons for mobile agents. Commun. ACM **42**(3), 88–89 (1999)
89. Luccio, F., Pagli, L., Santoro, N.: Network decontamination with local immunization. In: Proceedings of 8th Workshop on Advances in Parallel and Distributed Computational Models (2006)
90. Markou, E.: Identifying hostile nodes in networks using mobile agents. Bull. Eur. Assoc. Theor. Comput. Sci. **108**, 93–129 (2012)
91. Markou, E., Paquette, M.: Black hole search and exploration in unoriented tori with synchronous scattered finite automata. In: Baldoni, R., Flocchini, P., Binoy, R. (eds.) OPODIS 2012. LNCS, vol. 7702, pp. 239–253. Springer, Heidelberg (2012). https://doi.org/10.1007/978-3-642-35476-2_17
92. Miklík, S.: Exploration in faulty networks. Ph.D. thesis, Comenius University (2010)
93. Ng, S., Cheung, K.: Protecting mobile agents against malicious hosts by intention of spreading. In: Proceedings of International Conference on Parallel and Distributed Processing and Applications, pp. 725–729 (1999)
94. Oppliger, R.: Security issues related to mobile code and agent-based systems. Comput. Commun. **22**(12), 1165–1170 (1999)
95. Panaite, P., Pelc, A.: Exploring unknown undirected graphs. J. Algorithms **33**, 281–295 (1999)
96. Peng, M., Shi, W., Corriveau, J.-P.: Repairing faulty nodes and locating a dynamically spawned black hole search using tokens. In: IEEE Conference on Communications and Network Security, pp. 136–145. IEEE (2018)
97. Peng, M., Shi, W., Corriveau, J.-P., Pazzi, R., Wang, Y.: Black hole search in computer networks: state-of-the-art, challenges and future directions. J. Parallel Distrib. Comput. **88**, 1–15 (2016)
98. Prencipe, G.: Corda: distributed coordination of a set of autonomous mobile robots. In: Proceedings of 4th European Research Seminar on Advances in Distributed Systems, pp. 185–190 (2001)
99. Robins, G., Zelikovsky, A.: Improved Steiner tree approximation in graphs. In: Proceedings of 10th Annual ACM-SIAM Symposium on Discrete Algorithms, pp. 770–779 (2000)
100. Rollik, H.: Automaten in planaren graphen. Acta Informatica **13**, 287–298 (1980)
101. Sander, T., Tschudin, C.F.: Protecting mobile agents against malicious hosts. In: Vigna, G. (ed.) Mobile Agents and Security. LNCS, vol. 1419, pp. 44–60. Springer, Heidelberg (1998). https://doi.org/10.1007/3-540-68671-1_4
102. Schelderup, K., Ølnes, J.: Mobile agent security—issues and directions. In: Zuidweg, H., Campolargo, M., Delgado, J. (eds.) IS&N 1999. LNCS, vol. 1597, pp. 155–167. Springer, Heidelberg (1999). https://doi.org/10.1007/3-540-48888-X_16

103. Sen, J., Chandra, M.G., Harihara, S., Reddy, H., Balamuralidhar, P.: A mechanism for detection of gray hole attack in mobile ad hoc networks. In: 2007 6th International Conference on Information, Communications & Signal Processing, pp. 1–5. IEEE (2007)
104. Shannon, C.E.: Presentation of a maze-solving machine. In: Proceedings of 8th Conference of the Josiah Macy Jr. Foundation (Cybernetics), pp. 173–180 (1951)
105. Shi, W.: Black hole search with tokens in interconnected networks. In: Guerraoui, R., Petit, F. (eds.) SSS 2009. LNCS, vol. 5873, pp. 670–682. Springer, Heidelberg (2009). https://doi.org/10.1007/978-3-642-05118-0_46
106. Shi, W., Garcia-Alfaro, J., Corriveau, J.-P.: Searching for a black hole in interconnected networks using mobile agents and tokens. J. Parallel Distrib. Comput. **74**(1), 1945–1958 (2014)
107. Shi, W., Peng, M., Corriveau, J.-P., Croft, W.L.: Faulty node repair and dynamically spawned black hole search. In: Deng, R., Weng, J., Ren, K., Yegneswaran, V. (eds.) SecureComm 2016. LNICST, vol. 198, pp. 144–162. Springer, Cham (2017). https://doi.org/10.1007/978-3-319-59608-2_8
108. Spafford, E.H., Zamboni, D.: Intrusion detection using autonomous agents. Comput. Netw. **34**(4), 547–570 (2000)
109. Su, M.-Y.: Prevention of selective black hole attacks on mobile ad hoc networks through intrusion detection systems. Comput. Commun. **34**(1), 107–117 (2011)
110. Suzuki, I., Yamashita, M.: Distributed anonymous mobile robots: formation of geometric patterns. SIAM J. Comput. **28**(4), 1347–1363 (1999)
111. Vitek, J., Castagna, G.: Mobile computations and hostile hosts. In: Tsichritzis, D. (ed.) Mobile Objects, pp. 241–261. University of Geneva (1999)
112. Zarrad, A., Daadaa, Y.: A review of computation solutions by mobile agents in an unsafe environment. Int. J. Adv. Comput. Sci. Appl. **4**(4), 87–92 (2013)

# Network Decontamination

Nicolas Nisse(✉)

Université Côte d'Azur, Inria, CNRS, I3S, Sophia Antipolis, France
nicolas.nisse@inria.fr

**Abstract.** The Network Decontamination problem consists of coordinating a team of mobile agents in order to clean a contaminated network. The problem is actually equivalent to tracking and capturing an invisible and arbitrarily fast fugitive. This problem has natural applications in network security in computer science or in robotics for search or pursuit-evasion missions. Many different objectives have been studied: the main one being the minimization of the number of mobile agents necessary to clean a contaminated network.

Many environments (continuous or discrete) have also been considered. In this Chapter, we focus on networks modeled by graphs. In this context, the optimization problem that consists of minimizing the number of agents has a deep graph-theoretical interpretation. Network decontamination and, more precisely, *graph searching* models, provide nice algorithmic interpretations of fundamental concepts in the Graph Minors theory by Robertson and Seymour.

For all these reasons, graph searching variants have been widely studied since their introduction by Breish (1967) and mathematical formalizations by Parsons (1978) and Petrov (1982). This chapter consists of an overview of the algorithmic results on graph decontamination and graph searching.

**Keywords:** Graph searching · Path- and tree-decompositions (Distributed) graph algorithms · Computational complexity

## 1 Introduction

*Network Decontamination* is a problem in which a team of mobile agents, called *searchers*, aims at *clearing* the links and nodes of an *infected* network. Alternatively, it can be defined as a *pursuit-evasion* game between a malicious intruder, called the *fugitive*, and a team of searchers that must *capture* the fugitive.

Since its introduction by Breisch [Bre67], Parsons [Par78a] and Petrov [Pet82], this field has received a lot of attention due to its numerous applications in network security and distributed computing, in robotics and differential games, and in graph theory. Previous surveys on network decontamination have been proposed [Bie91, Als04, FS06, FT08, CHI11, Bre12]. Most of them mainly

This work has been partially supported by ANR program "Investments for the Future" under reference ANR-11-LABX-0031-01, the Inria Associated Team AlDyNet.


P. Flocchini et al. (Eds.): Distributed Computing by Mobile Entities, LNCS 11340, pp. 516–548, 2019.
https://doi.org/10.1007/978-3-030-11072-7_19

focus either on a centralized setting or on a distributed setting. This chapter aims at presenting an up-to-date (as exhaustive as possible) overview on graph decontamination both in distributed and centralized settings.

**Lost in a Cave.** In 1967, Breisch opened the field of network decontamination by asking the following question:

*"A person is lost in a particular cave and is wandering aimlessly. Is there an efficient way for the rescue party to search for the lost person? What is the minimum number of searchers required to explore a cave so that it is impossible to miss finding the victim if it is in the cave?"* [Bre67].

Parsons [Par78a] and Petrov [Pet82] independently formalized the problem in a continuous setting where the objective is, for a team of mobile agents, the *searchers*, to capture an invisible and arbitrarily fast *fugitive*, in an environment modeled by a continuous embedding of a graph $G$ on a surface. In this model, both the fugitive and the searchers move simultaneously in a continuous way from a point of $G$ (corresponding to a vertex or in the interior of an edge) to another. The searchers *capture* the fugitive if, at some time, the fugitive occupies the same point as a searcher. We recall this technical definition for completeness.

**Definition 1** [Par78a]. *For any $k \in \mathbb{N}^*$, let $\mathcal{C}_k(G)$ be the set of families $F = \{s_1, \cdots, s_k\}$ such that, for every $1 \leq i \leq k$, $s_i : [0, \infty[\to G$ is a continuous function. A* search plan *for $G$ is a family $F \in \mathcal{C}_k(G)$ such that, for every continuous function $f : [0, \infty[\to G$, there exists $t_f \in [0, \infty[$ and $i \in \{1, \cdots, k\}$ such that $s_i(t_f) = f(t_f)$.*

Intuitively, $k$ represents the number of searchers. A search plan of $\mathcal{C}_k(G)$ is therefore a set of trajectories determined for each searcher ($s_i$ represents the trajectory of searcher $i$), which ensures that, whatever be the trajectory $f$ of the fugitive in $G$, there is a searcher that will occupy the same point as the fugitive at some time $t_f$. In other words, a search plan ensures that whatever be the strategy used by the fugitive, it must eventually be captured. Note that this definition does not constrain the speeds of the searchers and of the fugitive. Note also that, this model of pursuit-evasion game is actually equivalent to the *network decontamination* problem where a team of searchers must clear an infected network (e.g., a system of tunnels contaminated by some toxic gas, a computer network infected by a virus, *etc.*).

The continuous model of Parsons and Petrov can equivalently be defined in a discrete setting where environments are modeled by graphs [Par78b, Gol89a, Gol89b]. This latter formulation (formal definitions and examples are postponed to Sect. 2.1) is often referred to as *Graph Searching*[1] in the literature. Besides its natural applications in robotics or network security, one of the reasons for the vast literature on graph searching is probably its close relationship with some of the cornerstones of the Graph Minors theory [RS83, RS04]. Precisely, graph

[1] We should emphasize that there is another different topic of graph theory, related to Depth/Breadth First Search, called *Graph Searching*, a.k.a. Graph Traversals (e.g. [CDH+16]).

searching provides an algorithmic interpretation of *tree- and path-decompositions* of graphs [RS90] that are important (algorithmic) tools of modern graph theory (see [CM93, DH08, BFL+09, FLS18]). This relationship led to numerous results common to graph searching and graph decompositions (see Sects. 2 and 4.1).

**Pursuit-Evasion Games.** Before starting our survey on graph searching, let us briefly mention different approaches for studying pursuit-evasion games. Roughly, the field may be divided into two main branches: pursuit-evasion games in continuous environments (polygonal environments, polyhedral surfaces, *etc.*) or in graphs. For the former approach, in a continuous setting, the reader is invited to see [GLL+99, CHI11, BKIS12, KS15, ABC+15] and references therein. In the case of discrete environments (i.e., in graphs), the field of pursuit-evasion games may also be divided into (at least) two different families of problems where results and tools are very different: *Cops and robber games* (see, e.g., [BN11] and Chap. 1 of [Nis14]) and Graph Searching. The main differences between these two approaches are (1) the different speeds of the fugitive, and (2) that Cops and Robber games are played turn-by-turn by two players while, in graph searching, the searchers and the fugitive move simultaneously. Roughly, both Graph Searching and Cops and Robber games are related to graph structural properties, but Cops and Robber games also rely on graph metric properties.

In this Chapter, we focus on graph environments, with searchers and an arbitrarily fast fugitive moving simultaneously, i.e., we speak about Graph Searching.

**Organization of the Chapter.** The main graph searching variants are formally defined in Sect. 2, where relationships with graph decompositions and algorithmic results are presented. In Sect. 3, we focus on the *connected* variant of graph searching and on distributed algorithms for graph decontamination. Finally, Sect. 4 is devoted to the study of several alternative graph searching models. In this paper, the network decontamination terminology will be mainly used in Sects. 2 and 3, and the pursuit-evasion terminology is used in Sect. 4.

We assume that the reader is familiar with graph terminology (see [Die12]). In particular, see [BLS99] for the definitions of the graph classes mentioned throughout the chapter.

## 2 Graph Searching

This section is devoted to the presentation of the basics of graph searching. We focus on computational complexity, algorithms in a centralized setting, and on the relationship between variants of graph searching and graph parameters and decompositions.

### 2.1 The Seminal Model: Edge-Search

*Graph searching* aims at clearing the vertices and edges of a graph using a team of mobile agents, called *searchers*. Let us now formally define the seminal variant of graph searching, *a.k.a. edge-searching* [Par78b].

**Network Decontamination Terminology.** Initially all vertices and edges of a graph[2] $G = (V, E)$ are *contaminated.* A vertex is *cleared* when it is occupied by a searcher (in particular, initially, no vertices are occupied). An edge $e \in E$ is *cleared* if a searcher slides along $e$. Once a vertex/edge has been cleared, it is said *clear.* However, an unoccupied clear vertex is *recontaminated* as soon as there is a path free of searchers from it to a contaminated vertex. Similarly, an edge is recontaminated as soon as one of its endpoints is recontaminated.

A *strategy* consists of a finite sequence of *steps*, or *moves*, where each step consists of either sliding a searcher along an edge, or placing a searcher at some vertex of the graph, or removing a searcher from a vertex of $G$. The number of searchers *used* by a strategy is the maximum number of searchers present in the graph among all its steps. A strategy is *winning* if, eventually, it results in a state where all vertices and edges are (simultaneously) clear.

**Simple Examples.** As a warm-up, let us consider the following examples (Fig. 1).

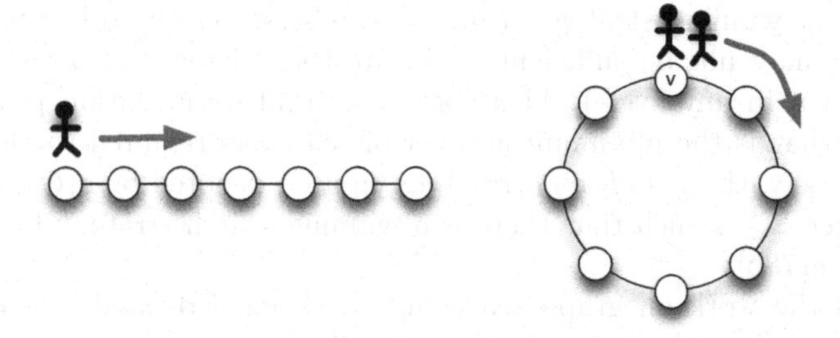


**Fig. 1.** Schematic overviews of optimal edge-search strategies in paths and cycles.

*Paths.* Let $P_n$ be an $n$-node path. A strategy in $P_n$ consists of, first, placing a searcher at one end of $P_n$ and, then, sequentially sliding this searcher along every edge until it reaches the other end of $P_n$. It is easy to see that, when the searcher reaches the second end, every node and edge of $P_n$ have been cleared and have never been recontaminated. Hence, such a strategy is winning.

*Cycles.* As a second example, let us consider the cycle $C_n$ on $n \geq 3$ vertices. The first step of any strategy can only consist of placing a searcher at some vertex $v \in V(C_n)$ which becomes clear. Now, removing this searcher from $v$ would result in the recontamination of $v$ by its neighbors (and so it would result in the initial state). On the other hand, sliding the searcher along an edge from $v$ to one of its neighbors $u$ would result in a symmetrical state where only $u$ is clear and occupied (since $v$ and the edge $uv$ would be recontaminated by the other neighbor of $v$). Hence, the only meaningful move is to place a second

[2] Unless stated otherwise, all graphs considered in this chapter are simple, undirected, and connected.

searcher at $v$ (actually we may imagine placing this second searcher at other vertices but it would lead to other recontaminations) and slide it "around" the cycle until it comes back to $v$. During every sliding step, the searcher at $v$ has *guarded* the vertex $v$, preventing the recontamination of all edges traversed by the second searcher. Therefore, the presented strategy using 2 searchers is winning.

*Universal strategy.* The last example is a *universal* strategy, i.e., which is winning in any (connected) graph $G = (V, E)$ with $n$ vertices. During the $n$ first steps, let us place one searcher at every vertex of $G$. Let $v \in V$ be any vertex of $G$ and consider the searcher $A$ at $v$. Sequentially, let us slide this searcher $A$ along the edges of $G$ until all edges of $G$ have been traversed at least once. At this step, all edges not incident to $v$ have both their ends occupied and have been cleared by $A$, therefore they are all clear. Finally, for every neighbor $u \in N(v)$[3] of $v$, let us slide the searcher at $u$ along the edge $uv$ from $u$ to $v$. Clearly, the presented strategy is winning and uses $n$ searchers.

**Search Number.** As illustrated by the above examples, in any $n$-node graph, there exists a winning strategy using $n$ searchers. On the other hand, a single searcher may not be sufficient to ensure the existence of a winning strategy (as shown in any cycle). Therefore, a natural optimization problem is to determine what is the minimum number of searchers required to clear a given graph. Precisely, the *search number* of a graph $G$, denoted by $\mathtt{s}(G)$, is the minimum integer $k \geq 1$ such that there is a winning search strategy for $G$ using $k$ searchers [Par78b].

Most of the work on graph searching has been dedicated to compute the search number of graphs and to design *optimal* strategies (i.e., winning strategies using the minimum number of searchers), both in centralized and distributed settings. However, several other objectives (see Sects. 3.2 and 4.4) have been considered such as minimizing the "cost" of a strategy, its "length", the number of moves of the searchers or the number of "rounds" of a strategy, *etc.*

**Monotonicity.** Before going further, let us define a crucial notion when dealing with search strategies. A search strategy is *monotone* if, when following it, no vertices nor edges are recontaminated. Said differently, in a monotone strategy, it is forbidden to remove a searcher from a vertex $v$ if $v$ has at least one incident contaminated edge and no other searcher is occupying $v$. Moreover, sliding a searcher from a vertex $v$ to one of its neighbors $u \in N(v)$ is allowed only: if $v$ is occupied by another searcher; or if all edges incident to $v$ are already clear; or if $vu \in E$ is the only edge incident to $v$ that is still contaminated.

One of the first challenges concerning Graph Searching has been to answer the following question: "does recontamination help?". In other words, does there always exist an optimal strategy that is monotone? This latter question was first asked (and conjectured to be true) by Megiddo *et al.* [MHG+88].

[3] Given a graph $G = (V, E)$ and $v \in V$, $N(v)$ denotes the set of neighbors of $v$, i.e., $N(v) = \{u \in V \mid uv \in E\}$.

At a first glance, this question looks "intuitively obviously true" (why would it be useful to let vertices be recontaminated?) but it is actually not obvious at all and, moreover, we will see (Sects. 3 and 4.2) that there are variants of graph searching where recontamination actually helps. The conjecture of Megiddo *et al.* has been first proved by LaPaugh [LaP93] and an elegant proof of it by Bienstock and Seymour [BS91] is sketched in the next sub-section.

**Theorem 1** [LaP93,BS91]. *"Recontamination does not help", i.e., in any graph $G$, there exists a winning monotone strategy using $\mathtt{s}(G)$ searchers.*

We refer to Theorem 1 by saying that the edge-search variant is *monotone*. To see the importance of this theorem, let us do the following remarks.

- First, there always exists a winning monotone strategy with a number of steps which is polynomial (actually linear) in the size of the graph (since each edge and vertex is cleared exactly once). Therefore, such a strategy constitutes a polynomial-size certificate for the search number, i.e., given a graph $G$ and an integer $k \geq 1$ as inputs, the problem of deciding if $\mathtt{s}(G) \leq k$ is in NP. We are not aware of another method to prove this fact.
- Second, monotone strategies are much easier to imagine and design, and they are much easier to manipulate to prove lower bounds. For instance, the universal strategy presented above allows to show that $\mathtt{s}(K_n) \leq n$ where $K_n$ is the complete graph with $n \geq 1$ vertices. For any $n > 3$, this result is tight, i.e., $\mathtt{s}(K_n) = n$ for every $n > 3$. The general technique to prove such a lower bound is to consider an optimal monotone strategy (which exists by Theorem 1) and assume, for the purpose of contradiction, that it uses less than $n$ searchers. Finally, it can be shown that because at most $n-1$ searchers are used, there must be a step with recontamination, leading to a contradiction.
- Last but not least, the monotonicity result allows to establish the equivalence between graph searching and other graph parameters such as the parameters related to graph decompositions that are the corner stone of the Graph Minor theory (see Sects. 2.2 and 4.1).

### 2.2 Mixed/Node-Search and Pathwidth

The edge-search model provides a natural way to describe the seminal problems of Breisch [Bre67] and Parsons [Par78a] and it is the main model studied in a distributed setting (see Sects. 3.2 and 3.3). Variants "close" to edge-search have been proposed because they are somehow easier to work with and, moreover, provide alternative definitions for graph parameters known in other contexts.

**Node-Search.** Kirousis and Papadimitriou defined *node graph searching* because of its relationship with *pebble games* [KP85,KP86,Bie91]. In this setting, a strategy is defined as a sequence of moves like in edge-searching with two main differences. First, only two moves are allowed: placement/removal of a searcher at/from a vertex (so searchers do not slide along edges). Second, an edge becomes clear as soon as both its ends are occupied. In this variant, recontamination and monotone strategies are defined as in edge-search. The corresponding graph invariant is the *node search number*, denoted by $\mathtt{ns}$. For any graph $G$, $\mathtt{ns}(G)-1 \leq \mathtt{s}(G) \leq \mathtt{ns}(G)+1$ [KP86]. Moreover, the three cases are possible since $\mathtt{s}(P_n) = 1 < \mathtt{ns}(P_n) = 2$ (where $P_n$ is a path on $n$ nodes), $\mathtt{s}(G) = \mathtt{ns}(G) = 2$ if $G$ is a star with at least three leaves and $\mathtt{s}(K_{3,3}) = 5 > \mathtt{ns}(K_{3,3}) = 4$ (Fig. 2) [KP86]. Simple (and polynomial-time) transformations allow to "transpose" node-search to edge-search and *vice versa*. Indeed, Kirousis and Papadimitriou proved that, for any graph $G$, $\mathtt{s}(G^{//}) = \mathtt{ns}(G) + 1$ and that $\mathtt{s}(G) = \mathtt{ns}(G^{++}) - 1$ where $G^{//}$ (resp., $G^{++}$) is obtained from $G$ by replacing each edge by two parallel edges (resp., by three edges in series) [KP86]. As we will see below, the node-search variant is important because monotone node-strategies provide an algorithmic interpretation of path-decompositions of graphs.

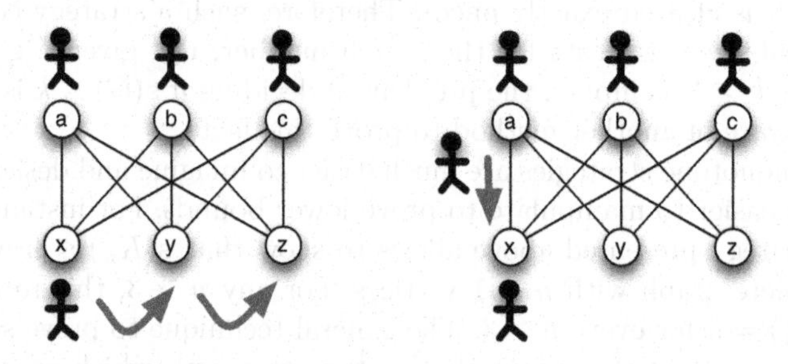


**Fig. 2.** Schematic overviews of optimal edge-search and node-search strategies in $K_{3,3}$. In both cases, one searcher remains at each of the vertices in $\{a, b, c\}$. In node-search (left), the fourth searcher goes sequentially to $x, y$ and then $z$. In edge-search (right), a fourth search first goes to $x$ while the fifth searcher sequentially clears the edges incident to $x$, then the fourth searcher goes to $y$ and the fifth searcher clears the edges incident to it, and so on.

**Mixed-Search.** To prove Theorem 1, Bienstock and Seymour defined the notion of *mixed-search strategy* [BS91] as an edge-search strategy with the difference that an edge is cleared either when it is crossed by a searcher or when both its ends are occupied by a searcher. The corresponding graph invariant is the *mixed search number*, denoted by $\mathtt{mixs}$. Again, there is a close relationship with edge-search. Precisely, for any graph $G$, $\mathtt{mixs}(G) \leq \mathtt{ns}(G) \leq \mathtt{mixs}(G)+1$ [BS91] and inequalities are tight. Moreover, $\mathtt{mixs}(G^{+}) = \mathtt{s}(G)$ for any graph $G$ where $G^{+}$ is obtained from $G$ by subdividing each edge once [BS91].

As mentioned above, mixed-searching has been introduced because it allows an elegant proof of Theorem 1. We aim at sketching this (a bit technical) proof because it is instructive since many studies on graph searching use a similar formalism, representing search-strategies by a sequence of tuples of sets of edges or vertices.

*Sketch of Proof of Theorem* 1 [BS91]. A *crusade* in a graph $G = (V, E)$ is a sequence $(E_0, \cdots, E_\ell)$ of subsets of $E$ such that $E_0 = \emptyset$, $E_\ell = E$ and $|E_i - E_{i-1}| \leq 1$ for every $1 \leq i \leq \ell$. The crusade uses $k$ searchers if $|\delta(E_i)| \leq k$ for every $0 \leq i \leq \ell$, where $\delta(E_i)$ is the set of vertices incident with an edge in $E_i$ and an edge in $E \setminus E_i$. A crusade is *progressive* if $E_{i-1} \subseteq E_i$ for every $1 \leq i \leq \ell$.

The proof consists of first easily showing that, if there is a mixed strategy using $k$ searchers, then there is a crusade using at most $k$ searchers. The second easy step of the proof is to show that, if there is a progressive crusade using $k$ searchers, then there is a monotone mixed strategy using at most $k$ searchers. The key of the proof is to prove that if there is a crusade using $k$ searchers then there exists a progressive crusade using at most $k$ searchers. The latter step is proved by considering a crusade $\mathcal{C}$ (using $k$ searchers) such that $\sum_{0 \leq i \leq \ell} (|\delta(E_i)|+1)$ is minimum and, under the previous assumption, $\sum_{0 \leq i \leq \ell} |E_i|$ is minimum. Then, using the submodularity of the function $\delta$ (i.e., for every $A, B \subseteq E$, $|\delta(A \cup B)| + |\delta(A \cap B)| \leq |\delta(A)| + |\delta(B)|$), it can be shown that $\mathcal{C}$ is progressive. ◇

This proves that mixed-searching is monotone. Noticing that the simple transformations presented above preserve monotonicity, this implies that both node-search and edge-search are monotone too.

Additionally, the monotonicity of mixed search allows to prove that, for any graph $G$, $\mathtt{mixs}(G) - 1$ actually equals the *proper pathwidth* of $G$ [TUK95].

**Path-Decomposition and Pathwidth.** *Pathwidth* is an important structural measure that appears in the Graph Minor theory [RS83,Bie91,Bod98] but also in other domains such as VLSI design [DKL87,Kin92,FL94]. Given a graph $G = (V, E)$, a *path-decomposition* is a sequence $P = (X_0, \cdots, X_\ell)$ of subsets of vertices of $G$, called *bags*, such that (1) $\bigcup_{0 \leq i \leq \ell} X_i = V$; (2) for every $uv \in E$, there exists $0 \leq i \leq \ell$ with $\{u, v\} \subseteq X_i$; and (3) for every $0 \leq i \leq j \leq k \leq \ell$, $X_i \cap X_k \subseteq X_j$. The *width* of $P$ is the size of its largest bag minus one, and the *pathwidth* of $G$, denoted by $\mathtt{pw}(G)$, is the minimum width of a path-decomposition of $G$ (see an example on Fig. 3).

From any path-decomposition $P = (X_0, \cdots, X_\ell)$ of a graph $G$, it is easy to derive a node-search strategy for $G$: for $i$ from 0 to $\ell$, sequentially place a searcher at every vertex of $X_i$ and then sequentially remove the searchers from the vertices in $X_i \setminus X_{i-1}$ (see an example on Fig. 3). From the properties of path-decompositions, for every $0 \leq i < \ell$, $S = X_i \cap X_{i+1}$ *separates* $A = \bigcup_{0 \leq j \leq i} X_j \setminus X_{i+1}$ from $B = \bigcup_{i < j \leq \ell} X_j \setminus X_i$ [Bod98] and therefore, the searchers at $S$ prevent $B$ from recontaminating $A$. Hence, it is easy to see that such a strategy is winning and monotone and that the number of searchers used equals the width of $P$ plus one. Reciprocally, from any monotone winning strategy using $k$ searchers, it is easy to derive a path-decomposition of width $k-1$ (where each bag corresponds to the set of vertices occupied by a searcher at each step). Therefore, by Theorem 1 (applied to node-search):

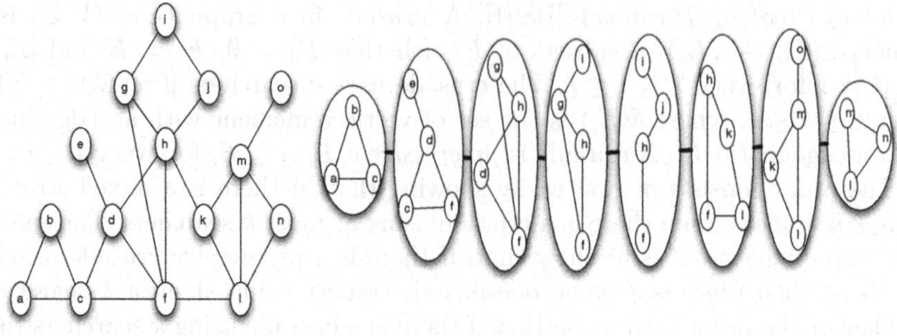

$P = (\{abc\}, \{cdef\}, \{dfgh\}, \{fghi\}, \{fhij\}, \{fhkl\}, \{klmo\}, \{lmn\})$

**Fig. 3.** Example of a graph $G$ (left) and one of its path-decompositions $P = (X_0, \cdots, X_7)$ of width 3 (right). Prove that it is optimal, i.e., that $\mathtt{pw}(G) = 3$. A (monotone winning) node-search strategy corresponding to $P$ is then $S = (a, b, c, \bar{a}, \bar{b}, d, e, f, \bar{c}, \bar{e}, h, g, \bar{d}, i, \bar{g}, j, \bar{i}, \bar{j}, k, l, \bar{f}, \bar{h}, m, o, \bar{k}, \bar{o}, n)$, where $x$ means to place a searcher at vertex $x$, while $\bar{x}$ means to remove the searcher at $x$.

**Theorem 2** [KP85,KP86,EST94]. *For any graph $G$, $\mathtt{ns}(G) = \mathtt{pw}(G) + 1$.*

Among other important consequences, Theorem 2 allows to transpose the numerous results concerning pathwidth to node-search and, sometimes, to edge-search and mixed-search by using the simple transformations seen above.

Note that pathwidth, and so node-search number, may be equivalently defined in terms of a measure, called *vertex-separation*, of linear layouts of vertices [Kin92]. Similarly, the mixed-search number of a graph $G$ can be defined in terms of *linear width* (some measure on the linear layouts of the edges) [Thi00].

## 2.3 Complexity and Algorithms

This subsection is devoted to the computational complexity of the edge-, node- and mixed graph searching problems. Algorithms to compute the search numbers (and corresponding strategies) in general graphs and particular graph classes are also presented.

**Hardness.** Given an $n$-node graph $G$ and an integer $k \geq 1$ as inputs, the problem of deciding if $\mathtt{s}(G) \leq k$ has first been proved to be NP-hard in [MHG+88]. Then, Monien and Sudborough proved that this problem is NP-hard in the class of planar graphs with maximum degree 3 [MS88]. This latter result also holds for both node-search and mixed-search. Later on, Gustedt proved that deciding the pathwidth (and so the node-search number) is NP-hard in the class of chordal graphs [Gus93]. Edge-search is also NP-hard in chordal graphs [PTK+00]. Moreover, assuming the *Small Set Expansion Conjecture*, the problem of deciding the pathwidth of a graph is NP-hard to approximate within a constant factor [WAPL14].

**Exact Generic Algorithms.** On the positive side, all graph searching variants mentioned so far are *closed under taking minor.* That is, for any minor[4] $H$ of a graph $G$, $\mathtt{s}(H) \leq \mathtt{s}(G)$ (resp., $\mathtt{ns}(H) \leq \mathtt{ns}(G)$ and $\mathtt{mixs}(H) \leq \mathtt{mixs}(G)$). Therefore, from the Graph Minor theory [RS04, Die12], it follows that, for any fixed $k \geq 1$, the set of *minimal obstructions* for having a search-number at most $k$ is finite, and so each search-number admits a *Fixed Parameter Tractable* (FPT) algorithm [RS95, CFK+15] (where the parameter is the size of the solution). In the case of node-search, Ellis *et al.* first designed an algorithm in time $O(n^{2k^2+4k+8})$ via dynamic programming [EST87]. They also gave structural characterizations of graphs with node-search number at most 3 [EST94]. Then, Bodlaender and Kloks gave the first constructive FPT algorithm for deciding whether $\mathtt{ns}(G) \leq k$ in time $k^{O(k^3)}n$ [BK96]. In the case of mixed-search (and linear width), Bodlaender and Thilikos gave a constructive FPT algorithm [BT04]. Thilikos also designed a linear-time algorithm to decide whether a graph has mixed-search (resp., edge-search) number at most two by fully characterizing the set of minimal obstructions [Thi00]. In addition to parameterized algorithms, other algorithms have been proposed to compute the pathwidth of general graphs. The best known exact exponential-time algorithm computing the pathwidth runs in time $O(1.89^n)$ [KKK+16] (see also [SV09]). Moreover, using the definition(s) of graph searching in terms of vertex-layout, several Integer Linear Programs solving these problems have been proposed [PSS13, CMN16, Cou16, Mal18].

**Graph Classes.** Search problems can be solved in various graph classes in polynomial time. The case of trees has been particularly studied [Par78a, MHG+88, EST94, PHH+00]. In particular, Skodinis designed a linear-time algorithm for computing the node-search number of trees and a corresponding strategy [Sko03]. A generic and distributed algorithm for computing, in time $O(n \log n)$, any of the search numbers in $n$-node trees (only the initial setting of the algorithm differ) has been designed in [CHM12], where the interesting notion of hierarchical decomposition of trees is introduced. The algorithms for trees are all based on the so-called Parsons' lemma. Since trees are particularly interesting in graph searching, we sketch its proof below. In the following, given a tree $T$, a vertex $v \in V(T)$ and a connected component $T'$ of $T \setminus v$, let $T' \cup v$ denote the subtree induced by the vertices of $T'$ and $v$.

**Lemma 1 (Parsons' lemma** [Par78b]**).** *For any $k \in \mathbb{N}^*$ and any tree $T$, $\mathtt{s}(T) \geq k+1$ if and only if $T$ has a vertex $v$ with at least three components $T_1, T_2, T_3$ of $T \setminus v$ such that $\mathtt{s}(T_i \cup v) \geq k$ for every $i \in \{1,2,3\}$.*

*Sketch of Proof.* The "if" part follows from monotonicity. Indeed, assume there is a monotone search strategy using at most $k$ searchers in $T$ and let $v, T_1, T_2$, and $T_3$ be as defined in the lemma. By monotonicity, we may assume that $T_1 \cup v$, $T_2 \cup v$ and $T_3 \cup v$ are cleared in this order. However, to clear $T_2 \cup v$, there must be a step at which all $k$ searchers are occupying vertices of $T_2$, which would imply a recontamination of $T_2 \cup v$ from $T_3$, a contradiction.

[4] A minor of a graph $G$ is any subgraph of any graph obtained from $G$ by contracting some edges.

On the other hand, if there exists no vertex $v$ as in the lemma, it is possible to find a subpath $P$, called an *avenue*, such that, for every connected component $T_u$ of $T \setminus P$ (where $u$ is the unique neighbor of $T_u$ in $P$), $\mathtt{s}(T_u \cup u) < k$ (see Fig. 4). Then, a strategy using $k$ searchers consists of sliding one searcher from one end of $P$ to its other end, while sequentially clearing the components of $T \setminus P$ using the $k-1$ remaining searchers. The avenues can be recursively computed by a dynamic programming algorithm. ◇

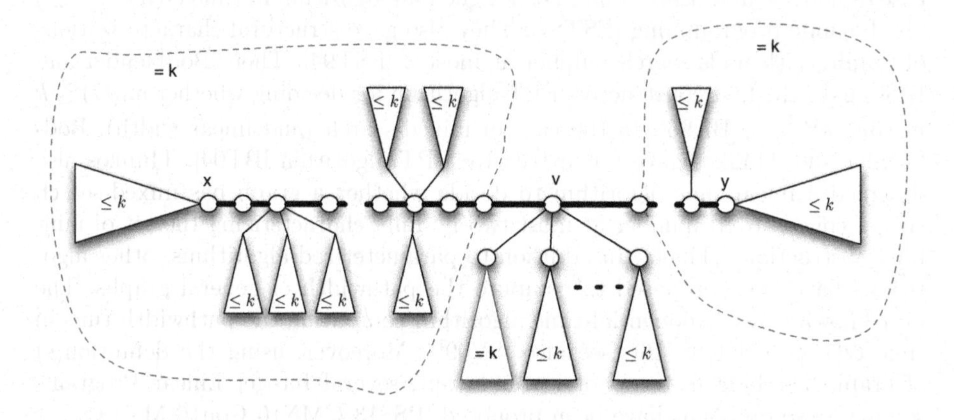


**Fig. 4.** Schematic overview of a tree $T$ with $\mathtt{s}(T) = k+1$. A triangle labelled with "$\leq k$" represents a subtree with search number at most $k$. The node $v$ has at least three components (in red) with search number $k$. The bold path (between $x$ and $y$) is an avenue in $T$. (Color figure online)

The above strategy implies that $\mathtt{s}(T) = O(\log n)$ in any $n$-node tree $T$. Moreover, this bound is tight (consider a rooted tree where all internal vertices have degree 3 and all leaves are at the same distance from the root).

In contrast, we should mention that graph searching is NP-hard in weighted trees, where weights on vertices represent the number of searchers required to preserve a vertex from recontamination, and edge-weights represent the number of searchers that must simultaneously traverse an edge to clear it [MT09].

Many other graph classes have been studied. The pathwidth is polynomial-time computable in *circular arc graphs* (in time $O(n^2)$) [ST07], *unicyclic graphs* [EM04, YZC07], *biconvex bipartite graphs* [PY07], in some subclasses of *chordal graphs* [Gus93], in *hypercubes* [CK06], in *cographs* [BM93] (in linear time), *etc.* The pathwidth can also be computed in time $n^{O(1)}$ (in the proposed algorithm, the exponent is larger than 11) in *outerplanar graphs* (using the fact that these graphs have bounded treewidth) and 2-approximation algorithms (using the *dual* of outerplanar graphs) running in time $O(n \log n)$ are designed in [BF02, CHS07].

The other variants have also been studied. The mixed search number can be computed in linear time in *interval graphs*, in polynomial time in *split graphs* [FHM10], and in linear time in *permutation graphs* [HM08]. The edge search number can be computed in linear time in *cographs* [GHM12] and in polynomial time in split graphs and interval graphs [PTK+00].

**Open Questions.** *We would like to conclude this section with some intriguing open questions. First, note that there are no known graph classes where the complexity of deciding the edge/node/mixed search number differs. Moreover, for any graph $G$ and any distinct $x, y \in \{\mathtt{s}, \mathtt{ns}, \mathtt{mixs}\}$ as inputs, the complexity of computing $x(G) - y(G)$ is not known.*

# 3 Connected Graph Searching and Distributed Setting

In all models defined in Sect. 2, removing a searcher and placing it at any vertex is allowed. Such a *jump* may however be unrealistic or even impossible in practical applications. Removing a searcher from a vertex $v$ and placing it at another vertex $u$ may be replaced by a sequence of slides along the edges of a path from $v$ to $u$. However, this would imply that the searcher travels in an unsafe environment (through a part that is still contaminated) and moreover, it may lead to strategies that are not monotone. To handle this problem, Barrière *et al.* proposed a new variant, called *connected graph searching*, where removing a searcher is not allowed and in which, at every step, the subgraph induced by the clear edges and vertices must be connected [BFFS02].

Two main questions were asked with the introduction of connected graph searching in [BFFS02,BFST03]. First, what is "the cost of connectivity" in terms of the number of searchers? That is, does there exist a constant $c$ such that any graph $G$ admits a connected search strategy using at most $c \cdot \mathtt{s}(G)$ searchers? Second, is connected graph searching monotone?

## 3.1 Cost of Connectivity

A *connected search strategy* $\mathcal{S}$ in a graph $G = (V, E)$ and using $k \geq 1$ searchers can be defined as follows. First, a vertex $v_0 \in V$, called *homebase*, is chosen and all the $k$ searchers are placed at it. Then, $\mathcal{S}$ is a sequence of moves, where each move consists of sliding a searcher at $u \in V$ along an edge $e = uv \in E$ and such a move is allowed only if, after the sliding, there is path of clear edges from $v_0$ to $v$, i.e., the clear part must always be connected (here we only consider the edge-search variant where an edge is cleared when a searcher slides along it). The *connected search number* of a graph $G$, denoted by $\mathtt{cs}(G)$, is the smallest $k$ such that there exists a connected search strategy that clears $G$ using $k$ searchers. Clearly, the choice of the homebase has an impact on the number of searchers (e.g., consider a path where the homebase is not one of its ends). Hence, the connected search number is defined with respect to the "best" possible homebase.

**(Non) Monotonicity.** Connected strategies clearly allow recontamination. Monotone connected search strategies are defined in a similar way: first, a vertex $v_0 \in V$ is chosen and all the $k$ searchers are placed at it, then, the strategy consists of a sequence of moves, where each move consists of sliding a searcher at $u \in V$ along an edge $e = uv \in E$ only if either $u$ is still occupied by a searcher after the move, or all incident edges of $u$ but possibly $e$ were already clear before the move. One important and surprising result is that, contrary to the classical graph searching, in the connected variant, recontamination may help [YDA09]. It is interesting to mention that their counter-example $G$ has about 400.000 vertices and is such that $\mathtt{cs}(G) = 281$ while any monotone connected strategy requires at least 290 searchers (see Fig. 5). We are not aware of a smaller example.

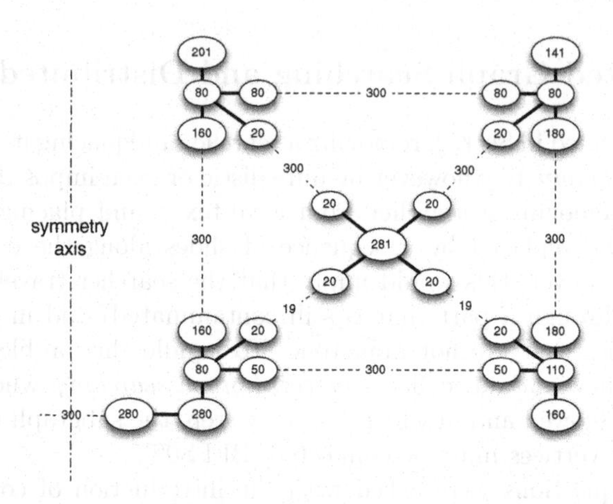


**Fig. 5.** Schematic overview of a graph $G$ such that $\mathtt{cs}(G) = 281 < \mathtt{mcs}(G) = 290$ [YDA09]. Any circle with label $x$ inside represents a clique of size $x$. A bold edge between two cliques $A$ and $B$ (with $|A| \leq |B|$) represents a perfect matching between the vertices of $A$ and a subset of size $|A|$ of the vertices of $B$. A dotted line with label $\ell$ between two cliques of same size $x$ represents a "path" of $\ell$ cliques of size $x$ where two consecutive cliques are joint by a perfect matching.

Hence, the *monotone connected search number* of a graph $G$, denoted by $\mathtt{mcs}(G)$, may be strictly larger than its connected search number $\mathtt{cs}(G)$. A consequence of this result is that it is not known whether the problem of computing the connected search number of a graph is in NP. As far as we know, there are no lower or (non-trivial) upper bounds on the number of steps of connected search strategies. Another difference between the search number and its (monotone or not) connected counter part is that mcs and cs are not minor-closed. Hence, it is not clear whether the problem of computing the (monotone) connected search number of graphs admits a fixed parameter tractable algorithm (nor even a polynomial-time algorithm when the number $k$ of searchers is fixed). However, both parameters are closed under taking contractions [BGTZ16].

Recontamination does not help for connected graph searching in trees, i.e., $\mathtt{mcs}(T) = \mathtt{cs}(T)$ in any tree $T$ [BFFS02, BFF+12] (proof *à la* Bienstock and Seymour). Besides, in any $n$-node tree $T$ and for any homebase $v_0$, there exists a monotone connected strategy, starting from $v_0$ and using at most $1 + \mathtt{cs}(T)$ searchers and, moreover, $\mathtt{cs}(T) = O(\log n)$ [BFFS02]. In the same paper, it is shown that computing the connected search number can be done in polynomial time in trees. Recently, it has been shown that recontamination does not help in the class of graphs with connected search number at most two [BGTZ16]. That is, for any graph $G$, $\mathtt{mcs}(G) = 2$ if and only if $\mathtt{cs}(G) = 2$. The result follows from the characterization of this class of graphs by exhibiting the family of 177 minimal-contraction obstructions.

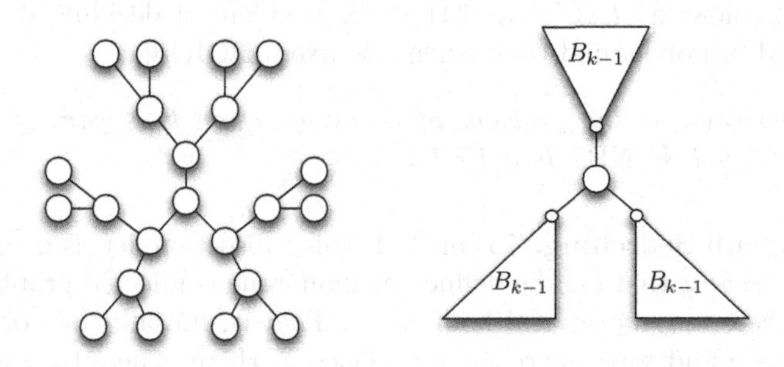


**Fig. 6.** On the left, the tree $D_3$ such that $\mathtt{s}(D_3) = 3 < \mathtt{cs}(D_3) = 4$. On the right, the recursive construction of the tree $D_k$ where $B_{k-1}$ is the complete binary tree of depth $k-1$ for any $k \geq 2$. For any tree $T$, $\mathtt{cs}(T) \geq k+1$ if and only if $T$ admits $D_k$ as a minor [BFST03, BFF+12].

**Cost in Number of Searchers.** Let us start with a simple example. Consider the complete rooted tree $D_3$ with all internal vertices with degree three and all leaves at distance 3 from the root. It is easy to see that $\mathtt{s}(D_3) = 3 < \mathtt{cs}(D_3) = 4$ (Fig. 6). Therefore, connectivity has some price in terms of minimum number of searchers. In any tree $T$, $\mathtt{cs}(T) \leq 2\mathtt{s}(T) - 2$ and this bound is tight [BFST03, BFF+12]. The proof relies on the fact that $\mathtt{cs}(T)$ is closed under taking minors in the class of trees and that $\mathtt{cs}(T) \geq k$ if and only if $T$ contains some specific tree $D_k$ as a minor (in contrast with the classical search number, the set of minimal obstructions for connected search number in trees is reduced to a single tree).

Therefore, the question of the cost of connectivity arises naturally: how far is the connected search number of a graph from its pathwidth? In other words, does there exist a constant $c \geq 2$ bounding the ratio between connected search number and search number in any graph? Several partial results have been proposed [Nis09, BFF+12] before Dereniowski closed the question:

**Theorem 3** [Der12b]. $\mathtt{mcs}(G) \leq 2\mathtt{s}(G) + 3$ *in any graph* $G$.

To prove Theorem 3, Dereniowski designed a polynomial time algorithm that transforms any monotone search strategy using $k$ searchers into a connected one using at most $2k+3$ searchers. His result shows that the ratio between monotone connected search number and connected search number is bounded by 2.

**Complexity and Algorithms.** On the complexity point of view, computing $\mathtt{cs}$ is NP-hard since $\mathtt{cs}(G^*) = \mathtt{s}(G) + 1$ for any graph $G$ where $G^*$ is obtained from $G$ by adding a universal vertex. Dereniowski proved that weighted connected graph searching is also NP-hard in weighted trees [Der11]. On the positive side, a polynomial time approximation with approximation ratio 3 is designed for this problem in trees [Der12a]. Similar results were proved while with different terminology (speaking about edge-width instead of weight) [BTK11]. The connected search number of outerplanar graphs has been investigated in [FTT05]. Very recently, Dereniowski *et al.* proved that the problem of deciding if $\mathtt{cs}(G) \leq k$ can be solved in polynomial-time when $k$ is fixed [DOR18].

**Open Questions.** *Is the problem of deciding (when $k$ is part of the input) whether* $\mathtt{cs}(G) \leq k$ *in* NP*? Is it FPT in $k$?*

**Internal Graph Searching.** To conclude this subsection, let us mention *internal graph searching* that can be defined as monotone connected graph searching but where there may be several homebases. That is, initially, one or more vertices are chosen and some searchers are placed at them. Then, the only allowed moves are to slide searchers if it does not create recontamination. This variant has been first introduced in [BFST03] and an interesting heuristic has been proposed in [FNS07]. In this paper, the initial vertices (the homebases) are chosen randomly and then the searchers grow the cleared part around these vertices in a BFS manner, then the best obtained strategies are used to generate next generations of strategies using a classical genetic algorithm.

### 3.2 Distributed (Monotone) Connected Graph Searching

A major reason for which the connectivity constraint has been introduced is that it ensures safe communications between the searchers during the execution of the strategy. For instance, when the searchers have to coordinate themselves but have no way to communicate when they are far from each other, possible solutions would be either to leave some messages on the vertices or to use a searcher for carrying instructions between other searchers. In both cases, the connectivity constraint helps since it allows to avoid that messages are left on contaminated vertices that the searcher crosses when moving in the contaminated area to transfer instructions.

In this subsection, we study the clearing of graphs in such environments where the searchers have only local vision of their environment and must communicate to coordinate the clearing.

**Distributed Model.** The $k$ searchers are modeled by synchronous autonomous mobile computing entities (automata) with distinct IDs from 1 to $k$. Otherwise searchers are all identical, run the same program, and use at most $O(\log n)$ bits of memory, where $n$ is the number of vertices of the network. A network is modeled by an undirected connected graph $G$. A priori, the network is asynchronous. However, as explained below, any synchronous algorithm can be transposed into an asynchronous environment by adding an extra searcher traveling in the (connected) clear part of the graph to synchronize the moves of all searchers. Moreover, the network is anonymous, that is, the vertices are not labelled. The edges incident to any vertex $u$ are labelled from 1 to its degree, so that the searchers can distinguish the different edges incident to a vertex. Every vertex of the network has a zone of local memory, the *whiteboard* in which searchers can read, erase, and write symbols (unless stated otherwise, whiteboards have size $O(\log n)$ bits and are only used for face-to-face communication between searchers occupying a same vertex). It is moreover assumed that searchers can access these whiteboards in fair mutual exclusion. The goal is then to design an algorithm, called a *search protocol*, such that the fewest number of searchers running this algorithm achieves the clearing of the graph in a connected way.

**Universal Algorithms.** In this section, we present several search protocols that have been designed to clear any graph $G = (V, E)$. In this setting, the searchers do not know in advance in which graph they are launched. That is, when occupying some vertex $u$, a searcher executes the algorithm only based on its current state (the *memory* of the searcher), on the content of the whiteboard at $u$, and on the degree of $u$. [BFNV08] designed a general algorithm allowing $\mathtt{mcs}(G)+1$ searchers to connectedly clear any graph $G$. Since the extra searcher (compared to the centralized case) cannot be avoided due to the asynchronicity of the network, this is optimal. Roughly, this algorithm orders all possible sequences of moves in some well-suited lexicographical order and tries them one after the other (sequentially increasing the number of searchers that are used) until the graph is clear. For this purpose, the searchers use whiteboards of size $O(|E| \log |V|)$ bits where they write all their moves. At the end, a description of the strategy is then stored in a distributed way on the whiteboards. This algorithm has however two drawbacks: it takes an exponential amount of time (which cannot be avoided unless $P = NP$) and the clearing is not monotone.

To deal with monotonicity, [NS09] proposed to address the problem by providing a small amount of information (*advice*) to the searchers, following the framework of [FIP06]. Precisely, it is shown that the minimum number of bits of information that must *a priori* be distributed in an $n$-node graph $G$ in order to clear it monotoneously with the optimal number of searchers is $\Theta(n \log n)$ [NS09]. Roughly, this piece of information encodes a spanning tree "along which" the clearing must be performed.

Another approach to handle monotonicity is to allow the use of more searchers. More precisely, the *cost* of a search protocol $\mathcal{P}$ in a graph $G$ with homebase $v_0$ is measured by the ratio between the number of searchers it uses to clear $G$ and the search number $\mathtt{mcs}(G)$ of $G$. This ratio, maximized over all graphs and all starting vertices, is called the *competitive ratio* of the protocol $\mathcal{P}$. [INS09] proved that monotonicity has an important cost (i.e., may increase significantly the minimum number of searchers) in a distributed setting since any search protocol (clearing any graph in a monotone connected way) has competitive ratio $\Omega(\frac{n}{\log n})$ and that this lower bound holds in the class of trees with maximum degree 3. On the positive side, this bound is tight: there exists a search protocol with competitive ratio $O(\frac{n}{\log n})$ [INS09]. The idea behind the algorithm is to "control" a (partial) spanning tree of the clear part and to determine the next edge to be cleared according to it in such a way that this tree does not contain a "high" ternary tree as a minor (since such a minor would lead to the use of many searchers).

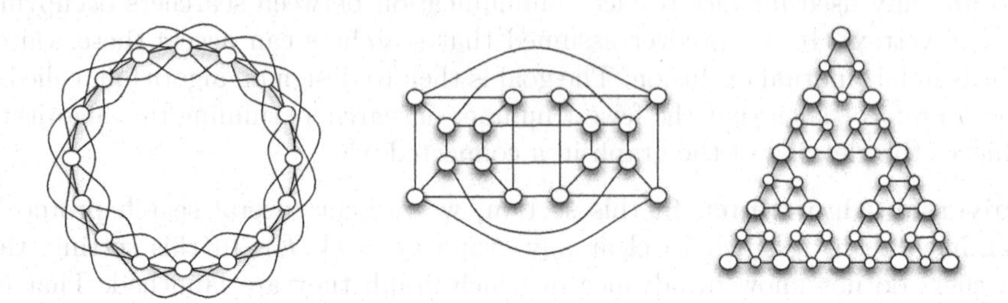

**Fig. 7.** Illustrations of a chordal ring (left), the hypercube of dimension 4 (center) and the Sierpiński graph built by 4 iterations (right).

**Specific Topologies.** Many distributed search protocols specialized for particular graph classes have also been designed. A distributed algorithm that computes the connected search number of trees has been proposed in [BFFS02,BFF+12]. Then, a *self-stabilizing* algorithm for clearing trees has been designed [MM09] and further improved in [BMM10]. The latter algorithm allows $1+\log n$ searchers to clear any $n$-node tree and stabilizes in $O(n \log n)$ moves after initialization. Moreover, it is a *non-silent* algorithm, meaning that it continues to clear the tree indefinitely.

Topologies that are commonly used for interconnection networks (see examples on Fig. 7) have been studied. Precisely, the following topologies have been considered: grids [FLS05], chordal rings and tori [FHL07], hypercubes [FHL08], and Sierpiński graphs [Luc09]. In this setting, the authors compare the number of searchers, moves, and the number of rounds of their algorithms in two models. In the first model, a particular searcher is used to coordinate the clearing while, in the latter one, the searchers are endowed with some visibility ability: they can see whether their neighbors are clear or contaminated, empty or occupied. All designed algorithms use the fact that all these graph classes admit relatively well-structured centralized strategies and, moreover, the symmetries of these topologies allow the searchers to benefit from some *sense of direction* (for instance, clockwise orientation in chordal rings or standard compass-labelling in grids). For instance, in a grid starting from one of its corners, the strategy first makes the searchers occupy all vertices of the first column and then move "in parallel" from one column to the next one until the grid is clear. In the case with visibility, the searchers can locally decide when they have to go to the next column without recontamination. Table 1 summarizes the obtained results (note that these results consider the clearing of vertices only or, said differently, an edge is cleared in the same way as in node-search).

**Open Questions.** *The studies of the tradeoffs between the number of searchers, moves, and time steps are left as open problems.*

**Table 1.** Monotone connected search in specific topologies. Results marked with a star (*) are known to be optimal.

| Topology | Model | # searchers | # moves | # time steps |
|---|---|---|---|---|
| $m \times n$ **Grids** [FLS05] ($m \leq n$) | Coordinator<br>Visibility | $m+1^*$<br>$m^*$ | $\frac{m^2+4mn-5m-2}{2}$<br>$\frac{m^2+2mn-3m}{2}$ | $mn-2$<br>$m+n-2^*$ |
| $m \times n$ **Tori** [FHL07] ($m \leq n$) | Coord.<br>Vis. | $2m+1^*$<br>$2m^*$ | $2mn-4m-1$<br>$mn-2m$ | $mn-2m$<br>$n-2$ |
| $n$-node **Chordal rings** [FHL07] with largest chord of length $\ell$ and $\ell'$ second largest chord | Coord.<br>Vis. | $2\ell+1^*$<br>$2\ell^*$ | $4n-6\ell-1$<br>$n-2\ell$ | $3n-4\ell-1$<br>$\lceil \frac{n-2\ell}{2(\ell-\ell')} \rceil$ |
| **Hypercubes** [FHL08] (dimension $n$) | Coord.<br>Vis. | $\Theta(\frac{n}{\sqrt{\log n}})^*$<br>$n/2$ | $O(n \log n)$<br>$O(n \log n)$ | $O(n \log n)$<br>$O(\log n)^*$ |
| **Sierpiński graphs** [Luc09] built by $n$ iterations | Coord.<br>Vis. | $n+1^*$<br>$n+2$ | $O(n3^n)$, $\Omega(3^n)$<br>– | $O(3^n)$, $\Omega(3^n/n)$<br>– |

A search protocol has also been designed for *partial grids* (i.e., connected subgraphs of $n \times n$ grids) that uses $O(\sqrt{n})$ searchers [DU16]. The algorithm strongly uses sense of direction and the algorithm in [BDK15] as a subprocedure. The algorithm is optimal since some partial grids require this amount of searchers and moreover, the authors prove that, for any search protocol, there are partial grids (with search number $O(\log n)$) that force the algorithm to use $\Omega(\sqrt{n})$ searchers [DU16].

To conclude this subsection, let us mention the *cloning* variant proposed in [FHL08]. In this model, the searchers may clone themselves, i.e., the searchers are not restricted to appear at the homebase but, at any step, a searcher at $v$ may create new searchers at $v$ (this essentially allows to decrease the number of moves and time steps). Various topologies have been studied in this setting: hypercubes [FHL08], graph products [ISZ07], grids and tori [ISZ08], and pyramids [SIS06].

### 3.3 Exclusive Graph Searching and Look/Compute/Move

*Exclusive graph searching* is defined as mixed graph searching with the extra *exclusivity constraint* (each vertex can be occupied by at most one searcher at a time) and such that searchers cannot jump from one vertex to another one, i.e., searchers can only slide along edges [BBN17].

Exclusive graph searching addresses two limitations of classical variants as far as practical applications are concerned. First, as in internal graph searching, the unrealistic assumption that searchers may jump is got rid of. Second, classical variants assume that any vertex can be simultaneously occupied by several searchers. This assumption may be unrealistic in several contexts. Typically, placing several searchers at the same vertex may simply be impossible in a physical environment in which, e.g., the searchers are modeling physical searchers moving in a network of pipes. In the case of software agents deployed in a computer network, maintaining several searchers at the same node may consume local resources (e.g., memory, computation cycles, etc.). The exclusivity constraint aims at dealing with this problem.

More formally, given a connected $n$-node graph $G$, an *exclusive search strategy* in $G$, using $k \leq n$ searchers consists of (1) placing the $k$ searchers at $k$ different vertices of $G$, and (2) performing a sequence of slidings ensuring the exclusivity constraint. An edge becomes clear whenever either a searcher slides along it, or one searcher is placed at each of its extremities (as in mixed-search). The exclusive-search number of $G$, denoted by $\mathtt{xs}(G)$ is the smallest $k$ for which there exists a winning search strategy in $G$. Exclusive graph searching behaves very differently from classical variants. For instance, $\mathtt{xs}(S_n) = n - 1$ for any star $S_n$ with $n \geq 3$ leaves. More important, it is not monotone even in trees and it is not closed under taking subgraphs [BBN17]. It has been proved that, for any graph $G$ with maximum degree $\Delta$, $\mathtt{ns}(G) \leq \mathtt{xs}(G) \leq (\Delta-1)(\mathtt{ns}(G)+1)$ [BBN17]. Surprisingly, computing the monotone exclusive search number is NP-hard in split graphs (where pathwidth can be polynomially computed) and can be solved in polynomial time in a subclass of star-like graphs (where pathwidth is NP-hard) [MNP17]. A linear-time algorithm in cographs is also proposed in [MNP17]. A polynomial-time algorithm that computes the monotone exclusive search number of trees has been designed in [BBN17]. It is based on a lemma in the same vein as Parsons' lemma (while more technical) and then follows the same principles as the algorithm of Ellis *et al.* for edge-search (see Lemma 1) but the proof is more technical due to the non-closedness under subgraph.

**Open Questions.** *Is the problem of deciding (when $k$ is part of the input) whether* $\mathtt{xs}(G) \leq k$ *in* NP*? Is it polynomial when $k$ is fixed? Is it FPT in $k$? Also the study of the exclusive search number in various graph classes is still open.*

Distributed exclusive graph searching has been studied in the `Look-Compute-Move` model where searchers have very weak abilities (they are anonymous and oblivious) but can "see" the whole network (see Chaps. 8 and 9 for more details). Algorithms for paths and trees, using the optimal number of searchers (or more), have been designed in [BBN12] and the case of cycles is studied in [DSN+15, DNN17]. The algorithm in cycles relies on a subprocedure that places the searchers in an adequate configuration that can also be used to solve other coordination problems such as *gathering* and *perpetual exploration.*

**Open Questions.** *One intriguing remaining question in the cycle is whether* 4 *searchers can exclusively clear any cycle with at least* 10 *vertices in the* `Look-Compute-Move` *model. Indeed, for any $n$-node cycle with $n > 10$, it is possible to clear it with $k \in \{5, \cdots, n-3\} \cup \{n\}$ searchers and not possible with* $\leq 3$ *searchers or $k \in \{n-2; n-1\}$ searchers* [DSN+15].

## 4 Plethora of Alternative Models

Recall that, in the Introduction, it was mentioned that the network decontamination problem can be equivalently seen as a pursuit-evasion game between a team of searchers and a *fugitive* (an intruder, a lost spelunker...). Variants of graph searching that have been described so far can all be stated in terms of capturing a lucky invisible and arbitrarily fast fugitive in a graph. By "lucky" (or "omniscient"), we mean that the objective is the design of a strategy that captures the fugitive in the worst case, i.e., whatever be the behavior of the fugitive. From now on, we use the pursuit-evasion terminology (except in Subsect. 4.3) because it fits the proposed models better.

### 4.1 Visible/Inert Fugitive and Tree-Like Structures

**Visible Fugitive.** A first natural extension of node-search concerns the case of a visible fugitive. In this variant, the fugitive occupies a vertex that is known by the searchers but may move at any step to another (known) vertex. In particular, if a step of a strategy consists of placing a searcher at the vertex occupied by the fugitive, the latter may simultaneously (before the searcher "lands") move to any vertex it can access (through a path free of searchers). The *visible search number* of a graph $G$, denoted by $\mathtt{vns}(G)$, is the minimum number of searchers required to catch a visible fugitive in this setting. For instance, $\mathtt{vns}(T) = 2$ for any tree $T$ (not reduced to a single vertex) while, for any $n \in \mathbb{N}^*$, there are $n$-node trees $T$ such that $\mathtt{ns}(T) = \Theta(\log n)$ (see Sect. 2.3). The visible search number shares a relationship with *treewidth*[5], denoted by $\mathtt{tw}(G)$, that is similar to the relationship between pathwidth and node-search number. Precisely:

**Theorem 4** [ST93]. *For any graph $G$, $\mathtt{vns}(G) = \mathtt{tw}(G) + 1$.*

As in the case of pathwidth and node-search, it is easy to show that monotone visible node-search strategies are equivalent to tree-decompositions. Again, the difficulty is to prove that there always exists an optimal strategy that is monotone. Seymour and Thomas proved the monotonicity of visible graph searching by defining a dual structure for the tree-decompositions, namely the *brambles* (initially called *screens*) [ST93], that actually corresponds to a winning strategy for the fugitive. Given a graph $G = (V, E)$, a *bramble* is a family $\mathcal{B} = (B_i)_{0 \leq i \leq \ell}$ of subsets of vertices such that (1) $B_i$ induces a connected subgraph of $G$ for each $i$ and (2) the sets $B_i$ are pairwise *touching* (i.e., any two sets intersect or there exists an edge linking them). The *order* of $\mathcal{B}$ is the minimum size of a hitting set, i.e., the smallest number of vertices in $V$ that intersect each set in $\mathcal{B}$. The treewidth of a graph $G$ is at most $k - 1 \in \mathbb{N}$ (and so $\mathtt{vns}(G) \leq k$) if and only if the maximum order of a bramble of $G$ is $k$ [ST93]. Given a graph $G$ with a bramble $\mathcal{B}$ of order $k+1$, it is easy to describe a winning strategy for the fugitive against $\leq k$ searchers. Indeed, at every step, the fugitive can move (since the sets are connected and pairwise touching) to a set of $\mathcal{B}$ whose vertices are occupied by no searcher. The notion of bramble is very useful to prove lower bounds on the visible search number of graphs. For instance, it is easy to show that $\mathtt{vns}(G_{n\times n}) \leq n+1$ in any $n \times n$ grid $G_{n\times n}$ and a bramble of order $n$ in

[5] Due to the huge number of works on treewidth, we have decided not to detail them (nor the definition of treewidth) and refer the reader to [Bod98, Die12, CFK+15].

$G_{n\times n}$ can also easily be found. Altogether, this proves that $\mathtt{vns}(G_{n\times n}) = n+1$ (such a result is rather technical without the help of brambles).

The connected capture of a visible fugitive has been studied in [FN08]. As its invisible counterpart, it is not monotone. However, in contrast with the invisible case (Theorem 3), this variant may require $\Omega(\log n * \mathtt{vns}(G))$ searchers in some $n$-node graphs $G$ and this is asymptotically tight [FN08].

**Non-deterministic Graph Searching.** *Non-deterministic graph searching* generalizes both node-search and visible node-search [FFN09]. Given a fixed integer $q \geq 0$, a non-deterministic strategy aims at catching an invisible fugitive with the additional ability that the fugitive is visible during at most $q$ steps of the game (the choice of when to see the fugitive is left to the searchers dynamically during the strategy). The minimum number of searchers required to catch the fugitive in this setting is denoted by $\mathtt{ns}_q(G)$. By definition, $\mathtt{ns}_0(G) = \mathtt{ns}(G)$ (the fugitive is always invisible) and $\mathtt{ns}_\infty(G) = \mathtt{vns}(G)$ (the fugitive is always visible). Computing $\mathtt{ns}_q(G)$ is NP-hard for any $q \geq 0$ and an exponential-time algorithm to compute it is presented in [FFN09]. The monotonicity of this variant is proved in [MN08] and a constructive FPT algorithm is designed in [BBM+13]. A polynomial-time dynamic programming algorithm to compute a 2-approximation of $\mathtt{ns}_q(T)$ in the class of trees $T$ (exact for $q \leq 1$) is designed in [ACN15].

**Open Questions.** *The existence of an exact polynomial-time algorithm that computes* $\mathtt{ns}_q(T)$ *in any tree* $T$ *and for any* $q > 1$ *is still open.*

*Another interesting open question is the definition of a dual structure (similar to brambles for visible node-search* [ST93] *or to blockage for node-search* [BRST91]*) for non-deterministic graph searching.*

**Inert Fugitive.** Another variant of node-search is related to tree-decompositions. A fugitive is *inert (a.k.a., lazy)* if it is invisible but can only move if a searcher is landing at the vertex it is currently occupying. In any graph $G$, the minimum number of searchers required to catch the fugitive in this setting also equals the treewidth of $G$ plus one [RT11].

**LIFO-Search.** Last but not least, let us mention a variant of graph searching related to another tree-like parameter of graphs. Namely, *LIFO-search* is a variant of node-search where the searchers are labelled with distinct integers and with the extra constraint that a searcher can be removed only if no searcher with smaller label is present in the graph [GHT12]. In [GHT12], this variant is proved to be monotone and equivalent to the *tree-depth* of graphs [NdM08].

### 4.2 Directed Graphs

During the last decade, several digraph decompositions have been proposed in order to try to bring to directed graphs the same algorithmic power as tree-decompositions provide for undirected graphs [GHK+16]. Interestingly, most of these attempts have been defined through graph searching games. An important difference between directed graph searching games and undirected ones arises via the notion of monotonicity. In the directed case, there are two distinct definitions of monotonicity: a game is *cop-monotone* if each vertex is occupied at most once by a searcher, it is *robber-monotone* if the area reachable by the fugitive never increases. Clearly a cop-monotone game is robber-monotone. However, as shown below, the converse is not always true.

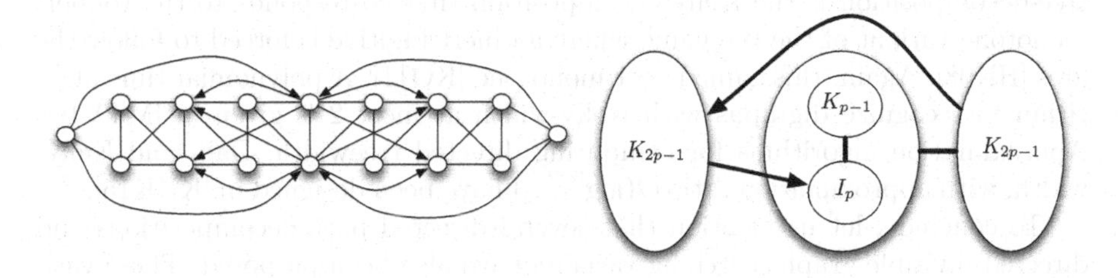


**Fig. 8.** On the left, a graph in which 4 searchers can capture a visible fugitive constrained to move in strongly connected components free of searchers (directed treewidth variant) but every robber-monotone strategy requires 5 searchers [Adl07]. On the right, the schematic overview of a graph where $3p-1$ searchers can capture a visible fugitive constrained to follow the orientation of the arcs (DAG-width variant) but every monotone strategy requires $4p-2$ searchers, for any $p \geq 2$ [KO11]. $K_x$ denotes a clique on $x$ vertices and $I_x$ is an independent set on $x$ vertices. A directed edge between two parts $A$ and $B$ means that there are edges from every vertex in $A$ to every vertex in $B$. Undirected edges mean that there are edges between $A$ and $B$ in both directions.

Johnson *et al.* first defined the *directed tree-decomposition* which roughly "translates" the connectivity properties of tree-decomposition into strong connectivity properties in directed graphs [JRST01]. Their variant is closely related to the graph searching game where a visible fugitive has the extra constraint that it can move only in strongly connected components free of searchers. That is, the fugitive can go from vertices $u$ to $v$ if there is a directed path from $u$ to $v$ free of searchers and a directed path from $v$ to $u$ free of searchers. It has been shown that, in this game, the non-monotone, the cop-monotone and the robber-monotone variants may differ [JRST01, Adl07] (see Fig. 8, left). Because of the non-monotonicity result, no min-max theorem can be expected via graph searching. However, [JRST01] proved a weaker result: if $k$ searchers have a winning strategy in a digraph $D$, then $3k-1$ searchers have a robber-monotone winning strategy in $D$, which leads to a min-max theorem up to a constant ratio between directed treewidth and so-called *havens* [JRST01]. In [EHS13], it is proved that the cop-monotone version of this game is actually equivalent to the D-width defined by Safari [Saf05] leading to an exact algorithm for computing this variant. Moreover, [EHS13] showed that D-width and directed treewidth are actually equivalent (in the sense that one is bounded if and only if the other is bounded).

The DAG-decomposition is weaker than directed tree-decomposition (bounded DAG-width implies bounded directed treewidth) [BDH+12].

It corresponds to the cop-monotone version of the game where the visible fugitive is constrained to follow the direction of arcs. While robber-monotone and cop-monotone variants coincide [BDH+12], this game is not monotone [KO11] (see Fig. 8, right). However, a drawback of DAG-decomposition is that the best known upper bound of the size of such a decomposition with width $k$ in an $n$-node digraph is $O(n^k)$ and that the corresponding optimization problem is PSPACE-complete [AKR16]. Another decomposition weaker than directed tree-decomposition is the Kelly-decomposition that corresponds to the robber-monotone variant of the the game where an inert fugitive is forced to follow the arcs [HK08]. Again, this game is not monotone [KO11]. A polynomial-time algorithm to recognize digraphs with Kelly-width at most 2 is given in [MTV10]. Approximation algorithms for computing directed treewidth, Dag- and Kelly-width, with approximation ratio $O(\log^{3/2} n)$ have been designed in [KKK15].

To conclude, let us mention that several directed path-decompositions and directed invisible graph searching variants have also been proposed. These variants mainly differ (1) in the abilities of the searchers and the fugitive: either both have to follow the direction of arcs, or only one of them, and (2) in the variant of graph searching that is considered: edge, node or mixed. More details can be found in [Bar06, YC07b, ADHY07, YC07a, Yan07, YC08c, YC08b, YC08a, YC09]. Contrary to their visible counterparts, all these directed variants are monotone.

### 4.3 Recontamination Alternatives

In classical graph searching, a vertex is recontaminated instantaneously if it is not occupied and adjacent to a contaminated vertex. Flocchini *et al.* proposed several alternative definitions for recontamination. In a first variant, with *threshold immunity* or *local immunity*, a vertex can be recontaminated only if a sufficient number of its neighbors are contaminated [LPS06]. In a second model, with *temporal immunity*, a vertex can be recontaminated only $t$ steps after it has been left by a searcher, where $t \geq 0$ is a parameter [FMS08].

**Local Immunity.** In [LPS06], a clear (and unoccupied) vertex is recontaminated if more than half of its neighbors are contaminated. In this setting, any graph with maximum degree three can be cleared by at most 2 searchers and by a single one if moreover there is a pendant vertex. [LPS06] gave search protocols for tori that are optimal in terms of number of searchers and asymptotically tight for the number of moves. They also considered the case of trees. Their results have been extended by the generalization proposed in [FLPS16].

In [FLPS16], the parameterized version of this problem is considered where the parameter $m \geq 1$ represents the minimum number of neighbors of a vertex $v$ that must be contaminated to recontaminate $v$. In $n_1 \times \cdots \times n_d$ $d$-dimensional grids, one searcher is sufficient for any $m \geq d$, $\Pi_{j=1}^{d-m} n_j$ searchers are sufficient otherwise. In the case of the $n_1 \times \cdots \times n_d$ $d$-dimensional torus, $2^d \cdot \Pi_{j=1}^{d-m} n_j$ searchers are sufficient for $m \leq d-1$ and $2^{2d-m}$ searchers are sufficient for $d \leq m \leq 2d$.

**Open Questions.** *It is not known whether these upper bounds are tight.*

Finally, a dynamic programming algorithm that computes optimal monotone search strategies in trees has also been designed [FLPS16].

**Temporal Immunity.** A graph $G$ has *temporal immunity* $t \geq 0$ if a vertex becomes recontaminated after having been exposed (i.e., unoccupied and with a contaminated neighbor) during more than $t$ steps [FMS08]. As an example, assume that $t = 2$, then an $n$-node cycle can be cleared by a single searcher moving clockwise during $2n$ steps (note that the strategy is not monotone). [DJS16] defined the *immunity number* $\iota_k(G)$ of $G$ as the minimum integer $t \geq 0$ such that $k \geq 1$ searchers can clear $G$ with temporal immunity $t$.

A distributed algorithm for computing the minimum number of searchers needed to clear a tree with temporal immunity $t \geq 0$ has been designed and a structural characterization of trees with $\iota_k(T) = t$ is provided [FMS08]. Roughly, $\iota_k(T) = t$ if and only if $T$ does not contain a subtree obtained from the complete ternary tree of height $k$ whose all edges have been subdivided $\lceil \frac{t}{2} \rceil + 1$ times. Finally, an algorithm for clearing any tree of height at most $h$ with $\lfloor \frac{2h}{t+2} \rfloor$ searchers is presented in [FMS08].

For any $k \in \{1, 2, 4\}$, any $n \times n$ grid with temporal immunity at least $(4 - k)(n - 1) - 1$ can be cleared by $k$ searchers [DFZ10], no results for other number of searchers are known. In the case of *strong grids*, $k$ searchers are sufficient to clear them when the immunity is at least $\lfloor \frac{2(2n-1)}{k} \rfloor$ [DFZ10].

Finally, $\iota_1(G)$ has been studied in several classes of graphs [DJS16] such as paths: $\iota_1(P_n) = 0$ for every $n$; cycles: $\iota_1(C_n) = 2$ for every $n$, and equals $n - 1$ if monotonicity is required; complete graphs: $\iota_1(K_n) = n - 1$ for every $n$; complete bipartite graphs: $\iota_1(K_{n,m}) = 2m - 1$ for $3 \leq m \leq n$; $n$-node trees: $\iota_1(T) \leq 30\sqrt{n}$; $p \times q$ grids: $p/2 \leq \iota_1 \leq p$ for $p \leq q$, *etc.* It can be shown that there are $n$-node trees $T$ for which $\iota_1(T) = \Omega(n^{1/3+\epsilon})$ for some constant $\epsilon > 0$ [DJS16].

**Open Questions.** *A challenge would be to close the gap with the upper bound $30\sqrt{n}$ in trees. The question of general planar graphs is also open.*

## 4.4 Other Models and Objectives

To conclude this chapter, we would like to mention some variants of graph searching that differ from previous ones by: the objective that must be optimized, the way the fugitive is captured, the speed of the fugitive, etc. We only mention some of these variants, others may be found in [FT08].

**Different Objectives.** The *cost* of a strategy is the sum of the number of occupied vertices over all steps of a strategy. This parameter (in the node-search variant) appears to equal the *profile* of the graph $G$ (minimum number of edges of an interval supergraph of $G$) [FG00,Fom04], while, in the visible (or inert) variant, it equals the minimum *fill in* of $G$ (minimum number of edges of a chordal supergraph). The cost in the case of edge-search has been studied in [DD13]. The *maximum occupation time* is the maximum over all vertices of the number of steps during which a vertex is occupied. This parameter coincides with the *bandwidth* of graphs [FHT05,Der09]. The *capture time* (minimum number of bags of a path-decomposition with a given width) has also been considered in [BH06,DKZ15].

**Different Speeds.** The case of a fugitive with bounded *speed* (the fugitive has speed $s$ if, at every step, it can move through a path of length at most $s$) has been considered in [Fom98,Fom99], and the case of an inert fugitive with bounded speed is considered in [DKT97].

**Different Rules.** [DYY08] introduced the *fast searching game* in which the searchers cannot be removed and every edge can be traversed only once. This variant has been studied in [SY09,Yan11,SY11,Yan13,DDD13,XYZZ16,XY17]. See also [MNP08,KP16] (and references therein) for the so-called *brush game.*

**Different Applications.** Surprisingly, a variant of graph searching has been defined to model the problem of routing reconfiguration in optical (WDM) networks [CHM+09]. In this variant, an invisible fugitive is moving following the orientation of the arcs in a directed graph and it is captured as soon as the searchers constrain the moves of the fugitive to a component that is not strongly connected. Monotonicity [NS16], computational complexity [CS11,CHM12] as well as trade-off between the number of searchers and the cost of strategies [CCM+11] have been studied.

# References

[ABC+15] Ames, B.P.W., et al.: A leapfrog strategy for pursuit-evasion in a polygonal environment. Int. J. Comput. Geom. Appl. **25**(2), 77–100 (2015)

[ACN15] Amini, O., Coudert, D., Nisse, N.: Non-deterministic graph searching in trees. Theor. Comput. Sci. **580**, 101–121 (2015)

[ADHY07] Alspach, B., Dyer, D., Hanson, D., Yang, B.: Arc searching digraphs without jumping. In: Dress, A., Xu, Y., Zhu, B. (eds.) COCOA 2007. LNCS, vol. 4616, pp. 354–365. Springer, Heidelberg (2007). https://doi.org/10.1007/978-3-540-73556-4_37

[Adl07] Adler, I.: Directed tree-width examples. J. Comb. Theory Ser. B **97**(5), 718–725 (2007)

[AKR16] Amiri, S.A., Kreutzer, S., Rabinovich, R.: Dag-width is PSPACE-complete. Theor. Comput. Sci. **655**, 78–89 (2016)

[Als04] Alspach, B.: Searching and sweeping graphs: a brief survey. Mathematiche **59**, 5–37 (2004)

[Bar06] Barát, J.: Directed path-width and monotonicity in digraph searching. Graphs Comb. **22**(2), 161–172 (2006)

[BBM+13] Berthomé, P., Bouvier, T., Mazoit, F., Nisse, N., Soares, R.P.: An unified FPT algorithm for width of partition functions. Research Report RR-8372, INRIA, September 2013

[BBN12] Blin, L., Burman, J., Nisse, N.: Brief announcement: distributed exclusive and perpetual tree searching. In: Aguilera, M.K. (ed.) DISC 2012. LNCS, vol. 7611, pp. 403–404. Springer, Heidelberg (2012). https://doi.org/10.1007/978-3-642-33651-5_29

[BBN17] Blin, L., Burman, J., Nisse, N.: Exclusive graph searching. Algorithmica **77**(3), 942–969 (2017)

[BDH+12] Berwanger, D., Dawar, A., Hunter, P., Kreutzer, S., Obdržálek, J.: The DAG-width of directed graphs. J. Comb. Theory Ser. B **102**(4), 900–923 (2012)

[BDK15] Borowiecki, P., Dereniowski, D., Kuszner, L.: Distributed graph searching with a sense of direction. Distrib. Comput. **28**(3), 155–170 (2015)

[BF02] Bodlaender, H.L., Fomin, F.V.: Approximation of pathwidth of outerplanar graphs. J. Algorithms **43**(2), 190–200 (2002)

[BFF+12] Barrière, L., et al.: Connected graph searching. Inf. Comput. **219**, 1–16 (2012)

[BFFS02] Barrière, L., Flocchini, P., Fraigniaud, P., Santoro, N.: Capture of an intruder by mobile agents. In: Proceedings of the 14th Annual ACM Symposium on Parallel Algorithms and Architectures (SPAA), pp. 200–209 (2002)

[BFL+09] Bodlaender, H.L., Fomin, F.V., Lokshtanov, D., Penninkx, E., Saurabh, S., Thilikos, D.M.: (Meta) kernelization. In: 50th Annual IEEE Symposium on Foundations of Computer Science (FOCS), pp. 629–638. IEEE Computer Society (2009)

[BFNV08] Blin, L., Fraigniaud, P., Nisse, N., Vial, S.: Distributed chasing of network intruders. Theor. Comput. Sci. **399**(1–2), 12–37 (2008)

[BFST03] Barrière, L., Fraigniaud, P., Santoro, N., Thilikos, D.M.: Searching is not jumping. In: Bodlaender, H.L. (ed.) WG 2003. LNCS, vol. 2880, pp. 34–45. Springer, Heidelberg (2003). https://doi.org/10.1007/978-3-540-39890-5_4

[BGTZ16] Best, M.J., Gupta, A., Thilikos, D.M., Zoros, D.: Contraction obstructions for connected graph searching. Discrete Appl. Math. **209**, 27–47 (2016)

[BH06] Brandenburg, F.J., Herrmann, S.: Graph searching and search time. In: Wiedermann, J., Tel, G., Pokorný, J., Bieliková, M., Štuller, J. (eds.) SOFSEM 2006. LNCS, vol. 3831, pp. 197–206. Springer, Heidelberg (2006). https://doi.org/10.1007/11611257_17

[Bie91] Bienstock, D.: Graph searching, path-width, tree-width and related problems (a survey). In: Proceedings of Reliability Of Computer And Communication Networks, a DIMACS Workshop. DIMACS Series in Discrete Mathematics and Theoretical Computer Science, vol. 5, pp. 33–50. DIMACS/AMS (1991)

[BK96] Bodlaender, H.L., Kloks, T.: Efficient and constructive algorithms for the pathwidth and treewidth of graphs. J. Algorithms **21**(2), 358–402 (1996)

[BKIS12] Bhadauria, D., Klein, K., Isler, V., Suri, S.: Capturing an evader in polygonal environments with obstacles: the full visibility case. Int. J. Robot. Res. **31**(10), 1176–1189 (2012)

[BLS99] Brandstädt, A., Le, V.B., Spinrad, J.P.: Graph Classes: A Survey. Society for Industrial and Applied Mathematics, Philadelphia (1999)

[BM93] Bodlaender, H.L., Möhring, R.H.: The pathwidth and treewidth of cographs. SIAM J. Discrete Math. **6**(2), 181–188 (1993)

[BMM10] Blair, J., Manne, F., Mihai, R.: Efficient self-stabilizing graph searching in tree networks. In: Dolev, S., Cobb, J., Fischer, M., Yung, M. (eds.) SSS 2010. LNCS, vol. 6366, pp. 111–125. Springer, Heidelberg (2010). https://doi.org/10.1007/978-3-642-16023-3_11

[BN11] Bonato, A., Nowakovski, R.J.: The Game of Cops and Robber on Graphs. American Mathematical Society, Providence (2011)

[Bod98] Bodlaender, H.L.: A partial k-arboretum of graphs with bounded treewidth. Theor. Comput. Sci. **209**(1–2), 1–45 (1998)

[Bre67] Breisch, R.L.: An intuitive approach to speleotopology. Southwest. Cavers **6**, 72–78 (1967)

[Bre12] Breish, R.L.: Lost in a Cave: Applying Graph Theory to Cave Exploration. Greyhound Press, Dallas (2012)

[BRST91] Bienstock, D., Robertson, N., Seymour, P.D., Thomas, R.: Quickly excluding a forest. J. Comb. Theory Ser. B **52**(2), 274–283 (1991)

[BS91] Bienstock, D., Seymour, P.D.: Monotonicity in graph searching. J. Algorithms **12**(2), 239–245 (1991)

[BT04] Bodlaender, H.L., Thilikos, D.M.: Computing small search numbers in linear time. In: Downey, R., Fellows, M., Dehne, F. (eds.) IWPEC 2004. LNCS, vol. 3162, pp. 37–48. Springer, Heidelberg (2004). https://doi.org/10.1007/978-3-540-28639-4_4

[BTK11] Borie, R.B., Tovey, C.A., Koenig, S.: Algorithms and complexity results for graph-based pursuit evasion. Auton. Robots **31**(4), 317–332 (2011)

[CCM+11] Cohen, N., Coudert, D., Mazauric, D., Nepomuceno, N., Nisse, N.: Tradeoffs in process strategy games with application in the WDM reconfiguration problem. Theor. Comput. Sci. **412**(35), 4675–4687 (2011)

[CDH+16] Corneil, D.G., Dusart, J., Habib, M., Mamcarz, A., de Montgolfier, F.: A tie-break model for graph search. Discrete Appl. Math. **199**, 89–100 (2016)

[CFK+15] Cygan, M., et al.: Parameterized Algorithms. Springer, Cham (2015)

[CHI11] Chung, T.H., Hollinger, G.A., Isler, V.: Search and pursuit-evasion in mobile robotics - a survey. Auton. Robots **31**(4), 299–316 (2011)

[CHM+09] Coudert, D., Huc, F., Mazauric, D., Nisse, N., Sereni, J.-S.: Reconfiguration of the routing in WDM networks with two classes of services. In: Conference on Optical Network Design and Modeling (ONDM), Braunschweig, Germany (2009)

[CHM12] Coudert, D., Huc, F., Mazauric, D.: A distributed algorithm for computing the node search number in trees. Algorithmica **63**(1–2), 158–190 (2012)

[CHS07] Coudert, D., Huc, F., Sereni, J.-S.: Pathwidth of outerplanar graphs. J. Graph Theory **55**(1), 27–41 (2007)

[CK06] Chandran, L.S., Kavitha, T.: The treewidth and pathwidth of hypercubes. Discrete Math. **306**(3), 359–365 (2006)

[CM93] Courcelle, B., Mosbah, M.: Monadic second-order evaluations on tree-decomposable graphs. Theor. Comput. Sci. **109**(1&2), 49–82 (1993)

[CMN16] Coudert, D., Mazauric, D., Nisse, N.: Experimental evaluation of a branch-and-bound algorithm for computing pathwidth and directed pathwidth. ACM J. Exp. Algorithmics **21**(1), 1.3:1–1.3:23 (2016)

[Cou16] Coudert, D.: A note on integer linear programming formulations for linear ordering problems on graphs. Research report, Inria, I3S, Universite Nice Sophia Antipolis, CNRS, February 2016

[CS11] Coudert, D., Sereni, J.-S.: Characterization of graphs and digraphs with small process numbers. Discrete Appl. Math. **159**(11), 1094–1109 (2011)

[DD13] Dereniowski, D., Dyer, D.: On minimum cost edge searching. Theor. Comput. Sci. **495**, 37–49 (2013)

[DDD13] Dereniowski, D., Diner, Ö.Y., Dyer, D.: Three-fast-searchable graphs. Discrete Appl. Math. **161**(13–14), 1950–1958 (2013)

[Der09] Dereniowski, D.: Maximum vertex occupation time and inert fugitive: recontamination does help. Inf. Process. Lett. **109**(9), 422–426 (2009)

[Der11] Dereniowski, D.: Connected searching of weighted trees. Theor. Comput. Sci. **412**(41), 5700–5713 (2011)

[Der12a] Dereniowski, D.: Approximate search strategies for weighted trees. Theor. Comput. Sci. **463**, 96–113 (2012)

[Der12b] Dereniowski, D.: From pathwidth to connected pathwidth. SIAM J. Discrete Math. **26**(4), 1709–1732 (2012)

[DFZ10] Daadaa, Y., Flocchini, P., Zaguia, N.: Network decontamination with temporal immunity by cellular automata. In: Bandini, S., Manzoni, S., Umeo, H., Vizzari, G. (eds.) ACRI 2010. LNCS, vol. 6350, pp. 287–299. Springer, Heidelberg (2010). https://doi.org/10.1007/978-3-642-15979-4_31

[DH08] Demaine, E.D., Hajiaghayi, M.T.: The bidimensionality theory and its algorithmic applications. Comput. J. **51**(3), 292–302 (2008)

[Die12] Diestel, R.: Graph Theory. Graduate Texts in Mathematics, vol. 173, 4th edn. Springer, Heidelberg (2012)

[DJS16] Daadaa, Y., Jamshed, A., Shabbir, M.: Network decontamination with a single agent. Graphs Comb. **32**(2), 559–581 (2016)

[DKL87] Deo, N., Krishnamoorthy, M.S., Langston, M.A.: Exact and approximate solutions for the gate matrix layout problem. IEEE Trans. CAD Integr. Circuits Syst. **6**(1), 79–84 (1987)

[DKT97] Dendris, N.D., Kirousis, L.M., Thilikos, D.M.: Fugitive-search games on graphs and related parameters. Theor. Comput. Sci. **172**(1–2), 233–254 (1997)

[DKZ15] Dereniowski, D., Kubiak, W., Zwols, Y.: The complexity of minimum-length path decompositions. J. Comput. Syst. Sci. **81**(8), 1715–1747 (2015)

[DNN17] D'Angelo, G., Navarra, A., Nisse, N.: A unified approach for gathering and exclusive searching on rings under weak assumptions. Distrib. Comput. **30**(1), 17–48 (2017)

[DOR18] Dereniowski, D., Osula, D., Rzazewski, P.: Finding small-width connected path decompositions in polynomial time. CoRR, abs/1802.05501 (2018)

[DSN+15] D'Angelo, G., Di Stefano, G., Navarra, A., Nisse, N., Suchan, K.: Computing on rings by oblivious robots: a unified approach for different tasks. Algorithmica **72**(4), 1055–1096 (2015)

[DU16] Dereniowski, D., Urbanska, D.: Distributed searching of partial grids. CoRR, abs/1610.01458 (2016)

[DYY08] Dyer, D., Yang, B., Yaşar, Ö.: On the fast searching problem. In: Fleischer, R., Xu, J. (eds.) AAIM 2008. LNCS, vol. 5034, pp. 143–154. Springer, Heidelberg (2008). https://doi.org/10.1007/978-3-540-68880-8_15

[EHS13] Evans, W., Hunter, P., Safari, M.A.: D-width and cops and robbers. Research report (2013, unpublished)

[EM04] Ellis, J.A., Markov, M.: Computing the vertex separation of unicyclic graphs. Inf. Comput. **192**(2), 123–161 (2004)

[EST87] Ellis, J.A., Sudborough, I.H., Turner, J.S.: Graph separation and search number. Technical report, Report Number: WUCS-87-11 (1987)

[EST94] Ellis, J.A., Sudborough, I.H., Turner, J.S.: The vertex separation and search number of a graph. Inf. Comput. **113**(1), 50–79 (1994)

[FFN09] Fomin, F.V., Fraigniaud, P., Nisse, N.: Nondeterministic graph searching: from pathwidth to treewidth. Algorithmica **53**(3), 358–373 (2009)

[FG00] Fomin, F.V., Golovach, P.A.: Graph searching and interval completion. SIAM J. Discrete Math. **13**(4), 454–464 (2000)

[FHL07] Flocchini, P., Huang, M.J., Luccio, F.L.: Decontaminating chordal rings and tori using mobile agents. Int. J. Found. Comput. Sci. **18**(3), 547–563 (2007)

[FHL08] Flocchini, P., Huang, M.J., Luccio, F.L.: Decontamination of hypercubes by mobile agents. Networks **52**(3), 167–178 (2008)

[FHM10] Fomin, F.V., Heggernes, P., Mihai, R.: Mixed search number and linear-width of interval and split graphs. Networks **56**(3), 207–214 (2010)

[FHT05] Fomin, F.V., Heggernes, P., Telle, J.A.: Graph searching, elimination trees, and a generalization of bandwidth. Algorithmica **41**(2), 73–87 (2005)

[FIP06] Fraigniaud, P., Ilcinkas, D., Pelc, A.: Oracle size: a new measure of difficulty for communication tasks. In: Proceedings of the Twenty-Fifth Annual ACM Symposium on Principles of Distributed Computing (PODC), pp. 179–187. ACM (2006)

[FL94] Fellows, M.R., Langston, M.A.: On search, decision, and the efficiency of polynomial-time algorithms. J. Comput. Syst. Sci. **49**(3), 769–779 (1994)

[FLPS16] Flocchini, P., Luccio, F., Pagli, L., Santoro, N.: Network decontamination under m-immunity. Discrete Appl. Math. **201**, 114–129 (2016)

[FLS05] Flocchini, P., Luccio, F.L., Song, L.X.: Size optimal strategies for capturing an intruder in mesh networks. In: Proceedings of the International Conference on Communications in Computing (CIC), pp. 200–206. CSREA Press (2005)

[FLS18] Fomin, F.V., Lokshtanov, D., Saurabh, S.: Excluded grid minors and efficient polynomial-time approximation schemes. J. ACM **65**(2), 10:1–10:44 (2018)

[FMS08] Flocchini, P., Mans, B., Santoro, N.: Tree decontamination with temporary immunity. In: Hong, S.-H., Nagamochi, H., Fukunaga, T. (eds.) ISAAC 2008. LNCS, vol. 5369, pp. 330–341. Springer, Heidelberg (2008). https://doi.org/10.1007/978-3-540-92182-0_31

[FN08] Fraigniaud, P., Nisse, N.: Monotony properties of connected visible graph searching. Inf. Comput. **206**(12), 1383–1393 (2008)

[FNS07] Flocchini, P., Nayak, A., Schulz, A.: Decontamination of arbitrary networks using a team of mobile agents with limited visibility. In: 6th Annual IEEE/ACIS International Conference on Computer and Information Science (ICIS), pp. 469–474. IEEE Computer Society (2007)

[Fom98] Fomin, F.V.: Helicopter search problems, bandwidth and pathwidth. Discrete Appl. Math. **85**(1), 59–70 (1998)

[Fom99] Fomin, F.V.: Note on a helicopter search problem on graphs. Discrete Appl. Math. **95**(1–3), 241–249 (1999)

[Fom04] Fomin, F.V.: Searching expenditure and interval graphs. Discrete Appl. Math. **135**(1–3), 97–104 (2004)

[FS06] Flocchini, P., Santoro, N.: Distributed security algorithms by mobile agents. In: Chaudhuri, S., Das, S.R., Paul, H.S., Tirthapura, S. (eds.) ICDCN 2006. LNCS, vol. 4308, pp. 1–14. Springer, Heidelberg (2006). https://doi.org/10.1007/11947950_1

[FT08] Fomin, F.V., Thilikos, D.M.: An annotated bibliography on guaranteed graph searching. Theor. Comput. Sci. **399**(3), 236–245 (2008)

[FTT05] Fomin, F.V., Thilikos, D.M., Todinca, I.: Connected graph searching in outerplanar graphs. Electron. Notes Discrete Math. **22**, 213–216 (2005)

[GHK+16] Ganian, R., et al.: Are there any good digraph width measures? J. Comb. Theory Ser. B **116**, 250–286 (2016)

[GHM12] Golovach, P.A., Heggernes, P., Mihai, R.: Edge search number of cographs. Discrete Appl. Math. **160**(6), 734–743 (2012)

[GHT12] Giannopoulou, A.C., Hunter, P., Thilikos, D.M.: LIFO-search: a min-max theorem and a searching game for cycle-rank and tree-depth. Discrete Appl. Math. **160**(15), 2089–2097 (2012)

[GLL+99] Guibas, L.J., Latombe, J.-C., LaValle, S.M., Lin, D., Motwani, R.: A visibility-based pursuit-evasion problem. Int. J. Comput. Geometry Appl. **9**(4/5), 471–494 (1999)

[Gol89a] Golovach, P.A.: Equivalence of two formalizations of a search problem on a graph. Vestnik Leningrad Univ. Math **22**, 13–19 (1989)

[Gol89b] Golovach, P.A.: A topological invariant in pursuit problems. Differ. Equ. **25**, 657–661 (1989)

[Gus93] Gustedt, J.: On the pathwidth of chordal graphs. Discrete Appl. Math. **45**(3), 233–248 (1993)

[HK08] Hunter, P., Kreutzer, S.: Digraph measures: kelly decompositions, games, and orderings. Theor. Comput. Sci. **399**(3), 206–219 (2008)

[HM08] Heggernes, P., Mihai, R.: Mixed search number of permutation graphs. In: Preparata, F.P., Wu, X., Yin, J. (eds.) FAW 2008. LNCS, vol. 5059, pp. 196–207. Springer, Heidelberg (2008). https://doi.org/10.1007/978-3-540-69311-6_22

[INS09] Ilcinkas, D., Nisse, N., Soguet, D.: The cost of monotonicity in distributed graph searching. Distrib. Comput. **22**(2), 117–127 (2009)

[ISZ07] Imani, N., Sarbazi-Azad, H., Zomaya, A.Y.: Capturing an intruder in product networks. J. Parallel Distrib. Comput. **67**(9), 1018–1028 (2007)

[ISZ08] Imani, N., Sarbazi-Azad, H., Zomaya, A.Y.: Intruder capturing in mesh and torus networks. Int. J. Found. Comput. Sci. **19**(4), 1049–1071 (2008)

[JRST01] Johnson, T., Robertson, N., Seymour, P.D., Thomas, R.: Directed tree-width. J. Comb. Theory Ser. B **82**(1), 138–154 (2001)

[Kin92] Kinnersley, N.G.: The vertex separation number of a graph equals its path-width. Inf. Process. Lett. **42**, 345–350 (1992)

[KKK15] Kintali, S., Kothari, N., Kumar, A.: Approximation algorithms for digraph width parameters. Theor. Comput. Sci. **562**, 365–376 (2015)

[KKK+16] Kitsunai, K., Kobayashi, Y., Komuro, K., Tamaki, H., Tano, T.: Computing directed pathwidth in $o(1.89^{n})$ time. Algorithmica **75**(1), 138–157 (2016)

[KO11] Kreutzer, S., Ordyniak, S.: Digraph decompositions and monotonicity in digraph searching. Theor. Comput. Sci. **412**(35), 4688–4703 (2011)

[KP85] Kirousis, L.M., Papadimitriou, C.H.: Interval graphs and searching. Discrete Math. **55**(2), 181–184 (1985)

[KP86] Kirousis, L.M., Papadimitriou, C.H.: Searching and pebbling. Theor. Comput. Sci. **47**(3), 205–218 (1986)

[KP16] Kinnersley, W.B., Pralat, P.: Game brush number. Discrete Appl. Math. **207**, 1–14 (2016)

[KS15] Klein, K., Suri, S.: Pursuit evasion on polyhedral surfaces. Algorithmica **73**(4), 730–747 (2015)

[LaP93] LaPaugh, A.S.: Recontamination does not help to search a graph. J. ACM **40**(2), 224–245 (1993)

[LPS06] Fabrizio, L., Pagli, L., Santoro, N.: Network decontamination with local immunization. In: Proceedings of 20th International Parallel and Distributed Processing Symposium (IPDPS). IEEE (2006)

[Luc09] Luccio, F.L.: Contiguous search problem in Sierpinski graphs. Theory Comput. Syst. **44**(2), 186–204 (2009)

[Mal18] Mallach, S.: Linear ordering based MIP formulations for the vertex separation or pathwidth problem. In: Brankovic, L., Ryan, J., Smyth, W.F. (eds.) IWOCA 2017. LNCS, vol. 10765, pp. 327–340. Springer, Cham (2018). https://doi.org/10.1007/978-3-319-78825-8_27

[MHG+88] Megiddo, N., Hakimi, S.L., Garey, M.R., Johnson, D.S., Papadimitriou, C.H.: The complexity of searching a graph. J. ACM **35**(1), 18–44 (1988)

[MM09] Mihai, R., Mjelde, M.: A self-stabilizing algorithm for graph searching in trees. In: Guerraoui, R., Petit, F. (eds.) SSS 2009. LNCS, vol. 5873, pp. 563–577. Springer, Heidelberg (2009). https://doi.org/10.1007/978-3-642-05118-0_39

[MN08] Mazoit, F., Nisse, N.: Monotonicity of non-deterministic graph searching. Theor. Comput. Sci. **399**(3), 169–178 (2008)

[MNP08] Messinger, M.-E., Nowakowski, R.J., Pralat, P.: Cleaning a network with brushes. Theor. Comput. Sci. **399**(3), 191–205 (2008)

[MNP17] Markou, E., Nisse, N., Pérennes, S.: Exclusive graph searching vs. pathwidth. Inf. Comput. **252**, 243–260 (2017)

[MS88] Monien, B., Sudborough, I.H.: Min cut is NP-complete for edge weighted trees. Theor. Comput. Sci. **58**(1), 209–229 (1988)

[MT09] Mihai, R., Todinca, I.: Pathwidth is NP-hard for weighted trees. In: Deng, X., Hopcroft, J.E., Xue, J. (eds.) FAW 2009. LNCS, vol. 5598, pp. 181–195. Springer, Heidelberg (2009). https://doi.org/10.1007/978-3-642-02270-8_20

[MTV10] Meister, D., Telle, J.A., Vatshelle, M.: Recognizing digraphs of Kelly-width 2. Discrete Appl. Math. **158**(7), 741–746 (2010)

[NdM08] Nesetril, J., de Mendez, P.O.: Grad and classes with bounded expansion i. Decompositions. Eur. J. Comb. **29**(3), 760–776 (2008)

[Nis09] Nisse, N.: Connected graph searching in chordal graphs. Discrete Appl. Math. **157**(12), 2603–2610 (2009)

[Nis14] Nisse, N.: Algorithmic complexity: between structure and knowledge how pursuit-evasion games help. Habilitation à Diriger des Recherches, Université Nice Sophia-Antipolis (2014). https://tel.archives-ouvertes.fr/tel-00998854

[NS09] Nisse, N., Soguet, D.: Graph searching with advice. Theor. Comput. Sci. **410**(14), 1307–1318 (2009)

[NS16] Nisse, N., Soares, R.P.: On the monotonicity of process number. Discrete Appl. Math. **210**, 103–111 (2016)

[Par78a] Parsons, T.D.: Pursuit-evasion in a graph. In: Alavi, Y., Lick, D.R. (eds.) Theory and Applications of Graphs. LNM, vol. 642, pp. 426–441. Springer, Berlin (1978). https://doi.org/10.1007/BFb0070400

[Par78b] Parsons, T.D.: The search number of a connected graph. In: 9th Southeastern Conference on Combinatorics, Graph Theory and Computing, Congress. Numer., vol. XXI, pp. 549–554. Utilitas Mathematica (1978)

[Pet82] Petrov, N.N.: A problem of pursuit in the absence of information on the pursued. Differ. Uravn. **18**, 1345–1352 (1982)

[PHH+00] Peng, S.-L., Ho, C.-W., Hsu, T., Ko, M.-T., Tang, C.Y.: Edge and node searching problems on trees. Theor. Comput. Sci. **240**(2), 429–446 (2000)

[PSS13] Penuel, J., Cole Smith, J., Shen, S.: Integer programming models and algorithms for the graph decontamination problem with mobile agents. Networks **61**(1), 1–19 (2013)

[PTK+00] Peng, S.-L., Tang, C.Y., Ko, M.-T., Ho, C.-W., Hsu, T.: Graph searching on some subclasses of chordal graphs. Algorithmica **27**(3), 395–426 (2000)

[PY07] Peng, S.-L., Yang, Y.-C.: On the treewidth and pathwidth of biconvex bipartite graphs. In: Cai, J.-Y., Cooper, S.B., Zhu, H. (eds.) TAMC 2007. LNCS, vol. 4484, pp. 244–255. Springer, Heidelberg (2007). https://doi.org/10.1007/978-3-540-72504-6_22

[RS83] Robertson, N., Seymour, P.D.: Graph minors. I. Excluding a forest. J. Comb. Theory Ser. B **35**(1), 39–61 (1983)

[RS90] Robertson, N., Seymour, P.D.: Graph minors. IV. Tree-width and well-quasi-ordering. J. Comb. Theory Ser. B **48**(2), 227–254 (1990)

[RS95] Robertson, N., Seymour, P.D.: Graph minors. XIII. The disjoint paths problem. J. Comb. Theory Ser. B **63**(1), 65–110 (1995)

[RS04] Robertson, N., Seymour, P.D.: Graph minors. XX. Wagner's conjecture. J. Comb. Theory Ser. B **92**(2), 325–357 (2004)

[RT11] Richerby, D., Thilikos, D.M.: Searching for a visible, lazy fugitive. SIAM J. Discrete Math. **25**(2), 497–513 (2011)

[Saf05] Safari, M.A.: D-width: a more natural measure for directed tree width. In: Jędrzejowicz, J., Szepietowski, A. (eds.) MFCS 2005. LNCS, vol. 3618, pp. 745–756. Springer, Heidelberg (2005). https://doi.org/10.1007/11549345_64

[SIS06] Shareghi, P., Imani, N., Sarbazi-Azad, H.: Capturing an intruder in the pyramid. In: Grigoriev, D., Harrison, J., Hirsch, E.A. (eds.) CSR 2006. LNCS, vol. 3967, pp. 580–590. Springer, Heidelberg (2006). https://doi.org/10.1007/11753728_58

[Sko03] Skodinis, K.: Construction of linear tree-layouts which are optimal with respect to vertex separation in linear time. J. Algorithms **47**(1), 40–59 (2003)

[ST93] Seymour, P.D., Thomas, R.: Graph searching and a min-max theorem for tree-width. J. Comb. Theory Ser. B **58**(1), 22–33 (1993)

[ST07] Suchan, K., Todinca, I.: Pathwidth of circular-arc graphs. In: Brandstädt, A., Kratsch, D., Müller, H. (eds.) WG 2007. LNCS, vol. 4769, pp. 258–269. Springer, Heidelberg (2007). https://doi.org/10.1007/978-3-540-74839-7_25

[SV09] Suchan, K., Villanger, Y.: Computing pathwidth faster than $2^n$. In: Chen, J., Fomin, F.V. (eds.) IWPEC 2009. LNCS, vol. 5917, pp. 324–335. Springer, Heidelberg (2009). https://doi.org/10.1007/978-3-642-11269-0_27

[SY09] Stanley, D., Yang, B.: Lower bounds on fast searching. In: Dong, Y., Du, D.-Z., Ibarra, O. (eds.) ISAAC 2009. LNCS, vol. 5878, pp. 964–973. Springer, Heidelberg (2009). https://doi.org/10.1007/978-3-642-10631-6_97

[SY11] Stanley, D., Yang, B.: Fast searching games on graphs. J. Comb. Optim. **22**(4), 763–777 (2011)

[Thi00] Thilikos, D.M.: Algorithms and obstructions for linear-width and related search parameters. Discrete Appl. Math. **105**(1–3), 239–271 (2000)

[TUK95] Takahashi, A., Ueno, S., Kajitani, Y.: Mixed searching and proper-path-width. Theor. Comput. Sci. **137**(2), 253–268 (1995)

[WAPL14] Yu, W., Austrin, P., Pitassi, T., Liu, D.: Inapproximability of treewidth and related problems. J. Artif. Intell. Res. **49**, 569–600 (2014)

[XY17] Xue, Y., Yang, B.: The fast search number of a Cartesian product of graphs. Discrete Appl. Math. **224**, 106–119 (2017)

[XYZZ16] Xue, Y., Yang, B., Zhong, F., Zilles, S.: Fast searching on complete $k$-partite graphs. In: Chan, T.-H.H., Li, M., Wang, L. (eds.) COCOA 2016. LNCS, vol. 10043, pp. 159–174. Springer, Cham (2016). https://doi.org/10.1007/978-3-319-48749-6_12

[Yan07] Yang, B.: Strong-mixed searching and pathwidth. J. Comb. Optim. **13**(1), 47–59 (2007)

[Yan11] Yang, B.: Fast edge searching and fast searching on graphs. Theor. Comput. Sci. **412**(12–14), 1208–1219 (2011)

[Yan13] Yang, B.: Fast-mixed searching and related problems on graphs. Theor. Comput. Sci. **507**, 100–113 (2013)

[YC07a] Yang, B., Cao, Y.: Directed searching digraphs: monotonicity and complexity. In: Cai, J.-Y., Cooper, S.B., Zhu, H. (eds.) TAMC 2007. LNCS, vol. 4484, pp. 136–147. Springer, Heidelberg (2007). https://doi.org/10.1007/978-3-540-72504-6_12

[YC07b] Yang, B., Cao, Y.: Monotonicity of strong searching on digraphs. J. Comb. Optim. **14**(4), 411–425 (2007)

[YC08a] Yang, B., Cao, Y.: Digraph searching, directed vertex separation and directed pathwidth. Discrete Appl. Math. **156**(10), 1822–1837 (2008)

[YC08b] Yang, B., Cao, Y.: Monotonicity in digraph search problems. Theor. Comput. Sci. **407**(1–3), 532–544 (2008)

[YC08c] Yang, B., Cao, Y.: On the monotonicity of weak searching. In: Hu, X., Wang, J. (eds.) COCOON 2008. LNCS, vol. 5092, pp. 52–61. Springer, Heidelberg (2008). https://doi.org/10.1007/978-3-540-69733-6_6

[YC09] Yang, B., Cao, Y.: Standard directed search strategies and their applications. J. Comb. Optim. **17**(4), 378–399 (2009)

[YDA09] Yang, B., Dyer, D., Alspach, B.: Sweeping graphs with large clique number. Discrete Math. **309**(18), 5770–5780 (2009)

[YZC07] Yang, B., Zhang, R., Cao, Y.: Searching cycle-disjoint graphs. In: Dress, A., Xu, Y., Zhu, B. (eds.) COCOA 2007. LNCS, vol. 4616, pp. 32–43. Springer, Heidelberg (2007). https://doi.org/10.1007/978-3-540-73556-4_6

# Mobile Agents on Dynamic Graphs

Giuseppe Antonio Di Luna(✉)

Aix-Marseille Université, LIS, CNRS, Université de Toulon, Toulon, France
g.a.diluna@gmail.com, diluna@lis-lab.fr

**Abstract.** At the core of distributed computing there is the necessity to coordinate a group of entities in face of the uncertainty present in the environment. Classically, such uncertainty was mainly the one introduced by the loss or the delay of messages (*asynchrony* and *failures*).

In this chapter we focus on the uncertainty introduced by the dynamism of the communication topology. We use the paradigm of mobile agents. In such paradigm the computational entities are *intelligent messages* circulating on top of a dynamic graph. We consider the problems of Exploration, Gathering and Deployment. We survey the most recent results in this interesting and relatively new field.

**Keywords:** Mobile agents · Dynamic graphs

## 1 Introduction

*Mobile Agents.* In the classic mobile agents paradigm, a set of computational entities also called *agents*, move on top of a static graph to solve a specific task. This paradigm abstracts a real world scenario of a computer network on which a set of *intelligent messages* circulates. Each one of these messages contains a set of instructions that, upon reception, are locally executed. These instructions interact with the local environment of the current node, changing certain memory locations or doing other local actions, and they decide the next destination of the message. Among the plethora of tasks that have been studied the most common are: Exploration [5,21,26,33,49], Gathering [11,20,27,31,38,40–42,55] and Deployment [28,50,53].

In the Exploration problem a set of mobile agents located on an unknown or known graph has to move in such a way to eventually visit each node.

In the Gathering problem the mobile agents, initially starting on distinct locations, have to reach the same node, or a subset of connected nodes.

The Deployment can be seen as the dual of Gathering: in this problem the agents have to scatter reaching a final configuration in which certain constraints on inter-agent distances are satisfied. As an example, on a ring graph the Uniform Deployment is solved when all distances between pairs of consecutive agents are equal [52], supposing that the number of agents divides the size of the ring.


P. Flocchini et al. (Eds.): Distributed Computing by Mobile Entities, LNCS 11340, pp. 549–584, 2019.
https://doi.org/10.1007/978-3-030-11072-7_20

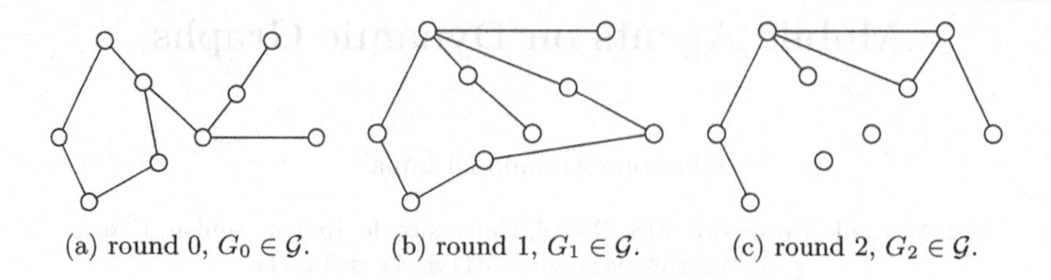

(a) round 0, $G_0 \in \mathcal{G}$. (b) round 1, $G_1 \in \mathcal{G}$. (c) round 2, $G_2 \in \mathcal{G}$.

**Fig. 1.** Example of evolving graph $\mathcal{G}$.

*Dynamic Networks.* Furthermore, it is evident that a real computer network cannot be seen as an immutable entity. On one hand, there are many events that physically prevent communication from working correctly; on the other hand, new links can be available creating new connections. The rates at which the network changes can be fast. Additionally, changes are mostly unpredictable, and failures or other disruptive events cannot be accurately and timely predicted.

This inherent dynamicity of wired computer networks is exacerbated when the communication is based on wireless technology. Wireless connections are plagued by temporary and frequent failures. Furthermore, nodes may physical move continuously changing the induced communication graph. With the virulent spread of wireless communication the design of sound software for a modern world can be done only when a perpetually and frequently changing topology is considered.

For the aforementioned reasons, there has a been a strong and fertile interest in studying classical and new computer science problems on top of *dynamic graphs*, such as: information dissemination [12,17,19,43,46,48], counting [22,23, 39], consensus and approximate agreement [6,13,44].

In the effort of capturing the several peculiarities of the dynamic setting, there has been the proposal of several dynamic graphs models, see [18,34]. One model, famous for its simplicity and manageability, is the one of *evolving graphs*, and it is the one that we will focus on.

Note that here we will discuss graphs where only edges, and thus connections among entities, are changing. Nevertheless, there are many settings in which the set $V$ varies along time. In a real world network we may have entities suddenly joining or leaving the computation, as example people may enter or exit a building while their smartphones are part of a local p2p network. This phenomenon is known as *churn* [4,8,45], and there has been a longstanding and deep investigation of it in the context of p2p systems [9,10,54]. Despite being a really interesting and challenging setting, we will neglect the churn and we will focus only on networks where connections are dynamics and the set of computational entities fixed.

*Mobile Agents on Dynamic Networks.* We will bring our attention to papers that assume a continuously dynamic system (also known as highly dynamic) and that investigate the three fundamental problems of Exploration [7,29,32], Gathering [25] and Uniform Deployment [3].

An important separation is the one between papers that assume to have the entire knowledge on the network topology and on the future actions of the adversary [1,2,29,35,47], and the ones that do not have such assumption [3,24,25,32]. When future changes are known, the challenges are due to the combinatorial aspects of the specific task investigated. As an example, in [47] is studied the problem of exploring an evolving graph, authors show that, when $\mathcal{G}$ has a limited lifetime, deciding if such exploration is doable or not is **NP**-complete. This setting has been defined as *Postmortem* in [51].

From the perspective of the distributed algorithms designer, we could argue that the main property of dynamic networks is the *uncertainty* of changes. It is this uncertainty that agents have to overcome when they solve tasks in a real communication network. We will see that this lack of knowledge greatly complicate solutions of tasks, even when we restrict ourselves to topologies that have a relatively easy structure. As an example, Terminating exploration of a constantly connected ring is far from trivial. The main subject of this chapter is this uncertain world, and, as in [51], we will call this setting *Live*.

*Chapter Outline.* The structure of the chapter is the following. We formalize the model in Sect. 2. After, we will have a brief look on works for the *Postmortem* setting in Sect. 3. In Sect. 4 we will show results for the *Live* Exploration. We will discuss the problem of Gathering in Sect. 5. Finally, we will show some really recent results for the problem of Uniform Deployment, see Sect. 6.

## 2 Model

### 2.1 Dynamic Networks

In our model the time is discrete, and it is divided in time units called *rounds*. Each round $r$ is a number in $\mathbb{N}$. An *evolving graph* $\mathcal{G} = (G_0 = (V, E_0), G_1 = (V, E_1), \ldots)$, is a sequence of static graphs. The element at index $r$ in the sequence corresponds to the graph at round $r$, we indicate such graph as $G_r = (V, E_r)$ where $E_r \subseteq \{\{v_1, v_2\} \mid v_1, v_2 \in V \wedge v_1 \neq v_2\}$. Each $G_r$ is a simple graph. If the sequence of graphs composing $\mathcal{G}$ is finite, we say that $\mathcal{G}$ has a finite lifetime, when not specified we assume that $\mathcal{G}$ has an infinite lifetime. With a small abuse of notations we write $G_1 \in \mathcal{G}$ to indicate that the static graph $G_1$ belongs to the dynamic graph $\mathcal{G}$. The set of nodes $V$ is fixed and does not change among rounds, what changes is the set of edges $E_r$, see Fig. 1.

Given an evolving graph $\mathcal{G}$, the *footprint* $F(\mathcal{G})$ of $\mathcal{G}$ is defined as $F(\mathcal{G}) = (V, \bigcup_{r=0}^{r=+\infty} E_r)$ that is the union of all graphs in the sequence $\mathcal{G}$, see Fig. 2. It is obvious, that the most general footprint is the complete graphs, other footprints implicitly limit the possible elements appearing in the evolving graph. Given two distinct nodes $v_s, v_t \in V$ and a round $r$, we say that there exists a *journey* from $v_s$ to $v_t$, $J_{v_s,v_t} = ((e_1, r_1), \ldots, (e_k, r_f))$, starting at round $r$, if $r \leq r_1 \leq r_2 \ldots \leq r_f$ and for any round $r_j$ in the journey, $e_{r_j} \in G_{r_j}$, see [18]. A *journey* $J_{v_s,v_t}$ starting at round $r$ and terminating at round $r_f$ is foremost [16], if there is no other journey starting at round $r$ from $v_s$ that reaches $v_t$ before $r_f$. That is a foremost journey is the fastest way for the source node to reach the target.

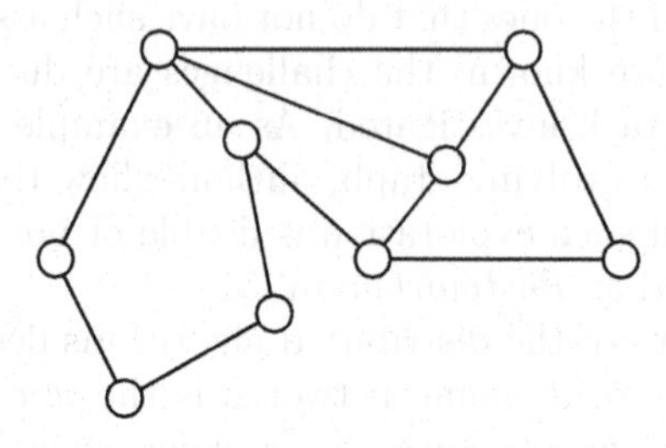

**Fig. 2.** Footprint of the dynamic graph in Fig. 1.

An evolving graph $\mathcal{G}$ is 1-*interval connected*, see [43], if for any round $r$ the graph $G_r$ is connected. An evolving graph $\mathcal{G}$ is $T$-*interval connected*, see [43], if for any round $r$ the graph $G_{[r,r+T-1]} = (V, \bigcap_{i=r}^{i=r+T-1} E_i)$ is connected. Intuitively, for any round $r$ there exists a stable backbone connecting nodes in $G$ that lasts at least $T$ rounds.

An evolving graph $\mathcal{G}$ is *connected over time* if for any pair of nodes $v_1, v_2 \in V$ and for any round $r$ there exists a journey from $v_1$ to $v_2$ starting at round $r$ in $\mathcal{G}$.

Given an evolving graph $\mathcal{G}$ its temporal diameter $D(\mathcal{G})$ is the maximum number of rounds needed for a foremost journey to go from any source node $v_s$ to any target node $v_t$ starting at any round $r$.

Apart from the assumptions on the connectivity of the graph, we may assume a pattern on the presence of edges. An evolving graph $\mathcal{G}$ is *recurrent* if for any edge $e \in F(\mathcal{G})$ and for any round $r$ there exists $r' > r$ such that $e \in G_{r'}$. An evolving graph $\mathcal{G}$ is $\delta$-*recurrent* if for any edge $e \in F(\mathcal{G})$ and for any round $r$ there exists $r' \in [r, r+\delta-1]$ with $e \in G_{r'}$.

Finally, an evolving graph $\mathcal{G}$ is *periodic* if for any edge $e \in F(\mathcal{G})$ there exists a $P_e$, such that, for any two rounds $r', r$ such that $r' - r \equiv 0 \mod P_e$ it holds $e \in G_{r'}$ if and only if $e \in G_r$. It is clear that if an evolving graph is periodic, and the footprint is connected, then there is connectivity over time.

### 2.2 Mobile Agents on Dynamic Networks

We consider a set of agents $A = \{a_0, a_1, \ldots, a_{k-1}\}$. They are mobile entities moving on top of an evolving graph $\mathcal{G}$. Each agent $a_j$ starts in a node $v \in V$. At each round $r$, agent $a_j$ may traverse only one edge of $G_r$. By traversing edges the agent moves in the evolving graph $\mathcal{G}$. Usually, agents are assumed to be *anonymous*, that is they do not have distinguishable identities and each of them executes the exact same algorithm as the others. When agents can be distinguished and they execute different algorithms we say that they have *IDs*.

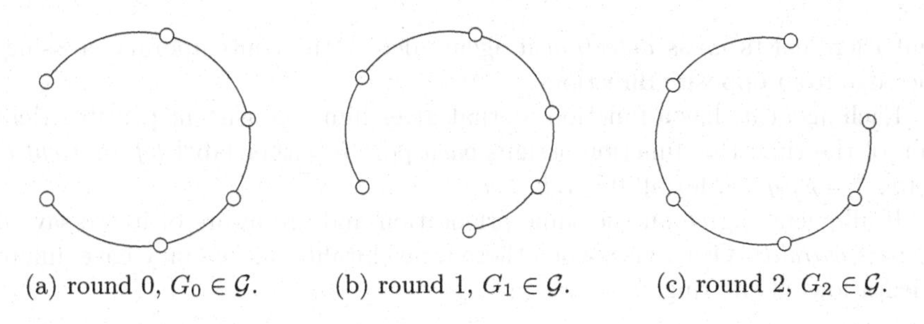

(a) round 0, $G_0 \in \mathcal{G}$. (b) round 1, $G_1 \in \mathcal{G}$. (c) round 2, $G_2 \in \mathcal{G}$.

**Fig. 3.** Dynamic 1-interval connected ring $\mathcal{R}$.

Please note that anonymous agents and agents with IDs are different only in the distributed case. In the case in which there is a single agent, or in which multiple agents are controlled by a centralised entity such distinction is not meaningful.

As detailed before, there are two main settings that can be found in the literature: the *Postmortem* and the *Live* setting.

**Postmortem model.** In the *Postmortem* model agents have complete knowledge of the evolving graph $\mathcal{G}$: agents can be seen as passive tokens moved by a centralised entity that has knowledge of $\mathcal{G}$. As said before, there is the restriction that moving one agent from a node to a neighbour takes one round: agents have to move using journeys over $\mathcal{G}$.

**Live model.** In the *Live* model agents dot have complete knowledge of $\mathcal{G}$. We will use as reference the model defined in [24]. Since the majority of works for the live setting solve problems on dynamic graph whose footprint is a ring, we will present the model for this specific case.

*Dynamic Ring.* A dynamic graph $\mathcal{G}$ is a dynamic ring if

$$F(\mathcal{G}) = (V = \{v_0, v_1, \dots, v_{n-1}\}, E),$$

with $(v_x, v_y) \in E \iff x = i \mod n \wedge y = (i+1) \mod n$. We will indicate a dynamic ring with the letter $\mathcal{R}$ (see Fig. 3 for an example of 1-interval connected ring). We say that the ring is *anonymous* if the nodes have no distinguishable identifiers, and *with landmark* if there is a node (the landmark) which is different from all others.

*Ports.* Each node $v_i$, is connected to its neighbours $v_{i-1}$ and $v_{i+1}$ via two distinctly labeled ports, $q_{i-}$ and $q_{i+}$. These ports will be used to enforce the movement of agents on $G$. To move on edge $(v_i, v_{i+1})$, an agent currently in $v_i$ has to position itself on the port $q_{i+}$.

*Cross Detection and Chirality.* Two agents moving in opposite directions on the same edge in the same round might or might not be able to detect this. We say that there exists *cross detection* if agents detect the contemporary crossing of the edge from opposite directions.

Each agent $a_j$ has a function $\lambda_j$ that gives him a consistent private orientation of the ring; this function assigns each port to either label *left* or *right* and $\lambda_j(q_{i-}) = \lambda_j(q_{k-})$, for all $0 \leq i, k < n$.

If all agents agree on the same orientation and are aware of it, we say that there is *chirality*. Otherwise, when there is no chirality, agents may have different orientations of the ring.

*The Agents' Life Cycle.* Initially, all agents are *inactive*. Each round starts with a non-empty subset of the agents becoming *active*. When an agent is activated it performs a sequence of operations: Look, Compute, and (possibly) Move.

Look: During the look operation each agent gets a local *snapshot* of the node where it resides. Specifically, it sees if there are other agents at the node and in which positions (i.e. whether they are or not on a port, and if so on which ones). Moreover the agent also determines its own position inside the node.

Compute: Using the content of its local memory, and the snapshot acquired in the Look Phase, each agent executes its algorithm to determine whether or not to move and, if so, in what direction; the result will be $direction \in \{left, right, nil\}$, where *left* and *right* are given according to the local orientation of the agent. If $direction = nil$, the agent becomes inactive. If $direction \neq nil$, $a_j$ accesses the appropriate port. The agent also determines if it has to updates its local memory.

Move: The move phase only involves agents inside ports. Therefore, let the agent $a_j$ be positioned on port $q_{i-}$ (resp., $q_{i+}$) after computing. If the link between $v_i$ and $v_{i-1}$ (resp., $v_{i+1}$) is present in this round, then agent $a_j$ will move to $v_{i-1}$ (resp., $v_{i+1}$), and reach it before the end of current round. If the link between $v_i$ and $v_{i-1}$ (resp., $v_{i+1}$) is not present, then agent $a_j$ will remain in the port and become inactive.

By definition of round, all active agents of round $r$ become inactive by the end of round $r$; when all agent are inactive, the system starts the new round $r+1$.

*Synchronous and Semi-synchronous Scheduler.* We will study two main activation schedulers: the *fully synchronous* ($\mathcal{FSYNC}$), in which at any round $r$ all agents are activated; and the *semi-synchronous* ($\mathcal{SSYNC}$), where a subset of agents is activated (the non activated agents are said to be *sleeping* or *passive* at round $r$). The fairness condition of $\mathcal{SSYNC}$ is that each agent is activated infinitely often. When an agent is activated, it does not know whether or not it was active in the previous round.

Note that in $\mathcal{SSYNC}$ it is possible for an agent to go inside a port, to remain there because the edge is not present, and then to be forced to sleep by the adversary. Therefore, it is interesting to specify what is the behaviour of an agent that is sleeping on a port while the corresponding edge is present. What may happen leads to three different models. These models have been defined for the first time in [24] and are:

- Passive Transport (PT): If at round $r$ an edge is present and agent is inside a port of the edge, then the agent moves to its destination (even if the agent is sleeping).
- Eventual Transport (ET): A sleeping agent cannot move. However, there is the following fairness condition: If an agent is sleeping on a port at round $r$ and the corresponding edge is present infinitely many times, then the agent will eventually become active at a round $r' > r$ when the corresponding edge is present.
- No Simultaneity (NS): A sleeping agent cannot move, and there is no other guarantee.

Let us remark that the ET condition does not prescribe that each edge has to be present an infinite number of rounds, it just says that if an edge $e$ is not perpetually removed, and an agent is waiting to traverse $e$, it will eventually succeed. However, there is no bound on the time that the agent has to wait in order to cross $e$. Neither is forbidden to have an edge perpetually removed.

### 2.3 Exploration, Gathering and Deployment

We define the three problems that we investigate in this chapter.

*Exploration.* Given a dynamic graph $\mathcal{G}$ and a set of agents $A$, algorithm $\mathcal{A}$ is an Exploration algorithm if there exists a round $r_f$, such that for each $v \in V$, $\exists r_v \in [0, r_f]$ and node $v$ is visited by one agent in round $r_v$. The previous specification says that all nodes in the graph has to be visited by at least one agent, within finite time.

However, it does not specify what has to be the behaviour of the algorithm when the exploration is completed. One may ask for Terminating exploration: Algorithm $\mathcal{A}$ has an *explicit termination* if at round $r_f$, each agent in $A$ is in a special terminal state, that is all agents are aware of the end of the exploration. An algorithm $\mathcal{A}$ has a *partial termination* if after round $r_f$ at least one agent in $A$ is in a special terminal state, that is at least one agent is aware of the termination.

A version of Exploration studied in the Postmortem setting is the one in which given $\mathcal{G}$ and an arrangement of agents, the algorithm $\mathcal{A}$ has to explore the ring in such a way that $r_f$ is minimised. When we consider a single agent such minimum $r_f$ is the optimal exploration time of $\mathcal{G}$. Following the terminology used in the literature of dynamic graphs [16], we call this version Foremost exploration.

Finally, a non terminating version is the Perpetual exploration, in this case algorithm $\mathcal{A}$ has to ensure that for any node $v \in V$ there exists an infinite sequence of rounds $S_v$, such that for each $r \in S_v$ there is at least one agent in $v$.

*Gathering.* Given a dynamic graph $\mathcal{G}$ and a set of agents $A$ we say that an algorithm $\mathcal{A}$ is a Gathering algorithm if there exists a round $r_f$, such that for any $r' \geq r_f$ all agents $A$ are on the same node $v$. Also in this case an algorithm is terminating if at round $r_f$, each agent in $A$ is in a special terminal state, that is all agents are aware that the gathering ended. Unfortunately, it has been shown that this version of Gathering is not solvable [25]. We will call the unsolvable version strict gathering. In the following when we use the term gathering we will consider a weaker version: An algorithm $\mathcal{A}$ solves gathering if there exists a round $r_f$ such that for any round $r' \geq r_f$ all agents $A$ are on two nodes $v_1, v_2$ such that $v_1$ and $v_2$ are neighbours in $F(G)$.

*Deployment.* Given a dynamic ring $\mathcal{R}$ and a set of agents $A$ we say that an algorithm $\mathcal{A}$ is an Uniform Deployment algorithm if there exists a round $r_f$, such that for any $r' \geq r_f$ the minimum distance between any two agents is $\lfloor \frac{|V|}{|A|} \rfloor$ [28].

## 3 Postmortem Exploration

In this section agents know the graph $\mathcal{G}$, if not specified otherwise, such knowledge is assumed by all the results presented in this chapter. Moreover, we assume a unique agent, that is $A = \{a\}$.

In the Postmortem setting the majority of the effort has been devoted to Foremost exploration. Before delving in the technical details, we give some basic definitions.

A foremost exploration schedule is one of the feasible solutions of a correct Foremost exploration algorithm. Specifically, given a graph $\mathcal{G}$, and a starting position $v_s$ of agent $a$, a foremost exploration schedule is a journey that allows $a$ to visit all nodes of $\mathcal{G}$ within the optimal exploration time of $\mathcal{G}$.

### 3.1 Arbitrary Graphs

Given an undirected static graph $G = (V, E)$, a single agent $a$, knowing the graph topology, can explore the graph in linear time: $a$ builds a spanning tree of $G$ and visits the tree using a DFS algorithm. The exploration time of this strategy is at most $2|V| - 1$ rounds.

In the dynamic case the optimal exploration time is far from linear, and computing a foremost exploration schedule is hard. It is even hard to compute an approximation of the foremost exploration schedule.

The first work examining the complexity of computing a foremost exploration schedule, has been [47]. Interestingly, and in strike contrast with the static case, if $\mathcal{G}$ has finite lifetime deciding if a single agent can explore it or not is **NP**-complete.

**Theorem 1** [47]. *Given a* 1*-interval connected graph* $\mathcal{G}$*, with finite lifetime, and a single agent* $a$ *placed on a node* $v_s \in V$*, deciding whether* Exploration *is feasible or not is* **NP***-complete.*

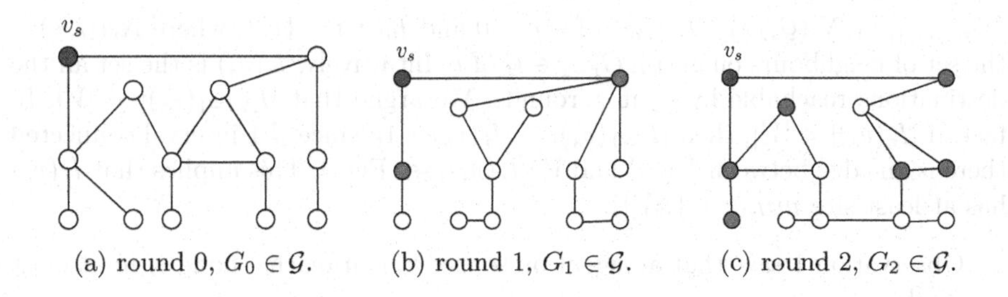


(a) round 0, $G_0 \in \mathcal{G}$. (b) round 1, $G_1 \in \mathcal{G}$. (c) round 2, $G_2 \in \mathcal{G}$.

**Fig. 4.** Temporal diameter of an 1-interval connected graph $\mathcal{G}$: The red nodes represent the set $I_r(v_s)$ (Color figure online)

*Proof.* The problem is clearly in **NP**: the certificate that $a$ can explore $\mathcal{G}$ is the union of at most $|V| - 1$ journeys and each of them contains at most $|V|$ edges. Thus it has size that is polynomial – at most quadratic – in $|V|$. To show that the problem is **NP**-complete, [47] constructs a reduction from Hamiltonian Path: let $H$ be a static graph, and let $\mathcal{G}_H : (G_0 = H, \ldots, G_{|V|-2} = H)$, an evolving graph constituted by a sequence of $|V| - 1$ instances all equal to graph $H$. If there exists an hamiltonian path in $H$ starting from $v_s$, then there must exist a journey that visit all nodes in $\mathcal{G}$ in $|V| - 1$ steps. On the other hand if $\mathcal{G}_H$ can be explored, then the exploration schedule is an hamiltonian path on $H$ starting from $v_s$, since there are only $|V| - 1$ instances agent $a$ cannot visit the same node twice, otherwise it will necessarily leave a node unexplored. □

As observed by [47], the previous proof can be adapted to show the difficulty of finding an approximation of the foremost exploration schedule for graphs that are not interval connected. Essentially, finding an exploration schedule with an exploration time that is within a polynomial factor from the optimal is equivalent to solve the hamiltonian path. The idea is to modify the $\mathcal{G}_H$ used in the proof of Theorem 1 by appending to it, first a sequence of $|V|^k$ graphs without edges, and then an infinite sequence constituted by instances of $H$. Any algorithm that does not visit all nodes in the first $|V| - 1$ rounds, using the hamiltonian path on $H$, has to wait at least $|V|^k$ rounds to complete the exploration. Since the optimal exploration time is $|V| - 1$ the inaproximability follows.

A key point of Theorem 1 is the finite, and very limited, lifetime of the graph. When $\mathcal{G}$ is 1-interval connected and its lifetime is big enough, Exploration is always feasible. To prove this, let us first do a simple observation on the temporal diameter of a 1-interval connected graph.

**Observation 1** [43,48]. *Given an 1-interval connected graph $\mathcal{G}$, with infinite lifetime, its temporal diameter is at most $|V| - 1$.*

*Proof.* Let us consider a source $v_s$ and a target $v_t$, we will show that in any $\mathcal{G}$ and for any round $r$ there exists a journey $J_{v_s,v_t}$ starting at round $r$ with duration upper bounded by $|V| - 1$ rounds. Let us define the set $I_r(v_s) =$

$(\cup_{\forall v \in I_{v_s}(r-1)} N_r(\mathcal{G}, v)) \cup I_{r-1}(v_s)$ for $r > 0$ and $I_0(v_s) = \{v_s\}$; where $N_r(\mathcal{G}, v)$ is the set of neighbours on graph $G_{r-1} \in \mathcal{G}$ of $v$. Intuitively, $I_r(v_s)$ is the set all the destinations reachable by $v_s$ in $r$ rounds. We argue that $|I_{|V|-1}(v_s)| = |V|$. In fact, if $|I_r(v_s)| < |V|$, then $|I_{r+1}(v_s)| \geq |I_r(v_s) + 1|$: since $\mathcal{G}$ 1-interval connected there is an edge between $I_r(v_s)$ and $V \setminus I_r(v_s)$, see Fig. 4. This implies that $I_r(v_s)$ has at least size $min(r+1, |V|)$. □

Observation 1 says that an agent on source $v_s$ can reach a node $v_t$ in at most $n$ rounds.

This observation allows [47] to give a simple algorithm (Simple Exploration) that explores any interval connected graph $\mathcal{G} = (V, E_r)$ in $\mathcal{O}(|V|^2)$ rounds. Simple Exploration starts with a set of unexplored node, initially equal to $V$, and it proceeds in iterations: at each iteration we pick one node $v_t$ from the set of unexplored nodes. From the current position of $a$ we reach $v_t$ using a foremost journey (see [16] for a polynomial algorithm to compute foremost journeys on dynamic graphs). Once $a$ is in $v_t$, we update the set of unexplored nodes by removing all nodes visited during the journey. We keep iterating until there are no more unexplored nodes. The Simple Exploration has a polynomial running time.

The reader would have noticed a large gap between the exploration time of Simple Exploration, that is $\mathcal{O}(|V|^2)$ rounds, and temporal diameter $|V| - 1$. However, it is not possible to do better than that:

**Theorem 2** [29]. *For every $n \geq 1$, there exists a 1-interval connected dynamic graph $\mathcal{G}$ with $|V| = 2n$, such that the optimal exploration time of $\mathcal{G}$ is $\Omega(|V|^2)$ rounds.*

*Proof.* The set of nodes is $V = C \cup L$, where $C = (c_0, \ldots, c_{n-1})$ and $L = (\ell_0, \ldots, \ell_{n-1})$ The graph $\mathcal{G} = (G_1, G_2, \ldots)$ is an infinite sequence of star graphs where each $G_r$ has center $c_r \mod n$. Let the agent $a$ be on node $\ell_i$. The fastest way to visit a node $\ell_j$ with $j \neq i$, is to first go to the center $c_r$ of the current graph $G_r$, and then to wait until $c_r$ is again the center of the star, thus $n-1$ round. Once $c_r$ is again the center, we can finally move to node $\ell_j$. This implies that to visit all nodes in $L$ at least $\Omega(|V|^2)$ rounds are needed. □

A similar Theorem can be found in [7], where the same idea is used to shown that a random walk has an exponential hitting time on dynamic graphs. According to Theorem 2, the optimal exploration time is $\Omega(|V|^2)$ in the worst case, while it is linear for static graphs. To make things worse, [29] shows that is not possible, unless $\mathbf{P} = \mathbf{NP}$, to find an exploration schedule that approximate the optimal one with a ratio that is better than $\mathcal{O}(|V|^{1-\epsilon})$. This holds even in 1-interval connected graphs:

**Theorem 3** [29]. *Given an 1-interval connected graph $\mathcal{G}$ finding an exploration schedule with an approximation ratio that is better than $\mathcal{O}(|V|^{1-\epsilon})$ is **NP**-hard.*

The dynamic graph used to prove Theorem 3 is built extending the one used in the proof of Theorem 2. The idea is to substitute nodes in $L$ with copies of the same static graph $G$. Inside graph $G$ two nodes are selected $s$ and $t$. The several copies of $G$ are all connected by paths. Each of these paths go from the $t$ node of copy $i$ to the $s$ node of copy $i+1$. In the last copy, the node $t$ is connected to nodes in $C$. The dynamic of the graph is built in such a way that, if in $G$ there exists an hamiltonian $s$-$t$ path, then an optimal exploration algorithm explores the graph in $\mathcal{O}(|V|)$, and viceversa. On the contrary, exploring the graph without using an hamiltonian $s$-$t$ path in $G$ takes $\mathcal{O}(|V|^2)$ rounds.

*Recurrent Graphs.* In the above we only discussed 1-interval connected graphs. An investigation on the complexity of finding a foremost exploration schedule for recurrent dynamic graphs can be found in [1]. Where there is an exponential time algorithm to compute the Foremost exploration for general recurrent graphs. In a follow-up paper [2], the same authors present a $\frac{12\delta}{5}$ approximation for general $\delta$-recurrent graphs.

### 3.2 Dynamic Rings

As we have seen in Sect. 3.1, exploring a 1-interval connected graph in less than $\mathcal{O}(|V|^2)$ rounds is not always possible. It is also **NP**-hard to decide whether $\mathcal{G}$ can be explored with less than a quadratic number of rounds.

Luckily, if we restrict ourselves to dynamic rings, the optimal exploration time is linear and a foremost exploration schedule can be computed by a polynomial algorithm. See [1] for an algorithm to compute a foremost exploration schedule on recurrent rings, that works also on 1-interval connected. The investigation of bounds on the optimal exploration time for 1-interval connected rings has been done independently in [29,37]. However, [37] is the only one that focus on the case of a $T$-interval connected $\mathcal{R}$. We will first describe the elegant proof in [37] for an upper bound of $2|V|-2$ rounds on the exploration time of an 1-interval connected ring $\mathcal{R}$.

**Theorem 4** [37]. *Given an 1-interval connected ring $\mathcal{R} = ((V, E_r))_{r=0}^{+\infty}$, an agent $a$ starting from an arbitrary source node $v_s \in V$ explores $\mathcal{R}$ in at most $2|V|-2$ rounds.*

*Proof.* The proof uses $n = |V|$ virtual agents. The first part of the proof shows that, starting from any round $r$, there exists a source node $v_{s_r} \in V$ such that an agent located in $v_{s_r}$ is able to move for $n-1$ consecutive rounds in clockwise direction, exploring the ring. Node $v_{s_r}$ exists despite the dynamic of $\mathcal{R}$.

Let us consider the case where each node is initially occupied by one agent. Each agent moves clockwise for $n-1$ consecutive rounds. At each round $r \in [0, n-2]$ the adversary removes only one edge of $\mathcal{R}$. With each edge removal the adversary blocks at most one new virtual agent. For the proof purpose we can imagine that a blocked agent disappears. However, there are at most $n-1$ removals and the number of agents is $n$. Thus, it must exist one agent that cannot be blocked by the adversary. This agent is the one that at round $r$ was located in $v_{s_r}$, and it moves clockwise for $n-1$ rounds exploring the ring.

To complete the proof we have to show how to use a single agent to explore the ring. Let $a$ be this agent. First of all $a$ computes the node $v_{s_{n-1}}$ using the simulation with virtual agents as discussed above. Now agent $a$ has to reach node $v_{s_{n-1}}$ by round $n-1$. However, by Observation 1 we know that $a$ can reach any node in $v \in V$ in at most $n-1$ rounds. □

The bound given by Theorem 4 is not tight. In [37] there are better upper bounds.

**Theorem 5** [37]. *Given a $T$-interval connected ring $\mathcal{R} = ((V, E_r))_{r=0}^{+\infty}$, with $|V| = n$, an agent $a$ starting from an arbitrary source node $v_s \in V$ explores $\mathcal{R}$ in at most $f(T)$ rounds, where:*

$$f(T) = \begin{cases} 2n-3 & \text{if } T = 1 \\ 2n-T-1, & \text{if } 2 \leq T \leq \frac{n+1}{2} \\ \lfloor \frac{3(n-1)}{2} \rfloor, & \text{if } T > \frac{n+1}{2} \end{cases}$$

And such bounds are tight:

**Theorem 6** [37]. *For any $n \geq 3$, there exists a $T$-interval connected ring $\mathcal{R} = ((V, E_r))_{r=0}^{+\infty}$, with $|V| = n$, such that an agent $a$ needs at least $f(T)$ rounds to explore $\mathcal{R}$. Where $f(T)$ is*

$$\begin{cases} 2n-3 & \text{if } T = 1 \\ 2n-T-1, & \text{if } 2 \leq T \leq \frac{n+1}{2} \\ \lfloor \frac{3(n-1)}{2} \rfloor, & \text{if } T > \frac{n+1}{2} \end{cases}$$

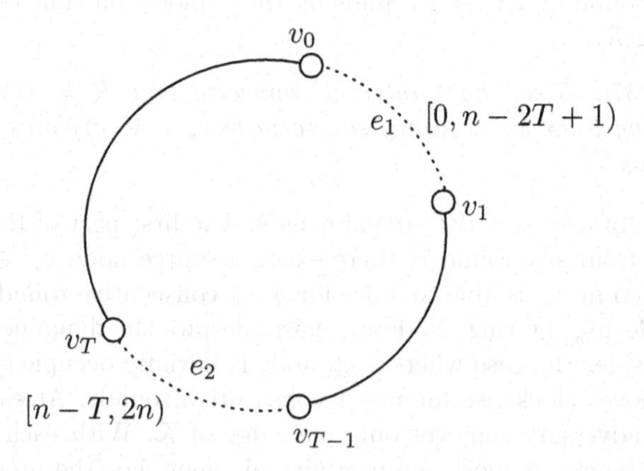


**Fig. 5.** Lower bound graph, from [37]. An interval is the set of rounds in which the corresponding edge is missing.

*Proof.* The proof is first given for $T \in [2, \lceil \frac{n+1}{2} \rceil]$; at the end we will show that this implies the correctness also for other values of $T$.

The graph $\mathcal{R} = ((V, E_r))_{r=0}^{+\infty}$ is constructed as follows. $V = \{v_0, v_1, \ldots, v_{n-1}\}$. In $E_r$ all the edges are always present with two exceptions: edge $e_1 = (v_0, v_1)$ and edge $e_2 = (v_{T-1}, v_T)$. Specifically, for any $r \in [0, n - 2T + 1)$ edge $e_1 \notin E_r$; for any $r \in [n - T, 2n)$ the edge $e_2 \notin E_r$, see Fig. 5. The graph $\mathcal{R}$ is $T$ interval connected: between the last round in which $e_1$ is removed and the first round in which $e_2$ is removed, there are $T - 1$ rounds where all edges are present.

Agent $a$ starts in node $v_0$, we now show that $a$ needs $2n - T - 1$ rounds to complete the exploration. We focus on the order in which $a$ explores node $v_{T-1}$ and node $v_T$.

**Case 1.** Agent $a$ explores node $v_{T-1}$ before $v_T$. In this case $a$ has to necessarily traverse edge $e_1$. However, the edge is absent until round $n - 2T - 1$, and the clockwise distance, according to our figure, between $v_0$ and $v_{T-1}$ is $T - 1$ edges. This means that, agent $a$ needs $n - T$ rounds to reach $v_{T-1}$. But, at time $n - T$ the edge $e_2$ is missing, and it will be removed until round $2n$. This forces agent $a$ to reach $v_T$ going in counter-clockwise direction all the way from $v_{T-1}$, paying $n - 1$ rounds. The total time is $n - T + n - 1 = 2n - T + 1$.

**Case 2.** Agent $a$ explores node $v_T$ before $v_{T-1}$. Note that the counter-clockwise distance between $v_0$ and $v_T$ is $n - T$ edges. Thus agent $a$ reaches node $v_T$ at time $n - T$. But, at time $n - T$ the edge $e_2$ is removed, and it will be removed until round $2n$. At this point agent $a$ has to still explore node $v_{T-1}$. So the only things it can do is to go counter-clockwise direction all the way from $v_T$, paying $n - 1$ rounds. Also in this case the total time is $n - T + n - 1 = 2n - T - 1$.

To conclude the proof we have to show that considering $T \in [2, \lceil \frac{n+1}{2} \rceil]$ is enough to prove the claim of the theorem. The case of $T = 1$ is trivially included: a 2-interval connected ring is also a 1-interval connected ring. The case of $T > \lceil \frac{n+1}{2} \rceil$ is implied by the corner case $T = \lceil \frac{n+1}{2} \rceil$. However, when $T = \lceil \frac{n+1}{2} \rceil$ the edge $e_1$ is always present. The only removed edge is $e_2$ and a dynamic graph where only one edge is removed is $T$-interval connected for any $T \geq 1$. □

### 3.3 Other Topologies

For 1-interval connected graphs, the study of several specific topologies can be found in [29]. Briefly, a 1-interval connected $2 \times n$ grid can be explored in time $\mathcal{O}(n \log^3 n)$; a graph with treewidth bounded by $k$ can be explored in $\mathcal{O}(n^{\frac{3}{2}} k^{\frac{3}{2}} \log n)$ rounds, and there are planar graphs with maximum degree 4 where the optimal exploration time is $\Omega(n \log n)$ rounds.

The case of 1-interval connected cactus has been studied in [35], where an algorithm with exploration time of $n \cdot 2^{\Theta(\sqrt{\log n})}$ rounds is given.

Regarding the complexity of computing a foremost exploration schedule in recurrent, $\delta$-recurrent and periodic dynamic graph, the reader can look at [1,2]. Where the authors prove that on recurrent graphs the foremost exploration schedule is not approximable, even when the footprint of $F(\mathcal{G})$ is a star or a tree with maximum degree 3. Interestingly, if the graph is $\delta$-*recurrent*, and the footprint is a tree there exists a $\delta$-approximation of the foremost exploration schedule, and such approximation is tight. Polynomial algorithms are given for the case of rings and paths.

# 4 Live Exploration

In this section we switch our focus to a setting where $\mathcal{G}$ is unknown, or partially unknown, and where agents have only a local view of the system. In the Live setting, concepts that were neglected in the Postmortem setting (such as the anonymity of the agents, the anonymity of the graph and the presence or not of common chirality) become key aspects, and they have a strong impact on the feasibility and design of distributed algorithm for Exploration.

At the time of writing, we are aware of only few papers [14,15,24,37] that tackle the Exploration in the non-periodic dynamic graphs, and all of them deals with the case of the dynamic ring. Other works investigated more general topologies, but assuming a special kind of periodic graphs, the so called carrier networks [30,32,36].

## 4.1 Interval Connected Rings

The study of Exploration in 1-interval connected ring $\mathcal{R} = ((V, E_r))_{r=0}^{+\infty}$ has been done in [24], where the main focus is on Terminating exploration. The paper that the access to ports is mutual exclusive. However, such assumption can be dropped if we assume that agents start from distinct nodes, as we do in this chapter. We use $n$ to indicate the number of nodes.

As expected, in the Live setting one agent is not enough to explore the ring:

**Observation 2.** *Given an agent $a$ on a node $v_0$, the adversary can block agent $a$ on node $v_0$ forever.*

*Proof.* If agent $a$ tries to leave $v_0$ using port $q_{0-}$, then the adversary removes edge $(v_{n-1}, v_0)$ and keeps it removed while $a$ is in the port $q_{0-}$. If agent $a$ leaves port $q_{0-}$ and enters port $q_{0+}$ the adversary removes edge $(v_0, v_1)$. □

Therefore, 2 or more agents are necessary. Now, we will show that Terminating exploration with 2 agents requires either some kind of knowledge on $\mathcal{R}$ (such as, an upper bound on its size), or that $\mathcal{R}$ is not an anonymous ring (there must exist a landmark). Before proving it we observe that two agents cannot meet.

**Observation 3** [24]. *Let us consider a $\mathcal{F}$SYNC scheduler, and two agents $a_0, a_1$ starting at different location. The adversary can always prevent them to meet at the same node, and it would do so by never blocking both agents in the same round.*

*Proof.* There are only two cases in which $a_0$ and $a_1$ meet: (1) they both try to move on the same node $v$, traversing two different edges $e, e'$, the adversary removes one of edge, let us say $e'$ preventing the meeting; or (2) one agent remains still, say $a_1$, in node $v$ and $a_0$ tries to move on $v$ using edge $e'$, the adversary removes edge $e'$. □

Note that the previous observation holds also if agents are moved by a centralised entity, therefore also if agents have IDs, communications, etc.

**Theorem 7** [24]. *There does not exist any partially terminating deterministic exploration algorithm of anonymous rings of unknown size by two agents, even with distinct IDs, common chirality, and when the scheduler is* $\mathcal{F}$SYNC.

*Proof.* The proof is by contradiction. Let $\mathcal{A}$ be a partially Terminating exploration algorithm. Let $E$ be an execution of $\mathcal{A}$ on a dynamic ring of size $n$, where agents $a$ and $b$ start in two distinct locations and where the adversary always prevents the meeting of the agents, never blocking the agents at the same round. By Observation 3, this run exists. We assume, w.l.o.g., that $a$ is the first agent terminating in $E$, and that it terminates at round $r(E)$. Using execution $E$, until round $r(E)$, we construct an execution $E'$ of $\mathcal{A}$ on a dynamic ring of size $n' = 8r(E)$. In execution $E'$ the agents start at two distinct locations at distance $4r(E)$. The execution $E'$ is constructed in such a way that, until round $r(E)$ neither agent can distinguish this execution from $E$. We argue that this is possible, since in execution $E$ the adversary never blocks the two agents at the same time, and the agents do not meet. This implies that $a$ terminates in $E'$ at round $r(E)$. However, the size of the ring is $8r(E)$ and the agents started $4r(E)$ apart. Therefore, at round $r(E)$ there are at least $6r(E)$ unexplored nodes. Since $a$ does not move anymore, it remains a unique active agent $b$ in the system. Execution $E'$ is completed by having the adversary blocking agent $b$ from round $r(E)$ on (see Observation 2). The existence of execution $E'$ contradicts the correctness of $\mathcal{A}$. □

If we have 3 agents with IDs, it is unknown whether it is possible or not to have a Terminating exploration for an anonymous unknown $\mathcal{R}$. However, if we restrict ourselves to anonymous agents, then increasing their quantity, to any constant number, does not help. Being the ring anonymous, if edges are not removed, they will move in a symmetric way, never meeting, and therefore never terminating.

**Fully-Synchronous Scheduler.** In this section we discuss algorithms for the $\mathcal{F}$SYNC scheduler. They will use as a building block procedure EXPLORE ($dir$ | $p_1{:}s_1$; $p_2{:}s_2$; ...; $p_k{:}s_k$), where $dir$ is either *left* or *right*, $p_i$ is a predicate, and $s_i$ is a state.

In Procedure EXPLORE, the agent performs Look, then evaluates the predicates $p_1, \ldots, p_k$ in order; as soon as a predicate is satisfied, say $p_i$, the procedure exits and the agent does a transition to the specified state, say $s_i$. If no predicate is satisfied, the agent tries to Move in the specified direction $dir$ and the procedure is executed again in the next round. Predicates that we use are, *catches*: the agent observes the other agent on the port corresponding to its moving direction; *caught*: the agent is on the port after a failed move, the other agent is observed in the node. Variables are: $Tsteps$: the total number of rounds since the beginning of the algorithm; *Btime*: the number of consecutive rounds the agent has been currently waiting in a port.

*Known Upper Bound on Ring Size.* When a bound $N$ on the ring size is known, [24] presents an algorithm that terminates in $3N - 6$ rounds. The Algorithm KnownNNoChirality is in Algorithm 1, and it does not assume neither chirality nor cross detection. Let us briefly explain the algorithm: Agents start in the Init state. Initially, each agent moves in its *left* direction. Note that the left direction is not necessarily the same for both agents. If an agent enters in the state Forward it keeps moving *left* until termination. When an agent enters in state Bounce it changes direction to *right*, and it keeps moving *right* until termination. When an agent terminates it enters the state Terminate, and it remains still forever. An agent goes from Init state to Bounce if: it *catches* the second agent; or it reaches round $2N-4$ and $Btime = N-1$, that is it has been blocked for the last $N - 1$ rounds of the first $2N - 4$ rounds. An agent goes from Init state to Forward if: it is *caught* by the second agent; or it reaches round $2N - 4$ and $Btime \neq N - 1$.

---

**Algorithm 1.** Algorithm KnownNNoChirality

---

States: {Init, Bounce, Forward, Terminate}.

<u>In state Init:</u>

Explore(*left* | ($Ttime \geq 2N - 4 \wedge Btime \geq N - 1$): Bounce; *catches*: Bounce; *caught*: Forward; $Ttime \geq 2N - 4$: Forward)

<u>In state Bounce:</u>

Explore(*right* | $Ttime \geq 3N - 6$: Terminate)

<u>In state Forward:</u>

Explore(*left* | $Ttime \geq 3N - 6$: Terminate)

---

**Theorem 8** [24]. *Algorithm* KnownNNoChirality *allows two anonymous agents without chirality to explore a* 1*-interval connected anonymous ring and to explicitly terminate in time* $3N - 6$*, where* $N$ *is a known upper bound on the ring size.*

*Proof.* First of all, each agent enters in Terminate state, and it terminates at round $3N - 6$. We have to show that when $r = 3N - 6$ the ring has been explored. Let us first notice that if the two agents cross each other and keep moving in opposite directions, then they explore the ring in $N-1$ rounds. This is obvious, since the adversary at each round can block only one of them, by removing one edge, then the agent that is not blocked makes progress.

We have two cases according to the chirality of agents:

**Case 1.** Agents disagree on the *left* direction: After $N-3$ rounds, the agents are at minimum distance 2, or are at minimum distance 1, or crossed each other.

In case the distance between the agents is 2, then they were initially on two neighbouring nodes; therefore, in the next round, the ring will be explored. In case they are at distance 1 and the edge between them is not missing for the successive $N-1$ rounds, then they will cross each other and the ring will be explored by at most round $3N-6$. Otherwise, in case the edge between them is missing for the successive $N-1$ rounds, then at round $2N-4$ they will both change direction and the ring will be explored in the next $N-2$ rounds. If they crossed each other, then they will not change direction for the successive $N-1$ rounds exploring the ring.

**Case 2.** Agents agree on the *left* direction: If they catch each other before round $2N-4$, then they will move in opposite directions starting from the same node. Therefore, the ring will be explored in the next $N-1$ rounds terminating the exploration by round $3N-6$. Let us suppose they do not catch each other, then at each round at least one of them will traverse an edge: they cannot be blocked at the same time. Let us suppose that $a$ traverses $N-k$ edges with $k>2$; then $b$ traversed at least $x=2N-4-(N-k)$ edges, that is $x=N-4+k \geq N-1$, exploring the ring. It remains the case when agent $a$ traverses exactly $N-2$ edges; but this implies that also $b$ has traversed at least $N-2$ edges and, since they start from different nodes, the ring has been explored also in this case. □

It is easy to see that Algorithm KnownNNoChirality has a termination time that is asymptotically optimal. The only knowledge on $\mathcal{R}$ is the upper bound $N$. Thus, if we consider a dynamic ring that has always all edges, let it be $R$ ($R$ is essentially a static ring), the only way for a single agent to terminate is to do $N-1$ steps in some direction. Having two agents, as in our case, does not help, since they are anonymous they make symmetric movements, never meeting on $R$. This means that each of them has to terminate on its own, and therefore it has to do $\mathcal{O}(N)$ steps.

*Unknown Ring.* In Theorem 7, we have seen that, if $\mathcal{R}$ is unknown and anonymous, two agents cannot terminate. However, it is possible for them to explore $\mathcal{R}$ *unconsciously*: they explore the ring but they are not able to terminate. The algorithm that we discuss is the Unconscious Exploration of [24] (see Algorithm 2, where variable *Etime* is the number of rounds since the last call of procedure Explore).

The main technique is to try to guess the ring size. The guess $G$, initially $G=2$, is repetitively increased until it is big enough to allow ring exploration. More precisely, the algorithm implicitly divides the time in phases. In each phase an agent moves in its direction for a number of rounds that is twice the current guess. At the start of a new phase, each agent doubles its guess $G$, and it decides if it has to keep, or change, the direction of the previous phase: it changes direction if it has been blocked for the last $G$ rounds, otherwise it keeps its direction.

**Algorithm 2.** Algorithm UNCONSCIOUS EXPLORATION

States: {Init, Bounce, Reverse, Forward, Keep}.
In state Init:
  $G \leftarrow 2$, $dir \leftarrow left$
  EXPLORE($dir$; $Etime \geq 2G \wedge Btime > G$: Reverse, $Etime \geq 2G$: Keep, *catches*: Bounce, *caught*:Forward)
In state Reverse:
  $G \leftarrow 2 \cdot G$, $dir \leftarrow opposite(dir)$
  EXPLORE($dir$; $Etime \geq 2G \wedge Btime > G$: Reverse, $Etime \geq 2G$: Keep, *catches*: Bounce, *caught*:Forward)
In state Keep:
  $G \leftarrow 2 \cdot G$
  EXPLORE($dir$; $Etime \geq 2G \wedge Btime > G$: Reverse, $Etime \geq 2G$: Keep, *catches*: Bounce, *caught*:Forward)
In state Bounce:
  EXPLORE($opposite(dir)$)
In state Forward:
  EXPLORE($dir$)

**Theorem 9** [24]. *Algorithm* UNCONSCIOUS EXPLORATION *allows two anonymous agents without chirality to explore, without terminating, an* 1*-interval connected anonymous ring; the exploration is completed in* $\mathcal{O}(n)$ *time.*

*Proof.* Once agents catch each other, they will have opposite direction forever. Thus, if agents catch each other, then they will explore the ring in the successive $n-1$ rounds. So we have to consider the case in which agents do not catch each other. Let a *phase* be the period of time when the guess remains the same. The guess $G$ is doubled after $2G$ time steps, this means that at time $t_n \leq 4n$, $G \geq n$. Let $P$ be the phase in which $G \geq n$ and $r$ be the first round of this phase. In case agents are moving in the same direction, at round $r$, in each time step at least one of them makes progress (our hypothesis is that they do not catch each other). Therefore in the next $2G$ time steps the ring will be explored.

Consider now the case when agents are moving in opposite directions, and let us examine what is the configuration at round $r_{endP} = r + 2G - 1$, that is the last round before phase $P+1$: If at time $r_{endP}$ neither of them is blocked on an edge, then they must have crossed each other. In phase $P+1$ they will both keep different directions, notice that at the end of round $r_{endP}$ both agent decides to not reverse direction. The ring will be explored by the end of phase $P+1$.

Otherwise, at time $r_{endP}$ at least one of them has to be blocked on an edge $e$. At the beginning of the next phase, round $r_{start(P+1)} = r_{endP} + 1$, we have three options:

- Only one agent decides to change direction; in this case, they will have the same direction in phase $P+1$, exploring in $O(n)$ rounds.

- Both agents decide to change direction; then both have been blocked for the last $G$ rounds of phase $P$. Being both blocked at the same time, then they have to be waiting on two opposite endpoints of the same edge $e$. When they both reverse direction they will explore the ring starting from $e$ by at most round $r_{start(P+1)} + n - 1$.
- No one changes direction; this means that no one of them has been blocked consecutively for the last $G \geq n$ rounds. Agents are going on opposite directions in phase $P$, therefore after the first $G \geq n$ rounds of phase $P$ they either are at distance 1 from each other or they crossed. In case they have not crossed, since they are not both blocked for the last $G$ rounds of phase $P$, then they must cross by at most round $r_{start(P+1)}$. This means that they explore the ring in phase $P+1$.

We have shown that the agents explore by the end of phase $P+1$. Phase $P+1$ is reached after a number of rounds that is linear in $n$, this means that the claimed bound of the theorem follows. □

In case the ring has a landmark, the termination is possible: there exists a $\mathcal{O}(n)$ algorithm when agents share the same chirality, in this case the landmark is used as reference to understand when agents loop around the ring. When there is no chirality [24] gives a $\mathcal{O}(n \log n)$ algorithm. The case without chirality it is quite complex. In that algorithm, agents exploits the absence of edges to break the symmetry. Once the symmetry is broken, each agent acquires a unique ID. The IDs and a doubling strategy, as for Algorithm Unconscious Exploration, are used to orchestrate the movements of agents in such a way that they either meet, establishing a common orientation for the ring, or they loop around the ring enough time to ensure the termination of both.

**Semi-synchronous Scheduler.** We have seen that under the $\mathcal{F}$SYNC scheduler, Terminating exploration and Exploration are possible also in the Live setting. In this part of the chapter, we consider a $\mathcal{S}$SYNC scheduler. In this case we have three different models NS, ET, PT, see Sect. 2.2. Between these models there is the following relationship $P(\mathsf{NS}) \subseteq P(\mathsf{ET}) \subseteq P(\mathsf{PT})$, where $P(\mathsf{X})$ is the set of problems solvable in model X. First of all in the weakest of the three models, NS, is not possible to explore.

**Theorem 10** [24]. *In the* NS *model, exploring the ring is impossible with any number of agents, even if the ring and the agents are not anonymous and there is chirality.*

*Proof* (Sketch). Let us imagine to have a ring R where all nodes but one, let it be $v$, are occupied by agents. The only way to visit $v$ is by having agents that go in $v$ by both its incident edges $e_1, e_2$. Unfortunately, this is not possible in NS. The adversary forces to sleep the agents that are in the port of $e_2$, and removes edge $e_1$. By alternating this strategy with a bit of carefulness, the adversary creates a fair scheduler and it prevents the exploration of $v$. □

Interestingly, two anonymous agents are not enough to explore a ring even in the strongest PT.

**Theorem 11** [24]. *In the* PT *model without chirality two anonymous agents are not sufficient to explore a ring of size* $n \geq 5$*. The result holds even if there is a distinguished landmark node and the exact network size is known to the agents.*

*Proof* (Sketch). The proof is by contradiction. Let $\mathcal{A}$ be a correct exploration algorithm for agents $a, b$. The adversary is able to decide the orientation of the agents, their starting position and the topology of the ring $\mathcal{R}$. The adversary picks 4 nodes $v, v'$ and $u, u'$, these nodes are disjoint and none of them is the landmark. It also partially fixes the topology of $\mathcal{R}$ by creating edge $(v, v')$ and $(u, u')$ and it puts agent $a$ on $v$ and agent $b$ on $u$. Note that as long as one agent does not go outside these nodes, the adversary does not have to specify the remaining topology of the ring. Initially, the adversary keeps agent $b$ sleeping and activates $a$. Agent $a$ is activated and it is prevented to go in other nodes but node $v'$. Eventually, $a$ either decides to visit node $v'$ or it decides to wait perpetually on the port of $v$ that goes outside the portion $\{v, v'\}$.

Let us assume that $a$ goes to $v'$. At this point the adversary activates agent $b$, and it uses the same scheduler that used for $a$. Now if $a$, resp. $b$, switches between nodes $v, v'$, resp. $u, u'$, then they will never explore the ring, and thus $\mathcal{A}$ is not correct. This means that eventually $a$ has to decide to perpetually wait on a port $p$ that connects the portion $\{v, v'\}$ to other nodes of the ring. But being agent $b$ anonymous also $b$ decides to do the same. At this point the adversary fixes the topology of the ring, and also orientation of the agents, in such a way that $a$ and $b$ are perpetually waiting on the two endpoints of the same edge $e$. The adversary removes edge $e$ forever keeping the agents stuck and preventing the exploration. □

Note that the above proof requires the absence of chirality. If chirality is present, 2 agents can explore and partially terminate both in PT and ET.

Without chirality three agents are enough to explore and partially terminate in PT and ET. In this case the number of steps required is quadratic in $n$, while linear steps were sufficient in the $\mathcal{F}$SYNC case. In $\mathcal{S}$SYNC we can have only partial termination, as shown in [24], it is impossible to build an algorithm in PT with chirality where two agents correctly terminate. Finally, in ET partial termination can be achieved only if there is exact knowledge of $n$, an upper bound or a landmark are not enough, this is in contrast with PT. The reader that is interested in these results can read [24].

### 4.2 Connected over Time Rings

In [14,15] the authors investigate the distributed Perpetual exploration of rings that are connected over time in $\mathcal{F}$SYNC. In their model there are no ports on nodes, that is an agent does not have to enter inside a port to move on an edge, but it directly goes on the edge. An agent during the Look sees if there is a missing edge incident in the current node.

This means that an agent cannot be trapped on a single node: if agent $a$ sees a missing *left* edge then it leaves the node taking the *right* edge. However, a variant of Observation 2 holds: a single agent can be trapped on two adjacent nodes.

A significative difference is that when edges are not accessed using ports an agent does not know the direction of other agents in its same node. However, [14,15] show that such knowledge is not required to solve Perpetual exploration. In a model without ports, predicate *catches* has the following equivalent: *In the previous round I was alone, I moved to a new node v, and now I am not alone in v.*

**Algorithm 3.** Algorithm PEF_3+

States: {Init, Reverse}.
<u>In state Init:</u>
  dir=*left*
  EXPLORE(dir | *catches*: Reverse)
<u>In state Reverse:</u>
  dir = opposite(dir)
  EXPLORE(dir | *catches*: Reverse)

In [15] the authors propose the PEF_3+ algorithm (Algorithm 3). PEF_3+ solves Perpetual exploration when there are 3 or more agents and the ring is connected over time. Note that agents do not have to share the same chirality.

**Theorem 12** [15]. *Given a connected over time ring $\mathcal{R}$,* 3 *or more anonymous agents executing algorithm* PEF_3+ *solve* Perpetual exploration.

We give the intuition behind the correctness of PEF_3+ for the case of 3 agents. Suppose first that all robots have the same direction. If two of them never meet then, the robots perpetually explore the ring. This is obvious since for the robots to not meet they have to keep spinning around the ring. To prevent this the adversary has to make at least two robots meet. However, when this happens one agent *catches* the other. Therefore at this point we have at least two agents with opposite directions, so we enter in the case in which not all agents have the same direction. If there are such two agents, we eventually reach a configuration in which there is a segment of the ring containing all robots, and two robots $a$ and $b$ at the two extremes of this segment, one agent going *left* and the other going *right*. These agents are the *extreme agents*. Note that the extreme agent is not fixed: during the execution other agents may switch position with $a$ or $b$. The key point is that no matter of how agents catch each other, there will always be two extreme agents. Since the extreme agents are going in opposite directions they either explore the ring perpetually, or they will be blocked forever on the two endpoints of the same edge. In this last case the presence of the third agent $c$ ensures the perpetual exploration: agent $c$ will ping-pong among $a$ and $b$ perpetually visiting all nodes.

Interestingly, the authors show that 2 agents are not enough to explore a connected over time ring of arbitrary size.

**Theorem 13** [15]. *Given a connected over time ring $\mathcal{R}$ of arbitrary size, there exists no algorithm solving* Exploration *with* 2 *anonymous agents.*

The idea used in the proof is similar to the one of Theorem 11, and it draws an interesting parallelism between 1-interval connected rings with a $\mathcal{S}$SYNC scheduler and connected over time rings with $\mathcal{F}$SYNC scheduler.

*Self-stabilizing Algorithm.* Paper [14] tackles the challenging case of designing a self-stabilizing algorithm for Exploration. They prove that 3 agents with IDs are enough to solve Exploration on rings of any size. The task is impossible if there are only two agents.

**Theorem 14** [14]. *Three non-anonymous agents are necessary and sufficient to solve the self-stabilising* Exploration *of a connected over time ring $\mathcal{R}$ of arbitrary size.*

The idea of the algorithm is similar to the one of PEF_3+. However, in this case there is the need to solve the situation in which two robots are on the same node and with the same state. To break this symmetry the IDs of the robots are used. They are scanned bit by bit in a circular fashion, and the bit is used to decide the direction of the agent in the current round.

### 4.3 Recurrent Rings

When the ring is $\delta$-recurrent, Observation 2 is not valid and a single agent is enough to explore the ring. [37] shows an algorithm, namely STUBBORN TRAVERSAL, for $\delta$-recurrent $T$-interval connected rings. The idea of STUBBORN TRAVERSAL is quite simple: the unique agent $a$ picks one direction and stubbornly moves in that way. Being the graph recurrent each edge eventually appears, thus $a$ visits all nodes. The upper bound of the algorithm is

$$n - 1 + \lceil \frac{n-1}{\max\{1, T-1\}} \rceil (\delta - 1)$$

rounds. The idea behind the bound is immediate. The agent has to traverse $n-1$ edges. For the $T$-interval property when a removed edge reappears there cannot be other removals for $T-1$ rounds. Therefore, the agent encounters at most $\lceil \frac{n-1}{\max\{1,T-1\}} \rceil$ blocked edges. Being the graph $\delta$-recurrent, an edge can be missing for at most $\delta - 1$ rounds.

Surprisingly, the algorithm is almost optimal. [37] shows a lower bound of

$$n - 1 + \lfloor \frac{n-3}{\max\{1, T-1\}} \rfloor (\delta - 1)$$

rounds.

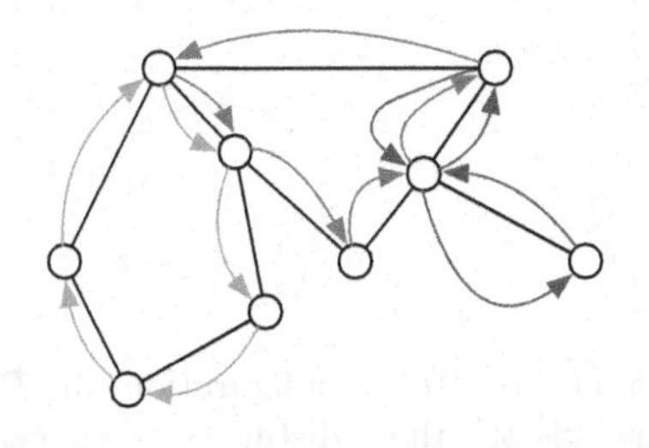

(a) On the static graph $G$ there are three carriers: red, blue and green.

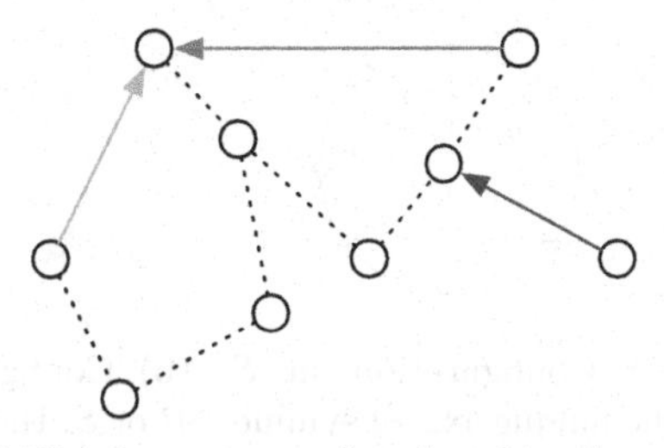

(b) The dynamic graph induced by the carries at round 0.

**Fig. 6.** Example of carrier network (Color figure online)

### 4.4 Periodic Graphs

The Terminating exploration of periodic graphs has been studied in [30,32,36]. [32] has been the first paper to look at the problem of exploration in the dynamic setting, if we exclude the random walk studied in [7].

The model of [32] is different from the one that we are analyzing. In their case the dynamic graph is induced by a set of carriers that are moving on top of a static graph $G$. Each carrier performs a periodic visit of certain nodes moving at constant speed; each node is covered by at least one carrier (see Fig. 6).

The unique agent $a$ has to visit the graph, but it is restricted to instantaneously move from one carrier to the other. Agent $a$ cannot disembark a carrier and wait for another one. In Fig. 6 the agent may move from the green carrier to the red one. Each carrier has a unique ID. This model is known as *carrier network*.

When $G$ is anonymous, and the carriers have all the same period (the carriers are *homogenous*), the authors propose a simple algorithm, namely HITCH-A-RIDE. The agent is required to know an upper bound $B$ on the maximum path of each carrier. The algorithm performs a DFS on a spanning tree of the carriers.

Initially, the agent starts in the *root* carrier. When an agent is inside a new carrier $c$ it stays there for $B$ rounds, and memorises the IDs of all other carriers it meets, increasing the set $ToVisit$ and the set $reachable(c)$. After $B$ rounds, carrier $c$ is considered visited, so it is removed from $ToVisit$. At this point the agent goes to a new carrier $c' \in ToVisit \cap reachable$ and set $parent(c') = c$. If $c'$ does not exist, the agent backtracks going to $parent(c')$.

**Theorem 15** [32]. *Given a carrier network $\mathcal{G}$, with $k$ carriers, and an upper bound $B$ on the maximum cycle of each carrier, an agent executing algorithm* HITCH-A-RIDE *explores the graph in at most $(3k-2)B$ rounds.*

Notice that $B$ has to be known. Given any algorithm $\mathcal{A}$ that does not need the knowledge of $B$, we can show that $\mathcal{A}$ is not correct. Let us consider a graph with nodes $\{v_0, v_1, v_2\}$ and a carrier that does a loop on the three nodes, in this case $B = 3$. On this graph $\mathcal{A}$ terminates after $t$ rounds. Now we construct an

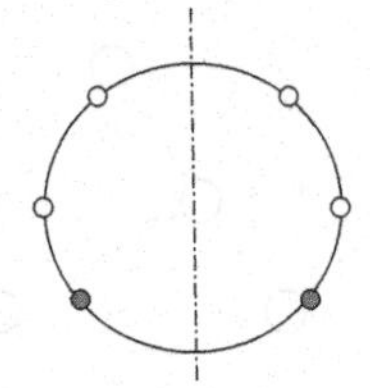
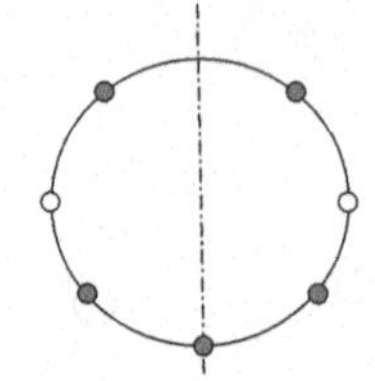
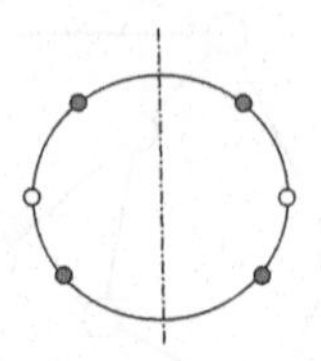
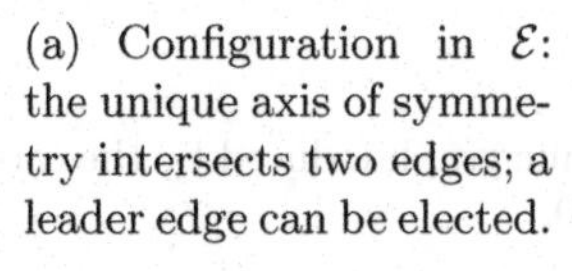

(a) Configuration in $\mathcal{E}$: the unique axis of symmetry intersects two edges; a leader edge can be elected.

(b) Configuration not in $\mathcal{P}$ or $\mathcal{E}$: the unique axis of symmetry intersects one node; a leader node can be elected.

(c) Configuration in $\mathcal{P}$: the distances between homebases are periodics; it is not possible to elect either a leader node or a leader edge.

**Fig. 7.** Examples of homebases configurations. The homebases are the red nodes. (Color figure online)

undistinguishable graph where the nodes are the same but the carrier performs a non-simple cycle doing $t$ loops on $v_0, v_1$ and then going to $v_2$, in this case $B = t + 1$. Being the graph anonymous $\mathcal{A}$ terminates also in this last graph, and it terminates after $t$ rounds, missing the visit of node $v_2$. A more formal proof of the necessity of $B$ knowledge is contained in [32].

The work of [36] builds on top of [32], allowing the agent to wait. While [30] considers the exploration of carrier networks with *black-hole*. A *black-hole* is a node, or an edge, that destroys visiting agents.

## 5 Live Gathering in Dynamic Rings

In this section we will focus on Gathering. Specifically, the results of [25] for anonymous agents on 1-interval connected rings under the $\mathcal{F}$SYNC scheduler. On a dynamic ring $\mathcal{R}$, Gathering is solved when agents are located on nodes that are neighbours in $F(\mathcal{R})$. The impossibility of strict gathering (i.e., all agents on the same node) is immediate from Observation 3.

*Homebases and Configurations.* We assume that agents starts in distinct nodes, the starting nodes are marked in such a way to be distinguishable from other nodes. These nodes are the homebases. Besides the homebases the ring is anonymous, and the homebases are all equal. It is well known that if the agents are anonymous, the ring is anonymous and there are no homebases, gathering is impossible even in static rings.

The locations of the $k$ homebases in $F(\mathcal{R})$ define a *configuration* $C$ (refer to Fig. 7). Intuitively, gathering can be solved only when the configuration is such that all agents ca uniquely identify a distinguished edge (leader edge) or a distinguished node (leader node). It is impossible to elect such references when the configuration of homebases is periodic, this impossibility holds also for the static case. Clearly, to elect a leader edge the disposition of homebases on the ring has to admit a unique axis of symmetry going through two edges; to elect a leader node the configuration has to either have a unique symmetry axis that goes through a node or no symmetry axes, see Fig. 7. Formal definitions are in [25]. We indicate with $\mathcal{P}$ the set of periodic configurations, and with $\mathcal{E}$ the set of configurations where a leader edge can be elected.

### 5.1 Chirality and Cross Detection

When there is cross detection and chirality is possible to solve Gathering in $\mathcal{O}(|V|)$ rounds; however, the knowledge of either $n = |V|$ or $k$ is required. When there is chirality is possible to elect a leader node also when the configuration is $\mathcal{E}$: the leader node is the left endpoint of the leader edge.

We present the Simple Gathering algorithm, the algorithm is obtained by adapting the techniques proposed in [25] for the special case of cross detection and chirality. Simple Gathering is divided in two phases: *Discovery Phase* and *Gathering Phase.*

In the Discovery Phase, agents try to learn the configuration by doing a loop of the ring in clockwise direction. If the loop is prevented, all agents gather on the same node and they are aware of that. Otherwise, Discovery Phase ends, all agents know the current configuration, and they agree on an elected leader node. Note that agents have memory, therefore even if they have only local view, after a loop of the ring, they are able to construct the current configuration in their memory.

The Gathering Phase starts after the Discovery Phase. In this phase, agents first try to gather on the elected node. If this does not happen, agents are partitioned in two groups. These two groups move in such a way that agents gather; also, agents are able to recognise the gathering and terminate correctly.

In the following, we will provide more details on these two phases.

*Discovery Phase.* Let us first introduce a basic observation that will be used in the Discovery Phase.

**Observation 4** [25]. *Let $A$ be a set of agents moving clockwise during an interval of time $I$ lasting at least $3n - 4$ rounds. If an agent $a^* \in A$ moves less than $n - 1$ steps in $I$, then there exists a round $r \in I$ where all agents in $A$ are on the same node.*

*Proof.* At each round only one edge could be missing. Therefore if at round $r \in I$ $a^*$ is blocked, then every $a \in A$ that at round $r'$ is not at the same node of $a^*$ moves. Agent $a^*$ moves less than $n-1$ steps in an interval lasting at least $3n-4$ rounds, this means that the number of rounds in which $a^*$ is blocked is at least $2n - 2$. For the above, the agents in $A$ that are not in the same node as $a^*$ have moved towards $a^*$ of at least $2n - 2$ steps. It is also true that, every time $a^*$ moves, the other agents might be blocked; however, by hypothesis, this has happened less than $n-1$ times. The initial distance between $a^*$ and an agent in $A$ is at most $n - 1$, this distance increases less than $n - 1$ (due to $a^*$ moving), but it decreases by $2n-2$ (due to $a^*$ being blocked); thus the distance is zero, (the agents reach the same node) by the end of interval $I$. □

**Algorithm 4.** Discovery Phase of Algorithm SIMPLE GATHERING when $n$ is known.

States: {Init, GatheringPhase, Terminate}.
<u>In state Init:</u>
EXPLORE ($left \mid Ttime = 3n-1 \wedge$ ( $Esteps < n-1 \vee \#Agents = k$): Terminate; $Ttime = 3n-1$: GatheringPhase)

**Algorithm 5.** Discovery Phase of Algorithm SIMPLE GATHERING when $k$ is known.

States: {Init, GatheringPhase, Terminate}.
<u>In state Init:</u>
EXPLORE ($left \mid \#Agents = k$: Terminate; $Ttime = 3n-1$: GatheringPhase)

There are two variants of the Discovery Phase: one is used when agents know $k$ and the other when they know $n$. In Algorithm 4 there is the variant for known $n$ and in Algorithm 5 the variant for known $k$. The variable $\#Agents$ indicates the number of agents in the current node. The variable $Esteps$ the number of steps that an agent did since the start of procedure EXPLORE.

- Variant for known $n$: Each agent goes clockwise (direction *left*). The only check is done at time $3n-1$. Where an agent terminates if $Esteps < n-1$ or if it sees $k$ agents.
  Note that $k$ is initially unknown. Nevertheless, an agent computes $k$ after a loop of the ring, this can be done by counting the homebases. By Observation 4, at time $3n-1$ if an agent does not terminate for $Esteps < n-1$, then it knows $k$, so it can perform the check $\#Agents = k$. If an agent does not terminate it goes to state GatheringPhase.
- Variant for known $k$: Each agent goes clockwise. An agent terminates if $\#Agents = k$. Otherwise, once $r = 3n-1$ it goes to state GatheringPhase. Note that $n$ is initially unknown. Nevertheless, an agent computes $n$ after a loop of the ring, the loop is detected by seeing $k+1$ homebases.

**Lemma 1.** *Given a set of agents, with common chirality, knowing n (resp. k) and executing the Algorithm 4 (resp. 5) on a 1-interval connected ring. The following hold:*

- *If an agent enters state* Terminate, *every agent does so. Moreover, all agents are on the same node.*
- *If an agent enters state* GatheringPhase, *every agent does so. Moreover, all agents have done a loop of the ring.*

*Proof.* We divide the proof in two parts.

- Agents know $n$ and execute the Algorithm of Fig. 4. We partition the agents in two sets $L$ and $NL$. The set $L$ is composed by agents that have done a loop of the ring ($Esteps \geq n-1$). The set $NL$ by the remaining agents, the ones with $Esteps < n-1$.
  It is immediate that if $NL = \emptyset$ the statement of the lemma is correct: all agents know $k$, so they either all terminate, on the same node, by seeing $\#Agents = k$ or they all go to GatheringPhase.
  In case $NL \neq \emptyset$, we have that agents in $NL$ terminates at round $3n-1$: by definition for them hold $Esteps < n-1$. However, by Observation 4 we know that all agents are on the same node. Therefore, at round $3n-1$ for agents in $L$ the predicate $\#Agents = k$ is true. This implies that all agents go to Terminate.
- Agents know $k$ and execute the Algorithm of Fig. 5. The correctness when $\#Agents = k$ is immediate. We have to show that when all agents enter in state GatheringPhase they have done a loop of the ring. By Observation 4 if at round $3n-1$ exists an agent that has not done a loop around the ring ($Esteps < n-1$), then all agents are on the same node, and they terminate seeing $\#Agents = k$. Therefore, if agents do not terminate by round $3n-1$, then they all know $n$. □

*Gathering Phase.* Thanks to Lemma 1 we know that at the start of Gathering Phase all agents know the configuration of the ring, and thus $k$ and $n$. Moreover, they agreed on a leader node $v_l$, recall that the configuration is not in $\mathcal{P}$.

The Gathering Phase is in Algorithm 6, for simplicity we assume it starts at round 0. The algorithm uses the boolean variable *Crossed*, the value of this variable is true when an agent crossed someone on an edge in the previous round.

Initially, all agents move *left*; an agent keeps walking until it either reaches $v_l$ (predicate $onNode(v_l)$) or $Ttime = 3n-1$. If an agent reaches $v_l$ it enters state Wait, otherwise it enters state ReachingElected. An agent in Wait state stays still until round $3n-1$ and then it enters state ReachedElected. At round $3n-1$ agents are partitioned in two sets: ReachingElected and ReachedElected. Agents in ReachingElected walks *right*, while agents ReachedElected walks *left*. In both states agents terminate using the same conditions: $\#Agents = k$ or $Crossed = true$ or $Btime = n-1$.

**Algorithm 6.** Gathering Phase of Algorithm SIMPLE GATHERING

States: {ReachedElected, ReachingElected, Terminate}.
In state GatheringPhase:
  dir=*left*
  EXPLORE (*dir* | *onNode*($v_l$): Wait; *Ttime* = $3n - 1$: ReachingElected)
In state Wait:
  EXPLORE (*nil* | *Ttime* = $3n - 1$: ReachedElected)
In state ReachedElected:
  EXPLORE (*dir* | $\#Agents = k \vee Crossed = true \vee Btime = n - 1$: Terminate)
In state ReachingElected:
  $dir = right$
  EXPLORE (*dir* | $\#Agents = k \vee Crossed = true \vee Btime = n - 1$: Terminate)

**Theorem 16.** *Given a set of agents knowing the configuration $C$ of homebases. If $C \notin \mathcal{P}$ and the agents execute Algorithm 6, then* Gathering *is solved in $O(n)$ rounds.*

*Proof.* Since $C \notin \mathcal{P}$ and there is chirality, it is possible to elect the leader node $v_l$. First of all, observe that at time $3n - 1$ all agents are partitioned in two set $R$ and $NR$. The set $NR$ is the one of agents in state ReachingElected, the set $R$ is the set of agents in state ReachedElected.

By using an analogous argument of the one in Observation 4, we can easily show that all agents in set $R$ (resp. $NR$) are on the same node. Thus at round $3n - 1$ agents are partitioned in two nodes forming two groups. A group is a set of agents in the same state, on the same node, and moving in the same direction. Note that a group behaves as a single agent.

The two groups move with opposite directions. After at most $n - 1$ rounds the groups either meet on the same node, or they are at distance 1, or they cross on an edge. They clearly terminate if they cross each other or if they go on the same node, predicates $\#Agents = k$ or $Crossed = true$. If they are at distance 1, waiting on the two endpoints of the same missing edge, then eventually one group terminates, predicate $Btime = n - 1$. The other group either terminates for the same reason, or it crosses the edge gathering on a single node and terminating for $\#Agents = k$. □

*Removing the Cross Detection.* Removing the cross detection from SIMPLE GATHERING is not immediate. While the Discovery Phase is not impacted by the cross detection, the Gathering Phase uses it.

A technique to remove the cross detection is the *Logic Ring* [25]. Intuitively, a *Logic Ring* assigns a sequence of rounds to each pair edge-direction. An agent is allowed to cross an edge in a specific direction only when the current round is in the sequence assigned to that edge-direction. In this way agents synchronise their movements such that they never cross the same edge, from two different directions, at the same round. The construction of the *Logic Ring* is possible when the symmetry of the ring is broken by the election of a node or an edge. The *Logic Ring* increases the cost of solving Gathering to $\mathcal{O}(n \log n)$ rounds, it is unknown if such increase is necessary.

### 5.2 No Chirality

The absence of chirality breaks SIMPLE GATHERING. More clever strategies have to be designed for both the Discovery Phase and the Gathering Phase. Interestingly, when $n$ is known, the absence of chirality does not impact the solvability of the problem, that can be still solved for any configuration $C \notin \mathcal{P}$. While when $k$ is unknown, the absence of chirality makes the problem unsolvable when the configuration is in $\mathcal{E}$.

**Theorem 17** [25]. *In rings with no chirality,* Gathering *is impossible without knowledge of $n$ when starting from a configuration $C \in \mathcal{E}$. This holds even if there is cross detection and $k$ is known.*

Interestingly, also the cross detection is necessary when the configuration is in $\mathcal{E}$.

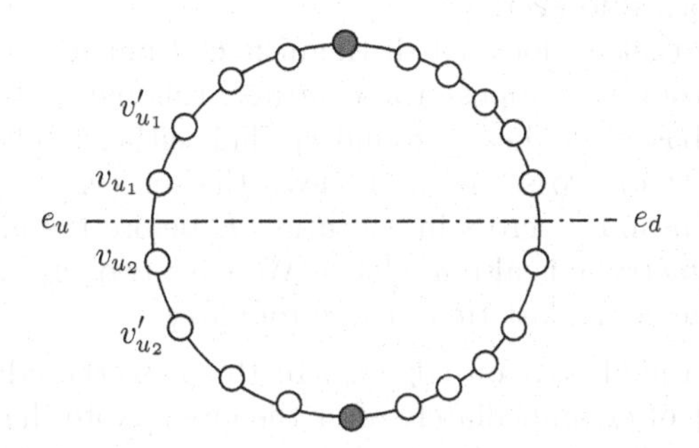


**Fig. 8.** Configuration where Gathering is impossible when there is no cross detection. Homebases are in red. (Color figure online)

**Theorem 18** [25]. *Without chirality and without cross detection,* Gathering *is impossible when starting from a configuration $C \in \mathcal{E}$. This holds even if the agents know $C$ (which implies knowledge of $n$ and $k$).*

*Proof* (Sketch). The proof is by contradiction. Let us consider an initial configuration $C$ with two agents $a_l, a_r$, a unique axis of symmetry passing through edges $e_u, e_d$, and where the two homebases $h_l, h_r$ are at distance at least 4 from $e_u$ and 5 from $e_d$ (see an example in Fig. 8). Let $\mathcal{A}$ be an algorithm that solves Gathering starting from configuration $C$.

We consider an execution $E$ where the adversary does not remove any edge. For the symmetry of $C$, in $E$ the agents cross each other only over $e_u$ or $e_d$ never meeting in the same node at the same time. Furthermore, agents can only gather at the two endpoints of one of these edges. W.l.o.g, we assume that algorithm $\mathcal{A}$ terminates when the agents are on the endpoints of $e_u = (v_{u_1}, v_{u_2})$. Let $v'_{u_2}$ be the neighbour of $v_{u_2}$ different from $v_{u_1}$ (resp. $v'_{u_1}$ the neighbour of $v_{u_1}$ different from $v_{u_2}$). Let $r_f$ be the round in which $a_l$ reaches $v_{u_1}$ and terminates (note that $a_l$ could have passed by $v_{u_1}$ several times before, without terminating) and let $r_1$, possibly equal to $r_f$, be the first round when $a_l$ reaches a node in the set $\{v_{u_1}, v_{u_2}, v'_{u_2}, v'_{u_1}\}$ and such that $a_l$ does not leave that set after $r_1$.

Agent $a_l$ may reach $v_{u_1}$ at round $r_f$ in several ways, but note that since the agents anonymous, $a_r$ does exactly the symmetric moves of $a_l$ with respect to the axis.

We now construct, using the same configuration $C$, a new execution $E'$. In execution $E'$ agents behave like in execution $E$ until they possibly find themselves blocked by an edge removal. Moreover, $E'$ is constructed in such a way that agent $a_l$ is never blocked. Thus $a_l$ behaves exactly as in execution $E$ terminating in node $v_{u_1}$ at round $r_f$. However, in execution $E'$ gathering is not achieved because we will block agent $a_r$ away from $a_l$.

Let us suppose that in execution $E$, agents $a_l$ and $a_r$ never cross before round $r_1$. In this case in execution $E'$, we block $a_r$ in the first round on its homebase, and we keep $a$ blocked forever.

We now prove that $a_l$ does not distinguish $E'$ from $E$, terminating at least two edges away from $a_r$. Suppose the contrary: this means that in $E$ agent $a_l$ reaches the homebase of $a_r$ before round $r_1$. This either implies that $a_r$ crosses the edge $e_d$ or that $a_r$ crosses $e_u$ and leaves the set $\{v_{u_1}, v_{u_2}, v'_{u_2}, v'_{u_1}\}$. Both cases imply that $a_l$ and $a_r$ cross in execution $E$ before round $r_1$ (each time $a_l$ traverses the symmetry axis also $a_r$ does). We have two cases, according to the edge where they cross the last time before round $r_1$.

- They cross at round $r_{last}$ on edge $e_d$. In this case the adversary blocks $a_r$ on an endpoint of $e_d$ immediately after the cross. Note that agent $a_l$ cannot reach the node where $a_r$ is blocked, it would contradict the fact that in $E$ round $r_{last}$ is the last round in which agents cross outside $\{v_{u_1}, v_{u_2}, v'_{u_2}, v'_{u_1}\}$.
- They cross at round $r_{last}$ on edge $e_u$. Since the cross is before round $r_1$, there exists a round $r' < r_1$ where both agents leave the set $\{v_{u_1}, v_{u_2}, v'_{u_2}, v'_{u_1}\}$, and round $r'$ is after the cross. In this case the adversary blocks $a_r$ right after $r'$, when $a_r$ is outside the set $\{v_{u_1}, v_{u_2}, v'_{u_2}, v'_{u_1}\}$. Also in this case, if $a_l$ reaches the node where $a_r$ is blocked, it contradicts the fact that $r_{last}$ was the last round in which they cross before $r_1$: to reach the node where $a_r$ is blocked, $a_l$ has to either cross edge $e_u$ and then leave $\{v_{u_1}, v_{u_2}, v'_{u_2}, v'_{u_1}\}$, or it has to cross edge $e_d$. Both cases contradict the fact that in execution $E$ the round $r_{last}$ is the round of the last cross between agents.

We have shown that in all cases agent $a_l$ cannot distinguish execution $E'$ from $E$. Thus in $E'$, $a_l$ terminates on $v_{u_1}$, while agent $a_r$ is not on a neighbour node of $v_{u_1}$. This contradicts the correctness of $\mathcal{A}$. □

When Gathering is solvable the known algorithms give the following upper bounds: with cross detection and without chirality Gathering can be solved in

$\mathcal{O}(n)$ rounds; the bound is $\mathcal{O}(n^2)$ when also cross detection is absent. It is an open problem to determine if the quadratic bound is tight.

# 6 Live Uniform Deployment in Dynamic Rings

The problem of Uniform Deployment has been studied in [3] (the authors call the problem Dispersion but we will use the term Uniform Deployment). The authors assume that the number of agents is equal to the number of nodes, that is $|A| = |V|$, we will see in the following why this assumption is somehow needed.

The model is as follows. There are no ports. Each agent has a visibility radius $\rho$. At each round, an agent sees the current topology up to $\rho$ hops from its position. Agents have IDs, and each agent is able to see the IDs of other agents on its same node. Moreover, agents have a global weak multiplicity detection, that is they can see if a certain node, inside their visibility radius, is occupied by more than one agent.

The dynamic is interesting, the adversary is allowed to permute the nodes of the ring. More precisely, the dynamic graph $\mathcal{VP} = (G_1, G_2, \dots)$ is such that each $G_j \in \mathcal{VP}$ is a ring, but we may have $G_j \neq G_i$[1]. We call this kind of dynamic graphs node permuting rings (VP). It is also possible to consider the union of node permutation and 1-interval connectivity (VP-1): in this case a graph $\mathcal{VP}$-1 $= (G_1, G_2, \dots)$ is such that each $G_j \in \mathcal{VP}$-1 is either a path or a ring.

It is easy to see that if the dynamic graph is VP any constant number of agents $k$ can be confined in at most $3k$ nodes. Let the $k$ agents be initially placed in different nodes; at round $r$ the agent $a_i$ is on node $v_i$, node $v_i$ has two neighbours $v_{i-1}, v_{i+1}$. Agent $a_i$ can move either to $v_{i-1}$ or $v_{i+1}$. Let us suppose w.l.o.g. that it goes to $v_{i-1}$. At round $r+1$ the adversary shuffle the ring such that $v_{i-1}$ has neighbours $v_{i+1}, v_i$.

The previous observation gives an hint on the necessity of non constant number of agents to solve non trivial problems in VP. An interesting open question is to characterise the non-trivial problems that are solvable in VP with a constant number of agents.

To solve the Uniform Deployment in VP-1, [3] proposes the algorithm VP-1-INTERVAL-CHAIN. The algorithm assumes full-visibility (i.e., the agents can see the entire ring so we have $\rho = \lceil \frac{|V|-1}{2} \rceil$) and chirality. Before introducing the VP-1-INTERVAL-CHAIN, we give few definitions.

A *hole* is an empty node, a *multinode* is a node with multiple agents, and a *singleton* is a node with only one agent. A *chain* is a sequence of consecutive nodes, starting with a multinode, followed by zero or more singletons and terminating with an empty node. A *good chain* is a chain such that there is no missing edge between nodes in the chain. A *bad chain* is a chain with a missing edge between two nodes of the chain. If in clockwise direction the multinode precedes the hole, then the chain is a clockwise chain. Otherwise, the chain is a counter-clockwise one. See Fig. 9.

[1] The footprint of the graph is not necessarily a ring.

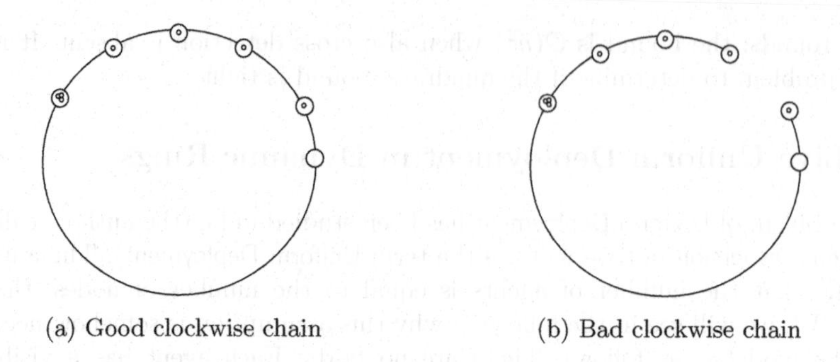

(a) Good clockwise chain (b) Bad clockwise chain

**Fig. 9.** Examples of chains, from [3].

The idea of VP-1-INTERVAL-CHAIN is as follows. If agent $a$ is in a singleton node in a clockwise *good chain*, then $a$ moves towards the hole. If agent $a$ is in a singleton node in a counter-clockwise *good chain*, then $a$ moves towards the hole if and only if the multinode of the chain is not part of another good chain. Regarding the agents in the multinode only the one with lowest id, let it be $u$, moves. If $u$ is in a multinode that is part of only one good chain, then $u$ moves towards the hole. If $u$ is part of two good chains, then $u$ moves clockwise.

**Theorem 19** [3]. *Algorithm* VP-1-INTERVAL-CHAIN *solves* Uniform Deployment *on* VP-*1 dynamic rings in* $\mathcal{O}(|V|)$ *rounds.*

*Proof.* It is immediate to see that as long as the dispersion is not solved there are multinodes, recall that $|A| = |V|$. It is also clear that there are at least two chains: take a sequence of empty nodes, let it be *empty*, and consider the two sequences of occupied nodes at left and right of such empty space. Let them be $chain_{left}$ and $chain_{right}$. Since only one edge can be absent at each round, at least one between $chain_{left}$ or $chain_{right}$ is a good chain. If one chain expands in *empty*, then the number of empty nodes decreased by at least 1.

Otherwise, by construction of the algorithm, $chain_{right}$ has to be a bad chain and $chain_{left}$ a good chain. Moreover $chain_{left}$ did not expand in *empty*. For $chain_{left}$ to not expand in *empty*, it means that its multinode is also part of a clockwise good chain, $chain'$. By the rule of the algorithm $chain'$ will expand. Thus, also in this case, the number of empty nodes decreased by at least 1.

The previous implies that the number of empty nodes decreases at each round. □

When chirality is not present authors solve Uniform Deployment under more restricted assumptions. If robots start all in the same node they solve the problem in $\mathcal{O}(|V|)$ rounds in VP-1; otherwise, they give solution algorithms for rings of odd size or for rings of size 4.

The authors show that Uniform Deployment is not solvable when the visibility of each robot is limited to its current node. The impossibility holds both in VP and in interval connected rings. It is an open problem to determine the solvability of Uniform Deployment for other visibility radiuses.

## 7 Conclusion

The study of agents on dynamic graph is a challenging topic. Few works have investigated the design of distributed algorithms in the agent model. For now, they considered a very limited set of problems and, most of the time, strong restrictive assumptions.

A future work could be to extend the study of the exploration on topologies that are more complex than rings, another one could be to investigate the impact of $\mathcal{S}$SYNC on the problems of Gathering and Uniform Deployment. The field is still in its early stage and there are many new directions worth considering.

**Acknowledgment.** The author thanks Giuseppe Prencipe and the anonymous reviewer for their invaluable feedbacks.

## References

1. Aaron, E., Krizanc, D., Meyerson, E.: DMVP: foremost waypoint coverage of time-varying graphs. In: Kratsch, D., Todinca, I. (eds.) WG 2014. LNCS, vol. 8747, pp. 29–41. Springer, Cham (2014). https://doi.org/10.1007/978-3-319-12340-0_3
2. Aaron, E., Krizanc, D., Meyerson, E.: Multi-robot foremost coverage of time-varying graphs. In: Gao, J., Efrat, A., Fekete, S.P., Zhang, Y. (eds.) ALGOSENSORS 2014. LNCS, vol. 8847, pp. 22–38. Springer, Heidelberg (2015). https://doi.org/10.1007/978-3-662-46018-4_2
3. Agarwalla, A., Augustine, J., Moses, W., Madhav, S., Sridhar, A.K.: Deterministic dispersion of mobile robots in dynamic rings. In: Proceedings of the 19th International Conference on Distributed Computing and Networking (ICDCN), pp. 19:1–19:4 (2018)
4. Aguilera, M.K.: A pleasant stroll through the land of infinitely many creatures. SIGACT News **35**(2), 36–59 (2004)
5. Albers, S., Henzinger, M.: Exploring unknown environments. SIAM J. Comput. **29**(4), 1164–1188 (2000)
6. Augustine, J., Pandurangan, G., Robinson, P.: Fast byzantine agreement in dynamic networks. In: Proceedings of the 32th Symposium on Principles of Distributed Computing (PODC), pp. 74–83 (2013)
7. Avin, C., Koucký, M., Lotker, Z.: Cover time and mixing time of random walks on dynamic graphs. Random Struct. Algorithms **52**(4), 576–596 (2018)
8. Baldoni, R., Bonomi, S., Raynal, M.: Regular register: an implementation in a churn prone environment. In: Kutten, S., Žerovnik, J. (eds.) SIROCCO 2009. LNCS, vol. 5869, pp. 15–29. Springer, Heidelberg (2010). https://doi.org/10.1007/978-3-642-11476-2_3
9. Baldoni, R., Platania, M., Querzoni, L., Scipioni, S.: Practical uniform peer sampling under churn. In: Proceedings of the 9th International Symposium on Parallel and Distributed Computing (IPDC), pp. 93–100 (2010)
10. Baldoni, R., Querzoni, L., Virgillito, A., Jiménez-Peris, R., Patiño-Martínez, M.: Dynamic quorums for DHT-based P2P networks. In: Proceedings of the 4th IEEE International Symposium on Network Computing and Applications (NCA), pp. 91–100 (2005)

11. Barrière, L., Flocchini, P., Fraigniaud, P., Santoro, N.: Rendezvous and election of mobile agents: impact of sense of direction. Theory Comput. Syst. **44**(3), 143–162 (2007)
12. Baumann, H., Crescenzi, P., Fraigniaud, P.: Parsimonious flooding in dynamic graphs. Distrib. Comput. **24**(1), 31–44 (2011)
13. Biely, M., Robinson, P., Schmid, U.: Agreement in directed dynamic networks. In: Even, G., Halldórsson, M.M. (eds.) SIROCCO 2012. LNCS, vol. 7355, pp. 73–84. Springer, Heidelberg (2012). https://doi.org/10.1007/978-3-642-31104-8_7
14. Bournat, M., Datta, A.K., Dubois, S.: Self-stabilizing robots in highly dynamic environments. In: Proceedings of the 18th International Symposium on Stabilization, Safety, and Security of Distributed Systems (SSS), pp. 54–69 (2016)
15. Bournat, M., Dubois, S., Petit, F.: Computability of perpetual exploration in highly dynamic rings. In: Proceedings of the 37th IEEE International Conference on Distributed Computing Systems (ICDCS), pp. 794–804 (2017)
16. Bui-Xuan, B., Ferreira, A., Jarry, A.: Computing shortest, fastest, and foremost journeys in dynamic networks. Int. J. Found. Comput. Sci. **14**(2), 267–285 (2003)
17. Casteigts, A., Flocchini, P., Mans, B., Santoro, N.: Shortest, fastest, and foremost broadcast in dynamic networks. Int. J. Found. Comput. Sci. **25**(4), 499–522 (2015)
18. Casteigts, A., Flocchini, P., Quattrociocchi, W., Santoro, N.: Time-varying graphs and dynamic networks. Int. J. Parallel Emergent Distrib. Syst. **27**(5), 387–408 (2012)
19. Clementi, A., Monti, A., Pasquale, F., Silvestri, R.: Information spreading in stationary markovian evolving graphs. IEEE Trans. Parallel Distrib. Syst. **22**(9), 1425–1432 (2011)
20. De Marco, G., Gargano, L., Kranakis, E., Krizanc, D., Pelc, A., Vaccaro, U.: Asynchronous deterministic rendezvous in graphs. Theor. Comput. Sci. **355**, 315–326 (2006)
21. Deng, X., Papadimitriou, C.H.: Exploring an unknown graph. J. Graph Theory **32**(3), 265–297 (1999)
22. Di Luna, G.A., Baldoni, R.: Brief announcement: investigating the cost of anonymity on dynamic networks. In: Proceedings of the 34th Symposium on Principles of Distributed Computing (PODC), pp. 339–341 (2015)
23. Di Luna, G.A., Baldoni, R., Bonomi, S., Chatzigiannakis, I.: Counting in anonymous dynamic networks under worst-case adversary. In: Proceedings of the IEEE 34th International Conference on Distributed Computing Systems (ICDCS), pp. 338–347 (2014)
24. Di Luna, G.A., Dobrev, S., Flocchini, P., Santoro, N.: Live exploration of dynamic rings. In: Proceedings of the 36th IEEE International Conference on Distributed Computing Systems (ICDCS), pp. 570–579 (2016)
25. Di Luna, G.A., Flocchini, P., Pagli, L., Prencipe, G., Santoro, N., Viglietta, G.: Gathering in dynamic rings. In: Proceedings of the 24th International Colloquium Structural Information and Communication Complexity (SIROCCO), pp. 339–355 (2017)
26. Dieudonn, Y., Pelc, A.: Deterministic network exploration by anonymous silent agents with local traffic reports. ACM Trans. Algorithms **11**(2) (2014). Article No. 10
27. Dobrev, S., Flocchini, P., Prencipe, G., Santoro, N.: Multiple agents rendezvous in a ring in spite of a black hole. In: Papatriantafilou, M., Hunel, P. (eds.) OPODIS 2003. LNCS, vol. 3144, pp. 34–46. Springer, Heidelberg (2004). https://doi.org/10.1007/978-3-540-27860-3_6

28. Elor, Y., Bruckstein, A.M.: Uniform multi-agent deployment on a ring. Theor. Comput. Sci. **412**(8), 783–795 (2011)
29. Erlebach, T., Hoffmann, M., Kammer, F.: On temporal graph exploration. In: Halldórsson, M.M., Iwama, K., Kobayashi, N., Speckmann, B. (eds.) ICALP 2015. LNCS, vol. 9134, pp. 444–455. Springer, Heidelberg (2015). https://doi.org/10.1007/978-3-662-47672-7_36
30. Flocchini, P., Kellett, M., Mason, P.C., Santoro, N.: Searching for black holes in subways. Theory Comput. Syst. **50**(1), 158–184 (2012)
31. Flocchini, P., Kranakis, E., Krizanc, D., Santoro, N., Sawchuk, C.: Multiple mobile agent rendezvous in a ring. In: Farach-Colton, M. (ed.) LATIN 2004. LNCS, vol. 2976, pp. 599–608. Springer, Heidelberg (2004). https://doi.org/10.1007/978-3-540-24698-5_62
32. Flocchini, P., Mans, B., Santoro, N.: On the exploration of time-varying networks. Theor. Comput. Sci. **469**, 53–68 (2013)
33. Fraigniaud, P., Ilcinkas, D., Peer, G., Pelc, A., Peleg, D.: Graph exploration by a finite automaton. Theor. Comput. Sci. **345**(2–3), 331–344 (2005)
34. Harary, F., Gupta, G.: Dynamic graph models. Math. Comput. Model. **25**(7), 79–88 (1997)
35. Ilcinkas, D., Klasing, R., Wade, A.M.: Exploration of constantly connected dynamic graphs based on cactuses. In: Halldórsson, M.M. (ed.) SIROCCO 2014. LNCS, vol. 8576, pp. 250–262. Springer, Cham (2014). https://doi.org/10.1007/978-3-319-09620-9_20
36. Ilcinkas, D., Wade, A.M.: On the power of waiting when exploring public transportation systems. In: Fernàndez Anta, A., Lipari, G., Roy, M. (eds.) OPODIS 2011. LNCS, vol. 7109, pp. 451–464. Springer, Heidelberg (2011). https://doi.org/10.1007/978-3-642-25873-2_31
37. Ilcinkas, D., Wade, A.M.: Exploration of the T-interval-connected dynamic graphs: the case of the ring. Theory Comput. Syst. **62**(5), 1144–1160 (2018)
38. Klasing, R., Markou, E., Pelc, A.: Gathering asynchronous oblivious mobile robots in a ring. Theor. Comput. Sci. **390**(1), 27–39 (2008)
39. Kowalski, D., Mosteiro, M.: Polynomial counting in anonymous dynamic networks with applications to anonymous dynamic algebraic computations. In: Proceedings of the 45th International Colloquium on Automata, Languages, and Programming (ICALP), pp. 156:1–156:14 (2018)
40. Kranakis, E., Krizanc, D., Markou, E.: Mobile agent rendezvous in a synchronous torus. In: Correa, J.R., Hevia, A., Kiwi, M. (eds.) LATIN 2006. LNCS, vol. 3887, pp. 653–664. Springer, Heidelberg (2006). https://doi.org/10.1007/11682462_60
41. Kranakis, E., Krizanc, D., Markou, E.: The Mobile Agent Rendezvous Problem in the Ring. Morgan & Claypool, San Rafael (2010)
42. Kranakis, E., Krizanc, D., Santoro, N., Sawchuk, C.: Mobile agent rendezvous problem in the ring. In: Proceedings of the 23rd International Conference on Distributed Computing Systems (ICDCS), pp. 592–599 (2003)
43. Kuhn, F., Lynch, N., Oshman, R.: Distributed computation in dynamic networks. In: Proceedings of the 42nd Symposium on Theory of Computing (STOC), pp. 513–522 (2010)
44. Kuhn, F., Moses, Y., Oshman, R.: Coordinated consensus in dynamic networks. In: Proceedings of the 30th Symposium on Principles of Distributed Computing (PODC), pp. 1–10 (2011)
45. Merritt, M., Taubenfeld, G.: Computing with infinitely many processes. Inf. Comput. **233**, 12–31 (2013)

46. Michail, O.: An introduction to temporal graphs: An algorithmic perspective. Internet Math. **12**(4), 239–280 (2016)
47. Michail, O., Spirakis, P.G.: Traveling salesman problems in temporal graphs. Theor. Comput. Sci. **634**, 1–23 (2016)
48. O'Dell, R., Wattenhofer, R.: Information dissemination in highly dynamic graphs. In: Proceedings of the Joint Workshop on Foundations of Mobile Computing (DIALM-POMC), pp. 104–110 (2005)
49. Panaite, P., Pelc, A.: Exploring unknown undirected graphs. J. Algorithms **33**, 281–295 (1999)
50. Proskurnikov, A.V., Parsegov, S.E.: Problem of uniform deployment on a line segment for second-order agents. Autom. Remote Control **77**(7), 1248–1258 (2016)
51. Santoro, N.: Time to change: on distributed computing in dynamic networks. In: Proceedings of the 19th International Conference on Principles of Distributed Systems (OPODIS), pp. 1–14 (2015)
52. Shibata, M., Kakugawa, H., Masuzawa, T.: Brief announcement: space-efficient uniform deployment of mobile agents in asynchronous unidirectional rings. In: Spirakis, P., Tsigas, P. (eds.) SSS 2017. LNCS, vol. 10616, pp. 489–493. Springer, Cham (2017). https://doi.org/10.1007/978-3-319-69084-1_37
53. Shibata, M., Mega, T., Ooshita, F., Kakugawa, H., Masuzawa, T.: Uniform deployment of mobile agents in asynchronous rings. In: Proceedings of the 35th ACM Symposium on Principles of Distributed Computing (PODC), pp. 415–424 (2016)
54. Stutzbach, D., Rejaie, R.: Understanding churn in peer-to-peer networks. In: Proceedings of the 6th ACM SIGCOMM Conference on Internet Measurement (IMC), pp. 189–202 (2006)
55. Ta-Shma, A., Zwick, U.: Deterministic rendezvous, treasure hunts, and strongly universal exploration sequences. ACM Trans. Algorithms **10**(3), 12:1–12:15 (2014)

# Other Computational Settings

# Geometric Aspects of Robot Navigation: From Individual Robots to Massive Particle Swarms

Sándor P. Fekete(✉)

Department of Computer Science, TU Braunschweig, Braunschweig, Germany
s.fekete@tu-bs.de

**Abstract.** We describe a spectrum of challenges and results related to geometric aspects of robot navigation, advancing from centralized methods for difficult offline problems (such as the *Art Gallery Problem*), to online tasks for many robots (as in *online exploration by a swarm of robots*), locally managing the connectivity and shape of a large swarm (i.e., *cohesive control*), all the way to *controlling massive swarms of particles by global forces.*

## 1 Introduction

Ever since the first work on autonomous robots, algorithmic aspects of robot navigation have played an important role, with new theoretical insights making it possible to expand the practical possibilities, and new real-world challenges motivate algorithmic innovation. Particularly important roles in these developments were played by geometry and by the advances of distributed models and methods. In the following, we provide a number of highlights: The *Art Gallery Problem* described in Sect. 2 deals with localizing stationary viewpoints for mapping all of a given, known region. Section 3 describes how to *explore and triangulate an unknown region* by a swarm of simple robots with only weak sensor and navigation capabilities. Section 4 deals with the challenge of *cohesive control*, i.e., organizing a swarm of simple agents by local interaction, such that connectivity is maintained, even in the presence of external forces and agent failures. The final Sect. 5 describes how to *control a massive swarm of particles by using uniform external forces.*

## 2 Art Gallery Problems

### 2.1 Motivation

Consider a robot platform that can produce high-resolution, virtual environments, based on a limited number of laser scans. For mapping all of a given region in the presence of obstacles, we need to compute an optimal set of scan positions. This is closely related to one of the classic problems of computational geometry: The Art Gallery Problem (AGP) asks for illuminating or surveying all of a given polygonal region $P$ from as few positions ("guards") as possible.


P. Flocchini et al. (Eds.): Distributed Computing by Mobile Entities, LNCS 11340, pp. 587–614, 2019.
https://doi.org/10.1007/978-3-030-11072-7_21

## 2.2 Formal Aspects

Consider a given polygonal region $P$. For any point $g \in P$, the *visibility region* $\mathcal{V}(g)$ is the set of all positions $p \in P$ for which there is a straightline connection between $G$ and $p$ that lies completely in $P$. The Art Gallery Problem (AGP) asks for a minimum cardinality guard set $G \subset P$ that sees all of $P$, i.e., such that $\bigcup_{g \in G} \mathcal{V}(g) = P$. An important distinction arises from the possible positions of guards: a *vertex guard* must be placed at a vertex of $P$, while a *point guard* can be located anywhere in $P$.

## 2.3 Context

As first proven by Chvátal [15] and shown by Fisk [32] in a beautiful and concise proof, $\lfloor \frac{n}{3} \rfloor$ guards are sometimes necessary and always sufficient for guarding a simple polygon $P$ with $n$ vertices. See O'Rourke [64] for an early overview.

Algorithmically, the AGP is NP-hard, even for a simply connected polygonal region $P$ [51]. Eidenbenz et al. [20] showed that for a region with holes, finding an optimal set of vertex guards is at least as hard as the problem *Set Cover*, so there is little hope of achieving a better approximation guarantee than $\Omega(\log n)$. It seems unlikely that this gets any easier when allowing general point guards, as there is no known simple characterization of a discrete candidate set of guard locations. Furthermore, recent work by Abrahamsen et al. [1] proves that the AGP is complete for the existential theory of the reals, implying that it is unlikely to even belong to the class NP.

All this shows the difficulty of the AGP, but it does not rule out methods that combine structural insights with powerful mathematical tools to achieve provably optimal solutions for instances of interesting size.

Computing optimal solutions for general AGP instances is not only relevant from a theoretical point of view, but has also gained in practical importance in the context of modeling, mapping, and surveying complex environments, such as in the fields of architecture, robotics and medicine.

## 2.4 Application Scenario

One particular real-world platform giving rise to instances of the Art Gallery Problem is Irma3D (**I**ntelligent **R**obot for **M**apping **A**pplications in 3D), an autonomous robot; see Fig. 1. Its main sensor is a Riegl VZ-400 laser scanner. A typical 3D laser scan needs 3 min, producing up to 20 million highly precise 3D measurements of the surrounding. A globally consistent scan matching is used to merge the 3D scans to a single scene [11]. Irma3D is built on a Volksbot RT-3 chassis; it uses the Xsens MTi IMU and odometry to sense its own position. See Fig. 2 for a real-world image (Top) and the schematic view with an optimal set of scan points (Bottom). The algorithmic challenge is to plan number and positioning of these scans.

**Fig. 1.** IRMA3D in front of the town hall of Bremen, scanning the city square.

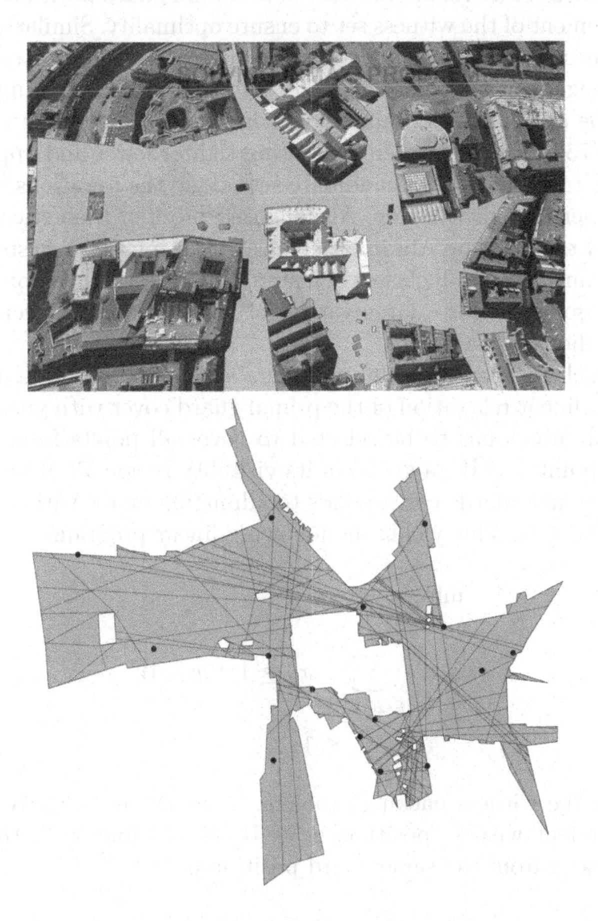

**Fig. 2.** (Top) part of a real-life AGP instance with 15 holes and 332 vertices: a square in the city center of Bremen. (Bottom) a corresponding extracted polygonal region, with an optimal set of scan positions, shown as 15 black dots. The white holes correspond to obstacles formed by buildings in the square.

## 2.5 Algorithmic Insights

Independently, different groups (Braunschweig and Campinas) have combined methods from integer (IP) and linear programming (LP) with non-discrete geometry in order to obtain optimal solutions; first for the discrete case of vertex guards [17], but later also for general point guards [46, 73].

The algorithm in [73] computes lower and upper bounds for the AGP, based on computing finite set cover instances with the help of a state-of-the-art IP solver. To generate a lower bound, a finite set of witness candidates is chosen and a restricted AGP is solved, in which only the witnesses have to be covered. For this, it suffices to extract a finite set of potential guard positions from the visibility arrangement of the witness set to ensure optimality. Similarly, finite sets of potential witness positions for a given finite guard set can be extracted from the visibility arrangement of the guards. This allows it to compute upper and lower bounds for the optimal AGP value by solving discrete set cover instances. The algorithm in [73] iterates between generating tighter lower and upper bounds by refining the witness and guard candidate sets along the iterations. It stops when lower and upper bounds coincide. Although no proof of theoretical convergence is known (and the work by Abrahamsen et al. [1] strongly suggests that no such convergence can exist for all classes of instances), in tests, the approach is able to yield optimal solutions for a large variety of instance classes, even for polygons with up to a thousand vertices.

An approach presented in [46] considers a similar primal-dual scheme, but focuses on the linear relaxation of the primal guard cover with guard set $G$, from which a small subset has to be selected to cover all points from a witness set $W$: For each point $w \in W$, a guard in its visibility region $\mathcal{V}(w)$ must be chosen. Allowing fractional guards corresponds to admitting guard variables $0 \leq x_g \leq 1$ for any guard $g \in G$. This yields the following linear program.

$$\min \sum_{g \in G} x_g \tag{1}$$

$$\text{s.t.} \sum_{g \in G \cap \mathcal{V}(w)} x_g \geq 1 \quad \forall w \in W \tag{2}$$

$$0 \leq x_g \leq 1 \qquad \forall g \in G \tag{3}$$

Its dual is the witness packing problem, in which the objective is to find as many independent witness positions $w \in W$ as possible, such that no two of them can be seen from the same guard position $g \in G$.

$$\max \sum_{w \in W} y_w \tag{4}$$

$$\text{s.t.} \sum_{g \in G \cap \mathcal{V}(w)} y_w \leq 1 \; \forall g \in G \tag{5}$$

$$0 \leq y_w \leq 1 \qquad \forall w \in W \tag{6}$$

Because of strong duality of linear programming, considering these fractional guard and witness values leads to optimal primal and dual solutions with identical objective values. To eliminate fractional solutions, we can apply appropriate cutting planes derived from the set cover polytope, based on specific subsets $J_1 \cap G$ and $J_2 \cap G$.

$$\sum_{g \in J_2 \cap G} 2x_g + \sum_{g \in J_1 \cap G} x_g \geq 2 \tag{7}$$

As it turns out [24], only a small subset of these inequalities matter in the context of AGP instances. Together with a similar primal-dual iteration scheme such as the one in [73], we can find optimal integral solutions for a large range of benchmark instances, including the one shown in the scenario above; see Fig. 3 for a pair of primal and dual solutions.

### 2.6 Extensions

There are various extensions and related questions. In the work described above, the robot has unlimited viewing distance, only the number of scans is to be minimized, and the given region is known in advance. We have also studied this problem for the case of limited viewing distance and an objective function that is a linear combination of the number of scans and distance traveled by the robot. See [29] and the cited related work. We have also studied the context of exploring an unknown region by a robot with discrete vision; see [31].

### 2.7 Acknowledgments

The content of this section is based on the abstract [10] and paraphrases the joint work with Dorit Borrmann, Pedro de Rezende, Cid de Souza, Stephan Friedrichs, Alexander Kröller, Andreas Nüchter and Christiane Schmidt contained in the papers [24,46]. For a visualization, see the video that accompanies [10], to be found at the website http://www.computational-geometry.org/SoCG-videos/socg13video/#Borrmann-etal.

## 3 Online Exploration and Triangulation by a Swarm of Simple Robots

### 3.1 Motivation

Consider a swarm of inexpensive robots without explicit mapping capabilities in an unknown area. Each robot has a limited visibility range, but can move around to get a more complete picture of the environment. Once the region has been fully covered, the robots can also stay around so that we can get live updates. How can we organize this exploration and make sure that we can continue to observe all parts of the environment after they are discovered?

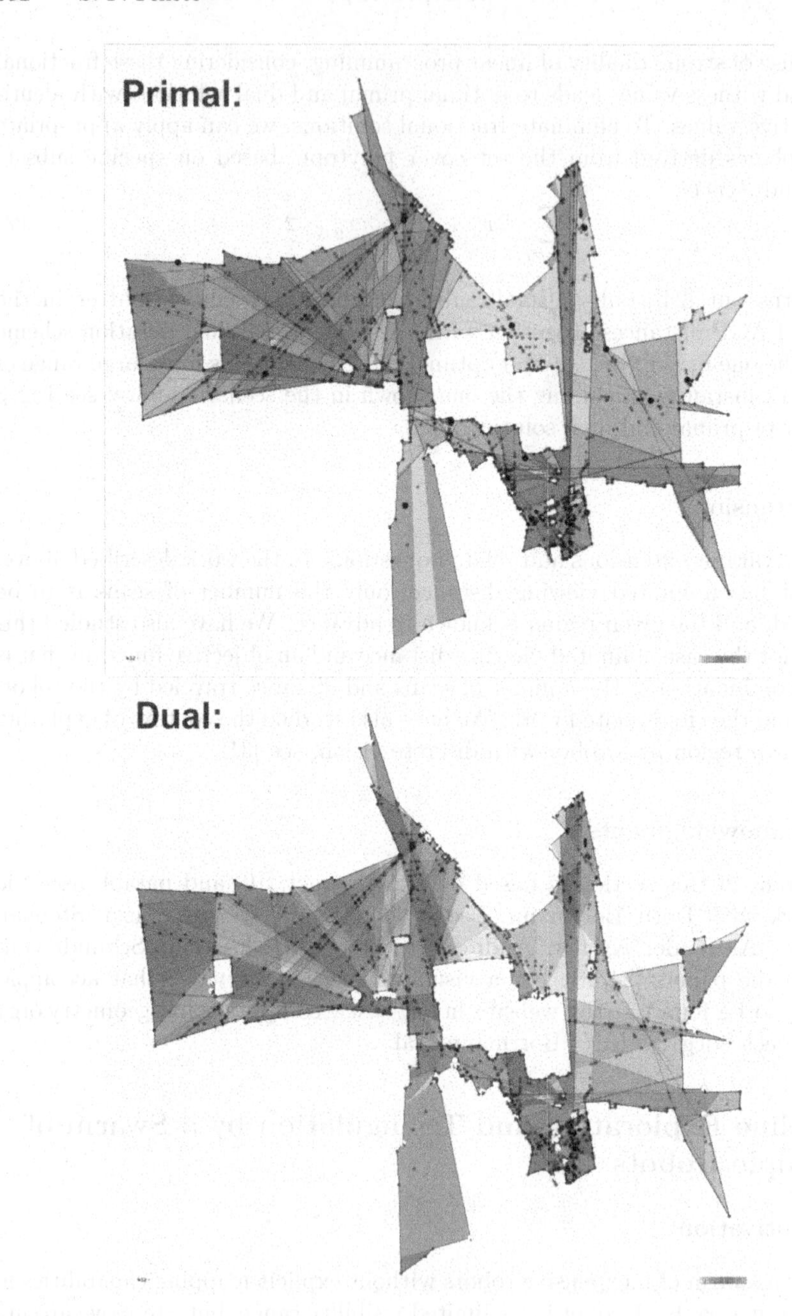


**Fig. 3.** A pair of primal and dual solutions to the fractional linear programming relaxation.

### 3.2 Formal Aspects

We are given a polygonal region $P$, and a point $z \in P$ on the boundary of $P$. In addition, we are given a supply of robots with limited (circular) communication range $r$; for ease of description, we normalize to $r = 1$. Within this range, perception of and communication with other robots is possible. In the Minimum Relay Triangulation Problem (MRTP), the goal is to compute a set $R$ of robot positions within $P$ (with $z \in R$ and $V \subseteq R$ for the vertex set $V$ of $P$), such that there is a (unit) triangulation of $P$ whose vertex set is exactly the set $R$ and whose edges stay within $P$ and have length at most 1. The objective is to minimize the number of robots. In the Maximum Area Triangulation Problem (MATP), the number of available robots is bounded by a number $k$; the goal is to determine a set $R$ of at most $k$ robot positions, with a unit triangulation covering a maximum possible area. For the online versions (OMRTP and OMATP), the polygon $P$ is unknown. Each robot may move through the area, and has to decide on a new location for a triangulation vertex, while still being within reach of the previously placed relays. Once it has stopped, it becomes part of the static triangulation, allowing other relays to extend the exploration.

### 3.3 Context

In recent years, the field of robotics has seen two diverging trends. One has been to achieve progress by increasing the capabilities of individual robots, keeping the cost of state-of-the-art machines relatively high. An opposite direction has been to develop simpler and cheaper platforms, at the expense of reducing the capabilities per robot. The latter raises new challenges for developing new principles and algorithms, such as coordinating many robots with limited capabilities into a swarm that can carry out difficult tasks, such as exploration, surveillance, and guidance.

### 3.4 Application Scenario

A real-life example of an advanced, low-cost, swarm robot design with limited sensor capabilities is the r-one [61], shown in Fig. 4. Its estimated unit cost is about US $250. Measuring only 11 cm in diameter, it has a 32-bit ARM-based microcontroller, running at 50 MHz with no floating point unit. The local infrared (IR) communication system is used for inter-robot communication and localization. Each robot has eight IR transmitters and eight receivers. The transmitters broadcast in unison and emit a radially uniform energy pattern. The robot's eight IR receivers are radially spaced to produce 16 distinct detection regions (shown in Fig. 4 (Right)). By monitoring the overlapping regions, the bearing of neighbors can be estimated to within $\approx\pi/8$. Thus, it has limited capabilities for measurement, which is intertwined with local communication. The IR receivers have a maximum bit rate of 1250 bits per second. Each robot transmits $(\Delta + 1)$ 4-byte messages during each round, one being a system announce message, the others containing the bearing measurements to that robot's neighbors.

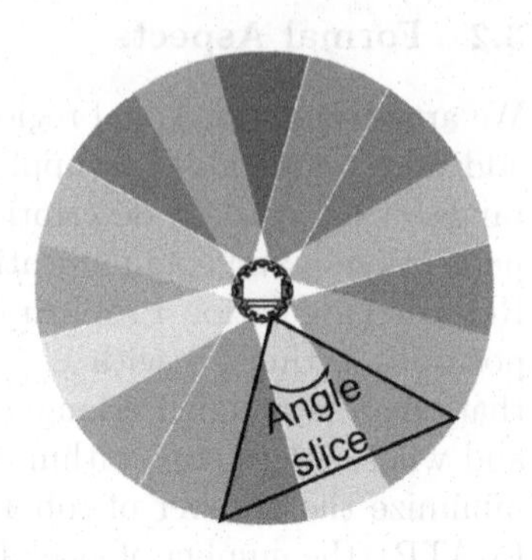


**Fig. 4.** (Left) the r-one for multi-robot research, designed by the MRSL group at Rice University. (Right) IR receiver detection regions. Each receiver detects an overlapping 68°, allowing to determine angles within about 22.5°.

The system supports a maximum of $\Delta = 10$. For more on experimental work on coordination and navigation of r-ones, see [61,62].

The algorithmic challenge is to exploit the capabilities of a swarm to overcome the limitations of the individual robots, and achieve overall behavior with provable performance guarantees that are rooted in solid algorithmic theory.

### 3.5 Algorithmic Insights

The problems MRTP and MATP were introduced in [26]; the currently best results for the online versions OMRTP and OMATP were presented in [25,68]. Both problems share their decision problem, which is known to be NP-hard. For the OMRTP, there is a lower bound of 6/5 on the competitive factor of any deterministic strategy, as well as a 3-competitive algorithm for general polygons. This strategy is shown in Fig. 5: We place robots at unit intervals along the boundary (green) and fill the interior with a regular triangular grid (blue). The space between the two is patched together using a third class (red). One can prove that the size of each of the three classes is bounded by the number of robots in an optimal solution. For polyominoes, algorithms with better competitive factors exist [30].

On the other hand, the OMATP does not admit a deterministic strategy with a constant competitive factor, if the polygon may have small corridors. If these can be excluded, greedy strategies perform well [30].

These strategies have been used on the real robots described above; see Fig. 6 for a snapshot.

### 3.6 Extensions

Once a well-formed triangulation (with lower bounds on minimum edge length and minimum internal angle) is established, it can be employed for a number

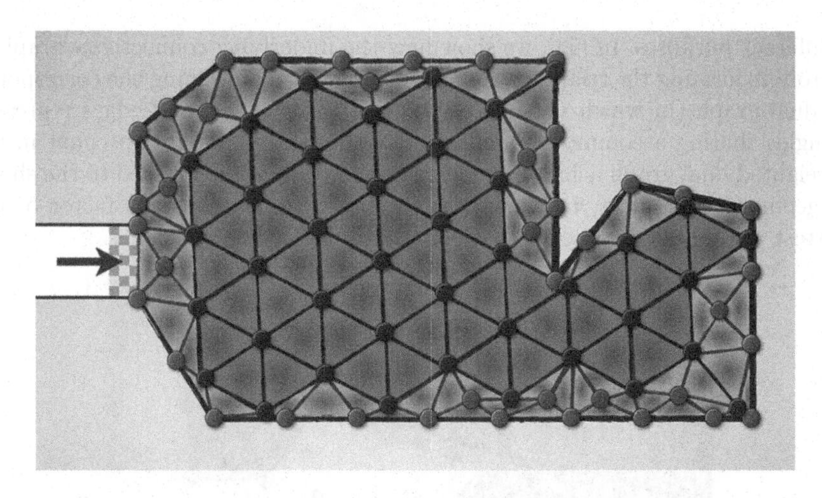

**Fig. 5.** The 3-competitive strategy for the OMRTP, consisting of three sets of robots: green robots are placed along the boundary at vertices and at unit distance along edges; blue robots fill the interior by a grid; red robots complete the triangulation by connecting boundary interior robots. (Color figure online)

**Fig. 6.** A swarm of r-ones executing the online algorithm for the OMATP in the real world.

of different purposes. In [53], we show how the underlying connectivity graph of the robots forming the triangulation can be used for maintaining the corresponding dual graph (in which vertices correspond to triangles, and edges represent triangles sharing a common boundary), and how a minimum hop count in the unweighted dual graph achieves a constant stretch factor compared to the shortest geometric distances, i.e., manages to stay within a constant factor of the shortest achievable distance with global information; see Fig. 7.

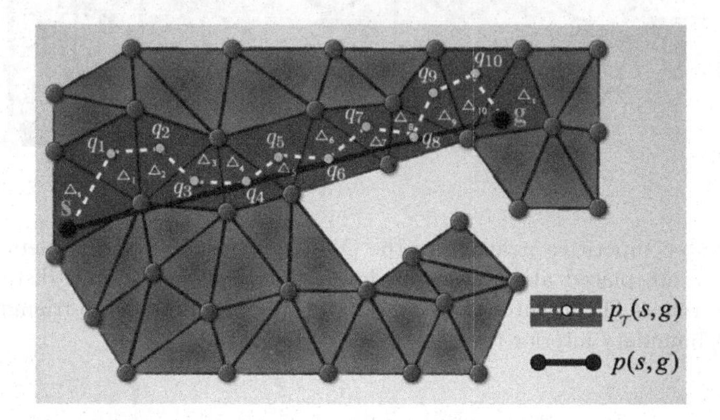


**Fig. 7.** Using a dual path for routing in a triangulated environment: a shortest path (shown in red) is approximated by a minimum-hop path (shown in yellow), achieving constant stretch. (Color figure online)

Another application of a triangulation is to use it for surveying the underlying region by additional, mobile robots, again based on the dual graph of the stationary triangulation. In [2,54,57], we discuss various aspects of local policies for patrolling the vertices of such a graph.

### 3.7 Acknowledgments

The content of this section is based on the abstract [28] and paraphrases the joint work with Aaron Becker, Tom Kamphans, Alexander Kröller, Seoung Kyou Lee, James McLurkin, Joe Mitchell, Christiane Schmidt, described in the papers [25,52,54]. For a visualization, see the video http://www.computational-geometry.org/SoCG-videos/socg13video/#Becker-etal that accompanies [28]. See the journal paper [54] for full technical details of the robotics side and its extensions.

## 4 Distributed Cohesive Control

### 4.1 Motivation

Consider a swarm of robots that needs to remain connected. There is no central control and no knowledge of the overall environment. This environment is hostile: The swarm is being pulled apart by external forces, stretching it into a number of different directions, so it is in danger of breaking up. Individual robots are weak, with limited sensing, limited communication, and limited connectivity; even worse, each robot's expected lifetime is limited by random, permanent failures, which may destroy connectedness and functioning of the swarm as a whole. How can we achieve coordinated dynamic swarm behavior without centralized coordination? How can we employ each robot as much as possible, without depending on it if it fails? How can we balance overall flexibility and robustness to deal with the hostile environment?

The challenge is to develop local self-stabilizing mechanisms that allow the swarm to stay locally well connected (forcing swarm members to stay close to each other), even when it is being pulled apart by several distant and mobile sites (forcing swarm members to spread out).

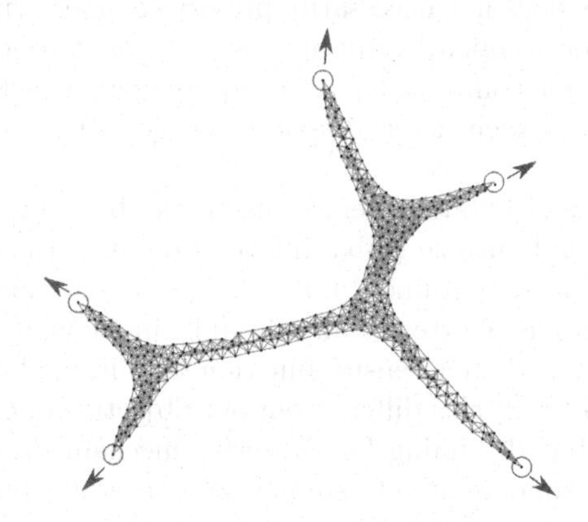

**Fig. 8.** A robust robot swarm emulating a Steiner tree between five diverging leader robots.

## 4.2 Formal Aspects

We consider a finite set of robots $\mathcal{R}$ with an externally controlled subset of *leader robots* $\mathcal{L} \subsetneq \mathcal{R}, |\mathcal{L}| \ll |\mathcal{R}|$. We want the remaining robots $\mathcal{R} \setminus \mathcal{L}$ to maintain a dynamic and robust network that keeps the swarm connected, even in the presence of random robot failures and arbitrary leader movements. Thus, the overall shape of the swarm should form a "thick" Steiner tree among the leaders with the robots $\mathcal{R} \setminus \mathcal{L}$ evenly distributed along the edges, as shown in Fig. 8.

Robots have the shape of circles; two of them are connected when within a maximum distance and with an unobstructed line of sight. Robots know the relative positions and orientations of their neighbors and can communicate asynchronously. Each robot has a unique ID; leader IDs are known by all others. Robot's translations and rotations are limited in velocity and acceleration. Communication is possible by broadcasting to immediate neighbors.

The perception of all robots is local; however, due to the known position and orientation difference, each robot can transform vectors of its neighbors to its own coordinate system. We avoid multi-hop transformations to keep errors small.

### 4.3 Context

One of the earliest works on flocking is Reynold's pioneering work [66]. In recent years, a considerable number of aspects and objectives have extended this perspective. We highlight only some of the ensuing papers, showing how they differ from our perspective.

A basic component of flocking is volumetric control, as it was presented by Spears [71]: Robots use local potential field controllers (with attractive and repulsive forces) for constructing a regular lattice with a corresponding base density [40,63]. This does not necessarily preserve connectivity [3,36,71]. While the latter can be side-stepped by simply assuming that robots are always connected [70], we aim for connectivity as a requirement, which is vital in a fully distributed setting in which deterministic recovery from disconnectedness may be impossible.

Some of the ideas of Olfati-Saber [63] form the basis of our work; however, in that model, robots do utilize global information, e.g., the position of a guide robot in a shared coordinate frame [14,49,50,63] or environmental potential [34]. Instead of the potentials, Cortes et al. [16] and Lindhé et al. [56] used Voronoi tessellation. This is based on a density function, requiring global information for covering a region. Overall, this differs from our objective of developing methods that are fully distributed, aiming for collective mechanisms for complex group behavior that go beyond relatively simple objectives [9], but also for systems that are robust against partial hardware failures [43].

The final property is *cohesiveness* of the overall swarm: all robots should maintain a unified state, such as desired distance or orientation; see [63] for a formal definition. As described in [60], detecting and maintaining a swarm boundary is of particular importance for maintaining swarm cohesiveness and connectedness. This is based on and related to work in the field of wireless sensor networks (WSNs), which has considered many geometric settings in which a large swarm of stationary nodes is faced with the task of achieving a large-scale overall goal, while the individual components can only operate locally, based on limited individual capabilities and information; refer to Fekete et al. [27,47] for a detailed description. In addition to the work on swarm robotics described above, there is a large body of theoretical work on geometric swarm behavior; here we only mention Chazelle [13] for flocking behavior, and Fekete et al. [27,47] for geometric algorithms for static sensor networks, including distributed boundary detection.

Beyond the involved properties and paradigm, the overall goal for the swarm can also be described as a distributed optimization problem: Maintain a generalized Steiner tree with limited edge lengths that connects a moving set of terminals. To the best of our knowledge, only Hamann and Wörn [35] have explicitly considered the construction of Steiner trees by a robot swarm. For static terminals, they start with an exploratory network; as soon as all terminals are connected, only best paths are kept and locally optimized.

Even in a centralized and static setting with full information, we must deal with a generalization of the well-known NP-hard problem of finding a good Steiner tree [33]; see the books by Hwang et al. [41] and Prömel and Steger [65] for further introduction. More specifically, we are faced with the relay placement problem: The input is a set of sensors and a number $r \geq 1$, the communication range of a relay. The objective is to place a minimum number of relays so that between every pair of sensors is connected by a path through sensors and/or relays. The best known theoretical performance bound for this NP-hard problem was given by Efrat et al. [19], who presented a 3.11-approximation algorithm; they also showed a worst-case lower bound of 3 for a large class of approximation algorithms. For a fixed number of available relays, this turns into our problem of maximizing the achievable networks size, with matching approximation factor.

### 4.4 Algorithmic Insights

A key insight is that achieving complex overall behavior can be based not only on local interaction that resembled physical forces, but also on principles of distributed algorithms that build more complex structures. To this end, we have developed a number of powerful local mechanisms for maintaining a dynamic swarm of robots with limited capabilities and information, in the presence of external forces and permanent node failures. These mechanisms consist of a set of local, self-stabilizing, *continuous* algorithms that together produce a generalization of a Euclidean Steiner tree, maintain a dynamic and robust network between leader robots. At any stage, the resulting overall shape achieves a good compromise between local thickness, global connectivity, and flexibility to further continuous motion of the terminals, adopting the directions of multiple leaders, while preserving a uniform thickness along the edges of the Steiner tree. The resulting swarm behavior scales well, is robust against node failures, and performs close to the best known approximation bound for a corresponding centralized static optimization problem.

We first sketch the base behavior of the robots, inducing an almost convex swarm shape. This is subsequently improved by leader forces, and stability improvement and thickness contraction.

#### 4.4.1 Base Behavior

Our base behavior consists of three components that result in a swarm shape of a droplet. (i) The *flocking algorithm* of Olfati-Saber [63] considers regular distribution and movement consensus. The algorithm is a stateless equation based on potential fields and is proven to converge. It uses three rules: Attraction to neighbors, repulsion from too close neighbors, and adaption to the velocity of neighbors. We slightly modified the algorithm for better response to additional forces. (ii) An extended version of the *boundary detection* algorithm of McLurkin and Demaine [60], which determines if a robot lies on the boundary and also identifies small holes by using the average angle. (iii) The *boundary tension* of Lee and McLurkin [55], which straightens and minimizes the boundary of the swarm. This is done by simply pushing boundary robots to the middle of its two boundary neighbors.

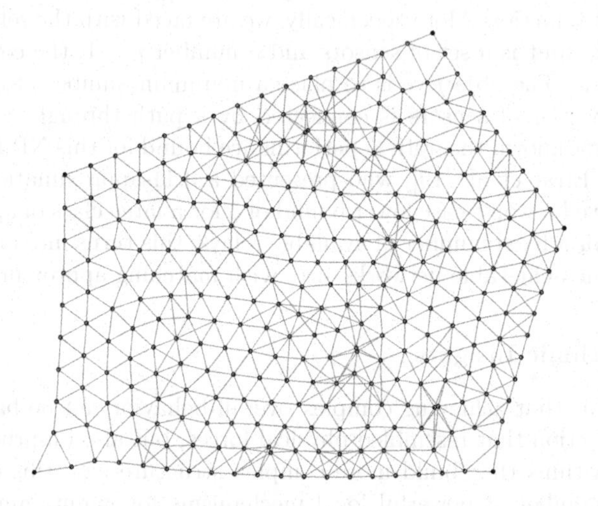

**Fig. 9.** A swarm configuration in which a purely physics-based mechanism lead to disconnection.

The base behavior without any other forces results in at most convex shapes before losing connectivity. Figure 9 shows a situation in which the swarm is about to lose connectivity. For stronger control and more variable shapes, leader forces are introduced.

#### 4.4.2 Leader Forces

A single leader constitutes the simplest form of swarm control. In this case the swarm motion is determined by the leader's velocity. With multiple (possibly antagonistic) leaders, the swarm is not just steered, but may be stretched to the limit until connectivity is lost. To this end, each robot needs to find an appropriate balance between the influence of different leaders. See top of Fig. 10 for an illustration. We therefore combine both methods by a smooth transition between velocity matching close to the leaders and leader pursuit when further away; see the bottom of Fig. 10.

#### 4.4.3 Stability Improvement and Thickness

Near Steiner points, connections along concave swarm boundaries may be stretched by boundary forces. When the involved edges approach the upper bound for communication, connections may be disrupted, to the point where the

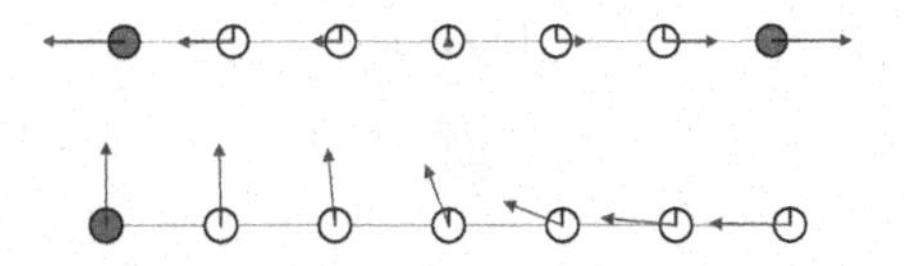

**Fig. 10.** (Top) A one-dimensional scenario with two leaders (red) moving in opposite directions. (Bottom) With increasing distance to the leader, the effect shifts from velocity matching to leader pursuit. (Color figure online)

swarm loses connectivity. By adding a thickness-dependent compression force, we reduce neighbor distances without influencing the Steiner-tree shape of the swarm; in effect, this works similar to compression stockings. Algorithmically, the involved mechanisms resemble methods that have been studied in the context of sensor networks, such as local methods for boundary detection and hop distance from the boundary. This gives rise to notions such as the hop circle of radius $h$ with robot $r$ as circle center: This is the set of all robots with a hop count $\leq h$ to $c$; only robots with hop distance at least $h$ may be on the boundary, so a hop circle of maximal radius around a given robot gives an indication of the local thickness in its neighborhood. (For an example, see Fig. 11.)

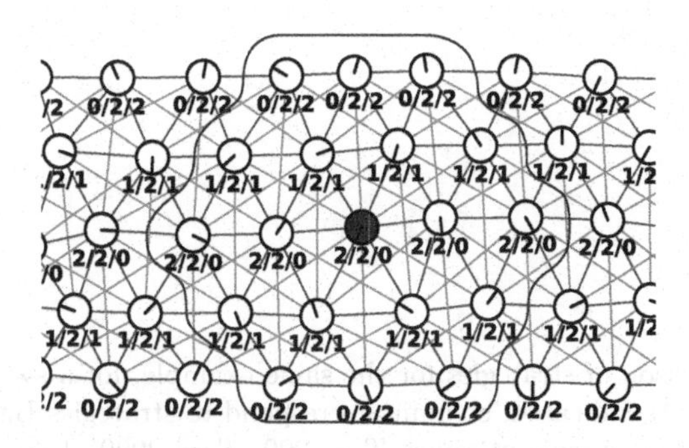


**Fig. 11.** Thickness determination for a limb part of a swarm. The indicated triples of numbers at each node $r$ correspond to $(b(r)/t(r)/h(r))$, where $b(r)$ is the hop count from the boundary, $t(r)$ is the local thickness, and $h(r)$ is the circle center distance. A largest *hop circle* is marked in blue. (Color figure online)

This allows local approaches to keep track of and maintain local thickness and connectivity, even in the presence of external forces and robot failures. See our paper [48] for more technical details.

### 4.5 Extensions

There are numerous possible extensions, most notably for dealing with a cohesive swarm in the presence of obstacles. At this time, these are still under development.

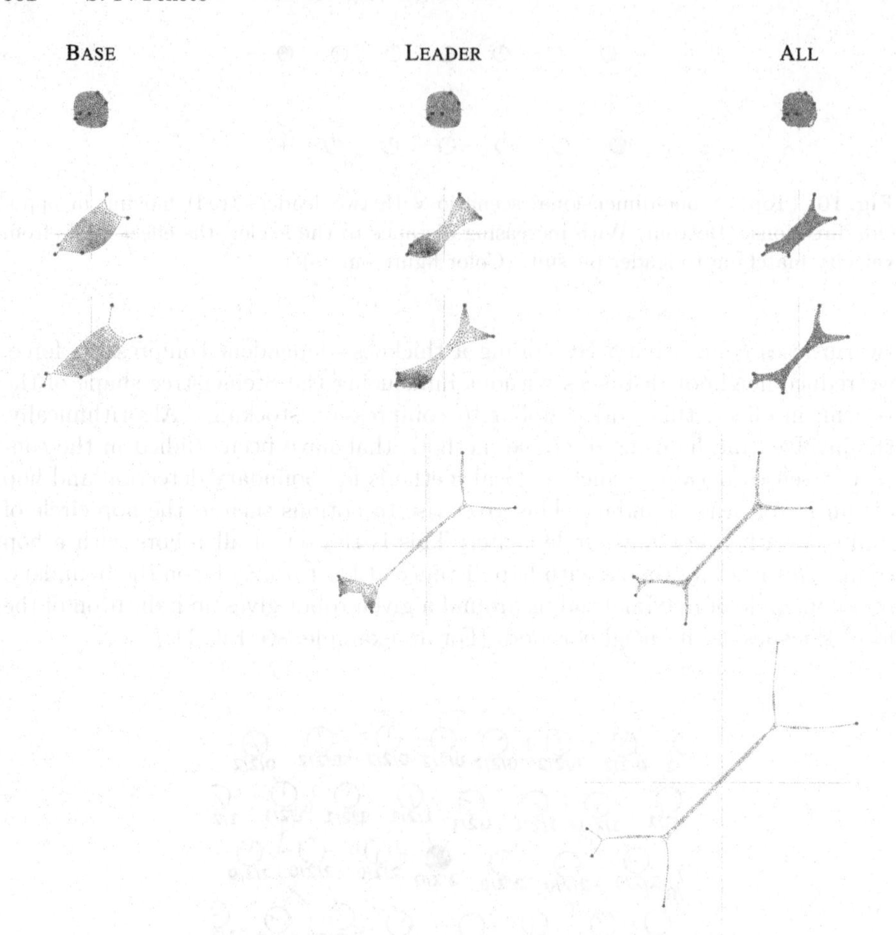


**Fig. 12.** A comparison of strategies for the same example, for a swarm with $n = 400$ and failure rate 0. As indicated, columns correspond to strategies BASE, LEADER, and ALL. Rows show the swarms at times $T = 200$, $T = 2000$, $T = 3000$, $T = 7600$, $T = 12,000$, with 60 steps per simulated second. When a swarm is no longer shown, it has become disconnected right after the previous time step.

### 4.6 Acknowledgments

The content of this section is based on the abstract [22] and paraphrases the joint work with Maximilian Ernestus, Michael Hemmer and Dominik Krupke contained in the paper [48] and the thesis [23] of Maximilian Ernestus and Dominik Krupke.

# 5 Controlling Swarms by Global Forces

## 5.1 Motivation

One of the exciting new directions of robotics is the design and development of micro- and nanorobot systems, with the goal of letting a huge population of robots perform complex operations in a complicated environment. Due to scaling issues, individual control of the involved robots becomes physically impossible: While energy storage capacity drops with the third power of robot size, medium resistance decreases much slower. A possible answer lies in applying a global, external force to all particles in the swarm. This is what many current micro- and nanorobot systems with many robots do: The whole swarm is steered and directed by an external force that acts as a common control signal. These common control signals include global magnetic or electric fields, chemical gradients, and turning a light source on and off.

## 5.2 Formal Aspects

We consider a two-dimensional grid world, with some cells occupied and others free. Initially, the planar square grid is filled with some unit-square particles (each occupying a cell of the grid) and some fixed unit-square blocks. All particles are commanded in unison: a valid command is "Go Up" ($u$),"Go Right" ($r$),"Go Down" ($d$), or "Go Left" ($l$). All particles move in the commanded direction until they hit an obstacle or another particle. A representative command sequence is $\langle u, r, d, l, d, r, u, \ldots \rangle$. We call these global commands *force-field moves.* We assume that we can bound the minimum particle speed and that we can guarantee that all particles have moved to their maximum extent.

## 5.3 Application Scenario

Becker et al. [8] demonstrate how to apply a magnetic field to simultaneously move cells containing iron particles in a specific direction within a fabricated workspace; see Fig. 13a. Other recent examples include using the global magnetic field from an MRI to guide magneto-tactic bacteria through a vascular network to deliver payloads at specific locations [12], and using electromagnets to steer a magneto-tactic bacterium through a micro-fabricated maze [45]; however, this still involves only individual particles at a time, not the parallel motion of a whole, massive swarm. How can we manipulate the overall swarm with coarse global control, such that individual particles arrive at multiple different destinations in a (known) complex vascular network such as the one in Fig. 13b?

## 5.4 Context

The problem resembles the logic puzzle *Tilt* [72], and dexterity ball-in-a-maze puzzles such as *Pigs in Clover* and *Labyrinth*, which involve tilting a board to cause all mobile pieces to roll or slide in a desired direction. Problems of this type

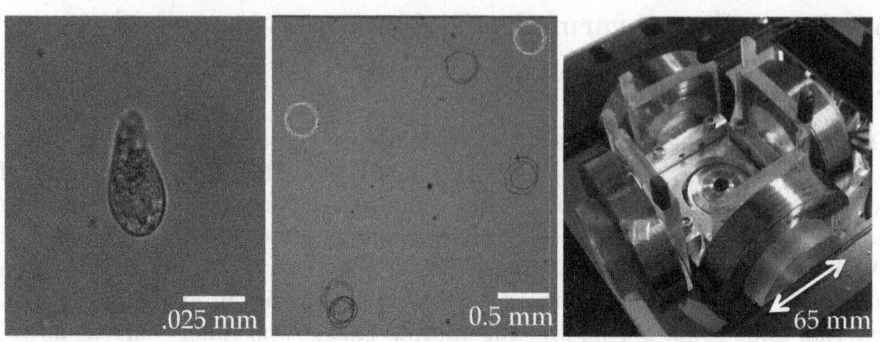


(a) (Left) After feeding iron particles to ciliate eukaryon (*Tetrahymena pyriformis*) and magnetizing the particles with a permanent magnet, the cells can be turned by changing the orientation of an external magnetic field (see colored paths in the center image). (Right) Using two orthogonal Helmholz electromagnets, Becker et al. [8] demonstrated steering many living magnetized *T. pyriformis* cells. All cells are steered by the same global field.

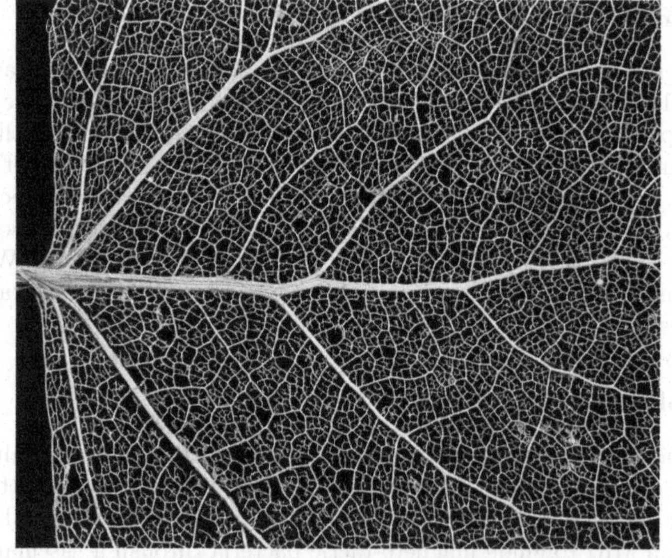

(b) Biological vascular network (cottonwood leaf). (Photo: Royce Bair/Flickr/Getty Images.) Given such a network along with initial and goal positions of $N$ particles, is it possible to bring each particle to its goal position using a global control signal? Note that this arrangement is *not* a tree, but a graph structure with many cycles. MATLAB code for driving $N$ particles through this network is available at `http://www.mathworks.com/matlabcentral/fileexchange/42892`.

**Fig. 13.** (Top) State of the art in controlling small objects by force fields. (Bottom) A complex vascular network, forming a typical environment for the parallel navigation of small objects. This section investigates parallel navigation in discretized 2D environments. (Color figure online)

are also similar to sliding-block puzzles with fixed obstacles [18,37–39], except that all particles receive the same control inputs, as in the Tilt puzzle. Another connection is to *Randolph's Ricochet Robots* [21], a game that allows individual and independent control of the involved particles.

In the real world, driving ferromagnetic particles with a magnetic resonance imaging (MRI) scanner gives examples of this challenge, from nano- to micro-scales; see [74].

## 5.5 Algorithmic Insights

Clearly, having only one global signal that uniformly affects all robots at once poses a strong restriction on the ability of the swarm to perform complex operations. The only hope for breaking symmetry is to use interactions between the robot swarm and obstacles in the environment. The key challenge is to establish if interactions with obstacles are sufficient to perform complex operations, ideally by analyzing the complexity of possible logical operations.

It is important to note that there are two fundementally different classes of algorithmic problems, which we denote by *External Computation* and *Internal Computation.*

### 5.5.1 External Computation

Considering the particle swarm as input for a given algorithmic problem, we are faced with a number of questions that need to be resolved by a computing device "outside" of the particle system, such as the following.

1. Given a map of an environment, such as the vascular network shown in Fig. 13b, along with initial and goal positions for each particle, does there *exist* a sequence of inputs that will bring each particle to its goal position?
2. Given a map of an environment, such as the vascular network shown in Fig. 13b, along with initial and goal positions for each particle, what is the *shortest* sequence of moves that will bring each particle to its goal position?
3. Given initial and goal positions for each particle in a swarm, how can we *design* a set of obstacles and a sequence of moves, such that each particle reaches its goal position?

Deliberate use of existing stationary obstacles leads to a wide range of possible particle configurations. In our work [4–6,69], we address all these issues. For the first two questions, we show that they may lead to computationally difficult situations. We also develop several positive results for the third question. The underlying idea is to construct artificial obstacles (such as walls) that allow arbitrary rearrangements of a given two-dimensional particle swarm.

**Theorem 1.** *Given a specified goal location and an initial configuration of movable particles and fixed obstacles, it is NP-hard to decide if a move sequence exists that ends with some particle at the goal location.*

The proof relies on a reduction from 3SAT. Suppose we are given $n$ Boolean variables $x_1, x_2, \ldots, x_n$, and $m$ disjunctive clauses $C_j = U_j \vee V_j \vee W_j$, where each literal $U_j, V_j, W_j$ is of the form $x_i$ or $\neg x_i$. We construct a problem instance that has a solution if and only if all clauses can be satisfied by a truth assignment to the variables. This instance is composed of *variable gadgets* for setting individual variables TRUE or FALSE, *clause gadgets* that construct the logical OR of groupings of three variables, and a *check gadget* that constructs the logical AND of all the clauses. A particle is only delivered to the goal location if the variables have been set in such a way that the formula evaluates to TRUE. See Fig. 14 for an overview of the whole construction.

On the positive side, we can show that for a given labeled arrangement of particles, arbitrary permutations can be achieved with appropriate sets of obstacles. To this end, we consider a 2D array of particles, as shown in Fig. 15. For an $a_r \times a_c$ matrix $A$ and a $b_r \times b_c$ matrix $B$, of equal total size $N = a_r a_c = b_r b_c$, a *matrix permutation* assigns each element in $A$ a unique position in $B$.

**Theorem 2.** *Let $A$ and $B$ be matrices with dimensions as above. Any matrix permutation that transforms $A$ into $B$ can be executed by a set of obstacles in just four moves. For $N$ particles, the constructed arrangement of obstacles requires $(3N+1)^2$ space and $4N+1$ obstacles. If particles move with a speed of $v$, the required time for those four moves is $12N/v$.*

This can be employed to realize larger sets of permutations all at once, as shown in Fig. 16.

The previous construction is efficient with respect to the number of required moves, at the expense of a possibly higher number of obstacles. By allowing a larger number of moves, we can limit the number of obstacles for achieving *any* permutation, as shown in Fig. 17.

**Theorem 3.** *We can construct a set of $O(N)$ obstacles such that any $a_r \times a_c$ arrangement of $N$ particles can be rearranged into any other $a_r \times a_c$ arrangement $\pi$ of the same particles, using at most $O(N^2)$ force-field moves.*

*Proof.* See Fig. 17. Use Theorem 2 to build two sets of obstacles, one each for $p$ and $q$, such that $p$ is realized by the sequence $\langle u, r, d, \ell \rangle$ (clockwise) and $q$ is realized by $\langle r, u, \ell, d \rangle$ (counterclockwise). Then we use the appropriate sequence for generating $\pi$ in $O(N^2)$ moves.

On the other hand, minimizing the number of moves for achieving a desired goal configuration for *all* particles turns out to be PSPACE-complete.

**Theorem 4.** *Given an initial configuration of (labeled) movable particles and fixed obstacles, it is* PSPACE*-complete to compute a shortest sequence of force-field moves to achieve another (labeled) configuration.*

The proof is largely based on a complexity result by Jerrum [42], who considered the following problem: Given a permutation group specified by a set of generators and a single target permutation $\pi$, which is a member of the group,

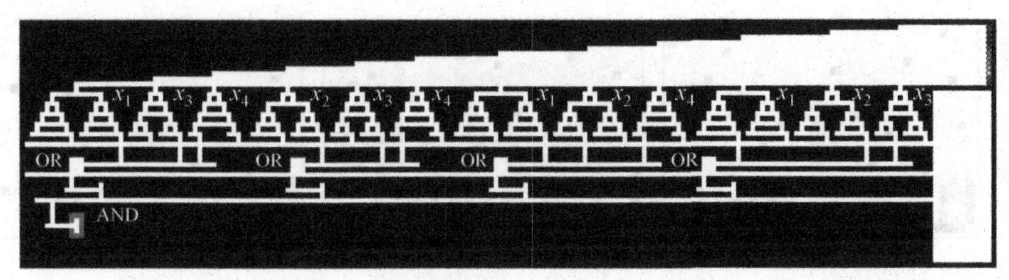


(a) Initial state with particles (colored) on the upper right. The objective is to move one particle into the grey target rectangle at lower left.

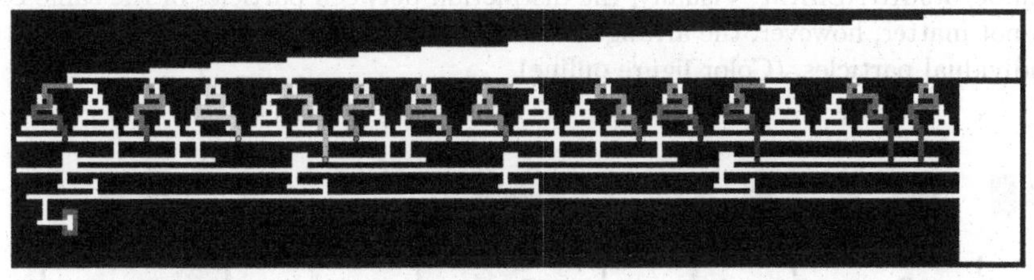

(b) Setting variables to (False, True, False, True) does not satisfy this 3SAT instance.

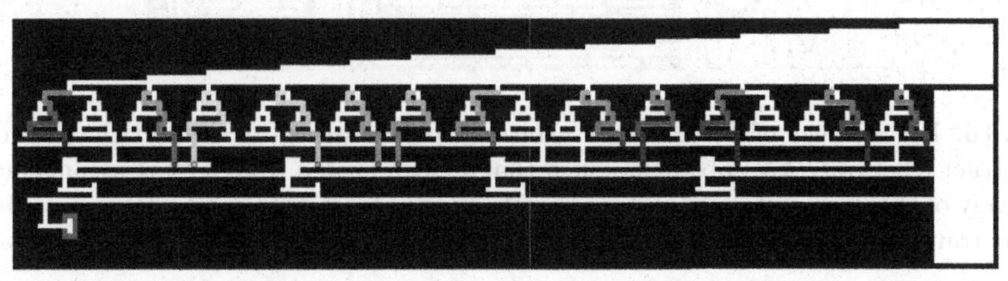

(c) Setting the variables (True, False, False, True) satisfies this 3SAT instance.

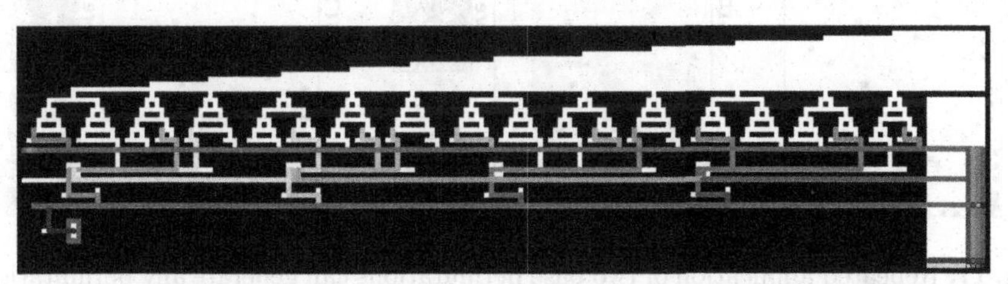

(d) Successful outcome. (True, False, False, True) moves a single particle into the target region.

**Fig. 14.** Combining twelve variable gadgets, four 3-input OR gadgets, and a 4-input AND gadget to realize the 3SAT expression $(\neg x_1 \vee \neg x_3 \vee x_4) \wedge (\neg x_2 \vee \neg x_3 \vee x_4) \wedge (\neg x_1 \vee x_2 \vee x_4) \wedge (x_1 \vee \neg x_2 \vee x_3)$. (Color figure online)

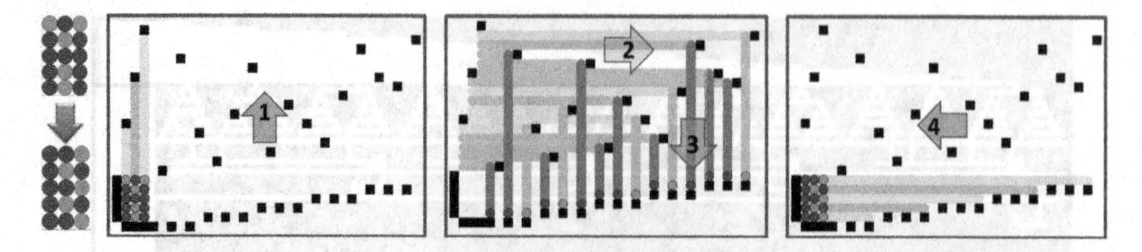


**Fig. 15.** In this image for $N = 15$, black cells are obstacles, white cells are free, and colored discs are individual particles. The world has been designed to permute the particles between 'A' into 'B' every four steps: $\langle u, r, d, \ell \rangle$. See the video at http://youtu.be/3tJdRrNShXM. Visually, the distinction between particles of the same color does not matter; however, the arrangement of obstacles induces a specific permutation of individual particles. (Color figure online)

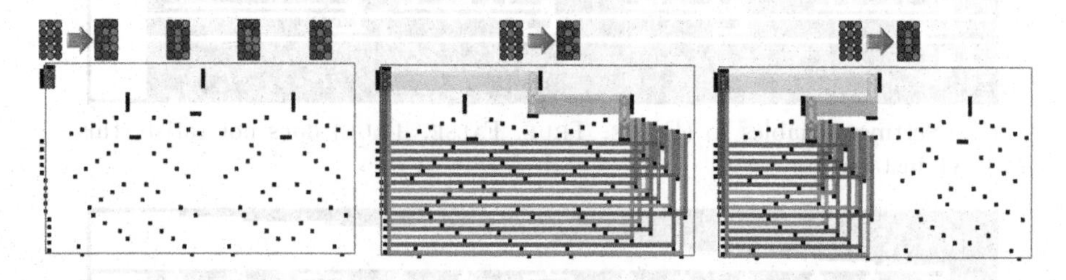

**Fig. 16.** For any set of $k$ fixed, but arbitrary permutations of $N$ particles, we can construct a set of $O(kN)$ obstacles, such that we can switch from a start arrangement into any of the $k$ permutations using at most $O(\log k)$ force-field moves. Here $k = 4$ and 'A' is transformed into 'B', 'C', 'D', or 'E' in eight moves: $\langle r, d, (r/\ell), d, (r/\ell), d, \ell, u \rangle$.

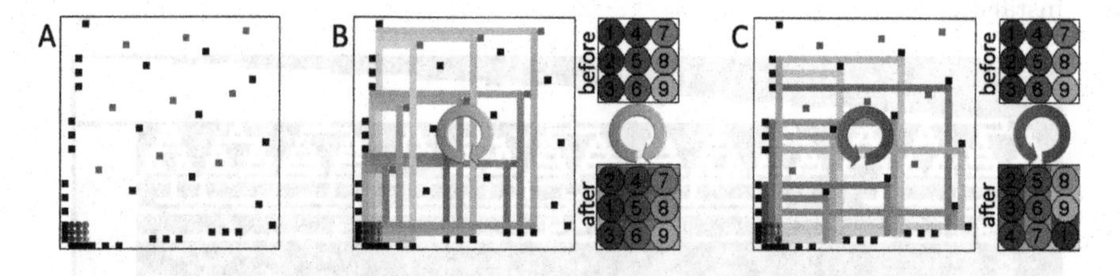


**Fig. 17.** Repeated application of two base permutations can generate any permutation, when used in a manner similar to BUBBLE SORT. The obstacles in (A) generate the base permutation $p = (1, 2)$ in the clockwise direction $\langle u, r, d, \ell \rangle$ (B) and $q = (1, 2, \ldots, N)$ in the counterclockwise direction $\langle r, u, \ell, d \rangle$ (C).

what is the shortest expression for the target permutation in terms of the generator? This problem was shown to be PSPACE-complete in [42], even when the generator set consists of only two permutations $\pi_1$ and $\pi_2$. Combining this with the idea of our construction of Theorem 3 yields the claimed result: Use two sets of obstacles for realizing $\pi_1$ by a sequence of four clockwise moves and $\pi_2$ by a sequence of four counterclockwise moves; then a shortest sequence of force-field moves for achieving a desired target permutation $\pi_3$ corresponds to a minimum generation of $\pi_3$ by $\pi_1$ and $\pi_2$.

### 5.5.2 Internal Computation

Considering the particle swarm as a complex system that can be reconfigured in various ways, we are faced with issues of the computational power of the swarm itself (as opposed to that of an external device), such as the following.

1. Can the complexity of particle interaction be exploited to model logical operations?
2. Are there limits to the computational power of the particle swarm?
3. How can we achieve computational universality with particle computation?

In [4,5,69], we give precise answers to all of these questions. In particular, we show that the logical operations AND, NAND, NOR, and OR can be implemented in our model using dual-rail logic. Using terminology from electrical engineering, we call these components that calculate logical operations *gates*. We establish a fundamental limitation for particle interactions: We cannot duplicate the output of a chain of gates without also duplicating the chain of gates. This means that a so-called FAN-OUT gate cannot be generated. We resolve this missing component with the help of $2 \times 1$ particles, which can be used to create FAN-OUT gates that produce multiple copies of the inputs without needing duplicate gates, as shown in Fig. 18 for a physical prototype. Using these FAN-OUT gates, we provide rules for replicating arbitrary digital circuits, allowing us to establish the full range of computational universality as presented by complex digital circuits.

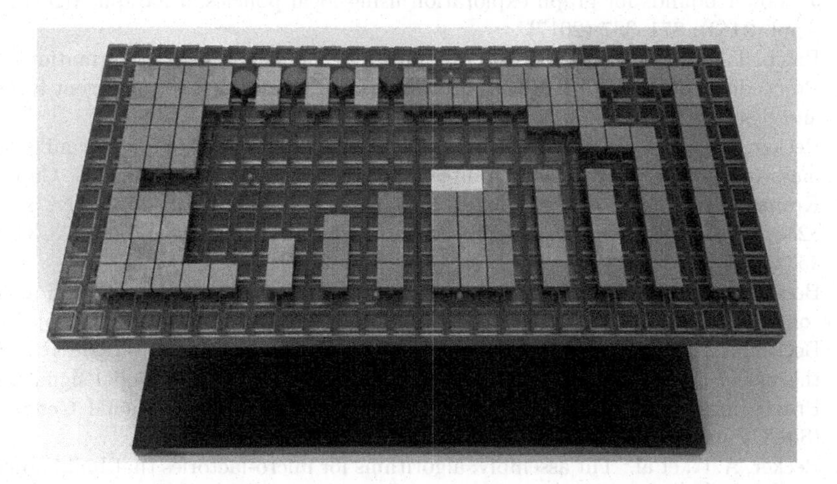

**Fig. 18.** Gravity-fed hardware implementation of particle computation. The reconfigurable prototype is set up as a FAN-OUT gate using a $2 \times 1$ robot (white)

### 5.6 Extensions

There are numerous extensions to the basic framework of control by uniform global forces. In [58] we develop methods for mapping, foraging, and coverage with a particle swarm. In [59] we show how to collect a particle swarm. In [44] we show how to use uniform global forces in combination with appropriate obstacles to efficiently sort and classify polyomino shapes.

In combination with "sticky" particles that bind together when brought into contact, we can use the basic setup to build production lines for assembling given shapes; see our paper [7] for a basic algorithmic and complexity analysis, and [67] for more efficient methods that proceed in a hierarchical fashion.

### 5.7 Acknowledgments

The content of this section is based on the abstract [6] and paraphrases the joint work with Aaron Becker, Erik Demaine, Golnaz Habibi, Jarrett Lonsford, James McLurkin, Hahmed Mohtasham Shad, Rose Morris-Wright contained in the papers [4,5,69]. For an animated visualization, see the video at https://youtu.be/H6o9DTIfkn0.

## References

1. Abrahamsen, M., Adamaszek, A., Miltzow, T.: The art gallery problem is $\exists\mathbb{R}$-complete. In: Proceedings of the 50th Annual ACM Symposium on Theory of Computing (STOC), pp. 65–73 (2018)
2. Akash, A.K., Fekete, S.P., Lee, S.K., López-Ortiz, A., Maftuleac, D., McLurkin, J.: Lower bounds for graph exploration using local policies. J. Graph Algorithms Appl. **21**(3), 371–387 (2017)
3. Balch, T., Hybinette, M.: Social potentials for scalable multi-robot formations. In: Proceedings of the 17th IEEE/RSJ International Conference on Intelligent Robots and Systems (ICRA), pp. 73–80 (2000)
4. Becker, A., Demaine, E.D., Fekete, S.P., Habibi, G., McLurkin, J.: Reconfiguring massive particle swarms with limited, global control. In: Flocchini, P., Gao, J., Kranakis, E., Meyer auf der Heide, F. (eds.) ALGOSENSORS 2013. LNCS, vol. 8243, pp. 51–66. Springer, Heidelberg (2014). https://doi.org/10.1007/978-3-642-45346-5_5
5. Becker, A.T., Demaine, E.D., Fekete, S.P., Lonsford, J., Morris-Wright, R.: Particle computation: complexity, algorithms, and logic. Natural Computing (to appear)
6. Becker, A.T., Demaine, E.D., Fekete, S.P., Shad, S.H.M., Morris-Wright, R.: Tilt: the video. Designing worlds to control robot swarms with only global signals. In: Proceedings of the 31st International Symposium on Computational Geometry (SoCG), pp. 16–18 (2015). https://youtu.be/H6o9DTIfkn0
7. Becker, A.T., et al.: Tilt assembly: algorithms for micro-factories that build objects with uniform external forces. In: Proceedings of the 28th International Symposium on Algorithms and Computation (ISAAC 2017), pp. 11:1–11:13 (2017). Full version to appear in: Algorithmica

8. Becker, A.T., Ou, Y., Kim, P., Kim, M.J., Julius, A.: Feedback control of many magnetized: tetrahymena pyriformis cells by exploiting phase inhomogeneity. In: Proceedings of the 26th IEEE/RSJ International Conference on Intelligent Robots and Systems (IROS), pp. 3317–3323 (2013)
9. Bonabeau, E., Meyer, C.: Swarm intelligence: a whole new way to think about business. Harvard Bus. Rev. **79**, 106–114 (2001)
10. Borrmann, D., et al.: Point guards and point clouds: solving general art gallery problems. In: Proceedings of the 29th Symposium on Computational Geometry (SoCG), pp. 347–348 (2013)
11. Borrmann, D., Elseberg, J., Lingemann, K., Nüchter, A., Hertzberg, J.: Globally consistent 3D mapping with scan matching. Robot. Auton. Syst. **56**(2), 130–142 (2008)
12. Chanu, A., Felfoul, O., Beaudoin, G., Martel, S.: Adapting the clinical MRI software environment for real-time navigation of an endovascular untethered ferromagnetic bead for future endovascular interventions. Magn. Reson. Med. **59**(6), 1287–1297 (2008)
13. Chazelle, B.: The convergence of bird flocking. J. ACM **61**(4), 21 (2014)
14. Chuang, Y.-L., Huang, Y.R., D'Orsogna, M.R., Bertozzi, A.L.: Multi-vehicle flocking: scalability of cooperative control algorithms using pairwise potentials. In: Proceedings of the 24th IEEE/RSJ International Conference on Intelligent Robots and Systems (ICRA), pp. 2292–2299 (2007)
15. Chvátal, V.: A combinatorial theorem in plane geometry. J. Comb. Theory Ser. B **18**, 39–41 (1974)
16. Cortes, J., Martinez, S., Karatas, T., Bullo, F.: Coverage control for mobile sensing networks. IEEE Trans. Robot. Autom. **20**(2), 243–255 (2004)
17. Couto, M.C., de Souza, C.C., de Rezende, P.J.: An exact and efficient algorithm for the orthogonal art gallery problem. In: Proceedings of the XX Brazilian Symposium on Computer Graphics and Image Processing (SIBGRAPI), pp. 87–94 (2007)
18. Demaine, E.D., Demaine, M.L., O'Rourke, J.: PushPush and Push-1 are NP-hard in 2D. In: Proceedings of the 12th Canadian Conference on Computational Geometry (CCCG), pp. 211–219 (2000)
19. Efrat, A., Fekete, S.P., Mitchell, J.S., Polishchuk, V., Suomela, J.: Improved approximation algorithms for relay placement. ACM Trans. Algorithms **12**, 20:1–20:28 (2016)
20. Eidenbenz, S., Stamm, C., Widmayer, P.: Inapproximability results for guarding polygons and terrains. Algorithmica **31**(1), 79–113 (2001)
21. Engels, B., Kamphans, T.: Randolphs robot game is NP-hard. Electron. Notes Discrete Math. **25**, 49–53 (2006)
22. Ernestus, M., Fekete, S.P., Hemmer, M., Krupke, D.: Continuous geometric algorithms for robot swarms with multiple leaders. In: Proceedings of the 31st European Workshop on Computational Geometry (EuroCG), pp. 69–72 (2015)
23. Ernestus, M., Krupke, D.: Distributed, scalable algorithmic methods for swarms with multiple leader robots. Bachelor thesis, TU Braunschweig (2014)
24. Fekete, S.P., Friedrichs, S., Kröller, A., Schmidt, C.: Facets for art gallery problems. Algorithmica **73**, 411–440 (2015)
25. Fekete, S.P., Kamphans, T., Kröller, A., Mitchell, J.S.B., Schmidt, C.: Exploring and triangulating a region by a swarm of robots. In: Goldberg, L.A., Jansen, K., Ravi, R., Rolim, J.D.P. (eds.) APPROX/RANDOM -2011. LNCS, vol. 6845, pp. 206–217. Springer, Heidelberg (2011). https://doi.org/10.1007/978-3-642-22935-0_18

26. Fekete, S.P., Kamphans, T., Kröller, A., Schmidt, C.: Robot swarms for exploration and triangulation of unknown environments. In: Proceedings of the 25th European Workshop on Computational Geometry, pp. 153–156 (2010)
27. Fekete, S.P., Kröller, A.: Geometry-based reasoning for a large sensor network. In: Proceedings of the 22nd Symposium on Computational Geometry (SoCG), pp. 475–476 (2006)
28. Fekete, S.P., Kröller, A., Lee, S., McLurkin, J., Schmidt, C.: Triangulating unknown environments using robot swarms. In: Proceedings of the 29th Symposium on Computational Geometry (SoCG), pp. 345–346 (2013)
29. Fekete, S.P., Mitchell, J.S.B., Schmidt, C.: Minimum covering with travel cost. J. Comb. Optim. **24**, 32–51 (2012)
30. Fekete, S.P., Rex, S., Schmidt, C.: Online exploration and triangulation in orthogonal polygonal regions. In: Ghosh, S.K., Tokuyama, T. (eds.) WALCOM 2013. LNCS, vol. 7748, pp. 29–40. Springer, Heidelberg (2013). https://doi.org/10.1007/978-3-642-36065-7_5
31. Fekete, S.P., Schmidt, C.: Polygon exploration with time-discrete vision. Comput. Geom.: Theory Appl. **43**(2), 148–168 (2010)
32. Fisk, S.: A short proof of Chvátal's watchman theorem. J. Comb. Theory Ser. B **24**, 374 (1978)
33. Garey, M.R., Graham, R.L., Johnson, D.S.: The complexity of computing Steiner minimal trees. SIAM J. Appl. Math. **32**(4), 835–859 (1977)
34. Gazi, V.: Swarm aggregations using artificial potentials and sliding-mode control. IEEE Trans. Robot. **21**(6), 1208–1214 (2005)
35. Hamann, H., Wörn, H.: Aggregating robots compute: an adaptive heuristic for the euclidean steiner tree problem. In: Asada, M., Hallam, J.C.T., Meyer, J.-A., Tani, J. (eds.) SAB 2008. LNCS (LNAI), vol. 5040, pp. 447–456. Springer, Heidelberg (2008). https://doi.org/10.1007/978-3-540-69134-1_44
36. Hayes, A.T., Dormiani-Tabatabaei, P.: Self-organized flocking with agent failure: off-line optimization and demonstration with real robots. In: Proceedings of the 19th IEEE/RSJ International Conference on Intelligent Robots and Systems (ICRA), pp. 3900–3905 (2002)
37. Hearn, R.A., Demaine, E.D.: PSPACE-completeness of sliding-block puzzles and other problems through the nondeterministic constraint logic model of computation. Theor. Comput. Sci. **343**(1–2), 72–96 (2005)
38. Hoffmann, M.: Motion planning amidst movable square blocks: Push-* is NP-hard. In: Proceedings of the 12th Canadian Conference on Computational Geometry (CCCG), pp. 205–210 (2000)
39. Holzer, M., Schwoon, S.: Assembling molecules in ATOMIX is hard. Theor. Comput. Sci. **313**(3), 447–462 (2004)
40. Howard, A., Matarić, M.J., Sukhatme, G.S.: Mobile sensor network deployment using potential fields: a distributed, scalable solution to the area coverage problem. In: Asama, H., Arai, T., Fukuda, T., Hasegawa, T. (eds.) Proceedings 6th International Symposium on Distributed Autonomous Robotics Systems (DARS), pp. 299–308. Springer, Tokyo (2002). https://doi.org/10.1007/978-4-431-65941-9_30
41. Hwang, F.K., Richards, D.S., Winter, P.: The Steiner Tree Problem. Annals of Discrete Mathematics, vol. 53. Elsevier, Amsterdam (1992)
42. Jerrum, M.R.: The complexity of finding minimum-length generator sequences. Theor. Comput. Sci. **36**, 265–289 (1985)

43. Kamimura, A., Murata, S., Yoshida, E., Kurokawa, H., Tomita, K., Kokaji, S.: Self-reconfigurable modular robot-experiments on reconfiguration and locomotion. In: Proceedings of the 14th IEEE/RSJ International Conference on Intelligent Robots and Systems (IROS), pp. 606–612 (2001)
44. Keldenich, P., et al.: On designing 2D discrete workspaces to sort or classify 2D polyominoes. In: Proceedings of the 35th IEEE/RSJ International Conference on Intelligent Robots and Systems (ICRA) (2018, to appear)
45. Khalil, I.S.M., Pichel, M.P., Reefman, B.A., Sukas, O.S., Abelmann, L., Misra, S.: Control of magnetotactic bacterium in a micro-fabricated maze. In: Proceedings of the 30th IEEE/RSJ International Conference on Intelligent Robots and Systems (ICRA), pp. 5488–5493 (2013)
46. Kröller, A., Baumgartner, T., Fekete, S.P., Schmidt, C.: Exact solutions and bounds for general art gallery problems. J. Exp. Algorithms **17**, 2–3 (2012)
47. Kröller, A., Fekete, S.P., Pfisterer, D., Fischer, S.: Deterministic boundary recognition and topology extraction for large sensor networks. In: Proceedings of the ACM/SIAM Symposium on Discrete Algorithms (SODA), pp. 1000–1009 (2006)
48. Krupke, D.M., Ernestus, M., Hemmer, M., Fekete, S.P.: Distributed cohesive control for robot swarms: maintaining good connectivity in the presence of exterior forces. In: Proceedings of the 28th IEEE/RSJ International Conference on Intelligent Robots and Systems (IROS), pp. 413–420 (2015)
49. La, H.M., Sheng, W.: Adaptive flocking control for dynamic target tracking in mobile sensor networks. In: Proceedings of the 22nd IEEE/RSJ International Conference on Intelligent Robots and Systems (IROS), pp. 4843–4848 (2009)
50. La, H.M., Sheng, W.: Flocking control of a mobile sensor network to track and observe a moving target. In: Proceedings of the 22nd IEEE/RSJ International Conference on Intelligent Robots and Systems (IROS), pp. 3129–3134 (2009)
51. Lee, D.-T., Lin, A.K.: Computational complexity of art gallery problems. IEEE Trans. Inf. Theory **32**(2), 276–282 (1986)
52. Lee, S.K., Becker, A.T., Fekete, S.P., Kröller, A., McLurkin, J.: Exploration via structured triangulation by a multi-robot system with bearing-only low-resolution sensors. In: Proceedings of the 31st IEEE/RSJ International Conference on Intelligent Robots and Systems (ICRA), pp. 2150–2157 (2014)
53. Lee, S.K., Fekete, S.P., McLurkin, J.: Virtual-agent coverage control in triangulated environments using geodesic Voronoi tessellation. In: Proceedings of the 27th IEEE/RSJ International Conference on Intelligent Robots and Systems (IROS), pp. 3858–3865 (2014)
54. Lee, S.K., Fekete, S.P., McLurkin, J.: Structured triangulation in multi-robot systems: coverage, patrolling, Voronoi partitions, and geodesic centers. Int. J. Robot. Res. **9**(35), 1234–1260 (2016)
55. Lee, S.K., McLurkin, J.: Distributed cohesive configuration control for swarm robots with boundary information and network sensing. In: Proceedings of the 27th IEEE/RSJ International Conference on Intelligent Robots and Systems (IROS), pp. 1161–1167. IEEE (2014)
56. Lindhé, M., Ögren, P., Johansson, K.H.: Flocking with obstacle avoidance: a new distributed coordination algorithm a new distributed coordination algorithm based on Voronoi partitions. In: Proceedings of the 22nd IEEE/RSJ International Conference on Intelligent Robots and Systems (ICRA), pp. 1785–1790 (2005)
57. Maftulec, D., Lee, S.K., Fekete, S.P., Akash, A.K., López-Ortiz, A., McLurkin, J.: Local policies for efficiently patrolling a triangulated region be a robot swarm. In: Proceedings of the 32nd IEEE/RSJ International Conference on Intelligent Robots and Systems (ICRA), pp. 1809–1815 (2015)

58. Mahadev, A.V., Krupke, D., Fekete, S.P., Becker, A.T.: Mapping, foraging, and coverage with a particle swarm controlled by uniform inputs. In: Proceedings of the 30th IEEE/RSJ International Conference on Intelligent Robots and Systems (IROS), pp. 1097–1104 (2017)
59. Mahadev, A.V., Krupke, D., Reinhardt, J.-M., Fekete, S.P., Becker, A.T.: Collecting a swarm in a 2D environment using shared, global inputs. In: Proceedings of the 13th Conference on Automation Science and Engineering (CASE 2016), pp. 1231–1236 (2016)
60. McLurkin, J., Demaine, E.D.: A distributed boundary detection algorithm for multi-robot systems. In: Proceedings of the 22th IEEE/RSJ International Conference on Intelligent Robots and Systems (IROS), pp. 4791–4798 (2009)
61. McLurkin, J., et al.: A low-cost multi-robot system for research, teaching, and outreach. In: Martinoli, A., et al. (eds.) Proceedings 10th International Symposium on Distributed Autonomous Robotics Systems (DARS), vol. 83, pp. 597–609. Springer, Heidelberg (2010). https://doi.org/10.1007/978-3-642-32723-0_43
62. McLurkin, J., et al.: A robot system design for low-cost multi-robot manipulation. In: Proceedings of the 27th IEEE/RSJ International Conference on Intelligent Robots and Systems (IROS), pp. 912–918 (2014)
63. Olfati-Saber, R.: Flocking for multi-agent dynamic systems: algorithms and theory. IEEE Trans. Autom. Control **51**, 401–420 (2006)
64. O'Rourke, J.: Art Gallery Theorems and Algorithms. International Series of Monographs on Computer Science. Oxford University Press, New York (1987)
65. Prömel, H.J., Steger, A.: The Steiner Tree Problem: A Tour Through Graphs, Algorithms, and Complexity. Vieweg, Braunschweig (2002)
66. Reynolds, C.W.: Flocking, herds, and schools: a distributed behavioral model. Comput. Graph. **21**(4), 25–34 (1987)
67. Schmidt, A., Manzoor, S., Huang, L., Becker, A.T., Fekete, S.P.: Efficient parallel self-assembly under uniform control inputs. IEEE Robot. Autom. Lett. **3**, 3521–3528 (2018)
68. Schmidt, C.: Algorithms for mobile agents with limited capabilities. Ph.D. thesis, TU Braunschweig (2011)
69. Shad, H.M., Moris-Wright, R., Demaine, E.D., Fekete, S.P., Becker, A.T.: Particle computation: device fan-out and binary memory. In: Proceedings of the 32nd IEEE/RSJ International Conference on Intelligent Robots and Systems (ICRA), pp. 5384–5389 (2015)
70. Shi, H., Wang, L., Chu, T., Xu, M.: Flocking coordination of multiple mobile autonomous agents with asymmetric interactions and switching topology. In: Proceedings of the 18th IEEE/RSJ International Conference on Intelligent Robots and Systems (IROS), pp. 935–940 (2005)
71. Spears, W.M., Spears, D.F., Hammann, J.C., Heil, R.: Distributed, physics-based control of swarms of vehicles. Auton. Robots **17**(2–3), 137–162 (2004)
72. ThinkFun: Tilt: gravity fed logic maze. http://www.thinkfun.com/tilt
73. Tozoni, D.C., de Rezende, P.J., de Souza, C.C.: The quest for optimal solutions for the art gallery problem: a practical iterative algorithm. In: Bonifaci, V., Demetrescu, C., Marchetti-Spaccamela, A. (eds.) SEA 2013. LNCS, vol. 7933, pp. 320–336. Springer, Heidelberg (2013). https://doi.org/10.1007/978-3-642-38527-8_29
74. Vartholomeos, P., Akhavan-Sharif, M., Dupont, P.E.: Motion planning for multiple millimeter-scale magnetic capsules in a fluid environment. In: Proceedings of the 29th IEEE/RSJ International Conference on Intelligent Robots and Systems (ICRA), pp. 1927–1932 (2012)

# Computing by Programmable Particles

Joshua J. Daymude[1(✉)], Kristian Hinnenthal[2], Andréa W. Richa[1], and Christian Scheideler[2]

[1] Computer Science, CIDSE, Arizona State University, Tempe, AZ, USA
{jdaymude,aricha}@asu.edu

[2] Department of Computer Science, Paderborn University, Paderborn, Germany
{krijan,scheidel}@mail.upb.de

**Abstract.** The vision for *programmable matter* is to realize a physical substance that is scalable, versatile, instantly reconfigurable, safe to handle, and robust to failures. Programmable matter could be deployed in a variety of domain spaces to address a wide gamut of problems, including applications in construction, environmental science, synthetic biology, and space exploration. However, there are considerable engineering and computational challenges that must be overcome before such a system could be implemented. Towards developing efficient algorithms for novel programmable matter behaviors, the *amoebot model* for self-organizing particle systems and its variant, *hybrid programmable matter*, provide formal computational frameworks that facilitate rigorous algorithmic research. In this chapter, we discuss distributed algorithms under these models for shape formation, shape recognition, object coating, compression, shortcut bridging, and separation in addition to some underlying algorithmic primitives.

**Keywords:** Programmable matter · Self-organizing particle systems
Distributed algorithms

## 1 Introduction

The idea of a robot that can transform into different shapes and sizes (e.g., Hasbro's *Transformers*) or multitudes of tiny mobile robots that collectively build structures, move objects, or even act as weapons or shields (e.g., Disney's *Big Hero 6* or Marvel's *Black Panther*) have become as ubiquitous and iconic in science fiction futurism as flying cars, holographic video conferencing, and teleportation. Yet this vision does not exist solely in fiction; since the 1990s, many researchers spanning across biology, chemistry, physics, mathematics, and computer science have contributed significant results towards realizing versatile, scalable robotic systems. In 1991, Toffoli and Margolus [54] defined *programmable matter* as a physical computing medium composed of simple, homogeneous nodes that can be (*i*) assembled into "lumps" of arbitrary size, (*ii*) dynamically reconfigured into any regular structure that grows at most polynomially, (*iii*) interactively controlled by user input or environmental stimuli, and (*iv*) accessed in real time for observation, analysis, or modification.


P. Flocchini et al. (Eds.): Distributed Computing by Mobile Entities, LNCS 11340, pp. 615–681, 2019.
https://doi.org/10.1007/978-3-030-11072-7_22

The vision for programmable matter is to realize a physical substance that is scalable, versatile, instantly reconfigurable, safe to handle, and robust to failures. Programmable matter could be deployed in numerous domain spaces to address a wide gamut of problems: in construction, it could be used as a self-repairing building material or as a dynamically reconfigurable support scaffolding; in environmental science, it could be used to locate and metabolize pollutants at the micro-scale; in biological processes, it could aid in the construction and maintenance of nano-scale structures or even boost healing by artificially transporting and applying medicine where it's most needed; in robotics, it could be used to sustain long-term missions in isolated or hazardous environments where it would be difficult for a human to intervene.

There are formidable challenges to realizing programmable matter both from the engineering and computational perspectives. In this chapter, we abstractly consider programmable matter as a collection of simple computational entities that must coordinate at the individual level to achieve useful behaviors at the system level. Using the amoebot model and hybrid programmable matter model as our computational frameworks, we will present a series of distributed algorithms for programmable matter and give rigorous theoretical results regarding their correctness and efficiency.

## 1.1 Related Work

When considering the various models and implementations of programmable matter, one can differentiate between *passive* and *active* systems. In passive systems, individual units of programmable matter cannot control their own movements, instead relying on their structural properties and interactions with their environment for locomotion. They may, in some cases, have limited computational abilities to make decisions and communicate. Prominent examples of passive systems include population protocols [2], molecular computing and tile self-assembly models [26,40], and slime molds [7,44].

Our focus is primarily on active systems, where the individual units can control their actions and movements to achieve some task. Examples of physical active systems include swarm robotics [47] as well as self-reconfigurable modular robotics [59]. These systems seek similar coordinated behaviors as those considered in the amoebot model, but use robots that often have significantly more powerful sensing and communication abilities. Among the theoretical models of active systems, the *nubot model* from molecular programming [58] and *metamorphic robots* [12,56] have the most similarities to the amoebot model, including their representations of space and their emphasis on simple, local computational units. However, they include some capabilities (e.g., rigid body movements in the nubot model) that prohibit a direct translation between the models.

The amoebot model for self-organizing particle systems (fully described in Sect. 2.1) envisions programmable matter as a system of simple, homogeneous *particles* that have only local communication and vision, constant-size memory, and no global sense of direction. This model was introduced to facilitate rigorous algorithmic research on programmable matter systems, and has since served as the computational framework for many theoretical investigations and even one experimental study [48].

Hybrid programmable matter (Sect. 6) combines the active and passive approaches by considering a passive structure of connected tiles that can be reconfigured by a collection of active robots. When considering only the passive tiles, this model shares many similarities with the tile self-assembly models mentioned before, where tiles bond to each other based on predefined "glues" (see, e.g., [26,40]). On the other hand, the active robots in the hybrid setting are very similar to the particles of the amoebot model, with the added capability of lifting and moving tiles. When considering only the robots' movements on a static tile structure, hybrid programmable matter reduces to an instance of the *mobile agents on graphs* model, where problems such as gathering/rendezvous [41], intruder caption [6], and graph searching and exploration [13,28] have been studied extensively. *DNA nanomachines* offer a promising realization of hybrid programmable matter, and are capable of walking on one- and two-dimensional surfaces [35,39,57], transporting cargo [51,53], and acting as the head of a finite automaton on an input tape [45].

### 1.2 Chapter Organization

In Sect. 2, we define the amoebot model for programmable matter, including the rationale behind its modeling choices and a list of common model extensions. Section 3 contains three algorithmic primitives under the amoebot model—leader election, the spanning forest primitive, and distributed binary counters—that are utilized by the algorithms of Sects. 4 and 5. The shape formation and object coating algorithms of Sect. 4 are largely deterministic, while the algorithms for compression, shortcut bridging, and separation in Sect. 5 are fully stochastic. Hybrid programmable matter is defined in Sect. 6, and algorithms for shape formation and shape recognition in this hybrid setting are described in Sect. 7. A summary of the chapter and an outline of future research are given in Sect. 8.

For clarity and brevity, we do not give any proofs of the theoretical results in this chapter and occasionally omit algorithm details that detract from a clear understanding of an algorithm's main ideas. However, we cite all underlying publications in their respective sections and encourage the interested reader to read further.

## 2 The Amoebot Model

The *amoebot model* is an abstract computational model of programmable matter intended to enable rigorous algorithmic analysis of collective systems at the nano-scale. Originally proposed as "amoeba-inspired self-organizing particle systems" in [25], the model was polished and formally announced as the *amoebot model* in [18]. It has since undergone many updates and changes over the years to support new settings and considerations, but has kept to the same core principles throughout. Here, we give a complete description of the current model in

Sect. 2.1, provide some intuition behind its details in Sect. 2.2, and describe its common extensions in Sect. 2.3.

## 2.1 Model Description

In the amoebot model, programmable matter consists of individual, homogeneous computational elements called *particles*. Any structure that a particle system $\mathcal{P}$ can form is represented as a subgraph of an infinite, undirected graph $G = (V, E)$ where $V$ represents all positions a particle can occupy relative to its structure and $E$ represents all atomic movements a particle can make. Each node in $V$ can be occupied by at most one particle at a time. In the *geometric amoebot model*, it is further assumed that $G = G_\Delta$, where $G_\Delta$ is the triangular lattice[1] (see Fig. 1a). Fixing the position of some particle, $G_\Delta$ represents the discretization of space relative to this particle and the possible atomic movements between these discrete positions. This discretization can be conceptualized as a tiling of two-dimensional space; $G_\Delta$ corresponds to the hexagonal tiling (Fig. 1a).

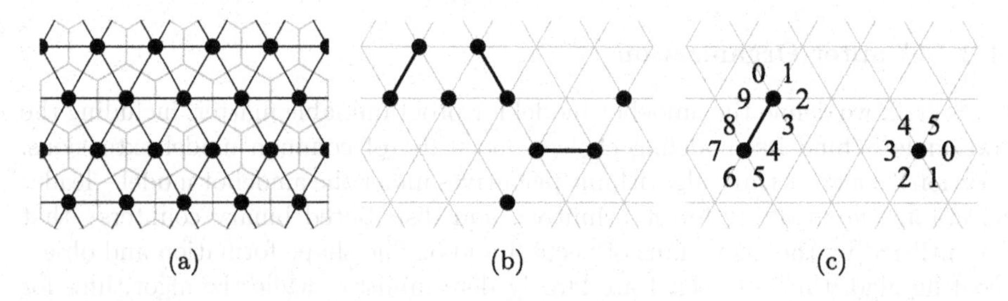


**Fig. 1.** (a) A section of the triangular lattice $G_\Delta$ (black) and its dual, the hexagonal tiling (gray). (b) Expanded and contracted particles (black dots) on $G_\Delta$ (gray lattice). Particles with a black line between their nodes are expanded. (c) Two particles with different offsets for their port labels.

Each particle occupies either a single node in $V$ (i.e., it is *contracted*) or a pair of adjacent nodes in $V$ (i.e., it is *expanded*), as in Fig. 1b. Particles move via a series of *expansions* and *contractions*: a contracted particle can expand into an unoccupied adjacent node to become expanded, and completes its movement by contracting to once again occupy a single node. An expanded particle's *head* is the node it last expanded into and the other node it occupies is its *tail*; a contracted particle's head and tail are both the single node it occupies.

Two particles occupying adjacent nodes are said to be *neighbors*. Neighboring particles can coordinate their movements in a *handover*, which can occur in one of two ways. A contracted particle $P$ can initiate a "push" handover with an expanded neighbor $Q$ by expanding into a node occupied by $Q$, forcing it to

[1] Some papers refer to $G_\Delta$ as the *equilateral triangular grid graph* $G_{\text{eqt}}$ or the triangular lattice $\Gamma$.

contract. Alternatively, an expanded particle $Q$ can initiate a "pull" handover with a contracted neighbor $P$ by contracting, forcing $P$ to expand into the node it is vacating.

Each particle keeps a collection of ports—one for each edge incident to the node(s) it occupies—that have unique labels from its own local perspective. Although each particle is *anonymous*, lacking a unique identifier, a particle can locally identify any given neighbor by its labeled port corresponding to the edge between them. The particles are assumed to have a common *chirality* (i.e., notion of clockwise direction), which allows each particle to label its ports in clockwise order. However, particles do not share a coordinate system or global compass and may have different offsets for their port labels, as in Fig. 1c.

Each particle has a constant-size local memory partitioned into internal memory and one addressed memory for each neighboring node. A particle can only write into its own memory, but can read all the addressed and internal memories of its neighbors for communication. A particle's internal memory stores whether it is expanded or contracted, its local port labeling (including which ports are incident to its head versus its tail), and any other application-specific information. Particles do not have any global information and—due to the limitation of constant-size memory—cannot know the total number of particles in the system nor any estimate of this value.

The system progresses through *atomic actions* according to the standard $\mathcal{A}$SYNC model of computation from distributed computing (see, e.g., [36]). A classical result under this model states that for any concurrent asynchronous execution of atomic actions, there exists a sequential ordering of actions producing the same end result, provided conflicts that arise in the concurrent execution are resolved. In the amoebot model, an atomic action corresponds to the activation of a single particle. Once activated, a particle can (*i*) perform an arbitrary, bounded amount of computation involving information it reads from its local memory and its neighbors' memories, (*ii*) write to its local memory, and (*iii*) perform at most one expansion or contraction. Conflicts involving simultaneous particle expansions into the same unoccupied node are assumed to be resolved arbitrarily such that at most one particle moves to some unoccupied node at any given time[2]. Thus, while in reality many particles may be active concurrently, it suffices when analyzing algorithms under the amoebot model to consider a sequence of activations where only one particle is active at a time. The resulting activation sequence is assumed to be *fair*: for any inactive particle $P$ at time $t$, $P$ will be activated again at some time $t' > t$. An *asynchronous round* is complete once every particle has been activated at least once. Unless otherwise specified, a *round* refers to an asynchronous round.

[2] A particle can only write into its own memory in the amoebot model's publishing-based communication, so no conflicts of concurrent writes to the same memory location are possible.

## 2.2 Rationale

We now provide some intuition behind the amoebot model and its details. At the highest level, we seek to answer the question: *what complex, collective behaviors are achievable by extremely simple, restricted programmable particles?* The amoebot model was designed to restrict the capabilities of the individual particles as much as possible in hope of developing algorithms that could be useful to many implemented systems across task domains and scales. For example, it may seem unnecessarily restrictive to a swarm robotics engineer to consider robots with only constant-size memory, since commodity hardware often supports $\mathcal{O}(\log n)$ or even $\mathcal{O}(n)$ memory, for reasonable swarm sizes $n$. However, this assumption makes algorithms under the amoebot model applicable both to swarm systems with extra memory as well as systems at the more restrictive micro- or nano-scales. Moreover, the resulting systems can be arbitrarily scalable to any number of units, a desirable property for programmable matter.

*Communication.* Restricting particle communication and vision to immediate neighbors captures the local nature of unit interactions in programmable matter. The amoebot model's communication scheme is a publishing-based version of standard message passing protocols in the $\mathcal{A}$SYNC model. If a particle $P$ wants to send some information $x$ to its neighbor $Q$, it writes $x$ to its addressed memory facing $Q$. Particle $P$ must wait for $Q$ to activate, read $x$, and acknowledge its receipt before $P$ can know the information was communicated. Situations where multiple neighbors try to send information to the same particle concurrently must be resolved by the recipient. (See Sect. 2.3 for a simpler variant of this publishing-based communication model).

*Chirality.* The chirality assumption, which states that particles have a common sense of clockwise direction, is reasonable in many settings. Having a shared chirality is essentially equivalent to the system's ability to break spatial symmetry, such as distinguishing between "up" and "down". This is usually fairly simple to decide; for example, if a particle system were deployed in any medium subject to gravity, the system's top and bottom would be trivially distinguishable. Recent results by Di Luna et al. suggest that this assumption may not be necessary for all applications [23, 24]. We will discuss this further at the end of Sect. 3.1.

*Connectivity.* A particle system is *connected* if the subgraph of $G$ induced by the occupied nodes of $V$ is also connected. This notion does not imply any particular kind of connectivity in a physical programmable matter system; connections could be physical bonds, points of contact between neighboring units, or even wireless communication links. Although the amoebot model does not require that a system remains connected, this is often a desirable property that its algorithms maintain. If a particle system disconnects, there is little hope the resulting components could ever reconnect. Since each particle can only see and communicate with its immediate neighbors and does not have a global compass, disconnected components have no way of knowing their relative positions and thus cannot intentionally move toward one another to reconnect.

*Space.* To aid system connectivity, we chose the triangular lattice $G_\Delta$ to represent space in the geometric amoebot model. In the other regular two-dimensional lattices (square and hexagonal), particles are often forced to momentarily disconnect from the rest of the system even to perform moves as simple as shifting "around" another particle by one position (see Fig. 2).

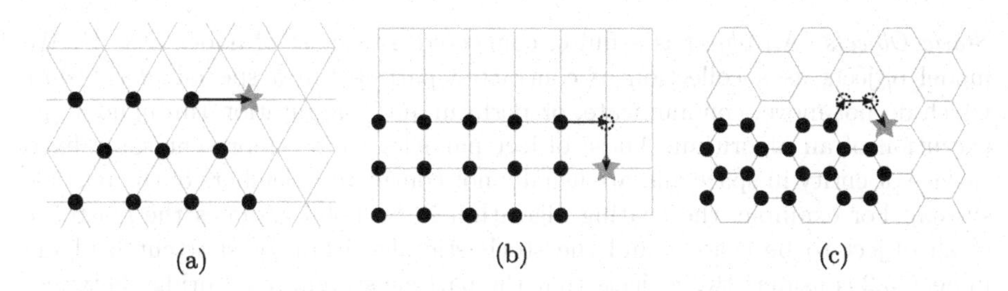


**Fig. 2.** Illustration of a particle moving "around" a neighboring particle to get to the next position on the surface, depicted as a gray star, on the (a) triangular lattice $G_\Delta$, (b) square lattice, and (c) hexagonal lattice.

*Movement.* Modeling movements as expansions, contractions, and handovers also has roots in connectivity. Splitting a particle's movement from one node to another into an expansion and a contraction can be thought of as a look-ahead mechanism in which the particle reserves a space and examines its new surroundings before deciding whether or not to go through with the movement. This is vaguely similar to human walking, where we put one foot forward before completely shifting our weight to take another step. By looking ahead, a particle can determine whether its move might break system connectivity before committing to it. Handovers, as described in Sect. 2.1, allow the system to maintain connectivity while moving. These movements were not simply included for convenience; there are tasks—such as moving through a static tunnel of width 1—which are impossible without handovers if connectivity must also be maintained at all times.

## 2.3 Extensions

Many papers on the amoebot model utilize techniques and assumptions that extend the core model described in Sect. 2.1. These extensions can be thought of as modules which combine and repackage core model features into useful, higher-level functionalities.

*Leader Particle.* Some algorithms under the amoebot model assume the existence of a unique *leader particle* (or *seed*) at initialization which can be used to coordinate the rest of the system. This assumption is reasonable, since the leader election algorithm in [16] can be used as a preprocessing step for obtaining this leader particle (see Sect. 3.1 for details). Notably, a leader can impose its labeling scheme on all other particles in the system to establish a global compass. In the other direction, if the system establishes a global compass without a leader, it becomes trivial to solve leader election (e.g., "elect the south-most, west-most particle"). Thus, any algorithm under the amoebot model which hopes to run in sublinear time—faster than the leader election algorithm of [16], which matches the worst-case lower bound—must restrict itself only to local compasses.

*Static Objects.* An *object* is a finite, connected, static set of nodes $O \subset V$. We model objects as a collection of contracted particles in a special *object state* which do not move, communicate, or perform any computation throughout the execution of an algorithm. These object particles simply represent some fixed surface or entity in space and are usually not considered members of the particle system. For example, the coating algorithm in Sect. 4.3 assumes the existence of an object to be coated, and the stochastic algorithm for shortcut bridging in Sect. 5.2 considers two objects that the particle system must bridge between. Particles can differentiate between object and non-object particles.

*Node Differentiation.* It is sometimes useful to consider physical spaces that have heterogeneous properties, such as a marsh with both land and water locations or a tree-lined path with some parts exposed to sunlight and others in shade. *Node differentiation* models these differences by considering an assignment $\Phi : V \to \{1, \ldots, k\}$ that maps each node of the graph $G$ to one of $k$ types, where $k$ is a constant. A contracted particle occupying a node $u \in V$ can read $\Phi(u)$, but cannot alter it. Analogously, an expanded particle occupying adjacent nodes $u, v \in V$ can read but not write $\Phi(u)$ and $\Phi(v)$. This assignment $\Phi$ need not be static, but any dynamics controlling its evolution should reflect changes in the environment and not the actions of the particle system. This extension should not be used to encode global information for the particles to utilize.

*Token Passing.* A *token* is a constant-size message that can be passed from particle to particle. More specifically, a particle $P$ can pass a token $t$ to a neighbor $Q$ by publishing $t$ to its addressed memory facing $Q$, wait for $Q$ to read, copy, and acknowledge $t$, and then delete $t$ from its own memory. Due to the constant-size memory constraint of the amoebot model, each particle can hold only a constant number of tokens at once. Rules on whether tokens must be passed in a pipelined fashion, merge together, or interact in more complex ways may vary by each algorithm's need.

Many algorithms under the amoebot model use token passing to relay information beyond a particle's immediate neighborhood. For example, the leader election algorithm in Sect. 3.1 uses tokens extensively to facilitate communication and competition between candidates which are often far from one another. Although token passing often makes algorithms more complex, especially when tokens of different protocols interact, its flexibility and direct compatibility with the model make it a viable tool for many applications.

*Direct Write Communication.* In the publishing-based communication scheme of the amoebot model, a particle only has write access to its own memory, but has read access to all addressed and internal memories of its neighbors. As described in Sect. 2.2, if a particle $P$ wants to send some information $x$ to its neighbor $Q$, a multi-step process of publishing and acknowledging $x$ is initiated. However, this can greatly complicate the presentation of algorithms that rely heavily on writes (e.g., any algorithm that uses token passing), as every write is split over multiple particle activations that must be locally synchronized.

One could greatly simplify communication descriptions by employing a variant of the publishing-based communication model. In the *direct write communication* model, a particle can do the following in one activation: (*i*) perform an arbitrary, bounded amount of computation involving information it reads from its local memory and its neighbors' memories, (*ii*) write to its local memory, (*iii*) directly write updates to at most one neighbor's memory, and (*iv*) perform at most one expansion or contraction. However, in the asynchronous setting of the amoebot model, this direct write communication allows for *write conflicts*, where multiple particles concurrently attempt to write to the internal memory of a common neighbor. These conflicts are assumed to be resolved arbitrarily such that each particle is involved in at most one write at any given time (i.e., at any given time, either a particle $P$ is writing to a neighbor, a neighbor is writing to $P$, or neither). This direct write communication model can be faithfully emulated by the publishing-based communication scheme of the amoebot model via a simple emulation primitive, described fully in [17].

*Random Number Generation.* It is often assumed that each particle has access to random bits with which it can generate random values. However, due to the constant-size memory constraint of the core model, each particle can only hold a constant number of random bits and thus can only store constant precision random values. It is left to the algorithm designer to ensure that constant precision is sufficient for their application; see [1,11] for examples of such arguments.

*Agent Emulation.* It can be useful for a particle to run multiple instances of an algorithm at once, especially in settings where it needs to participate in different phases of an algorithm concurrently. In the leader election algorithm in Sect. 3.1, for example, a particle executes up to three instances concurrently (one per boundary it is incident to). To accommodate this, a single particle can emulate up to a constant number of *agents*, each with its own memory, running its own instance of a given algorithm. This respects the constant-size memory constraint as each algorithm instance requires only constant memory and each particle emulates at most a constant number of agents.

## 3 Algorithmic Primitives for the Amoebot Model

Several algorithmic primitives exist under the amoebot model, acting as reusable building blocks for the algorithms of Sects. 4 and 5. These include *leader election*

(Sect. 3.1), which is used by other algorithms as a black box for obtaining a unique leader particle, the *spanning forest primitive* (Sect. 3.2), which is used to locally organize and move a particle system along a specified path, and *distributed binary counters* (Sect. 3.3), which enable $n$ particles to collectively emulate a binary counter storing unsigned values up to $2^n - 1$.

### 3.1 Leader Election

To date, there have been two approaches to the classical problem of leader election under the amoebot model: the token-based approach by Derakhshandeh and Daymude et al. [16,22], and the erosion-based approach by Di Luna et al. [23]. We will focus primarily on the algorithm in [16], which simplifies and extends the algorithm and analysis of [22] to a fully local, distributed, asynchronous setting. At a glance, the algorithm in [16] elects a leader in $\mathcal{O}(L)$ asynchronous rounds with high probability[3], where $L$ is the length of the outer boundary of the system and w.h.p. applies to both correctness and runtime. A brief comparison to the erosion-based approach is made at the end of this section.

**Problem Description.** An algorithm is said to solve the leader election problem if for any connected particle system of initially contracted particles, eventually a single particle *irreversibly* declares itself the *leader* (e.g., by setting a dedicated bit in its memory) and no other particle ever declares itself to be the leader. The running time of a leader election algorithm is defined to be the number of asynchronous rounds until a leader is declared. The algorithm is not required to terminate for particles other than the leader, though a leader could broadcast its existence to the rest of the system to trigger termination, if desired.

**Algorithm.** We begin with a high-level overview of the algorithm's six *phases.* These phases are not strictly synchronized among each other, i.e., at any point in time, different parts of the particle system may execute different phases. Furthermore, a particle can be involved in the execution of multiple phases at the same time. The first phase is *boundary setup.* In this phase, each particle locally checks whether it is part of a *boundary* of the particle system. Only particles on a boundary participate in leader election. Particles occupying a common boundary organize themselves into a directed cycle. The remaining phases operate on each boundary independently. In the *segment setup* phase, the boundaries are divided into *segments.* Each particle flips a fair coin: particles that flip heads become *candidates* and compete for leadership whereas particles that flip tails become *non-candidates* and assist the candidates in their competition. A segment consists of a candidate and all subsequent non-candidates along the boundary up to the next candidate. The *identifier setup* phase assigns a random identifier to each candidate. The identifier of a candidate is stored distributedly among the

[3] An event occurs *with high probability (w.h.p.)* if the probability of success is at least $1 - 1/n^c$, where $c > 0$ is a constant; in our setting, $n$ is the number of particles.

particles of its segment. In the *identifier comparison* phase, the candidates compete for leadership by comparing their identifiers using a token passing scheme. Whenever a candidate sees an identifier higher than its own, it revokes its candidacy. Whenever a candidate sees its own identifier, the *solitude verification* phase is triggered. In this phase, the candidate checks whether it is the last remaining candidate on the boundary. If so, it initiates the *boundary identification* phase to check if it occupies the unique *outer boundary* of the system. In that case, it becomes the leader; otherwise, it revokes its candidacy.

*Boundary Setup.* The boundary setup phase organizes the particle system into a set of *boundaries*, as in Fig. 3a. Let $A$ be the set of nodes in $G_\Delta$ that are occupied by particles, and consider the graph $G_\Delta|_{V \setminus A}$ induced by the unoccupied nodes in $G_\Delta$. An *empty region* is a maximal connected component of $G_\Delta|_{V \setminus A}$. Let $N(R)$ be the neighborhood of an empty region $R$ in $G_\Delta$; that is, $N(R) = \{u \in V \setminus R : \exists v \in R \text{ such that } (u, v) \in E\}$. Note that by definition, all nodes in $N(R)$ are occupied by particles. We refer to $N(R)$ as the *boundary* of the particle system corresponding to $R$. Since $A$ corresponds to a finite set of particles, exactly one empty region has infinite size while any others have finite size. The boundary corresponding to the infinite empty region is the unique *outer boundary*, and any boundary corresponding to a finite empty region is an *inner boundary*.

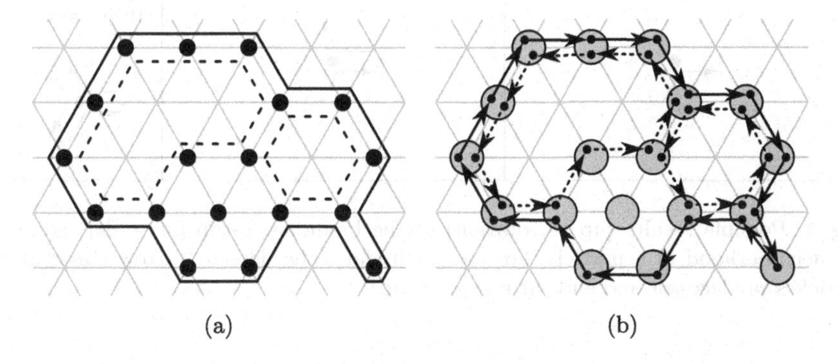


**Fig. 3.** (a) Boundaries of a particle system. The solid line represents the unique outer boundary and the dashed lines represent the inner boundaries. (b) Agents (black dots) of particles (gray circles) organized into directed cycles along the boundaries of (a).

Next, the particles of each boundary organize into a directed cycle. Upon its first activation, each particle $P$ determines its place in these cycles using only local information as follows. If $P$ has no neighbors, then since the particle system is connected, $P$ must be the only particle. So $P$ immediately declares itself the leader and terminates. If $P$ is surrounded (i.e., it has six neighbors), $P$ is not part of any boundary and simply terminates.

Otherwise, the neighborhood of $P$ must contain at least one occupied and one unoccupied node. For each maximal, connected sequence of unoccupied nodes

$S$ in the neighborhood of $P$ (of which there can be at most three; see Fig. 4), let $P$ act as a distinct *agent* $a_S$ that independently executes the remainder of the leader election algorithm. This ensures that the leader election algorithm runs on each boundary independently, since $P$ cannot locally decide which such sequences belong to which boundary. Each agent $a_S$ chooses the particle immediately clockwise (resp., counterclockwise) of $S$ to be its *successor* (resp., *predecessor*).

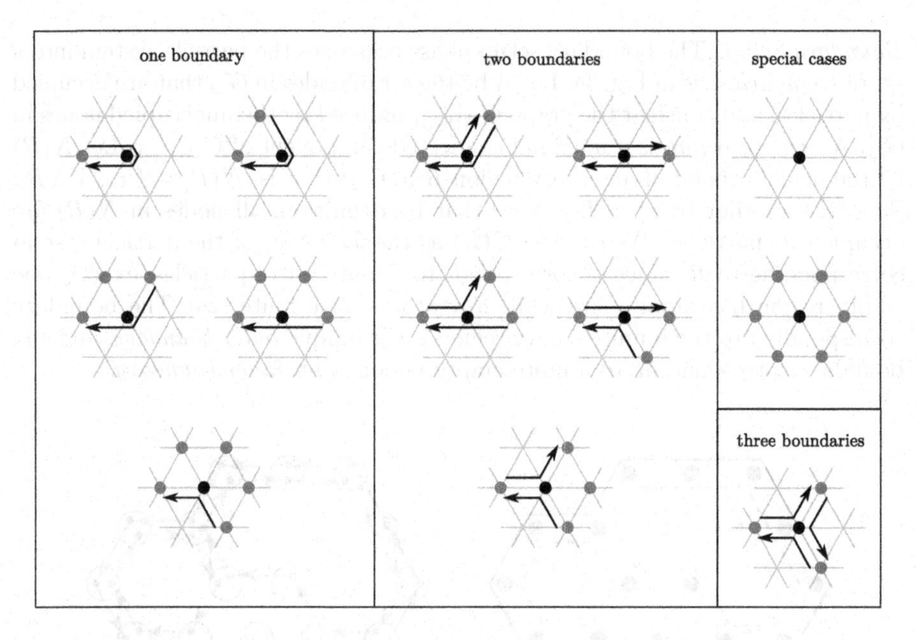


**Fig. 4.** Possible results (up to rotation) of the boundary setup phase depending on the neighborhood of a particle. For each boundary, the depicted arrow starts at the particle's predecessor and ends at its successor.

Figure 4 shows all possible neighborhoods of a particle (up to rotation) and the corresponding predecessor and successor assignments of its agents. These assignments organize the set of all agents into disjoint cycles spanning the boundaries of the particle system (see Fig. 3b). It is possible that a particle can occur up to three times on the same boundary as different agents. While this property can be ignored for most of the remaining phases, it will remain a cause for special consideration in the solitude verification phase.

*Segment Setup.* All remaining phases (including this one) execute exclusively on each boundary independently. Therefore, we only consider a single boundary for the remainder of the algorithm description. The segment setup phase divides the boundary into disjoint "segments" as follows. Each agent flips a fair coin;

those that flip heads become *candidates* and those that flip tails become *non-candidates.* In the following phases, candidates compete for leadership while non-candidates assist this competition. A *segment* is a maximal sequence of agents $(a_1, a_2, \ldots, a_k)$ such that $a_1$ is a candidate, $a_i$ is a non-candidate for $i > 1$, and $a_i$ is the successor of $a_{i-1}$ for $i > 1$. We refer to the segment starting at a candidate $c$ as the segment of $c$ (denoted $c$.seg) and denote its length as $|c.\text{seg}|$. In the following phases, each candidate uses its segment as a distributed memory.

*Identifier Setup.* After the segments have been set up, each candidate generates a random *identifier* for use in the competition of the next phase by assigning a random digit to each agent in its segment. Note that the term identifier is slightly misleading in that two distinct candidates can have the same identifier.

To generate its random identifier $c$.id, a candidate $c$ sends a token (recall token passing from Sect. 2.3) along its segment in the direction of the boundary. As the token traverses the segment, it assigns a value chosen uniformly at random from $\{0, 1\}$ to each visited agent[4]. The resulting identifier is a binary number consisting of $|c.\text{seg}|$ bits where $c$ holds the most significant bit and the last agent of $c$.seg holds the least significant bit.

After generating $c$.id, each candidate $c$ creates a copy of $c$.id that is stored in *reverse digit order* in its segment. This copy is used in the next phase to compare against the identifiers of other candidates. The token that generated $c$.id is reused in creating the reversed copy as follows. It first reads the digit of the last agent of the segment $c$.seg. It is then passed to the beginning of $c$.seg (to candidate $c$) and stores a copy of that digit. It then reads the digit of $c$ and is passed back to the end of the segment where it stores a copy of that digit. This continues in a similar fashion with the second to last and second agent and so on until $c$.id is completely copied. Finally, the token is passed to $c$ to signal the end of this identifier setup phase.

*Identifier Comparison.* During the identifier comparison phase, the candidate agents use their identifiers to compete with each other. When comparing identifiers of different lengths, longer identifiers are defined to be higher than shorter ones; otherwise, the identifiers are compared directly. A candidate with the highest identifier eventually progresses to the solitude verification phase, described in the next section, while any candidate with a lower identifier withdraws its candidacy. To achieve the comparison, the non-reversed copies of the identifiers remain stored in their respective segments while the reversed copies move backwards along the boundary as a sequence of tokens. More specifically, a *digit token* is created for each digit of a reversed identifier. A digit token created by the last agent of a segment is marked as a *delimiter token.* We define the *token sequence* of a candidate $c$ as the sequence of digit tokens created by the agents in $c$.seg. Once created, digit tokens traverse the boundary against the direction of the cycle spanning it. Each agent is allowed to hold at most two tokens

[4] In [16], the digits are chosen uniformly at random from $[0, r-1]$ where $r$ is a fixed constant. The resulting identifiers are numbers with radix $r$.

at a time, and can forward at most one token per activation. Tokens are not allowed to overtake each other. Furthermore, an agent can only receive a token after it creates its own digit token. This ensures that token sequences of distinct candidates remain separated and the tokens within a token sequence maintain their relative order along the boundary.

We give a high level description of the token passing scheme for identifier comparison, illustrated in Fig. 5, using two successive candidates $c$ and $c'$. Here, the token sequence of $c'$ is compared with $c$.id. Initially, agents are *active* and tokens are *inactive*, as in Fig. 5a. Whenever a token is forwarded by a candidate into a new segment, the token becomes active, as in Fig. 5b. When an active agent receives an active token, they *match*, storing the result of their digit comparison ($<$, $>$, or $=$) in the agent and both becoming inactive (Fig. 5c). A matched (inactive) token is then simply passed on without incident until reaching $c$, who reactivates it when forwarding it into the next segment (Fig. 5d).

The delimiter token of $c'$, say $d_{c'}$, eventually enters $c$.seg (Fig. 5d). As $d_{c'}$ traverses $c$.seg, it sees the results of the previous digit comparisons from least to most significant and updates its record of the overall comparison accordingly (Fig. 5e–f). When candidate $c$ eventually receives $d_{c'}$, it locally compares identifier lengths as follows. If $c$ already matched with a non-delimiter token of $c'$, then $|c.\text{seg}| < |c'.\text{seg}|$ and $c$ withdraws its candidacy. If the delimiter token $d_{c'}$ already matched with some agent before $c$, then $|c.\text{seg}| > |c'.\text{seg}|$ and $c$ remains a candidate. Finally, if $c$ matches with $d_{c'}$ (as in Fig. 5g), we have $|c.\text{seg}| = |c'.\text{seg}|$.

In this last case, $c$ must use the record of the overall comparison stored in $d_{c'}$ in combination with its own digit comparison with $d_{c'}$ to decide the comparison result. If $c.\text{id} < c'.\text{id}$, $c$ withdraws its candidacy. If $c.\text{id} > c'.\text{id}$, $c$ remains a candidate. Finally, if $c.\text{id} = c'.\text{id}$, $c$ may have just compared against its own identifier and thus initiates the solitude verification phase to determine if it is the only remaining candidate on the boundary.

As an aside, candidates who withdraw candidacy still reactivate inactive tokens when forwarding them. The delimiter token also resets inactive agents as it passes over them, preparing them for future identifier comparisons (Fig. 5e–g).

*Solitude Verification.* The goal of the solitude verification phase is for a candidate $c$ to check whether it is the last remaining candidate on its boundary. Solitude verification is triggered during the identifier comparison phase whenever a candidate detects equality between its own identifier and the identifier of a token sequence that traversed its segment. Such a token sequence can either be a candidate's own or that of another candidate with the same identifier. Once the solitude verification phase is started, it runs in parallel to the identifier comparison phase and does not interfere with it.

A necessary (but insufficient) condition for candidate $c$ to be the only remaining candidate on its boundary is if the next candidate along the boundary occupies the same node as $c$. The following algorithm checks this condition. Treat the directed edges of the boundary cycles as vectors in the two-dimensional Euclidean plane. The next candidate along the boundary, say $c'$, occupies the same node as $c$ if and only if the sum of the vectors corresponding to boundary

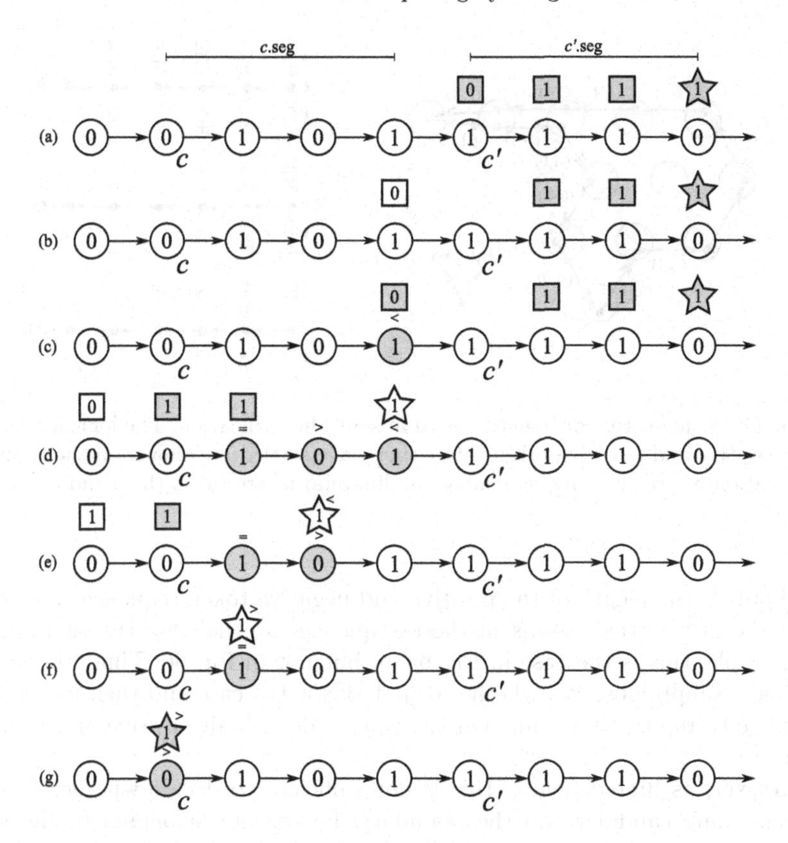


**Fig. 5.** Illustration of identifier comparison between $c$.id = 0101 and $c'$.id = 1110. Active elements are white while inactive elements are gray. The digit tokens are depicted as squares, while the star depicts the special delimiter token.

edges from $c$ to $c'$ is 0. To decide if this is the case in a local manner, $c$ defines a local two-dimensional coordinate system (e.g., as in Fig. 6) and uses a token passing scheme to check whether the $x$-components of the vectors sum to 0. An analogous scheme is used for the $y$-components, which runs in parallel.

First, $c$ sends an *activation token* in the direction of the cycle towards the next candidate. Whenever the token moves in the positive (resp., negative) direction of the locally defined $x$-axis, it creates a *positive token* (resp., *negative token*). These tokens are sent back towards $c$. Positive and negative tokens move independently of each other, but cannot overtake tokens of the same type. Once these tokens either reach $c$ or cannot move any closer to $c$, they become *settled.* Note that once all positive (or negative) tokens are settled, they form a consecutive sequence whose length corresponds to the number of tokens, as in Fig. 6a–c.

When the activation token reaches the next candidate, it reverses its movement back towards $c$, staying behind any positive or negative tokens that have not settled. Once they have settled, deciding whether the vectors sum to 0 can be done in a local manner: the vectors from $c$ to the next candidate sum to 0

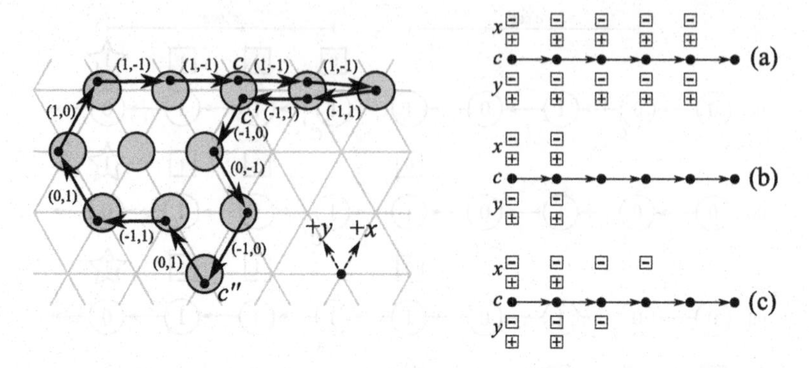


**Fig. 6.** The local vector construction used in solitude verification. The logical positions of the positive and negative tokens after they have settled are shown on the right, for the situation where the only remaining candidate(s) is/are (a) $c$, (b) $c$ and $c'$, and (c) $c$ and $c''$.

if and only if the length of the positive and negative token sequences are equal; i.e., if the last settled tokens in these sequences are held by the same agent. For example, this is the case in Fig. 6a–b, but not in Fig. 6c. Thus, the activation token simply observes whether or not this is the case and then moves back towards $c$ to report the result. On the way, it deletes all positive and negative tokens.

However, as hinted before, this is not sufficient to decide whether $c$ is the last remaining candidate on the boundary. For agents belonging to the same particle on the same boundary, as with $c$ and $c'$ in Fig. 6b, the vectors will sum to 0 despite there being at least two agents remaining. To handle this case, each particle assigns a locally unique identifier from $\{1, 2, 3\}$ to each of its agents in an arbitrary way. When the activation token reaches the next candidate, it reads its agent identifier and carries this information back to $c$. It is not hard to see that $c$ is the last remaining candidate on the boundary if and only if the vectors sum to 0 and the agent identifier stored in the activation token equals the agent identifier of $c$.

Finally, we address the interaction between the solitude verification and identifier comparison phases. If solitude verification is triggered for a candidate $c$ while $c$ is still performing a previously triggered execution of solitude verification, it ignores this trigger and simply continues with the already ongoing execution. Candidate $c$ may also be eliminated by the identifier comparison phase while it is performing solitude verification. In this case, $c$ waits for the ongoing solitude verification to finish and only then withdraws its candidacy.

*Boundary Identification.* Once a candidate $c$ determines that it is the only remaining candidate on its boundary, it initiates the boundary identification phase to check if it lies on the unique outer boundary. If so, the particle acting as $c$ declares itself the leader; otherwise, $c$ revokes its candidacy. This phase uses the fact that, due to the boundary setup phase, the outer boundary is oriented clockwise while any inner boundary is oriented counterclockwise (see Fig. 3b).

To distinguish between clockwise and counterclockwise oriented boundaries, a candidate $c$ sends a token along its boundary that sums the angles of the turns it takes according to Fig. 7, storing the results in a counter $\alpha$. When the token returns to $c$, there are two cases: $\alpha = 360°$ for the unique outer boundary, and $\alpha = -360°$ for any inner boundary. We encode $\alpha$ as $k \in \mathbb{Z}$ such that $\alpha = k \cdot 60°$. It is sufficient to store $k$ modulo 5 so that we have $k = 1$ for the outer boundary and $k = 4$ for an inner boundary, requiring only three bits of memory.

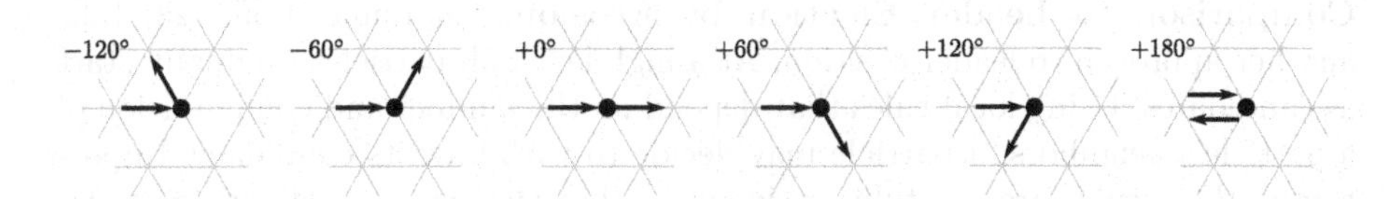


**Fig. 7.** Determining $\alpha$. The incoming and outgoing arrows represent the directions the token enters and leaves an agent, respectively, up to rotation.

**Analysis.** We now briefly discuss the correctness and runtime of the leader election algorithm, stating the main results.

*Correctness.* Recall that for a leader election algorithm to be correct, a single particle must irreversibly declare itself the leader and no other particle can ever do so. The boundary identification phase ensures that no candidate on an inner boundary can ever declare itself the leader, so the analysis focuses only on the outer boundary. For a single candidate agent $c^*$ on the outer boundary to become the leader, it must have the highest identifier on the boundary; i.e., $c^*.\text{id} > c.\text{id}$ for every other candidate $c \neq c^*$ on the outer boundary. The analysis upper bounds the probability of another candidate having the same highest identifier by an inverse polynomial in $n$, the number of particles in the system. Barring this event, a unique candidate $c^*$ with the highest identifier emerges. It eventually compares its own identifier with itself, triggering solitude verification. When solitude verification succeeds, boundary identification will indicate that $c^*$ is on the unique outer boundary. Thus, we can state the following result.

**Theorem 1.** *The algorithm correctly solves the leader election problem, w.h.p.*

*Runtime.* The first two phases of the algorithm (boundary setup and segment setup) are performed by each particle in their first activation, and thus are completed in the first round. The identifier setup phase takes $\mathcal{O}(\ell^2)$ rounds for a segment of length $\ell$, and the length of a segment on the outer boundary is $\mathcal{O}(\log n)$, w.h.p. Combining these results, the identifier setup phase completes for all candidates on the outer boundary in $\mathcal{O}(\log^2 n)$ rounds. Token sequences of the identifier comparison phase are then shown to traverse the outer boundary in $\mathcal{O}(L)$ rounds, where $L$ is the number of agents on the outer boundary. By a similar argument, the solitude verification phase is shown to take $\mathcal{O}(\ell)$ rounds, where $\ell$ is the number of agents between the current and next candidates. Finally, the boundary identification phase on the outer boundary it proven to take $\mathcal{O}(L)$ rounds. Therefore, all together, we obtain the following.

**Theorem 2.** *The algorithm solves the leader election problem in $\mathcal{O}(L) = \mathcal{O}(n)$ rounds, w.h.p., where $L$ is the number of agents on the outer boundary and $n$ is the number of particles in the system.*

**Comparison to Leader Election by Erosion.** Di Luna et al. [23] take another approach to leader election. At a high level, all particles originally start as candidates. Using local rules that depend on the number and configuration of a particle's neighbors, a particle may decide to withdraw its candidacy. Importantly, these rules are carefully designed so that the set of candidate particles always forms a connected component. At the end of this *erosion* process, there are one to three remaining candidates in a symmetric configuration. They give a deterministic protocol for attempting to break this symmetry, but it is possible that it may fail and a simple coin-flipping scheme must be used instead.

In comparison to the token-based algorithm described above, the erosion-based algorithm does not need long-range communication in the form of token passing, nor does it require that the system has a common chirality (i.e., notion of clockwise direction). However, it cannot handle particle systems which contain empty regions ("holes"), and was not shown to be extensible to situations where the particle system is moving, as is shown for the token-based algorithm in [21] (see Sect. 4.3 for more details). Their algorithm achieves the same runtime bound of $\mathcal{O}(n)$ rounds w.h.p., where $n$ is the number of particles in the system.

### 3.2 Spanning Forest Primitive

Without a global compass or shared coordinate system, particles of a particle system must use some local mechanisms to coordinate their movements. Many algorithms under the amoebot model solve this problem using the *spanning forest primitive*, originally introduced in [22]. This primitive organizes the particle system into one or more "trees", each of which is composed of *follower* particles following a single *root* particle. The root is responsible for directing the movement of its tree; the followers simply perform a follow-the-leader protocol to trail along behind the root. For simplicity, we will present this primitive with respect to a single spanning tree; in general, this primitive executes on each tree of the spanning forest concurrently and independently.

**Problem Description.** Consider an initially connected particle system $\mathcal{P}$ of contracted, *idle* particles, a designated *root* particle $R$, and a (simple) path $\mathcal{L} \subset G_\Delta$ beginning at the node occupied by $R$. We desire for the particle system to traverse the path $\mathcal{L}$ exactly without becoming disconnected.

Two caveats are needed for considering this problem in practice. First, the root particle $R$ is not usually predetermined. A root can either arises as the result of some local mechanism (e.g., "if adjacent to an object, become a root") or can be elected using leader election (Sect. 3.1) as a subprimitive. Second, $\mathcal{L}$ is usually not given explicitly; it is more often the path the root particle $R$ traverses as it executes some local algorithm. Nevertheless, for this standalone presentation of the primitive, we assume $R$ and $\mathcal{L}$ are given.

**Algorithm.** Particles can be in one of three states: idle, follower, or root. All particles (except the unique root $R$), are initially idle. When an idle particle $P$ is activated, it checks if it has a follower or root neighbor $Q$. If so, $P$ sets its parent pointer $P$.parent $\leftarrow Q$ and becomes a follower; otherwise, $P$ does nothing.

When a follower particle $P$ is activated, it first checks whether it is contracted or expanded. If $P$ is contracted and its parent $Q$ (pointed at by $P$.parent) is expanded, $P$ expands in a push handover with $Q$, forcing $Q$ to contract. In doing so, it may need to update $P$.parent so it still points to $Q$. Otherwise, if $P$ is expanded, there are two cases. First, if $P$ has no idle neighbors and no children—i.e., no neighbors $Q$ such that $Q$.parent points to the tail of $P$—$P$ simply contracts. Otherwise, if $P$ has a child $Q$ that is contracted, $P$ contracts in a pull handover with $Q$, forcing $Q$ to expand. Similar to the push handover, $P$ may need to update $Q$.parent so it still points to $P$.

When the root particle $R$ is activated, it also checks whether it is contracted or expanded. The rules for when it is expanded are the same as for followers: if it has no idle neighbors and no children, it contracts; otherwise, if it has a contracted child, it performs a pull handover. If $R$ is contracted, on the other hand, it checks if it has reached the end of the given path $\mathcal{L}$. If so, it does nothing; otherwise, it simply expands into the next node of $\mathcal{L}$.

*Example.* Consider an example run of the spanning forest primitive, illustrated in Fig. 8. All particles except the root $R$ are initially idle (Fig. 8a). Particle $P_1$ has the root $R$ as a neighbor and becomes a follower, setting $P_1$.parent $\leftarrow R$, while $R$ expands into the next node of the path $\mathcal{L}$ (Fig. 8b). Eventually, all idle particles become followers; a handover occurs between $R$ and $P_1$ (Fig. 8c). Root $R$ moves into the final node of $\mathcal{L}$ via expansions and handovers that propagate out through the rest of the tree (Fig. 8d–f). Eventually, all particles become contracted, and $\mathcal{L}$ has been traversed by the particle system (Fig. 8g).

*Root Swaps.* One caveat is necessary: it is possible that as $R$ traverses $\mathcal{L}$, it may be blocked by another part of the particle system, as in Fig. 9a. In this case, it will not be able to expand into the next node of $\mathcal{L}$ since this node is occupied by another particle $Q$. Instead, $R$ performs a *root swap* with $Q$ (if $Q$ is contracted), in which $R$ transfers the contents of its memory to $Q$, promotes $Q$ to become the new root, demotes itself to become a follower, and sets $R$.parent $\leftarrow Q$ (see Fig. 9b). Note that this does not disrupt the tree structure of the particle system, even if $Q$ was idle or has idle neighbors.

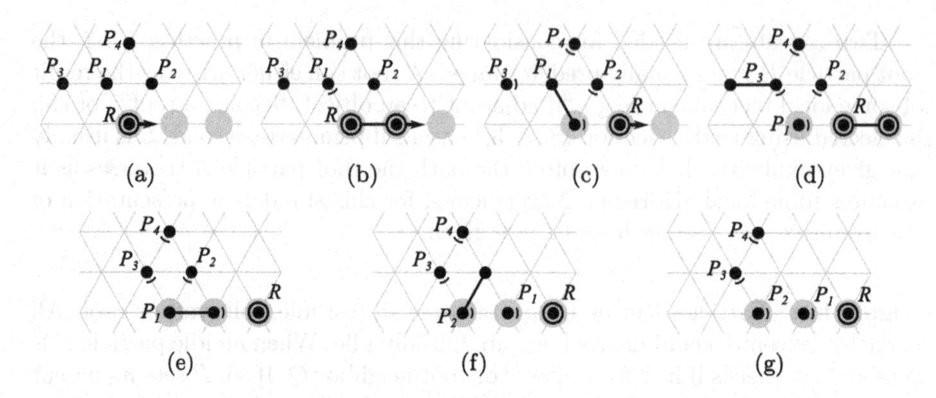


**Fig. 8.** Example of the spanning forest primitive. The root $R$ is shown as a black dot with a black circle, and the path $\mathcal{L}$ it follows is shown in gray. The black arcs point from a follower's head to its parent, and expanded particles have a black line connecting their head and tail.

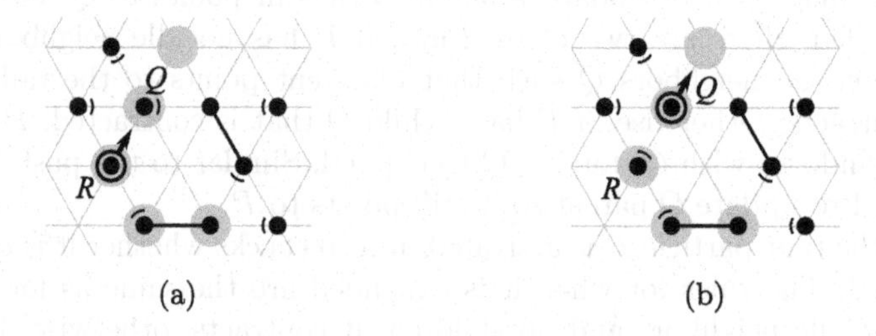


**Fig. 9.** Example of using a root swap to overcome a blocked path.

**Analysis.** We now state the main correctness and runtime results for the spanning forest primitive.

*Correctness.* Two properties must hold for the particle system $\mathcal{P}$ to correctly traverse path $\mathcal{L}$: *(i)* $\mathcal{P}$ always makes eventual progress along $\mathcal{L}$ until $\mathcal{L}$ has been entirely traversed, and *(ii)* $\mathcal{P}$ never becomes disconnected. These are the *liveness* and *safety* conditions for the spanning forest primitive, respectively. Liveness depends on the root's ability to continue expanding and performing handovers along $\mathcal{L}$. In [21,22], Derakhshandeh et al. prove that every expanded particle (including the root) eventually contracts. Thus, a contracted root will always be able to either expand into the next node of $\mathcal{L}$ or perform a root swap with the particle blocking it, since this blocking particle is also guaranteed to eventually be contracted. Safety follows more immediately. The only way $\mathcal{P}$ can become disconnected is if a particle with children or idle neighbors contracts outside a handover, breaking their connectivity to the tree. However, the protocol explicitly disallows this, so we have the following.

**Theorem 3.** *A particle system $\mathcal{P}$ following the spanning forest primitive will correctly traverse any given simple path $\mathcal{L}$.*

*Runtime.* A *dominance argument* [14,20] is used to analyze the runtime of the spanning forest primitive. These arguments first consider a parallel execution of an algorithm where all particles make progress in lock-step. This is often much easier to reason about. For the spanning forest primitive, it is shown that in $\mathcal{O}(n)$ parallel rounds, all $n$ particles are the root or its followers and the resulting tree forms a pattern of alternating expanded and contracted particles. Once in this configuration, the tree makes exactly one node of progress along $\mathcal{L}$ every 2 parallel rounds. Together, this implies that the spanning forest primitive traverses the entire path $\mathcal{L}$ in $\mathcal{O}(|\mathcal{L}|)$ parallel rounds.

Using careful case analysis, the concurrent, asynchronous execution of an algorithm is then shown to always make at least as much progress per round as its parallel counterpart. This implies that the runtime bound for the parallel execution is an upper bound on the runtime of the asynchronous execution. For the spanning forest primitive, this argument yields the following bound.

**Theorem 4.** *A particle system $\mathcal{P}$ following the spanning forest primitive will traverse a given path $\mathcal{L}$ in $\mathcal{O}(|\mathcal{L}|)$ rounds, where $|\mathcal{L}|$ is the length of path $\mathcal{L}$.*

### 3.3 Distributed Binary Counters

Many behaviors for programmable matter are realized by efficient algorithms under the amoebot model that only use constant-size state variables and messages. However, it can be useful in many applications to work with values that are on the order of $\mathcal{O}(\log n)$ or even $\mathcal{O}(n)$, where $n$ is the number of particles in the system. An example of such values appeared in leader election (Sect. 3.1), where candidates generated identifiers of logarithmic length to compete for leadership. Other applications could include, for example, measuring the size of an object with a particle system in order to replicate its shape.

Due to the constant-size memory constraint of the amoebot model (Sect. 2.1), individual particles cannot keep these larger values in memory by themselves. However, with the help of a leader particle, a system of $n$ particles can be organized into *distributed memory* in the form of a *distributed binary counter* that stores unsigned values up to $2^n - 1$ and supports increments and decrements by one as well as zero-testing. Porter and Richa first introduced an increment-only binary counter under the amoebot model in [42]; Daymude et al. extended this work to support decrements and zero-tests in [15].

**Problem Description.** Consider an initially connected particle system $\mathcal{P}$ of contracted particles organized in a simple path $P_0, P_1, \ldots, P_{n-1}$ with a leader particle $\ell = P_0$ at its start. We desire for the particle system to self-organize into a distributed binary counter that supports increments and decrements by one (initiated by the leader $\ell$). Additionally, the distributed binary counter should support reliable zero-testing, i.e., the leader $\ell$ should reliably be able to determine whether or not the counter's value is equal to zero.

Although the particles could move while maintaining the binary counter (e.g., according to the spanning forest primitive of Sect. 3.2), for ease of presentation we will assume they are static. It is also assumed that the leader $\ell$ never causes the counter to reach negative values; i.e., at any given time, $\ell$ has initiated at least as many increments as decrements. Presumably, $\ell$ could perform a zero-test before initiating a decrement to ensure the counter will not go negative, but for our presentation this nonnegativity assumption will suffice.

**Algorithm.** Each particle $P_i$ (for $0 \leq i < n$) has a bit value $P_i.\text{bit} \in \{\emptyset, 0, 1\}$, where $P_i.\text{bit} = \emptyset$ implies $P_i$ is not part of the counter; i.e., it is beyond the most significant bit. A *final token* $f$ represents the end of the counter. If a particle $P_i$ holds $f$, the counter value is represented by the bits of each particle from the leader $\ell$ (holding the least significant bit) up to and including $P_{i-1}$ (holding the most significant bit). Although not necessary for increments and decrements, utilizing $f$ will allow the leader to zero-test the counter locally and efficiently.

The leader $\ell$ is responsible for initiating counter operations, while the other particles carry these operations out using only local information. To increment the counter, the leader $\ell$ simply generates an increment token $c^+$ (assuming it was not already holding a token). Now consider this operation from the perspective of any particle $P_i$ holding a $c^+$ token, where $0 \leq i < n$. If $P_i.\text{bit} = 0$, $P_i$ can simply consume $c^+$ and set $P_i.\text{bit} \leftarrow 1$. Otherwise, if $P_i.\text{bit} = 1$, this increment needs to be carried over to the next most significant bit. As long as $P_{i+1}$ is not already holding a token, $P_i$ can forward $c^+$ to $P_{i+1}$ and set $P_i.\text{bit} \leftarrow 0$. Finally, if $P_i.\text{bit} = \emptyset$, this increment has been carried over past the counter's end, so $P_i$ must also be holding the final token $f$. In this case, $P_i$ simply forwards $f$ to $P_{i+1}$ and sets $P_i.\text{bit} \leftarrow 1$.

Decrements are similar; when considering this operation from the perspective of any particle $P_i$ holding a $c^-$ token, where $0 \leq i < n$, the cases for $P_i.\text{bit} \in \{0, 1\}$ are anti-symmetric to those for the increment, with two exceptions. First, we only allow $P_i$ to consume $c^-$ and set $P_i.\text{bit} \leftarrow 0$ if $P_{i+1}$ is not also holding a $c^-$. While not necessary for the correctness of the decrement operation, this will enable conclusive zero-testing. Second, if $P_i.\text{bit} = 1$ and $P_{i+1}$ is holding $f$, then $P_i$ is the most significant bit. So this decrement shrinks the counter by one bit; thus, $P_i$ consumes $c^-$, takes $f$ from $P_{i+1}$, and sets $P_i.\text{bit} \leftarrow \emptyset$.

Finally, the zero-test operation: if $P_1$ is holding a decrement token $c^-$ and $P_1.\text{bit} = 1$, $\ell$ cannot perform the zero-test conclusively. Otherwise, the counter value is 0 if and only if $\ell.\text{bit} = 0$, $P_1$ is holding the final token $f$, and $P_1$ is not holding an increment token $c^+$.

**Analysis.** We only present the correctness and runtime results for the distributed binary counter primitive from [15], as these subsume those in [42].

*Correctness.* For the distributed binary counter primitive to be correct, it must eventually yield the same values as a centralized counter, assuming both are given the same sequence of operations as input. Because the algorithm ensures that the increment and decrement tokens are processed in the same order that their respective operations are initiated, the resulting distributed counter will correctly process an arbitrary number of increment and decrement operations (assuming the counter value remains in $\{0, 1, \ldots, 2^n - 1\}$). The zero-test operation is shown to always eventually be available (i.e., $P_1$ is not holding a decrement token $c^-$ or $P_1$.bit $\neq 1$) and is reliable whenever it is available. Together, this yields:

**Theorem 5.** *A particle system $\mathcal{P}$ running the distributed binary counter primitive will maintain a distributed counter that eventually yields the same values as a centralized counter, given the same sequence of increment, decrement, and zero-test operations.*

*Runtime.* The following runtime bound is proven using a careful dominance argument (see, e.g., Sect. 3.2) applied to the progress of the increment and decrement tokens in the counter.

**Theorem 6.** *Given any nonnegative sequence of $m$ operations, a particle system $\mathcal{P}$ running the distributed binary counter primitive processes all operations in $\mathcal{O}(m)$ rounds.*

**Applications.** We briefly discuss some applications of the distributed binary counter primitive whose details are beyond the scope of this chapter. In [42], Porter and Richa give an algorithm using the increment-only counter for matrix-vector multiplication. For an $a \times b$ matrix (i.e., $a$ rows and $b$ columns), this algorithm requires $\mathcal{O}(ab)$ rounds to set up and an additional $\mathcal{O}(a + b)$ rounds to perform the multiplication. By performing a sequence of these matrix-vector multiplications, an $a \times b$ matrix can be multiplied by an $b \times c$ matrix in $\mathcal{O}(ab + c(a + b))$ rounds. Applications of these matrix multiplication algorithms to edge detection and color transformation in image processing are described in [42].

In [15], Daymude et al. use the distributed binary counter primitive to aid in *convex hull formation*, in which a particle system must seal an object using as few particles as possible. At a very high level, a leader particle traverses the surface of the given object, updating its estimate of the object's convex hull as it goes. Using its followers as a distributed binary counter, it stores the distances from its current position to each of the six half-planes constituting the convex hull. Once it completes its estimation, it can use these stored distances to lead its followers along the convex hull itself, eventually sealing the object as desired.

## 4 Deterministic Algorithms Under the Amoebot Model

This section is devoted to the deterministic algorithms under the amoebot model. These algorithms use the full capabilities of the amoebot model, including several of its extensions (Sect. 2.3) and algorithmic primitives (Sect. 3). We focus on two basic problems for programmable matter: forming a shape using all the particles in the system, and coating an object. In *basic shape formation* (Sect. 4.1), a follow-the-leader type protocol is used to construct regular shapes such as lines and hexagons. In *general shape formation* (Sect. 4.2), a more complex algorithm is given for constructing a much broader range of shapes. Finally, in *object coating* (Sect. 4.3), an algorithm is given for coating surfaces in one or more layers of particles as evenly as possible.

### 4.1 Basic Shape Formation

Shape formation is one of the most immediate and natural applications of programmable matter. A "lump" of programmable matter should be able to reconfigure into new shapes based on user input or autonomous sensing of its environment. Moreover, the final shape should scale with the size of the initial "lump". In *basic shape formation*, a particle system self-organizes to form regular, geometric shapes. Here, we focus on lines, hexagons, and triangles; however, the general framework can be applied to obtain other shapes as well. Line formation was first investigated in [22], while hexagon and triangle formation were introduced in [19]. Detailed versions of all three algorithms can be found in [52].

**Problem Description.** An instance of a *shape formation* problem has the form $(I, \mathcal{G})$, where $I$ is the initial configuration of the particle system and $\mathcal{G}$ is a set of goal configurations. An instance is *valid* if (*i*) $I$ and all configurations of $\mathcal{G}$ are each connected, and (*ii*) $I$ is composed of all contracted, idle particles and a unique *seed* particle. An algorithm solves a valid instance of the shape formation problem if, starting from initial configuration $I$, the algorithm terminates in a configuration of $\mathcal{G}$, after which all particles are contracted and no longer move.

The specific *line formation*, *hexagon formation*, and *triangle formation* problems simply define the desired set of goal configurations $\mathcal{G}$ as all configurations of straight lines, (almost) regular hexagons, and (almost) regular triangles, respectively. Note that, depending on the number of particles in $I$, the outermost layer of a hexagon or triangle may not be complete.

**Algorithm.** Particles can be in one of four states: idle, follower, root, and retired. All particles are initially idle except the unique seed particle, which is always retired. At a high level, this algorithm constructs the desired shape one particle at a time in a snake-like fashion, starting at the seed. Thus, at any point in time before termination, the structure of retired particles partially forms the goal configuration.

The spanning forest primitive (Sect. 3.2) is used to organize the system. In basic shape formation, root particles always traverse the structure of retired particles in a clockwise direction, trailing their followers behind them. The most recent particle to retire (starting with the seed), say $P$, keeps a pointer $P$.retireDir to the next node to be filled by a retired particle. When a contracted root $Q$ finds that it occupies this node (by seeing $P$.retireDir pointing to it), it retires, locally calculates the next node to be filled by a retired particle, and sets $Q$.retireDir to point to it. As other roots continue their traversal and their followers become roots when they touch the surface of retired particles, eventually all particles retire, forming the desired shape.

It remains to specify how a newly retired particle calculates the next node to add to the structure. Recall from Sect. 2.1 that each particle keeps a set of ports (one for each edge incident to the node(s) it occupies) labeled in clockwise order. Suppose a particle $P$ has already retired and set $P$.retireDir, and another particle $Q$ has just retired in the position referenced by $P$.retireDir. Let $i$ be the port label of $Q$ pointing to $P$. For line formation, $Q$ sets $Q$.retireDir $\leftarrow (i+3) \bmod 6$; this specifies that the next node to join the line should be opposite the direction of the existing structure, resulting in a straight line (see Fig. 10). For hexagon formation, $Q$ sets $Q$.retireDir to the label of the first port clockwise from $i$ that does not point to another retired particle; this causes the hexagon to be formed in counterclockwise order (see Fig. 11).

Finally, for triangle formation, a two-step mechanism is used[5]. Particle $Q$ first checks the position immediately clockwise from $i$, say $j = (i+1) \bmod 6$. If $j$ does not point to a retired particle (as in Fig. 12c), $Q$ is on a new side of the triangle and sets $Q$.retireDir $\leftarrow j$ to grow a new layer on this side. Otherwise, if $j$ points to a retired particle (as in Fig. 12d), $Q$ is simply extending an existing layer. So it sets $Q$.retireDir $\leftarrow (j+2) \bmod 6$.

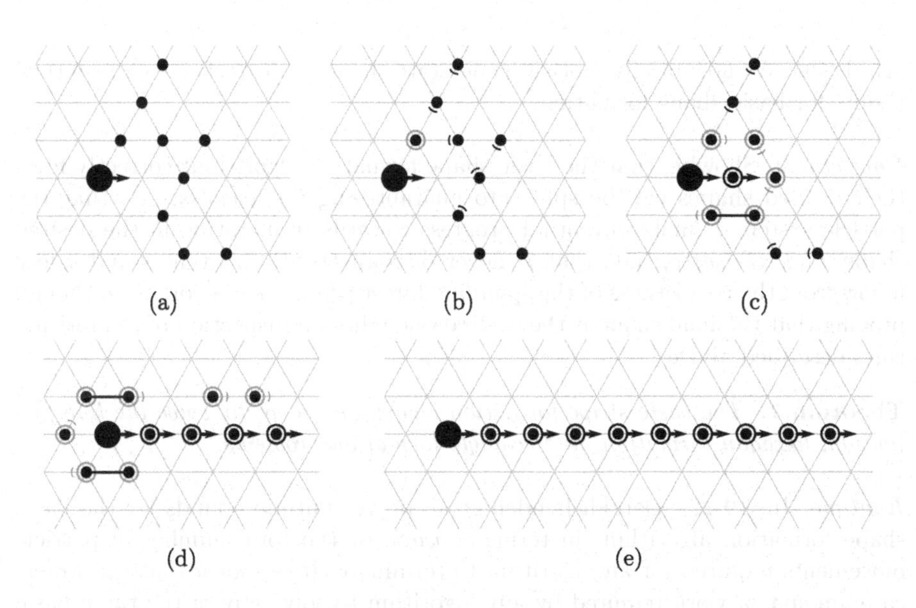


**Fig. 10.** Example of line formation with 10 particles, where the seed is depicted as a large black circle. Followers are shown with black arcs pointing to their parents, roots are shown with a gray circle, and retired particles are shown with a black circle. A retired particle's retireDir is shown as a black arrow.

[5] For this presentation, we use a simplified scheme that results in a triangle with the seed at its center; the original scheme given in [19,52] is significantly more complex and results in a triangle with the seed at one vertex.

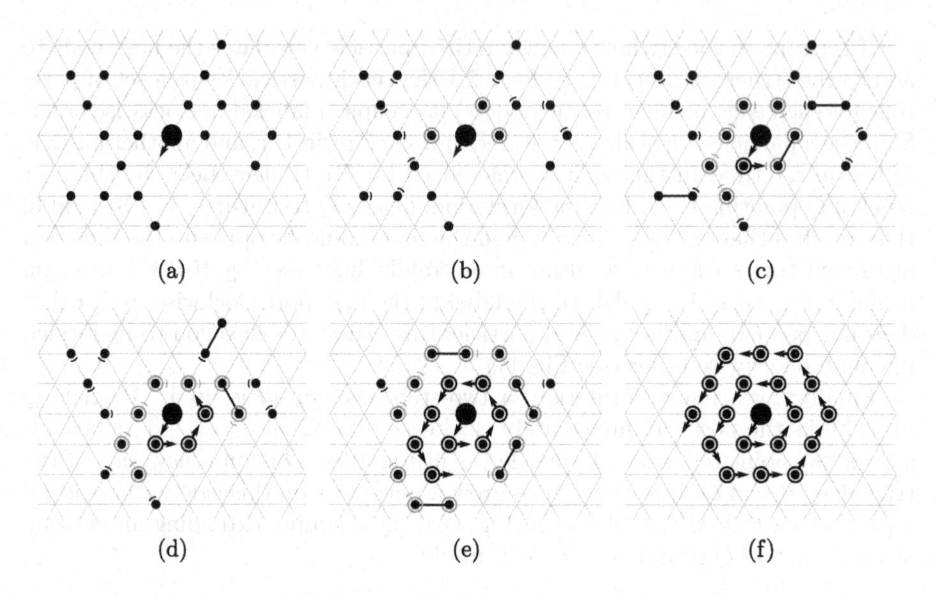

**Fig. 11.** Example of hexagon formation with 18 particles.

**Analysis.** We now briefly motivate and state the main correctness and runtime results for basic shape formation.

*Correctness.* Showing that the basic shape formation algorithms correctly form their desired shapes can be split into guaranteeing (*i*) *liveness*; i.e., that the particle system $\mathcal{P}$ makes eventual progress towards terminating in the desired shape, and (*ii*) *safety*; i.e., that $\mathcal{P}$ never disconnects. Both liveness and safety follow from the correctness of the spanning forest primitive (Theorem 3), though proving that the final shape is the desired one relies on inspection of the retiring rules described above.

**Theorem 7.** *The basic shape formation algorithms correctly solve the line formation, hexagon formation, and triangle formation problems.*

*Runtime.* In [19,22], Derakhshandeh et al. prove runtime bounds for the basic shape formation algorithms in terms of *work*, or the total number of particle movements required for an algorithm to terminate. It is shown that the worst-case amount of work required by any algorithm to solve any of the three basic shape formation problems (line formation, hexagon formation, or triangle formation) is $\Omega(n^2)$, where $n$ is the number of particles in the system. The basic shape formation algorithm described above is shown to match this worst-case bound in all three cases, terminating in $\mathcal{O}(n^2)$ work.

In his Ph.D. thesis, Strothmann uses a dominance argument (see, e.g., Sect. 3.2) to prove runtime bounds for these three algorithms in terms of asynchronous rounds [52]. Formally, his thesis gives the following theorem.

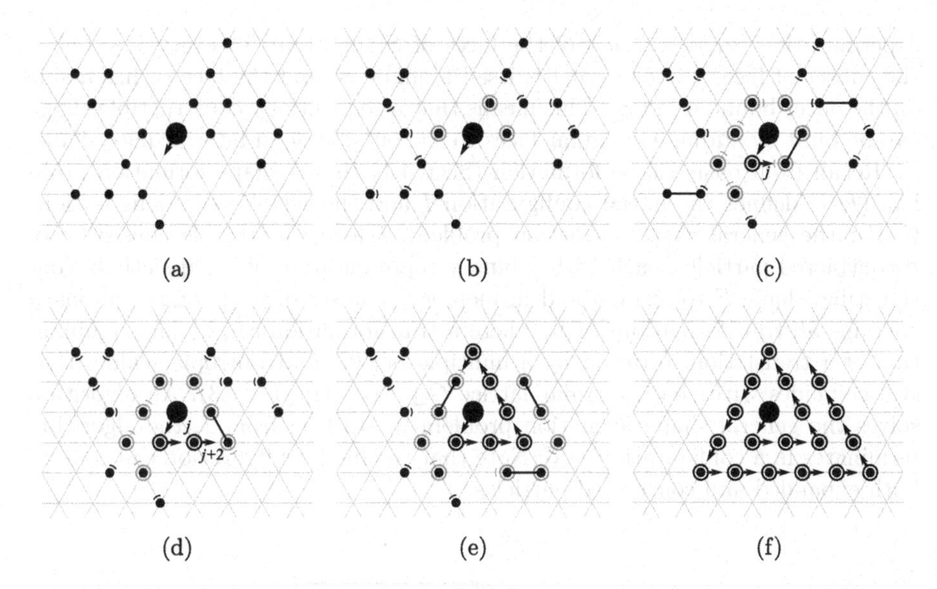


**Fig. 12.** Example of triangle formation with 18 particles. Unlike in Fig. 11, the resulting shape in (f) is not fully regular due to the number of particles. Additionally, note that (a)–(c) are the same as Fig. 11a–c; the basic shape formation algorithms only differ in the way retired particles set their pointers.

**Theorem 8.** *The total number of rounds required by the basic shape formation algorithms to solve the line formation, hexagon formation, or triangle formation problem is $\mathcal{O}(n)$, where $n$ is the number of particles in the system.*

## 4.2 General Shape Formation

The basic shape formation algorithm of Sect. 4.1 has two major disadvantages that the *general shape formation algorithm* of this section seeks to alleviate. First, it can only construct simple shapes that are amenable to being built one particle at a time along a continuous path. Second, as stated in Theorem 8, it generally requires $\mathcal{O}(n)$ rounds to construct a shape of $n$ particles, even when the shape could be formed more efficiently. The general shape formation algorithm [20, 29] can form a much broader class of shapes and, assuming the particle system is well initialized, can do so as fast as any other local-control algorithm.

**Problem Description.** Let $S$ be a finite set of faces in the triangular lattice $G_\Delta$. $S$ is a *shape* if the faces of $S$ are connected—i.e., there exists a path from any face in $S$ to any other via pairs of faces that share a side—and the number of faces $s = |S|$ is constant. A shape $S$ is *sequentially constructible*[6] if there exists

[6] The original publication on "universal" shape formation [20] claimed the algorithm could construct any shape with a constant number of faces. However, Gmyr corrected an oversight in this paper's analysis in his Ph.D. thesis [29] and, as a result, the class of shapes had to be restricted to sequentially constructible shapes.

a permutation $(a_1, a_2, \ldots, a_s)$ of the faces of $S$ such that for every $1 \leq i \leq s$, the subset of faces $(a_1, a_2, \ldots a_i)$ is itself a shape $S_i$ and the face $a_i$ has a side on the outer boundary of $S_i$. This means that a sequentially constructible shape can be built by adding triangular faces to the outside of an intermediate shape.

Recall from basic shape formation (Sect. 4.1) that a shape formation problem $(I, \mathcal{G})$ defines the initial configuration $I$ and the set of goal configurations $\mathcal{G}$. For the general shape formation problem, $I$ forms a triangle consisting of $n$ contracted particles, each with a binary representation of a sequentially constructible shape $S$ to form stored in memory. The set of goal configurations $\mathcal{G}$ contains all transformations of $S$, where a transformation can be any combination of a translation, rotation by a multiple of 60°, and isotropic scaling that still coincides with the triangular lattice $G_\Delta$ (see, e.g., Fig. 13). An algorithm solves the general shape formation problem if, starting from $I$, the algorithm terminates in a configuration of $\mathcal{G}$. Note that the final configuration can contain both expanded and contracted particles.

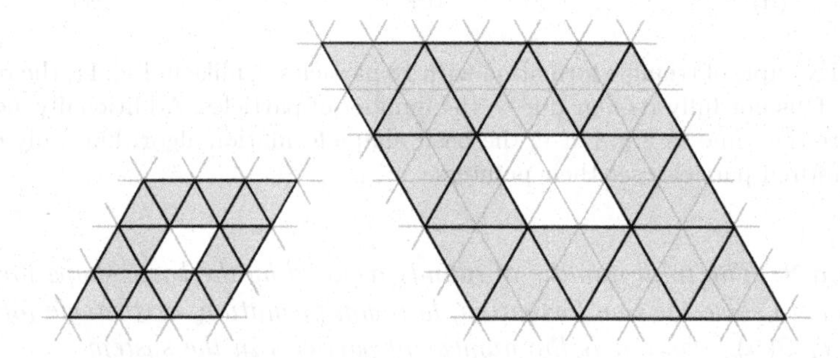

**Fig. 13.** A sequentially constructible shape consisting of 16 faces (left) and one of its transformations (right) involving a 120° rotation counterclockwise and an isotropic 2× scale-up.

**Algorithm.** We present the general shape formation algorithm in several parts. First, we describe several movement primitives that are used throughout the algorithm. Next, we give highlights of an algorithm that transforms the initial triangle of particles into a useful intermediate structure. Finally, we describe the actual formation process. Our presentation prioritizes the algorithm's main ideas over its details, and we refer the interested reader to [20, 29] for a more thorough description.

*Movement Primitives.* Several movement primitives are used throughout the algorithm to move large sets of particles in a coordinated and efficient manner. The first of these is *chain movement*, in which a "chain" of particles moves along a simple path $\mathcal{L}$ without disconnecting. This is essentially the spanning forest primitive described in Sect. 3.2, but the set of particles moving along $\mathcal{L}$ is already organized into a simple path instead of a tree.

The remaining movement primitives operate on a set of contracted particles that form a triangle. Four primitives are available to such triangles: *expansion*, *contraction*, *rotation*, and *shift*. Figure 14 depicts the first three of these primitives. An expansion of a triangle results in an *expanded triangle*, which is a rhombus composed of two triangles that share a side and are each the same size as the original triangle. A contraction of a triangle transforms an expanded triangle back into a triangle. Together, an expansion and contraction of a triangle rotates a triangle by 60° about one of its vertices. Finally, a shift of a triangle moves all of its particles by one node in a common direction.

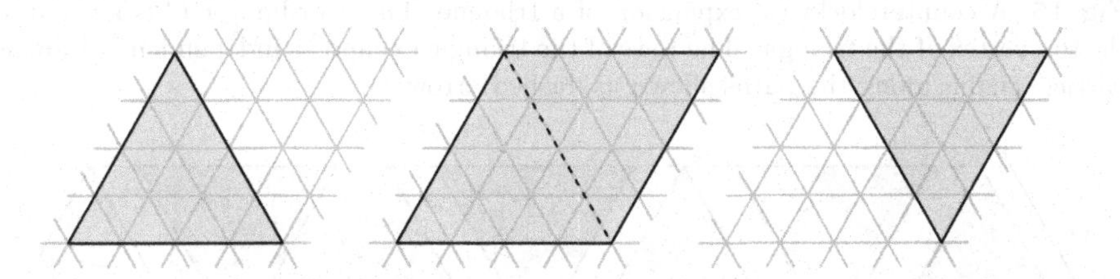

**Fig. 14.** Expansion, contraction, and rotation of a triangle. A triangle (left) can expand to form an expanded triangle (middle), which in turn can contract back to a triangle (right), rotating the original triangle by 60°.

At a high level, the triangle movement primitives are initiated by a *triangle coordinator* that occupies one of the triangle's three vertices. The coordinator organizes the particles in each row of its triangle into particle chains that can then be moved according to the chain movement primitive; for example, Fig. 15 depicts how the particle chains are moved in an expansion of a triangle. The coordinator initiates and completes these chain movements using a token passing scheme whose details we omit.

*Reaching the Intermediate Structure.* The first step of the general shape formation algorithm is to reconfigure the particle system from its initial triangle shape to the intermediate structure depicted in Fig. 16. This intermediate structure is composed of $\Delta$ equilateral triangles of side length $\ell$ arranged in a line and a remainder composed of too few particles to form an additional triangle. All particles in the intermediate structure should be contracted.

The side length $\ell$ must be chosen carefully so that the resulting number of triangles $\Delta$ in the intermediate structure is sufficient to construct the desired shape $S$. More specifically, $\Delta$ should satisfy $(3/4)s + 1 \leq \Delta \leq s - 3$, where $s = |S|$ is the number of faces in $S$; the choice of these particular bounds is explained in [29]. If $L$ is the side length of the initial triangle of all $n$ particles, then $\ell$ is chosen to be $\ell = \lceil L/\lfloor c \cdot \sqrt{s} \rfloor \rceil$, where $c < 1$ is a constant. It is shown in [29] that there is a $c$ which yields the desired bounds on $\Delta$, assuming $s$ and $L$ are sufficiently large.

To form this intermediate structure, the particle system first performs leader election (Sect. 3.1) to elect a unique leader particle on the boundary of the initial

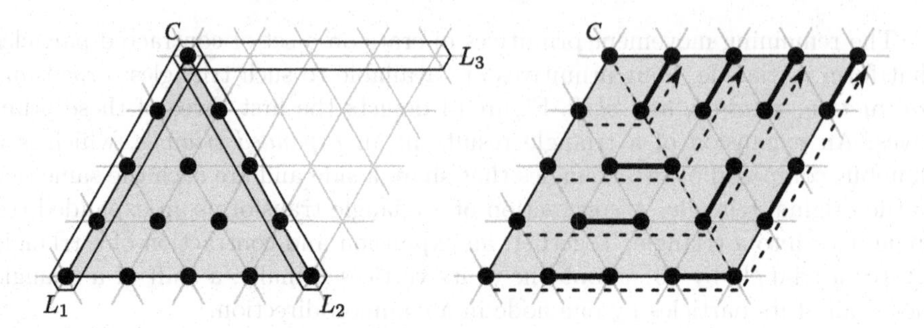


**Fig. 15.** A counterclockwise expansion of a triangle. The coordinator $C$ is located at the top vertex of the triangle. The rows of the triangle expand as independently moving particle chains along the paths shown as dashed arrows.

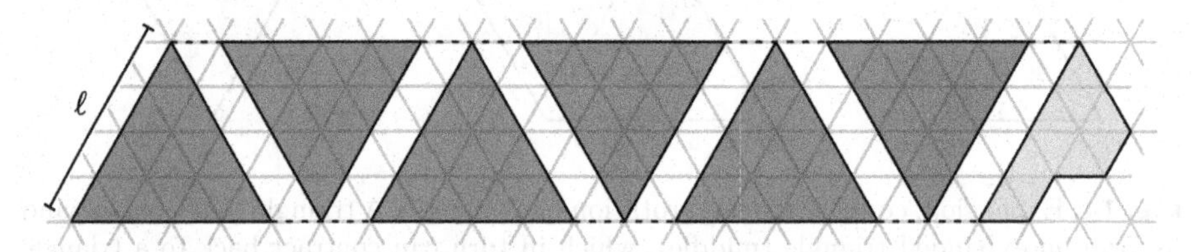


**Fig. 16.** The intermediate structure for general shape formation. The triangles (dark gray) form a straight line, and the remainder (light gray) is not large enough to form another triangle.

triangle. The leader then uses a token to transfer its leadership to a particle at one of the triangle's vertices. Next, the particle system determines the value of $\ell$. However, since $\ell = \Omega(\sqrt{n})$, a single particle cannot keep $\ell$ in its constant-size memory. So the leader initiates the following token passing scheme to store $\ell$ in a distributed fashion over multiple particles. The leader generates a *counter token* and sends it down one of the initial triangle's sides. The counter token stores the number of steps it has taken modulo $\lfloor c \cdot \sqrt{s} \rfloor$, which is a constant since $c$ and $s$ are. Whenever the token's count is 0 (starting at the leader), a *marker token* is generated that moves back towards the leader without overtaking other marker tokens. When the counter token is consumed by the particle at the end of the triangle's side, exactly $\ell$ marker tokens will have been generated. Thus, when all marker tokens have moved back towards the leader as far as possible—which can be detected by the leader using a process similar to "settling" in the solitude verification phase of leader election (Sect. 3.1)—exactly $\ell$ particles, starting with the leader, will be holding marker tokens.

Using $\ell$, the leader coordinates a recursive process to form the intermediate structure. The process first splits the current triangle into a smaller triangle and an isosceles trapezoid with legs of length $\ell$ (see Fig. 17). The trapezoid immediately becomes part of the intermediate structure, while the smaller triangle must be rotated, shifted twice, and rotated again to be placed in line with the next part of the intermediate structure. This process then recurs on the smaller triangle until a triangle with side length at most $\ell$ is moved, completing the intermediate structure.

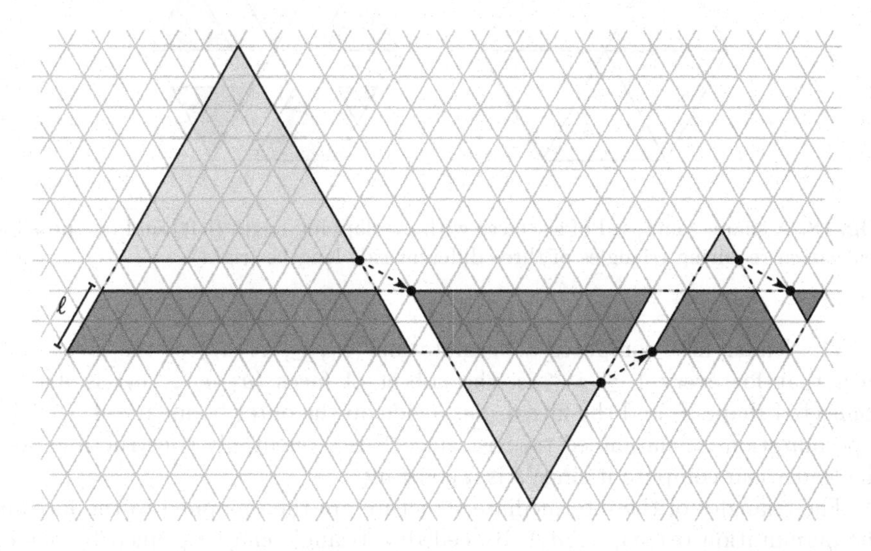


**Fig. 17.** Constructing the intermediate structure. The parts shown in dark gray form the intermediate structure, while the triangles that are moved are shown in light gray. The dashed arrows show these triangle movements. For example, the leftmost triangle is rotated 60° about the marked vertex, shifted once right and once down-right, and then rotated 120° about the next marked vertex.

*Forming the Final Shape.* To form the desired shape $S$, the leader coordinates a process that sequentially adds triangles from the intermediate structure to the outside of the shape under construction. In [29], Gmyr first describes a "simple algorithm" which captures the main ideas of the general shape formation algorithm. However, there are three issues the simple algorithm does not address. First, faces of $S$ have overlapping edges while triangles formed by particles do not, since each node of $G_\Delta$ can be occupied by at most one particle. Thus, realizing a shape may require pruning triangles to different side lengths (as in Fig. 18) and reincorporating the pruned particles elsewhere. Second, the remainder of particles in the intermediate structure that did not form a complete triangle must somehow be incorporated into the final shape. Finally, the algorithm for constructing the intermediate structure constructs at most $s - 3$ triangles, but the desired shape $S$ has $s$ faces. Thus, some expanded triangles need to make up for the missing triangles. We will only present the simple algorithm for general shape formation, and refer the interested reader to [29] for the full algorithm.

In the simplified setting, suppose that the particles pruned to create smaller triangles do not need to be reincorporated and that the intermediate structure consists of exactly $s$ triangles without a remainder. The algorithm constructs a scaled representation of the desired shape $S$, where each face has side length $\ell$.

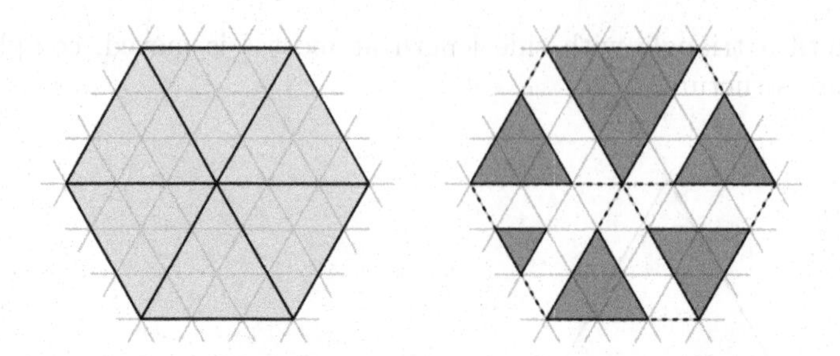

**Fig. 18.** A shape composed of six faces with overlapping edges (left) and one possible realization requiring triangles of three different side lengths (right).

The leader first computes a permutation $(a_1, a_2, \ldots, a_s)$ of the faces of $S$ such that for every $1 \leq i \leq s$, the subset of faces $(a_1, a_2, \ldots a_i)$ is itself a connected shape $S_i$ and the face $a_i$ has a side on the outer boundary of $S_i$. Such a permutation is guaranteed to exist since $S$ is sequentially constructible, and the leader can compute it since $s$ is a constant.

The triangles of the intermediate structure are then mapped to the faces in the permutation $(a_1, a_2, \ldots, a_s)$. If needed, a triangle can be pruned by one or two rows of particles; the pruned rows are moved out of the way using chain movements. The triangles of the intermediate structure are then added to the outer boundary of the shape in the order defined by the permutation using triangle rotations and shifts. If the intermediate structure or pruned rows obstruct the placement of a triangle, they are simply moved out of the way.

**Analysis.** As described earlier, the intermediate structure with a desired number of triangles is constructed correctly, assuming the number of faces $s$ in the goal shape and the number of particles in the system $n$ are sufficiently large. It is shown in [20,29] that this intermediate construction completes in $\mathcal{O}(\sqrt{n})$ rounds, w.h.p. The w.h.p. qualifier is inherited from leader election (Sect. 3.1).

A triangle can be moved from the intermediate structure to its goal position in the final shape using only a constant number of triangle rotations and shifts, each of which requires $\mathcal{O}(\ell)$ rounds, where $\ell = \mathcal{O}(\sqrt{n})$ is the side length of the triangle. The number of rows pruned from triangles is constant, and each has length $\mathcal{O}(\ell) = \mathcal{O}(\sqrt{n})$. These pruned rows are moved out of the way a distance $\mathcal{O}(\sqrt{n})$ at most a constant number of times, so moving the pruned rows requires $\mathcal{O}(\sqrt{n})$ rounds in total, by Theorem 4. Therefore, all together, the following runtime bound is obtained.

**Theorem 9.** *The general shape formation algorithm constructs any sequentially constructible shape in $\mathcal{O}(\sqrt{n})$ rounds, w.h.p.*

We conclude with the following theorem, which shows that the general shape formation algorithm achieves the optimal bound.

**Theorem 10.** *Any local-control algorithm for forming a non-triangular shape $S$ requires $\Omega(\sqrt{n})$ rounds in the worst case.*

### 4.3 Object Coating

*Object coating* is another natural application of programmable matter. For example, one can imagine a particle system coating remote parts of a bridge to identify stress points, or coating a vehicle as a layer of smart paint. Instead of developing a class of coating algorithms that each coats a specific object, a more elegant approach might seek one general algorithm that dynamically adapts to any given object. We present such an algorithm in this section. The *universal coating algorithm* was defined and proven to be correct in [21], and its runtime analysis and proofs of worst-case optimality appeared in [14].

**Problem Description.** We begin with some terminology. *Layer $i$* of an object $O$ is the set of unoccupied nodes in $G_\Delta$ whose shortest path to $O$ has length $i$. Let $B_i$ denote the number of nodes in layer $i$. An object $O$ has a *tunnel of width $k$* if the subgraph of $G_\Delta$ induced by the non-object nodes is not $k$-connected; for example, Fig. 19 depicts an object with a tunnel of width 1.

An instance of the object coating problem has the form $(\mathcal{P}, O)$, where $\mathcal{P}$ is a system of $n$ particles and $O$ is a static object to be coated. An instance is *valid* if: (*i*) all particles in $\mathcal{P}$ are initially contracted and in an idle state, (*ii*) the nodes occupied by particles of $\mathcal{P}$ and the object $O$ induce a connected subgraph of $G_\Delta$, (*iii*) $O$ does not contain holes, and (*iv*) $O$ does not contain any tunnels of width $2(\lceil n/B_1 \rceil + 1)$. Coating an object with narrow tunnels requires technical mechanisms that complicate the protocol without contributing to the main idea of coating, so these types of objects are not considered. An algorithm solves a valid instance $(\mathcal{P}, O)$ of the object coating problem if it terminates in a configuration where all particles of $\mathcal{P}$ are as close to the object $O$ as possible, after which no particle ever moves or changes state. Intuitively, this means that $\mathcal{P}$ coats $O$ as evenly as possible.

**Fig. 19.** An object with a tunnel of width 1.

**Algorithm.** The universal coating algorithm is composed of several algorithmic primitives that run concurrently without any underlying synchronization. The first of these is the spanning forest primitive described in Sect. 3.2, which orients the particles towards the object. The *complaint-based coating primitive* is responsible for coating layer 1 of the object (also called the *surface layer*). The *node-based leader election primitive* is a variant of the leader election algorithm described in Sect. 3.1 that, instead of directly electing a leader particle, elects a leader node on the surface layer. After the surface layer is completely coated, the particle occupying the leader node becomes the leader particle. This leader particle triggers the *general layering primitive*, which allows each layer $i \geq 2$ to form once layer $i-1$ is complete. We describe each of these primitives in detail after introducing some preliminaries.

*Preliminaries.* Particles can be in one of five states: idle, follower, root, marker, and retired. Throughout the coating algorithm, a particle $P$ keeps track of its current layer number, denoted $P$.layer. However, to respect the constant-size memory constraint, $P$.layer is stored modulo 4 in a particle's memory. A layer is said to be *filled* if all nodes in that layer are occupied by retired particles.

*Spanning Forest Primitive.* The spanning forest primitive for coating extends the spanning forest primitive in Sect. 3.2. Instead of assuming the root particles are predetermined, idle particles become roots if they are adjacent to the object or a retired particle. Additionally, if the new root was adjacent to the object, it makes the node it occupies a *leader candidate node* and begins to assist in leader election, described below. As usual, the root is responsible for leading its tree of followers; the path it traverses is defined by the complaint-based coating and general layering primitives, described below. It is possible that a root may be blocked by particles of another tree during its path traversal. Instead of using the root swap operation defined in Sect. 3.2, blocked roots in the coating algorithm simply wait. An unblocked root is called a *super-root.*

*Complaint-Based Coating.* The complaint based-coating primitive is responsible for coating the surface layer with particles. When an idle particle becomes a follower according to the spanning forest primitive, it generates a *complaint flag.* Complaint flags are forwarded from particles to their parents through the spanning forest. In more detail, each particle can hold at most two complaint flags. Whenever a particle $P$ is holding at least one complaint flag, it forwards one flag to its parent as long as its parent is holding less than two flags.

These complaint flags eventually accumulate at and behind a super-root of the spanning forest. A super-root can only expand along its traversal path if it is holding a complaint flag and, when it expands, it consumes one complaint flag (see Fig. 20b–c). No other particles (i.e., roots or followers) need complaint flags to perform their movements. When a root is in a situation where it could perform a handover with either another root on the object or a follower not yet on the object, it gives preference to the follower (see Fig. 20d–e). These movement rules ensure that a particle not yet in the surface layer will eventually join, provided the surface layer is not already filled.

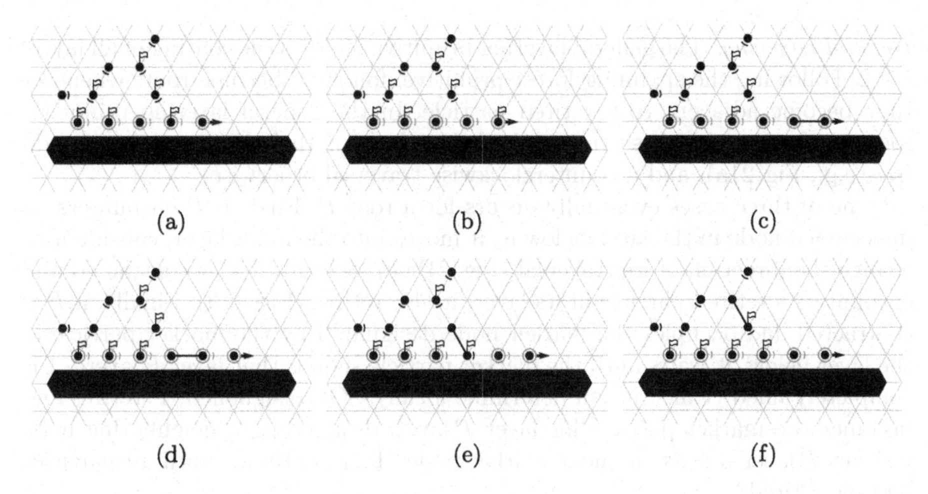


**Fig. 20.** Example of the complaint-based coating primitive. The roots (black dots with gray circles) are coating the object (black polygon) behind the super-root (gray circle with black arrow). Complaint flags are forwarded to the super-root to enable movement and allow more followers to join the surface layer.

*Node-Based Leader Election.* The node-based leader election primitive runs in parallel with the complaint-based coating primitive to elect a leader node on the surface layer. This primitive uses a variant of the algorithm presented in Sect. 3.1, where the leader candidates are nodes instead of static particles. The roots in the surface layer facilitate the competition between leader node candidates by storing and transferring all the tokens and state information used by the leader election algorithm for the node(s) they occupy. Root movements are handled carefully so that no leader election information is lost or moved to a different node; for example, during a handover between an expanded particle $P$ and a contracted particle $Q$, $P$ transfers all leader election information about the node occupied by its tail to $Q$.

The node-based leader election primitive will not successfully elect a leader node until all nodes of the surface layer are occupied. Once a leader node emerges, the first contracted particle to occupy it becomes both retired and a marker. This marker particle designates a neighboring node in layer 2 as a *marker node*, which will act as the starting point for the next layer. If a contracted root is following a retired particle, it also becomes retired, causing the surface layer to fill with retired particles in counterclockwise order. Once the surface layer is completely filled with retired particles, the general layering primitive is activated.

One caveat is necessary before presenting the final primitive. If the surface layer is longer than the number of particles in the system (i.e., $B_1 > n$), a leader node will never be elected. However, the complaint-based coating primitive will still continue until all complaint flags are consumed by the super-root, bringing all particles in the system into the surface layer. This results in a successful coating of the object; all particles are as close to the object as possible.

*General Layering.* The general layering primitive handles the coating of all layers $i \geq 2$. Following the spanning forest primitive, followers become roots whenever they become adjacent to a retired particle. In the general layering primitive, roots perform a clockwise traversal of their layer if their layer number is odd (see, e.g., Fig. 21a), and a counterclockwise traversal otherwise.

One of three cases eventually occurs for a root $P$. First, if $P$ encounters an unoccupied node in the layer below it, it moves into the lower layer, causing it to change the direction of its traversal (Fig. 21b–c). Second, if $P$ is contracted and encounters a retired particle in its layer, it also retires (Fig. 21d). Finally, if $P$ is contracted and occupies the marker node designated by the marker particle in the layer below, it waits until the marker particle signals that layer $P.\text{layer} - 1$ is completely filled (which it can determine locally). Once signalled, $P$ retires and becomes the marker particle for layer $P.\text{layer}$, designating a neighboring node in layer $P.\text{layer} + 1$ as the next marker node. This continues until all particles become retired.

**Analysis.** The correctness of the universal coating algorithm follows from the *safety* condition, i.e., that the set nodes occupied by particles and objects remains connected at all times, and the *liveness* condition, i.e., that the system eventually makes progress. Progress is made if an idle particle becomes active (i.e., a root or follower), a movement is executed, or an active particle retires. Both safety and liveness are proven by Derakshandeh et al. in [21].

In [14], Daymude et al. use a careful dominance argument to bound the runtime of the universal coating algorithm. To coat the surface layer, the particle system first organizes into a spanning forest. By Theorem 4, this is achieved in $\mathcal{O}(n)$ rounds. The surface layer is then coated in $\mathcal{O}(B_1) = \mathcal{O}(n)$ rounds, where $B_1$ is the length of the surface layer, assuming there are enough particles to do so. By Theorem 2, a leader node is elected in an additional $\mathcal{O}(B_1)$ rounds, w.h.p., allowing the particles of the surface layer to retire in $\mathcal{O}(B_1)$ more rounds. By a similar argument, coating the higher layers is shown to take another $\mathcal{O}(n)$ rounds in the worst case. All together, we have the following theorem.

**Theorem 11.** *The universal coating algorithm correctly solves a valid instance of the object coating problem $(\mathcal{P}, O)$ in $\mathcal{O}(n)$ rounds in the worst case, w.h.p., where $n$ is the number of particles in $\mathcal{P}$.*

We conclude with the following theorem, which shows that the universal coating algorithm achieves the optimal bound.

**Theorem 12.** *The worst-case runtime required by any local-control algorithm to solve the object coating problem for a system of $n$ particles is $\Omega(n)$ rounds.*

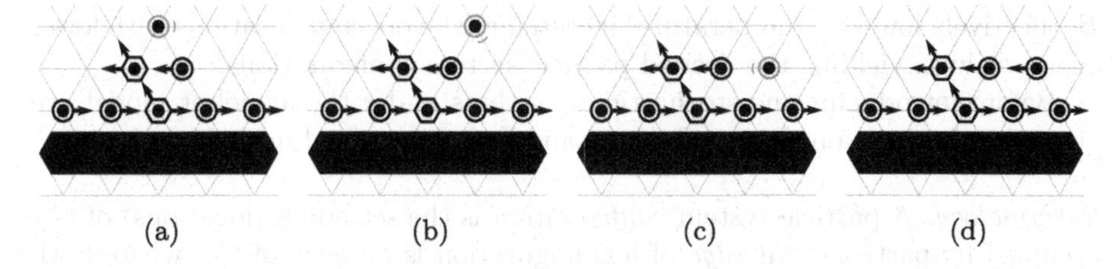


**Fig. 21.** Example of the general layering primitive. Retired particles are shown in black circles, and marker particles are shown in black hexagons. A root (gray circle) (a) traverses layer 3 in a clockwise direction, (b) encounters an unoccupied location in layer 2, (c) enters layer 2 and changes directions, and (d) retires.

# 5 Stochastic Algorithms Under the Amoebot Model

In the *stochastic approach to self-organizing particle systems*, algorithms under the amoebot model are designed and analyzed using concepts from statistical physics and stochastic processes. Instead of using careful state management and communication to drive particle computation and movement as in the algorithms of Sect. 4, these stochastic algorithms are stateless, use almost no communication, and depend only on probabilistic decisions. Designing these algorithms begins by defining an energy function that captures the objectives for the particle system. One then designs *Markov chains* that, in the long run, favor particle system configurations with desirable energy values. Although Markov chains are usually defined at a global, system level, these must be designed carefully so that they can be translated into fully distributed, local, asynchronous algorithms run by each particle individually.

The motivation underlying the design of these Markov chains comes from statistical physics, where ensembles of particles similar to those considered in the amoebot model represent physical systems. Previous studies on these systems have shown that local micro-behaviors can induce global, macro-scale changes to a system [3,4,46], yielding the kind of emergent phenomena desirable for programmable matter. Like a spring relaxing, physical systems favor configurations that minimize energy. Each system configuration $\sigma$ is assigned an energy value by a Hamiltonian $H(\sigma)$ and a corresponding weight $w(\sigma) = e^{-B \cdot H(\sigma)}$, where $B = 1/T$ is inverse temperature. Markov chains have been well-studied as a tool for sampling system configurations with probabilities proportional to their weight $w(\sigma)$, where configurations with the least energy $H(\sigma)$ are the most likely to be sampled.

The stochastic approach to self-organizing particle systems utilizes a Hamiltonian $H(\sigma)$ over particle system configurations $\sigma$ that assigns the lowest values to desirable configurations; a corresponding Markov chain algorithm is then designed to favor these configurations with small $H(\sigma)$. Each problem uses a different Hamiltonian. Each problem also uses a bias parameter $\lambda = e^B$. The weight of a configuration then becomes $w(\sigma) = \lambda^{-H(\sigma)}$. Thus, raising $\lambda$ (by increasing

$B$, effectively lowering temperature) increasingly favors configurations with lower energy values, yielding the desired particle system configurations.

Before introducing the stochastic algorithms under the amoebot model, we give a brief primer on the terminology and techniques used in their design.

*Terminology.* A particle system *configuration* is the set nodes (locations) of $G_\Delta$ occupied by particles. An *edge* of a configuration is an edge of $G_\Delta$ where both endpoints are occupied by particles. When referring to a *path*, we mean a path of such edges. Two particles are *connected* if there exists a path between them, and a configuration is *connected* if all pairs of particles are[7]. A *hole* in a configuration is a maximal finite connected component of unoccupied locations.

*Markov Chains.* A *Markov chain* $\mathcal{M}$ is a memoryless stochastic process defined on a state space $\Omega$. Here, $\Omega$ is a set of particle system configurations; thus, we only consider state spaces that are finite and discrete. The transition matrix $Q : \Omega \times \Omega \rightarrow [0, 1]$ is defined so that $Q(\sigma, \tau)$ is the probability of moving from state $\sigma$ to state $\tau$ in one step, for any pair of states $\sigma, \tau \in \Omega$. In Markov chains for particle systems, transitions correspond to one particle moving one unit in one direction, and the transition probabilities are chosen carefully to achieve some objective. The $t$-step transition probability $Q^t(\sigma, \tau)$ is the probability of moving from $\sigma$ to $\tau$ in exactly $t$ steps.

A Markov chain is *irreducible*, or its state space is *connected*, if there is a sequence of valid transitions from any state to any other state; i.e., for all $\sigma, \tau \in \Omega$, there is a $t$ such that $Q^t(\sigma, \tau) > 0$. A Markov chain is *aperiodic* if for all $\sigma, \tau \in \Omega$ we have $\gcd\{t : Q^t(\sigma, \tau) > 0\} = 1$. A Markov chain is *ergodic* if it is both irreducible and aperiodic. Any finite, ergodic Markov chain converges to a unique *stationary distribution* $\pi$ defined as $\lim_{t\rightarrow\infty} Q^t(\sigma, \tau) = \pi(\tau)$, for all $\sigma, \tau \in \Omega$. Any distribution $\pi'$ satisfying the *detailed balance condition*—$\pi'(\sigma)Q(\sigma, \tau) = \pi'(\tau)Q(\tau, \sigma)$ for all $\sigma, \tau \in \Omega$—must be this unique stationary distribution (see, e.g., [27]).

Given a state space $\Omega$, a set of allowable state transitions, and a desired stationary distribution $\pi$ on $\Omega$, the celebrated Metropolis-Hastings algorithm [32] defines a Markov chain on $\Omega$ that uses only allowable transitions and has stationary distribution $\pi$. This algorithm sets the probabilities of state transitions as follows. Starting at a state $\sigma \in \Omega$, choose a neighboring state $\tau \in \Omega$ (one with an allowable transition from $\sigma$ to $\tau$) uniformly with probability $1/(2\Delta)$, where $\Delta$ is the maximum number of neighbors of any state. Move from $\sigma$ to $\tau$ with probability $\min\{1, \pi(\tau)/\pi(\sigma)\}$; with the remaining probability stay at $\sigma$ and repeat. Assuming the allowable transitions connect the state space, then detailed balance will verify that this algorithm yields a Markov chain with $\pi$ as its stationary distribution. Although calculating $\pi(\tau)/\pi(\sigma)$ seems to require global knowledge, this ratio can often be calculated with only local information when many terms cancel out, as is the case for the algorithms presented here.

[7] This definition of configuration connectivity is equivalent to that of system connectivity given in Sect. 2.2.

*Obliviousness and Robustness.* Compared to the mostly deterministic algorithms in Sect. 4, these stochastic algorithms are *nearly oblivious*, requiring very little memory. Markov chains are naturally memoryless, and thus the distributed algorithms derived from them also have very little dependence on memory and communication. As we will see in Sects. 5.1, 5.2 and 5.3, only 1–2 bits of memory per particle are required in the resulting distributed algorithms. This is an artifact of translating the Markov chain algorithms—where a particle moves from one node to a neighboring node in a single step—into ones that fully respect the constraints of the amoebot model, where a particle can perform at most one expansion or contraction per activation.

Our stochastic algorithms are also significantly more *robust* (i.e., have a higher tolerance to failures) than those in Sect. 4. A distributed algorithm's *fault tolerance* is its ability to correctly solve a problem in spite of potential failures, a highly desirable and relevant property for programmable matter. One can imagine that in a system of thousands of particles, a small number of them may die and cease to move, compute, or communicate (i.e., a *crash failure*) or become corrupted and move, compute, or communicate erroneously (i.e., a *Byzantine failure*). Even a single fault of either type would cause complete failure in the algorithms in Sect. 4. In contrast, the stochastic algorithms have robustness built in. Because these algorithms are stateless and do not rely on communication, crashed particles and particles attempting to communicate arbitrary or malicious information have no real effect on the behavior of non-faulty particles. However, crashed particles act as fixed points that will affect the resulting particle system configurations.

### 5.1 Compression

The original publication on compression was the first to introduce the stochastic approach to self-organizing particle systems [11]. This line of work continued with shortcut bridging (Sect. 5.2) and separation (Sect. 5.3), both of which extend the original algorithm design of compression to achieve more complex behaviors. We will thus give more motivation and details for compression, while focusing more on the distinguishing features of the later algorithms.

In the *compression* problem, a particle system gathers together as tightly as possible, as in a sphere or its equivalent in the presence of some underlying geometry. There are many metrics that capture this behavior; e.g., system diameter or average particle distance to the system's center of mass. Here, a particle system must reorganize to minimize its *perimeter*, where a configuration's perimeter is measured by the length of the walk along its boundary. Several examples of this behavior exist in nature, particularly in social insects: fire ants gather to form floating rafts [38], cockroach larvae perform self-organizing aggregation [34], and honeybees choose hive locations based on decentralized swarming and recruitment [9]. While no individual insect can view the whole group when making decisions and soliciting information, it can take cues from its immediate neighbors to achieve cooperation.

**Problem Description.** Let $p(\sigma)$ denote the perimeter of a particle system configuration $\sigma$. For a system of $n$ particles, the minimum possible perimeter is $p_{min} := p_{min}(n) = \Theta(\sqrt{n})$. A configuration $\sigma$ with no holes is said to be *$\alpha$-compressed* if $p(\sigma) < \alpha \cdot p_{min}$, for any $\alpha > 1$. An algorithm solves the compression problem if, given any particle system in an initially connected configuration and any $\alpha > 1$, eventually the system reaches and remains in a set of $\alpha$-compressed configurations with all but a probability exponentially small in $n$.

Analogously, the maximum possible perimeter for a system of $n$ particles is $p_{max} := p_{max}(n) = 2n - 2$. A configuration $\sigma$ with no holes is said to be *$\beta$-expanded* if $p(\sigma) > \beta \cdot p_{max}$, for any $0 < \beta < 1$. An algorithm solves the expansion problem if, given any particle system in an initially connected configuration and any $0 < \beta < 1$, eventually the system reaches and remains in a set of $\beta$-expanded configurations with all but a probability exponentially small in $n$.

**Markov Chain $\mathcal{M}$.** We first present the Markov chain $\mathcal{M}$ for compression and will later show its translation into a distributed, local, asynchronous algorithm $\mathcal{A}$ that can be run by individual particles. $\mathcal{M}$ is defined on the state space $\Omega$ of all connected particle system configurations of $n$ contracted particles. Both $\mathcal{M}$ and $\mathcal{A}$ start in an arbitrary configuration $\sigma_0 \in \Omega$ and take a bias parameter $\lambda > 1$ as input, where $\lambda$ controls the preference for having small perimeter.

Recall that, for each stochastic algorithm, a Hamiltonian $H(\sigma)$ is defined which assigns the lowest energy values—and thus the largest weight $w(\sigma) = e^{-B \cdot H(\sigma)} = \lambda^{-H(\sigma)}$—to desirable configurations. To achieve compression, the lowest energy values should be assigned to the configurations with the smallest perimeter. In [11], Cannon et al. prove that minimizing configuration perimeter $p(\sigma)$ is equivalent to maximizing the number of configuration edges $e(\sigma)$. Thus, by setting $H(\sigma) = -e(\sigma)$, we obtain $w(\sigma) = \lambda^{e(\sigma)}$. This implies that $\lambda > 1$ corresponds to particles favoring having more neighbors while $\lambda < 1$ corresponds to particles favoring having fewer neighbors.

Markov chain $\mathcal{M}$ is carefully designed to maintain several critical properties that are necessary for its correctness and for applying certain tools from Markov chain analysis. First, $\mathcal{M}$ must keep the particle system connected and hole-free throughout its execution, assuming it starts in a connected, hole-free configuration. Next, $\mathcal{M}$ should be ergodic and, after a move is made, there should be a nonzero probability that it is undone in the next step. Finally, in order to solve the compression problem, $\mathcal{M}$ must eventually reach a stationary distribution that favors system configurations proportional to their weight $w(\sigma) = \lambda^{e(\sigma)}$ using only local information and local particle moves.

*Local Properties for Simple Connectivity.* Two local properties are used to ensure the particle system remains connected and hole-free throughout the execution of $\mathcal{M}$. Together, these properties ensure that a particle's local connectivity with respect to its neighbors does not change as a result of its move. Moreover, they ensure that every move can be undone, as desired.

We use the following notation. For a location $\ell$ of $G_\Delta$, let $N(\ell)$ denote the set of particles adjacent to $\ell$. For adjacent locations $\ell$ and $\ell'$, we use $N(\ell \cup \ell')$ to denote the set $N(\ell) \cup N(\ell')$, excluding particles occupying $\ell$ or $\ell'$. Let $\mathbb{S} = N(\ell) \cap N(\ell')$ be the particles adjacent to both locations; note that $|\mathbb{S}| \in \{0, 1, 2\}$. The following properties can be locally checked by an expanded particle occupying both $\ell$ and $\ell'$, and are symmetric with respect to these locations (see Fig. 22).

*Property 1.* $|\mathbb{S}| \in \{1, 2\}$ and every particle in $N(\ell \cup \ell')$ is connected to a particle in $\mathbb{S}$ by a path through $N(\ell \cup \ell')$.

*Property 2.* $|\mathbb{S}| = 0$, $\ell$ and $\ell'$ each have at least one neighbor, all particles in $N(\ell) \setminus \{\ell'\}$ are connected by paths within this set, and all particles in $N(\ell') \setminus \{\ell\}$ are connected by paths within this set.

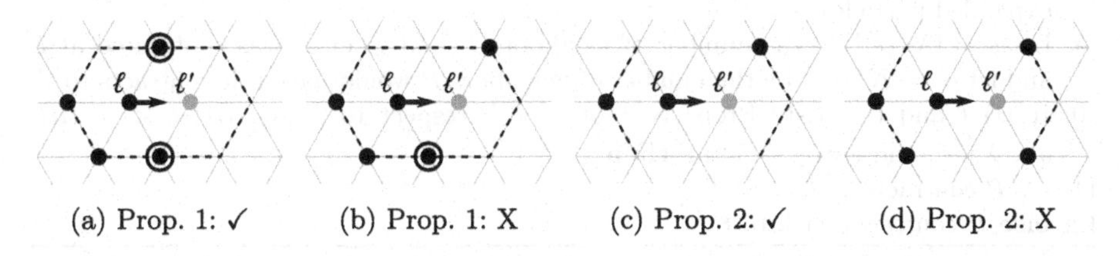

(a) Prop. 1: ✓ (b) Prop. 1: X (c) Prop. 2: ✓ (d) Prop. 2: X

**Fig. 22.** Examples of particle neighborhoods with respect to Properties 1 and 2. Particles of $\mathbb{S}$ are drawn with black circles around them.

We can now present the Markov chain $\mathcal{M}$ for compression (Algorithm 1).

---

**Algorithm 1.** Markov Chain $\mathcal{M}$ for Compression

Beginning at any connected configuration $\sigma_0$ of $n$ contracted particles, repeat:
1: Select particle $P$ uniformly at random from among all particles; let $\ell$ be its location.
2: Choose a neighboring location $\ell'$ and $q \in (0, 1)$ uniformly at random.
3: **if** $\ell'$ is unoccupied **then**
4: $P$ expands to occupy both $\ell$ and $\ell'$.
5: Let $e = |N(\ell)|$ be the number of neighbors $P$ had when it was contracted at $\ell$, and let $e' = |N(\ell')|$ be the number of neighbors $P$ would have if it contracts to $\ell'$.
6: **if** (*i*) $\ell$ and $\ell'$ satisfy Property 1 or 2, (*ii*) $e < 5$, and (*iii*) $q < \lambda^{e'-e}$ **then**
7: $P$ contracts to $\ell'$.
8: **else** $P$ contracts back to $\ell$.

---

**Distributed, Local, Asynchronous Algorithm $\mathcal{A}$.** We now present the local, distributed, asynchronous algorithm $\mathcal{A}$ that each particle runs. Recall from Sect. 2.1 that during a single activation of a particle $P$, $P$ can perform an arbitrary amount of computation and at most one expansion or contraction. In particular, $P$ cannot do both an expansion and a contraction in one activation as $\mathcal{M}$ does in a single step. Thus, $\mathcal{A}$ must decouple a single step of $\mathcal{M}$ into two (not necessarily consecutive) particle activations and carefully handle the way

**Algorithm 2.** Distributed Algorithm $\mathcal{A}$ for Compression run by Particle $P$

If $P$ is contracted:
1: Let $\ell$ denote $P$'s current location.
2: Particle $P$ chooses neighboring location $\ell'$ uniformly at random from the six possible choices, and generates a random number $q \in (0, 1)$.
3: **if** $\ell'$ is unoccupied and $P$ has no expanded neighboring particle at $\ell$ **then**
4: $P$ expands to occupy both $\ell$ and $\ell'$.
5: **if** there are no expanded particles adjacent to $\ell$ or $\ell'$ **then**
6: $P$ sets a flag $f =$ TRUE in its local memory.
7: **else** $P$ sets $f =$ FALSE.

If $P$ is expanded:
8: Let $N^*(\cdot) \subseteq N(\cdot)$ be the set of neighboring particles excluding any heads of expanded particles.
9: Let $e = |N^*(\ell)|$ be the number of neighbors $P$ had when it was contracted at $\ell$, and let $e' = |N^*(\ell')|$ be the number of neighbors $P$ would have if it contracts to $\ell'$.
10: **if** $(i)$ $\ell$ and $\ell'$ satisfy Property 1 or 2 with respect to $N^*(\cdot)$, $(ii)$ $e < 5$, $(iii)$ $q < \lambda^{e'-e}$, and $(iv)$ $f =$ TRUE **then**
11: $P$ contracts to $\ell'$.
12: **else** $P$ contracts back to $\ell$.

in which the particle's neighborhood may change between its two activations (see [11] for full details of this decoupling).

Each particle $P$ continuously runs Algorithm $\mathcal{A}$, executing Steps 1–7 if $P$ is contracted and Steps 8–12 if $P$ is expanded. Conditions $(i)$–$(iii)$ in Step 10 of $\mathcal{A}$ are the same as those in Step 6 of $\mathcal{M}$. The additional Condition $(iv)$ ensures $P$ is the only particle in its neighborhood potentially moving to a new position since it last expanded. Any conflicts arising from two particles concurrently attempting to expand into the same location are assumed to be resolved arbitrarily (Sect. 2.1). Hence, any concurrent movements will cover pairwise disjoint neighborhoods and the respective actions will be mutually independent.

However, in an asynchronous setting, one cannot typically assume the next particle to be activated is equally likely to be any particle, as in Step 1 of $\mathcal{M}$. This assumption is made in order to explicitly calculate the stationary distribution of $\mathcal{M}$ (Lemma 1) and rigorously analyze it (Theorems 13 and 14), but the system's behavior is not expected to differ substantially if this requirement was relaxed.

These random sequences of particle activations can be approximated using Poisson clocks with mean 1. That is, each particle can activate and execute Algorithm $\mathcal{A}$ at a random real time drawn from the exponential distribution $e^{-t}$. After each action, a particle could then compute another random time drawn from the same distribution $e^{-t}$ and activate again after that amount of time has elapsed. The exponential distribution is unique in that, if particle $P$ has just activated, it is equally likely that any particle will be the next particle to activate, including particle $P$ (see, e.g., [27]). Moreover, a particle updates without requiring knowledge of any other particle's clock. As an aside, the analysis can be modified to accommodate each clock having its own constant mean; however, for ease of presentation, we assume here that they are all i.i.d.

**Results.** Cannon et al. give a detailed analysis of Markov chain $\mathcal{M}$ in [11]; here, we briefly present the major results on some properties of $\mathcal{M}$, its stationary distribution $\pi$, and its correctness in solving compression. We conclude with results on using $\mathcal{M}$ to also solve expansion. Note that all results for $\mathcal{M}$ extend to the local algorithm $\mathcal{A}$.

*Invariants of $\mathcal{M}$.* Cannon et al. prove that if the particle system is initially connected, $\mathcal{M}$ maintains system connectivity. Moreover, once a connected configuration with no holes is reached, $\mathcal{M}$ will never introduce new holes to the system. Together, these imply that once $\mathcal{M}$ reaches the subspace $\Omega^* \subset \Omega$ of all connected, hole-free configurations of $n$ particles, it will remain in $\Omega^*$ forever. Since $\mathcal{M}$ is finite and ergodic on $\Omega^*$ (a result shown in [11]), it converges to a unique stationary distribution $\pi$ on $\Omega^*$ that can be verified by detailed balance.

**Lemma 1.** *Markov chain $\mathcal{M}$ for compression has a unique stationary distribution $\pi$ given by:*

$$\pi(\sigma) = \begin{cases} \lambda^{e(\sigma)}/Z & \text{if } \sigma \in \Omega^*; \\ 0 & \text{otherwise,} \end{cases}$$

*where $Z = \sum_{\tau \in \Omega^*} \lambda^{e(\tau)}$ is the normalizing constant or partition function.*

*Achieving Compression and Expansion.* Markov chain $\mathcal{M}$—and, by extension, local algorithm $\mathcal{A}$—solve both the compression and expansion problems depending on the value of bias parameter $\lambda$. Although compression or expansion could occur before $\mathcal{M}$ even converges to its stationary distribution $\pi$, the proofs in [11] rely on analyzing $\pi$. First, it is shown that for any $\alpha > 1$ and provided $\lambda$ and $n$ are large enough, a configuration chosen at random according to $\pi$ is $\alpha$-compressed with all but a probability that is exponentially small in $n$.

**Theorem 13.** *For any $\alpha > 1$, let $\lambda^* = (2+\sqrt{2})^{\alpha/(\alpha-1)}$. There exists $n^* \geq 0$ and $\zeta < 1$ such that for all $\lambda > \lambda^*$ and $n > n^*$, the probability that a random sample $\sigma$ drawn according to the stationary distribution $\pi$ of $\mathcal{M}$ is not $\alpha$-compressed is exponentially small:*

$$\Pr_{\sigma \sim \pi}\left[p(\sigma) \geq \alpha \cdot p_{min}\right] < \zeta^{\sqrt{n}}.$$

It can also be shown that there is some constant $\alpha$ for which $\alpha$-compression occurs when $\lambda > 2 + \sqrt{2}$ is fixed. However, there is a tradeoff: smaller values of $\lambda$ require larger values of $\alpha$ and vice versa.

The algorithm also provably achieves expansion for different values of the bias parameter $\lambda$. It is shown that, for all $0 < \lambda < 2.17$ and provided $n$ is large enough, there is a constant $0 < \beta < 1$ such that a configuration chosen at random according to the stationary distribution of $\mathcal{M}$ is $\beta$-expanded with all but exponentially small probability in $n$. This is counterintuitive, since it implies that $\lambda > 1$ (i.e., favoring more neighbors) is not sufficient to guarantee particle compression.

**Theorem 14.** *For any $0 < \beta < 1$, let $\lambda^* = (\sqrt{2}/(2+\sqrt{2})^{\beta})^{1/(1-\beta)}$. There exists $n^* \geq 1$ and $\zeta < 1$ such that for all $\lambda < \lambda^*$ and $n \geq n^*$, the probability that a random sample $\sigma$ drawn according to the stationary distribution $\pi$ of $\mathcal{M}$ is not $\beta$-expanded is exponentially small:*

$$\Pr_{\sigma \sim \pi}[p(\sigma) \leq \beta \cdot p_{max}] < \zeta^{\sqrt{n}}.$$

Similar to compression, it is also shown that there is a constant $\beta$ for which $\beta$-expansion occurs when $0 < \lambda < 2.17$ is fixed. Again, there is a tradeoff in larger values of $\lambda$ requiring smaller values of $\beta$ and vice versa.

*Convergence Time of $\mathcal{M}$.* Although compression provably occurs with all but exponentially small probability once $\mathcal{M}$ converges to its stationary distribution, no explicit bounds are given on the time required for this to occur. However, simulations support the conjecture that the worst-case number of steps of $\mathcal{M}$ needed to observe compression is $\Omega(n^3)$ and $\mathcal{O}(n^4)$, or $\mathcal{O}(n^3)$ rounds of $\mathcal{A}$.

**Simulations.** In practice, Markov chain $\mathcal{M}$ yields good compression. Figure 23 depicts a simulation of $\mathcal{M}$ for $\lambda = 4$ on 100 particles that begin in a line and become compressed. In contrast, $\lambda = 2$ (which still favors having more particle neighbors), does not yield compression; see Fig. 24, where even after 20 million steps of $\mathcal{M}$, the particles have not compressed. Cannon et al. conjecture there is a phase transition in $\lambda$, i.e., a critical value $\lambda_c$ such that for all $\lambda > \lambda_c$ the particles compress and for all $\lambda < \lambda_c$ they do not. Such phase transitions exist for similar statistical physics models (e.g., [8]). Theorems 13 and 14 indicate that if $\lambda_c$ exists, then $2.17 \leq \lambda_c \leq 2 + \sqrt{2}$.

### 5.2 Shortcut Bridging

Andrés Arroyo et al. further validated the stochastic approach to self-organizing particle systems in an investigation of *shortcut bridging* [1]. This work is inspired by an entomological study [43] that found army ants of the genus *Eciton* continuously modify the shape and position of foraging bridges—constructed and maintained by their own bodies—across holes and uneven surfaces on the forest floor. These bridges appear to stabilize in a structural formation that balances the "benefit of increased foraging trail efficiency" with the "cost of removing workers from the foraging pool to form the structure" [43]. Shortcut bridging is an attractive goal for programmable matter, which may need to make similar tradeoffs when maintaining bridges over terrain with structural irregularities.

To consider this problem in the amoebot model, two model extensions (Sect. 2.3) are employed. First, static objects are used to anchor the particle system to certain fixed sites. Second, the locations of $G_\Delta$ are considered either *gap* (unsupported) or *land* (supported) using node differentiation. The notion of configuration perimeter used in compression (Sect. 5.1) is extended to address this new land/gap setting as follows. The *weighted perimeter* $\overline{p}(\sigma, c)$ of a particle system configuration $\sigma$ is the summed weight of the edges on the boundary of $\sigma$,

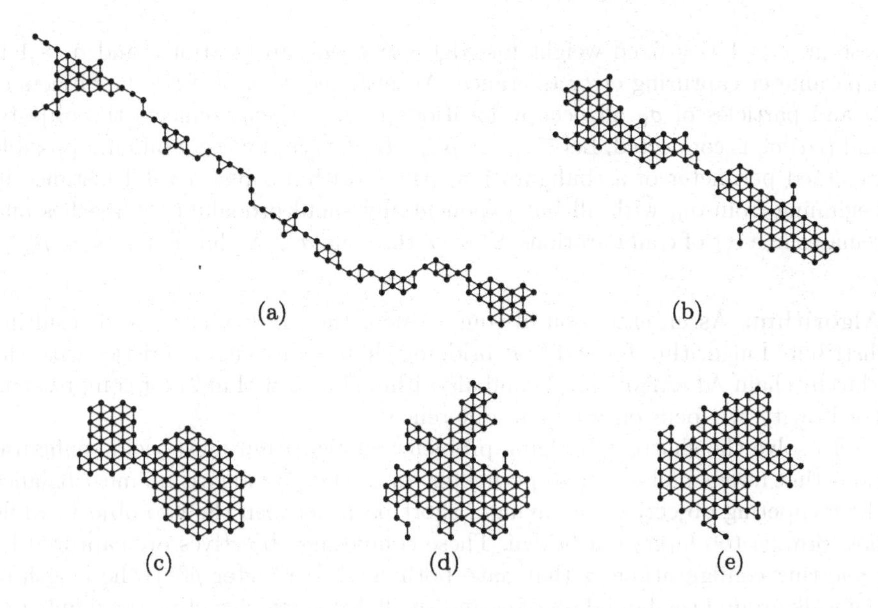


**Fig. 23.** 100 particles in a line with occupied edges drawn, after (a) 1 million, (b) 2 million, (c) 3 million, (d) 4 million, and (e) 5 million steps of $\mathcal{M}$ with bias $\lambda = 4$.

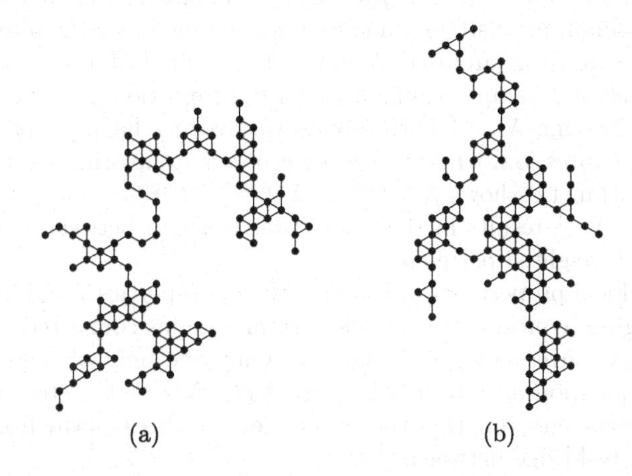


**Fig. 24.** 100 particles in a line with occupied edges drawn, after (a) 10 million and (b) 20 million steps of $\mathcal{M}$ with bias $\lambda = 2$.

where edges between land locations have weight 1, edges between gap locations have weight $c \geq 1$, and edges with one endpoint on land and one endpoint in the gap have weight $(1 + c)/2$.

**Problem Description.** An instance of the shortcut bridging problem has the form $(L, O, \sigma_0, c, \alpha)$, where $L \subseteq V$ is the set of land locations, $O$ is the set of (two) objects to bridge between, $\sigma_0$ is the initial configuration for the particle system, $c \geq 1$ is a fixed weight for edges between gap locations, and $\alpha > 1$ is a parameter capturing error tolerance. An instance is *valid* if (*i*) the objects of $O$ and particles of $\sigma_0$ all occupy locations in $L$, (*ii*) $\sigma_0$ connects the objects, and (*iii*) $\sigma_0$ is connected. Let $\overline{p}_{min} := \overline{p}_{min}(L, O, \sigma_0, c)$ be the minimum possible weighted perimeter of a configuration. An algorithm solves a valid instance if, beginning from $\sigma_0$, with all but exponentially small probability it reaches and remains in a set of configurations $\Sigma^*$ such that any $\sigma \in \Sigma^*$ has $\overline{p}(\sigma, c) < \alpha \cdot \overline{p}_{min}$.

**Algorithm.** As an extension of compression, the Markov chain and resulting distributed algorithm for shortcut bridging share many characteristics with the Markov chain $\mathcal{M}$ (Algorithm 1) and algorithm $\mathcal{A}$ (Algorithm 2) for compression. For brevity, we focus on the main differences.

To solve the shortcut bridging problem, an algorithm must yield configurations that have small weighted perimeter. In essence, an algorithm must balance the competing objectives of having a short path between the two objects while not forming too large of a bridge. These competing objectives are captured by preferring configurations $\sigma$ that have both small perimeter $p(\sigma)$, the length of the walk around the boundary of $\sigma$, and small *gap perimeter* $g(\sigma)$, the number of perimeter edges in the gap. Using the weights defined above, weighted perimeter becomes $\overline{p}(\sigma, c) = p(\sigma) + (c - 1)g(\sigma)$; thus, minimizing weighted perimeter is equivalent to simultaneously minimizing both perimeter and gap perimeter.

Two bias parameters are used: $\lambda$ and $\gamma$. The desired Markov chain for shortcut bridging should sample configurations $\sigma$ proportional to weight $w(\sigma) = \lambda^{-p(\sigma)}\gamma^{-g(\sigma)}$. Setting $\lambda > 1$ corresponds to favoring having small perimeter, as it did for compression, while $\gamma > 1$ corresponds to favoring having small gap perimeter. Arithmetic shows $\lambda^{-\overline{p}(\sigma,c)} = \lambda^{-p(\sigma)-(c-1)g(\sigma)} = \lambda^{-p(\sigma)}(\lambda^{c-1})^{-g(\sigma)}$, so setting $\gamma = \lambda^{c-1}$ results in the desired relationship between perimeter, gap perimeter, and weighted perimeter.

The same local properties from compression (Properties 1 and 2) are used in shortcut bridging to ensure the particle system remains connected, no new holes form, and every move made can be undone. Only one small change is made: the definitions of location neighborhoods (e.g., $N(\ell)$, $N(\ell \cup \ell')$, etc.) are extended to include objects, ensuring that the system does not break away from the points it is supposed to bridge between.

We can now present the Markov chain $\mathcal{M}_B$ for shortcut bridging. It follows the same procedure as Markov chain $\mathcal{M}$ for compression (Algorithm 1), with two differences. First, instead of counting neighbors as in Step 5 of $\mathcal{M}$, $\mathcal{M}_B$ simply refers to the configuration with particle $P$ at its original location $\ell$ as $\sigma$ and the configuration with $P$ at its new location $\ell'$ as $\sigma'$. Then, $\mathcal{M}_B$ replaces the probability condition in Step 6 of $\mathcal{M}$ with $q < \lambda^{p(\sigma)-p(\sigma')}\gamma^{g(\sigma)-g(\sigma')}$. One might observe that this probability is not defined locally; however, an argument in [1] shows that it can be calculated by an expanded particle occupying $\ell$ and $\ell'$ using only local information from its neighborhood.

Markov chain $\mathcal{M}_B$ can be directly translated to a fully distributed, local, asynchronous algorithm $\mathcal{A}_B$ for shortcut bridging using the same decoupling and Poisson clock mechanisms described for compression.

**Results.** In [1], Andrés Arroyo et al. give a detailed analysis of the Markov chain $\mathcal{M}_B$ and resulting distributed algorithm $\mathcal{A}_B$ for shortcut bridging. Here, we will only present the highlights. As for compression, $\mathcal{M}_B$ is shown to maintain system connectivity. It is also shown to eventually reach the set of configurations with no holes, after which all configurations will remain hole-free.

The resulting stationary distribution $\pi_B$ over $\Omega_B$ (the set of all configurations reachable from $\sigma_0$, the initial configuration, via valid transitions of $\mathcal{M}_B$) is shown to be $\pi_B(\sigma) = \lambda^{-p(\sigma)}\gamma^{-g(\sigma)}/Z$, where $Z = \sum_{\tau \in \Omega_B} \lambda^{-p(\tau)}\gamma^{-g(\tau)}$ is the normalizing constant. Analyzing this stationary distribution with a careful Peierls argument results in the following theorem, which shows $\mathcal{M}_B$ correctly solves the shortcut bridging problem.

**Theorem 15.** *Consider any $\alpha > 1$ and let $\lambda^* = (2+\sqrt{2})^{\alpha/(\alpha-1)}$. There exists $n^* > 0$ such that for all $\lambda > \lambda^*$ and $n > n^*$, if $\gamma = \lambda^{c-1}$, the probability that a random sample $\sigma$ drawn from the stationary distribution of $\mathcal{M}_B$ has weighted perimeter $\overline{p}(\sigma, c) \geq \alpha \cdot \overline{p}_{min}$ is exponentially small in $n$, where $n$ is the number of particles in the system.*

Simulation results supporting the findings of Theorem 15 can be seen in Figs. 25 and 26, where $\mathcal{M}_B$ was run with biases $\lambda = 4$ and $\gamma = 2$ (i.e., $c = 3/2$) on a V-shaped and N-shaped land mass, respectively.

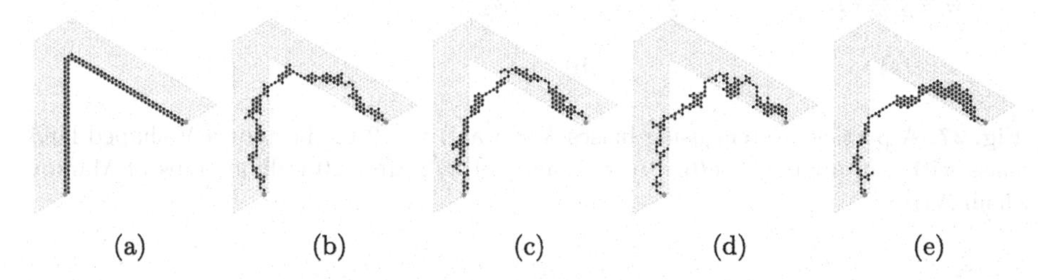


**Fig. 25.** A particle system, beginning in configuration (a), using biases $\lambda = 4$ and $\gamma = 2$ to shortcut a V-shaped land mass after (b) 2 million, (c) 4 million, (d) 6 million, and (e) 8 million steps of Markov chain $\mathcal{M}_B$. The two objects (large, dark gray) anchor the particle system (black) to land (gray).

*Dependence on Gap Angle.* The shortcut bridging algorithm is also shown in [1] to have a dependence on the internal angle $\theta$ of the gap similar to that of the army ant bridges studied in [43]. Informally, it is shown that when $\theta$ is sufficiently small, with all but exponentially small probability the bridge constructed by the particles stays close to the bottom of the gap (away from the apex of angle $\theta$).

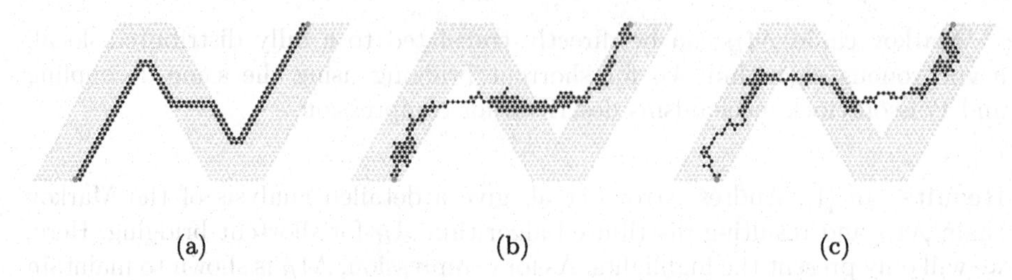

**Fig. 26.** A particle system, beginning in configuration (a), using biases $\lambda = 4$ and $\gamma = 2$ to shortcut an N-shaped land mass after (b) 10 million and (c) 20 million steps of Markov chain $\mathcal{M}_B$.

For some large values of $\theta$, it is shown that the bridge constructed by the particles stays close to the top of the gap (nearer to land) with all but exponentially small probability. Although the bounds obtained in [1] are relatively narrow (e.g., the proofs for small $\theta$ only hold for gap angles less than $\theta_1 = 0.0879$–$5.03°$), simulations suggest these bounds are far from tight. Figure 27 depicts simulation results that are consistent with the proven angle dependence behaviors, but were obtained using angles and bias parameter values outside the proven bounds.

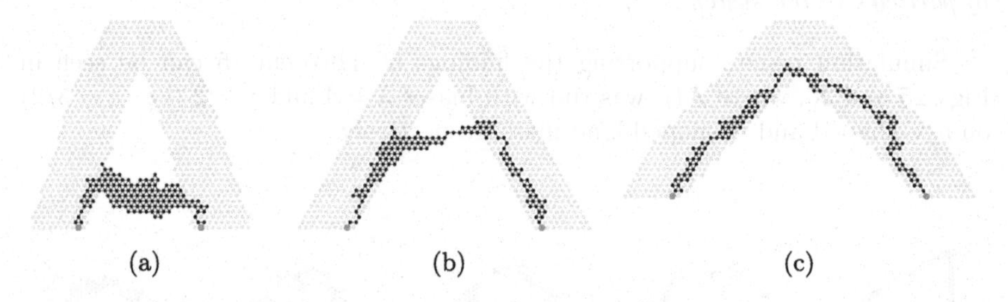

**Fig. 27.** A particle system using biases $\lambda = 4$ and $\gamma = 2$ to shortcut a V-shaped land mass with gap angle (a) $\pi/6$, (b) $\pi/3$, and (c) $\pi/2$ after 20 million steps of Markov chain $\mathcal{M}_B$.

### 5.3 Separation

Examples of heterogeneous entities separating and integrating exist at many scales, from molecules exhibiting attractive and repulsive forces, to mixed solutions of varying viscosities, to inherent human biases that influence how we form and maintain social groups. This fundamental behavior of heterogeneous entities separating or integrating in response to environmental stimuli spans remarkably diverse areas of study. Of particular relevance to the stochastic approach to self-organizing particle systems are the Schelling model [49, 50]—which explores how micro-motives of individuals can induce macro-phenomena such as racial segregation in residential neighborhoods—and the Ising model of ferromagnetism from statistical physics [55].

In the *separation* problem, a *heterogeneous particle system*—i.e., one in which particles have different immutable *colors*—must self-organize to form monochromatic clusters, resulting in observable color class separation. Each particle $P$ keeps a color $c(P) \in \{c_1, \ldots, c_k\}$ in its memory that is visible to itself and its neighbors, where $k < n$ is some small constant. An edge between two neighboring particles $P$ and $Q$ is *homogeneous* if $c(P) = c(Q)$ and is *heterogeneous* otherwise. Cannon et al. give a Markov chain algorithm $\mathcal{M}_S$ for separation in a two-color particle system in [10], though the algorithm has also been shown to generalize to $k$-color particle systems in simulations (for a constant $k$). We will focus on the two-color case for the problem statement, results, and simulations, but will describe the Markov chain $\mathcal{M}_S$ and corresponding distributed algorithm $\mathcal{A}_S$ for separation in full $k$-color generality.

**Problem Description.** Informally, a two-color particle system configuration is *separated* if there is a set $R$ of particles such that $R$ mostly contains particles of color $c_1$, its complement $\overline{R}$ mostly contains particles of color $c_2$, and the boundary between $R$ and $\overline{R}$ is small. If this is the case, $R$ and $\overline{R}$ are called *clusters.* More formally, a configuration is $(\beta, \delta)$*-clustered*, for $\beta > 0$ and $\delta < 1/2$, if there are at most $\delta|R|$ particles of color $c_2$ in $R$, at most $\delta|\overline{R}|$ particles of color $c_1$ in $\overline{R}$, and the boundary between $R$ and $\overline{R}$ is of size at most $\beta\sqrt{n}$, where $n$ is the number of particles in the system.

An instance of the separation problem has the form $(\sigma_0, \beta, \delta)$ where $\sigma_0$ is a connected initial configuration of colored particles and $\beta > 0$ and $0 < \delta < 1/2$ are constants. An algorithm solves an instance if, beginning from configuration $\sigma_0$, with all but exponentially small probability it reaches and remains in a set of configurations that are $(\beta, \delta)$-clustered.

**Algorithm.** As was the case for shortcut bridging (Sect. 5.2), the Markov chain $\mathcal{M}_S$ and corresponding distributed algorithm $\mathcal{A}_S$ for separation are also extensions of the compression algorithm (Sect. 5.1) and follow the stochastic approach to self-organizing particle systems. To achieve separation, an algorithm should favor configurations with many edges (inducing small perimeter, as in Sect. 5.1) and large monochromatic clusters. These objectives are achieved by sampling configurations $\sigma$ proportional to their weight $w(\sigma) = \lambda^{e(\sigma)}\kappa^{a(\sigma)}$ where $e(\sigma)$ is the number of edges and $a(\sigma)$ is the number of homogeneous edges in $\sigma$. Bias parameter $\lambda$ controls the system's preference for having small perimeter, as in compression and shortcut bridging; larger values of $\lambda$ increasingly favor compressed configurations while for small $\lambda$ the opposite is true. The additional bias parameter $\kappa$ controls separation, favoring clustered/separated configurations when $\kappa$ is large and well-integrated configurations when $\kappa$ is small.

A new *swap move* is introduced to enable adjacent particles of different colors to switch places. For two neighboring contracted particles $P$ and $Q$, either $P$ or

$Q$ can initiate a swap to exchange colors, which can be implemented as follows: $P$ reads $x \leftarrow c(Q)$ from the memory of $Q$, overwrites $c(Q) \leftarrow c(P)$ in the memory of $Q$, and finally updates its own color $c(P) \leftarrow x$. Adding this swap move enables faster convergence of the separation algorithm in practice, but is not necessary for any of its formal results.

In addition to the usual location neighborhood definitions (e.g., $N(\ell)$, $N(\ell \cup \ell')$) used in compression and shortcut bridging, separation also uses color-specific location neighborhoods. More precisely, for a location $\ell$, $N_i(\ell)$ denotes the set of particles of color $c_i$ adjacent to location $\ell$. For neighboring locations $\ell$ and $\ell'$, $N_i(\ell \cup \ell')$ denotes the set $N_i(\ell) \cup N_i(\ell')$, excluding particles occupying $\ell$ and $\ell'$. These color-specific neighborhoods are used when calculating the difference in the number of homogeneous edges between a particle's new and old position.

Separation also uses local Properties 1 and 2 from compression to ensure the particle system remains connected and no new holes form. However, these properties need not be verified for swap moves, which do not change the set of occupied nodes and thus cannot disconnect the system or create a hole. We can now present the Markov chain $\mathcal{M}_S$ for separation (Algorithm 3).

---

**Algorithm 3.** Markov Chain $\mathcal{M}_S$ for Separation

---

Beginning at any connected configuration $\sigma_0$ of $n$ contracted particles, repeat:

1: Select particle $P$ uniformly at random from among all particles; let $c_i$ be its color and $\ell$ its location.
2: Choose a neighboring location $\ell'$ and $q \in (0,1)$ uniformly at random.
3: **if** $\ell'$ is unoccupied **then**
4: $P$ expands to occupy both $\ell$ and $\ell'$.
5: **if** (*i*) $\ell$ and $\ell'$ satisfy Property 1 or 2, (*ii*) $|N(\ell)| < 5$, and (*iii*) $q < \lambda^{|N(\ell')|-|N(\ell)|}\kappa^{|N_i(\ell')|-|N_i(\ell)|}$ **then**
6: $P$ contracts to $\ell'$.
7: **else** $P$ contracts back to $\ell$.
8: **else if** $\ell'$ is occupied by particle $Q$ of color $c_j$ **then**
9: $P$ calculates $|N_i(\ell)|$ and $|N_j(\ell) \setminus \{Q\}|$ and sends these values to $Q$.
10: $Q$ calculates $|N_i(\ell') \setminus \{P\}|$ and $|N_j(\ell')|$.
11: **if** $q < \gamma^{|N_i(\ell')\setminus\{P\}|-|N_i(\ell)|+|N_j(\ell)\setminus\{Q\}|-|N_j(\ell')|}$ **then** $Q$ swaps with $P$.

---

While $\mathcal{M}_S$ has much in common with the Markov chain $\mathcal{M}$ for compression (Algorithm 1), it also has important differences. In particular, condition (*iii*) in Step 5 of $\mathcal{M}_S$ specifically addresses homogeneous edges and Steps 8–11 of $\mathcal{M}_S$ implement the new swap move. Moreover, the translation of $\mathcal{M}_S$ to a fully distributed, local, asynchronous algorithm $\mathcal{A}_S$ for separation does not follow trivially from the translation given for compression. Although decoupling a particle's expansion and contraction in Steps 3–7 of $\mathcal{M}_S$ can be done in a similar manner to that of compression, locally synchronizing the swap move of Steps 8–11 of $\mathcal{M}_S$ requires additional mechanisms. Details can be found in [10].

**Results.** Cannon et al. rigorously analyze the Markov chain $\mathcal{M}_S$ and corresponding distributed algorithm $\mathcal{A}_S$ for $k$-color separation in [10]. As for compression and shortcut bridging, $\mathcal{M}_S$ is shown to maintain system connectivity and remain hole-free throughout its execution, assuming it begins at a connected and hole-free configuration $\sigma_0$. Additionally, its stationary distribution $\pi_S$ over $\Omega_S$ (the set of all connected, hole-free configurations with the same number of particles of each color as $\sigma_0$) is shown to be $\pi_S(\sigma) = \lambda^{e(\sigma)}\kappa^{a(\sigma)}/Z$, where $Z = \sum_{\tau \in \Omega_S} \lambda^{e(\tau)}\kappa^{a(\tau)}$ is the normalizing constant.

While these initial results follow from standard techniques, proving that $\mathcal{M}_S$ achieves separation requires significantly heavier machinery. Using a Markov chain analysis technique known as *bridging* [37] (not to be confused with the shortcut bridging behavior described in Sect. 5.2), it is shown that, among two-color configurations with the same small external boundary, a configuration sampled according to the stationary distribution $\pi_S$ of $\mathcal{M}_S$ will be clustered/separated as desired with all but exponentially small probability. We present the corresponding theorem in its full formality below, but refer the reader to [10] for a more detailed explanation.

**Theorem 16.** *For any $\alpha > 1$, $\beta > 4\alpha$, and $\delta < 1/2$, there exists $\kappa^*$ and $n_0$ (which depend on $\alpha$, $\beta$, and $\delta$) such that for all $\kappa > \kappa^*$ and $n > n_0$, for any $\alpha$-compressed boundary $\mathcal{B}$, the probability that a configuration sampled according to $\pi_S$ from among two-color configurations with $n$ particles and boundary $\mathcal{B}$ is not $(\beta, \delta)$-clustered is at most $\zeta^{\sqrt{n}}$ for some constant $\zeta < 1$.*

**Simulations.** In simulation, $\mathcal{M}_S$ exhibits the expected separation behavior for large $\lambda$ and $\kappa$, as well as integration behaviors for other parameter values. Figure 28 shows a simulation of $\mathcal{M}_S$ on a two-color system with 50 particles of each color using biases $\lambda = 4$ and $\kappa = 4$, the regime in which the system should compress and individual color classes should separate. Much of the progress towards a compressed and separated system occurs in the first million steps, though the simulation runs for much longer. Figure 29 compares the resulting configurations after running $\mathcal{M}_S$ from the same initial configuration for the same number of steps, varying only the values of $\lambda$ and $\kappa$. Four distinct phases appear: expanded-integrated, expanded-separated, compressed-integrated, and compressed-separated. Thus, although Theorem 16 only proves separation for two-color systems with small, fixed boundaries, $\mathcal{M}_S$ appears to be capable of a diverse set of dynamic behaviors in practice.

## 6 Hybrid Programmable Matter

In this section, we discuss a variant of the amoebot model known as *hybrid programmable matter*, in which a collection of *active robots* operate on a connected system of *passive tiles*. The robots have similar capabilities to the particles of the amoebot model, but the tiles—which are uniform and stateless—cannot move themselves nor perform any computation. To change its (relative) position, a

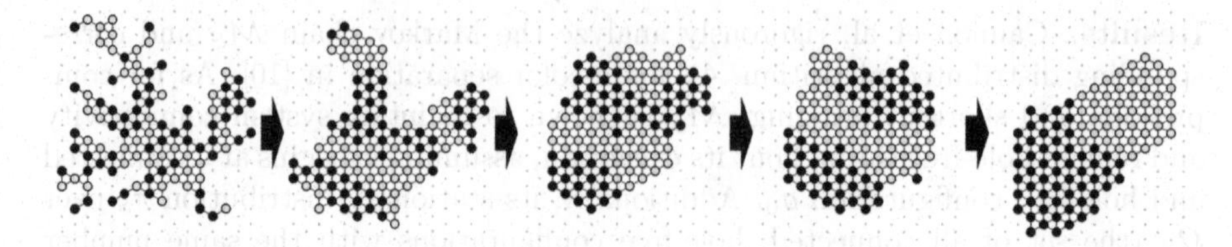

**Fig. 28.** A two-color heterogeneous particle system starting in an arbitrary state after—from left to right—0; 50,000; 1,050,000; 17,050,000; and 68,250,000 steps of $\mathcal{M}_S$ with $\lambda = 4$ and $\kappa = 4$.

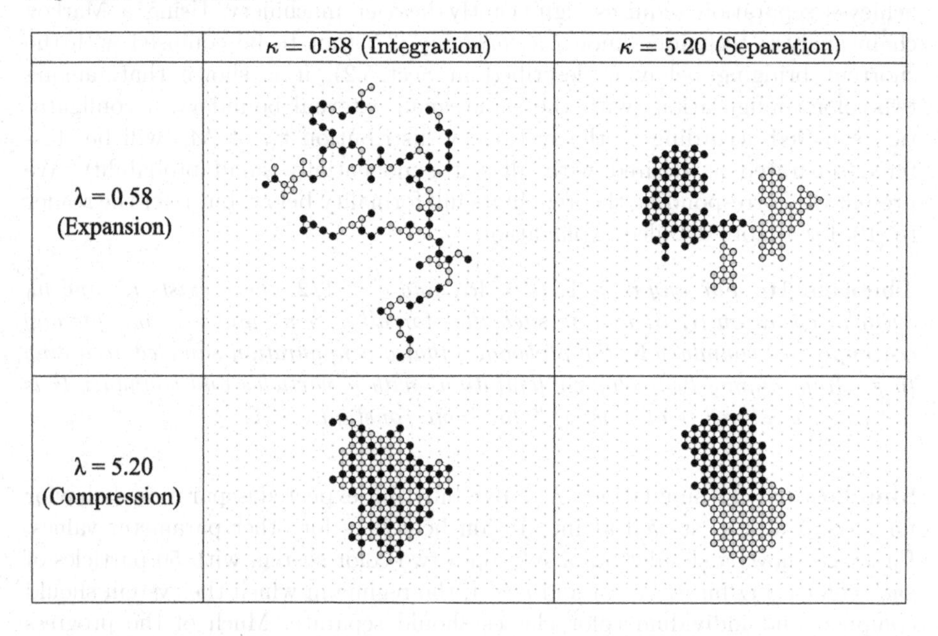


**Fig. 29.** A two-color heterogeneous particle system starting in the leftmost configuration of Fig. 28 after 50 million steps of $\mathcal{M}_S$ for various values of the parameters $\lambda$ and $\kappa$.

tile must be lifted and moved by a robot. Unlike in the amoebot model, system connectivity is defined with respect to the structure of tiles (including tiles being carried by robots) and must be maintained at all times. Since the set of robots does not need to stay connected, we can abstract away from any specific locomotion primitive such as the amoebot model's expansions and contractions.

Although most of the algorithms presented in this section focus on systems with only one robot, we present the complete model with respect to multiple robots, as it is the basis of ongoing research. A system of hybrid programmable matter is composed of $k$ active robots operating on a set of $n$ passive hexagonal tiles. Each tile occupies exactly one node of the triangular lattice $G_\Delta = (V, E)$ (see, e.g., Fig. 30a). A *configuration* $(T, P)$ consists of a set $T \subset V$ of all nodes occupied by tiles, and the robots' positions $P \subset V$. Each node can be occupied

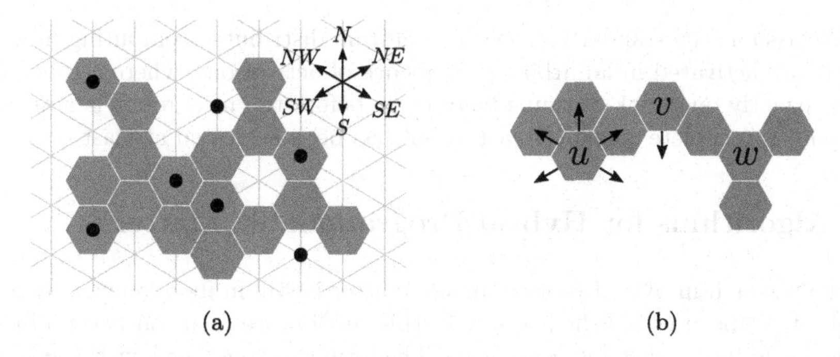


**Fig. 30.** (a) A connected set of tiles positioned on the triangular lattice $G_\Delta$. The black dots indicate robot positions. (b) Possible movements of tiles $u$, $v$, and $w$. Tile $w$ cannot be moved anywhere without violating connectivity.

by at most one robot. We describe the relative positions of adjacent nodes by the six compass directions *N*, *NE*, *SE*, *S*, *SW*, and *NW* (see Fig. 30a). As in the amoebot model, we assume that the robots have a common chirality, but do not share a coordinate system or global compass. For this chapter, we will present the algorithms using a single robot as if the robot's orientation was a global one; for ease of presentation, we will also assume a common orientation for the distributed algorithms utilizing multiple robots.

Each robot must occupy or be adjacent to a node occupied by a tile. Additionally, the subgraph of $G_\Delta$ induced by $T \cup P_c$ must stay connected, where $P_c \subseteq P$ is the set of positions occupied by robots carrying a tile. In a scenario where a tile structure swims in a liquid, for example, this restriction prevents the robots or parts of the tile structure from floating apart. Some examples of possible tile-moving steps are shown in Fig. 30b.

The robots act as finite automata and operate in rounds of `Look-Compute-Move` cycles. In the `Look` phase, a robot observes the node it occupies, say $p$, and the six nodes adjacent to $p$. For each of these nodes, the robot can determine if it is occupied by a tile, if it is occupied by a robot, and, in the latter case, the state of that robot. In the `Compute` phase, a robot potentially changes its state and determines its next move according to the observed information. Furthermore, it may change the state of any robots occupying adjacent nodes. In the `Move` phase, a robot can either (*i*) lift a tile from $p$ if $p \in T$, (*ii*) place a tile it is carrying on $p$ if $p \notin T$, or (*iii*) move to an adjacent node while possibly carrying a tile with it. Each robot can carry at most one tile.

A robot is additionally allowed to carry a constant number of *pebbles*, which can be placed on tiles in order to *mark* them. More specifically, in the `Look` phase, a robot can additionally observe whether any tile in its neighborhood is marked by a pebble. In the `Move` phase, in addition to its other options, a robot can either pick up a pebble (if its current tile is marked by a pebble) or place a pebble (if the current tile is not already marked and the robot has a pebble at its disposal). Pebbles are stateless and indistinguishable.

We assume the standard $\mathcal{A}$SYNC model from distributed computing in which robots are activated in an arbitrary sequence of activations, where a robot performs exactly one `Look-Compute-Move` cycle before the next robot is activated. A *round* is complete whenever each robot has been activated at least once.

## 7 Algorithms for Hybrid Programmable Matter

As the research in hybrid programmable matter is still in its infancy, only a few results have been established so far. In this section, we focus on two problems: *shape formation* and *shape recognition*. The former is concerned with *transforming* the tile structure into some desired shape, while the latter considers the problem of *recognizing* a given shape. As in the amoebot model, the main difficulty in designing algorithmic solutions for these problems lies in the robot's limited memory and visibility. Here, we investigate how these challenges can be overcome for hybrid programmable matter.

### 7.1 Algorithmic Primitives

Before presenting the shape formation and recognition algorithms, we first outline some basic results and present a set of helpful primitives.

*Exploring the Tile Structure.* The hybrid programmable matter model essentially reduces to a set of robots moving in a possibly dynamic but discrete environment. A natural question to ask is whether the robots are able to gather information about the environment. For example, one may ask whether a single robot is able to explore the tile structure, i.e., visit each node at least once, and then halt. It is known that, if the tile structure is assumed to be compact, a single robot can fully traverse the structure by visiting the adjacent *columns* (i.e., consecutive connected tiles from $N$ to $S$) of any column in clockwise order. An additional pebble (or a second robot) suffices to let the algorithm terminate. However, when considering arbitrary tile structures, even a single pebble does not suffice; this follows from the observation that, when the tile structure is static, the hybrid model reduces to Finite Automata in Labyrinths [33]. On the other hand, this problem can be solved with two pebbles [5].

The exploration problem relates to many other practical problems: *tile movement safety*, i.e., deciding whether the removal of a tile would disconnect the tile set, *hole detection*, i.e., deciding whether a structure is simply connected, and *boundary detection*, i.e., reaching the structure's outer boundary. Similar to the construction given by Gmyr et al. [31], it can be shown that none of these problems can be solved by a single robot without using a pebble. The same authors show that while tile movement safety can be decided with a single pebble, the best known solutions for hole and boundary detection require two pebbles. However, these results only hold if the robots are not allowed to move tiles. Once this is allowed, these problems can be easily solved using approaches similar to the algorithms we present in Sects. 7.2 and 7.3.

*Safe Tile Movements.* Given the previous discussion on the difficulty of deciding tile movement safety, it may seem that a complex, multi-robot strategy for identifying tiles to be moved may be necessary for ensuring tile structure connectivity. However, a different—and surprisingly simple—strategy to maintain connectivity under tile movements is the following. If the tiles in the neighborhood of a tile $t$ satisfy *local connectivity*—i.e., they form a connected component around $t$—$t$ can be safely moved, as in Fig. 31a. If the neighboring tiles form two connected components separated by a single empty node $u$ in the neighborhood of $t$, then $t$ can be safely placed onto $u$ without violating connectivity (Fig. 31b). Otherwise, tile movement safety cannot be locally decided (Fig. 31c). Using these local rules, a tile $t$ that is safe to move can always be found by moving *NW*, *SW*, or *N* (in that precedence) until reaching a tile that has no adjacent tile in any of these directions. If $t$ satisfies local connectivity, it can be lifted and moved anywhere; otherwise, it must have adjacent tiles to the northeast and south but no adjacent tile to the southeast, so $t$ can be placed onto the empty node to the southeast. By repeating this strategy, a robot is guaranteed to eventually find a tile that is locally connected, and therefore safely removable.

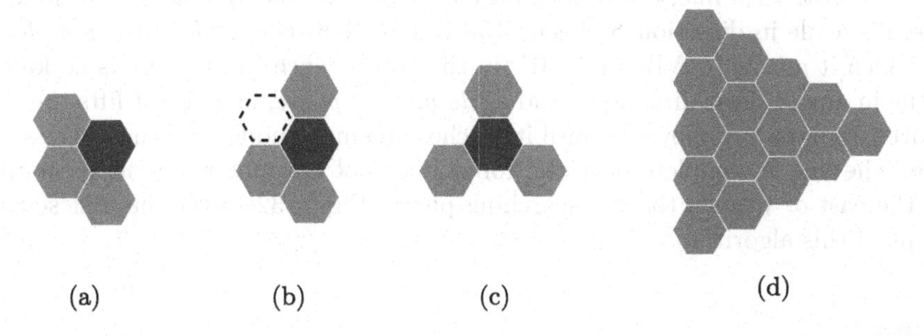


**Fig. 31.** The black tile can (a) be safely removed, (b) only be moved to the adjacent node marked by a dashed outline, or (c) not be moved at all. (d) The triangle to be constructed by the shape formation algorithm in Sect. 7.2.

### 7.2 Shape Formation

Arguably one of the most interesting problems for hybrid programmable matter is shape formation. Shape formation problems may consider different shapes to form (e.g., a hexagon as in Sect. 4.1, or a sequentially constructible shape as in Sect. 4.2), different optimization goals (e.g., runtime, or distance of moved tiles), and side conditions (e.g., avoiding moving tiles beyond the initial structure's convex hull). In this section, we consider the problem of forming a triangle. We first present an algorithm for triangle formation by a single robot, and then show how it can be modified to handle a multi-robot setting. Finally, we present two variants of the algorithm that aim at minimizing the algorithm's running time and only moving tiles within the initial structure's convex hull, respectively.

The section is based on the work of Gmyr et al. [31], where proofs and more detailed descriptions of the algorithms can be found.

**Problem Description.** Consider a connected set of tiles and a single robot that is initially placed on some tile. An algorithm solves the triangle formation problem for hybrid programmable matter if it transforms the initial tile structure into a triangle that is axis-aligned along the robot's north-south axis and grows from east to west (see Fig. 31d).

**Algorithm.** In a naive approach to shape formation, the robot could iteratively search for a tile that can be removed without disconnecting the tile structure and then move that tile to some position such that the shape under construction is extended. Although a safely removable tile always exists, a robot may not be able to find it as previously discussed. Instead, the safe tile movements discussed in Sect. 7.1 are used to first transform the structure into a *line* (i.e., a sequence of connected tiles from north to south). From this intermediate structure, a triangle can easily be constructed in a second stage.

To construct a line, the robot first moves $S$ as far as possible, i.e., as long as there is a tile in direction $S$. Then, it alternates between a *tile searching phase*, in which it moves $N$, $NW$, and $SW$ (in that precedence) until there is no longer a tile in any of these directions; and a *tile moving phase*, in which it lifts the tile, moves one step $SE$, moves $S$ until it reaches an empty node, and then places the tile. The line is complete once the robot does not encounter any adjacent tiles to the east or west in the tile searching phase. Figure 32 shows the first several steps of this algorithm.

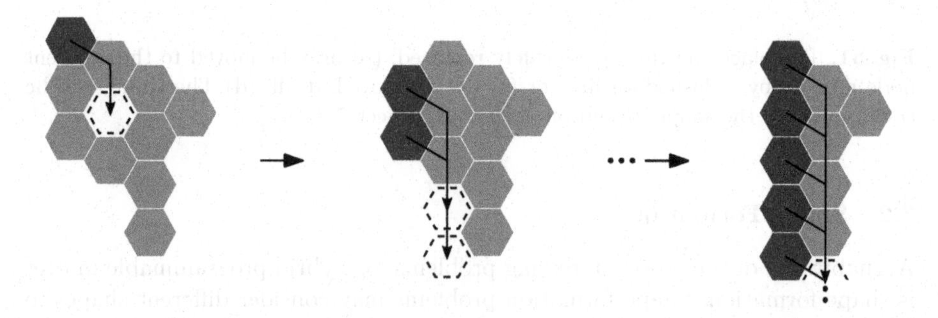

**Fig. 32.** First several steps of line formation. The black tiles are moved to the positions marked by dashed outlines.

In the second stage, the triangle is built by repeatedly taking the northernmost tile of the line, carrying it south to the *vertex* of the forming triangle, and adding it to the westernmost *layer* of the triangle (see Fig. 33). More specifically, the robot places the first tile $NW$ of the triangle's vertex. Every other tile of the triangle is then placed as follows. The robot first takes the northernmost tile of the line and brings it to the vertex. It then walks *NW* and *S* (in that precedence) until an empty node is reached. If there is a tile to the southeast, the robot moves one step *S* and places the tile. Otherwise, the robot moves *N* to the top of the layer, takes one step *NW*, and places the tile. In this manner, the robot continues to extend the triangle tile by tile until the line only consists of the triangle's vertex.

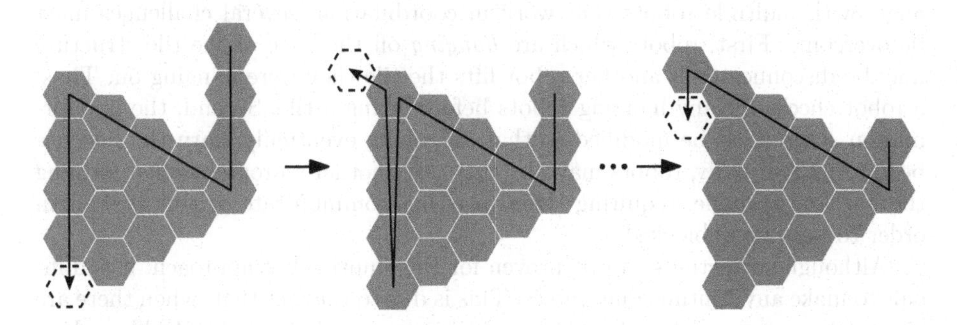

**Fig. 33.** Snapshots of triangle formation. If the number of tiles is not triangular, the final layer will not be completely filled.

**Analysis.** We now state the main results for the triangle formation algorithm.

*Correctness.* The two stages of the algorithm can be analyzed separately. The correctness of the line formation stage follows from (*i*) the tile searching phase always leads the robot to discover a safe tile to move, (*ii*) the tile moving phase never disconnects the tile structure, and (*iii*) the algorithm terminates when a line is formed. The correctness of the triangle formation stage is easily established, resulting in the following theorem.

**Theorem 17.** *The algorithm correctly transforms any connected tile structure into a triangle.*

*Runtime.* The two stages are again analyzed separately. In the first stage, each tile is moved by at most $n$ steps. This is due to the fact that the easternmost column of the initial configuration is never moved, and since tiles are only moved *S* and *SE*. As there are $n$ tiles, the total number of rounds that account to moving tiles is bounded by $\mathcal{O}(n^2)$. Additionally, the robot has to move through the structure in order to search for tiles. By assigning coordinates to the nodes and using the coordinates of the robot as a potential function, it can be shown that the total number of rounds that account to searching for tiles is also bounded by $\mathcal{O}(n^2)$. In the second stage, each tile is carried by a distance of at most $n$, and moving to the next tile takes an additional $n$ steps at most. Therefore, we have the following theorem.

**Theorem 18.** *The algorithm constructs a triangle within $\mathcal{O}(n^2)$ rounds.*

It is not hard to see that $\Omega(n^2)$ rounds are necessary to rearrange an arbitrary initial tile configuration into a triangle using a single robot. If the initial configuration is a line, then a constant fraction of the tiles must be moved by a distance linear in $n$ and thus, in total, $\Omega(n^2)$ move steps are necessary.

**Distributed Algorithm.** In order to extend the algorithm to construct a triangle with multiple robots that work in coordination, several challenges must be overcome. First, robots which are *hanging* off the edge of the tile structure may be disconnected if another robot lifts the tile they were hanging on. Thus, a robot checks for any hanging robots before lifting a tile. Second, the line formation stage must be modified so that all robots eventually learn the line has been formed. Finally, robots may obstruct one another's progress when forming the line and triangle, requiring them to either communicate or simply wait in order to become unblocked.

Although correctness can be proven for this multi-robot approach, it is difficult to make any runtime guarantees. This is due to the fact that, when there are many robots compared to the number of tiles, many robots are blocked by others and must wait to make progress. However, simulation results for distributed line formation [31] suggest a reasonable speedup for few robots in randomly generated initial configurations. In order to guarantee a speedup for multiple robots, we believe different formation strategies that better utilize coordination between the robots will be useful.

**Alternative Intermediate Structures.** Although a line can be constructed efficiently, its linear *diameter* (i.e., the maximal length of a shortest path between any two tiles) may make it an undesirable intermediate structure. In fact, if both the initial diameter $D$ and the diameter of the desired shape are small, moving tiles by a linear distance seems to be an excessive effort. Therefore, we briefly describe how to construct two alternate intermediate structures, namely a *block* and a *tree*, noting their advantages and disadvantages.

*Block Formation.* In a block, all tiles except those farthest to the west have a neighbor to the northwest. A block has only one westernmost column, and every row begins with a tile from that column (see Fig. 34a).

As in the line formation algorithm, constructing a block alternates between searching and moving phases. The robot first searches for a locally northwesternmost tile by repeatedly moving *NW*, *SW*, or *N* (in that precedence). The robot then lifts the tile, moves *SE* until it reaches an empty node, and places the tile there. Although this simple algorithm correctly constructs a block, detecting its completion requires a series of more complex tests performed alongside the block's construction.

**Theorem 19.** *The algorithm constructs a block within $\mathcal{O}(nD)$ rounds and ensures that no tile is ever moved by more than a distance $D$.*

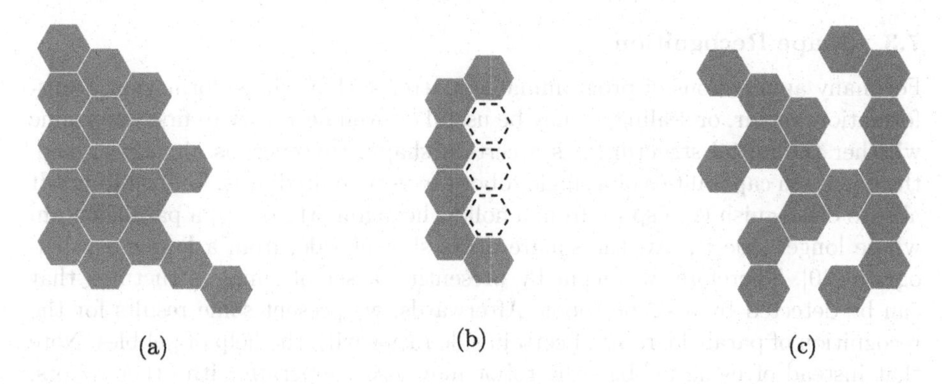


**Fig. 34.** A (a) block, (b) overhang, and (c) tree.

Note that since tiles are exclusively moved *SE*, the resulting block has at most $D$ rows consisting of at most $D$ tiles each, and therefore has diameter $\mathcal{O}(D)$. Therefore, by using a block as an intermediate structure, a triangle can be constructed in $\mathcal{O}(nD)$ rounds. When the initial configuration's diameter is low, i.e., $D = \mathcal{O}(\sqrt{n})$, a triangle can thus be formed in $\mathcal{O}(n^{3/2})$ rounds.

*Tree Formation.* While the block-based approach focuses on quickly constructing a suitable intermediate structure, it may also be desirable to minimize the required work space. Both the line and block are, in many cases, built almost completely outside the initial configuration's convex hull (where we refer to the convex hull of the corresponding set of hexagonal tiles in the Euclidean plane). We briefly describe an algorithm that builds a tree in time $\mathcal{O}(n^2)$ by exclusively moving tiles inside the structure's convex hull. A *tree* is a connected tile configuration that does not contain an *overhang*, i.e., a set of vertically adjacent empty nodes bounded by tiles to the north, west, and south. Examples of an overhang and a tree can be found in Fig. 34b and c, respectively.

We describe the tree formation algorithm at a high level and refer the interested reader to [31] for a detailed description. Roughly speaking, the robot traverses the columns of the tile structure from west to east until it encounters an overhang. It then fills the overhang by retrieving tiles from western columns. Here, the robot exploits the property that the western columns no longer have overhangs, which allows it to find safely removable tiles efficiently and to bring them back to the overhang. After filling the overhang, the algorithm recurs. Correctness of this approach is established by proving that the algorithm (*i*) fills an overhang if one exists and (*ii*) terminates after a complete traversal. Together with a detailed runtime analysis, we have the following theorem.

**Theorem 20.** *The algorithm constructs a tree within* $\mathcal{O}(n^2)$ *rounds without ever placing a tile outside the initial structure's convex hull.*

### 7.3 Shape Recognition

For many applications of programmable matter such as shape formation, transformation, repair, or sealing, it may be useful or even necessary to first determine whether the initial structure has a certain shape. However, as already argued, the detection capabilities of a single robot are very limited: it is, for example, not able to distinguish (*i*) a spiral from a hollow hexagon [31], or (*ii*) a parallelogram whose longer side is, say, the square of its shorter side, from a larger parallelogram [30]. Therefore, we begin by presenting a set of simple structures that can be detected by a single robot. Afterwards, we present some results for the recognition of parallelograms of certain side ratios with the help of pebbles. Note that instead of using pebbles the robot may also cooperate with other robots, each mimicking the behaviour of a pebble. The details of the results presented in this chapter can be found in [30].

**Problem Description.** Consider a single robot that is placed on some tile of an arbitrary initial tile configuration. The shape recognition problem tasks the robot with deciding if the tile structure is of a certain shape, e.g., a line, triangle, hexagon, or parallelogram. If the shape is a parallelogram, the robot must additionally decide if its longer side has length $\ell = f(h)$, for a given function $f(\cdot)$, where $h$ is the length of its shorter side. In this section, we ask whether $\ell = ah + b$ for some constants $a$ and $b$.

**Algorithm.** All four types of shapes can be recognized using a similar strategy. For example, to test if a given tile shape is a line, the robot first chooses a direction in which there is a tile (say, w.l.o.g., $N$), walks in that direction as far as possible, and then traverses the structure in the opposite direction until no longer possible. If it ever encounters a tile to the east or west of any traversed tile, the structure is not a line.

To test if a given tile shape is a (filled) parallelogram axis-aligned along the robot's $N$ and $NE$ directions, the robot first moves to a locally southernmost tile by moving $S$ and $SW$ as long as there is a tile in either of these directions. It then traverses the shape column by column in a snake-like fashion (see Fig. 35a) by repeating the following movements: move $N$ as far as possible; move one step $NE$; move $S$ as far as possible, and finally move one step $NE$. The above procedure is repeated until a $NE$ movement is impossible. The robot can decide whether the structure is a parallelogram by performing a sequence of checks alongside these movements. Any other axis-aligned parallelogram, and all other aforementioned shapes, can be tested in a similar fashion.

Now consider the problem of determining whether $\ell = ah + b$. W.l.o.g., assume that the longer side of the parallelogram is in the northeast direction. The longer side can be determined by moving to the northernmost tile of column 0 (where the columns are numbered from 0 to $\ell - 1$ from west to east), and then moving $SE$ as far as possible; if there is a tile to the southeast (resp., to the south), then the longer side is in the northeastern (resp., southern) direction. If there is no tile to the northeast or south, then the parallelogram is a rhombus.

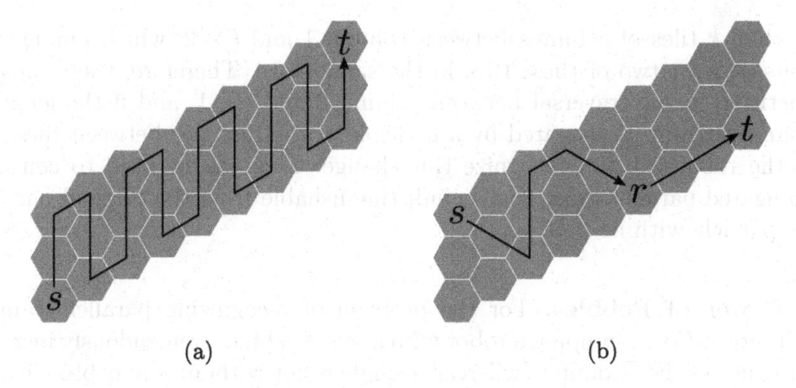


**Fig. 35.** (a) The snake-like traversal to detect a parallelogram. (b) The traversal to decide whether $\ell = ah + b$ (where in this case $a = 2$ and $b = 3$).

To decide whether $\ell = ah + b$, the robot first moves to the northernmost tile of column 0. It then traverses the tile structure in two stages to verify the ratio of the sides. In the first stage, the robot "measures" the distance $ah$ along the length of the parallelogram by moving in a zig-zag fashion as depicted in Fig. 35b. In the second stage, the robot measures $b$. More specifically, in the first stage, the robot repeats the following movements in a loop: (*i*) move *SE* as far as possible, (*ii*) move *N* as far as possible, and (*iii*) move one step *NE*. After having performed the complete sequence of *SE* movements $a$ times, the robot moves on to the second stage, in which it makes an additional $b$ *NE* steps.

If the robot reaches the easternmost column before completing the above procedure, or halts on a tile with a neighboring tile to the northeast, it terminates with a negative result. Otherwise, it terminates with a positive result.

**Analysis.** We have the following theorem for shape recognition.

**Theorem 21.** *A single robot can detect whether the structure is a line, a triangle, a hexagon, or a parallelogram. Furthermore, it can detect whether the structure is a parallelogram with* $\ell = ah + b$ *for any constants* $a, b \in \mathbb{N}$.

However, the following theorem shows that a single robot by itself cannot hope to verify any side ratios of a parallelogram that are not linear.

**Theorem 22.** *A single robot without any pebbles cannot decide whether the tile configuration is a parallelogram with* $\ell = f(h)$, *where* $f(x) = \omega(x)$.

Theorem 22 can be proven using the following observation: to correctly detect the length of a parallelogram of length $f(h)$, the robot must move from the westernmost to the easternmost column (or vice versa) at least once (otherwise it would be unaware of any elongation of the parallelogram). Therefore, if $h$ is chosen such that $\lfloor (f(h) - 2)/h \rfloor > k$, where $k$ is the given number of states of the robot, there must be a row of the parallelogram in which the robot steps on more than $k$ tiles of columns between column 1 and $\ell - 2$, which implies that it visits at least two of these tiles in the same state. Therefore, there must be a repetition in the traversal between column 0 and $\ell - 1$, and if the length of the parallelogram is elongated by a multiple of the distance between these two tiles, the robot will not recognize this change. This can be used to construct an elongated parallelogram that is indistinguishable from the original one by a single particle with no pebbles.

**The Power of Pebbles.** For the problem of recognizing parallelograms of certain side ratios, equipping a robot with a set of pebbles tremendously increases its capabilities. For example, whereas a single robot without any pebbles cannot decide any superlinear function, by Theorem 22, a single pebble suffices to decide any polynomial of constant degree and with constant coefficients. Furthermore, two pebbles suffice to decide certain exponential functions. An overview of some results for a robot with pebbles, which can be found in full detail in the work of Gmyr et al. [30], is shown in Table 1.

**Table 1.** A summary of results for recognizing whether a given parallelogram has height $h$ and length $\ell = f(h)$ given a certain number of pebbles.

| Pebbles | Possible | Impossible | Remarks |
|---|---|---|---|
| 1 | $f(x) = a_n x^n + \ldots + a_0$ | $f(x) = \omega(x^{6k+2})$ | $n, a_i$ constant for all $i$; $k$ is the number of the robot's states |
| 2 | $f(x) = \underbrace{2^{2^{\cdot^{\cdot^{2^x}}}}}_{s+1}$ | — | $s$ constant; asymptotic lower bound does not exist |
| $n$ | $f_n(x)$ | $f_{n+1}(x)$ | For some function family $f_n$ |

## 8 Conclusion and Future Work

This chapter presented a comprehensive review of the distributed algorithms for programmable matter defined under the amoebot model, and surveyed some initial results in the budding new model of hybrid programmable matter. For the amoebot model, we presented two distinct types of algorithms: those that are (mostly) deterministic and heavily utilize state management and particle communication, and those that are fully stochastic, keeping little to no state and requiring very little communication between particles. There is an inherent tradeoff between these two approaches. The former often yields algorithms that provably terminate within a linear number of rounds or better, but are more complex to design, more difficult to implement, and have single points of failure. On the other hand, the stochastic algorithms are very difficult to analyze in terms of convergence times and are observed to be relatively slow in simulations; however, they are inherently robust and are relatively easy to design.

Looking forward, there are several intriguing research directions for the amoebot model. First, any physical implementation of programmable matter would need to meaningfully address the challenge of power management as it expends energy moving and computing. Extending the amoebot model to incorporate energy costs for particle actions could lead to interesting modeling questions, such as how individual particles obtain energy and share it among the collective. Moreover, considering energy usage as an alternative to time complexity when analyzing algorithm efficiency could yield new paradigms for algorithm design. Second, developing a general framework for fault tolerant algorithms under the amoebot model would be a huge step towards realizing programmable matter systems that can handle unpredictable and potentially hazardous application domains. Third, generalizing the amoebot model to three-dimensional space is an exciting goal and a current research direction that may bring our theoretical investigations closer to physically realizable systems.

This is also of interest for hybrid programmable matter; in particular, considering three-dimensional tiles that are hollow (e.g., skeletal polyhedra) would allow active robots to move through the resulting structure, generalizing the movement primitives used in two-dimensional space. One could additionally imagine that the hollow tiles could fold into a condensed shape, enabling robots to transport tiles through other tiles. Other directions for future work on hybrid programmable matter include developing algorithms for multiple robots that provably benefit from the power of coordination, considering settings where tiles may break or need to be repaired, and settings where tiles may attach and detach at random based on environmental changes outside of the robots' control.

**Acknowledgements.** Our warmest gratitude belongs to all of our wonderful collaborators, both past and present, without whom this research would not have been possible. We would like to thank Robert Gmyr, Thim Strothmann, and Zahra Derakhshandeh for their trailblazing work on self-organizing particle systems during their Ph.D. studies. We would especially like to thank Robert for letting us use materials from his Ph.D. thesis for this chapter (in particular, his excellent images). To Dana Randall and Sarah Cannon, thank you for leading us into a new paradigm by showing us just how much one can do with a whole lot of randomness. To Irina Kostitsyna and Dorian Rudolph, thank you for all your work in developing hybrid programmable matter. Finally, to our undergraduate research assistants, especially Alexandra Porter: thank you for your enthusiasm, energy, and effort.

## References

1. Andrés Arroyo, M., Cannon, S., Daymude, J.J., Randall, D., Richa, A.W.: A stochastic approach to shortcut bridging in programmable matter. Nat. Comput. **17**(4), 723–741 (2018)
2. Angluin, D., Aspnes, J., Diamadi, Z., Fischer, M.J., Peralta, R.: Computation in networks of passively mobile finite-state sensors. Distrib. Comput. **18**(4), 235–253 (2006)
3. Baxter, R.J., Enting, I.G., Tsang, S.K.: Hard-square lattice gas. J. Stat. Phys. **22**, 465–489 (1980)
4. Blanca, A., Chen, Y., Galvin, D., Randall, D., Tetali, P.: Phase coexistence for the hard-core model on $\mathbb{Z}^2$. Comb. Probab. Comput. 1–22 (2018). https://www.cambridge.org/core/journals/combinatorics-probability-and-computing/article/phase-coexistence-for-the-hardcore-model-on-2/9B652165B36865C568285FD7A37D8B59
5. Blum, M., Kozen, D.: On the power of the compass (or, why mazes are easier to search than graphs). In: 19th Annual Symposium on Foundations of Computer Science, SFCS 1978, pp. 132–142 (1978)
6. Bonato, A., Nowakowski, R.J.: The Game of Cops and Robbers on Graphs. AMS (2011)
7. Bonifaci, V., Mehlhorn, K., Varma, G.: Physarum can compute shortest paths. J. Theor. Biol. **309**, 121–133 (2012)
8. Borgs, C., et al.: Torpid mixing of some Monte Carlo Markov chain algorithms in statistical physics. In: Proceedings of the 40th Annual Symposium on Foundations of Computer Science, FOCS 1999, pp. 218–229 (1999)
9. Camazine, S., Visscher, P.K., Finley, J., Vetter, R.S.: House-hunting by honey bee swarms: collective decisions and individual behaviors. Insectes Soc. **46**(4), 348–360 (1999)
10. Cannon, S., Daymude, J.J., Gokmen, C., Randall, D., Richa, A.W.: Brief announcement: a local stochastic algorithm for separation in heterogeneous self-organizing particle systems. In: Proceedings of the 2018 ACM Symposium on Principles of Distributed Computing, PODC 2018, pp. 483–485 (2018). https://arxiv.org/abs/1805.04599
11. Cannon, S., Daymude, J.J., Randall, D., Richa, A.W.: A Markov chain algorithm for compression in self-organizing particle systems. In: Proceedings of the 2016 ACM Symposium on Principles of Distributed Computing, PODC 2016, pp. 279–288 (2016). A significantly updated journal version is in preparation. https://arxiv.org/abs/1603.07991
12. Chirikjian, G.S.: Kinematics of a metamorphic robotic system. In: Proceedings of the 1994 IEEE International Conference on Robotics and Automation, ICRA 1994, vol. 1, pp. 449–455 (1994)
13. Das, S.: Mobile agents in distributed computing: network exploration. Bull. Eur. Assoc. Theor. Comput. Sci. **109**, 54–69 (2013)
14. Daymude, J.J., et al.: On the runtime of universal coating for programmable matter. Natural Comput. **17**(1), 81–96 (2018)
15. Daymude, J.J., Gmyr, R., Hinnenthal, K., Kostitsyna, I., Scheideler, C., Richa, A.W.: Convex hull formation for programmable matter (2018). https://arxiv.org/abs/1805.06149

16. Daymude, J.J., Gmyr, R., Richa, A.W., Scheideler, C., Strothmann, T.: Improved leader election for self-organizing programmable matter. In: Fernández Anta, A., Jurdzinski, T., Mosteiro, M.A., Zhang, Y. (eds.) ALGOSENSORS 2017. LNCS, vol. 10718, pp. 127–140. Springer, Cham (2017). https://doi.org/10.1007/978-3-319-72751-6_10
17. Daymude, J.J., Richa, A.W., Scheideler, C.: The amoebot model (2018). https://sops.engineering.asu.edu/sops/amoebot
18. Derakhshandeh, Z., Dolev, S., Gmyr, R., Richa, A.W., Scheideler, C., Strothmann, T.: Brief announcement: amoebot - a new model for programmable matter. In: Proceedings of the 26th ACM Symposium on Parallelism in Algorithms and Architectures, SPAA 2014, pp. 220–222 (2014)
19. Derakhshandeh, Z., Gmyr, R., Richa, A.W., Scheideler, C., Strothmann, T.: An algorithmic framework for shape formation problems in self-organizing particle systems. In: Proceedings of the Second Annual International Conference on Nanoscale Computing and Communication, NANOCOM 2015, pp. 21:1–21:2 (2015)
20. Derakhshandeh, Z., Gmyr, R., Richa, A.W., Scheideler, C., Strothmann, T.: Universal shape formation for programmable matter. In: Proceedings of the 28th ACM Symposium on Parallelism in Algorithms and Architectures, SPAA 2016, pp. 289–299 (2016)
21. Derakhshandeh, Z., Gmyr, R., Richa, A.W., Scheideler, C., Strothmann, T.: Universal coating for programmable matter. Theor. Comput. Sci. **671**, 56–68 (2017)
22. Derakhshandeh, Z., Gmyr, R., Strothmann, T., Bazzi, R., Richa, A.W., Scheideler, C.: Leader election and shape formation with self-organizing programmable matter. In: Phillips, A., Yin, P. (eds.) DNA 2015. LNCS, vol. 9211, pp. 117–132. Springer, Cham (2015). https://doi.org/10.1007/978-3-319-21999-8_8
23. Di Luna, G.A., Flocchini, P., Santoro, N., Viglietta, G., Yamauchi, Y.: Shape formation by programmable particles. In: 21st International Conference on Principles of Distributed Systems, OPODIS 2017, vol. 95, pp. 31:1–31:16 (2018)
24. Di Luna, G.A., Flocchini, P., Prencipe, G., Santoro, N., Viglietta, G.: Line recovery by programmable particles. In: Proceedings of the 19th International Conference on Distributed Computing and Networking, ICDCN 2018, pp. 4:1–4:10 (2018)
25. Dolev, S., Gmyr, R., Richa, A.W., Scheideler, C.: Ameba-inspired self-organizing particle systems (2013). Workshop paper at Biological Distributed Algorithms (BDA) (2013). https://arxiv.org/abs/1307.4259
26. Doty, D.: Theory of algorithmic self-assembly. Commun. ACM **55**(12), 78–88 (2012)
27. Feller, W.: An Introduction to Probability Theory and Its Applications, vol. 1. Wiley, New York (1968)
28. Fomin, F.V., Thilikos, D.M.: An annotated bibliography on guaranteed graph searching. Theor. Comput. Sci. **399**(3), 236–245 (2008)
29. Gmyr, R.: Distributed algorithms for overlay networks and programmable matter. Ph.D. thesis, Paderborn University (2017)
30. Gmyr, R., Hinnenthal, K., Kostitsyna, I., Kuhn, F., Rudolph, D., Scheideler, C.: Shape recognition by a finite automaton robot. In: 43rd International Symposium on Mathematical Foundations of Computer Science, MFCS 2018, pp. 52:1–52:15 (2018)
31. Gmyr, R., et al.: Forming tile shapes with simple robots. In: DNA Computing and Molecular Programming. DNA24, pp. 122–138 (2018)
32. Hastings, W.K.: Monte Carlo sampling methods using Markov chains and their applications. Biometrika **57**(1), 97–109 (1970)

33. Hoffmann, F.: One pebble does not suffice to search plane labyrinths. In: Gécseg, F. (ed.) FCT 1981. LNCS, vol. 117, pp. 433–444. Springer, Heidelberg (1981). https://doi.org/10.1007/3-540-10854-8_47
34. Jeanson, R., et al.: Self-organized aggregation in cockroaches. Anim. Behav. **69**(1), 169–180 (2005)
35. Lund, K., et al.: Molecular robots guided by prescriptive landscapes. Nature **465**(7295), 206–210 (2010)
36. Lynch, N.: Distributed Algorithms. Morgan Kauffman, Burlington (1996)
37. Miracle, S., Randall, D., Streib, A.P.: Clustering in interfering binary mixtures. In: Goldberg, L.A., Jansen, K., Ravi, R., Rolim, J.D.P. (eds.) APPROX/RANDOM -2011. LNCS, vol. 6845, pp. 652–663. Springer, Heidelberg (2011). https://doi.org/10.1007/978-3-642-22935-0_55
38. Mlot, N.J., Tovey, C.A., Hu, D.L.: Fire ants self-assemble into waterproof rafts to survive floods. Proc. Natl Acad. Sci. **108**(19), 7669–7673 (2011)
39. Omabegho, T., Sha, R., Seeman, N.C.: A bipedal DNA Brownian motor with coordinated legs. Science **324**(5923), 67–71 (2009)
40. Patitz, M.J.: An introduction to tile-based self-assembly and a survey of recent results. Natural Comput. **13**(2), 195–224 (2014)
41. Pelc, A.: Deterministic rendezvous in networks: a comprehensive survey. Networks **59**(3), 331–347 (2012)
42. Porter, A., Richa, A.: Collaborative computation in self-organizing particle systems. In: Stepney, S., Verlan, S. (eds.) UCNC 2018. LNCS, vol. 10867, pp. 188–203. Springer, Cham (2018). https://doi.org/10.1007/978-3-319-92435-9_14
43. Reid, C.R., Lutz, M.J., Powell, S., Kao, A.B., Couzin, I.D., Garnier, S.: Army ants dynamically adjust living bridges in response to a cost-benefit trade-off. Proc. Natl Acad. Sci. **112**(49), 15113–15118 (2015)
44. Reid, C.R., Latty, T.: Collective behaviour and swarm intelligence in slime moulds. FEMS Microbiol. Rev. **40**(6), 798–806 (2016)
45. Reif, J.H., Sahu, S.: Autonomous programmable DNA nanorobotic devices using dnazymes. Theor. Comput. Sci. **410**, 1428–1439 (2009)
46. Restrepo, R., Shin, J., Tetali, P., Vigoda, E., Yang, L.: Improving mixing conditions on the grid for counting and sampling independent sets. Probab. Theory Relat. Fields **156**, 75–99 (2013)
47. Şahin, E.: Swarm robotics: from sources of inspiration to domains of application. In: Şahin, E., Spears, W.M. (eds.) SR 2004. LNCS, vol. 3342, pp. 10–20. Springer, Heidelberg (2005). https://doi.org/10.1007/978-3-540-30552-1_2
48. Savoie, W., et al.: Phototactic supersmarticles. Artif. Life Robot. **23**(4), 459–468 (2018)
49. Schelling, T.C.: Models of segregation. Am. Econ. Rev. **59**(2), 488–493 (1969)
50. Schelling, T.C.: Dynamic models of segregation. J. Math. Sociol. **1**(2), 143–186 (1971)
51. Shin, J.S., Pierce, N.A.: A synthetic DNA walker for molecular transport. J. Am. Chem. Soc. **126**(35), 10834–10835 (2004)
52. Strothmann, T.F.: Self-* algorithms for distributed systems: programmable matter & overlay networks. Ph.D. thesis, Paderborn University (2017)
53. Thubagere, A.J., et al.: A cargo-sorting DNA robot. Science **357**(6356), eaan6558 (2017)
54. Toffoli, T., Margolus, N.: Programmable matter: concepts and realization. Phys. D: Nonlinear Phenom. **47**(1), 263–272 (1991)
55. Vinković, D., Kirman, A.: A physical analogue of the Schelling model. Proc. Natl Acad. Sci. **103**(51), 19261–19265 (2006)

56. Walter, J.E., Tsai, E.M., Amato, N.M.: Algorithms for fast concurrent reconfiguration of hexagonal metamorphic robots. IEEE Trans. Robot. **21**(4), 621–631 (2005)
57. Wickham, S.F., et al.: A DNA-based molecular motor that can navigate a network of tracks. Nat. Nanotechnol. **7**(3), 169–173 (2012)
58. Woods, D., Chen, H.L., Goodfriend, S., Dabby, N., Winfree, E., Yin, P.: Active self-assembly of algorithmic shapes and patterns in polylogarithmic time. In: Proceedings of the 4th Conference on Innovations in Theoretical Computer Science. pp. 353–354 (2013)
59. Yim, M., et al.: Modular self-reconfigurable robot systems [grand challenges of robotics]. IEEE Robotics Automation Magazine **14**(1), 43–52 (2007)

# Author Index